Recent Advances in Horticultural Plant Genomics

Recent Advances in Horticultural Plant Genomics

Guest Editors

Zhongxiong Lai
Yuling Lin
Yuqiong Guo

MDPI

Basel • Beijing • Wuhan • Barcelona • Belgrade • Novi Sad • Cluj • Manchester

Guest Editors
Zhongxiong Lai
Institute of Horticultural Biotechnology
Fujian Agriculture and Forestry University
Fuzhou
China

Yuling Lin
Institute of Horticultural Biotechnology
Fujian Agriculture and Forestry University
Fuzhou
China

Yuqiong Guo
College of Horticulture
Fujian Agriculture and Forestry University
Fuzhou
China

Editorial Office
MDPI AG
Grosspeteranlage 5
4052 Basel, Switzerland

This is a reprint of the Special Issue, published open access by the journal *Plants* (ISSN 2223-7747), freely accessible at: https://www.mdpi.com/journal/plants/special_issues/K9BWGGRV76.

For citation purposes, cite each article independently as indicated on the article page online and as indicated below:

Lastname, A.A.; Lastname, B.B. Article Title. *Journal Name* **Year**, *Volume Number*, Page Range.

ISBN 978-3-7258-6276-4 (Hbk)
ISBN 978-3-7258-6277-1 (PDF)
https://doi.org/10.3390/books978-3-7258-6277-1

Contents

About the Editors

Zhongxiong Lai

Zhongxiong Lai, Ph.D., is a professor and doctoral supervisor at Fujian Agriculture and Forestry University. In business, he focuses on long-term teaching and scientific research. In terms of teaching, he has been teaching at both the graduate and undergraduate level for many years, covering topics such as horticulture, landscape architecture, biotechnology, and other specialties. He has trained more than 130 post-doctoral students and graduates. In terms of scientific research, he has been committed to researching biotechnology and the genetic resources of horticultural plants for a long time, aiming to decipher the genome of Longan for the first time. For more than a decade, he has presided over more than 20 important scientific research projects, such as the National Natural Science Foundation of China, the National Science and Technology Support Program, the 948 Program, the National Modern Agricultural Industry Technology System Special Fund, the Fok Yingdong Education Foundation of the Ministry of Education for Young Teachers of Colleges and Universities, the first batch of funding programs for the backbone of the Teachers of Colleges and Universities of the Ministry of Education, the Special Scientific Research Fund for Doctoral Programs of the Ministry of Education, and the Provincial Major Science and Technology Special Project (chief expert).

Yuling Lin

Yuling Lin, Ph.D., is a professor and doctoral supervisor at Fujian Agriculture and Forestry University. She is the director of the Institute of Horticultural Biotechnology of Fujian Agriculture and Forestry University. She is mainly engaged in researching horticultural plant genomics, non-coding RNA, and germplasm resource identification. She deciphered the genome of Longan for the first time and drew the single-cell transcription map of the embryonic callus. She has presided over three projects from the National Natural Science Foundation of China and three projects of the Fujian Natural Science Foundation of China, including one project from the Fujian Distinguished Youth Fund. She has also participated in one project related to the national industrial system and one project focusing on key research and development.

Yuqiong Guo

Yuqiong Guo, Ph.D, is a professor and doctoral supervisor at Fujian Agriculture and Forestry University. Her research focuses on tea plant resources, as well as processing and quality chemistry. She aims to form an effective linkage in the application of basic research, technology promotion, and industrial services, carry out technical research on the processing of rainwater green tea, and help the high-quality development of Wuyi rock tea. She focused on the drought-resistance mechanism of tea plants to provide effective improvement targets for drought-tolerant varieties. She also aims to explore new models of understory tea cultivation and promote the economic development and innovation of the tea industry.

plants

MDPI

Article

Integrated Transcriptomic and Developmental Analyses Provide Insights into the Intrafloral Stamen Differentiation in *Cassia fistula* L.

Zhonglai Luo [1,*], Tingting Duan [2], Xiaoyuan Li [1], Jianxuan Zhou [1], Qiankun Liu [1] and Libo Jiang [1,*]

[1] School of Life Sciences and Medicine, Shandong University of Technology, Zibo 255000, China
[2] College of Coastal Agricultural Sciences, Guangdong Ocean University, Zhanjiang 524088, China
* Correspondence: luozhongl@sdut.edu.cn (Z.L.); libojiang@sdut.edu.cn (L.J.)

Abstract

Selective pressure targeting male functions plays a crucial role in the evolution of floral morphological traits. In some angiosperm groups, flowers contain two or more sets of stamens that vary in size, color, and morphology, a phenomenon known as heteranthery. This reflects an evolutionary adaptation of stamens. However, the developmental basis and molecular mechanisms remain poorly understood. This study integrates transcriptomic and developmental approaches to elucidate the molecular and morphological mechanisms underlying intra-floral stamen differentiation in *Cassia fistula* L., an economic leguminous tree exhibiting heteranthery with three distinct stamen types: long stamens (LS), short stamens (SS), and degenerated stamens (St). We documented asynchronous stamen primordia initiation and development trajectories across stamen types. Transcriptomic profiling and protein–protein interaction analysis identified differentially expressed genes (DEGs) between filaments of the three stamen sets, with significant enrichment in brassinosteroid (BR) related pathways. *CYP90D1* (Cf_f19903) and *CYP90C1* (Cf_f56973) emerged as candidate genes related to stamen length differentiation in *C. fistula*. This study not only helped elucidate the developmental and genetic framework of heteranthery in *C. fistula* but also provided new insights for exploring floral organ evolution in leguminous plants.

Keywords: *Cassia fistula*; flower development; heteranthery; stamen differentiation; transcriptomics

1. Introduction

During the evolution of angiosperms, diverse morphological traits of flowers have evolved under selective pressure acting on male functions, specifically the dispersal of pollen to receptive stigmas [1–4]. In most angiosperms, the morphology, size, and color of stamens within the same flower are generally uniform, which is referred to as isomorphic stamens. As direct participants in male function, the morphology and structure of stamens exhibit certain variations under this selective pressure. Significant differentiation is observed among the stamens within a single flower of some angiosperms, a phenomenon known as heteranthery, which has intrigued many researchers ever since the age of Darwin.

Heteranthery has independently arisen on numerous occasions across diverse angiosperm lineages. This floral trait polymorphism has been documented in at least 20 families spanning 12 orders [5,6]. This form of somatic polymorphism is a floral adaptation that functions to mitigate fitness costs associated with pollen consumption by pollinators in species lacking nectar, and in which pollen is the primary floral reward for bees [7–9].

Early studies on heteranthery primarily focused on describing floral characteristics and pollinator behaviors (e.g., [10,11]). Based on observations of stamens in Melastomataceae and communications with the German naturalists Fritz and Hermann Müller, Darwin proposed the "division-of-labor" hypothesis for heteranthery [12], that one set of stamens serves to meet pollinators' nutritional needs by providing pollen as food ("feeding stamens"), while another set ensures the safe transfer of pollen to the stigma ("pollination stamens"). Despite Darwin's "enormous labor" [12,13] and the extensive work by subsequent researchers on multiple heterantherous groups (e.g., *Commelina coelestis*; *Solanum rostratum*; *Tripogandra grandiflora*; *Senna didymobotrya*) [10,14–16], this hypothesis long remained without effective validation until the 21st century. Making use of differentiation in pollen exine ornamentation, Luo et al. (2008) provided the first unequivocal experimental support for Darwin's division-of-labor hypothesis explaining stamen dimorphism in *Melastoma malabathricum* [8]. In the following years, the functional and evolutionary adaptation of heteranthery has been explored in various plant taxa [7,17–19]. However, the developmental mechanisms and molecular regulation underpinning this floral trait remain poorly characterized.

The structure of an organ is the basis of its function. Although the "division of labor" for heteranthery has been experimentally confirmed in species such as *M. malabathricum* [8] and *Solanum rostratum* [9], the developmental mechanisms behind stamen differentiation are still inadequately studied. Tucker's (1996) research on *S. didymobotrya* proposed that morphological and size differences between "pollination" and "feeding" stamens resulted from asynchronous development of stamen primordia [20]. In *Melastoma* and *Senna bicapsularis*, variations in the initiation time and developmental rates among different stamen groups have also been observed [21]. Nevertheless, key developmental stages determining stamen differentiation are still unclear, and whether different heterantherous taxa share common developmental patterns requires further validation in more species.

Heteranthery serves as an excellent research system for investigating the adaptive evolution of male organs in angiosperms. Despite more than a century of research in this field, critical scientific questions, particularly the mechanistic basis of stamen differentiation, remain unresolved. With advances in sequencing technology, research on model breeding systems such as distyly (e.g., Primulaceae, Turneraceae, Menyanthaceae, Rubiaceae) [22–26] and dioecy (e.g., Ebenaceae, Cucurbitaceae) [27,28] has entered the genomics era. The developmental and molecular regulation of these systems has been elucidated and experimentally validated. In contrast, most studies on heteranthery still focus on macro-level analyses. Recently, Yang et al. (2025) conducted whole-genome sequencing of *Monochoria elata* (Pontederiaceae) and identified genes from the LEAFY, MADS-box, and TCP families as putative regulators governing anther pigmentation, filament elongation, and spatial arrangement, which revealed gene networks underlying heteranthery [29]. However, integrative studies dissecting both developmental trajectories and molecular regulatory pathways of intra-floral stamen differentiation remain limited.

In the large family Leguminosae, heteranthery mainly distributes in several genera of the subfamily Caesalpinioideae, with particularly typical stamen differentiation observed in flowers of *Cassia* and *Senna* [30–36]. In the recent study, we investigated the development process and genetic basis of heteranthery in *Cassia fistula*, the Golden shower tree, a horticultural species exhibiting significant intra-floral stamen differentiation. This work integrated micromorphological, developmental, metabolomic, and comparative transcriptomic approaches, aiming at elucidating the mechanisms underlying heteranthery formation, thereby providing novel insights into the evolutionary dynamics of floral structure in economic leguminous plants and related angiosperms.

2. Results

2.1. Floral Biology of *Cassia fistula*

The golden shower tree is an important horticultural plant that produces pendulous racemes with bright yellow flowers (Figure 1A). In South China, *Cassia fistula* typically initiates flowering in early May, with the flowering period extending until September–October. The peak flowering time lasts from June to August. Flowers open at dawn (05:00–06:00), with petals unfurling and coiled stamens/stigmas gradually extending. By 18:00, stamens typically detach. Individual flower longevity is approximately 12 h.

At the study site, primary floral visitors were carpenter bees (*Xylocopa* spp.), including the gray-breasted Carpenter Bee (*X. phalothorax*) and Bamboo Carpenter Bee (*X. dissimilis*). Bees landed on the central short stamens (SS) of the flower when foraging, clasping the anthers tightly against their thorax with their legs while gripping the SS and St filaments with mandibles (Figure S1A,B). Rapid contraction of indirect flight muscles generated vibrations transmitted through the bee's body to the stamens and pistil, triggering pollen expulsion from the apical pores of the short-stamen anthers, which was later collected by the bees for the larvae's food. LS pollen dispersed through anther apical slits, adhering to the bee's dorsal thorax and dorsal abdomen. Concurrently, the stigma at the tip of the curved style contacted the bee's dorsal part, thereby completing pollination (Figure S1C).

2.2. Morphological Differentiation of Androecium

The androecium of *C. fistula* consists of ten stamens arranged in two whorls (Figure 1B,C). The outer whorl is episepalous, comprising three long stamens (LS) and two degenerated stamens (outer staminodes, out-St); while the inner whorl is epipetalous, with four short stamens (SS) and one degenerated stamen (inner staminode, in-St). LS have the longest filaments (30–40 mm), curved in an S-shape, bearing yellow, crescent-shaped anthers that flatten inward and slightly below the stigma (Figure 1B,F,I). Filaments of SS are shorter (6–10 mm long) than those of LS but longer than staminodes, with yellow, elliptical anthers (Figure 1E,I). The outer and inner staminodes have the shortest and straightest filaments, with degenerated anthers (Figure 1D,G).

Another difference in the dehiscence patterns between the two whorls of stamens. For the outer whorl stamens (three LS and two out-St), anthers dehisced through discontinuous longitudinal slits at both apex and base (Figure 1G,I), while anthers of the inner-whorl (four SS and one in-St) dehisced poricidally at the distal end (Figure 1G,H).

2.3. Developmental Process of the Three Stamen Sets

The initiation and development of floral organs constitute a continuous process. Following Tucker (1996) [20], the developmental process of floral organs in *C. fistula* can be divided into three stages: (1) Early stage: from the initiation of the floral primordium to the onset of carpel primordium invagination. (2) Mid-stage: from carpel primordium invagination to carpel elongation prior to fusion. (3) Late stage: encompassing the gradual fusion of carpels until the flower is fully matured.

When all petal primordia had enlarged, the primordium of the middle LS emerged first (Figure 2A), followed sequentially by the primordia of the two lateral LS (Figure 2B,C). Then, primordia of the two out-St form in close succession to complete the whorl (Figure 2D). So, the outer whorl stamen primordia initiated unidirectionally from the abaxial to adaxial side.

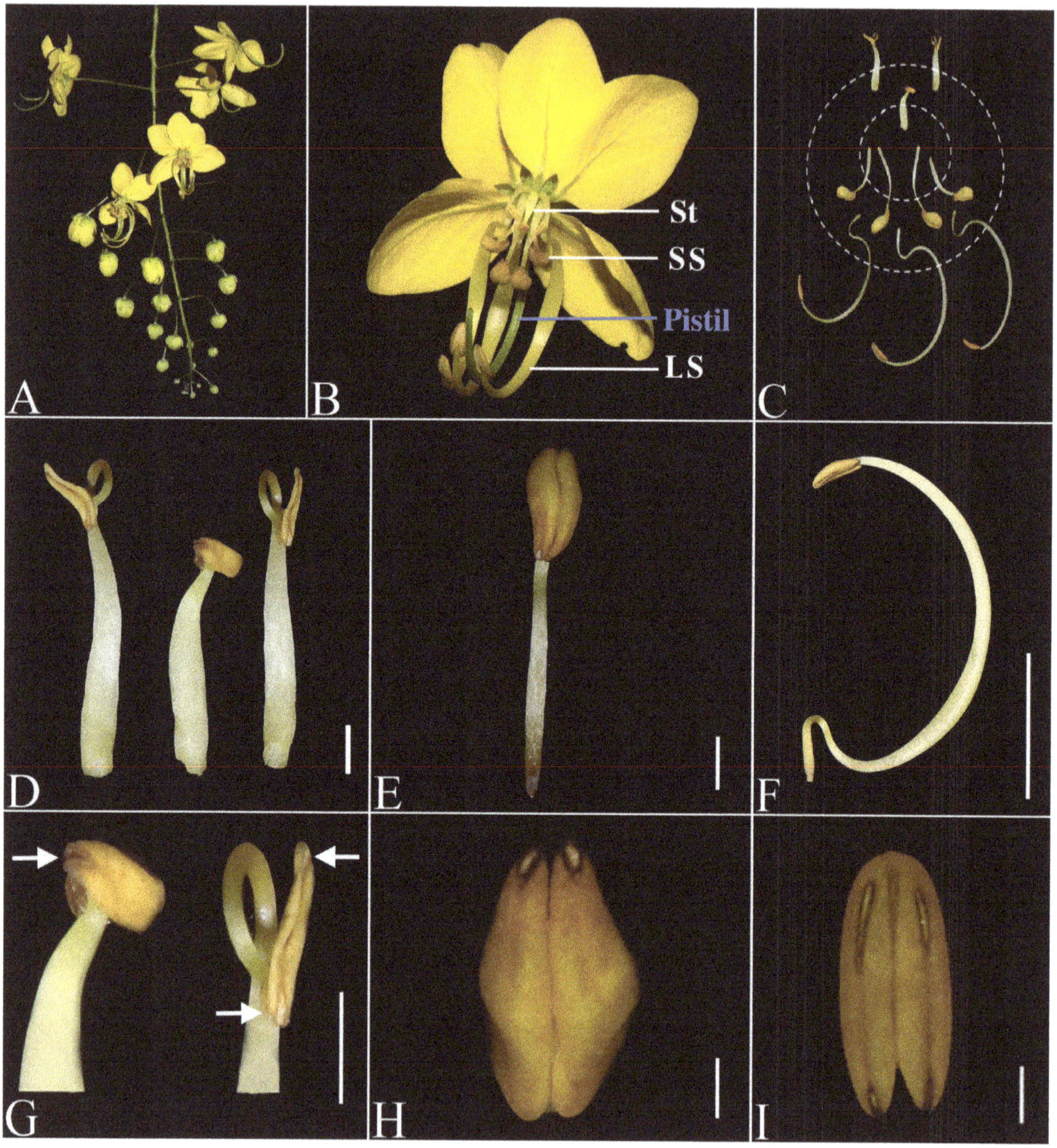

Figure 1. The inflorescence, floral morphology, and stamen differentiation pattern of *Cassia fistula*. (**A**) Raceme; (**B**) Flower, showing the long stamens (LS), short stamens (SS), staminodes (St), and Pistil; (**C**) Distribution of inner and outer whorls of stamens. Dashed circles highlight the two whorls of stamens; (**D**) Staminodes; (**E**) Short stamen; (**F**) Long stamen. (**G**) The anther of the inner staminode (left) with terminal pores (white arrow). The anther of outer staminodes (right) with incomplete longitudinal dehiscence at top and bottom (white arrows); (**H**) The anther of SS showing terminal pores; (**I**) The anther of LS showing incomplete longitudinal dehiscence at top and bottom. Bar = 1 cm in (**F**); 1 mm in (**D**,**G**–**I**); 2 mm in (**E**).

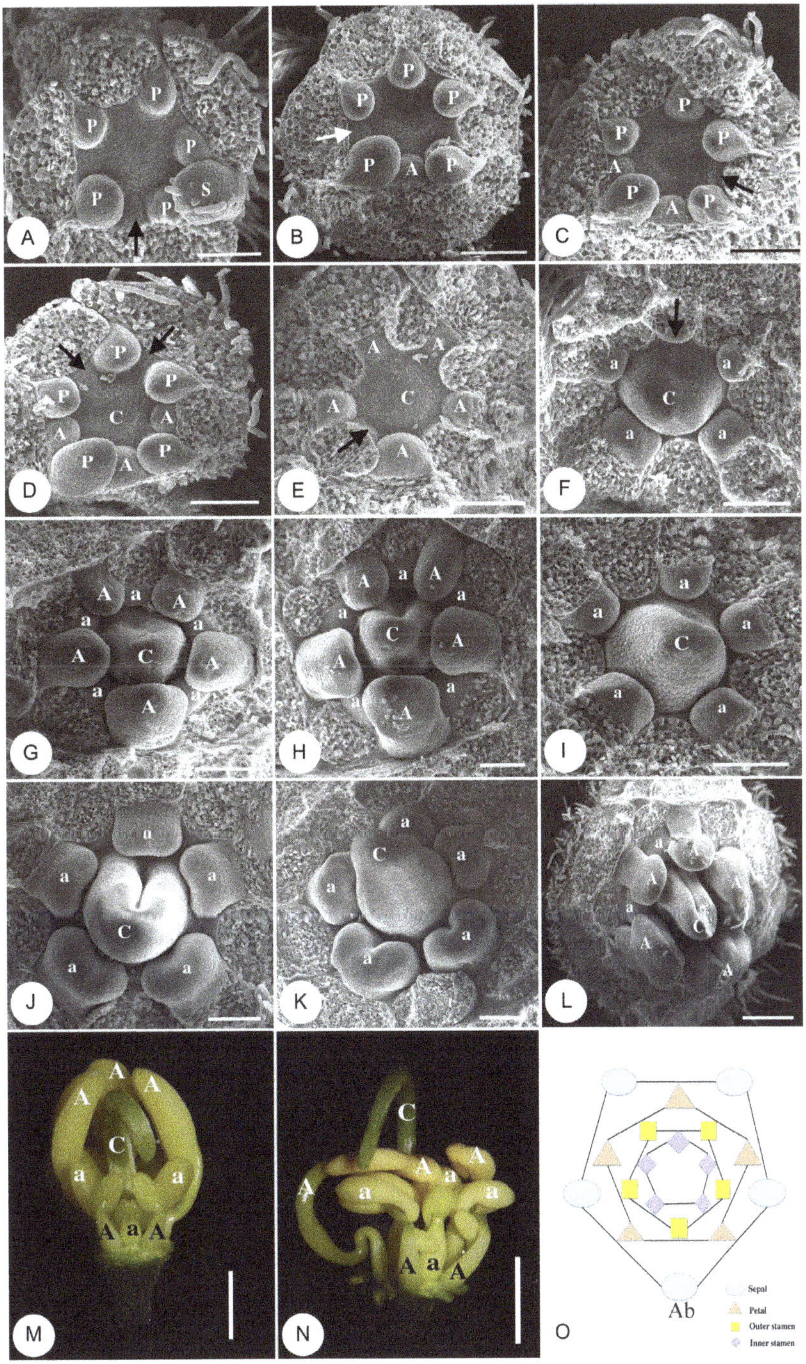

Figure 2. Floral organogenesis and development of *Cassia fistula*. Bracts and bracteoles removed in all figures; sepals removed in (**B–N**); petals removed in (**E–N**). (**A**) The middle LS primordium arises at the midpoint of the abaxial side, positioned precisely between the first and second petal primordia ("P"), as indicated by the arrow. "S", sepal primordium. (**B**) The second LS primordium emerges on the left lateral side, as marked by the arrow. (**C**) The third LS primordium emerges on the right lateral

side, as marked by the arrow. (**D**) Primordia of the two out-St arise on the adaxial side (arrows). Carpel primordia ("C") initiate development. (**E**) All five outer whorl (antesepalous) stamen primordia are established ("A"). The arrow points to the initiation site of the first SS primordium. (**F**) Four SS primordia ("a") are established, and the arrow indicates the primordium of in-St at the central midpoint of the adaxial side. (**G**) The outer whorl (antesepalous, "A") and inner-whorl (antepetalous, "a") stamens developed alternately, with the three LS significantly larger than the two out-St. (**H**) Primordia of the three LS become horseshoe-shaped, and epidermal trichomes begin to develop. (**I**). Five stamen primordia of the inner whorl are of similar size. (**J**,**K**) Differentiation between the four SS primordia and the in-St primordium increased; (**L**) The differentiation of anthers and filaments in the outer-whorl stamens has become pronounced, while the development of the three adaxial staminodes (two out-St and one in-St) is markedly delayed. (**M**) Stamen differentiation in the flower bud is ca. 7 mm in diameter. (**N**) Stamen differentiation in a flower bud, ca. 15 mm in diameter. (**O**) Diagram illustrating the arrangement of sepals, petals, outer whorl stamens, and inner whorl stamens. "Ab" indicates abaxial side. Flower buds in (**A–L**) are arranged with the abaxial side at the lower side of each figure. Bar = 100 μm in (**A–K**); 200 μm in (**L**); 2 mm in (**M**); 5 mm in (**N**).

The initiation of the inner-whorl (opposite the petals) stamen primordia closely follows that of the outer whorl stamens, also exhibiting a unidirectional developmental sequence. The two abaxial SS primordia sequentially emerge (Figure 2E), followed by the two lateral short stamens, and finally, the adaxial in-St primordium is initiated (Figure 2F).

By the time all inner-whorl stamen primordia have formed, the outer-whorl stamens already exhibit distinct differentiation, with the three abaxial LS primordia significantly larger than the two adaxial out-St primordia (Figure 2G). Subsequently, the three LS primordia initiate differentiation into anthers and filaments. When the apex of the primordia becomes horseshoe-shaped, anther formation commences, accompanied by the growth of epidermal trichomes (Figure 2H). During the development and expansion of the microsporangia, longitudinal grooves have emerged on both sides and the center of the three LS anthers (Figure 2L). In contrast, the two adaxial out-St primordia develop significantly slower than the three abaxial LS primordia, ultimately differentiating into two staminodes with markedly shorter filaments at maturity (Figure 2L–N).

The anther differentiation of inner-whorl stamens initiates relatively late, commencing after the anther differentiation of outer-whorl stamens. Initially, the five inner-whorl stamen primordia develop at a consistent pace with minimal size variation (Figure 2I). As development progresses, the differentiation between the four SS primordia and the single adaxial in-St primordium becomes increasingly distinct (Figure 2J). A prominent groove has emerged along the midline of the SS anthers (Figure 2K), while the development of the in-St is notably delayed (Figure 2K,L, arrows).

From the developmental process, it can be observed that aside from the differences in the initiation timing of development, the LS filaments exhibited a faster overall elongation rate compared to the filaments of SS and St. In the later stages of development, the distinctions among the three stamen groups become increasingly apparent (Figure 2M,N).

The initiation of carpel primordium occurs almost simultaneously with the emergence of outer whorl (antisepalous) stamen primordia. The central region of the floral primordium, initially flat, gradually develops into a dome-shaped structure (Figure 2C–E). Upon the emergence of all five inner-whorl (antipetalous) stamen primordia, the circular carpel primordium is already clearly demarcated (Figure 2F). As carpel depression deepens with stamen differentiation, its incurving margins form a profound cleft (Figure 2J–L), and the initially erect carpel becomes curved within the floral bud (Figure 2M,N). The arrangements of the two whorls of stamens, as well as the sepals and petals in the flower, are illustrated in Figure 2O.

During late developmental stages, from the large floral bud (ca. 12 mm in diameter) to anthesis, filament growth in both St and SS of *C. fistula* is largely arrested. In contrast, the filaments of LS and the style continue elongating, forming an S-shaped curvature at the base of the LS filaments. At full bloom, the apex of LS anthers and the stigma are positioned at nearly the same level (Figure 3A,B and Figure S1C), facilitating LS in exercising its function as the "pollination stamen" [37].

Figure 3. The process of flower opening (**A**) and stamen elongation (**B**), petals removed from *Cassia fistula*. Bar = 1 cm.

2.4. Phytohormone Quantification and Comparison

Crucial hormones involved in plant organ growth were quantified in the filaments of different stamen types. Brassinolides (BR) promote cell elongation by modulating cytoskeletal organization and cell wall loosening (26). The mean concentration of BR in LS filaments (0.032 $\pm$ 0.007 ng/g, FW) was significantly higher than that in SS (0.019 $\pm$ 0.002 ng/g, FW) or St (0.016 $\pm$ 0.003 ng/g, FW) filaments (Figure 4A). In contrast, similar concentrations were found between LS, SS, and St for IAA (3-indoleacetic acid), GA1 (gibberellic acid 1), GA7 (gibberellic acid 7), and methyl jasmonate (Figure 4B–D).

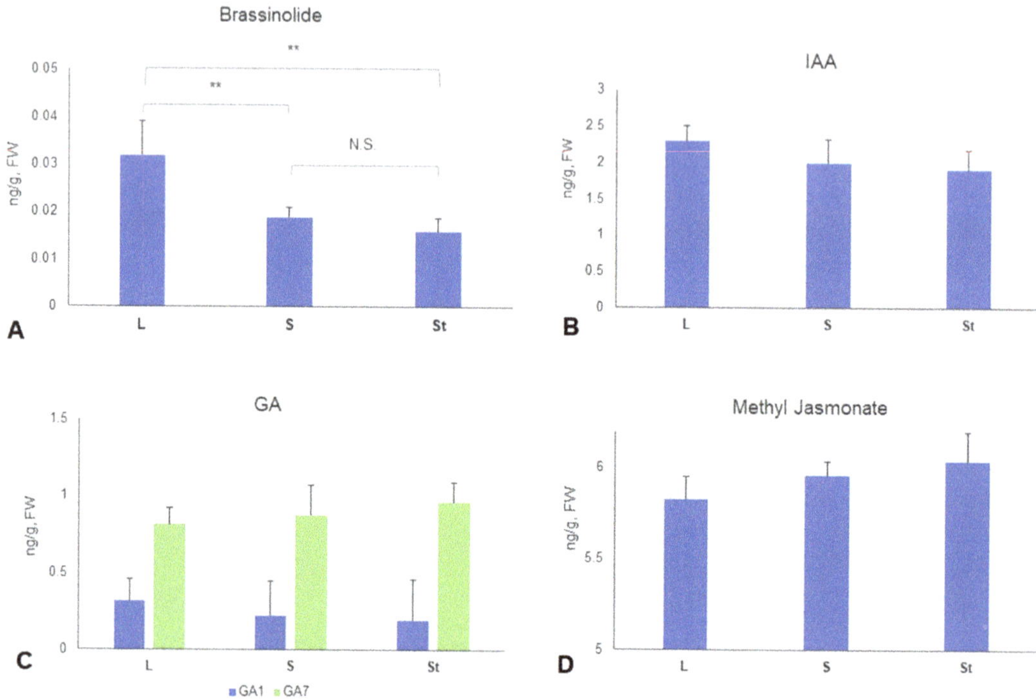

Figure 4. Quantification of important phytohormones in the three types of staminal filaments of *C. fistula*. (**A**) Briassinolide; (**B**) IAA; (**C**) GA; (**D**) Methyl jasmonate. ** $p < 0.01$ in Wilcoxon rank-sum test. N.S. Difference not significant.

2.5. RNA-Sequencing, Functional Annotation, and Classification

Staminal filaments from LS, SS, and St were subjected to transcriptomic sequencing. We achieved at least 8 Gb clean data for each sample and obtained approximately 293 million pair-end reads with Q30 > 90.0.

After assembly, 145,223 unigenes were obtained for the staminal filament samples. With the aid of the eggNOG-mapper (v.2) pipeline [38], protein sequences of unigenes were searched against different databases, including Swissprot, PFAM, eggNOG, GO, and KEGG for functional annotations. There are 72,815 unigenes that have been annotated in one or more databases (Table 1).

Table 1. Number of unigenes annotated in different databases.

Anno_Database	Annotated_Number	300 ≤ Length < 1000	Length ≥ 1000
COG	19,179	7519	6677
GO	42,783	19,494	12,304
KEGG	27,609	12,872	7230
KOG	39,141	17,660	11,647
Pfam	38,903	16,280	14,562
Swissprot	42,912	19,325	13,496
eggNOG	64,952	29,132	18,657
nr	70,647	32,164	19,286
All_Annotated	72,815	32,544	19,325

A total of 27,609 unigenes were matched to 128 KEGG pathways for the filament samples. Among them, the pathways "Carbon metabolism (ko01200)", "Biosynthesis of amino acids (ko01230)", and "Ribosome (ko03010)" ranked among the top three, having the largest number of annotated genes. Additionally, numerous unigenes were mapped to pathways such as "starch and sucrose metabolism (ko00500)", "Protein processing in endoplasmic reticulum (ko04141)", and "Plant hormone signal transduction (ko04075)", indicating active biosynthesis and metabolic activities (Figure S2).

For the filament samples, 42,783 unigenes were assigned one or more GO terms. Regarding the BP category (Biological Processes), a significant number of genes were associated with the terms "oxidation-reduction process" (GO: 0055114; 3351 unigenes) and "protein phosphorylation" (GO: 0006468; 1956 unigenes). "ATP binding" (GO:0005524; 5625 unigenes) holds the highest representation in the Molecular Function (MF) category, and "integral component of membrane" (GO:0016021; 3740 unigenes) is the most frequently occurring GO term in the Cellular Components (CC) category (Figure S3).

2.6. Differential Expression and Enrichment Analysis of Unigenes

Gene expression in filaments and anthers between different stamen types was compared to screen for genes potentially associated with the phenotypic differences during stamen development. For filaments, the largest number of DEGs (168) is found between LS and St, followed by the comparison between LS and SS (49). Only 16 DEGs were identified between SS and St filaments. The number of DEGs was a bit greater than that reported in Yang et al. (2025) [29], which identified 15 DEGs between pollinating and feeding anthers across three developmental stages.

Between LS and St filaments, KEGG pathways "Plant hormone signal transduction" (ko04075), "Sesquiterpenoid and triterpenoid biosynthesis" (ko00909), and "Metabolic pathways" (ko01100) were significantly enriched (Q-value < 0.05). Many genes involved in these pathways were related to BR biosynthesis or auxin metabolism, including *CYP90D1* (Cf_f49903), *CYP90C1* (Cf_f56973), *YUCCA* (Cf_f57541), and *AUX28* (Cf_f36862). "Sesquiterpenoid and triterpenoid biosynthesis" (ko00909) was the only significantly enriched pathway between LS and SS filaments, while no significantly enriched pathway was identified for the SS-St comparison.

Results of the KEGG enrichment analysis are shown in Figures 5 and S4.

2.7. Screening of Genes Potentially Associated with Heteranthery

Based on the gene enrichment and phytohormone analysis, we primarily focused on the expression patterns of BR-related genes.

DEGs *CYP90D1* (Cf_f49903) and *CYP90C1* (Cf_f56973) are key genes involved in BR signaling. *CYP90D1* plays a critical role in the brassinosteroid (BR) biosynthetic pathway and growth regulation. *CYP90D1* (Cf_f49903) was expressed significantly higher in LS filaments compared to SS and St filaments (Figure 6A). The expression values (FPKM) of Cf_f49903 were lower than one in SS filaments, and <0.1 in the St filaments. *CYP90C1* acts downstream of *CYP90D1*. Its homolog, Cf_f56973, showed much higher expression in LS filaments than St filaments (Figure 6B). qRT-PCR was performed to verify the expression levels of candidate DGEs. The expression pattern of *CYP90D1* (Cf_f49903) and *CYP90C1* (Cf_f56973) in LS, SS, and St filaments is shown in Figure 6C, which is consistent with the RNA-Seq data in general.

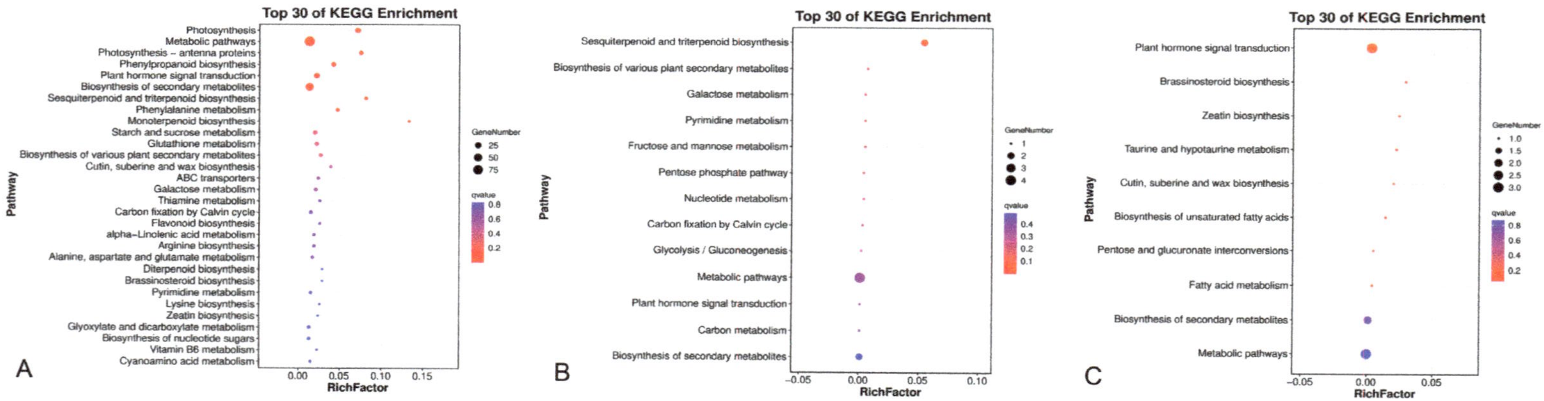

Figure 5. KEGG pathway enrichment analysis for DEGs in the comparisons of LS vs. St filaments (**A**), LS vs. SS filaments (**B**), and SS vs. St filaments (**C**).

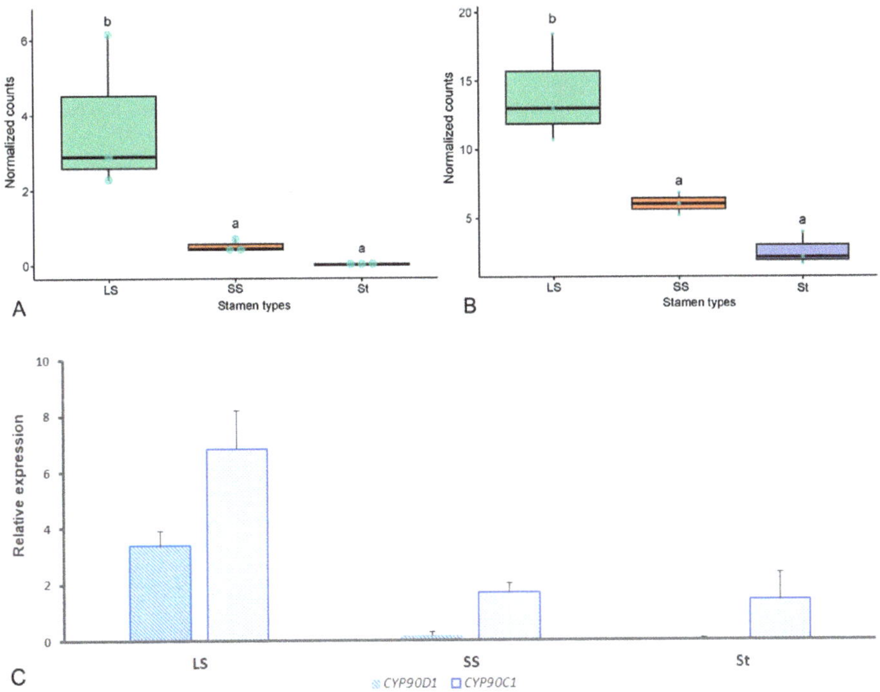

Figure 6. Differential expression of unigenes *CYP90D1* (Cf_f49903) (**A**), *CYP90C1* (Cf_f56973) (**B**), and their expression levels tested by qRT-PCR in three types of stamens (LS, SS, St) of *Cassia fistula*. Different letters (a and b) represent significant differences with Log2(FC) > 1 and FDR < 0.05. Error bars in (**C**) indicate the S.E. of three biological replicates.

2.8. Protein–Protein Interaction Network Analysis

PPI networks were constructed to illustrate the biological interactions of the genes of interest in greater detail. Expressions were compared between different stamen types, and unigenes with FoldChange (FC) ≥ 2 in any replicates were selected to create a gene set (Supplementary Materials). Then, protein sequences of these unigenes were mapped to the STRING database, and the generated network was exported to Cytoscape for further analysis. The PPI subnetwork was composed of 1084 nodes and 2420 edges (Figure S5).

BiNGO (v.3.0.5) implemented in Cytoscape was used to assess over-representation of GO categories in the network. "Hormone biosynthetic process" (GO: 0042446), "brassinosteroid biosynthetic process" (GO: 0016132), "brassinosteroid metabolic process" (GO: 0016131), and "flower development" (GO: 0009908) were all significantly enriched (FDR < 0.05), in which *CYP90D1* (Cf_f49903) and *CYP90C1* (Cf_f56973) were included. The related GO term network is illustrated in Figure S6.

3. Discussion

As a complex floral trait, heteranthery reflects the high adaptability and evolutionary potential of male organs in angiosperms, holding significant importance for exploring the evolution of breeding systems and floral structural diversity. Although this special floral adaptation has intrigued scientists for over one hundred years, the developmental basis and molecular mechanisms are still underexplored for most heterantherous groups. In the present study, we investigated the development process of different stamen types in *C. fistula*, identified potential gene networks associated with the intra-floral stamen differentiation through transcriptomic and metabolomic analyses.

Morphological differentiation is a hallmark of heteranthery. In *C. fistula*, the androecium displays pronounced structural divergence among its three stamen sets, primarily represented by filament length difference (Figure 1B,C). This phenomenon aligns with observations in other heterantherous taxa, such as *Lagerstroemia indica*, *Senna bicapsularis*, *S. wislizeni*, and *Commelina coelestis* [7,14,35,36,39]. However, the developmental mechanisms underlying filament length variation remain poorly understood across these systems.

Endress (1997) hypothesized that temporal shifts in stamen initiation—specifically, the "centrifugal" developmental pattern in *Dillenia alata*—could lead to asynchronous maturation between outer and inner stamen whorls, thereby contributing to heteranthery [40]. This model provides a plausible framework.

Our data demonstrated that, in *C. fistula*, asynchronous development not only exists between the outer and inner whorls of the stamen but was also detected within the same whorl. The outer whorl stamen preceded the inner whorl stamen in primordia initiation (Figure 2B–H). In the outer whorl, the middle LS initiated first, and then the two lateral LS primordia emerged (Figure 2B–D). The two out-St formed last (Figure 2G,H). In the inner whorl, the two abaxial SS primordia sequentially emerged, followed by the two lateral short stamens, and finally, the adaxial in-St primordium initiated (Figure 2E,F). Therefore, primordia development in both whorls exhibited unidirectional patterns, from the abaxial side to the adaxial side. Tucker (1996) found that in the congenera species *C. javanica*, the abaxial stamens emerged first, showing a similar pattern with *C. fistula* [20]. Marazzi and Endress (2008) also observed the unidirectional initiation order for both stamen whorls in *Senna* species [35].

But Singh and Sharma (1978) have reported the stamens of *C. fistula* initiated "alternatively" in the two whorls [41]. Our result is consistent with Tucker (1996) [20], suggesting the descriptions in Singh and Sharma (1978) [41] might be inaccurate.

The differentiation of LS and St started early at the primordia stage. The three abaxial LS primordia were significantly larger than the two adaxial out-St primordia when carpel depression deepens (Figure 2F–H). Elongation of the LS and the style continued during the development of the floral buds, resulting in comparable lengths in open flowers. This arrangement facilitated the LS anthers and the stigma contacting the dorsal part of the bee's body and completing pollination (Figure S1C). A similar pattern could also be observed in heterantherous clades *Melastoma* [7,8], *Senna bicapsularis* [7], and *Solanum rostratum* [9], where anther tips of the long stamens ("pollination stamens") and the stigma contacted the bee's abdomen when they visited the flowers, in good agreement with the "division of labor" hypothesis [8,9].

The morphological differentiation and "division of labor" for heteranthery are primarily determined by the length and shape of filaments. Yang et al. (2025) suggested that in *M. elata*, jasmonic acid (JA) and lignin might contribute to filament length and morphological differentiation between feeding and pollinating stamens [29]. Luo et al. (2016) reported *ARFs* (Auxin Response Factors) and *GA20ox* (Gibberellin 20-oxidase) potentially regulated stamen length in *S. bicapsularis* [42]. In *C. fistula*, either the IAA, GA, or methyl jasmonate levels differed little among the filaments of long stamens, short stamens, and staminodes (Figure 4). Similar trends have been reported in *M. elata*, where the IAA content was nearly identical in the two anther types [29]. Our analysis reveals that brassinolides (BRs) showed significantly higher concentration in LS filaments compared to SS or St filaments (Figure 4), suggesting a potential correlation of BR content with filament length differentiation in *C. fistula*.

BRs play a pivotal role in regulating cell elongation and organ growth in plants. In *Arabidopsis*, BR-deficient mutants (e.g., *bri1*, *det2-1*) show shortened hypocotyl due to reduced cell size. Additionally, application of BRs to the mutants significantly restored organ elongation by inducing *CESA* genes for cellulose synthesis [43]. In distylous species, the formation of short-styled morphs has been demonstrated to be associated with S-locus genes that degrade BRs in many lineages (e.g., Primula, Gelsemium, Linum, and *Nymphoides*) [26,29,44,45]. Wang et al. (2023) [46] revealed that BRs promoted stamen filament elongation by activating BZR1-family transcription factors (BFTFs), which directly upregulated katanin-encoding genes (e.g., *KTN1, KTN80s*). Mutants that lack functional katanin (e.g., *lue1*, *ktn80.1234*) or BFTFs (e.g., *qui-2*) exhibit stunted filament growth due to impaired microtubule dynamics.

Although BRs have important functions in regulating organ elongation across plant species, their potential involvement in the development of heteranthery remains unexplored in the current literature [29,42]. BR concentration was significantly higher in LS filaments than in SS or St filaments of *C. fistula*. The endogenous BR content often exhibits considerable variation across different plant species and tissue types. Although the BR concentration detected in the filaments of *C. fistula* was relatively low, it falls within the range (0.01–0.1 ng/g, FW) reported in previous studies [47,48], potentially implying tissue-specific regulatory mechanisms. Our transcriptomic data revealed that several phytohormone-related KEGG pathways, such as "Plant hormone signal transduction" (ko04075) and "Sesquiterpenoid and triterpenoid biosynthesis" (ko00909), were significantly enriched in the comparison of LS and St filaments in C. fistula (Figures 5, S4 and S5). GO category analysis further demonstrated that "brassinosteroid metabolic process" (GO: 0016131) and "brassinosteroid biosynthetic process" (GO: 0016132) were both overrepresented in the PPI network (Figure S6). *CYP90D1* (*Cf_f49903*) and *CYP90C1* (*Cf_f56973*) were included in these pathways and GO terms. *Cf_f49903* was expressed significantly higher in LS filaments compared to St (log2FC = 8.25, FDR < 0.05) or SS (log2FC = 2.96, FDR < 0.05) filaments (Figure 6), i.e., showed an expression ratio of LS:SS:St = 378:51:1. *CYP90D1* plays a critical role in the BR biosynthetic pathway and growth regulation. As a cytochrome P450 enzyme, it is essential for maintaining BR homeostasis, as demonstrated by severe little person phenotypes in *CYP90D1* mutants due to BR deficiency [49,50]. *CYP90C1* acts downstream of early BR biosynthetic enzymes such as *CYP90D1*, ensuring the progression of BR synthesis from sterol intermediates to fully active brassinolide. Both CYP90C1 and CYP90D1 catalyze C-23 hydroxylation of 22-hydroxylated BR intermediates, indicating their roles as functionally redundant C-23 hydroxylases [51,52]. Genetic evidence has been found in *Arabidopsis* that single mutants (*rot3*/*cyp90c1* or *cyp90d1*) showed mild phenotypes (e.g., reduced leaf elongation), while the *cyp90c1 cyp90d1* double mutant exhibited severe dwarfism and male sterility due to rudimentary filaments [50]. The homolog of *CYP90C1* in *C. fistula*, Cf_f56973, shows much higher expression in LS filaments than St filaments (log2FC = 4.82, FDR < 0.05), while only slightly higher than SS filaments (log2FC = 1.00, FDR = 0.9) (Figure 6). This expression pattern suggests *CYP90D1* takes the primary role in regulating filament length, and the low expression of *CYP90C1* might have contributed to the degeneration of St.

4. Materials and Methods

4.1. Plant Material and Floral Biology Study

Cassia fistula, commonly known as the golden shower tree, is native to Southeast Asia and widely distributed in the tropical and subtropical regions of Asia. It is the national flower of Thailand, with great horticultural and economic values. The wood is dense and durable, making it suitable for use as posts or bridges. Additionally, the fruits (pods), bark, and leaves are utilized for medicinal purposes in certain regions.

Samples were collected from mature *C. fistula* trees in Guangdong province. Normally developed and fully opened flowers were randomly selected, and floral characteristics such as flower symmetry, style and stigma morphology, and nectar secretion were examined and recorded. Specifically, attentions were paid to the differentiation pattern of stamens, including the structure, length, and color variations among stamen groups.

4.2. Stamen Morphogenesis and Development

Flower buds of different development stages were collected. Some fresh flower buds were dissected and photographed under a stereoscope (SMZ25, Nikon, Tokyo, Japan), and others were fixed in 4% glutaraldehyde solution, vacuumed, and then stored at 4 °C.

Samples for SEM studies were prepared following Luo et al. (2009) [7] and observed with the scanning electron microscope (JSM-6360LV; JEOL, Tokyo, Japan). Images were digitally recorded, and morphological data were analyzed with the aid of software Image Pro Plus (v. 6.0).

4.3. Phytohormone Measurements and Analysis

To examine phytohormones potentially involved in filament elongation, fresh filaments from the long, short, and degenerated stamens were collected from three trees and frozen immediately in liquid nitrogen. Mixed-stage samples were weighed and freeze-dried [26]. Extraction, purification, and quantification of BR and other important hormones were performed by Biomarker Technologies Corporation (Beijing, China) according to the company's protocol, employing the UPLC-MS (Ultra Performance Liquid Chromatography-Mass Spectrometry) system (Thermo, Waltham, MA, USA).

4.4. RNA Extraction, Library Construction, and Sequencing

Comparative transcriptomic profiling was conducted to examine the genes and metabolic pathways potentially involved in the differentiation of filaments, anthers, and pollen grains. As the elongation of filaments is a continuous process, filaments from small flower buds (ca. 6 mm in diameter) to fully opened flowers were collected and mixed for each stamen type. There are nine filament samples in all.

Total RNA was extracted with the HiPure Universal RNA Mini Kit (Magen, Guangzhou, China) following the manufacturer's instructions, then stored at −80 °C until use. RNA quality was assessed by the Agilent Bioanalyzer 2100 system (Agilent Technologies, Santa Clara, CA, USA) before sequencing. RNA samples with RIN (RNA Integrity Number) > 7.0 and OD 260/280 > 2.0 were used for PacBio library preparation, Illumina library construction, and sequencing.

RNA libraries were prepared by the NEBNext® Ultra™ RNA Library Prep Kit for Illumina® (NEB, Ipswich, MA, USA) according to the manufacturer's instructions. Sequencing was performed on an Illumina Hiseq 2000 platform, and paired-end reads (raw reads) were generated with a length of 150 bp. Raw reads with adaptors or undetermined bases (poly-N) were removed, and reads with low quality (containing more than 50% bases with a Q-value $\leq$ 20) were also discarded.

4.5. Gene Functional Annotation

Unigene sequences were aligned by blastx to protein databases NR (http://www.ncbi.nlm.nih.gov/), Swiss-Prot (http://www.expasy.ch/sprot), KEGG (http://www.genome.ad.jp/kegg/), and COG (http://www.ncbi.nlm.nih.gov/COG/) and aligned by blastn to nucleotide databases NT with an e-value < 1.0×10^{-5}. GO (Gene Ontology) annotation was performed with AmiGO [53] in three ontology categories: Biological Process (BP), Cellular

Components (CC), and Molecular Function (MF). KEGG annotation was performed using the KEGG Automatic Annotation Server (KAAS) [54], providing additional functional information showing the pathways in which the transcript isoforms are involved.

4.6. Gene Expression and Enrichment Analysis

To identify potential transcripts involved in stamen differentiation, clean reads obtained from Illumina sequencing were mapped to the consensus isoforms using Bowtie2 [55]. The RSEM program was employed to estimate the expression abundances of unigenes [56]. FPKM (the number of fragments per kilobase of exon per million mapped fragments) was used to represent the relative expression levels of each transcript. Differential expression analysis was conducted in R with the DESeq2 package (ver. 1.26.0), and the statistical test results (p-values) were corrected for multiple testing with the Benjamini–Hochberg false discovery rate (FDR). Sequences were assigned as DEGs if the FDR ≤ 0.05 and FC (fold change) ≥ 2 (log2FC ≥ 1 or log2FC ≤ -1) between two transcriptomes. qRT-PCR (Real-Time quantitative PCR) was employed to verify the expression of certain important DGEs following standard procedures. Relative abundance of transcripts was determined with the $2^{-\Delta\Delta Ct}$ method, using the actin gene as the reference.

GO and KEGG enrichment analyses of DEGs for different stamen types were conducted using the GSEA (Gene Set Enrichment Analysis) pipeline (v. 4.2.3) [57], a second-generation pathway analysis tool providing more accurate and less biased results by utilizing the entire expression data sets [58], implemented on the Omicshare Bioinformatics Cloud Platform (GENEDENOVO, Guangzhou, China).

4.7. Protein–Protein Interaction Network Analysis

Protein–protein interactions (PPIs) are fundamental to most cellular biological processes. Recent studies increasingly integrate gene expression profiles with PPI networks to uncover functional insights into candidate genes. DEGs were analyzed using STRING (https://string-db.org/), a comprehensive database cataloging experimentally validated and predicted PPIs across >5000 organisms. Protein sequences of DEGs were aligned against STRING to construct PPI networks, followed by the removal of isolated nodes and self-interactions.

The resulting networks were exported to Cytoscape (v3.9.1) for topological visualization and further analysis. Subcellular localization annotations for network proteins were retrieved from UniProt (https://www.uniprot.org/) and integrated into Cytoscape. Functional enrichment analysis of Gene Ontology (GO) terms was performed using the BiNGO plugin [59], with statistical significance evaluated via the Benjamini–Hochberg false discovery rate (FDR) correction.

5. Conclusions

This study provides an integrated analysis of both developmental processes and molecular mechanisms underlying stamen differentiation in *Cassia fistula*. Our findings reveal that asynchronous development serves as the primary driver for the morphological divergence among the three stamen sets. The initiation of LS preceded both the SS and St. The filaments of LS continued elongation in the following stages until the flowers fully opened, while the development of St and SS filaments was markedly retarded. LS filaments exhibited an overall faster elongation process compared to the SS and St filaments. The cellular basis of filament elongation is an important direction for future studies. Transcriptomic profiling highlighted the potential role of brassinosteroid (BR) biosynthesis and metabolism, with two key genes, *CYP90D1* (Cf_f49903) and *CYP90C1* (Cf_f56973), emerging as candidate regulators related to stamen length differentiation in

C. fistula. These results help shed light on the developmental and genetic framework of heteranthery in *C. fistula*, and suggest a methodology for exploring floral organ evolution in leguminous plants.

Supplementary Materials: The following supporting information can be downloaded at: https://www.mdpi.com/article/10.3390/plants14223490/s1, Figure S1: (A) Carpenter bee *Xycolopa phalothorax* is visiting the flower of *Cassia fistula*. (B) Filaments of SS and St have been bitten by *Xycolopa*, showing the dark necrotic marks. (C) A carpenter bee is buzzing the anthers of SS, solid arrow showing the bee grasping the anthers of short stamens with its appendages, and dashed arrow showing the bee gripping the filaments of SS and St with its mandibles. Circle shows anthers of LS and the stigma touching the dorsal surface of bee's abdomen; Figure S2: Number of unigenes in the top 15 Annotated KEGG pathways for the filament of *Cassia fistula*; Figure S3: Top ten annotated GO terms for filaments in the Biological Processes (BP), Cellular Components (CC) and Molecular Function (MF) categories, respectively; Figure S4: Circular plots showing enriched KEGG pathways in the comparisons of LS vs. St filaments (A), LS vs. SS filaments (B) and SS vs. St filaments (C); Figure S5. PPI network constructed from selected unigenes of *Cassis fistula*. *CYP90D1* (Cf_f49903), *CYP90C1* (Cf_f56973) (dark pink) and directly related unigenes (light pink) were highlighted; Figure S6: Biological network illustrating phytohormone and floral development related GO terms. Colored circles showing over-represented GO categories (FDR < 0.05) in filament comparisons. "Brassinosteroid biosynthetic process" (GO: 0016132) (dark pink), "brassinosteroid metabolic process" (GO: 0016131) (light pink) are highlighted.

Author Contributions: Z.L., T.D. and X.L. carried out the experiments and data analysis. J.Z. and Q.L. assisted in data processing. Z.L. drafted the manuscript. Z.L., T.D. and L.J. conceived the study and participated in its design. Z.L. and L.J. critically revised the manuscript. All authors have read and agreed to the published version of the manuscript.

Funding: This work was supported by the National Natural Science Foundation of China (31970252), and the Initial Funding for Doctoral Research from Shandong University of Technology (Grant No. 4041/422047).

Data Availability Statement: Data are contained within the article and Supplementary Materials.

Acknowledgments: We are grateful to Hongdan Ren, Jia Wang, and Lina Mei for material collection and lab assistance.

Conflicts of Interest: The authors declare no conflicts of interest.

References

1. Barrett, S.C.H. Mating strategies in flowering plants: The outcrossing-selfing paradigm and beyond. *Philos. Trans. R. Soc. B Biol. Sci.* **2003**, *358*, 991–1004. [CrossRef]
2. Barrett, S.C.H. Sexual interference of the floral kind. *Heredity* **2002**, *88*, 154–159. [CrossRef]
3. Stanton, M.L.; Snow, A.A.; Handel, S.N. Floral evolution-attractiveness to pollinators increases male fitness. *Science* **1986**, *232*, 1625–1627. [CrossRef]
4. Willson, M.F. Sexual selection in plants. *Am. Nat.* **1979**, *113*, 777–790. [CrossRef]
5. Vallejo-Marín, M.; Da Silva, E.M.; Sargent, R.D.; Barrett, S.C.H. Trait correlates and function of heteranthery. *New Phytol.* **2010**, *188*, 418–425. [CrossRef] [PubMed]
6. Barrett, S.C.H. Heteranthery. *Curr. Biol.* **2021**, *31*, R774–R776. [CrossRef] [PubMed]
7. Luo, Z.L.; Gu, L.; Zhang, D.X. Intrafloral differentiation of stamens in heterantherous flowers. *J. Syst. Evol.* **2009**, *47*, 43–56. [CrossRef]
8. Luo, Z.L.; Zhang, D.X.; Renner, S.S. Why two kinds of stamens in buzz-pollinated flowers? Experimental support for Darwin's division-of-labour hypothesis. *Funct. Ecol.* **2008**, *22*, 794–800. [CrossRef]
9. Vallejo-Marín, M.; Manson, M.; Thomson, J.D.; Barrett, S.C.H. Division of labour within flowers: Heteranthery, a floral strategy to reconcile contrasting pollen fates. *J. Evol. Biol.* **2009**, *22*, 828–839. [CrossRef]
10. Todd, J.E. On the flowers of *Solanum rostratum* and *Cassia chamaecrista*. *Am. Nat.* **1882**, *16*, 281–287. [CrossRef]

11. Hooker, J.D. Letter to Darwin C.R. 1 January 1862. Available online: https://www.darwinproject.ac.uk/DCP-LETT-3373 (accessed on 7 November 2024).
12. Darwin, C.R. Letter to Asa Gray. 22 January 1862. Available online: https://www.darwinproject.ac.uk/DCP-LETT-3404 (accessed on 7 November 2024).
13. Darwin, C.R. *The Various Contrivances by Which Orchids Are Fertilised by Insects*; John Murray: London, UK, 1899.
14. Müller, H. Two kinds of stamens with different functions in the same flower. *Nature* **1882**, *24*, 307–308. [CrossRef]
15. Lee, R.E. Pollen dimorphism in *Tripogandra grandiflora*. *Baileya* **1961**, *9*, 53–56.
16. Dulberger, R. The floral biology of *Cassia didymobotrya* and *C. auriculata* (Caesalpiniaceae). *Am. J. Bot.* **1981**, *68*, 1350–1360. [CrossRef]
17. Luo, Z.L.; Chen, S.; Zhang, D.X. Floral reward presentation favored the expression of male function in the pollen-only flower *Melastoma malabathricum*. *J. Syst. Evol.* **2012**, *50*, 488–495. [CrossRef]
18. Mesquita-Neto, J.N.; Costa, B.K.P.; Schlindwein, C. Heteranthery as a solution to the demand for pollen as food and for pollination-Legitimate flower visitors reject flowers without feeding anthers. *Plant Biol.* **2017**, *19*, 942–950. [CrossRef]
19. Dellinger, A.S.; Artuso, S.; Pamperl, S.; Michelangeli, F.A.; Penneys, D.S.; Fernández-Fernández, D.M.; Alvear, M.; Almeda, F.; Armbruster, W.S.; Staedler, Y.; et al. Modularity increases rate of floral evolution and adaptive success for functionally specialized pollination systems. *Commun. Biol.* **2021**, *4*, 453.
20. Tucker, S.C. Trends in evolution of floral ontogeny in *Cassia* sensu stricto, *Senna*, and *Chamaecrista* (Leguminosae: Caesalpinioideae: Cassieae: Cassiinae); A study in convergence. *Am. J. Bot.* **1996**, *83*, 687–711. [CrossRef]
21. Wang, J. A Comparative Study on Pollen Heteromorphism in Heterantherous Plants. Master's Dissertation, Chinese Academy of Sciences, Guangzhou, China, 2021.
22. Luo, Z.L.; Barrett, S.C.H.; Tu, T.Y.; Zhao, Z.T.; Jia, S.S.; Gu, S.R.; Duan, T.T.; Zhang, Y.; Xu, B.Q.; Gu, L.; et al. Genetic architecture of the *S*-locus supergene revealed in a tetraploid distylous species. *New Phytol.* **2025**, *248*, 1973–1988. [CrossRef]
23. Yuan, S.; Barrett, S.C.H.; Tang, C.Q.; Zhang, Y.M.; Sun, Q.L.; Zhao, Z.T.; Zhang, Y.; Zhang, D.X.; Luo, S.X. Genomic evidence unveils the genetic architecture and evolution of the *S*-locus controlling heterostyly in Rubiaceae. *New Phytol.* **2025**, *247*, 1925–1941. [CrossRef] [PubMed]
24. Yang, J.S.; Xue, H.R.; Li, Z.Z.; Zhang, Y.; Shi, T.; He, X.Y.; Barrett, S.C.H.; Wang, Q.F.; Chen, J.M. Haplotype-resolved genome assembly provides insights into the evolution of S-locus supergene in distylous *Nymphoides indica*. *New Phytol.* **2023**, *240*, 2058–2071. [CrossRef]
25. Shore, J.S.; Hamam, H.J.; Chafe, P.D.J.; Labonne, J.D.J.; Henning, P.M.; McCubbin, A.G. The long and short of the S-locus in *Turnera* (Passifloraceae). *New Phytol.* **2019**, *224*, 1316–1329. [CrossRef]
26. Huu, C.N.; Kappel, C.; Keller, B.; Sicard, A.; Takebayashi, Y.; Breuninger, H.; Nowak, M.D.; Bäurle, I.; Himmelbach, A.; Burkart, M.; et al. Presence versus absence of *CYP734A50* underlies the style-length dimorphism in primroses. *eLife* **2016**, *5*, e17956. [CrossRef] [PubMed]
27. Boualem, A.; Troadec, C.; Kovalski, I.; Sari, M.A.; Perl-Treves, R.; Bendahmane, A. A conserved mutation in an ethylene biosynthesis enzyme leads to andromonoecy in melons. *Science* **2016**, *353*, 398–400. [CrossRef] [PubMed]
28. Hileman, L.C. Trends in flower symmetry evolution revealed through phylogenetic and developmental genetic advances. *Philos. Trans. R. Soc. Lond. B Biol. Sci.* **2014**, *369*, 20130348. [CrossRef]
29. Yang, J.S.; Chen, J.M.; He, X.Y.; Wang, G.X.; Barrett, S.C.H.; Li, Z.Z. The *Monochoria* genome provides insights into the molecular mechanisms underlying floral heteranthery. *J. Genet. Genomics.* **2025**, *52*, 826–838. [CrossRef] [PubMed]
30. Müller, F. Two kinds of stamens with different functions in the same flower. *Nature* **1883**, *27*, 364–365. [CrossRef]
31. Meehan, T. On the Fertilization of *Cassia marilandica*. *Proc. Acad. Nat. Sci. Phila.* **1886**, *38*, 314–318.
32. Tucker, S.C. Floral development in Legumes. *Plant Physio.* **2003**, *131*, 911–926. [CrossRef]
33. Buchmann, S.L. Buzz pollination of *Cassia quiedondilla* (Leguminosae) by bees of the genera *Centris* and *Melipona*. *Bull. South. Calif. Acad. Sci.* **1974**, *73*, 171–173.
34. Marazzi, B.; Conti, E.; Endress, P.K. Diversity in anthers and stigmas in the buzz-pollinated genus *Senna* (Leguminosae, Cassiinae). *Int. J. Plant Sci.* **2007**, *168*, 371–391. [CrossRef]
35. Marazzi, B.; Endress, P.K. Patterns and development of floral asymmetry in *Senna* (Leguminosae, Cassiinae). *Am. J. Bot.* **2008**, *95*, 22–40. [CrossRef]
36. Marazzi, B.; Endress, P.K.; Queiroz, L.P.; Conti, E. Phylogenetic relationships within *Senna* (Leguminosae, Cassiinae) based on three chloroplast DNA regions: Patterns in the evolution of floral symmetry and extrafloral nectaries. *Am. J. Bot.* **2006**, *93*, 288–303. [CrossRef]
37. Ren, H.D. Intrafloral Stamen Differentiation and Its Ddaptive Significance in *Cassia fistula* L. Master's Dissertation, Chinese Academy of Sciences, Guangzhou, China, 2017.
38. Cantalapiedra, C.P.; Hernández-Plaza, A.; Letunic, I.; Bork, P.; Huerta-Cepas, J. eggNOG-mapper v2: Functional annotation, orthology assignments, and domain prediction at the metagenomic scale. *Mol. Biol. Evol.* **2021**, *38*, 5825–5829. [CrossRef]

39. Nepi, M.; Guarnieri, M.; Pacini, E. "Real" and feed pollen of *Lagerstroemia indica*: Ecophysiological differences. *Plant Biol.* **2003**, *5*, 311–314. [CrossRef]
40. Endress, P.K. Relationships between floral organization, architecture, and pollination mode in *Dillenia* (Dilleniaceae). *Plant Syst. Evol.* **1997**, *206*, 99–118. [CrossRef]
41. Singh, V.; Sharma, S. Floral organogenesis in *Cassia fistula* L. (Caesalpiniaceae). *Proc. Indian Acad. Sci.* **1978**, *87B*, 215–221. [CrossRef]
42. Luo, Z.L.; Hu, J.; Zhao, Z.T.; Zhang, D.X. Transcriptomic analysis of heteromorphic stamens in *Cassia biscapsularis* L. *Sci. Rep.* **2016**, *6*, 31600.
43. Xie, L.; Yang, C.; Wang, X. Brassinosteroids can regulate cellulose biosynthesis by controlling the expression of CESA genes in *Arabidopsis*. *J. Exp. Bot.* **2011**, *62*, 4495–4506. [CrossRef]
44. Gutiérrez-Valencia, J.; Fracassetti, M.; Berdan, E.L.; Bunikis, I.; Soler, L.; Dainat, J.; Kutschera, V.E.; Losvik, A.; Désamoré, A.; Hughes, P.W.; et al. Genomic analyses of the *Linum* distyly supergene reveal convergent evolution at the molecular level. *Curr. Biol.* **2022**, *32*, 4360–4371. [CrossRef]
45. Zhao, Z.T.; Zhang, Y.; Shi, M.M.; Liu, Z.Y.; Xu, Y.Q.; Luo, Z.L.; Yuan, S.; Tu, T.Y.; Sun, Z.L.; Zhang, D.X.; et al. Genomic evidence supports the genetic convergence of a supergene controlling the distylous floral syndrome. *New Phytol.* **2023**, *237*, 601–614. [CrossRef]
46. Wang, J.; Wang, G.; Liu, W.; Yang, H.; Wang, C.; Chen, W.; Zhang, X.; Tian, J.; Yu, Y.; Li, J.; et al. Brassinosteroid signals cooperate with katanin-mediated microtubule severing to control stamen filament elongation. *EMBO J.* **2023**, *42*, e111883. [CrossRef]
47. Manghwar, H.; Hussain, A.; Ali, Q.; Liu, F. Brassinosteroids (BRs) role in plant development and coping with different stresses. *Int. J. Mol. Sci.* **2022**, *23*, 1012. [CrossRef]
48. Kanwar, M.K.; Bajguz, A.; Zhou, J.; Bhardwaj, R. Analysis of Brassinosteroids in Plants. *J. Plant Growth Regul.* **2017**, *36*, 1002–1030. [CrossRef]
49. Nolan, T.M.; Vukašinović, N.; Liu, D.; Russinova, E.; Yin, Y. Brassinosteroids: Multidimensional regulators of plant growth, development, and stress responses. *Plant Cell* **2020**, *32*, 295–318. [CrossRef]
50. Ohnishi, T.; Szatmari, A.M.; Watanabe, B.; Fujita, S.; Bancos, S.; Koncz, C.; Lafos, M.; Shibata, K.; Yokota, T.; Sakata, K.; et al. C-23 hydroxylation by *Arabidopsis CYP90C1* and *CYP90D1* reveals a novel shortcut in brassinosteroid biosynthesis. *Plant Cell* **2006**, *18*, 3275–3288. [CrossRef]
51. Zhan, H.; Lu, M.; Luo, Q.; Tan, F.; Zhao, Z.; Liu, M.; He, Y. *OsCPD1* and *OsCPD2* are functional brassinosteroid biosynthesis genes in rice. *Plant Sci.* **2022**, *325*, 111482. [CrossRef] [PubMed]
52. Sakamoto, T.; Ohnishi, T.; Fujioka, S.; Watanabe, B.; Mizutani, M. Rice CYP90D2 and CYP90D3 catalyze C-23 hydroxylation of brassinosteroids in vitro. *Plant Physiol. Biochem.* **2012**, *58*, 220–226. [CrossRef]
53. Carbon, S.; Ireland, A.; Mungall, C.J.; Shu, S.; Marshall, B.; Lewis, S.; The AmiGO Hub; Web Presence Working Group. AmiGO: Online access to ontology and annotation data. *Bioinformatics* **2009**, *25*, 288–289. [CrossRef]
54. Moriya, Y.; Itoh, M.; Okuda, S.; Yoshizawa, A.C.; Kanehisa, M. KAAS: An automatic genome annotation and pathway reconstruction server. *Nucleic Acids Res.* **2007**, *35*, W182–W185. [CrossRef] [PubMed]
55. Langmead, B.; Salzberg, S.L. Fast gapped-read alignment with bowtie 2. *Nat. Methods* **2012**, *9*, 357–359. [CrossRef] [PubMed]
56. Li, B.; Dewey, C.N. RSEM: Accurate transcript quantification from RNA-Seq data with or without a reference genome. *BMC Bioinform.* **2011**, *12*, 323. [CrossRef] [PubMed]
57. Subramanian, A.; Tamayo, P.; Mootha, V.K.; Mukherjee, S.; Ebert, B.L.; Gillette, M.A.; Paulovich, A.; Pomeroy, S.L.; Golub, T.R.; Lander, E.S.; et al. Gene set enrichment analysis: A knowledge-based approach for interpreting genome-wide expression profiles. *Proc. Natl. Acad. Sci. USA* **2005**, *102*, 15545–15550. [CrossRef]
58. Khatri, P.; Sirota, M.; Butte, A.J. Ten years of pathway analysis: Current approaches and outstanding challenges. *PLoS Comput. Biol.* **2012**, *8*, e1002375. [CrossRef] [PubMed]
59. Maere, S.; Heymans, K.; Kuiper, M. BiNGO: A Cytoscape plugin to assess overrepresentation of gene ontology categories in biological networks. *Bioinformatics* **2005**, *21*, 3448–3449. [CrossRef] [PubMed]

plants

MDPI

Article

Comprehensive Genome-Wide Investigation and Transcriptional Regulation of the bZIP Gene Family in Litchi Fruit Development

Jiaxuan Liu [1,2,3,†], Saimire Silaiyiman [1,2,3,†], Jiaxin Wu [1,2,3], Lejun Ouyang [1,2], Zheng Cao [4] and Chao Shen [1,*]

1 Guangdong Provincial Key Laboratory for Green Agricultural Production and Intelligent Equipment, College of Biological and Food Engineering, Guangdong University of Petrochemical Technology, Maoming 525000, China; 18375432965@163.com (J.L.); 19990885586@163.com (S.S.); 15077668902@163.com (J.W.); ouyanglejun@gdupt.edu.cn (L.O.)

2 College of Life and Geographic Sciences, Kashi University, Kashi 844000, China

3 Key Laboratory of Biological Resources and Ecology of Pamirs Plateau in Xinjiang Uygur Autonomous Region, Kashi 844000, China

4 Maoming Agricultural Science and Technology Extension Center, Maoming 525000, China; caozheng5192@163.com

* Correspondence: cshen@gdupt.edu.cn

† These authors contributed equally to this work.

Abstract

Litchi chinensis, a crucial tropical and subtropical fruit tree in southern China, is widely appreciated for its distinctive flavor, high nutritional value, and significant economic impact. The bZIP (basic leucine zipper) gene family plays an essential role in regulating key biological functions during plant growth and development. In this study, we performed a comprehensive bioinformatics analysis of the bZIP gene family in litchi to systematically elucidate its molecular characteristics and functional properties. A total of 55 bZIP gene family members were identified, with the encoded proteins containing between 129 and 845 amino acid residues and theoretical isoelectric points (pI) ranging from 4.85 to 10.23. Protein–protein interaction network analysis revealed that 46 proteins exhibited interaction relationships. Phylogenetic analysis classified these genes into 13 distinct subgroups (A–K, M, and S). Chromosomal localization analysis indicated that bZIP gene family members were successfully mapped to 15 chromosomes. Intraspecific collinearity analysis identified 39 segmental duplication events, while interspecific and single-gene collinearity analyses suggested evolutionary conservation, with only a few genes exhibiting duplication or loss events. Cis-acting element analysis revealed a total of 213 elements associated with growth and development, which may play an important role in fruit development regulation. The results of differential gene expression, related to fruit development across different *litchi* cultivars, tissues, and flowering stages, combined with qRT-PCR validation, suggest that *LITCHI017015.m1* and *LITCHI004463.m1* may be involved in the early regulation of fruit development, while *LITCHI018843.m1* may play a regulatory role during the later stages of fruit development. These findings provide a strong theoretical foundation for understanding the roles of bZIP genes in litchi fruit growth and development, and lay the groundwork for further functional studies. This study has potential application value in litchi fruit development and genetic improvement.

Keywords: litchi chinensis; bZIP family; transcriptional regulation; fruit development

1. Introduction

Transcription factors (TFs) are key regulatory proteins that control gene expression, typically by binding to cis-acting elements in the promoter regions of target genes to either activate or repress their transcription. Basic leucine zipper (bZIP) transcription factors are widely distributed in eukaryotes and represent one of the largest TF families in plants [1]. Structurally, bZIP proteins consist of two functional domains: a basic region at the N-terminus composed of 16–20 basic amino acids responsible for recognizing and binding to the ACGT core sequence [2,3], and a leucine zipper region at the C-terminus composed of heptad repeat units that facilitate dimer formation. These dimers exhibit high specificity, with members of the same subfamily tending to form homodimers, while those from different subfamilies may form heterodimers. The formation and function of these dimers can vary under different environmental conditions and are also influenced by post-translational modifications such as phosphorylation [4]. Additionally, some plant bZIP proteins contain transcriptional activation domains enriched in proline, glutamine, or acidic amino acids [5]. The "Y"-shaped structure of the bZIP dimers enables precise binding to cis-elements in the promoters of target genes, thereby regulating gene expression and participating in various biological processes [6]. bZIP transcription factors have been identified in numerous plant species, including *Arabidopsis thaliana* [7], rice (*Oryza sativa*) [8], sorghum (*Sorghum bicolor*) [9], melon (*Cucumis melo*) [10], poplar (*Populus tremula*) [11], and rapeseed (*Brassica napus*) [12], providing a theoretical foundation for understanding their functions and the evolutionary processes of plants.

Studies have found that bZIP transcription factors play an important role in plant growth and development, regulating responses to abiotic stresses such as drought, high salinity, and low temperature in plants like *Arabidopsis thaliana* [13], rice (*Oryza sativa*) [14], wheat (*Triticum aestivum*) [15], tomato (*Solanum lycopersicum*) [16], pepper (*Capsicum annuum*) [17], soybean (*Glycine max*) [18], and maize (*Zea mays*) [19]. They are also involved in organ development and formation, covering critical stages such as seed maturation and germination, embryo development, flowering, and photomorphogenesis [20]. For example, in wheat, candidate bZIP genes regulate starch biosynthesis [21], while in rice, the OsbZIP76 mutant exhibits a reduced seed size and lower amylose content [22]. Additionally, bZIP transcription factors are involved in plant adaptation to extreme environmental conditions [23], such as AabZIP1, which activates ADS and CYP71AV1 to promote artemisinin synthesis in *Artemisia annua* [24].

In various signal transduction pathways, bZIP proteins regulate responses to ABA, light, osmotic stress, and pathogen infection, which are crucial for plant adaptation to adversity [25,26]. bZIP proteins also play a role in organ differentiation, cell elongation, carbon and nitrogen metabolism, unfolded protein response, and seed gene expression regulation [27]. For instance, AaAPK1 can phosphorylate AabZIP1 to enhance its transcriptional activation function [28], and the overexpression of AabZIP9 promotes artemisinin accumulation [29]. Furthermore, LV et al. found that AaTGA6 mediates salicylic acid signaling through AaERF1, regulating artemisinin synthesis [30], further highlighting the multifunctionality of bZIP factors.

In fruit ripening, bZIP transcription factors also play a central role. In grape (*Vitis vinifera*),, the *VvABF2* is significantly correlated with grape berry ripening, acting as a regulatory factor of ABA-dependent [31]. In peaches, certain bZIP genes may function as "rhythmizers" in the metabolic pathways regulating ripening [32]. Moreover, bZIP genes can influence amino acid and sugar metabolism, improving fruit flavor and quality [33]. For example, S1-bZIPs (such as tbz17 and AtbZIP11 mORF) have been shown to increase sucrose concentration in transgenic plants [34]. In tomatoes, SlbZIP1 and SlbZIP2 also enhance fruit sweetness without negative effects on the plant [35].

Litchi chinensis, a tropical evergreen tree in the Sapindaceae family, is widely distributed in subtropical regions of southern China and is popular for its unique flavor and rich nutritional value [36,37]. Its cultivation history dates back over two thousand years to the Han Dynasty. In addition to its high culinary value, litchi also exhibits antioxidant, anti-inflammatory, and health-promoting properties due to its polyphenolic and flavonoid compounds [38], and it has widespread use in traditional Chinese medicine.

Although bZIP transcription factors have been extensively studied in model plants, research on their roles in *Litchi chinensis* remains relatively scarce. Litchi growth depends on warm and humid climates, and its fruit development is highly sensitive to abiotic stresses such as low temperature, high temperature, and drought, which pose significant challenges to the litchi industry. Therefore, an in-depth study of the structure and function of the bZIP gene family in litchi, and especially its expression patterns during fruit development, holds both theoretical and practical significance. Through systematic identification and expression analysis, this research can provide solid scientific support for improving litchi quality and ensuring the sustainable development of the industry.

2. Results

2.1. Identification, Physicochemical Property Analysis, and Subcellular Localization of Litchi bZIP Transcription Factors

In *Litchi chinensis*, a total of 55 bZIP gene family members containing a leucine zipper domain were identified (Supplementary Table S1). Physicochemical property analysis revealed that the amino acid count of bZIP transcription factors varied from 129 in *LITCHI014383.m1* to 845 in *LITCHI020028.m1*. The relative molecular weight (Mr) had a large range, from 14,967.32 (*LITCHI014383.m1*) to 92,177.38 (*LITCHI020028.m1*). The isoelectric points (pIs) of litchi bZIP transcription factors were found to range from 4.85 (*LITCHI009858.m1*) to 10.23 (*LITCHI018144.m1*), with 29 classified as acidic proteins and 25 as basic proteins. Only *LITCHI020193.m1* had a neutral pI of 7, indicating its electrical neutrality. Additionally, the instability index (II) varied between 31.61 (*LITCHI025169.m1*) and 82.1 (*LITCHI006674.m1*). The aliphatic index ranged from 46.18 (*LITCHI010080.m1*) to 91.45 (*LITCHI023292.m1*), while the grand average of hydropathicity (GRAVY) values were between −1.211 (*LITCHI014217.m1*) and −0.331 (*LITCHI005494.m1*), indicating that all identified proteins exhibited hydrophilic properties (Supplementary Table S1). Furthermore, subcellular localization analysis indicated that all litchi bZIP proteins were localized in the nucleus (Supplementary Table S1).

2.2. Phylogenetic Analysis of the Litchi bZIP Gene Family

A phylogenetic tree was constructed by aligning litchi bZIP protein sequences with previously reported *Arabidopsis thaliana* bZIP proteins. Based on the classification of the *Arabidopsis* bZIP family, the litchi bZIP proteins were divided into 13 subfamilies: A, B, C, D, E, F, G, H, I, J, K, S, and M (Figure 1). Among them, subfamily A contained the highest number of litchi bZIP members (11), while subfamily B had only 1 member (*LITCHI008625.m1*). Subfamily C included 4 members and D had 6. E contained 2 (*LITCHI013041.m1* and *LITCHI020965.m1*). F had 2 (*LITCHI002032.m1* and *LITCHI004463.m1*) and G contained 3 (*LITCHI006674.m1*, *LITCHI020028.m1*, and *LITCHI025816.m1*). H had 2 (*LITCHI014217.m1* and *LITCHI017015.m1*), I included 8, J had 1 (*LITCHI018815.m1*), K had 1 (*LITCHI009858.m1*), S contained 10, and M had 2 (*LITCHI018144.m1* and *LITCHI030316.m1*). This classification lays a foundation for further functional studies on the role of bZIP transcription factors in litchi.

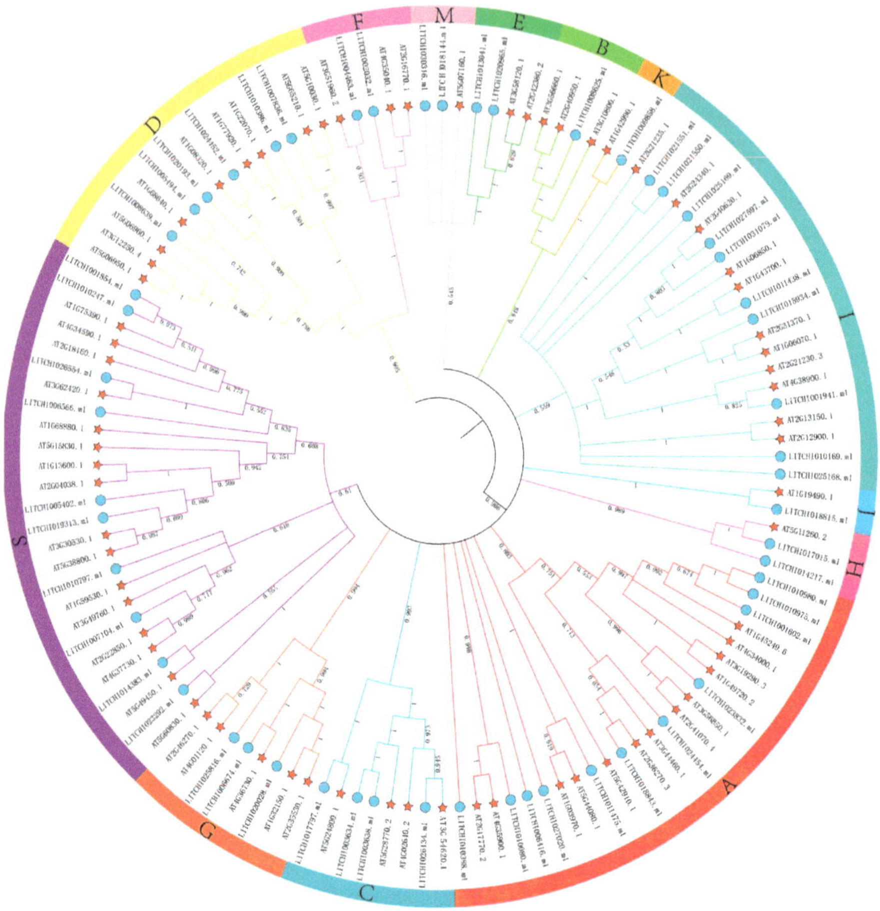

Figure 1. Classification and phylogenetic analysis of the bZIP gene family in *Arabidopsis thaliana* (pentagram) and *litchi* (circle). All bZIP genes were categorized into 13 classes, each class represented by a different color.

2.3. Prediction of Secondary and Tertiary Structures of Litchi bZIP Transcription Factors

To investigate the functions of the 55 bZIP transcription factor proteins in *Litchi chinensis*, we conducted secondary structure predictions. The results revealed that random coils and α-helices dominate the secondary structures of litchi bZIP transcription factors, both of which play essential roles in protein functionality (Supplementary Figure S1). Notably, bZIP members of the same subgroup showed greater structural similarity than members from different subgroups, emphasizing the evolutionary conservation of their structural features (Supplementary Figure S2).

2.4. Conserved Motif and Gene Structure Analysis of the Litchi bZIP Gene Family

Through an in-depth analysis of the evolutionary relationships, gene structure (Figure 2), and conserved domain (Supplementary Figure S3) of the litchi bZIP gene family, we observed that members within the same subfamily exhibit a high degree of similarity in both gene structure and conserved motifs. Motif analysis revealed that most litchi bZIP genes contain motif 1, which appears to be a widely conserved sequence in this gene family and may play a crucial role in gene function and structural stability. Regarding gene structure, our analysis indicated that all litchi bZIP genes consist of both exons and introns, with exon numbers ranging from 1 (*LITCHI026554.m1*) to 18 (*LITCHI020028.m1*).

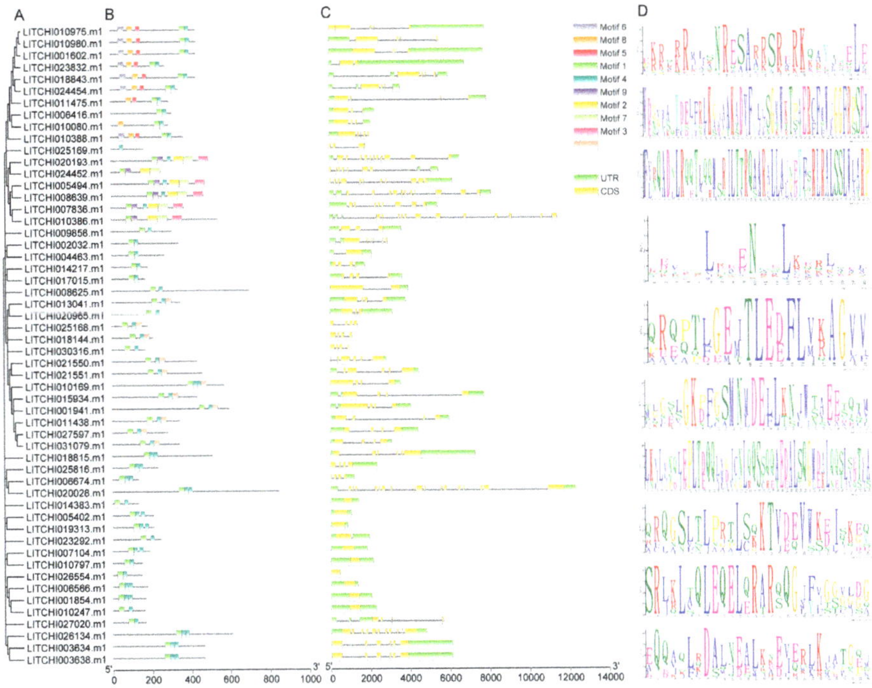

Figure 2. Analysis of conserved motifs and gene structure of bZIP protein in *litchi*. Phylogenetic tree of *litchi* bZIP gene family (**A**), conserved motifs (**B**), gene structure (**C**), and each amino acid sequence (**D**).

2.5. Chromosomal Localization and Interaction Analysis of Litchi bZIP Genes

The bZIP gene family members in *Litchi chinensis* have been successfully mapped to 15 chromosomes, with an uneven distribution pattern (Figure 3). Notably, Chr9 does not contain any bZIP genes, while Chr2 and Chr10 each harbor only one bZIP gene. By contrast, Chr4, Chr6, and Chr7 each contain two bZIP genes. Interestingly, most other chromosomes contain three or more bZIP genes. Among them, Chr8 harbors the highest number of bZIP transcription factors (11 genes), accounting for 20.0% of the total. Furthermore, the spatial distribution of bZIP genes on individual chromosomes varies. For instance, bZIP genes tend to cluster at the terminal regions of Chr5, Chr6, and Chr11, whereas their presence in the central regions is relatively sparse. By contrast, Chr3 exhibits a more evenly distributed pattern.

To investigate the functional relationships among proteins, we employed homology mapping to predict the interaction network of litchi bZIP transcription factors (Figure 4). Our study revealed interactions among 46 bZIP proteins. The number of protein interactions was mapped using a color gradient, with deeper colors indicating a higher number of interactions. Notably, *LITCHI020965.m1*, *LITCHI010247.m1*, *LITCHI014383.m1*, *LITCHI008625.m1*, and *LITCHI026554.m1* exhibited the highest interaction counts and were represented by the deepest colors. Meanwhile, other proteins also displayed interconnections within the network.

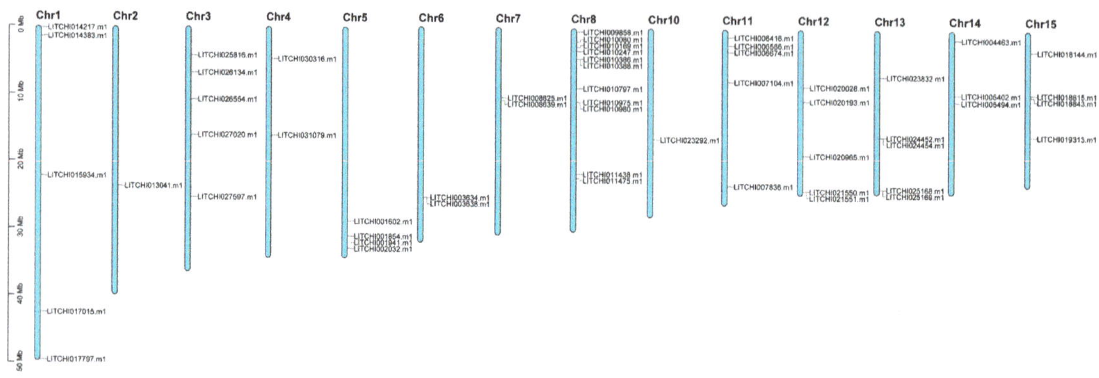

Figure 3. Chromosomal localization of bZIP genes in litchi.

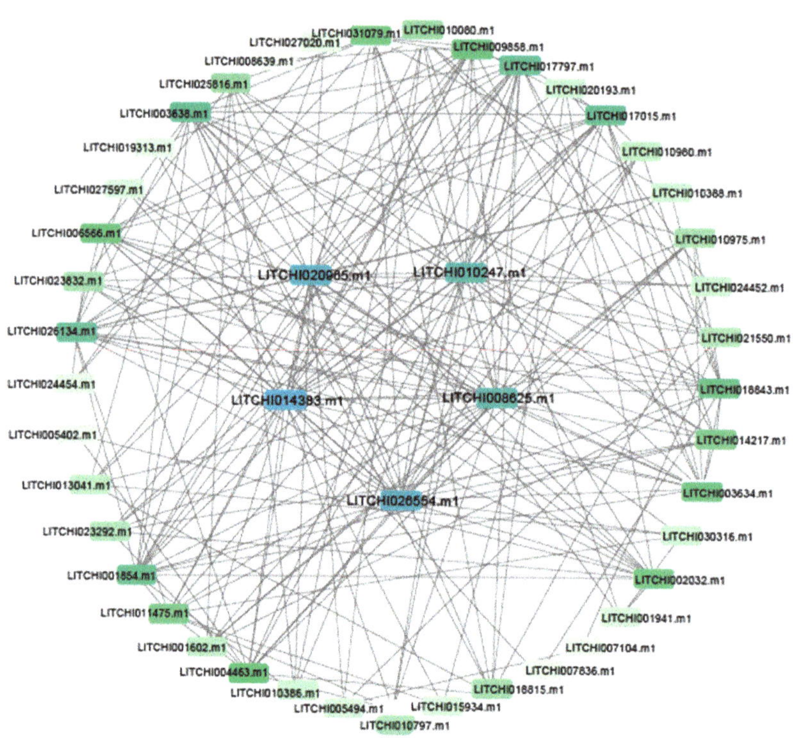

Figure 4. The litchi bZIP protein interaction network. Each box represents a protein. The color of the protein represents the strength of the interaction; the darker the color, the stronger the interaction. The five genes in the center are hub genes.

2.6. *Intra- and Interspecies Synteny Analysis, Gene Duplication, and Divergence of Litchi bZIP Genes*

The gene duplication events in the bZIP gene family revealed a high degree of homology among bZIP genes across different chromosomes. Among the 55 bZIP genes, 39 segmental duplication events were detected, with individual bZIP genes undergoing between 1 and 4 duplication events (Figure 5A). To further explore the evolutionary relationships, a synteny analysis of bZIP genes was conducted across six species within the *Sapindaceae* family, providing insights into the macro-level distribution patterns of bZIP genes in different species (Figure 5B). Focusing on the gene duplication and loss events of individual litchi bZIP genes within *Sapindaceae*, we found that *LITCHI003634.m1* and *LITCHI019313.m1* were duplicated in both litchi and its close relative *Dimocarpus longan* (Figure 6A and Supplementary Figure S4). Meanwhile, *LITCHI007104.m1* underwent duplication during the evolutionary transition from litchi to *Nephelium lappaceum*, but was subsequently lost in later evolutionary stages (Figure 6C and Supplementary Figure S5). Conversely, some bZIP genes, such as *LITCHI014217.m1* and *LITCHI026134.m1*, remained as single-copy genes, with no observed duplication or loss throughout evolution (Figure 6B and Supplementary Figure S6). By contrast, genes such as *LITCHI018815.m1* and *LITCHI005494.m1* were lost in litchi or its closely related species (Figure 6D,E and Supplementary Figure S7). These findings highlight the complex evolutionary history of bZIP genes in *Litchi chinensis* and its relatives.

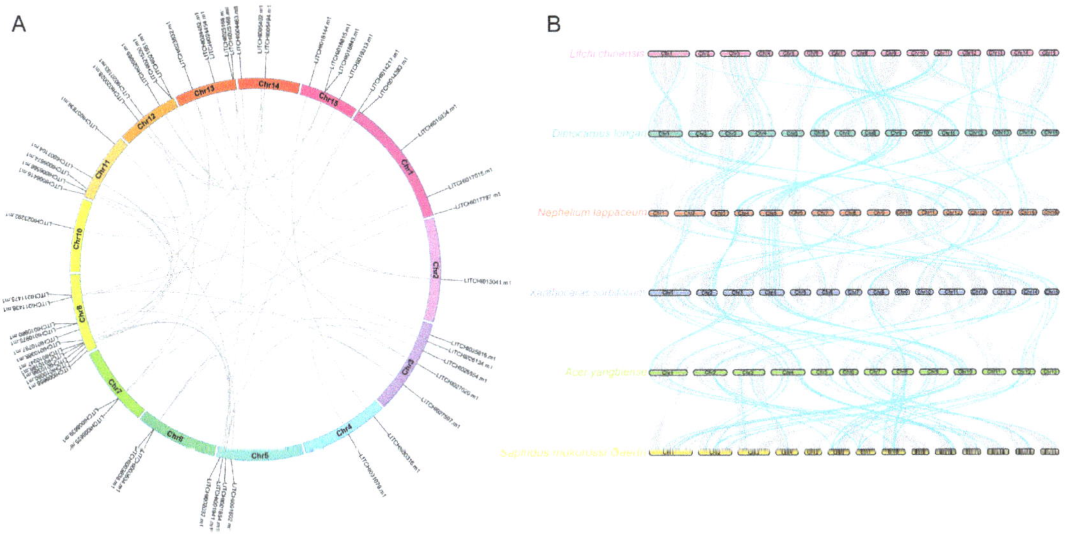

Figure 5. Synteny analysis of litchi bZIP genes within and across species. Collinearity analysis of *Litchi* bZIP genes (**A**) within the *Litchi* genome and (**B**) across species. The blue lines represent collinear bZIP genes.

2.7. *Prediction of Cis-Acting Elements in the Litchi bZIP Gene Family*

In this study, cis-acting elements were extracted from the 2000 bp upstream regions of the litchi bZIP gene coding sequences (Figure 7), and they were classified into four major categories: light-responsive elements, plant growth and developmental regulation elements, plant-hormone-responsive elements, and stress-responsive elements. Among the light-responsive elements, 369 elements were identified, including ACE and G-Box elements. The plant-hormone-responsive elements were the most diverse, with 514 elements detected, such as ARBE (abscisic acid response), AuxRR-core (auxin response), MBSI (flavonoid biosynthesis regulation), TATC-box (gibberellin response), and TCA-element (salicylic acid response). The stress-responsive elements were the most abundant, totaling 723 elements, with key elements including TC-rich repeats (defense and stress response), MBS (drought response), LTR (low-temperature response), ARE (anaerobic response), and WUN-motif (mechanical injury response). In terms of regulatory elements of plant growth and development, 213 elements were identified, including circadian (circadian rhythm regulation), GCN4_motif (endosperm expression), CAT-box (meristem expression), RY-element (seed-specific regulation), and O2-site (zein metabolism regulation). These results suggest that the *bZIP* gene family in litchi plays an essential role in environmental adaptation, hormone signaling, and plant developmental processes.

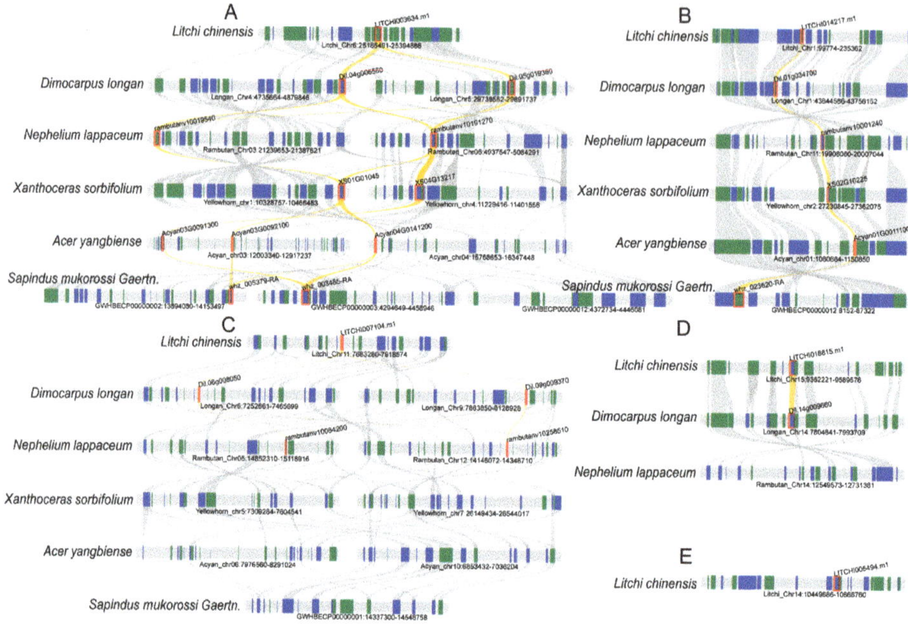

Figure 6. Duplication and loss events of *Sapindaceae* bZIP genes. The red boxes indicate the locations of bZIP genes, while the yellow lines represent gene collinearity. The gene is (**A**) duplicated among species; (**B**) conserved among species and is a single copy; (**C**) duplicated among some species and lost among others; (**D**) conserved among some species, is a single copy, and is lost among some species; and (**E**) unique to litchi.

2.8. Tissue Expression Patterns of Litchi bZIP Genes and Their Differential Expression in Various Fruit Developmental Stages

The tissue-specific expression patterns of bZIP genes in litchi are crucial for understanding their potential functions within the genome. In this study, we analyzed 616 GB of transcriptomic data from various litchi tissues, revealing widespread expression of the bZIP genes across different tissues.

Among them, *LITCHI001854.m1*, *LITCHI026554.m1*, *LITCHI010247.m1*, and *LITCHI001941.m1* exhibited higher expression levels compared to other genes (Figure 8A). Furthermore, during the fruit flesh and peel developmental stages at 55, 62, and 69 days after anthesis in the two cultivars *Nuomici* and *Huaizha*, the genes *LITCHI006654.m1*, *LITCHI028625.m1*, and *LITCHI001854.m1* showed significantly higher expression, while most other bZIP genes displayed relatively low expression levels (Figure 8B). During fruit peel development, *LITCHI006654.m1*, *LITCHI001854.m1*, and *LITCHI011438.m1* exhibited the highest expression among the bZIP genes (Figure 8C). In the study of the "sugar receding" phenomenon in *Feizixiao* litchi pulp, we analyzed the expression patterns of bZIP genes at 35, 63, and 69 days after anthesis. The results showed that *LITCHI006654.m1*, *LITCHI001854.m1*, and *LITCHI011438.m1* consistently exhibited high expression levels across all three time points (Figure 8D). A comprehensive analysis of bZIP gene expression patterns across different cultivars, developmental stages, and tissues revealed that *LITCHI006654.m1* maintained high expression throughout fruit development, followed by *LITCHI001854.m1*, suggesting that these two genes play crucial roles in litchi tissue differentiation and fruit development.

Figure 7. An analysis of the cis-regulatory elements of bZIP gene family promoters in litchi. The scale at the bottom indicates the length of the sequence. Different cis-acting elements are represented by different colors.

To further investigate the differential expression of bZIP genes during litchi fruit development, we analyzed the expression patterns of bZIP genes in the pericarp and aril of two litchi cultivars, *Nuomici* and *Huaizha*, in different post-anthesis stages (Figure 9A, Supplementary Figure S8A, Supplementary Table S2). We found that the genes *LITCHI017015.m1* and *LITCHI004463.m1* exhibited downregulated expression in the pericarp and aril of *Nuomici* and *Huaizha* at 55, 62, and 69 days after flowering (DAF), while *LITCHI006416.m1* also exhibited downregulated expression in the pericarp at 13, 26, and 40 DAF. Conversely, the genes *LITCHI014383.m1*, *LITCHI018843.m1*, *LITCHI013041.m1*, *LITCHI010797.m1*, and *LITCHI021550.m1* exhibited upregulated expression across different developmental stages in both the pericarp and aril of *Nuomici* and *Huaizha* (Figure 9A, Supplementary Figure S8A, Supplementary Table S2).

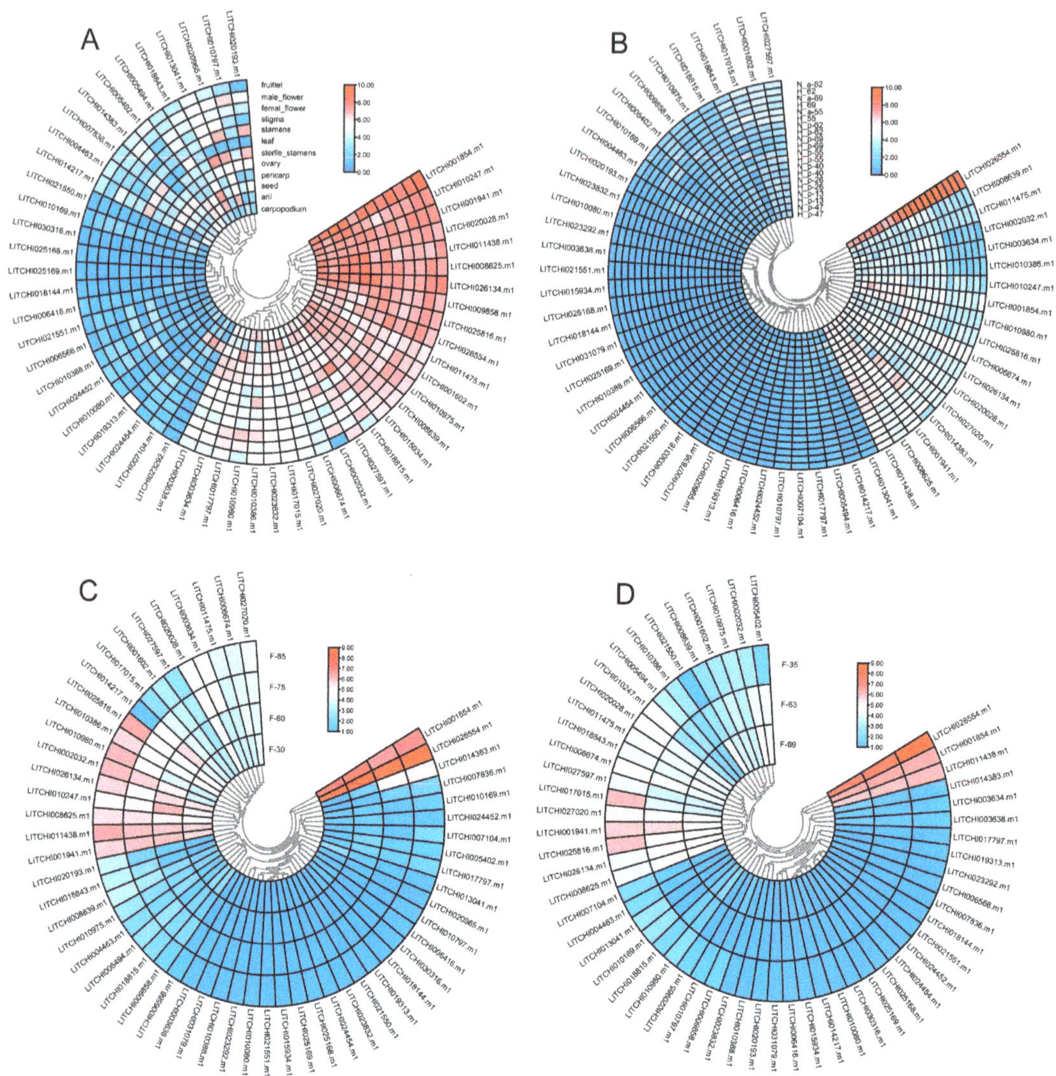

Figure 8. Tissue expression patterns of the litchi bZIP gene family across different tissues, species, and developmental stages after flowering. Expression analysis of bZIP genes (**A**) in different tissues of litchi, (**B**) in the pericarp and aril of Nuomici and Huaizha in different post-anthesis stages, (**C**) in litchi pericarp in different stages after flowering, and (**D**) in litchi "Feizixiao" in different stages after flowering.

During litchi pericarp development, we further analyzed the differential expression of bZIP genes in three distinct time periods: 30–60, 60–75, and 75–85 DAF (Figure 9B, Supplementary Figure S8B, Supplementary Table S3). In the 30–60 DAF period, five genes, including *LITCHI020193.m1* and *LITCHI014383.m1*, exhibited upregulated expression, while five genes, including *LITCHI006416.m1* and *LITCHI010797.m1*, exhibited downregulated expression. During the 60–75 DAF period, six genes, including *LITCHI020193.m1* and *LITCHI018843.m1*, exhibited upregulated expression, whereas five genes, including *LITCHI013041.m1* and *LITCHI024454.m1*, exhibited downregulated expression. In the 75–85 DAF period, six genes, including *LITCHI020193.m1* and *LITCHI017797.m1*, exhibited upregulated expression, while three genes, including *LITCHI020965.m1* and *LITCHI014217.m1*, exhibited downregulated expression. Notably, *LITCHI020193.m1* exhibited upregulated expression across all three periods, while *LITCHI020965.m1* exhibited downregulated expression consistently in all three periods.

Figure 9. Up- and downregulated expressions of bZIP genes in different litchi cultivars, tissues, and flowering stages. The differential expression of bZIP genes (**A**) in the pericarp and aril of *Nuomici* and *Huaizha* in different post-anthesis stages, (**B**) during litchi pericarp development at 30–60, 60–75, and 75–85 days after flowering, and (**C**) in the *Feizixiao* cultivar 35–63, 35–69, and 63–69 days after flowering. "N" represents the species *Nuomici*, "H" represents the species *Huaizha*, "a" represents aril, and "p" represents pericarp. The numbers indicate days after flowering. "Up" indicates upregulated expression and "Down" indicates downregulated expression.

In the study of the "sugar receding" phenomenon in the pulp of the *Feizixiao* cultivar, we analyzed gene expression changes during the 35–63, 35–69, and 63–69 DAF periods (Figure 9C, Supplementary Figure S8C, Supplementary Table S4). During the 35–63 DAF period, ten genes, including *LITCHI010169.m1* and *LITCHI010797.m1*, exhibited upregulated expression, while eight genes, including *LITCHI010388.m1* and *LITCHI007836.m1*, exhibited downregulated expression. In the 35–69 DAF period, nine genes, including *LITCHI005494.m1* and *LITCHI025816.m1*, exhibited upregulated expression, whereas seven genes, including *LITCHI005402.m1* and *LITCHI007836.m1*, exhibited downregulated expression. However, no differentially expressed bZIP genes were detected in the 63–69 DAF period.

2.9. qRT-PCR Expression Analysis of the bZIP Genes

Based on the differential expression analysis of bZIP genes across various litchi cultivars, tissues, and flowering stages during fruit development, we hypothesize that *LITCHI017015.m1* and *LITCHI004463.m1* may be involved in the regulation of early fruit development, whereas *LITCHI018843.m1* may play a role during the later stages of fruit maturation. To validate this hypothesis, we conducted quantitative real-time PCR (qRT-PCR) analysis on the three bZIP candidate genes (*LITCHI017015.m1*, *LITCHI004463.m1*, and *LITCHI018843.m1*) to examine their expression patterns throughout the fruit developmental process. The qRT-PCR results revealed that *LITCHI017015.m1* and *LITCHI004463.m1* were significantly upregulated between 20 and 40 days after flowering (DAF), corresponding to the early stage of fruit development, while their expression levels declined notably by 60 DAF. These results suggest that both genes may play important regulatory roles in the early phase of fruit development (Figure 10). By contrast, *LITCHI018843.m1* exhibited low expression levels at 20 and 40 DAF, but was significantly upregulated at 60 DAF, indicating its potential involvement in the regulation of late-stage fruit development (Figure 10).

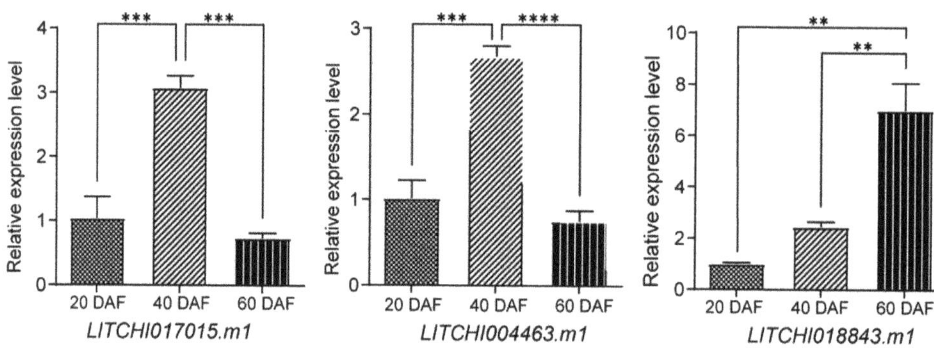

Figure 10. Fluorescence quantitative expressions of the bZIP gene. Asterisks indicate significant differences based on one-way ANOVA (** $p < 0.01$; *** $p < 0.001$; **** $p < 0.0001$).

3. Discussion

The bZIP transcription factor family is one of the most conserved and widely distributed transcription factor families in the plant kingdom. These factors play a crucial role in regulating key physiological processes such as plant growth and development, stress responses, seed germination, and senescence [39]. The bZIP gene family varies in size across different plant species [40], with the smallest family sizes found in lower plants [41]. For example, *Chlamydomonas reinhardtii* contains 17 bZIP genes, whereas *Physcomitrella patens*, one of the earliest known land mosses, has 43 bZIP genes [42]. In angiosperms, most plant genomes contain more than 40 bZIP genes [41]. In this study, based on whole-genome data for litchi, we identified a total of 55 bZIP transcription factors using bioinformatics approaches. In recent years, bZIP family members have been identified and reported in multiple species, including *Arabidopsis thaliana* (78 genes) [43], *Vitis vinifera* (55 genes) [44], *Phaseolus vulgaris* (84 genes) [45], *Malus domestica* (114 genes) [46], *Citrullus lanatus* (62 genes) [47], and *Zea mays* (54 genes) [48]. We found that the number of bZIP genes in litchi differs from that in other plant species, reflecting variations in gene family size among different plants. Subsequently, we conducted a comprehensive bioinformatics analysis of these gene family members to elucidate their characteristics.

Through the analysis of physicochemical properties, we found that all bZIP proteins in litchi are hydrophilic. Additionally, subcellular localization analysis revealed that all bZIP proteins in litchi are localized in the nucleus, consistent with findings in maize [48]. Furthermore, the secondary structure analysis of the 55 bZIP family members in litchi showed that their structures are primarily composed of random coils and α-helices. Phylogenetic analysis further classified these 55 bZIP genes into 13 subgroups, with the A and S subfamilies containing the highest number of bZIP genes. By comparison, the number of subgroups in other species varies, such as 10 in apple [46] and 11 in grape [44]. The increased number of subgroups in litchi suggests an expansion of the bZIP gene family in this species. The interspecies phylogenetic tree of litchi and Arabidopsis thaliana indicates that the bZIP gene family members in litchi share evolutionary similarities with those in Arabidopsis, suggesting their conserved nature.

Differences in gene structure organization may indicate distinct evolutionary trajectories of bZIP genes [49]. The number of exons and introns plays a crucial role in plant adaptation and developmental processes [50]. Further motif analysis revealed that most bZIP genes contain motif 1, which has also been observed in *Brassica napus* [12], suggesting its high conservation in protein structure and its crucial role in maintaining gene function and structural stability. Among the 55 identified bZIP genes, 39 segmental duplication events were detected, with duplication frequencies ranging from one to four occurrences per gene. By contrast, a study in maize [48] identified only 18 segmental duplications, significantly fewer than in litchi. However, no tandem duplication events were detected, indicating that segmental duplication is the primary driver of bZIP gene family expansion in litchi. This finding aligns with the analysis of bZIP genes in *Cannabis sativa* [51]. Similarly, in *Nephelium lappaceum* (rambutan), another species in the Sapindaceae family, segmental duplication has been demonstrated to play a crucial role in the expansion of resistance genes [52].

By analyzing the duplication events of individual bZIP genes in litchi, we found that only a few *bZIP* genes underwent duplication, while the majority remained as single-copy genes, such as *LITCHI014217.m1* and *LITCHI026134.m1*. This observation highlights the functional conservation of the *bZIP* gene family across species, which is consistent with previous litchi genome studies. These studies suggest that litchi, longan (*Dimocarpus longan*), and Xanthoceras sorbifolium share a γ-triplication event but have not undergone independent whole-genome duplication (WGD) events since then [53].

Gene expression patterns are one of the key indicators for elucidating gene function [54]. Tissue-specific expression analysis revealed significant differences in the expression levels of 55 bZIP genes across different tissues. For example, *LITCHI013041.m1* exhibited high expression in seeds and arils, whereas *LITCHI007104.m1* was predominantly expressed in seeds but showed low or nearly undetectable expression in other tissues. Notably, *LITCHI001854.m1* displayed consistently high expression levels across all tissues, suggesting its broad regulatory role in litchi growth and development. Further comprehensive analysis of bZIP gene expression across different litchi cultivars and developmental stages after anthesis revealed that *LITCHI006654.m1* maintained consistently high expression throughout fruit development. Additionally, *LITCHI006654.m1* exhibited the highest number of protein interactions, indicating its potential role as a key regulatory factor in litchi fruit development. This suggests that *LITCHI006654.m1* functions both as a highly expressed gene involved in fruit development and as a central hub in the protein interaction network, regulating multiple related biological processes.

The bZIP gene family has been demonstrated to play a crucial role in fruit development across various plant species. For instance, the bZIP gene *OebZIP22* has been shown to regulate fruit development in olives [55], while *PpbZIP44h* in pears controls the expression of genes related to fruit quality compounds [56]. However, research on the role of bZIP genes in litchi fruit development remains limited.

To further investigate the impact of bZIP genes on litchi fruit development, we integrated transcriptome data from different litchi tissues and analyzed the differential expression of *bZIP* genes in fruit and pericarp development across different post-anthesis stages in multiple litchi cultivars. Our results revealed that all 55 bZIP genes were expressed during fruit development and maturation, with 32 of them exhibiting differential expression across the three cultivars. These findings suggest that bZIP genes are extensively involved in litchi fruit development and maturation, potentially playing significant regulatory roles.

By analyzing the differential expression of bZIP genes during fruit and pericarp development across different post-anthesis stages in various litchi cultivars, we found that *LITCHI017015.m1* exhibited downregulated expression in the pericarp and aril of *Nuomici* and *Huaizha* at 55, 62, and 69 days after flowering (DAF). Similarly, in the cultivar *Feizixiao*, this gene also exhibited downregulated expression during the 75–85 DAF fruit development stage but exhibited upregulated expression during the 35–60 DAF stage. Meanwhile, *LITCHI004463.m1* exhibited upregulated expression during the 35–69 DAF fruit development stage but exhibited downregulated expression in the pericarp and aril of *Nuomici* and *Huaizha* at 55, 62, and 69 DAF. These findings suggest that both genes may play roles in the early stages of fruit development. The qRT-PCR results indicate that the expression levels of genes *LITCHI004463.m1* and *LITCHI017015.m1* are significantly upregulated during the early stages of fruit development, further confirming the above hypothesis.

Similar phenomena have been reported in other plant species. For example, the bZIP gene *npr1* in tomatoes exhibits upregulated expression during the development of locular tissue and mesocarp at 20 DAF [57], while the *MabZIP* gene in bananas plays an essential role in the early stages of fruit development [58]. In carrots, *CAREB1* is an important bZIP transcription factor that regulates gene expression during late embryonic development, dormancy, and maturation [59]. Likewise, our study found that *LITCHI018843.m1* exhibited upregulated expression during the late fruit development stage (40–85 DAF). Moreover, this gene exhibited upregulated expression in the pericarp and aril of *Nuomici* and *Huaizha* at 40, 55, 62, and 69 DAF. In addition, qRT-PCR validation results further confirmed that the expression level of *LITCHI018843.m1* is significantly upregulated during the later stages of fruit development, suggesting that *LITCHI018843.m1* may play a regulatory role in the later stages of litchi fruit development.

4. Materials and Methods

4.1. Identification of the bZIP Gene Family

In this study, the genome sequence data for litchi were obtained from the Litchi Genome Database [53]. We retrieved the gene IDs of the bZIP transcription factor family and downloaded the corresponding gene sequences, protein sequences, 2000 bp upstream promoter sequences, and genome annotation files. Additionally, the bZIP gene sequences for *Arabidopsis thaliana* were obtained from the Arabidopsis Information Resource (TAIR) (https://www.arabidopsis.org/, accessed on 18 April 2025).

4.2. Protein Physicochemical Property Analysis and Subcellular Localization

Using the ProtParam tool from the ExPASy online analysis platform (https://web.expasy.org/protparam/, accessed on 19 April 2025) [60], we conducted a detailed analysis of the physicochemical properties of litchi bZIP transcription factors, including amino acid count, molecular weight, isoelectric point (pI), hydrophilicity/hydrophobicity, aliphatic index, and instability index. Additionally, the subcellular localization was predicted using Plant-mPLoc from the Cell-PLoc 2.0 online tool (http://www.csbio.sjtu.edu.cn/bioinf/plant-multi/, accessed on 16 April 2025) [61], based on their amino acid sequences.

4.3. Protein Structure Prediction and Analysis and Protein–Protein Interaction Network Analysis

We employed the online service SOPMA (https://npsa-prabi.ibcp.fr/cgi-bin/npsa_automat.pl?page=npsa%20_sopma.html, accessed on 20 April 2025) to predict the secondary structure of proteins in the litchi bZIP gene family. The corresponding stacked graphs were plotted using Excel software for visualization. Additionally, the tertiary structure modeling of litchi bZIP transcription factors was performed using the SWISS-MODEL tool available on the ExPASy [60] online platform (https://beta.swissmodel.expasy.org/, accessed on 18 April 2025). The modeling results were further assembled and presented using PowerPoint software.

To investigate potential protein–protein interactions among bZIP gene family members, we conducted protein interaction network modeling using the STRING online database (https://cn.string-db.org/, accessed on 23 April 2025), referencing data from the Arabidopsis protein database. This analysis provided insights into the functional interactions among proteins within the litchi bZIP gene family.

4.4. Phylogenetic Analysis of the Gene Family

Multiple sequence alignments of the amino acid sequences and conserved domains of 78 *Arabidopsis thaliana* and 55 *Litchi* bZIP genes were performed using MEGA 7.0 software [62]. Phylogenetic clustering was conducted using the Neighbor-Joining (NJ) method, with 1000 bootstrap replicates to assess the reliability of the results. The evolutionary distances were calculated based on the Poisson correction model. Finally, the exported phylogenetic tree in nwk format was refined and visually enhanced using the ITOL online tool (https://itol.embl.de/, accessed on 25 April 2025).

4.5. Conserved Domain, Gene Structure, and Motif Analysis of the Gene Family

We utilized the Simple MEME Wrapper function in Tbtools v2.225 [63] software to predict and analyze the conserved domains of bZIP protein sequences, setting the number of motifs to 10 while keeping other parameters at their default values. Subsequently, gene structure analysis was performed using the Gene Structure View (Advanced) function in TBtools [63], incorporating the GFF gene structure file of *Litchi*.

4.6. Chromosomal Localization and Intra- and Interspecific Synteny Analysis of the Gene Family

The chromosomal localization of the Litchi bZIP gene family was determined using the Gene Location Visualize from GTF/GFF function in TBtools [63], by integrating gene IDs with the litchi GFF annotation file. Intraspecific synteny analysis was performed using BLASTp. The results were visualized using the Circle Gene View function in TBtools and further refined using Adobe Illustrator. For interspecific synteny analysis, we utilized JCVI v1.4.19 software [64] to compare bZIP genes among six species in the Sapindaceae family. Blue lines represent syntenic relationships among bZIP genes.

4.7. Promoter Analysis of the Gene Family

The 2000 bp upstream promoter sequences of *Litchi* bZIP genes were extracted from the *Litchi* genome dataset [53] and submitted to PlantCARE (https://bioinformatics.psb.ugent.be/webtools/plantcare/html/, accessed on 25 April 2025) for cis-regulatory element analysis. After filtering and integrating the analysis results, the Simple BioSequence Viewer function in TBtools [63] was used for visualization.

4.8. Tissue Expression Patterns of Litchi bZIP Genes and Differential Expression Analysis During Fruit Development

To analyze the expression patterns of bZIP genes in *Litchi chinensis*, we collected transcriptome data from different tissues (Project ID: PRJNA747875) [53]. Additionally, we integrated multiple transcriptome datasets, including transcriptome data from the pericarp and aril during fruit development and cracking (Project ID: PRJNA681070), with a total data volume of 483 Gb [65]; transcriptome data of litchi fruits in different developmental stages after flowering (Project ID: PRJNA883943), with a total data volume of 72 Gb [66]; and transcriptome data from the pulp of the "Feizixiao" cultivar at 35, 63, and 69 days after flowering (Project ID: PRJNA906072), with a total data volume of 61 Gb [67]. The raw sequencing data were subjected to quality control using Trimmomatic (v0.32) [68] and subsequently aligned to the reference genome with HISAT2 (v2.1.0) under default parameters [69]. Gene expression levels were quantified using Cufflinks, generating FPKM (Fragments Per Kilobase of transcript per Million mapped reads) values to assess gene expression and identify differentially expressed genes (DEGs) [70]. To visually represent the expression patterns, TBtools Heatmap Illustrator (v1.123) [63] was used to generate FPKM-based heatmaps, while EVenn was employed for the visualization of differentially expressed genes [71].

4.9. Fluorescence Quantitative Verification

Total RNA samples from litchi fruit, stored at −80 °C in our laboratory, were extracted at 20, 40, and 60 days after flowering, followed by cDNA synthesis and gDNA removal. The resulting cDNA was used as the template for qRT-PCR experiments with 3 biological repetitions. The reaction system consisted of 10 μL of ArtiCanATM SYBR qPCR Mix, 0.4 μL of forward and reverse primers each, 1 μL of the template, 0.4 μL of 50 × ROX Reference Dye I/II, and 7.8 μL of ddH2O. The reaction program was set as follows: 95 °C for 1 min for pre-denaturation, 95 °C for 10 s for denaturation, 60 °C for 20 s for annealing, and 72 °C for 15 s for extension, with a total of 40 cycles. After the experiment, the running data were collected and analyzed using the $2^{-\triangle\triangle Ct}$ method, followed by compiling and generating the data analysis report. The sequences of the fluorescence quantitative primers are listed in Supplementary Table S5.

5. Conclusions

This study systematically identified the molecular characteristics and functional properties of the bZIP gene family in *Litchi chinensis* through bioinformatics analysis. A total of 55 bZIP gene family members were identified in the *Litchi* genome and classified into 13 subfamilies. These genes were successfully mapped onto 15 chromosomes, demonstrating high evolutionary conservation and intraspecific collinearity. Furthermore, expression analysis across different cultivars, tissues, and fruit developmental stages, along with qRT-PCR validation, indicated that *LITCHI017015.m1* and *LITCHI004463.m1* may be involved in the early regulation of fruit development, while *LITCHI018843.m1* primarily functions during the later stages of fruit development. The findings of this study provide essential theoretical insights into the evolution, function, and regulatory roles of the bZIP gene family in fruit development. Moreover, they lay a foundation for genetic improvement and environmental adaptability research in *Litchi*, highlighting its potential applications.

Supplementary Materials: The following supporting information can be downloaded at: https://www.mdpi.com/article/10.3390/plants14101453/s1.

Author Contributions: Conceptualization, C.S.; methodology, C.S. and Z.C.; investigation, J.L. and S.S.; data curation, J.L., S.S. and J.W.; writing—original draft preparation, J.L. and S.S.; writing—review and editing, L.O., Z.C. and C.S.; funding acquisition, L.O. and C.S. All authors have read and agreed to the published version of the manuscript.

Funding: This work was financially supported by the Guangdong Provincial Science and Technology Plan Project (2021B0707010003), Guangdong Provincial Rural Revitalization Science and Technology Project "Maoming Tropical Characteristic Crop Germplasm Resource Exploration and Biobreeding Technology Innovation Center" (2021S0074), and the Projects of Talents Recruitment of Guangdong University of Petrochemical Technology (2019rc112).

Data Availability Statement: No new data were created or analyzed in this study.

Conflicts of Interest: The authors declare no conflicts of interest.

References

1. Pérez-Rodríguez, P.; Riano-Pachon, D.M.; Corrêa, L.G.G.; Rensing, S.A.; Kersten, B.; Mueller-Roeber, B. PlnTFDB: Updated content and new features of the plant transcription factor database. *Nucleic Acids Res.* **2010**, *38*, D822–D827. [CrossRef] [PubMed]
2. Foster, R.; Izawa, T.; Chua, N.H. Plant bZIP proteins gather at ACGT elements. *FASEB J.* **1994**, *8*, 192–200. [CrossRef]
3. Corrêa, L.G.; Riaño-Pachón, D.M.; Schrago, C.G.; dos Santos, R.V.; Mueller-Roeber, B.; Vincentz, M. The role of bZIP transcription factors in green plant evolution: Adaptive features emerging from four founder genes. *PLoS ONE* **2008**, *3*, e2944. [CrossRef] [PubMed]
4. Lee, S.; Shuman, J.D.; Guszczynski, T.; Sakchaisri, K.; Sebastian, T.; Copeland, T.D.; Miller, M.; Cohen, M.S.; Taunton, J.; Smart, R.C.; et al. RSK-mediated phosphorylation in the C/EBP{beta} leucine zipper regulates DNA binding, dimerization, and growth arrest activity. *Mol. Cell. Biol.* **2010**, *30*, 2621–2635. [CrossRef]
5. Hijazi, M.; Durand, J.; Pichereaux, C.; Pont, F.; Jamet, E.; Albenne, C. Characterization of the arabinogalactan protein 31 (AGP31) of Arabidopsis thaliana: New advances on the Hyp-O-glycosylation of the Pro-rich domain. *J. Biol. Chem.* **2012**, *287*, 9623–9632. [CrossRef]
6. Glover, J.N.; Harrison, S.C. Crystal structure of the heterodimeric bZIP transcription factor c-Fos-c-Jun bound to DNA. *Nature* **1995**, *373*, 257–261. [CrossRef]
7. Dröge-Laser, W.; Snoek, B.L.; Snel, B.; Weiste, C. The Arabidopsis bZIP transcription factor family—An update. *Curr. Opin. Plant Biol.* **2018**, *45*, 36–49. [CrossRef]
8. Nijhawan, A.; Jain, M.; Tyagi, A.K.; Khurana, J.P. Genomic survey and gene expression analysis of the basic leucine zipper transcription factor family in rice. *Plant Physiol.* **2008**, *146*, 333. [CrossRef]
9. Wang, J.; Zhou, J.; Zhang, B.; Vanitha, J.; Ramachandran, S.; Jiang, S.Y. Genome-wide expansion and expression divergence of the basic leucine zipper transcription factors in higher plants with an emphasis on *sorghum* F. *J. Integr. Plant Biol.* **2011**, *53*, 212–231. [CrossRef]
10. Liu, W.; Wang, M.; Zhong, M.; Luo, C.; Shi, S.; Qian, Y.; Kang, Y.; Jiang, B. Genome-wide identification of bZIP gene family and expression analysis of BhbZIP58 under heat stress in wax gourd. *BMC Plant Biol.* **2023**, *23*, 598. [CrossRef]
11. Zhao, K.; Chen, S.; Yao, W.; Cheng, Z.; Zhou, B.; Jiang, T. Genome-wide analysis and expression profile of the bZIP gene family in poplar. *BMC Plant Biol.* **2021**, *21*, 1–16. [CrossRef] [PubMed]
12. Zhou, Y.; Xu, D.; Jia, L.; Huang, X.; Ma, G.; Wang, S.; Zhu, M.; Zhang, A.; Guan, M.; Lu, K. Genome-wide identification and structural analysis of bZIP transcription factor genes in Brassica napus. *Genes* **2017**, *8*, 288. [CrossRef] [PubMed]
13. Uno, Y.; Furihata, T.; Abe, H.; Yoshida, R.; Shinozaki, K.; Yamaguchi-Shinozaki, K. Arabidopsis basic leucine zipper transcription factors involved in an abscisic acid-dependent signal transduction pathway under drought and high-salinity conditions. *Proc. Natl. Acad. Sci. USA* **2000**, *97*, 11632–11637. [CrossRef]
14. Xiang, Y.; Tang, N.; Du, H.; Ye, H.; Xiong, L. Characterization of OsbZIP23 as a key player of the basic leucine zipper transcription factor family for conferring abscisic acid sensitivity and salinity and drought tolerance in rice. *Plant Physiol.* **2008**, *148*, 1938–1952. [CrossRef]

15. Kobayashi, F.; Maeta, E.; Terashima, A.; Takumi, S. Positive role of a wheat HvABI5 ortholog in abiotic stress response of seedlings. *Physiol. Plant* **2008**, *134*, 74–86. [CrossRef]
16. Yáñez, M.; Cáceres, S.; Orellana, S.; Bastías, A.; Verdugo, I.; Ruiz-Lara, S.; Casaretto, J.A. An abiotic stress-responsive bZIP transcription factor from wild and cultivated tomatoes regulates stress-related genes. *Plant Cell Rep.* **2009**, *28*, 1497–1507. [CrossRef]
17. Lee, S.C.; Choi, H.W.; Hwang, I.S.; Choi, D.S.; Hwang, B.K. Functional roles of the pepper pathogen-induced bZIP transcription factor, CAbZIP1, in enhanced resistance to pathogen infection and environmental stresses. *Planta* **2006**, *224*, 1209–1225. [CrossRef]
18. Rodriguez-Uribe, L.; O'Connell, M.A. A root-specific bZIP transcription factor is responsive to water deficit stress in tepary bean (*Phaseolus acutifolius*) and common bean (*P. vulgaris*). *J. Exp. Bot.* **2006**, *57*, 1391–1398. [CrossRef] [PubMed]
19. Kusano, T.; Berberich, T.; Harada, M.; Suzuki, N.; Sugawara, K. A maize DNA-binding factor with a bZIP motif is induced by low temperature. *Mol. Gen. Genet.* **1995**, *248*, 507–517. [CrossRef]
20. Huang, X.; Ouyang, X.; Yang, P.; Lau, O.S.; Li, G.; Li, J.; Chen, H.; Deng, X.W. Arabidopsis FHY3 and HY5 positively mediate induction of COP1 transcription in response to photomorphogenic UV-B light. *Plant Cell* **2012**, *24*, 4590–4606. [CrossRef]
21. Zhong, X.; Feng, X.; Li, Y.; Guzmán, C.; Lin, N.; Xu, Q.; Zhang, Y.; Tang, H.; Qi, P.; Deng, M. Genome-wide identification of bZIP transcription factor genes related to starch synthesis in barley (*Hordeum vulgare* L.). *Genome* **2021**, *64*, 1067–1080. [CrossRef] [PubMed]
22. Niu, B.; Deng, H.; Li, T.; Sharma, S.; Yun, Q.; Li, Q.; Zhiguo, E.; Chen, C. OsbZIP76 interacts with OsNF-YBs and regulates endosperm cellularization in rice (*Oryza sativa*). *J. Integr. Plant Biol.* **2020**, *62*, 1983–1996. [CrossRef]
23. Sornaraj, P.; Luang, S.; Lopato, S.; Hrmova, M. Basic leucine zipper (bZIP) transcription factors involved in abiotic stresses: A molecular model of a wheat bZIP factor and implications of its structure in function. *Biochim. Biophys. Acta* **2016**, *1860*, 46–56. [CrossRef]
24. Zhang, F.; Fu, X.; Lv, Z.; Lu, X.; Shen, Q.; Zhang, L.; Zhu, M.; Wang, G.; Sun, X.; Liao, Z.; et al. A basic leucine zipper transcription factor, AabZIP1, connects abscisic acid signaling with artemisinin biosynthesis in *Artemisia annua*. *Mol. Plant* **2015**, *8*, 163–175. [CrossRef] [PubMed]
25. Zhiguo, E.; Zhang, Y.P.; Zhou, J.H.; Wang, L. Mini review roles of the bZIP gene family in rice. *Genet. Mol. Res.* **2014**, *13*, 3025–3036.
26. Guo, Z.; Dzinyela, R.; Yang, L.; Hwarari, D. bZIP transcription factors: Structure, modification, abiotic stress responses and application in plant improvement. *Plants* **2024**, *13*, 2058. [CrossRef]
27. Shen, H.; Cao, K.; Wang, X. A conserved proline residue in the leucine zipper region of AtbZIP34 and AtbZIP61 in Arabidopsis thaliana interferes with the formation of homodimer. *Biochem. Biophys. Res. Commun.* **2007**, *362*, 425–430. [CrossRef]
28. Zhang, F.; Xiang, L.; Yu, Q.; Zhang, H.; Zhang, T.; Zeng, J.; Geng, C.; Li, L.; Fu, X.; Shen, Q.; et al. ARTEMISININ BIOSYNTHESIS PROMOTING KINASE 1 positively regulates artemisinin biosynthesis through phosphorylating AabZIP1. *J. Exp. Bot.* **2018**, *69*, 1109–1123. [CrossRef]
29. Shen, Q.; Huang, H.; Zhao, Y.; Xie, L.; He, Q.; Zhong, Y.; Wang, Y.; Wang, Y.; Tang, K. The Transcription Factor Aabzip9 Positively Regulates the Biosynthesis of Artemisinin in *Artemisia annua*. *Front. Plant Sci.* **2019**, *10*, 1294. [CrossRef] [PubMed]
30. Lv, Z.; Guo, Z.; Zhang, L.; Zhang, F.; Jiang, W.; Shen, Q.; Fu, X.; Yan, T.; Shi, P.; Hao, X.; et al. Interaction of bZIP transcription factor TGA6 with salicylic acid signaling modulates artemisinin biosynthesis in *Artemisia annua*. *J. Exp. Bot.* **2019**, *70*, 3969–3979. [CrossRef]
31. Yu, F.-Y.; Jiang, L.-L.; Jia, J.-Q.; Zhang, Z.-Z.; Chen, S.-J.; Niu, X.-L.; Wang, M.-X.; Huang, S.-X. Comparative genomics analysis of bZIP family members in *Actinidia chinensis*. *Chin. J. Appl. Environ. Biol.* **2020**, 226–235.
32. Lovisetto, A.; Guzzo, F.; Tadiello, A.; Confortin, E.; Pavanello, A.; Botton, A.; Casadoro, G. Characterization of a bZIP gene highly expressed during ripening of the peach fruit. *Plant Physiol. Biochem.* **2013**, *70*, 462–470. [CrossRef]
33. Wang, H.; Zhang, Y.; Norris, A.; Jiang, C.-Z. S1-bZIP transcription factors play important roles in the regulation of fruit quality and stress response. *Front. Plant Sci.* **2022**, *12*, 802802. [CrossRef] [PubMed]
34. Ma, J.; Hanssen, M.; Lundgren, K.; Hernández, L.; Delatte, T.; Ehlert, A.; Liu, C.M.; Schluepmann, H.; Dröge-Laser, W.; Moritz, T. The sucrose-regulated Arabidopsis transcription factor bZIP11 reprograms metabolism and regulates trehalose metabolism. *New Phytol.* **2011**, *191*, 733–745. [CrossRef]
35. Sagor, G.; Berberich, T.; Tanaka, S.; Nishiyama, M.; Kanayama, Y.; Kojima, S.; Muramoto, K.; Kusano, T. A novel strategy to produce sweeter tomato fruits with high sugar contents by fruit-specific expression of a single bZIP transcription factor gene. *Plant Biotechnol. J.* **2016**, *14*, 1116–1126. [CrossRef]
36. Jiang, G.; Lin, S.; Wen, L.; Jiang, Y.; Zhao, M.; Chen, F.; Prasad, K.N.; Duan, X.; Yang, B. Identification of a novel phenolic compound in litchi (*Litchi chinensis Sonn.*) pericarp and bioactivity evaluation. *Food Chem.* **2013**, *136*, 563–568. [CrossRef]
37. Zhao, L.; Wang, K.; Wang, K.; Zhu, J.; Hu, Z. Nutrient components, health benefits, and safety of litchi (*Litchi chinensis Sonn.*): A review. *Compr. Rev. Food Sci. Food Saf.* **2020**, *19*, 2139–2163. [CrossRef]

38. Yao, P.; Gao, Y.; Simal-Gandara, J.; Farag, M.A.; Chen, W.; Yao, D.; Delmas, D.; Chen, Z.; Liu, K.; Hu, H. Litchi (*Litchi chinensis Sonn.*): A comprehensive review of phytochemistry, medicinal properties, and product development. *Food Funct.* **2021**, *12*, 9527–9548. [CrossRef]
39. Li, Y.-Y.; Meng, D.; Li, M.; Cheng, L. Genome-wide identification and expression analysis of the bZIP gene family in apple (*Malus domestica*). *Tree Genet. Genomes* **2016**, *12*, 1–17. [CrossRef]
40. Ma, H.; Liu, C.; Li, Z.; Ran, Q.; Xie, G.; Wang, B.; Fang, S.; Chu, J.; Zhang, J. ZmbZIP4 contributes to stress resistance in maize by regulating ABA synthesis and root development. *Plant Physiol.* **2018**, *178*, 753–770. [CrossRef]
41. Liu, X.; Chu, Z. Genome-wide evolutionary characterization and analysis of bZIP transcription factors and their expression profiles in response to multiple abiotic stresses in *Brachypodium distachyon*. *BMC Genom.* **2015**, *16*, 1–15. [CrossRef] [PubMed]
42. Ji, C.; Mao, X.; Hao, J.; Wang, X.; Xue, J.; Cui, H.; Li, R. Analysis of bZIP transcription factor family and their expressions under salt stress in *Chlamydomonas reinhardtii*. *Int. J. Mol. Sci.* **2018**, *19*, 2800. [CrossRef]
43. Jakoby, M.; Weisshaar, B.; Dröge-Laser, W.; Vicente-Carbajosa, J.; Tiedemann, J.; Kroj, T.; Parcy, F. bZIP transcription factors in Arabidopsis. *Trends Plant Sci.* **2002**, *7*, 106–111. [CrossRef]
44. Liu, J.; Chen, N.; Chen, F.; Cai, B.; Dal Santo, S.; Tornielli, G.B.; Pezzotti, M.; Cheng, Z.M. Genome-wide analysis and expression profile of the bZIP transcription factor gene family in grapevine (*Vitis vinifera*). *BMC Genom.* **2014**, *15*, 281. [CrossRef]
45. Zhang, Q.; Zhang, W.-J.; Yin, Z.-G.; Li, W.-J.; Xia, C.-Y.; Sun, H.-Y.; Yang, Y.-M.; Wu, H.-B.; Zhang, S.; Zhao, H.-H. Genome-wide identification reveals the potential functions of the bZIP gene family in common bean (*Phaseolus vulgaris*) in response to salt stress during the sprouting stage. *J. Plant Growth Regul.* **2021**, 1–16. [CrossRef]
46. Wang, S.; Zhang, R.; Zhang, Z.; Zhao, T.; Zhang, D.; Sofkova, S.; Wu, Y.; Wang, Y. Genome-wide analysis of the bZIP gene lineage in apple and functional analysis of *MhABF* in *Malus halliana*. *Planta* **2021**, *254*, 78. [CrossRef] [PubMed]
47. Yang, Y.; Li, J.; Li, H.; Yang, Y.; Guang, Y.; Zhou, Y. The bZIP gene family in watermelon: Genome-wide identification and expression analysis under cold stress and root-knot nematode infection. *PeerJ* **2019**, *7*, e7878. [CrossRef]
48. Cao, L.; Lu, X.; Zhang, P.; Wang, G.; Wei, L.; Wang, T. Systematic Analysis of Differentially Expressed Maize ZmbZIP Genes between Drought and Rewatering Transcriptome Reveals bZIP Family Members Involved in Abiotic Stress Responses. *Int. J. Mol. Sci.* **2019**, *20*, 4103. [CrossRef]
49. Azeem, F.; Tahir, H.; Ijaz, U.; Shaheen, T. A genome-wide comparative analysis of bZIP transcription factors in *G. arboreum* and *G. raimondii* (Diploid ancestors of present-day cotton). *Physiol. Mol. Biol. Plants* **2020**, *26*, 433–444. [CrossRef]
50. Zhu, M.; Yan, B.; Hu, Y.; Cui, Z.; Wang, X. Genome-wide identification and phylogenetic analysis of rice FTIP gene family. *Genomics* **2020**, *112*, 3803–3814. [CrossRef]
51. Huai, H.; Ning, K.; Hou, C.; Yang, S.-M.; Wang, J.-Z.; Chen, S.-L.; Dong, L.-L. The identification of bZIP gene family in *Cannabis sativa* L. and its preliminary research of the function in regulation of lipid metabolism. *Acta Pharm. Sin.* **2022**, *57*, 2528–2542.
52. Zheng, J.; Meinhardt, L.W.; Goenaga, R.; Matsumoto, T.; Zhang, D.; Yin, Y. The chromosome-level rambutan genome reveals a significant role of segmental duplication in the expansion of resistance genes. *Hortic. Res.* **2022**, *9*, uhac014. [CrossRef]
53. Hu, G.; Feng, J.; Xiang, X.; Wang, J.; Salojärvi, J.; Liu, C.; Wu, Z.; Zhang, J.; Liang, X.; Jiang, Z. Two divergent haplotypes from a highly heterozygous lychee genome suggest independent domestication events for early and late-maturing cultivars. *Nat. Genet.* **2022**, *54*, 73–83. [CrossRef]
54. Yamazaki, M.; Saito, K. Differential display analysis of gene expression in plants. *Cell. Mol. Life Sci. CMLS* **2002**, *59*, 1246–1255. [CrossRef]
55. Rong, S.; Wu, Z.; Cheng, Z.; Zhang, S.; Liu, H.; Huang, Q. Genome-wide identification, evolutionary patterns, and expression analysis of bZIP gene family in olive (*Olea europaea* L.). *Genes* **2020**, *11*, 510. [CrossRef]
56. Wang, H.; Xu, K.; Li, X.; Blanco-Ulate, B.; Yang, Q.; Yao, G.; Wei, Y.; Wu, J.; Sheng, B.; Chang, Y. A pear S1-bZIP transcription factor PpbZIP44 modulates carbohydrate metabolism, amino acid, and flavonoid accumulation in fruits. *Hortic. Res.* **2023**, *10*, uhad140. [CrossRef]
57. Mounet, F.; Moing, A.; Garcia, V.; Petit, J.; Maucourt, M.; Deborde, C.; Bernillon, S.; Le Gall, G.; Colquhoun, I.; Defernez, M. Gene and metabolite regulatory network analysis of early developing fruit tissues highlights new candidate genes for the control of tomato fruit composition and development. *Plant Physiol.* **2009**, *149*, 1505–1528. [CrossRef]
58. Hu, W.; Wang, L.; Tie, W.; Yan, Y.; Ding, Z.; Liu, J.; Li, M.; Peng, M.; Xu, B.; Jin, Z. Genome-wide analyses of the bZIP family reveal their involvement in the development, ripening and abiotic stress response in banana. *Sci. Rep.* **2016**, *6*, 30203. [CrossRef]
59. Guan, Y.; Ren, H.; Xie, H.; Ma, Z.; Chen, F. Identification and characterization of bZIP-type transcription factors involved in carrot (*Daucus carota* L.) somatic embryogenesis. *Plant J.* **2009**, *60*, 207–217. [CrossRef]
60. Artimo, P.; Jonnalagedda, M.; Arnold, K.; Baratin, D.; Csardi, G.; De Castro, E.; Duvaud, S.; Flegel, V.; Fortier, A.; Gasteiger, E. ExPASy: SIB bioinformatics resource portal. *Nucleic Acids Res.* **2012**, *40*, W597–W603. [CrossRef]
61. Chou, K.-C.; Shen, H.-B. Cell-PLoc 2.0: An improved package of web-servers for predicting subcellular localization of proteins in various organisms. *Nat. Sci.* **2010**, *2*, 1090–1103. [CrossRef]

62. Kumar, S.; Stecher, G.; Tamura, K. MEGA7: Molecular Evolutionary Genetics Analysis Version 7.0 for Bigger Datasets. *Mol. Biol. Evol.* **2016**, *33*, 1870–1874. [CrossRef]
63. Chen, C.; Chen, H.; Zhang, Y.; Thomas, H.R.; Frank, M.H.; He, Y.; Xia, R. TBtools: An Integrative Toolkit Developed for Interactive Analyses of Big Biological Data. *Mol. Plant* **2020**, *13*, 1194–1202. [CrossRef]
64. Tang, H.; Krishnakumar, V.; Zeng, X.; Xu, Z.; Taranto, A.; Lomas, J.S.; Zhang, Y.; Huang, Y.; Wang, Y.; Yim, W.C. JCVI: A versatile toolkit for comparative genomics analysis. *Imeta* **2024**, *3*, e211. [CrossRef]
65. Wang, J.; Wu, X.F.; Tang, Y.; Li, J.G.; Zhao, M.L. RNA-Seq provides new insights into the molecular events involved in "Ball-Skin versus Bladder Effect" on fruit cracking in litchi. *Int. J. Mol. Sci.* **2021**, *22*, 454. [CrossRef]
66. Zhong, R.; Wei, J.; Liu, B.; Luo, H.; Zhang, Z.; Pang, X.; Fang, F. Metabolite and transcriptome profiles of proanthocyanidin biosynthesis in the development of litchi fruit. *Int. J. Mol. Sci.* **2022**, *24*, 532. [CrossRef]
67. Peng, J.; Du, J.; Wuqiang, M.; Chen, T.; Shui, X.; Liao, H.; Lin, X.; Zhou, K. Transcriptomics-based analysis of the causes of sugar receding in Feizixiao litchi (*Litchi chinensis Sonn.*) pulp. *Front. Plant Sci.* **2022**, *13*, 1083753. [CrossRef]
68. Bolger, A.M.; Lohse, M.; Usadel, B. Trimmomatic: A flexible trimmer for Illumina sequence data. *Bioinformatics* **2014**, *30*, 2114–2120. [CrossRef]
69. Kim, D.; Langmead, B.; Salzberg, S.L. HISAT: A fast spliced aligner with low memory requirements. *Nat. Methods* **2015**, *12*, 357–360. [CrossRef]
70. Ghosh, S.; Chan, C.-K.K. Analysis of RNA-Seq data using TopHat and Cufflinks. *Plant Bioinform. Methods Protoc.* **2016**, *1374*, 339–361.
71. Yang, M.; Chen, T.; Liu, Y.X.; Huang, L. Visualizing set relationships: EVenn's comprehensive approach to Venn diagrams. *Imeta* **2024**, *3*, e184. [CrossRef] [PubMed]

plants

MDPI

Article

Comprehensive Genome-Wide Analysis of the Receptor-like Protein Gene Family and Functional Analysis of *PeRLP8* Associated with Crown Rot Resistance in *Passiflora edulis*

Weijun Yu [1,2,†], Fan Liang [1,†], Yue Li [1], Wenjie Jiang [1], Yongkang Li [1], Zitao Shen [1], Ting Fang [1,3] and Lihui Zeng [1,3,*]

1 Institute of Genetics and Breeding in Horticultural Plants, College of Horticulture, Fujian Agriculture and Forestry University, Fuzhou 350002, China; 15060537779@163.com (W.Y.); fanliang980430@126.com (F.L.); liyuehka@163.com (Y.L.); 15860691180@163.com (W.J.); leeykang97@gmail.com (Y.L.); ssszt08@163.com (Z.S.); fangting@fafu.edu.cn (T.F.)
2 Fujian Academy of Forestry Sciences, Fuzhou 350012, China
3 Key Laboratory of Ministry of Education for Genetics, Breeding and Multiple Utilization of Crops, Fujian Agriculture and Forestry University, Fuzhou 350002, China
* Correspondence: lhzeng@fafu.edu.cn
† These authors contributed equally to this work.

Abstract

Passion fruit (*Passiflora edulis* Sims) is a Passifloraceae plant with high economic value. Crown rot caused by *Rhizoctonia solani* is a major fungal disease, which can seriously reduce the yield and quality of passion fruit. Receptor-like proteins (RLPs), which act as pathogen recognition receptors, are widely involved in plant immune responses and developmental processes. However, the role of *RLP* family members of passion fruit in resistance to crown rot remains unclear. In this study, evolutionary dynamics analysis and comprehensive genomic characterization of the *RLP* genes family were performed on passion fruit. A total of 141 *PeRLPs* in the genome of the 'Zixiang' cultivar and 79 *PesRLPs* in the genome of the 'Tainong' cultivar were identified, respectively. Evolutionary analysis showed that proximal and dispersed duplication events were the primary drivers of *RLP* family expansion. RNA-seq data and RT-qPCR analysis showed that *PeRLPs* were constitutively expressed in different tissues and induced by low temperature, JA, MeJA, and SA treatments. The *PeRLP8* gene was identified as the hub gene by RNA-seq analysis of passion fruit seedlings infected by *Rhizoctonia solani*. The expression levels of *PeRLP8* of the resistant variety *Passiflora maliformis* (LG) were significantly higher than those of the sensitive variety *Passiflora edulis* f. *flavicarpa* (HG). Transient overexpression of *PeRLP8* tobacco and passion fruit leaves enhanced the resistance to *Rhizoctonia solani*, resulting in reduced lesion areas by 52.06% and 54.17%, respectively. In addition, it can increase reactive oxygen species levels and upregulated expression of genes related to active oxygen biosynthesis and JA metabolism in passion fruit leaves. Our research provides new insights into the molecular mechanism and breeding strategy of passion fruit resistance to crown rot.

Keywords: passion fruit; receptor-like protein (RLP); crown rot; pathogen resistance

1. Introduction

To ward off threats from diverse pathogens, plants have evolved a two-layer immune system, including pattern-triggered immunity (PTI) and effector-triggered immunity (ETI) [1,2]. PTI is the initial layer of plant immunity and plays a key role when the pattern recognition receptors (PRRs) on the surface of plant cells detect pathogen- or microbe-related molecular patterns (PAMPs or MAMPs). PRRs are an important part of complex immune signal networks, which can sense various pathogens [3]. Recently, plant PRRs have been effectively employed to provide broad-spectrum resistance in pepper [4], poplar [5] and *Arabidopsis* [6].

The plant PTI response is primarily initiated by the recognition or perception of apoplastic elicitors by plasma membrane-localized pattern recognition receptors (PRRs). This recognition activates downstream immune responses, including the production of reactive oxygen species (ROS), extracellular Ca^{2+} influx, activation of mitogen-activated protein kinase (MAPK) cascades, and the expression of immune-related genes [7]. Based on structural characteristics, PRRs comprise receptor-like kinases (RLKs) and receptor-like proteins (RLPs), which lack a cytoplasmic kinase domain. In *Arabidopsis*, 57 RLP family members were receptor proteins located on the cell surface and composed of three distinct domains, including a leucine-rich repeat (LRR) domain on the extracellular surface, a transmembrane domain, and a short intracellular peptide instead of a kinase domain [8]. The first *RLP* gene discovered was *Cf-9* in tomato, which is involved in the immune response of tomato leaves against *Cladosporium fulvum* [9]. Further research has shown that members of the *RLP* family play a role in immune defense. For example, proteins encoded by the *Cf* and *Ve* genes in tomato, as well as the HcrVf2 protein in apple, were involved in mediating plant resistance to fungal pathogens [10]. Additionally, when the *TaRLP1* gene was knocked out, wheat exhibited reduced resistance to *Puccinia striiformis* f. sp. *Tritici* (*Pst*) and became more susceptible to infection. In contrast, overexpression of the *TaRLP1* gene enhances wheat's resistance to *Pst*. These results suggested that *RLP* genes played a crucial role in plant response to pathogen infection.

Passion fruit (*Passiflora edulis* Sims) is a perennial evergreen vine-like fruit tree with high nutritional and economic value. However, passion fruit production is severely threatened by crown rot disease, which leads to significant yield losses and economic damage. When seedlings are infected with crown rot, dark brown lesions appear at the base of the stem, followed by gradual softening and rotting of the cortex, leading to brown rot development. Previous studies based on traditional morphological identification of fungal strains identified the primary pathogens responsible for stem rot in passion fruit as *Fusarium solani* and *Fusarium oxysporum*. In recent years, research on passion fruit has mainly focused on field symptoms, disease control measures, and cultivation techniques related to stem rot [11]. The chemical control methods were costly and harmful to the environment, making the development of disease-resistant cultivars one of the most effective strategies for managing passion fruit stem rot. However, there is limited in-depth research on the molecular mechanisms underlying crown rot resistance in passion fruit. In particular, the role of *RLP* genes in conferring resistance to crown rot is worth further studying.

In this study, *RLP* family genes were identified and characterized and a comparative evolutionary analysis was performed in the genome of the 'Zixiang' (ZX) and 'Tainong' (TN) cultivar. The orthologous and paralogous relationships were determined among these *RLP* genes. The chromosome distribution, gene structure, conserved motifs, and promoters of *PeRLP* genes were investigated. Moreover, a co-expression network with the *PeRLP* gene as the hub gene and positive response after being infected by *Rhizoctonia solani* was constructed. In addition, the *PeRLP8* gene was cloned from passion fruit, and its potential function was verified by RT-qPCR analysis and transient expression in wild tobacco and passion fruit leaves. Our study will provide potential candidate genes for the breeding of passion fruit resistant to crown rot through the genetic engineering method.

2. Results

2.1. Comprehensive Identification and Evolutionary Analysis of RLP Family Genes in Passion Fruit

Through excluding redundant sequences and genes encoding proteins with NB-ARC domains or kinase domains and verifying the *RLP* gene structures, 141 *RLP*-encoding genes in the *passiflra edulis* Sims (Zixiang, referred to as "ZX") genome and 79 *RLP*-encoding genes in the *passiflra edulis* Sims (Tainong, referred to as "TN") genome were identified. The number of two passion fruit *RLPs* had a significant expansion compared to the 57 *AtRLP* members in *Arabidopsis*. The *RLP* genes were renamed *PeRLP1-141* (ZX) and *PesRLP1-79* (TN) according to the location of the *RLP* genes to the respective chromosomes in the passion fruit genome. The 141 *PeRLPs* and 79 *PesRLPs* were distributed across nine chromosomes with varying numbers and uneven distribution (Figure S1). Many *PeRLP* genes were clustered in specific chromosomal regions in the ZX genome. For example, the highest number of *PeRLP* genes were found on Chr1 and Chr2, with 63 and 31 genes, respectively. Chr7 and Chr8 contain only 1 and 2 *PeRLP* genes, respectively. LG02 and LG05 had the highest number of *PesRLP* genes, with 13 and 31 genes in the TN genome, respectively. According to MCScanX results, dispersed and proximal duplications account for more than two-thirds of the duplication events in the ZX and TN genomes. These results indicated that the formation of these gene clusters and the expansion of *RLP* genes in passion fruit were likely due to these types of duplications.

To explore the evolutionary relationships among *RLPs* of eight plants, a phylogenetic analysis was conducted using Maximum Likelihood (ML) methods (Figure 1). The result showed that all *RLP* members cluster into nine distinct subgroups, labeled Group 1–9. This classification was supported by bootstrap values ranging from 60% to 100%, indicating the reliability of the *RLP* genes subfamily classification. A substantial number of *PeRLP* and *PesRLP* members were assigned to two groups (Group 5 and Group 8). Group 5 had the largest number of *RLP* members, including known disease resistance genes such as *HcrVf2*, *EIX2*, and homologous genes (35 *PeRLPs* and 21 *PesRLPs*) in passion fruit. Group 8 had the second largest number of *RLP* members, containing 50 *RLP* genes, but lacked homologs genes in *Arabidopsis* and tomato. Most *AtRLPs* were clustered together to form *Arabidopsis*-specific evolutionary branches (Group 3). These results suggested that *PeRLPs* and *PesRLPs* had undergone expansion following their divergence from a common ancestor with *Arabidopsis*.

2.2. Gene Structure Analysis of RLPs in Passion Fruit

Intron and exon structures play an important role in elucidating the phylogenetic relationships of gene families [12]. The exon-intron distribution pattern of *PeRLP* and *PesRLP* genes was further analyzed (Figure 2). The results showed that the exon number of *PeRLP* family genes ranged from 1 to 26, and the number of exons of *PesRLP* family genes ranged from 1 to 14. Most genes contain one or two exons. The exon number of *PeRLPs* and *PesRLPs* within the same group was also different in the evolutionary relationship. For example, *PeRLP1* had 12 exons, while *PeRLP119* had 9 exons in Group 1. These results provide a basis for the functional differentiation of the *PeRLP* and *PesRLP* gene families.

2.3. Conserved Domain Analysis of RLP Genes in Passion Fruit

To investigate the regulatory mechanisms of the *RLP* genes family in passion fruit, five highly conserved domains were identified (Figure 3). These domains were similar to those found in *RLP* genes of *Arabidopsis* and tomato [8,13]. Additionally, Multiple Em for Motif Elicitation (MEME) was used to further analyze the predicted RLP sequences to identify conserved motifs. A total of ten conserved motifs were identified among the predicted

RLPs and were designated as motifs 1–10 based on e-values (Figure 3). The most common motif at the N-terminus was motif 10. Among 141 *PeRLPs* and 79 *PesRLPs*, 133 *RLP* genes were found to contain motif 10, accounting for 60.5% of the total number of *RLP* genes. No subgroup-specific motifs were identified among the nine subfamilies, indicating that these gene subfamilies may have originated from a common ancestor. Interestingly, the types and numbers of conserved motifs were relatively similar within each subfamily, indicating that there were potential functional similarities among genes within the same family.

Certain subfamilies exhibited similar motifs and motif sequences. For example, motifs 2, 3, 6, 7, and 8 were commonly found in the Group 9 subfamily, and motifs 7 and 10 were more prevalent across all subfamilies, indicating close phylogenetic relationships. Conversely, some subfamilies lacked specific motifs; for instance, *RLP* genes in Group 1 were missing motifs 3, 5, 6, 8, and 9. These results further supported functional differences among the Group 1–9 subfamilies. The difference in domain structures among *RLP* family members may contribute to diverse pathogen recognition and immune function, highlighting the multifaceted roles of passion fruit *RLPs* in pathogen response.

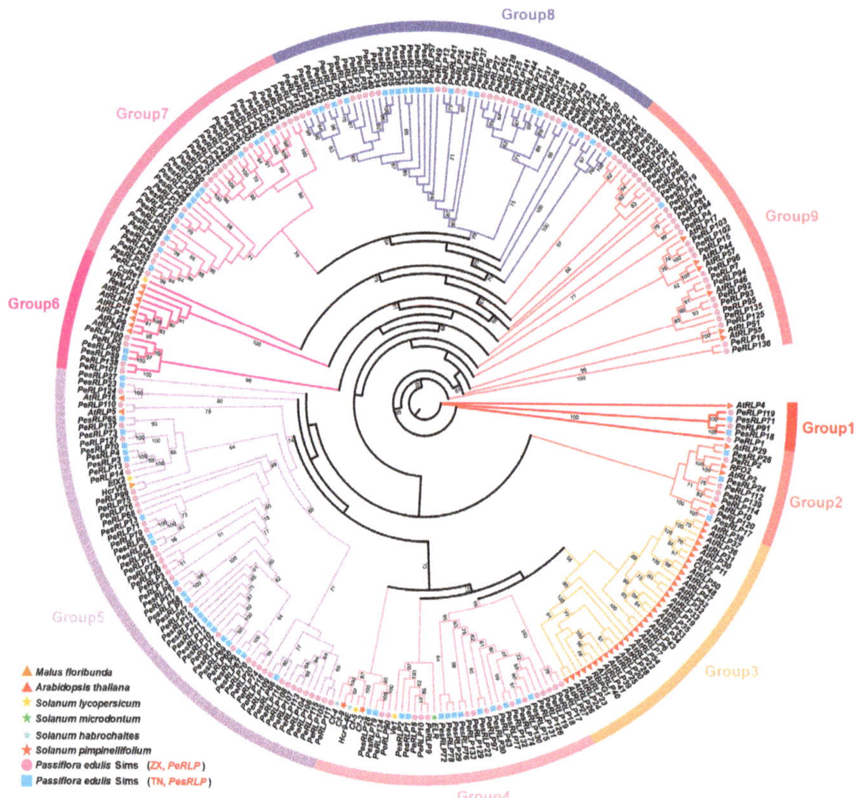

Figure 1. Identification and evolutionary analysis of *RLP* family genes in passion fruit. Phylogenetic analysis of the RLP homolog proteins from eight plant species, including *Malus floribunda*, *Arabidopsis thaliana*, *Solanum lycopersicum*, *Solanum microdontum*, *Solanum habrochaites*, *Solanum pimpinellifolium*, *passiflra edulis* Sims (ZX), and *passiflra edulis* Sims (TN). Each of the eight species was represented by a different shape in the evolutionary tree.

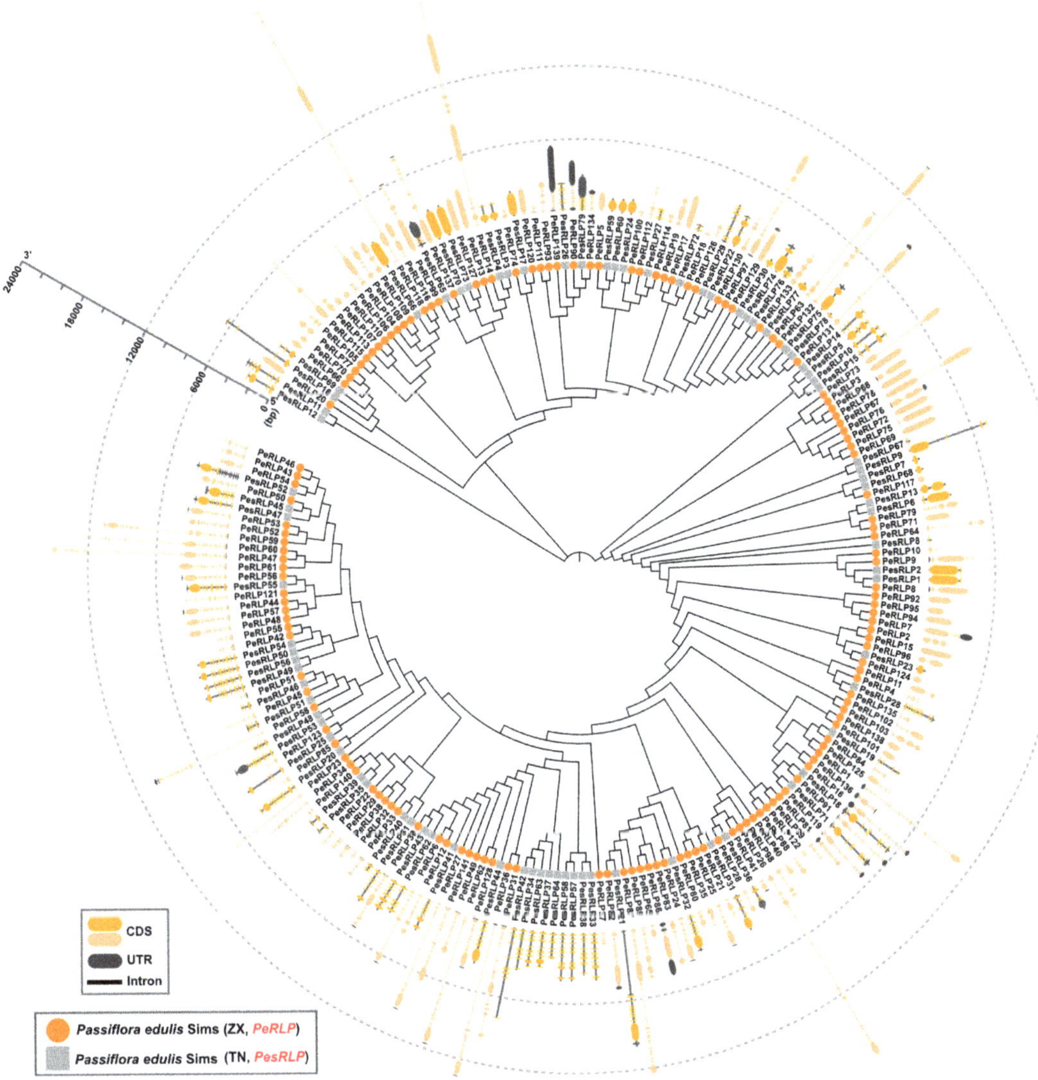

Figure 2. Gene structure of *RLP* genes in two passion fruit genomes. The circle represents 141 *PeRLP* genes in the ZX genome, and the square represents 79 *PesRLP* genes in the TN genome. The dotted line circles indicated the length of *RLP* genes.

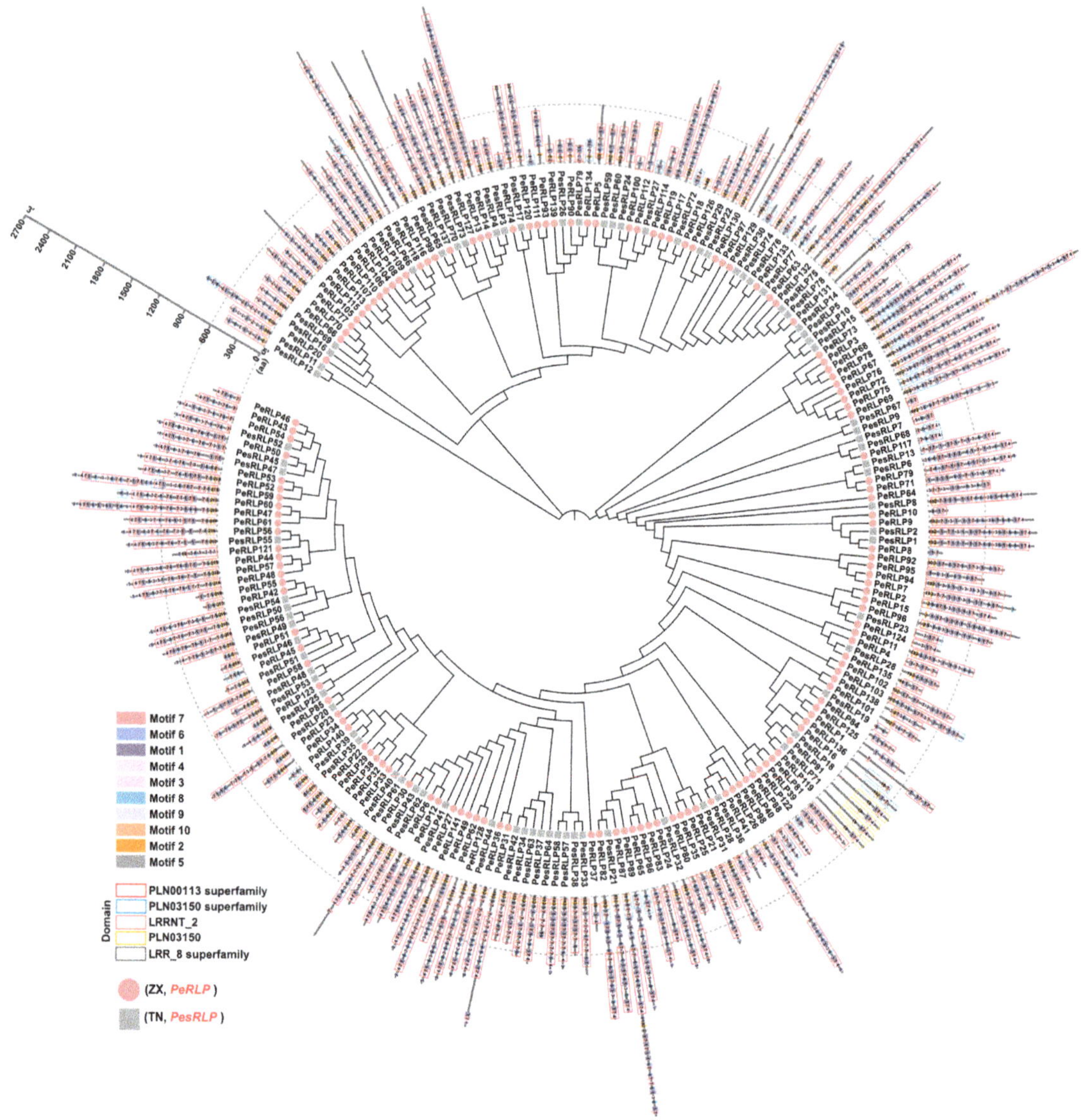

Figure 3. Conserved motif analysis of 141 PeRLP and 79 PesRLPs in the ZX and TN genomes. Different color modules indicated different motifs. Different color boxes indicate different conserved domains. The dotted line circles indicated the length of RLP proteins.

2.4. Collinearity Analysis of RLP Genes in Passion Fruit

A collinearity map was constructed to further explore the evolutionary trajectory of *RLP* members, and a list of collinear genes for *RLP* members across four species with chromosome-level genomes was generated (Figure 4). A number of highly conserved collinear blocks were shared among these genomes, with 74 of the 141 *PeRLP* genes (52.5%) mapped to these orthologous blocks. Clustering and identification of orthologous groups in ZX-TN and ZX-*Vitis vinifera* L. genomes revealed that their encoding genes exhibited similar intron distribution patterns and protein domain architectures. These results indicated that *RLP* orthologous groups among these different species may be functionally conserved.

Additionally, 32 paralogous gene pairs in the ZX genome and 12 paralogous gene pairs in the TN genome were identified (Figure 4). Approximately 94% of these paralogous *RLP* genes displayed identical protein domain structures. Except for the *PeRLP70-PeRLP76* paralogous pair, all paralogous *RLP* gene pairs had *Ka*/*Ks* values of less than 0.9, indicating that most paralogous *RLP* genes had undergone purifying selection. These paralogous genes might be the extant product of whole-genome duplications (WGDs) during the evolutionary history of the Passifloraceae family.

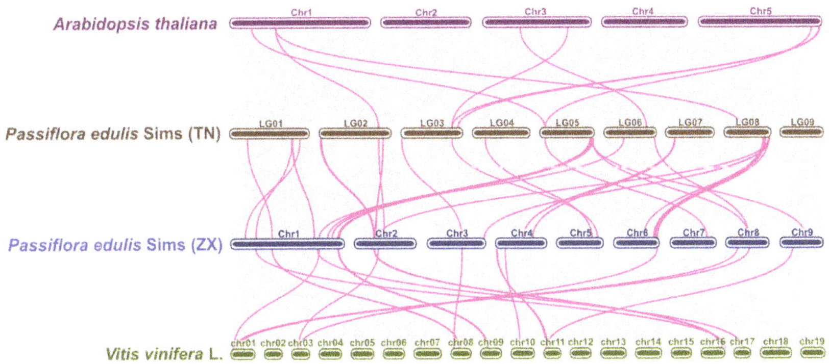

Figure 4. Synteny analysis of *RLP* genes in four plants, including *Arabidopsis*, *Passiflora edulis* Sims (ZX), *Passiflora edulis* Sims (TN), and *Vitis vinifera* L. genome. The purple lines display the collinear *RLP* genes among four plant genomes (*Arabidopsis*, TN, ZX and *Vitis vinifera* L.). The light gray lines represented collinear blocks.

2.5. Cis-Regulatory Element Assessment

RLP gene family members were involved in various developmental processes. To gain insight into their expression regulation, the 2000bp region upstream of each *PeRLPs* translation initiation site was analyzed using the PlantCARE database to identify potential transcription factor binding *cis*-elements. It was found that the promoter of *PeRLPs* contained *cis*-elements responsive to light, jasmonate (JA), methyl jasmonate (MeJA), abscisic acid (ABA), auxin, salicylic acid (SA), drought, defense, gibberellin, low temperature, circadian rhythms, and other factors (Figure S2). Among these, light-responsive *cis*-elements were the most common across all *PeRLPs* promoters. The *PeRLP126* promoter contained the highest number of *cis*-elements compared to other promoters. Notably, *PeRLP* promoters included *cis*-elements responsive to different plant hormones, especially had a higher number of MeJA (CGTCA and TGACG) and SA response elements, suggesting *PeRLPs* potentially respond to several hormones, such as SA, MeJA, and ABA. However, most *PeRLP* promoters contain single auxin (AuxREs, TGA-element, and TGA-box) and gibberellin response element. These findings indicated that *PeRLP* genes were more involved in stress responses, which were often primarily regulated by hormones, such as JA, SA, and ABA, rather than by growth hormones like auxin and gibberellin.

2.6. Expression Profile of PeRLP Genes

Gene expression profiles provide valuable resources for gene identification and functional analysis. The expression of 141 candidate *PeRLP* genes in flowers, leaves, roots, and seeds of passion fruit was analyzed by RNA-seq (Figure 5A). The TAU (τ) algorithm was used to determine the tissue-specific levels of 141 candidate *PeRLP* genes. It was found that 98 *PeRLP* genes had TAU values greater than 0.6, showing obvious tissue-biased expression patterns (Figure S3). Among them, 44 *PeRLP* family genes were preferentially expressed in passion fruit leaves, and 33 *PeRLP* family genes were preferentially expressed in the root

(Figure S3). *PeRLP2*, *PeRLP104*, *PeRLP13*, *PeRLP107*, *PeRLP113*, *PeRLP135*, and *PeRLP139* were specifically expressed in the root (TAU value was 1), and the expression of *PeRLP134* in the root was much higher than that in other tissues. In addition, six *PeRLP* genes were only specifically expressed in leaves (TAU value was 1) (Figure S3).

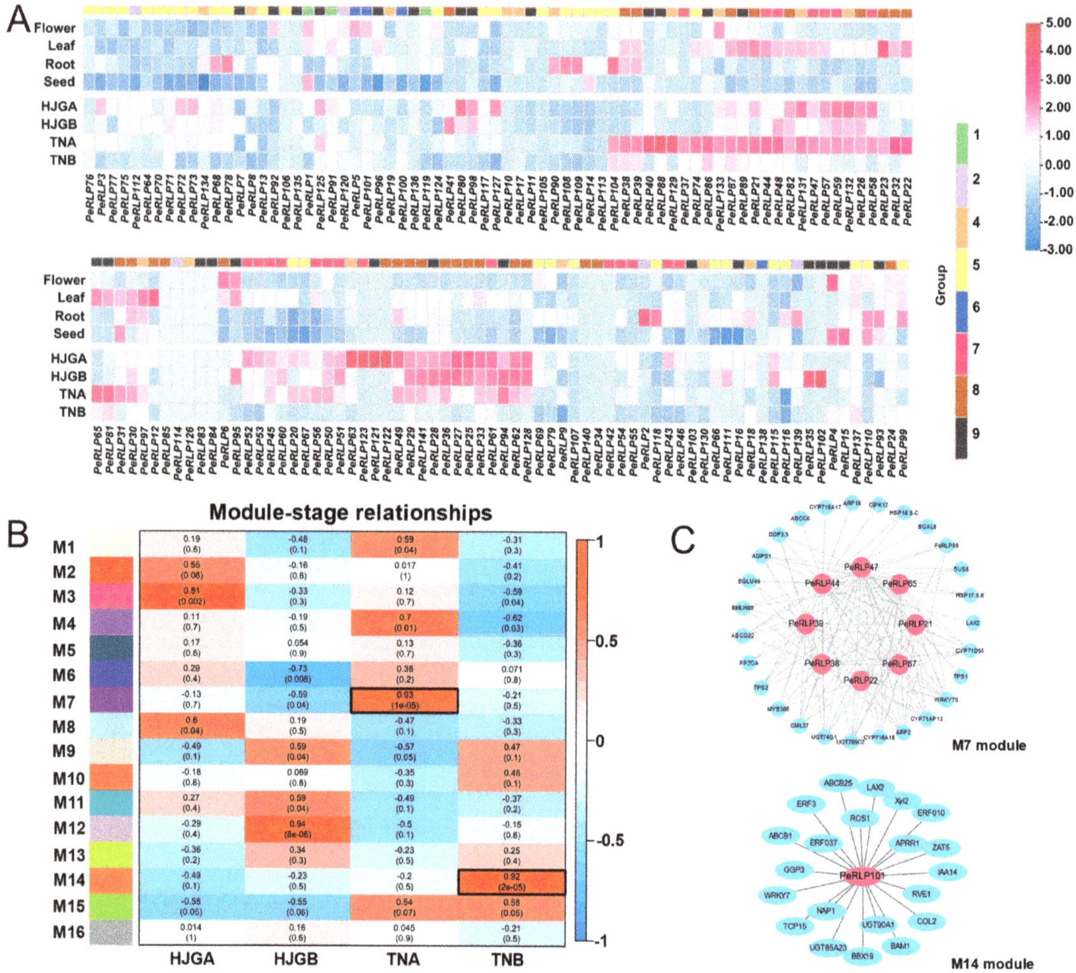

Figure 5. Expression profiles of 141 *PeRLP* genes. (**A**) Expression patterns of *PeRLP* genes in different tissues (root, leaf, seed, and flower) and seedlings of passion fruit with cold treatment. Three biological replicates of HJG (HJGA1, HJGA2, and HJGA3 were recorded as HJGA) and TN (TNA1, TNA2, and TNA3 were recorded as TNA) under normal temperature conditions. Three biological replicates of HJG (HJGB1, HJGB2, and HJGB3 were recorded as HJGB) and TN (TNB1, TNB2, and TNB3 were recorded as TNB) under cold stress. Differences in gene expression changes were shown in color as the scale, mediumvioletred for high expression, and steelblue for low expression. (**B**,**C**) Weighted gene co-expression network (WGCNA) was used to identify resistance genes associated with cold-tolerant variety TN. M1-M16 module indicated that the main branches constituted 16 merge modules (based on a threshold of 0.25) and were marked with different colors. Coexpression networks were constructed with eight and one *PeRLP* hub genes in the M7 and M14 modules, respectively.

The expression levels of all candidate *PeRLP* genes in the cold-sensitive variety 'Huangjinguo' (HJG) and cold-tolerant variety 'Tainong' (TN) under cold stress were calculated by reanalyzing the previously reported cold stress-free transcriptome data of passion fruit with reference to the ZX genome (Figure 5A). More than 10 *PeRLP* genes were found highly expressed under cold stress (Figure 5A). Five genes (*PeRLP83*, *PeRLP84*, *PeRLP85*, *PeRLP114*, and *PeRLP126*) did not have the expression. WGCNA was used to further identify the regulatory network of *PeRLP* genes involved in cold stress processes. A total of 16 co-expression modules were obtained. M7 and M14 modules were identified as highly correlated with cold tolerant of variety TN, suggesting the module members contained in them may be closely related to cold stress response (Figure 5B). A total of eight *PeRLP* hub genes belonging to the M7 module and one *PeRLP* hub gene belonging to the M14 module were identified by using WGCNA (Figure 5C). The co-expression network constructed with *PeRLP101* as the center included genes related to stress response, auxin signaling, and sucrose metabolism, indicating that *PeRLP101* may interact with these genes to actively respond to cold stress.

2.7. Expression Analysis of PeRLP Genes Under JA, MeJA, and SA Treatments

Given that most *PeRLP* family genes contain *cis*-acting elements related to hormone signaling, Six *PeRLP* gene promoter regions containing plant hormone response elements were selected to verify the transcriptional response to the treatments with JA, ABA, and MeJA hormones at 100 μmol/L concentrations (Figure S4). The results showed that six *PeRLP* family members differ in their specific responses to hormonal treatment. Notably, the expression levels of *PeRLP11* and *PeRLP70* changed most significantly under MeJA treatment, while *PeRLP8*, *PeRLP11*, and *PeRLP76* were significantly up-regulated within 0–8 h after JA treatment. The expression level of *PeRLP8* increased by nearly five times in 4 h compared with 0 h, indicating that it was more sensitive to JA treatment. In contrast, the transcription levels of *PeRLP17* and *PeRLP100* were significantly down-regulated under SA treatment. The expression patterns of *PeRLP8* and *PeRLP1* were opposite to those of *PeRLP70* and *PeRLP76* in response to SA treatment. The high sensitivity of these *PeRLP* genes under hormone treatments also validated the reliability of the predicted hormone-responsive *cis*-acting elements associated with them.

2.8. Expression Levels of PeRLP Genes in R. solani Infection

The crown rot of HG seedlings was identified as a disease caused by *R. solani*. The expression levels of all candidate *PeRLP* genes in the resistant cultivar *P. maliformis* (LG) *P. edulis* f. *flavicarpa* (HG) seedlings infected by *R. solani* were calculated using the ZX genome as a reference. Most of the *PeRLPs* belonging to Group 5 had high expression in the three stages of LG seedling infection with the bacteria (Figure 6A). WGCNA was used to further identify the regulatory network of *PeRLP* genes in response to the infection process of *R. solani*. For module-stage association, the MEblue module showed a high correlation with the L5 stage (Figure 6B). *PeRLP8* was identified as a hub gene in the MEblue module (Figure 6C). The co-expression networks of resistance-related genes include genes involved in ABA biosynthesis gene (*NCED2*), catabolism (*CYP707A*), and JA signal transduction pathway genes (*PP2C*), as well as other co-expressed genes involved in JA and ETH biosynthesis, signal transduction, sucrose metabolism and the key regulator of ROS (*RBOHD*). *PeRLP8* was also co-expressed with 16 *WRKY* family genes (Figure 6C). In addition, the expression pattern of PeRLP8 was supported by quantitative results (Figure S5). These results suggested that *PeRLP8* may interact with co-expressed genes to enhance LG's resistance to *R. solani*.

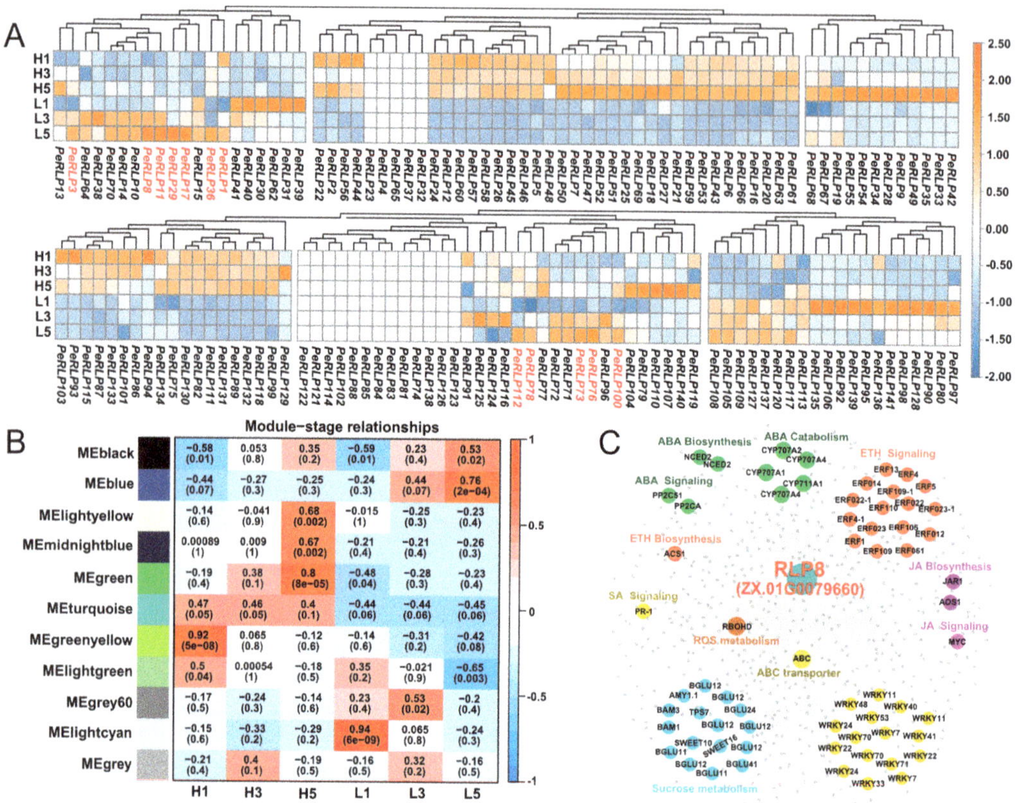

Figure 6. Analysis of gene co-expression network in RNA-seq data of passion fruit infected by *R. solani*. (**A**) Transcriptional expression analysis of 141 *PeRLP* during the infection of resistant variety LG and sensitive variety with *R. solani*. The heatmap was created based on the log2 (FPKM + 0.01) value of *PeRLP* genes and normalized by row. Differences in gene expression changes are shown in color as the scale, orange for high expression, and steelblue for low expression. L1, L3, and L5, respectively, represented the 1, 3, and 5 days after the resistant cultivar LG was infected by *R. solani*. H1, H3, and H5, respectively, represented the 1, 3, and 5 days after the sensitive cultivar HG was infected by *R. solani*. (**B**) The correlation analysis of modules and stages by using WGCNA, the heat map showed the correlation between modules and stages. Red and blue indicated positive and negative correlations, respectively. (**C**) The co-expression gene network was constructed with *PeRLP8* as the center in the MEblue module. SA signaling, Salicylic acid signaling; ETH biosynthesis and signaling, Ethene biosynthesis and signaling; ABA biosynthesis and catabolism, Abscisic acid biosynthesis and catabolism; JA biosynthesis and signaling, Jasmonic acid biosynthesis and signaling; ROS metabolism, reactive oxygen species metabolism.

2.9. Transient Expression of PeRLP8 Gene in Passion Fruit and Tobacco Leaves

To further study the role of *PeRLP8* in responding to *R. solani*, the CDS sequence of the *PeRLP8* gene was successfully cloned and sequenced. The constructed empty control 35S:: GFP and 35S:: *PeRLP8*-GFP were injected into passion fruit and tobacco leaves separately. Then, these leaves were inoculated with the fungal pathogen *R. solani*. It was observed that the passion fruit and tobacco disease leaves in the area of the leaves injected with 35S:: *PeRLP8*-GFP construct was significantly reduced by 54.17% and 52.06%, respectively (Figure 7A,B,D,E) and the expression of *PeRLP8* showed significantly up-regulated in injected areas compared to the control group (Figure 7C,F). These results suggested that the immune response was evoked after transient overexpression of the *PeRLP8* gene.

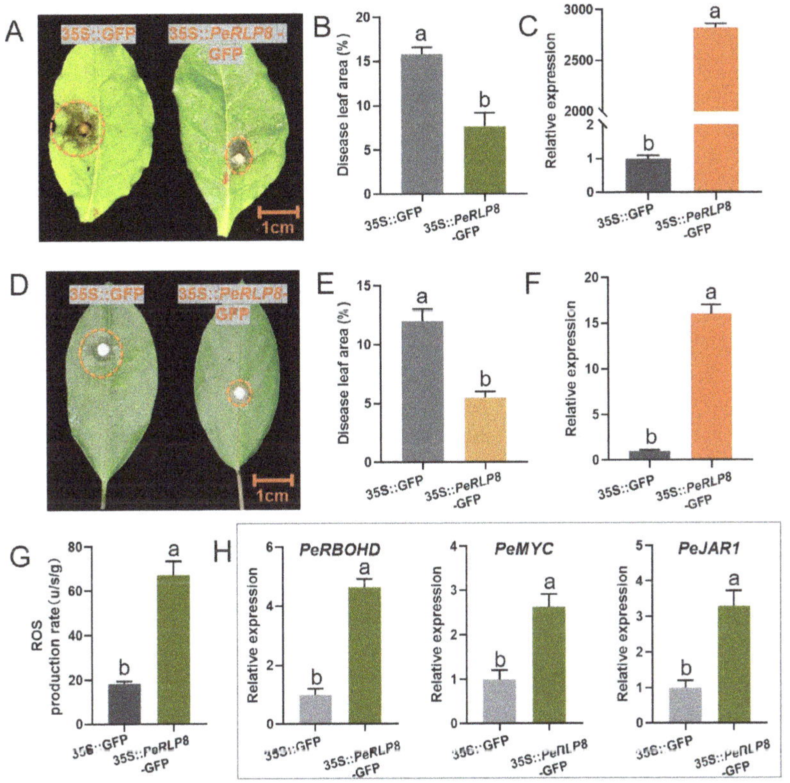

Figure 7. Resistance analysis of *PeRLP8*-overexpressed passion fruit and tobacco leaves against *R. solani*. The phenotype observation of transiently overexpressed passion fruit (**A**) and tobacco (**D**) leaves after inoculation with *R. solani*. The dotted lines on the leaves indicate the disease area. Bar = 1 cm. (**B**,**E**) The disease area statistics in *PeRLP8*-overexpressed passion fruit and tobacco leaves. (**C**,**F**) The expression levels of *PeRLP8* in transient-transformed tobacco and passion fruit leaves. (**G**) ROS content in transient-transformed passion fruit leaves. (**H**) Quantitative detection of 3 genes co-expressed with *PeRLP8* in *PeRLP8*-overexpressed passion fruit leaves. Different letters indicate statistically significant differences compared with the 35S:: GFP control (Student's *t*-test, $p < 0.05$). The error bars represented standard error.

The physiological analysis showed that the content of ROS in passion fruit leaves overexpressing 35S:: *PeRLP8*-GFP was significantly higher than that of the control (Figure 7G). This suggested an enhanced oxidation reaction may contribute to the resistance to pathogen attack. In addition, the expression levels of three co-expressed genes related to the ROS signaling pathway (*PeRBOHD* gene) and the JA biosynthesis and signal transduction pathway (*PeMYC* and *PeJAR1*) were significantly up-regulated in passion fruit leaves with transient expression of *PeRLP8* (Figure 7H), verifying that *PeRLP8* may activate the defense responses via ROS and JA metabolic pathways.

3. Discussion

Plants sense pathogen infection through cell surface and intracellular receptors. *RLPs* are the main defense layer of the natural immune system against pathogen infection. The

RLP gene family has been systematically studied in many plant species against pathogen infection [13–15]. However, *RLPs* in passion fruit have not been comprehensively analyzed. In this study, 141 *PeRLPs* and 79 *PesRLPs* genes were identified in the ZX and TN genomes, respectively. In terms of the number of genes, ZX and TN had significantly more genes than *Arabidopsis* which had 57 *RLP* genes [5,16]. This may be due to a WGD event in the Passiflora family [17,18]. Resistance gene families are often expanded through gene replication and diversification selection, which allows new pathogen recognition specificities to emerge [19]. In our study, the proximal and dispersed duplications were the main drivers of *RLP* family expansion in passion fruit. Furthermore, it has been reported that plants contain gene clusters of resistance gene analogs (RGAs), including *NLRs*, *RLKs*, and *RLPs* [13,20,21]. In tomatoes and peppers, several genes associated with RGAs were localized in gene clusters on different chromosomes [21–23]. Consistent with this, 141 *PeRLP* and 79 *PesRLP* members were found irregularly distributed across nine chromosomes and a large number of RLP genes formed gene clusters, which may be attributed to a single WGD event and proximal and dispersed gene duplication events. Additionally, the Group 4 branch of the *RLP* gene phylogenetic tree in different species contains several other *Cf* genes related to plant resistance to *C. fulvum* [24–27], Ve genes with race-specific resistance to tomato *Verticillium* [28], and 29 homologous *RLP* genes (*PeRLP8*, *PesRLP28*, *PesRLP79*, etc.) of passion fruit. This finding supported the view that the 29 homologous *RLP* genes that clustered with known *RLPs* involved in disease resistance may have a similar role in immunity.

Phylogenetic analysis showed that all *RLP* members in eight species clustered into nine different subgroups. Group 5 showed the major expansion and diversification of *PeRLPs* and *PesRLPs* from passion fruit, unlike *Arabidopsis*, which showed a sharp contraction of clade members with only one RLP. In contrast, all the genes of Group 3 belonged to *AtRLP* genes, which were greatly expanded and diversified. This phylogenetic analysis was consistent with the evolutionary analysis of *RLP* disease resistance genes in other plants [4,15,20]. The contraction or expansion of *RLP* genes in phylogenetic branches may be the result of plants coping with different selection pressures. Meanwhile, it is found that the conserved motifs of 141 *PeRLPs* and 79 *PesRLPs* members of ZX and TN were variable. It is reported that conserved motifs of genes were closely related to plants' response to biotic and abiotic stresses [29]. These results suggested that passion fruit had formed its own immune receptor diversity under different selection pressures after being separated from its common ancestor.

Expression profiles were regarded as important clues for gene function analysis. *RLP* gene families in many plant species exhibit different expression profiles in different tissues or organs and regulate various biological processes [8,13,15,20]. For example, *RLP* genes specifically expressed in roots may play a role as a root-specific regulator of tomato [13]. There were 18 *MaRLP* genes expressed in the Cavendish cultivar roots [15]. This is consistent with our results that 6 *PeRLP* genes were specifically expressed only in passion fruit leaves (TAU value 1) and 33 *PeRLP* genes were expressed in passion fruit root. These results suggested that *PeRLP* genes were widely involved in biological functions during plant development, and may play an important role in coping with biological and abiotic stresses in passion fruit.

The occurrence of crown rot has become an important limiting factor for the development of the passion fruit industry. *RLP* genes were the main genes involved in plant defense response, which were involved in many aspects of life activities, including plant growth and development and resistance to exogenous pathogens [30,31]. It has been reported that the *AtRLP1* gene participated in the MAMPS pathway and could recognize bacterial eMAX factor [30]. The tomato *CF-9* gene was involved in the immune defense response of tomato leaves to Xanthomonas [9]. All the genes resistant to *Xanthomonas campestris* in tomatoes were cloned, and it was found that all the genes belonged to the *RLP* family [8]. These results indicated that *RLP* family members play a certain role in immune defense. In our study, WGCNA was used to analyze transcriptome data of passion fruit seedlings infected with *R. solani* and identified *PeRLP8* as a hub gene that was significantly up-regulated in pathogen-resistant variety LG compared to the sensitive variety HG during the infection of *R. solani*. Moreover, the leaf lesion area of tobacco and passion fruit with transient expression of 35S:: *PeRLP8*-GFP were reduced after infection with *R. solani* compared with the control. These results indicated that the *PeRLP8* gene played a role in the resistance of passion fruit to *R. solani*.

As a stress signal molecule, JA accumulated rapidly when plant tissues were invaded by pathogens or insects [32], and played an important role in the defense response to vegetative pathogens [33]. Previous studies reported that overexpression of SlRLP6/10 in tomatoes could induce the accumulation of ROS and increase the contents of JA and ET, thus enhancing tomato resistance to *P. infestans* [34]. In this study, the expression levels of *PeMYC* and *PeJAR1* which were co-expressed with *PeRLP8* were significantly up-regulated in passion fruit leaves when transformed with 35S:: *PeRLP8*-GFP, and the transcription level of *PeRLP8* was significantly up-regulated under exogenous JA treatment. These results indicated that *PeRLP8* may interact with the JA pathway and enhance the resistance to *R. solani* in passion fruit. In addition, it is reported that the expression of the *RBOH* gene can produce superoxide and other reactive oxygen species, which play a role in the host defense process [35]. ROS can induce plants to produce substances that inhibit the infection of pathogens and also induce plant defense response in signal pathway [36]. In our study, overexpression of *PeRLP8* led to significant up-regulation of the *PeRBOHD* gene, which was the key gene in the ROS biosynthesis pathway. Therefore, it was suggested that *PeRLP8* may promote the production of ROS in passion fruit leaves by regulating the *PeRBOHD* gene to improve the resistance of passion fruit to *R. solani*, which needs more experiments to be verified. Overall, these results suggested that the *PeRLP8* gene may played a role in the defense against pathogen through the synergies of ROS and JA metabolism in passion fruit.

4. Materials and Methods

4.1. Plant Materials

The passion fruit seedlings of crown rot resistant cultivar *Passiflora maliformis* (LG) and crown rot susceptible cultivar *P. edulis* f. *flavicarpa* (HG) were provided by the Genetic Breeding Laboratory of Horticultural Plant Genetics and Breeding, College of Horticulture, Fujian Agriculture and Forestry University, China, and forty-five LG and HG seedlings were used to be infected with *Rhizoctonia solani* (*R. solani*), respectively. The stem base samples of the two cultivars were collected at 1, 3, and 5 days after the seedlings were infected by *R. solani*. The samples collected at each stage after infection had three biological repeats, and each biological repeat consisted of five infected stem base samples. The material used for transient gene expression was derived from the leaves of wild-type tobacco and the 'Qinguo No. 9' variety passion fruit. According to previous studies, three exogenous hormones (JA, SA, and MeJA) with concentrations of 100 μmol/L were applied to six passion fruit seedlings of 'Qinguo No. 9', respectively [37–39]. The above samples were quickly frozen and stored in liquid nitrogen.

4.2. Identification of RLP Gene in Passion Fruit

"Zixiang" (ZX) and "Tainong" (TN) genome and available genome annotation data were downloaded from the National Genomics Data Center website (PRJCA004251 and PRJCA003078). Multiple strategies were used to search and identify members of the

RLPs gene family, including keyword search, Hidden Markov Model search, and BLASTP search [4]. The BLASTP comparison was performed in the passion fruit genome by using 57 identified AtRLP sequences as reference. Next, the tBLASTn search was performed using the HMMER domain in the protein sequence of the passion fruit genome (threshold: 10^{-4}). Combining the comparison results of HMM and BLASTP, a hypothetical *RLP* gene set was formed, which was manually screened to remove redundant sequences. The structure of *RLPs* was annotated using the Pfam database, and the genes with kinase and NB-ARC domains were filtered out with Pfam IDs (PF07714.12, PF00069.20, and PF00931). Finally, 141 *PeRLP* and 79 *PesRLP* genes were identified in ZX and TN genomes, respectively.

4.3. Distribution and Duplication of RLP Genes on Chromosomes in Passion Fruit

The distribution of *RLP* genes in the ZX and TN genome pseudochromosomes were mapped by the Advanced Circos program in the TBtools comprehensive toolkit [40,41]. The repetitive pattern of each *RLP* gene in the ZX and TN genomes was analyzed using BLAST search with E-value $< 1 \times 10^{-5}$. The BLAST results import MCScanX software (https://github.com/wyp1125/MCScanX (accessed on 15 August 2024)) for tandem and WGD duplication identification with the default parameter [42]. In addition, MCScanX was used to identify the collinear modules of *RLP* genes between species and species [42], and these replicators were classified by fragment replication, tandem replication, and random replication events within the species. The previous independent replicators were spliced into linear expansion pathways.

4.4. Phylogenetic Analysis of RLP Gene Family in Passion Fruit

MUSCLE was used to perform multiple sequencing on RLP protein sequences [12]. A phylogenetic tree of RLPs was constructed using IQ-TREE software (v2.0.3) with the maximum-likelihood (ML) method, JTT model, and a bootstrap value of 1000 [43].

4.5. Collinearity and Ka/Ks Calculation Analysis of RLP Genes

The One Step MCScanX-Super Fast plug-in in TBtools-II comprehensive toolkit was used to generate interspecies collinearity files for *Arabidopsis*, ZX, TN, and grape genome [40]. The coding sequence of tandem duplications for *PeRLP* gene pairs was selected to calculate non-synonymous (*Ka*) and synonymous (*Ks*) substitution ratios by the Neigojobori method [44].

4.6. Promoter Cis-Elements Analysis of RLP Genes

The PlantCARE was used to predict cis elements contained in the promoter sequence with the 2000 bp sequence upstream of the translation initiation codon of the *PeRLP* genes family [45], and the results were visualized via the iTOL website [46].

4.7. Gene Structure and Conserved Motif Analysis of RLP Genes

The coding sequence and UTR structure of *PeRLPs* from the original gff3 annotation file of ZX and TN genomes were visualized with TBtools-II [40]. The conserved motifs of the *PeRLPs* were predicted and analyzed through the MEME online website with default parameters [47].

4.8. The Expression Pattern of PeRLP Genes Family in Different Transcriptome Datasets

RNA-seq raw data of four passion fruit tissue samples (roots, leaves, seeds, and flowers) and cold-treated passion fruit seedlings were obtained from previous reports [17,48,49]. All of the above samples for RNA-seq had 3 biological replicates. The filtered data were mapped to the ZX genome by using the HISAT2 software (v2.10) with default parameters [50]. The StringTie software (v1.3.4) was used to standardize the expression levels of all genes in passion fruit [51]. Differentially expressed genes (DEGs) were screened and identified by the R package DESeq2 (https://github.com/thelovelab/DESeq2 (accessed on 15 August 2024)) and $|\log_2 \text{foldchange}| \geq 1$ and p-value < 0.05 parameters [52]. The tissue specificity index (TAC Calc) program in TBtools-II was used to calculate the tissue specificity of each *PeRLP* gene in 4 tissues with filter values greater than 0.6 [40]. WGCNA was used to identify resistance genes associated with cold tolerant variety TN [53]. According to our previous report [54], the hub *PeRLP* genes were screened with the correlation degree of module members greater than 0.9, and the co-expression network was constructed based on this. Coexpression networks were visualized by the Cytoscape software (3.7.1) [55].

R. solani was provided by the Horticultural Plant Genetics and Breeding Laboratory of Fujian Agriculture and Forestry University. In total, 45 seedlings of resistant variety LG and sensitive variety HG were, respectively, selected, and each seedling was infected with *R. solani*. With the stems of five seedlings as a biological duplication, three biological duplication stem samples were collected for RNA-seq. The WGCNA tool was used to identify the modules associated with the highly resistant variety LG for screening *PeRLP* genes in response to the infection of *R. solani*. Then, the hub *PeRLP* genes were screened with the correlation degree of module members greater than 0.9, and the co-expression network was constructed based on this. The reliability of RNA-seq was verified by quantitative detection of candidate *PeRLP* genes, according to the previous RT-qPCR description method [54]. Primers used in qPCR were listed in Supplementary Table S1.

4.9. Transient Expression Analysis of PeRLP8 in Tobacco and Passion Fruit Leaves

Amplification of *PeRLP8* CDS sequence was carried out using 2× Taq Master Mix (Vazyme, Nanjing, China). The resulting fusion products were inserted into a GFP-tagged pCAMBIA1302-GFP vector under CaMV35S promoter control by using Phanta Super-Fidelity DNA Polymerase (Vazyme, Nanjing, China) to form 35S:: *PeRLP8*-GFP recombinant vector. The heavy suspension of 35S:: *PeRLP8*-GFP recombinant plasmid was injected on the back of the leaves of tobacco and passion fruit with empty vector 35S:: GFP as the control. A total of 6 tobacco leaves (3 seedlings) and 6 passion fruit leaves (6 seedlings) were injected, respectively. The injected tobacco and passion fruit plants were cultured for 24 h in the dark environment, and cultured for 24 h under normal conditions. Then, the *R. solani* block with a diameter of 0.5 cm was inoculated on the leaves of tobacco and passion fruit expressing 35S:: GFP and 35S:: *PeRLP8*-GFP instantaneously, and the phenotype of the infected leaves was observed. The leaves of tobacco and passion fruit were collected on the third day after infection. There were 3 biological replicates with 2 leaves as a replicate. Tobacco and passion fruit leaves with transient expressions of 35S:: GFP and 35S:: *PeRLP8*-GFP were collected to measure expression levels of *PeRLP8*. In addition, the relative expression levels of three genes (*PeRBOHD*, *PeMYC*, and *PeJAR1*) co-expressed with the *PeRLP8* gene were also detected using qPCR in the transiently transformed passion fruit leaves. Primers used in qPCR were listed in Supplementary Table S1.

The contents of reactive oxygen species (ROS) were determined using ROS production rate test kit (COMIN, Suzhou, China) with spectrophotometry, as described previously [56].

4.10. Statistical Analyses

Statistical analyses were performed by Student's t test or one-way ANOVA at a significance level of 0.05 using the SPSS 19.0. Figures were geminated by the GraphPad Prism 8.0.2 software.

Supplementary Materials: The following supporting information can be downloaded at https://www.mdpi.com/article/10.3390/plants13233264/s1, Table S1 Primers for real-time quantitative PCR, gene full length cloning, and vector construction. Figure S1 Identification and evolutionary analysis of *RLP* family genes in passion fruit. (A) Chromosomal localization and collinearity analysis of 141 *PeRLP* genes in the ZX genome. (B) Chromosomal localization and collinearity analysis of 79 *PesRLP* genes in the TN genome. Figure S2. The cis-regulatory elements in promoters of *PeRLP* family genes. The number of different cis-regulatory elements in the promoter was represented by square bars of different colors. Figure S3 Tissue specific calculations of 141 *PeRLP* genes in root, leaf, seed, and flower. Differences in gene expression changes were shown in color as the scale, mediumvio-letred for high expression and steelblue for low expression. Figure S4. The expression levels of 6 *PeRLP* genes at 0 h, 4 h and 8 h after JA, MeJA and SA treatment, respectively. Figure S5. Quantitative detection of PeRLP8 gene in resistant variety LG and sensitive variety HG infected by *R. solani*. The data presented are the mean values ± standard deviation (SD) from three independent replicates. Different letters denote significant differences (Student's t-test, $p < 0.05$).

Author Contributions: All the authors have contributed significantly. L.Z. conceived the study and designed the experiments. Y.L. (Yue Li), W.J., Y.L. (Yongkang Li), and Z.S. carried out the experiments. W.Y. and F.L. analyzed data and wrote the manuscript. W.Y. and F.L. contributed equally to the writing of the manuscript. T.F. revised the manuscript. L.Z. directly supervised the project. All authors have read and agreed to the published version of the manuscript.

Funding: This work was financially supported by the Central Government Guides Local Science and Technology Development Projects "Study on Key Genes and Mechanism of Resistance to crown rot in Passion Fruit" (2023L3004) and "14th Five-Year Plan" Fujian Province Seed Industry Innovation and Industrialization Project (zycxny2021010).

Data Availability Statement: The datasets generated during and/or analyzed during the current study are available from the corresponding author upon reasonable request.

Conflicts of Interest: The authors declare no conflicts of interest.

References

1. Lee, H.A.; Yeom, S.I. Plant NB-LRR proteins: Tightly regulated sensors in a complex manner. *Brief. Funct. Genom.* **2015**, *14*, 233–242. [CrossRef] [PubMed]
2. Wan, W.L.; Frohlich, K.; Pruitt, R.N.; Nurnberger, T.; Zhang, L. Plant cell surface immune receptor complex signaling. *Curr. Opin. Plant Biol.* **2019**, *50*, 18–28. [CrossRef]
3. Albert, I.; Hua, C.; Nurnberger, T.; Pruitt, R.N.; Zhang, L. Surface Sensor Systems in Plant Immunity. *Plant Physiol.* **2020**, *182*, 1582–1596. [CrossRef] [PubMed]
4. Kang, W.H.; Lee, J.; Koo, N.; Kwon, J.S.; Park, B.; Kim, Y.M.; Yeom, S.I. Universal gene co-expression network reveals receptor-like protein genes involved in broad-spectrum resistance in pepper (*Capsicum annuum* L.). *Hortic. Res.* **2022**, *9*, uhab003. [CrossRef]
5. Petre, B.; Hacquard, S.; Duplessis, S.; Rouhier, N. Genome analysis of poplar LRR-RLP gene clusters reveals RISP, a defense-related gene coding a candidate endogenous peptide elicitor. *Front. Plant Sci.* **2014**, *5*, 111. [CrossRef] [PubMed]
6. Tör, M.; Brown, D.; Cooper, A.; Woods-Tör, A.; Sjölander, K.; Jones, J.D.G.; Holub, E.B. Arabidopsis downy mildew resistance gene encodes a receptor-like protein similar to and tomato Cf-9. *Plant Physiol.* **2007**, *143*, 1079.
7. Tang, D.; Wang, G.; Zhou, J.M. Receptor Kinases in Plant-Pathogen Interactions: More Than Pattern Recognition. *Plant Cell* **2017**, *29*, 618–637. [CrossRef]
8. Wang, G.D.; Ellendorff, U.; Kemp, B.; Mansfield, J.W.; Forsyth, A.; Mitchell, K.; Bastas, K.; Liu, C.M.; Woods-Tör, A.; Zipfel, C.; et al. A genome-wide functional investigation into the roles of receptor-like proteins in Arabidopsis. *Plant Physiol.* **2008**, *147*, 503–517. [CrossRef]
9. Piedras, P.; Rivas, S.; Droge, S.; Hillmer, S.; Jones, J.D. Functional, c-myc-tagged Cf-9 resistance gene products are plasma-membrane localized and glycosylated. *Plant J.* **2000**, *21*, 529–536. [CrossRef]
10. Belfanti, E.; Silfverberg-Dilworth, E.; Tartarini, S.; Patocchi, A.; Barbieri, M.; Zhu, J.; Vinatzer, B.A.; Gianfranceschi, L.; Gessler, C.; Sansavini, S. The HcrVf2 gene from a wild apple confers scab resistance to a transgenic cultivated variety. *Proc. Natl. Acad. Sci. USA* **2004**, *101*, 886–890. [CrossRef]
11. Liu, C.L.; Liang, N. Integrated control techniques of passion fruit root rot. *Rural. Sci. Technol.* **2020**, *16*, 94–95.
12. Xu, Y.; Zhang, H.; Zhong, Y.; Jiang, N.; Zhong, X.; Zhang, Q.; Chai, S.; Li, H.; Zhang, Z. Comparative genomics analysis of bHLH genes in cucurbits identifies a novel gene regulating cucurbitacin biosynthesis. *Hortic. Res.* **2022**, *9*, uhac038. [CrossRef]
13. Kang, W.H.; Yeom, S.I. Genome-wide Identification, Classification, and Expression Analysis of the Receptor-Like Protein Family in Tomato. *Plant Pathol. J.* **2018**, *34*, 435–444. [CrossRef]

14. Jupe, F.; Pritchard, L.; Etherington, G.J.; MacKenzie, K.; Cock, P.J.A.; Wright, F.; Sharma, S.K.; Bolser, D.; Bryan, G.J.; Jones, J.D.G.; et al. Identification and localisation of the NB-LRR gene family within the potato genome. *BMC Genom.* **2012**, *13*, 75. [CrossRef]
15. Alvarez-Lopez, D.; Herrera-Valencia, V.A.; Gongora-Castillo, E.; Garcia-Laynes, S.; Puch-Hau, C.; Lopez-Ochoa, L.A.; Lizama-Uc, G.; Peraza-Echeverria, S. Genome-Wide Analysis of the LRR-RLP Gene Family in a Wild Banana (*Musa acuminata* ssp. malaccensis) Uncovers Multiple Fusarium Wilt Resistance Gene Candidates. *Genes* **2022**, *13*, 638. [CrossRef] [PubMed]
16. Fritz-Laylin, L.K.; Krishnamurthy, N.; Tor, M.; Sjolander, K.V.; Jones, J.D. Phylogenomic analysis of the receptor-like proteins of rice and Arabidopsis. *Plant Physiol.* **2005**, *138*, 611–623. [CrossRef]
17. Ma, D.N.; Dong, S.S.; Zhang, S.C.; Wei, X.Q.; Xie, Q.J.; Ding, Q.S.; Xia, R.; Zhang, X.T. Chromosome-level reference genome assembly provides insights into aroma biosynthesis in passion fruit. *Mol. Ecol. Resour.* **2021**, *21*, 955–968. [CrossRef]
18. Xia, Z.Q.; Huang, D.M.; Zhang, S.K.; Wang, W.Q.; Ma, F.N.; Wu, B.; Xu, Y.; Xu, B.Q.; Chen, D.; Zou, M.L.; et al. Chromosome-scale genome assembly provides insights into the evolution and flavor synthesis of passion fruit (*Passiflora edulis* Sims). *Hortic. Res.* **2021**, *8*, 14. [CrossRef] [PubMed]
19. Leister, D. Tandem and segmental gene duplication and recombination in the evolution of plant disease resistance genes. *Trends Genet.* **2004**, *20*, 116–122. [CrossRef]
20. Seo, E.; Kim, S.; Yeom, S.I.; Choi, D. Genome-Wide Comparative Analyses Reveal the Dynamic Evolution of Nucleotide-Binding Leucine-Rich Repeat Gene Family among Solanaceae Plants. *Front. Plant Sci.* **2016**, *7*, 1205. [CrossRef]
21. Andolfo, G.; Sanseverino, W.; Rombauts, S.; Van de Peer, Y.; Bradeen, J.M.; Carputo, D.; Frusciante, L.; Ercolano, M.R. Overview of tomato (*Solanum lycopersicum*) candidate pathogen recognition genes reveals important Solanum R locus dynamics. *New Phytol.* **2013**, *197*, 223–237. [CrossRef]
22. Kim, S.B.; Kang, W.H.; Huy, H.N.; Yeom, S.I.; An, J.T.; Kim, S.; Kang, M.Y.; Kim, H.J.; Jo, Y.D.; Ha, Y.; et al. Divergent evolution of multiple virus-resistance genes from a progenitor in *Capsicum* spp. *New Phytol.* **2017**, *213*, 886–899. [CrossRef] [PubMed]
23. Kruijt, M.; MJ, D.E.K.; de Wit, P.J. Receptor-like proteins involved in plant disease resistance. *Mol. Plant Pathol.* **2005**, *6*, 85–97. [CrossRef]
24. Dixon, M.S.; Jones, D.A.; Keddie, J.S.; Thomas, C.M.; Harrison, K.; Jones, J.D. The tomato Cf-2 disease resistance locus comprises two functional genes encoding leucine-rich repeat proteins. *Cell* **1996**, *84*, 451–459. [CrossRef]
25. Dixon, M.S.; Hatzixanthis, K.; Jones, D.A.; Harrison, K.; Jones, J.D. The tomato Cf-5 disease resistance gene and six homologs show pronounced allelic variation in leucine-rich repeat copy number. *Plant Cell* **1998**, *10*, 1915–1925. [CrossRef]
26. Takken, F.L.; Thomas, C.M.; Joosten, M.H.; Golstein, C.; Westerink, N.; Hille, J.; Nijkamp, H.J.; De Wit, P.J.; Jones, J.D. A second gene at the tomato Cf-4 locus confers resistance to cladosporium fulvum through recognition of a novel avirulence determinant. *Plant J.* **1999**, *20*, 279–288. [CrossRef]
27. Thomas, C.M.; Jones, D.A.; Parniske, M.; Harrison, K.; Balint-Kurti, P.J.; Hatzixanthis, K.; Jones, J.D. Characterization of the tomato Cf-4 gene for resistance to Cladosporium fulvum identifies sequences that determine recognitional specificity in Cf-4 and Cf-9. *Plant Cell* **1997**, *9*, 2209–2224.
28. Kawchuk, L.M.; Hachey, J.; Lynch, D.R.; Kulcsar, F.; van Rooijen, G.; Waterer, D.R.; Robertson, A.; Kokko, E.; Byers, R.; Howard, R.J.; et al. Tomato Ve disease resistance genes encode cell surface-like receptors. *Proc. Natl. Acad. Sci. USA* **2001**, *98*, 6511–6515. [CrossRef] [PubMed]
29. Zhao, Z.; Zhang, R.; Wang, D.; Zhang, J.; Zang, S.; Zou, W.; Feng, A.; You, C.; Su, Y.; Wu, Q.; et al. Dissecting the features of TGA gene family in Saccharum and the functions of ScTGA1 under biotic stresses. *Plant Physiol. Biochem.* **2023**, *200*, 107760. [CrossRef]
30. Jiang, Z.N.; Ge, S.; Xing, L.P.; Han, D.J.; Kang, Z.S.; Zhang, G.Q.; Wang, X.J.; Wang, X.U.; Chen, P.D.; Cao, A.Z. A novel wheat receptor-like protein gene, is involved in the defence response against f. sp. *J. Exp. Bot.* **2013**, *64*, 3735–3746. [CrossRef]
31. Bi, G.; Liebrand, T.W.; Cordewener, J.H.; America, A.H.; Xu, X.; Joosten, M.H. Arabidopsis thaliana receptor-like protein AtRLP23 associates with the receptor-like kinase AtSOBIR1. *Plant Signal. Behav.* **2014**, *9*, e27937. [CrossRef]
32. Pan, L.Y.; Zhou, J.; Sun, Y.; Qiao, B.X.; Wan, T.; Guo, R.Q.; Zhang, J.; Shan, D.Q.; Cai, Y.L. Comparative transcriptome and metabolome analyses of cherry leaves spot disease caused by Alternaria alternata. *Front. Plant Sci.* **2023**, *14*, 1129515. [CrossRef] [PubMed]
33. Hu, C.; Wei, C.; Ma, Q.; Dong, H.; Shi, K.; Zhou, Y.; Foyer, C.H.; Yu, J. Ethylene response factors 15 and 16 trigger jasmonate biosynthesis in tomato during herbivore resistance. *Plant Physiol.* **2021**, *185*, 1182–1197. [CrossRef] [PubMed]
34. Zhang, J.H.; Wei, H.B.; Hong, Y.H.; Yang, R.R.; Meng, J.; Luan, Y.S. The lncRNA20718-miR6022-RLPs module regulates tomato resistance to Phytophthora infestans. *Plant Cell Rep.* **2024**, *43*, 57. [CrossRef] [PubMed]
35. Bedard, K.; Krause, K.H. The NOX family of ROS-generating NADPH oxidases: Physiology and pathophysiology. *Physiol. Rev.* **2007**, *87*, 245–313. [CrossRef]
36. Yoshioka, H.; Numata, N.; Nakajima, K.; Katou, S.; Kawakita, K.; Rowland, O.; Jones, J.D.; Doke, N. Nicotiana benthamiana gp91phox homologs NbrbohA and NbrbohB participate in H_2O_2 accumulation and resistance to Phytophthora infestans. *Plant Cell* **2003**, *15*, 706–718. [CrossRef]

37. Zang, S.; Qin, L.; Zhao, Z.; Zhang, J.; Zou, W.; Wang, D.; Feng, A.; Yang, S.; Que, Y.; Su, Y. Characterization and Functional Implications of the Nonexpressor of Pathogenesis-Related Genes 1 (NPR1) in Saccharum. *Int. J. Mol. Sci.* **2022**, *23*, 7984. [CrossRef]
38. Wu, Q.B.; Pan, Y.B.; Su, Y.C.; Zou, W.H.; Xu, F.; Sun, T.T.; Grisham, M.P.; Yang, S.L.; Xu, L.P.; Que, Y.X. WGCNA Identifies a Comprehensive and Dynamic Gene Co-Expression Network That Associates with Smut Resistance in Sugarcane. *Int. J. Mol. Sci.* **2022**, *23*, 10770. [CrossRef]
39. Wang, D.J.; Qin, L.Q.; Wu, M.X.; Zou, W.H.; Zang, S.J.; Zhao, Z.N.; Lin, P.X.; Guo, J.L.; Wang, H.B.; Que, Y.X. Identification and characterization of WAK gene family in and the negative roles of under the pathogen stress. *Int. J. Biol. Macromol.* **2023**, *224*, 1–19. [CrossRef]
40. Chen, C.; Wu, Y.; Li, J.; Wang, X.; Zeng, Z.; Xu, J.; Liu, Y.; Feng, J.; Chen, H.; He, Y.; et al. TBtools-II: A "one for all, all for one" bioinformatics platform for biological big-data mining. *Mol. Plant* **2023**, *16*, 1733–1742. [CrossRef]
41. Chen, C.; Chen, H.; Zhang, Y.; Thomas, H.R.; Frank, M.H.; He, Y.; Xia, R. TBtools: An Integrative Toolkit Developed for Interactive Analyses of Big Biological Data. *Mol. Plant* **2020**, *13*, 1194–1202. [CrossRef] [PubMed]
42. Wang, Y.; Tang, H.; Debarry, J.D.; Tan, X.; Li, J.; Wang, X.; Lee, T.H.; Jin, H.; Marler, B.; Guo, H.; et al. MCScanX: A toolkit for detection and evolutionary analysis of gene synteny and collinearity. *Nucleic Acids Res.* **2012**, *40*, e49. [CrossRef] [PubMed]
43. Nguyen, L.T.; Schmidt, H.A.; von Haeseler, A.; Minh, B.Q. IQ-TREE: A fast and effective stochastic algorithm for estimating maximum-likelihood phylogenies. *Mol. Biol. Evol.* **2015**, *32*, 268–274. [CrossRef]
44. Kumar, S.; Stecher, G.; Tamura, K. MEGA7: Molecular Evolutionary Genetics Analysis Version 7.0 for Bigger Datasets. *Mol. Biol. Evol.* **2016**, *33*, 1870–1874. [CrossRef]
45. Lescot, M.; Dehais, P.; Thijs, G.; Marchal, K.; Moreau, Y.; Van de Peer, Y.; Rouze, P.; Rombauts, S. PlantCARE, a database of plant cis-acting regulatory elements and a portal to tools for in silico analysis of promoter sequences. *Nucleic Acids Res.* **2002**, *30*, 325–327. [CrossRef]
46. Letunic, I.; Bork, P. Interactive tree of life (iTOL) v3: An online tool for the display and annotation of phylogenetic and other trees. *Nucleic Acids Res.* **2016**, *44*, W242–W245. [CrossRef]
47. Bailey, T.L.; Boden, M.; Buske, F.A.; Frith, M.; Grant, C.E.; Clementi, L.; Ren, J.; Li, W.W.; Noble, W.S. MEME SUITE: Tools for motif discovery and searching. *Nucleic Acids Res.* **2009**, *37*, W202–W208. [CrossRef]
48. Xu, Y.; Huang, D.M.; Ma, F.N.; Yang, L.; Wu, B.; Xing, W.T.; Sun, P.G.; Chen, D.; Xu, B.Q.; Song, S. Identification of key genes involved in flavonoid and terpenoid biosynthesis and the pathway of triterpenoid biosynthesis in Passiflora edulis. *J. Integr. Agr.* **2023**, *22*, 1412–1423. [CrossRef]
49. Wu, Y.; Huang, W.; Tian, Q.; Liu, J.; Xia, X.; Yang, X.; Mou, H. Comparative transcriptomic analysis reveals the cold acclimation during chilling stress in sensitive and resistant passion fruit (*Passiflora edulis*) cultivars. *PeerJ* **2021**, *9*, e10977. [CrossRef]
50. Thakur, V. RNA-Seq Data Analysis for Differential Gene Expression Using HISAT2-StringTie-Ballgown Pipeline. *Methods Mol. Biol.* **2024**, *2812*, 101–113.
51. Kovaka, S.; Zimin, A.V.; Pertea, G.M.; Razaghi, R.; Salzberg, S.L.; Pertea, M. Transcriptome assembly from long-read RNA-seq alignments with StringTie2. *Genome Biol.* **2019**, *20*, 278. [CrossRef] [PubMed]
52. Liu, S.; Wang, Z.; Zhu, R.; Wang, F.; Cheng, Y.; Liu, Y. Three Differential Expression Analysis Methods for RNA Sequencing: Limma, EdgeR, DESeq2. *J. Vis. Exp.* **2021**, *175*, e62528. [CrossRef]
53. Zhu, M.; Xie, H.; Wei, X.; Dossa, K.; Yu, Y.; Hui, S.; Tang, G.; Zeng, X.; Yu, Y.; Hu, P.; et al. WGCNA Analysis of Salt-Responsive Core Transcriptome Identifies Novel Hub Genes in Rice. *Genes* **2019**, *10*, 719. [CrossRef] [PubMed]
54. Liang, F.; Zhang, Y.; Wang, X.; Yang, S.; Fang, T.; Zheng, S.; Zeng, L. Integrative mRNA and Long Noncoding RNA Analysis Reveals the Regulatory Network of Floral Bud Induction in Longan (*Dimocarpus longan* Lour.). *Front. Plant Sci.* **2022**, *13*, 923183. [CrossRef] [PubMed]
55. Shannon, P.; Markiel, A.; Ozier, O.; Baliga, N.S.; Wang, J.T.; Ramage, D.; Amin, N.; Schwikowski, B.; Ideker, T. Cytoscape: A software environment for integrated models of biomolecular interaction networks. *Genome Res.* **2003**, *13*, 2498–2504. [CrossRef]
56. Wang, N.; Fan, X.; He, M.; Hu, Z.; Tang, C.; Zhang, S.; Lin, D.; Gan, P.; Wang, J.; Huang, X.; et al. Transcriptional repression of TaNOX10 by TaWRKY19 compromises ROS generation and enhances wheat susceptibility to stripe rust. *Plant Cell* **2022**, *34*, 1784–1803. [CrossRef]

plants MDPI

Article

Emission and Transcriptional Regulation of Aroma Variation in *Oncidium* Twinkle 'Red Fantasy' Under Diel Rhythm

Yan Chen [1,2], Shengyuan Zhong [1,3], Lan Kong [1,2], Ronghui Fan [1,2], Yan Xu [4], Yiquan Chen [1,2] and Huaiqin Zhong [1,2,*]

1 Institute of Crop Sciences, Fujian Academy of Agricultural Sciences (Fujian Germplasm Resources Center), Fuzhou 350013, China; spencer_cy@163.com (Y.C.)
2 Fujian Engineering Research Center for Characteristic Floriculture, Fuzhou 350013, China
3 Fuzhou Sub-Center for New Plant Variety Tests, Ministry of Agriculture and Rural Affairs, Fuzhou 350013, China
4 College of Landscape Architecture and Art, Fujian Agriculture and Forestry University, Fuzhou 350002, China
* Correspondence: zhqeast@163.com

Abstract

Oncidium hybridum is one of the important cut-flowers in the world. However, the lack of aroma in its cut-flower varieties greatly limits the sustainable development of the *Oncidium hybridum* cut-flowers industry. This paper is an integral investigation of the diel pattern and influencing factors of the aroma release of *Oncidium* Twinkle 'Red Fantasy'. GC-MS analysis revealed that the release of 3-Carene peaked at 10:00, while Butyl tiglate and Prenyl senecioate did so at 14:00, with a diel rhythm. By analyzing the correlation network between aroma component synthesis and differentially expressed genes, 15 key structural genes were detected and regulated by multiple circadian rhythm-related transcription factors. *Cluster-17371.18_TPS*, *Cluster-65495.1_TPS*, *Cluster-46699.0_TPS*, *Cluster-60935.10_DXS*, *Cluster-47205.4_IDI*, and *Cluster-65313.7_LOX* were key genes in the terpenoid and fatty acid derivative biosynthetic pathway, which were co-expressed with aroma release. Constant light/dark treatments revealed that the diurnal release of 3-Carene may be influenced by light and the circadian clock, and Butyl tiglate and Prenyl senecioate may be mainly determined by endogenous circadian clock. Under constant light treatment, the *TPS*, *DXS*, *IDI*, and *LOX* genes seem to lose their regulatory role in the release of aroma compounds from *Oncidium* Twinkle 'Red Fantasy'. Under constant dark treatment, the *TPS* genes were consistent with the release pattern of 3-Carene, which may be a key factor in regulating the diel rhythm of 3-Carene biosynthesis. These results laid a theoretical foundation for the study of floral transcriptional regulation and genetic engineering technology breeding of *Oncidium hybridum*.

Keywords: *Oncidium hybridum*; floral scent; diel rhythm; transcriptome; light/dark treatment

1. Introduction

Floral scent is one of the important qualities and ornamental traits of ornamental plants, serving as a medium for communication between plant flowers and the external environment. It plays a crucial role in attracting pollinators, resisting pathogens and parasites, and serving as relevant signals in response to biotic and abiotic stress [1]. Meanwhile, floral volatiles are widely used in perfumes, scents, and cosmetics. Floral scent consists of a mixture of volatile organic compounds (VOCs), which are classified into terpenoids, phenylpropanoids/benzenoids, fatty acid derivatives, and amino acid derivatives according to biosynthetic pathways [2]. Terpenes are the most diverse and abundant secondary metabolites in plant floral scents, synthesized through two independent pathways: the mevalonic acid (MVA) pathway and the methylerythritol phosphate (MEP) pathway [3,4]. The MEP pathway mainly synthesizes sesquiterpenes, monoterpenes, and diterpenes, while the MVA pathway mainly synthesizes sesquiterpenes [5]. Among them, the MEP pathway uses pyruvate and glyceraldehyde 3-phosphate (G3P) as precursors, and undergoes seven enzymatic reactions to produce isopentenyl diphosphate (IPP) or dimethylallyl diphosphate (DMAPP) [6–8]. IPP and DMAPP generate sesquiterpenes, monoterpenes, diterpenes, and volatile carotenoid derivatives under the catalysis of geranylgeranyl diphosphate synthase (GGPPS), geranyl diphosphate synthase (GPPS), and terpene synthase (TPS) enzymes [6–8]. Fatty acid derivatives are another major class of VOCs in plants, and their biosynthesis mainly occurs through the lipoxygenase (LOX) pathway [9]. Linoleic and linolenic acid are the main substrates of LOX in plant tissues, which catalyze the production of unsaturated 9-hydroperoxides or 13-hydroperoxides through 9- and 13-LOX [5,10]. Unsaturated 9-hydroperoxides or 13-hydroperoxides are further oxidized and converted to volatile compounds by hydroperoxide lyase (HPL) and alcohol dehydrogenase (ADH) [5,10]. Up to now, there have been a lot of studies on the function, biosynthesis, and regulation of VOCs in ornamental plants [5,11]. However, the composition and abundance of VOCs in floral scents vary greatly among plants [12].

The synthesis of floral compounds in plants is a dynamic process throughout the entire plant cycle and is regulated by many factors, including light [13], flowering stages [14], diurnal patterns [15], and visiting insects [16]. The release of flower compounds usually exhibits a distinct diel rhythm, which has been demonstrated in many plants. These include snapdragon [17], rose [18], and Wisteria [19] flowers that release scent in the daytime, as well as *Petunia axillaris* [20] and *Nymphaea* subg. *Hydrocallis* [21] that release scent in the night. The release of the main active aroma compounds Myrcene and (E)-β-ocimene in snapdragon exhibited a diel rhythm variation with an increase during the day and a decrease at night [17]. The emission of the four major volatiles (E)-β-ocimene, benzaldehyde, geranyl isovalerate, and benzyl tiglate in *Globe amaranth* showed a diel rhythmic pattern, with the maximum emission occurring in the late afternoon and the lowest emission occurring at night [22]. The main floral components of *Petunia axillaris* were methyl benzoate, benzaldehyde, and isoeugenol, which exhibited a circadian rhythm [20]. The highest release occurred at 0 a.m. and lowest at 12 p.m. [20]. The release of floral compounds is regulated by genes, and the expression of many genes in the MEP pathway of terpenoid synthesis is rhythmic [17]. The monoterpene synthetase and deoxyoxyketose-5-phosphate synthetase (*DXPS*) genes of snapdragon showed a consistent change pattern of terpenoids, and this circadian rhythm was regulated by the endogenous circadian clock [17]. The expression levels of *DXS*, *DXR2*, *MCT*, and *HDR* genes in *Populus trichocarpa* followed a consistent rhythm with the release of monoterpenes [23]. The expression of terpene synthase gene *OfTPS5* in osmanthus flower showed a rhythm of increasing in the day and decreasing at night [24]. Many esters' metabolism was also regulated by the circadian clock. In *Rosa hybrida*, the release of geranyl acetate exhibited a consistent diel rhythm with the gene expression of the *RhAAT* in the LOX pathway [18]. In addition, the accumulation of *ZmLOX10* in the maize LOX pathway was strictly regulated by the diel rhythm, and was consistent with the highest photosynthetic activity over time [25]. At present, the research on the molecular mechanism of diel rhythm mainly focuses on Arabidopsis [26,27]. In petunias, the circadian rhythm-related gene *PhLHY* regulates the volatilization of aroma compounds by inhibiting the expression of genes on the floral volatile benzenoid/phenylpropanoid pathway during the day [28]. Yon et al. [29] silenced two key circadian rhythm-related genes, *NaLHY* and *NaZTL*, resulting in changes in the endogenous rhythm of volatile compound release. Many studies have found that genes involved in the synthesis and metabolism of different volatile compounds exhibit rhythmic expression. However, the specific metabolic mechanisms by which the diel rhythm affects floral scent are still unclear.

Oncidium hybridum belongs to the genus *Oncidium* of the Orchidaceae, also known as "dancing lady orchid", native to the Neotropical region. *Oncidium hybridum* is one of the most important cut-flower varieties in the world with its unique flower posture and shape, bright colors, long flowering period, and wide adaptability [30]. At present, there are few varieties of cut-flowers of *Oncidium hybridum* on the market, all of which are scentless varieties [30]. Floral scent is one of the key characteristics of many floricultural crops. Due to the poor affinity of interspecific and intergeneric hybridization, it is difficult to introduce aroma traits into cut-flower varieties by traditional breeding methods. This greatly limits the sustainable development of the cut-flower industry of *Oncidium hybridum* [31]. Therefore, a thorough analysis of the molecular mechanism underlying the scents formation of *Oncidium hybridum* is necessary to effectively regulate the breeding process of aroma varieties and improve breeding efficiency. At present, there are few aromatic varieties of *Oncidium hybridum*, and research on its aromatic volatile compounds has mainly focused on potted varieties such as *Oncidium* Sharry Baby [32,33] and *Oncidium* Rosy Sunset [34]. The characteristic aroma compounds of *Oncidium* Sharry Baby include 3,7-dimethyl-1,3,6-Octatriene, (E)-2-Butenoic acid, 2-(methylenecyclopropyl) prop-2-yl ester, 3,7-dimethyl-1,6-Octadien-3-ol, and (Z)-3,7-dimethyl-2,6-Octadien-1-ol, among which terpenes are important compounds affecting the aroma of *Oncidium* Sharry Baby [32,33]. Yeh et al. [35] validated the mechanism of the diel release of the scent of *Oncidium* Sharry Baby. It was also shown that *Circadian Clock Associated 1* (*CCA1*) transcription factor (TF) affected the diel rhythm release of scent by up-regulating *TPS* under constant light treatment. The main VOCs released by *Oncidium* Rosy Sunset was the Linalool belonging to terpenoid [34]. Its aroma release has a rhythmic pattern, with the highest concentration from 10:00 to 12:00, and the lowest concentration from 20:00 to 24:00 [34]. The *Oncidium* Twinkle series was known for its intense scent, but the type and rhythm regulation of its aroma have not been reported. Up to now, the biosynthesis of orchid scents is still unclear, and only a few studies have reported on this. Hsiao et al. [30] reported the monoterpenoid biosynthetic pathway and related genes in Phalaenopsis. The *PbGDPS* gene was specifically expressed in flowers, and aroma biosynthesis of Phalaenopsis was mainly due to the production of geranyl and Linalool by *DXPS* [30]. In *Cymbidium ensifolium*, the release of methyl jasmonate and its related genes *CeLOX*, *CeAOS*, *CeAOC*, and *CeJMT* showed the same trend in sepals and petals, and were all highly regulated [36]. These studies laid a foundation for exploring the formation and diel rhythm regulation mechanism of *Oncidium hybridum* scents. However, due to its long life cycle and large genome size, little was known about the way of producing its floral scents.

In this study, the potted aroma variety *Oncidium* Twinkle 'Red Fantasy' was used as the material, which has a strong floral aroma. We explored the mechanism of aroma release and diel rhythm-related mechanisms of *Oncidium* Twinkle 'Red Fantasy' based on the scientific phenomenon of the rhythmic release of its scent. The volatiles released by the *Oncidium* Twinkle 'Red Fantasy' were collected at different time points within 48 h (under 12 h light/12 h dark) using the dynamic headspace technique and analyzed using gas chromatography-mass spectrometry (GC-MS). Combined with transcriptome sequencing, the genes and TFs related to VOCs biosynthesis and the diel rhythm pathway in the transcriptome of *Oncidium* Twinkle 'Red Fantasy' were analyzed and excavated.

In order to further explore the relationship between aroma synthesis genes and the diel rhythm of aroma release, we detected the content of VOCs released and the expression of aroma synthesis genes in *Oncidium* Twinkle 'Red Fantasy' under constant light/dark treatment. These results will lay the foundation for analyzing the rhythmic release of aroma in *Oncidium hybridum*, and also provide a reference for the genetic engineering breeding of floral aroma.

2. Results

2.1. Oncidium Twinkle 'Red Fantasy' VOCs Were Emitted by Diel Rhythm Pattern Synchronizing

To identify the diel rhythm pattern of VOCs associated with *Oncidium* Twinkle 'Red Fantasy', GC−MS was employed for profiling volatiles at different time points within 48 h (under 12 h light/12 h dark) of flowering. We collected samples every 4 h over two days. GC-MS analysis of the headspace trapping revealed five volatile components. Three typical peaks were detected in *Oncidium* Twinkle 'Red Fantasy' flowers and identified as 3-Carene, Butyl tiglate, and Prenyl senecioate, which accounted for >96% of the total peak areas (Figure S1). 3-Carene belongs to monoterpenes, and its release began to increase and peak around 10:00 (Days 1 and 2) during diel oscillation, followed by a rapid decrease (Figure 1a,b). Butyl tiglate and Prenyl senecioate belong to ester compounds, and their diel oscillation displayed a peak around 14:00 (Day 1 and 2) (Figure 1c–f). The emission of Butyl tiglate increased from 2:00 until 14:00, and decreased from 14:00 until 22:00 (Days 1 and 2) (Figure 1c,d). The emission of Prenyl senecioate decreased from 2:00 to 6:00, while it showed an upward trend from 6:00 to 14:00, and decreased from 14:00 to 22:00 (Days 1 and 2) (Figure 1e,f). The diel release pattern of three volatile components were almost the same across all two days (Figure 1). This indicated that the floral emission pattern of *Oncidium* Twinkle 'Red Fantasy' was a diel control that may be regulated by diel rhythmic genes.

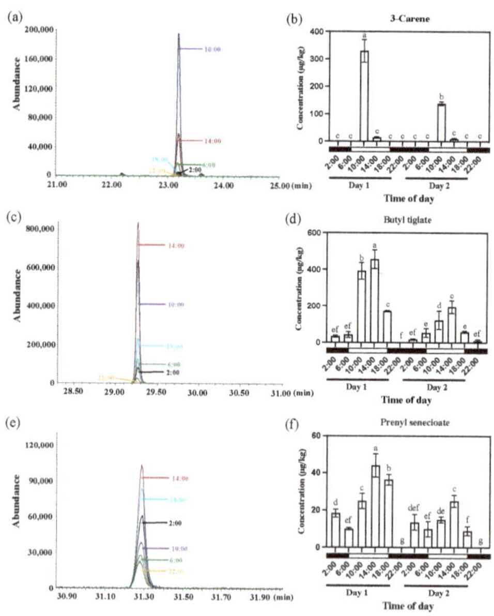

Figure 1. The emission patterns of three floral scent compounds—3-Carene, Butyl tiglate, and Prenyl senecioate—in *Oncidium* Twinkle 'Red Fantasy' under normal photoperiod (under 12 h light/12 h dark). (**a**) Overlapping analysis of 3-Carene ion current in samples at different time points within 24 h. The abscissa represents the retention time (min), and the ordinate represents the ion current intensity.

(**b**) The emission patterns of 3-Carene from *Oncidium* Twinkle 'Red Fantasy' flowers within 48 h. (**c**) Overlapping analysis of Butyl tiglate ion current in samples at different time points within 24 h. The abscissa represents the retention time (min), and the ordinate represents the ion current intensity. (**d**) The emission patterns of Butyl tiglate from *Oncidium* Twinkle 'Red Fantasy' flowers within 48 h. (**e**) Overlapping analysis of Prenyl senecioate ion current in samples at different time points within 24 h. The abscissa represents the retention time (min), and the ordinate represents the ion current intensity. (**f**) The emission patterns of Prenyl senecioate from *Oncidium* Twinkle 'Red Fantasy' flowers within 48 h. Each treatment was conducted in triplicate with three technical repeats. Values are mean ± SD. Different lowercase letters indicate a statistically significant difference ($p < 0.05$).

2.2. Overview of Transcriptome Sequencing

In order to obtain insights into the diel rhythm changes of floral VOCs biosynthesis of *Oncidium* Twinkle 'Red Fantasy', transcriptomic analysis at different time points within 24 h (under 12 h light/12 h dark) was performed. Principal component analysis (PCA) demonstrated that samples were well discriminated according to the classification of different time points, and clustered mainly according to three biological replicates (Figure 2a). The heatmap and hierarchical clustering showed different patterns of gene expression during the diel rhythm (from 2:00 to 22:00) (Figure 2b). Among them, the time points of ante meridiem (2:00, 6:00 and 10:00) and post meridiem (14:00, 18:00, and 22:00) were clustered, respectively. The results indicated that there were different gene expression patterns of *Oncidium* Twinkle 'Red Fantasy' between the ante meridiem and post meridiem.

To further explore the molecular regulatory mechanism of aroma formation under diel rhythm, RNA-seq data were analyzed by integrating aroma release patterns. We used the K-means clustering algorithm to classify the different characteristics of differentially expressed genes (DEGs). Eight different expression clusters (Cluster 1 to Cluster 8) were directly exhibited, with 1818, 958, 1111, 2341, 1773, 878, 815, and 1476 members, respectively (Figure 2c). Based on the change in the contents of key aroma compounds and gene expression profiles, we found that the gene expressions in Cluster 6 and Cluster 8 reached their peak at 14:00 and 10:00, respectively. This was similar to the trend of changes in the aroma content of *Oncidium* Twinkle 'Red Fantasy' (Figure 2c). In addition, the expression of genes regulating aroma synthesis may precede the release of aroma. Therefore, we focused on the functional annotation of genes in Cluster 4, Cluster 6, and Cluster 8 (Figure 2d). The content of 3-Carene peaked at 10:00, and the gene expression of Cluster 4 and Cluster 8 peaked at 6:00 and 10:00, respectively. The results of the Kyoto Encyclopedia of Genes and Genomes (KEGG) enrichment analysis showed that genes in Cluster 4 and Cluster 8 were significantly enriched in the pathways related to terpene synthesis, including "Terpenoid backbone biosynthesis", "Carotenoid biosynthesis", and "Sesquiterpenoid and triterpenoid biosynthesis". In addition, Cluster 8 genes were also significantly enriched in "Monoterpenoid biosynthesis" and "Zeatin biosynthesis" pathways. The Cluster 6 genes were significantly enriched in the "Linoleic acid metabolism" pathway related to ester metabolism, following the same trend as the content of Butyl tiglate and Prenyl senecioate, which reached its peak at 14:00. Meanwhile, Cluster 4, Cluster 6, and Cluster 8 genes were enriched in the "Circadian rhythm-plant" pathway. Therefore, we speculated that the genes in Cluster 4 and Cluster 8 were closely related to terpenoid biosynthesis, and Cluster 6 genes were related to ester volatile compound biosynthesis. These genes may be the reason for the formation of diel rhythm changes in the aroma of *Oncidium* Twinkle 'Red Fantasy'.

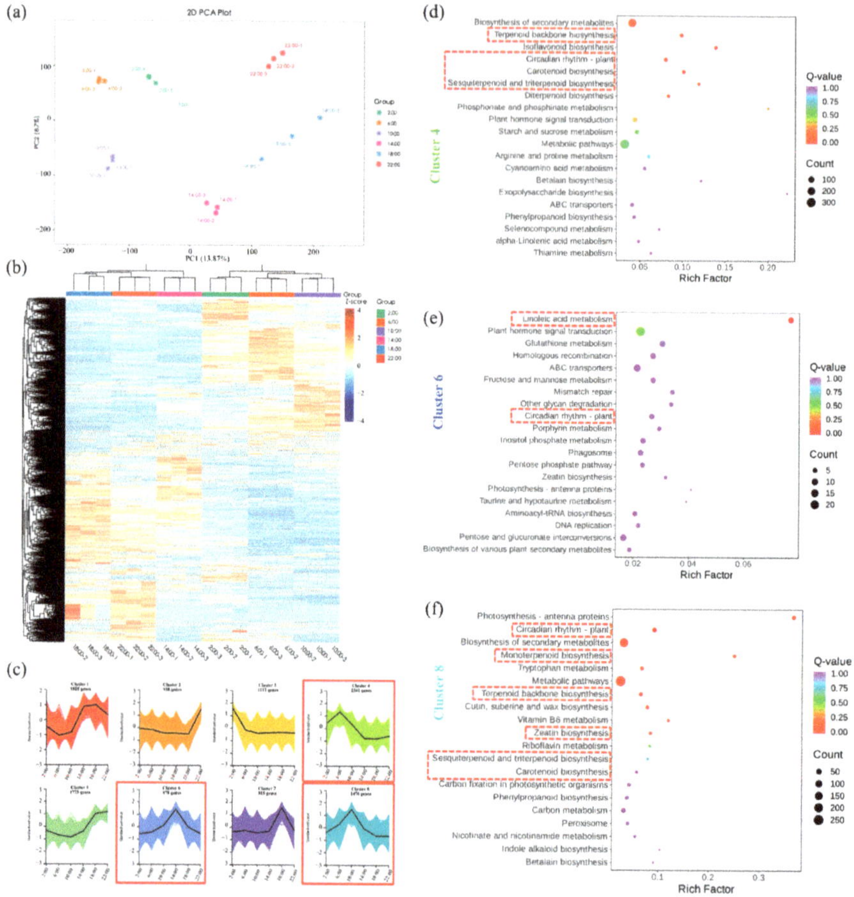

Figure 2. Transcriptomic analysis of *Oncidium* Twinkle 'Red Fantasy' at different time points within 24 h (under 12 h light/12 h dark). (**a**) Principal component analysis (PCA) plot showed overall differences among six groups (2:00, 6:00, 10:00, 14:00, 18:00, and 22:00) and the variability between intra-group samples. (**b**) Heatmap of differentially expressed genes (DEGs) sorted by K-means clustering across the samples collected at different time points. The numbers 1, 2, and 3 with each sample represented number of replicates. (**c**) Eight K-means clusters (Clusters 1–8) showed differential expression trends of DEGs at different time points. (**d**) KEGG enrichment analysis of DEGs in Cluster 4. The red boxes indicate metabolic pathways related to aroma rhythm release. (**e**) KEGG enrichment analysis of DEGs in Cluster 6. The red boxes indicate metabolic pathways related to aroma rhythm release. (**f**) KEGG enrichment analysis of DEGs in Cluster 8. The red boxes indicate metabolic pathways related to aroma rhythm release.

2.3. *Characterization of Fatty Acid Derivative and Terpenoid Biosynthetic Pathway Under Diel Rhythm*

To gain further insights into the biosynthesis of aroma compounds under diel rhythm, we outlined the biosynthesis pathways of fatty acid derivative and terpenoid based on the changes in the content of aroma compounds in *Oncidium* Twinkle 'Red Fantasy' and the KEGG database (Figure 3). Fatty acid derivatives are one of the main volatile organic compounds in plants, and lipoxygenase (LOX) was the crucial functional protein for the formation of the characteristic aroma compounds of the esters [9]. Polyunsaturated fatty acids are catabolized via the LOX pathway to form intermediates of 9-hydroperoxide and

13-hydroperoxide [5]. A total of 21 genes were found to be differentially expressed at 6:00 vs. 2:00, 10:00 vs. 2:00, 14:00 vs. 2:00, 18:00 vs. 2:00, 22:00 vs. 2:00, 10:00 vs. 6:00, 14:00 vs. 6:00, 18:00 vs. 6:00, 22:00 vs. 6:00, 14:00 vs. 10:00, 18:00 vs. 10:00, 22:00 vs. 10:00, 18:00 vs. 14:00, 22:00 vs. 14:00, and 22:00 vs. 18:00 at different steps of LOX pathway (Figure S2a). Among the 21 differentially enriched genes, the highly expressed genes were mainly concentrated at 14:00 (38.10%), while the remaining genes highly expressed at 2:00, 6:00, 10:00, and 18:00 were 9.52%, 23.80%, 14.29%, and 14.29%, respectively. Among them, eight *LOX* genes belong to Cluster 6, which was similar to the release diel rhythm of ester compounds in *Oncidium* Twinkle 'Red Fantasy', including two *13-LOX* genes (*Cluster-62377*.0 and *Cluster-62377.1*) and six *9-LOX* genes (*Cluster-65313.13*, *Cluster-65313.5*, *Cluster-65313.7*, *Cluster-65313.11*, *Cluster-65313.4*, and *Cluster-65313.9*) (Figure 3a). Consequently, it was reasonable to speculate that the eight differentially expressed *LOX* genes may lead to the peak emission of ester volatile compounds in *Oncidium* Twinkle 'Red Fantasy'.

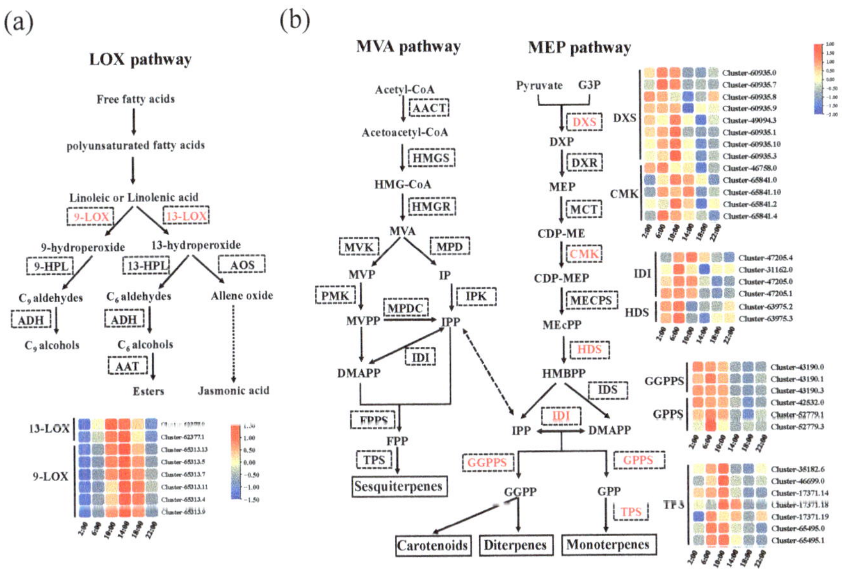

Figure 3. Overview of metabolites and DEGs in the biosynthesis pathways of fatty acid derivative and terpenoid in *Oncidium* Twinkle 'Red Fantasy'. (**a**) The DEGs of Cluster 6 were enriched in the fatty acid derivative biosynthesis pathway. *9-lipoxygenase* (*9-LOX*), *13-lipoxygenase* (*13-LOX*), *9-hydroperoxide lyase* (*9-HPL*), *13-hydroperoxide lyase* (*13-HPL*), *alcohol dehydrogenase* (*ADH*), *allene oxide synthase* (*AOS*), and *alcohol acyltransferase* (*AAT*). The black dashed boxes represent genes enriched in the LOX pathway, and the red fonts represent differentially expressed genes. (**b**) The DEGs of Cluster 4 and Cluster 8 were enriched in the terpenoid biosynthesis pathway. *Acetyl-CoA acetyltransferase* (*AACT*), *hydroxymethylglutaryl-CoA synthase* (*HMGS*), *hydroxymethylglutaryl-CoA reductase* (*HMGR*), *mevalonate kinase* (*MVK*), *mevalonate phosphate decarboxylase* (*MPD*), *phosphomevalonate kinase* (*PMK*), *isopentenyl phosphate kinase* (*IPK*), *mevalonate diphosphate decarboxylase* (*MPDC*), *isopentenyl diphosphate isomerase* (*IDI*), *farnesyl pyrophosphate synthase* (*FPPS*), *terpenoid synthase* (*TPS*), *1-deoxy-D-xylulose 5-phosphate synthase* (*DXS*), *1-deoxy-D-xylulose 5-phosphate reductoisomerase* (*DXR*), *2-C-methyl-D-erythritol 4-phosphate cytidylyltransferase* (*MCT*), *4-(cytidine 5′-diphospho)-2-C-methyl-D-erythritol kinase* (*CMK*), *2-C-methyl-D-erythritol 2,4-cyclodiphosphate synthase* (*MECPS*), *4-hydroxy-3-methylbut-2-en-1-yl diphosphate synthase* (*HDS*), *isoprenyl diphosphate synthase* (*IDS*), *geranylgeranyl pyrophosphate synthase* (*GGPPS*), and *geranyl diphosphate synthase* (*GPPS*). The black dashed boxes represent genes enriched in the MVA and MEP pathway, and the red fonts represent differentially expressed genes.

In plants, volatile terpenes are derived from two distinct pathways, designated as the MEP pathway and the MVA pathway [11]. In this study, structural genes encoding enzymes involved in these two pathways were isolated, and their expression levels were evaluated at six time points within 24 h (2:00, 6:00, 10:00, 14:00, 18:00, and 22:00). In the MVA pathway and MEP pathway, a total of 46 structural genes related to terpenoid synthesis were differentially enriched (Figure S2b). Among the 46 differentially enriched genes, most of them had the highest expression at 6:00 (34.78%) or 10:00 (28.26%). The genes highly expressed at 2:00, 14:00, 18:00, and 22:00 accounted for 10.87%, 8.7%, 8.7%, and 6.52% of the total, respectively. The results showed that structural genes related to terpenoid synthesis were highly expressed in ante meridiem (2:00, 6:00, and 10:00). Based on the previous analysis, the genes in Cluster 4 and Cluster 8 were closely related to the biosynthesis of terpenoids. We identified 32 DEGs in Cluster 4 and Cluster 8 that were enriched in the MEP pathway, including *DXS* (*Cluster-60935.0*, *Cluster-60935.7*, *Cluster-60935.8*, *Cluster-60935.9*, *Cluster-49094.3*, *Cluster-60935.1*, *Cluster-60935.10* and *Cluster-60935.3*), *CMK* (*Cluster-46758.0*, *Cluster-65841.0*, *Cluster-65841.10*, *Cluster-65841.2* and *Cluster-65841.4*), *IDI* (*Cluster-31162.0*, *Cluster-47205.0*, *Cluster-47205.1* and *Cluster-47205.4*), *HDS* (*Cluster-63975.2* and *Cluster-63975.3*), *GGPPS* (*Cluster-43190.0*, *Cluster-43190.1* and *Cluster-43190.3*), *GPPS* (*Cluster-42532.0*, *Cluster-52779.1* and *Cluster-52779.3*) and *TPS* (*Cluster-35182.6*, *Cluster-46699.0*, *Cluster-17371.14*, *Cluster-17371.18*, *Cluster-17371.19*, *Cluster-65495.0*, and *Cluster-65495.1*) (Figure 3b). Taken together, Cluster 4 and Cluster 8 DEGs, enriched in the MEP pathway, may play an important role in the biosynthesis of dominant terpenoid.

2.4. Fifteen Structural Genes and TF Genes Were Hub Genes Related to the Diel Rhythm of Aroma Synthesis in Oncidium Twinkle 'Red Fantasy'

To explore co-expression networks of genes and possible relationships between the changes in diel rhythm of aroma synthesis and genes, we performed a weighted gene co-expression network analysis (WGCNA), which revealed 10 modules (labeled Blue, Turquoise, Black, Green, Pink, Purple, Brown, Yellow, Magenta, and Red) (Figure 4a,b). The expression trends in each module were shown in Figure 4b and Figure S3. We searched for highly connected TFs related to DEGs and biological processes of diel rhythm of aroma synthesis. Four modules were further analyzed, including the Blue, Red, Yellow, and Brown modules (Figure 4b). In the Blue module, genes were generally up-regulated from 2:00 to 14:00, and down-regulated from 14:00 to 22:00 (Figure 4b). Genes co-expressed in this module were significantly enriched in "Fatty acid degradation", "α-Linolenic acid metabolism", "Linolenic acid metabolism", and "Fatty acid biosynthesis" (Figure S4a). In the Red module, genes were up-regulated from 2:00 to 10:00, and gradually down-regulated from 10:00 to 22:00 (Figure 4b). These genes were significantly enriched in "Circadian rhythm-plant" and "Monoterpenoid biosynthesis" (Figure S4b). In the Yellow module, genes were sharply up-regulated from 2:00 to 6:00, continuously up-regulated from 6:00 to 10:00, stayed at those levels, and then were down-regulated from 10:00 to 22:00 (Figure 4b). In the Brown module, genes were up-regulated from 2:00 to 6:00, and gradually down-regulated from 6:00 to 22:00 (Figure 4b). The Yellow and Brown module genes were significantly enriched in "Terpenoid backbone biosynthesis", "Circadian rhythm-plant", "Monoterpenoid biosynthesis", "Carotenoid biosynthesis", and "Zeatin biosynthesis", which were closely associated with the synthesis of terpenes (Figure S4c).

To identify highly connected TFs related to the diel rhythm of aroma synthesis in *Oncidium* Twinkle 'Red Fantasy', the selected modules were exported and visualized using Cytoscape (Figure 4c–e). Therefore, we discerned the TF genes from the network of Blue module, and a total of 44 TFs were identified and distributed across 25 families. Among them, *WRKY* genes with seven members presented the greatest number of TFs (Figure 4c). In the Red module, *Cluster-17371.18_TPS* was predicted to be regulated by

80 TFs, distributed across 25 families (Figure 4d). The nine structural genes in the Brown module were regulated by four transcription factors, including *Cluster-18175.1_HSF*, *Cluster-15990.4_EIL*, *Cluster-21183.2_C3H*, and *Cluster-18175.2_HSF* (Figure 4e). There were 243 pairs of TF-structural genes identified in the Yellow module, such as *MYB-related* (19), *C2H2* (11), *AP2/ERF* (9), *NAC* (9), and *WRKY* (9) (Figure 4f). These results suggest that TFs may participate in aroma-formation indirectly by modulating the expression of aroma-related structural genes. In the regulatory network of the four modules, we further screened TFs that were differentially enriched in the "Circadian rhythm-plant" pathway. In the Red module, three diel rhythm-related TFs *TCP* (*Cluster-63662.6_TCP*, *Cluster-64611.8_TCP*, and *Cluster-51358.0_TCP*) regulated *Cluster-17371.18_TPS* were identified, while none was identified in the Blue and Brown modules. Notably, we identified 10 diel rhythm-related TFs in the Yellow module, including eight *MYB-related* (*Cluster-53654.1_MYB-related*, *Cluster-53654.3_MYB-related*, *Cluster-53654.5_MYB-related*, *Cluster-59600.0_MYB-related*, *Cluster-59600.2_MYB-related*, *Cluster-59600.3_MYB-related*, *Cluster-59600.5_MYB-related*, and *Cluster-59600.6_MYB-related*), one *bHLH* (*Cluster-61618.3_bHLH*), and one *bZIP* (*Cluster-44312.0_bZIP*), all of which regulated the same structural gene (*Cluster-17371.14_TPS*) related to terpenoid synthesis. Therefore, it was speculated that *MYB*-related TFs and their related structural genes in the regulatory network were potential hub genes for the diel rhythm of aroma synthesis in *Oncidium* Twinkle 'Red Fantasy'.

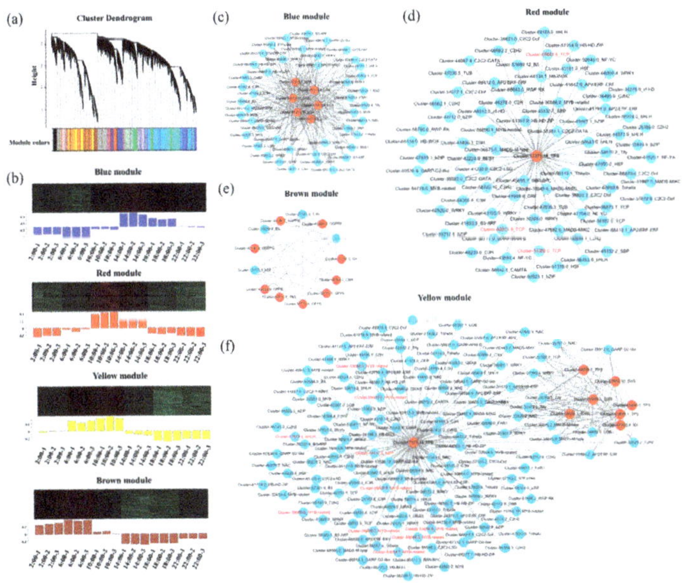

Figure 4. Establishment of weighted gene co-expression network analysis (WGCNA) modules of the differentially expressed genes (DEGs) at different time points. (**a**) Hierarchical clustering tree of the co-expression modules. The major tree branches constituted 10 distinct co-expression modules. (**b**) The gene expression patterns of the Blue, Red, Yellow, and Brown modules in WGCNA. The upper part was the clustering heatmap of genes within this module, with red indicating high expression and green indicating low expression. The lower part showed the expression patterns of module feature values in different samples. (**c**) Co-expression network of the genes from the Blue module. The red circles represent the key hub genes enriched in fatty acid derivative biosynthesis pathway, and the blue circles represent aroma synthesis related transcription factors (TFs). (**d**) Co-expression network

of the genes from the Red module. The red circles represent the key hub genes enriched in terpenoid biosynthesis pathway, and the blue circles represent aroma synthesis related TFs. The red font represents TFs that were differentially enriched in the "Circadian rhythm-plant" pathway. (**e**) Co-expression network of the genes from the Brown module. The red circles represent the key hub genes enriched in terpenoid biosynthesis pathway, and the blue circles represent aroma synthesis related TFs. (**f**) Co-expression network of the genes from the Yellow module. The red circles represent the key hub genes enriched in terpenoid biosynthesis pathway, and the blue circles represent aroma synthesis related TFs. The red font represents TFs that were differentially enriched in the "Circadian rhythm-plant" pathway. The networks were visualized by Cytoscape (v3.5.1) software.

2.5. Expression Analysis of Aroma Synthesis Genes in Oncidium Twinkle 'Red Fantasy' Under Diel Rhythm

To compare the transcription of aroma-related structural genes with the release patterns of VOCs under diel rhythm, the expression levels of six candidate structural genes (*Cluster-17371.18_TPS, Cluster-65495.1_TPS, Cluster-46699.0_TPS, Cluster-60935.10_DXS, Cluster-47205.4_IDI*, and *Cluster-65313.7_LOX*) (FPKM > 10) under diel rhythm were analyzed by quantitative real-time PCR (qRT-PCR) (Figure 5). Terpene synthases (TPS) constitute key enzymes in the biosynthetic pathway to produce monoterpene compounds. The GC–MS data suggested that the final product was 3-Carene (Figure 1). When we examined the expression patterns of *TPS* genes, the expression levels of *Cluster-17371.18_TPS* and *Cluster-65495.1_TPS* peaked at 10:00 (Figure 5). Among them, the expression of *Cluster-17371.18_TPS* was similar to the emission pattern of 3-Carene, mainly reflected on Day 1, while *Cluster-65495.1-TPS* showed similar peaks on both Day 1 and Day 2. However, *Cluster-46699.0_TPS* only showed a similar trend to 3-Carene on Day 2 (Figure 5). *DXS* and *IDI* genes are regarded as key rate-limiting enzymes in the MEP pathway, which directly affect the yield and efficiency of terpenoids [6–8]. The expression of *Cluster-60935.10_DXS* significantly increased from 2:00 to 6:00, maintained high expression from 6:00 to 10:00, and then significantly decreased (Day 1). At the same time, a peak was reached at 10:00 on Day 2 (Figure 5). Notably, *Cluster-47205.4_IDI* has a typical peak at 10:00 on both Day 1 and Day 2, which was consistent with the release trend of 3-Carene under diel rhythm (Figure 5). *Cluster-65313.7_LOX* was enriched in the fatty acid derivative biosynthesis pathway, and belonged to the Blue module in WGCNA analysis. During the diel oscillation, the expression of *Cluster-65313.7_LOX* was significantly up-regulated from 2:00 (Day 1) to 14:00 (Day 1), reaching its peak at 14:00 (Day 1), and then significantly down-regulated from 14:00 (Day 1) to 2:00 (Day 2). On Day 2, the expression level of *Cluster-65313.7_LOX* was significantly up-regulated from 2:00 to 10:00, and maintained high expression levels from 10:00 to 18:00 (Figure 5). The *Cluster-65313.7_LOX* expression trend was similar to the release pattern of Butyl tiglate and Prenyl senecioate in *Oncidium* Twinkle 'Red Fantasy', suggesting that *Cluster-65313.7_LOX* might be involved in the diel rhythmicity of ester volatile compound biosynthesis. In summary, the diel rhythm of the terpenoids in *Oncidium* Twinkle 'Red Fantasy' may be co-regulated by multiple enzymes in the synthetic pathway.

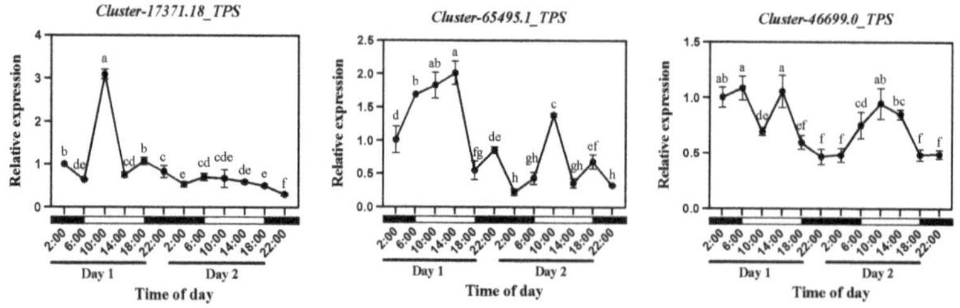

Figure 5. *Cont.*

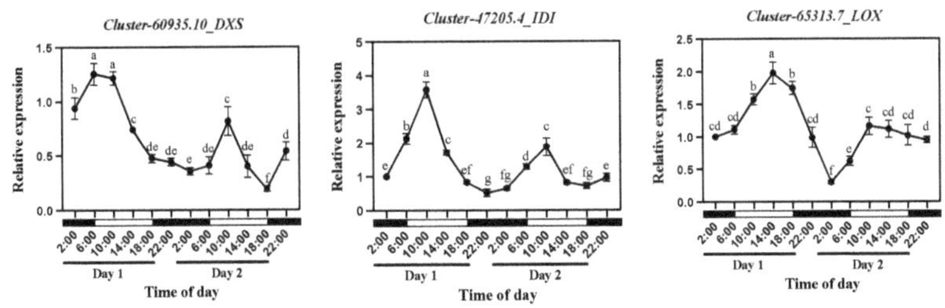

Figure 5. Relative expression of structural genes *Cluster-17371.18_TPS*, *Cluster-65495.1_TPS*, *Cluster-46699.0_TPS*, *Cluster-60935.10_DXS*, *Cluster-47205.4_IDI*, and *Cluster-65313.7_LOX* in *Oncidium* Twinkle 'Red Fantasy' flowers within 48 h (under 12 h light/12 h dark). Each treatment was conducted in triplicate with three technical repeats. Values are mean ± SD. Different lowercase letters indicate a statistically significant difference ($p < 0.05$).

2.6. Analysis of Aroma Release Pattern and Aroma Synthesis Genes Expression of Oncidium Twinkle 'Red Fantasy' Under Constant Light and Constant Dark Treatments

To further explore the relationship between aroma synthesis genes and the diel rhythm of aroma release, we first examined the content of VOCs released under constant light and constant dark treatments (Figure 6). The GC-MS data showed that the release of 3-Carene reached its peak at 6:00 (Day 1) under constant light treatment, and a peak also appeared at the same time point (6:00) on Day 2 (Figure 6a). Interestingly, both ester compounds (Butyl tiglate, and Prenyl senecioate) showed the same tendency to release under constant light treatment. The content of Butyl tiglate and Prenyl senecioate gradually increased from 2:00 to 14:00 (Day 1), rapidly decreased at 18:00 (Day 1), and remained stable from 10:00 to 22:00 on Day 2 (Figure 6a). These results indicated that the aroma release of *Oncidium* Twinkle 'Red Fantasy' was not only influenced by diel rhythm, but also by light signals. Under constant dark treatment, 3-Carene showed peaks at 10:00 on Day 1 and 6:00 on Day 2 (Figure 6b). Notably, the change trends of Butyl tiglate and Prenyll senecioate contents were similar to those under constant light treatment, with a peak appearing at 14:00 (Day 1) (Figure 6b). Thus, the aroma release of *Oncidium* Twinkle 'Red Fantasy' may be more regulated by diel rhythms than light signals under a constant dark condition.

When we explored the expression patterns of *TPS*, *DXS*, and *IDI* genes under constant light treatment, the expression of *TPS* genes (*Cluster-17371.18_TPS*, *Cluster-65495.1_TPS*, *Cluster-46699.0_TPS*) lost its rhythm and reached its peak at 10:00 (Day 2), 22:00 (Day 1), and 14:00 (Day 2), respectively (Figure 7a). The *IDI* gene was also in the same situation, and the expression reached its peak at 10:00. In response to the light signal, the expression of the *DXS* gene was gradually up-regulated with an increase in light duration from 14:00 (Day 1). For *Cluster-65313.7_LOX*, related to the biosynthesis of fatty acid derivatives, the expression peaks were mainly distributed at 10:00 (Day 1) and 6:00 (Day 2). Therefore, *TPS*, *DXS*, *IDI*, and *LOX* genes seem to lose their regulatory effect on the release of aroma compounds of *Oncidium* Twinkle 'Red Fantasy' under constant light treatment. We analyzed the expression of *TPS* genes under constant dark treatment and found that the expression trend of *Cluster-17371.18_TPS*, *Cluster-65495.1_TPS*, and *Cluster-46699.0_TPS* was similar to the release trend of 3-Carene (Figure 7b). It was up-regulated from 2:00 to 10:00 on Day 1, and from 2:00 to 6:00 on Day 2. The expression level of *Cluster-60935.10_DXS* was the highest at 6:00 on Day 1 and at 2:00 on Day 2, respectively, while the expression level of *Cluster-47205.4_IDI* was the highest at 10:00 (Day 1 and Day 2). The expression of *Cluster-65313.7_LOX* was up-regulated on Day 1 and rapidly down-regulated on Day 2. After removing the influence of light, *TPS* genes may be a key factor in regulating the diel rhythm of 3-Carene biosynthesis.

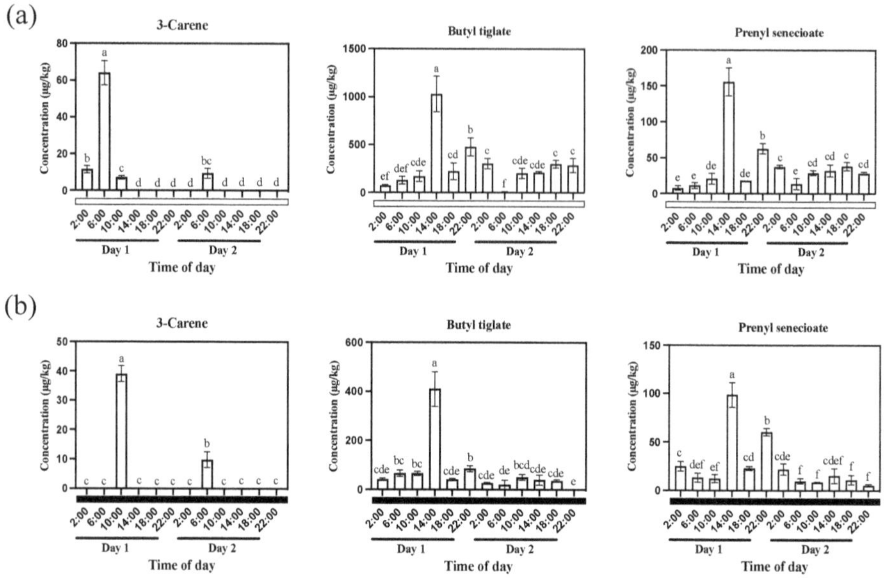

Figure 6. Analysis of aroma release pattern of *Oncidium* Twinkle 'Red Fantasy' under constant light and constant dark treatments. (**a**) The emission patterns of three floral scent compounds from *Oncidium* Twinkle 'Red Fantasy' flowers within 48 h under constant light. (**b**) The emission patterns of three floral scent compounds from *Oncidium* Twinkle 'Red Fantasy' flowers within 48 h under constant dark. Different lowercase letters indicate a statistically significant difference ($p < 0.05$).

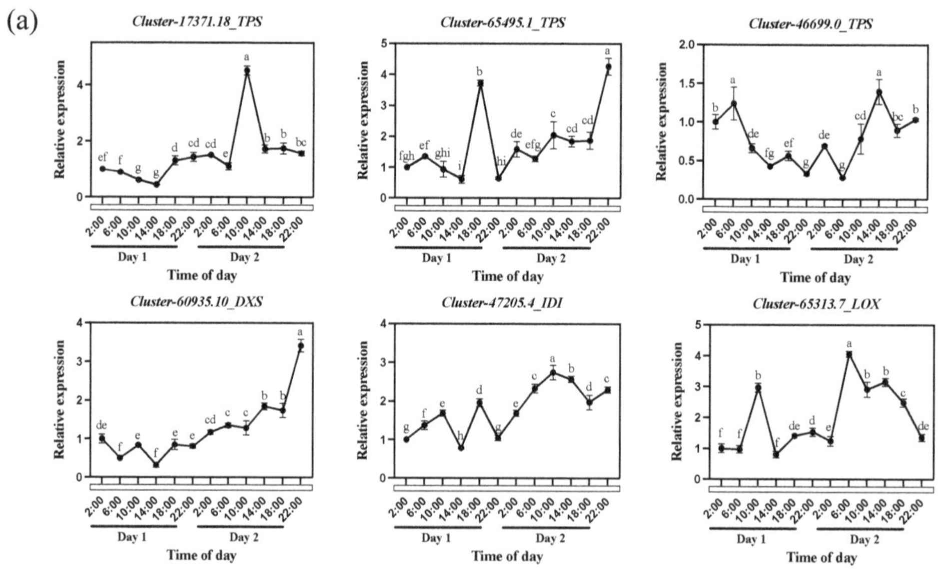

Figure 7. *Cont.*

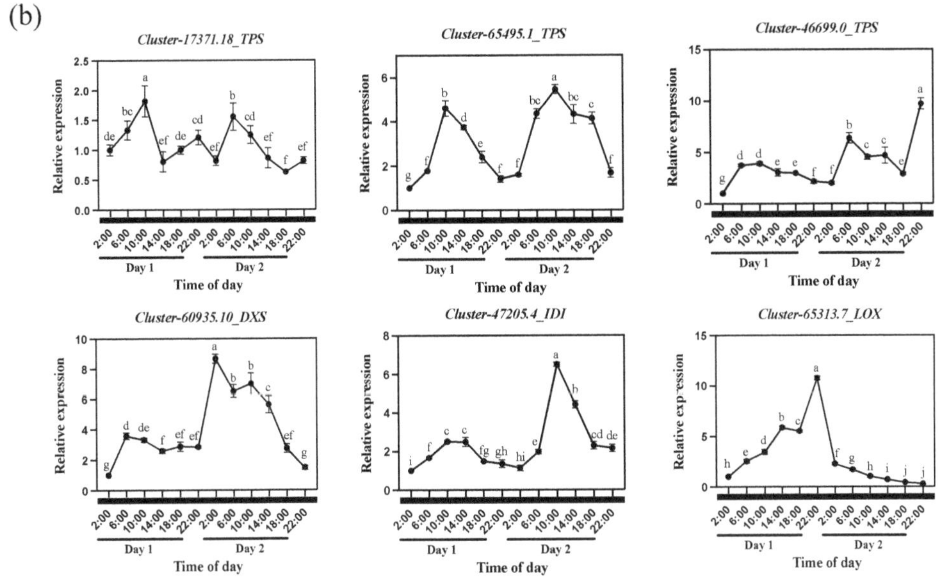

Figure 7. Analysis of aroma synthesis genes expression of *Oncidium* Twinkle 'Red Fantasy' under constant light and constant dark treatments. (**a**) Relative expression of structural genes *Cluster-17371.18_TPS, Cluster-65495.1_TPS, Cluster-46699.0_TPS, Cluster-60935.10_DXS, Cluster-47205.4_IDI*, and *Cluster-65313.7_LOX* in *Oncidium* Twinkle 'Red Fantasy' flowers within 48 h under constant light. (**b**) Relative expression of structural genes *Cluster-17371.18_TPS, Cluster-65495.1_TPS, Cluster-46699.0_TPS, Cluster-60935.10_DXS, Cluster-47205.4_IDI*, and *Cluster-65313.7_LOX* in *Oncidium* Twinkle 'Red Fantasy' flowers within 48 h under constant dark. Each treatment was conducted in triplicate with three technical repeats. Values are mean ± SD. Different lowercase letters indicate a statistically significant difference ($p < 0.05$).

3. Discussion

3.1. The Main VOCs of Oncidium Twinkle 'Red Fantasy' Were 3-Carene, Butyl Tiglate, and Prenyl Senecioate

Floral scent is one of the most important ornamental traits of plants, and is the main basis for evaluating flower quality. Compared with visual characteristics such as flower posture and color, the study of floral scent is relatively lagging behind. Due to the long life cycle and the complex and large genome of *Oncidium hybridum*, little was known about the synthesis and regulatory pathways of *Oncidium hybridum* floral scent. To date, the floral components of a variety of *Oncidium hybridum* have been identified. According to the types of floral scent component, terpenoids play a dominant role in the floral components of *Oncidium hybridum* [37]. Multiple studies have analyzed *Oncidium* Sharry Baby and identified its main volatile compounds as β-Ocimene, 3,7-dimethyl-1,3,6-Octatriene, Allo-ocimene, and 3-Carene, all of which belong to monoterpenes [32,38]. The main volatile compound of *Oncidium ornithorhynchum* and *Oncidium* Sweet Sugar was (*cis*)-3,7-dimethyl-1,3,6-Octatriene, while *Oncidium* 'Wils. Golden Afternoon Rich Yellow', *Oncidium*'Gower Ramsey Gold 2', *Oncidium* 'Gower Ramsey', and *Oncidium* 'Rosy Sunset' were 3,7-dimethyl-1,3,6-Octatriene, (-)-α-Cubebene, α-Copaene, and Linalool, respectively [37]. Monoterpenes are a class of important aromatic compounds, which are synthesized in plastids and are the main components of flower and fruit aromas [39]. We also identified 3-Carene, a monoterpene, as the main volatile component of *Oncidium* Twinkle 'Red Fantasy'. 3-Carene is a bicyclic monoterpene that has a strong rosin flavor and is an important natural scent. In the *Dendrobium* genus of the Orchidaceae family, *Dendrobium chrysotoxum* and *Dendrobium hancockii* detected 3-Carene as the main aromas. 3-Carene has also been identified as one of the main aromatic compounds in *Oncidium* Sharry Baby and *Oncidium* 'Red Fair'. Therefore, 3-Carene was the main component of the aroma compounds in *Oncidium* Twinkle 'Red Fantasy'.

In addition to terpenoids, two fatty acid derivatives (Butyl tiglate and Prenyl senecioate) were also identified in *Oncidium* Twinkle 'Red Fantasy'. Fatty acid derivatives are another major class of compounds that make up the floral components of plants. In *Rosa hybrida*, the main component was geranyl acetate, which contributed to its unique aroma [18]. Methyl jasmonate was a class of fatty acid derivative and also the main aroma component of *Cymbidium ensifolium* [40]. Meanwhile, methyl benzoate was one of the most abundant odor compounds in snapdragon flowers [41]. Chen et al. [42] detected the aroma compounds of *Oncidium* Twinkle 'Yellow Fantasy' and *Oncidium* Twinkle 'White Fantasy'. The results showed that 11 fatty acid derivatives were detected in *Oncidium* Twinkle 'Yellow Fantasy', including Isoamyl tiglate, Benzyl tiglate, 2-Methylbutyl acetate, Prenyl acetate, 2-Methylbutyl 2-methylbutyrate, 2-Methylbutyl isovalerate, Benzyl acetate, Methyl salicylate, Geranyl acetate, 2-methyl-butanoic aci phenylmethyl ester, and Benzyl valerate [42]. *Oncidium* Twinkle 'White Fantasy' detected five types of fatty acid derivative, including Benzyl tiglate, Methyl salicylate, Geranyl acetate, Methyl geranate, and Benzyl acetate [42]. Butyl tiglate and Prenyll senecioate were detected in *Oncidium* Twinkle 'Red Fantasy' of this study, but not in *Oncidium* Twinkle 'Yellow Fantasy' and *Oncidium* Twinkle 'White Fantasy'. In summary, there were differences in the relative content of volatile compounds and floral composition among different species of *Oncidium hybridum*.

3.2. Diurnal Rhythms of Aroma Synthesis Genes and VOCs Emissions in Oncidium Twinkle 'Red Fantasy'

Floral scents also have important biological significance to the plants. Floral scents can not only resist the harm of pests and diseases to flowers, but also attract insects to help them complete the reproduction process. The release of floral compounds usually exhibits a clear diel rhythm, which was generally consistent with the corresponding pollination insect habits of plants. *Petunia hybrida* was pollinated mainly by moths and releases its main floral compounds at night, while the aroma compounds of snapdragon pollinated by bees were mainly released during the day [17,20]. In this study, the main floral compounds 3-Carene, Butyl tiglate, and Prenyl senecioate in *Oncidium* Twinkle 'Red Fantasy' were mainly released during the day, which may be related to the activity habits of pollinating insects such as bees. As early as 1729, De Mairan observed the circadian rhythm and the folding and unfolding leaves of a sensitive heliotrope plant in complete darkness, and suggested that these movements were related to the internal rhythms of the plant [43]. Henri-Louis Duhamel du Monceau and Alphonse de Candolle further demonstrated the existence of endogenous circadian rhythms, which were independent of the environment [44]. Plants have a biological clock that predicts seasonal changes and adjusts their physiology and development accordingly [45,46]. As an internal molecular timer, the biological clock measures periodic changes in the environment, including daily or seasonal fluctuations, and resets daily [43,45,46]. The circadian rhythm is endogenous, autonomous, and independent of environmental changes. Light is not the cause of circadian rhythms, but it can regulate and reset the phase of the circadian biological clock and synchronize it [44,47]. In petunias, the release of β-Ionone showed a consistent circadian rhythm. After 24 h of constant light treatment, the normal circadian rhythm was no longer displayed, which indicated that this phenomenon was regulated by the photoperiod [15]. The release of methyl jasmonate, a floral compound in *Cymbidium ensifolium*, exhibited a diel rhythm within 48 h (12 h light/12 h dark) [40]. Under 48 h of constant light treatment, the floral scent was not released [40]. Under 48 h of constant dark treatment, the release of scent from flowers has a rhythm, but the phase and amplitude of the rhythm have changed [40]. In our study, the release of 3-Carene, Butyl tiglate, and Prenyl senecioate exhibited the diel rhythm. 3-Carene maintained the diel rhythm under constant light/dark, but the phase and amplitude of its rhythm changed. Butyl tiglate and Prenyl senecioate showed the same release tendency under constant light/dark, and maintained a diel rhythm of photoperiod (12 h light/12 h dark). Therefore, the diel rhythm release of 3-Carene may be influenced by light and the circadian clock, and Butyl tiglate and Prenyl senecioate may be mainly determined by the endogenous circadian clock.

Many physiological and biochemical activities of plants are regulated by genetic genes, and the release of floral compounds is closely related to gene expression. In the terpenoid and fatty acid derivative biosynthetic pathways, *DXPS*, *DXR*, *MCT*, *HDR*, *TPS*, *ATT*, and *LOX* genes have been shown to be associated with the diel rhythm of aroma release [17,18,23–25]. In *Oncidium* Twinkle 'Red Fantasy', we identified two *13-LOX* and six *9-LOX* genes in the biosynthesis pathway of fatty acid derivatives that exhibited a similar diel rhythm trend to the release of ester compounds. At the same time, 15 genes with the same trend as terpenoid release were identified, including *DXS*, *CMK*, *IDI*, and *TPS* genes. Therefore, the diel rhythm release of floral compounds in *Oncidium* Twinkle 'Red Fantasy' may be closely related to the expression of aroma synthesis-related genes. In general, the diel rhythm of aroma compounds was regulated by the circadian clock or photoperiod affecting gene expression. The circadian rhythm-related *bHLH* TF (*MYC2*) directly binds to the promoters of the terpenoid synthase genes *AtTPS11* and *AtTPS21*, and promotes their gene expression [48]. Zhou et al. [14] identified *MYB*-related TF *LATE ELONGATED HYPOCOTYL* (*JsLHY*) as a key gene regulating jasmine aroma formation, which can bind to the gene promoter regions of six aroma-related structural genes (*JsBEAT1*, *JsTPS34*, *JsCNL6*, *JsBPBT*, *JsAAAT5*, and *Js4CL7*) and directly promoted their expression. Yeh et al. [35] confirmed that CIRCADIAN CLOCK ASSOCIATED 1 (CCA1) in up-regulated *TPS* bound to the *TPS* promoter region, thereby affecting the circadian release of floral scent in *Oncidium* Sharry Baby. We identified 10 circadian rhythm-related TFs in the WGCNA network of aroma synthesis, including *TCP* (3), *MYB*-related (8), *bHLH* (1), and *bZIP* (1). Notably, eight *MYB*-related, one *bHLH*, and one *bZIP* genes predicted the regulation of the structural gene *Cluster-17371.14_TPS*, which was associated with terpenoid synthesis. Interestingly, the *TPS* genes were consistent with the release pattern of 3-Carene under constant dark treatment. This further indicated that *TPS* genes may play an important role in regulating the 3-Carene diel rhythm of *Oncidium* Twinkle 'Red Fantasy'.

4. Materials and Methods

4.1. Plant Materials

Oncidium Twinkle 'Red Fantasy' was planted in the Characteristic Orchid Germplasm Resources Preservation Nursery (Fuzhou, China). Before the experiment, 100 initial flowering plants were grown in a greenhouse at 30 °C/25 °C (14 h light/10 h dark). We selected *Oncidium* Twinkle 'Red Fantasy' at full flowering stage and cultured it at 25 °C for 12 h light/12 h dark (switching at 6:00 and 18:00) under constant light and constant dark treatment. The light intensity in this study was 20,000 Lx. Each treatment set up 30 plants to study the diel rhythm of VOCs emission and gene expression level.

4.2. VOCs Collection and GC–MS Analysis

The headspace solid-phase micro-extraction (HS-SPME) technique coupled with the GC–MS system was used to collect and analyze *Oncidium* Twinkle 'Red Fantasy' floral volatile compounds, following the method described by Yeh et al. [35]. Samples were collected at 2:00, 6:00, 10:00, 14:00, 18:00, and 22:00 over 2 days, with a total of 12 time points. Three replicates were set for each sample. The whole floret organ was cut (1 g) and immediately sealed in a 20-mL headspace sample bottle, following the injection of 0.6 μL (240 μg/mL) *n*-nonane as an internal standard solution. The bottle equilibrated in a laboratory stirrer/hot plate at 40 °C for 10 min. Then, a stainless steel needle, housing 65 μm SPME fiber (Sigma Aldrich, St. Louis, MO, USA), was placed through the hole to expose the fiber at the position of 1 cm over the liquid surface for 30 min with magnetic stirring at 100 rpm. After extraction, the SPME device was inserted into the injection port of an Agilent 7890A gas chromatograph coupled with an Agilent 5975C (Agilent Technologies, Santa Clara, CA, USA) series mass spectrometer and desorbed at 250 °C for 5 min in a split/splitless GC injection port. The analytical conditions were as follows: the GC column was DB-5MS (30 m × 0.25 mm × 0.25 μm); the temperature of the injector and the MS interface temperature were set at 250 and 260 °C, respectively; the oven temperature programs of the DB-5MS columns were 35 °C (1 min), 2 °C/min to 80 °C (1 min), 5 °C/min to 160 °C (1 min), and 10 °C/min to 220 °C (10 min); the carrier gas was helium (99.9995%) at 1 mL/min; the injector mode was splitless; the mass spectrum in the electron impact mode was generated at 70 eV; and temperatures for ion source and the quadrupole mass filter were 230 and 150 °C. The identification of the component of floral scent used the MS spectra from NIST 17. The concentration was calculated using the formula $C_{voc} = (A_{voc}/A_{std}) \times (m_{std}/m_s)$, where m_{std} was the known level of internal standard and m_s was the weight of the sample; A_{voc} and A_{std} were quantifying ion peak areas of identified compound and internal standard, respectively; and C_{voc} was the concentration of the identified compound.

4.3. RNA Extraction and Transcriptome Sequencing

In this study, RNA-seq sampling was performed at six time points (2:00, 6:00, 10:00, 14:00, 18:00, and 20:00) for 12 h light/12 h dark (switching at 6:00 and 18:00). Each sample was collected in three independent biological replicates. Total RNAs were isolated using ethanol precipitation and CTAB-PBIOZOL. The integrity and concentration of RNA were measured by Qubit fluorescence quantitative analyzer and Qsep400 high-throughput biological fragment analyzer, respectively. Illumina sequencing libraries were constructed and sequenced using the Illumina system NovaSeq6000. Due to the lack of a reference genome, we used Trinity (v2.13.2) [49] (https://github.com/trinityrnaseq/trinityrnaseq, accessed on 19 February 2024) to splice clean reads after obtaining them. --normalize_max_read_cov was used to set the maximum coverage of assembled reads (default: 50). --min_kmer_cov was used to set the minimum number of times (frequency) that k-mer must be observed. K-mers with frequencies below the threshold were excluded from the assembly (default: 5). --Min_gluemin number of reads needed to glue two inch contigs together (default: 10). We used the Corset (1.09) [50] (https://github.com/Oshlack/Corset, 19 February 2024) with default parameters compared to the expression of the number of reads on the transcript and model for hierarchical clustering transcript. We used Benchmarking Universal Single-Copy Orthologs (BUSCO) software (5.2.2) to evaluate the integrity of the assembled transcripts, and DIAMOND (v2.0.9) [51] software to compare the Unigene sequences with the Kyoto Encyclopedia of Genes and Genomes (KEGG), NCBI non-redundant proteins (NR), Swiss-Prot, gene ontology (GO), clusters of orthologous groups for eukaryotic complete genomes (KOG), and Trembl databases. We set the e-value to 1×10^{-5} to filter out results with poor alignment, and added the—max_hsps 1 parameter during alignment to only retain the hit that best aligned with the query sequence. After predicting the amino acid sequence of Unigene, HMMER software (3.1) (default parameter) was used to compare it with the Pfam database to obtain the annotation information of Unigenes.

4.4. Analysis of RNA-Seq Data

To estimate the expression levels of genes at different time points, clean transcriptome sequencing reads were aligned to the transcriptome obtained by Trinity splicing using Bowtie2 [52]. The number of reads mapped to each transcript was calculated using RNA sequencing by Expectation Maximization (RSEM) [53], and the fragments per kilobase of transcript per million mapped fragments (FPKM) was estimated to measure the transcript level of each gene [54]. DESeq2 (1.22.2) was used to analyze the difference of expression between sample groups. The Benjamini–Hochberg method was used to correct the hypothesis testing probability (p-value) by multiple hypothesis tests and obtain the False Discovery Rate (FDR). The screening criteria for differentially expressed genes (DEGs) were $|\log2\text{Fold Change}| >= 1$ and FDR < 0.05. Subsequently, Gene Ontology (GO) [55,56] and the Kyoto Encyclopedia of Genes and Genomes (KEGG) [57] enrichment analysis were conducted for the DEGs. A pathway with FDR ≤ 0.05 was defined as one in which DEGs were significantly enriched. The gene expression values (FPKM) of all samples were analyzed by principal component analysis (PCA) linear algebra to evaluate the differences between groups and the situation of samples within groups. Related-genes expression heat maps were visualized using Tbtools (v2.119) [58]. We used iTAK (1.7a) software for transcription factor (TFs) prediction with default parameters. The WGCNA R software (4.2.1) package [59] was used for weighted gene co-expression network analysis (WGCNA), and a cluster tree was constructed based on the correlation between gene expression levels, and divided into modules. The threshold for module merging was set at 0.25, and the minimum number of module genes was 50. We visualized the co-expression network using Cytoscape (v3.5.1) software.

4.5. RNA Extraction and qRT-PCR Analysis

Total RNA samples were extracted with the FastPure Universal Plant Total RNA Isolation Kit (Vazyme, Nanjing, China), and cDNA synthesis was carried out according to the instruction manual of PrimeScript^{TM} II 1st Strand cDNA Synthesis Kit (Takara Bio, Kusatsu, Shiga, Japan). Finally, through using gene-specific primers (Table S1), RT-qPCR analysis was conducted by a QuantStudio 1 Plus system with 2X SYBR Green *Pro Taq* HS Premix (Rox Plus) (Accurate Biology, Changsha, China). The relative expression was calculated by the $2^{-\Delta\Delta CT}$ method, and Actin was used as a reference gene to normalize the expression data.

4.6. Statistical Analysis

Data were analysed using SPSS statistics (v.24). One-way analysis of variance (ANOVA) with Duncan tests was performed for multiple comparisons in this study. The steps in carrying out the ANOVA are: 1. Calculate the variance between the samples. To calculate the variance between the samples, take the sum of the squared deviations of the means of various samples from the grand average, and divide this total by the degree of freedom k-1, where k = number of samples. 2. Calculate the variance within samples. To calculate the variance within the samples, take the sum of squares of the deviations of various items from the mean values of the respective samples and divide this total by the degree of freedom, n-k, where n = total number of all the observations and k = number of samples. 3. Calculate the total variance. The total variation is calculated by taking the squared deviation of each item from the grand average and dividing this total by the degree of freedom, n-1, where n = total number of observations. 4. Calculate the F ratio. This measures the ratio of between-column variance and within-column variance. If there is a real difference between the groups, the variance between groups will be significantly larger than the variance within the groups. 5. Decision Rule. At a given level of significance,

$\alpha = 0.05$, and at n-k and k-1 degrees of freedom, the value of F is tabulated from the table. On comparing the values, if the calculated value is greater than the tabulated value, reject the null hypothesis. That means the test is significant or there is a significant difference between the sample means. The different letters in the histogram and line chart showed that there was a statistical difference in the mean value at $p < 0.05$. GraphPad (v.9.1.1) software was used to make expression graphs.

5. Conclusions

In this study, volatile organic compounds and transcriptome analysis were used to clarify the mechanism related to the diel rhythm of aroma release in *Oncidium* Twinkle 'Red Fantasy'. The main aroma compounds of *Oncidium* Twinkle 'Red Fantasy' were 3-Carene, Butyl tiglate, and Prenyl senecioate, and the aroma release had a diel rhythm. Through the constant light and constant dark treatments, it was further verified that the diel release of 3-Carene may be influenced by light and the circadian clock, and Butyl tiglate and Prenyl senecioate may be mainly determined by the endogenous circadian clock. Meanwhile, the *TPS* genes were consistent with the release pattern of 3-Carene, which may be a key factor in regulating the diel rhythm of 3-Carene biosynthesis (Figure 8).

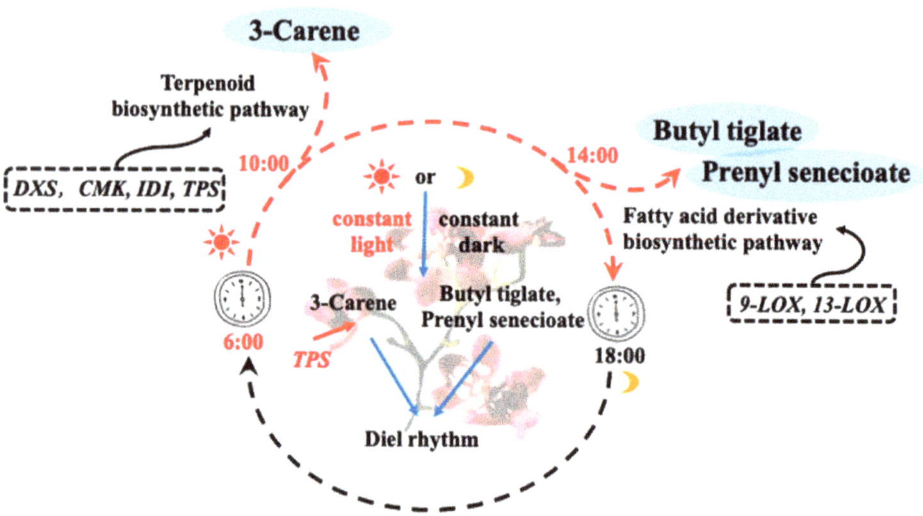

Figure 8. Schematic model of the mechanism by which the circadian rhythm regulates the aroma release of *Oncidium* Twinkle 'Red Fantasy'. The main aroma compounds of *Oncidium* Twinkle 'Red Fantasy' were 3-Carene, Butyl tiglate, and Prenyl senecioate. 3-Carene were mainly released at 10:00, while Butyl tiglate and Prenyl senecioate were mainly released at 14:00. *DXS*, *CMK*, *IDI*, *TPS*, and *LOX* were key genes in the terpenoid or fatty acid derivative biosynthetic pathway, which were co-expressed with aroma release. Under the treatment of constant light or dark, the aroma release maintained a circadian rhythm.

Supplementary Materials: The following supporting information can be downloaded at: https://www.mdpi.com/article/10.3390/plants13223232/s1, Figure S1: total ion chromatogram of mass spectrometry analysis of *Oncidium* Twinkle 'Red Fantasy'. ① Prenol; ② Nonane; ③ 3-Carene; ④ Butyl tiglate; ⑤ Prenyl senecioate. Figure S2: expression patterns of differentially expressed genes (DEGs) enriched in fatty acid derivative and terpenoid biosynthetic pathway. (a) Expression profiles of 21 DEGs enriched in fatty acid derivative biosynthetic pathway. (b) Expression profiles of 47 DEGs enriched in terpenoid biosynthetic pathway. Figure S3: the gene expression patterns of the Black, Green, Magenta, Pink, Purple, and Turquoise modules in WGCNA. The upper part was the clustering heatmap of genes within this module, with red indicating high expression and green indicating low expression. The lower part showed the expression patterns of module feature values in different samples. Figure S4: KEGG enrichment analysis of DEGs in modules. (a) KEGG enrichment analysis of DEGs in Blue module. (b) KEGG enrichment analysis of DEGs in Red module. (c) KEGG enrichment analysis of DEGs in Yellow module. (d) KEGG enrichment analysis of DEGs in Brown module. Table S1: primers used for real-time quantitative PCR.

Author Contributions: Conceptualization, H.Z.; validation, Y.C. (Yan Chen), L.K., R.F., Y.X., Y.C. (Yiquan Chen) and S.Z.; formal analysis, Y.C. (Yan Chen) and S.Z.; resources, H.Z.; writing—original draft preparation, Y.C. (Yan Chen); writing—review and editing, all authors. All authors have read and agreed to the published version of the manuscript.

Funding: This work was supported by the Natural Science Foundation of Fujian Province (2023J01198), the Free Exploration Science and Technology Innovation Project of the Fujian Academy of Agricultural Sciences (ZYTS202405), the National Key Research and Development Program Project (2023YFD1600505), the Youth Talent Project of the Fujian Academy of Agricultural Sciences (YC2021018), the High-quality Development of Agriculture in Fujian Province Exceeds the "5511" Collaborative Innovation Project (XTCXGC2021016), and the Science and Technology Innovation Team Construction Project of the Fujian Academy of Agricultural Sciences (CXTD2021010-2).

Data Availability Statement: The datasets generated for this study can be found in the National Centre for Biotechnology Information (NCBI) Sequence Read Archive under accession number PRJNA1128565.

Conflicts of Interest: The authors declare no conflicts of interest.

References

1. Schiestl, F.P. Ecology and evolution of floral volatile-mediated information transfer in plants. *New Phytol.* **2015**, *206*, 571–577. [CrossRef] [PubMed]
2. Pichersky, E.; Noel, J.P.; Dudareva, N. Biosynthesis of plant volatiles: Nature's diversity and ingenuity. *Science* **2006**, *311*, 808–811. [CrossRef] [PubMed]
3. Pulido, P.; Perello, C.; Rodriguez-Concepcion, M. New insights into plant isoprenoid metabolism. *Mol. Plant* **2012**, *5*, 964–967. [CrossRef] [PubMed]
4. Vranová, E.; Coman, D.; Gruissem, W. Network analysis of the MVA and MEP pathways for isoprenoid synthesis. *Annu. Rev. Plant Biol.* **2013**, *64*, 665–700. [CrossRef]
5. Mostafa, S.; Wang, Y.; Zeng, W.; Jin, B. Floral Scents and Fruit Aromas: Functions, Compositions, Biosynthesis, and Regulation. *Front. Plant Sci.* **2022**, *13*, 860157. [CrossRef]
6. Falara, V.; Akhtar, T.A.; Nguyen, T.T.; Spyropoulou, E.A.; Bleeker, P.M.; Schauvinhold, I.; Matsuba, Y.; Bonini, M.E.; Schilmiller, A.L.; Last, R.L.; et al. The tomato terpene synthase gene family. *Plant Physiol.* **2011**, *157*, 770–789. [CrossRef]
7. Tholl, D. Biosynthesis and biological functions of terpenoids in plants. *Adv. Biochem. Eng. Biotechnol.* **2015**, *148*, 63–106. [CrossRef]
8. Abbas, F.; Ke, Y.; Yu, R.; Yue, Y.; Amanullah, S.; Jahangir, M.M.; Fan, Y. Volatile terpenoids: Multiple functions, biosynthesis, modulation and manipulation by genetic engineering. *Planta* **2017**, *246*, 803–816. [CrossRef]
9. Espino-Díaz, M.; Sepúlveda, D.R.; González-Aguilar, G.; Olivas, G.I. Biochemistry of Apple Aroma: A Review. *Food Technol. Biotechnol.* **2016**, *54*, 375–397. [CrossRef]
10. Shalit, M.; Guterman, I.; Volpin, H.; Bar, E.; Tamari, T.; Menda, N.; Adam, Z.; Zamir, D.; Vainstein, A.; Weiss, D.; et al. Volatile ester formation in roses. Identification of an acetyl-coenzyme A. Geraniol/Citronellol acetyltransferase in developing rose petals. *Plant Physiol.* **2003**, *131*, 1868–1876. [CrossRef]
11. Qiao, Z.; Hu, H.; Shi, S.; Yuan, X.; Yan, B.; Chen, L. An Update on the Function, Biosynthesis and Regulation of Floral Volatile Terpenoids. *Horticulturae* **2021**, *7*, 451. [CrossRef]
12. Zhu, G.; Gou, J.; Klee, H.; Huang, S. Next-Gen Approaches to Flavor-Related Metabolism. *Annu. Rev. Plant Biol.* **2019**, *70*, 187–212. [CrossRef] [PubMed]
13. Hu, Z.; Zhang, H.; Leng, P.; Zhao, J.; Wang, W.; Wang, S. The emission of floral scent from Lilium 'siberia' in response to light intensity and temperature. *Acta Physiol. Plant.* **2013**, *35*, 1691–1700. [CrossRef]

14. Zhou, C.; Tian, C.; Wen, S.; Yang, N.; Zhang, C.; Zheng, A.; Tan, J.; Jiang, L.; Zhu, C.; Lai, Z.; et al. Multiomics Analysis Reveals the Involvement of JsLHY in Controlling Aroma Production in Jasmine Flowers. *J. Agric. Food Chem.* **2023**, *71*, 17455–17468. [CrossRef] [PubMed]
15. Simkin, A.J.; Underwood, B.A.; Auldridge, M.; Loucas, H.M.; Shibuya, K.; Schmelz, E.; Clark, D.G.; Klee, H.J. Circadian regulation of the PhCCD1 carotenoid cleavage dioxygenase controls emission of beta-ionone, a fragrance volatile of petunia flowers. *Plant Physiol.* **2004**, *136*, 3504–3514. [CrossRef]
16. Andrews, E.S.; Theis, N.; Adler, L.S. Pollinator and Herbivore Attraction to Cucurbita Floral Volatiles. *J. Chem. Ecol.* **2007**, *33*, 1682. [CrossRef]
17. Dudareva, N.; Martin, D.; Kish, C.M.; Kolosova, N.; Gorenstein, N.; Fäldt, J.; Miller, B.; Bohlmann, J. (E)-β-Ocimene and Myrcene Synthase Genes of Floral Scent Biosynthesis in Snapdragon: Function and Expression of Three Terpene Synthase Genes of a New Terpene Synthase Subfamily. *Plant Cell* **2003**, *15*, 1227–1241. [CrossRef]
18. Hendel-Rahmanim, K.; Masci, T.; Vainstein, A.; Weiss, D. Diurnal regulation of scent emission in rose flowers. *Planta* **2007**, *226*, 1491–1499. [CrossRef]
19. Jiang, Y.; Chen, X.; Lin, H.; Wang, F.; Chen, F. Floral Scent in Wisteria: Chemical Composition, Emission Pattern, and Regulation. *J. Am. Soc. Hortic. Sci. Am. Soc. Hortic. Sci.* **2011**, *136*, 873–884. [CrossRef]
20. Oyama-Okubo, N.; Ando, T.; Watanabe, N.; Marchesi, E.; Uchida, K.; Nakayama, M. Emission mechanism of floral scent in Petunia axillaris. *Biosci. Biotechnol. Biochem.* **2005**, *69*, 773–777. [CrossRef]
21. Maia, A.C.D.; de Lima, C.T.; Navarro, D.; Chartier, M.; Giulietti, A.M.; Machado, I.C. The floral scents of Nymphaea subg. Hydrocallis (Nymphaeaceae), the New World night-blooming water lilies, and their relation with putative pollinators. *Phytochemistry* **2014**, *103*, 67–75. [CrossRef] [PubMed]
22. Jiang, Y.F.; Zhao, N.; Wang, F.; Chen, F. Emission and Regulation of Volatile Chemicals from Globe Amaranth Flowers. *J. Am. Soc. Hortic. Sci.* **2011**, *136*, 16–22. [CrossRef]
23. Cinege, G.; Louis, S.; Hänsch, R.; Schnitzler, J.P. Regulation of isoprene synthase promoter by environmental and internal factors. *Plant Mol. Biol.* **2009**, *69*, 593–604. [CrossRef] [PubMed]
24. Liu, C.; Xi, W.; Yuan, J.; Zhu, L. Molecular Cloning and Functional Characterization of Linalool Synthase Gene OfTPS5 in Osmanthus fragrans'Lianzi Dangui'Flowers. *Acta Hortic. Sin.* **2020**, *47*, 11.
25. Christensen, S.A.; Nemchenko, A.; Borrego, E.; Murray, I.; Sobhy, I.S.; Bosak, L.; DeBlasio, S.; Erb, M.; Robert, C.A.; Vaughn, K.A.; et al. The maize lipoxygenase, ZmLOX10, mediates green leaf volatile, jasmonate and herbivore-induced plant volatile production for defense against insect attack. *Plant J.* **2013**, *74*, 59–73. [CrossRef]
26. McClung, C.R.; Gutierrez, R.A. Network news:prime time for systems biology of the plant circadian clock. *Curr. Opin. Genet. Dev.* **2010**, *20*, 588–598. [CrossRef]
27. Michael, T.P. Phase-Specific Circadian Clock Regulatory Elements in Arabidopsis. *Plant Physiol.* **2002**, *130*, 627–638. [CrossRef]
28. Fenske, M.P.; Hewett Hazelton, K.D.; Hempton, A.K.; Shim, J.S.; Yamamoto, B.M.; Riffell, J.A.; Imaizumi, T. Circadian clock gene LATE ELONGATED HYPOCOTYL directly regulates the timing of floral scent emission in Petunia. *Proc. Natl. Acad. Sci. USA* **2015**, *112*, 9775–9780. [CrossRef]
29. Yon, F.; Joo, Y.; Cortés Llorca, L.; Rothe, E.; Baldwin, I.T.; Kim, S.G. Silencing Nicotiana attenuata LHY and ZTL alters circadian rhythms in flowers. *New Phytol.* **2016**, *209*, 1058–1066. [CrossRef]
30. Hsiao, Y.Y.; Tsai, W.C.; Kuoh, C.S.; Huang, T.H.; Wang, H.C.; Wu, T.S.; Leu, Y.L.; Chen, W.H.; Chen, H.H. Comparison of transcripts in Phalaenopsis bellina and Phalaenopsis equestris (Orchidaceae) flowers to deduce monoterpene biosynthesis pathway. *BMC Plant Biol.* **2006**, *6*, 14. [CrossRef]
31. Luo, Y.; Huang, M.; Wu, J. Progress in Oncidium Breeding Study. *Acta Agric. Jiangxi* **2012**, *24*, 15–20.
32. Hua, C.; Yan-Ping, F. Analysis of Aroma Components of Oncidium. *Acta Agric. Univ. Jiangxiensis* **2012**, *34*, 692–698.
33. Zhang, Y.; Li, X.L.; Wang, Y.; Tian, M.; Fan, M.H. Changes of Aroma Components in Oncidium Sharry Baby in Different Florescence and Flower Parts. *Sci. Agric. Sin.* **2011**, *44*, 110–117.
34. Chiu, Y.T.; Chen, H.C.; Chang, C. The Variation of Oncidium Rosy Sunset Flower Volatiles with Daily Rhythm, Flowering Period, and Flower Parts. *Molecules* **2017**, *22*, 1468. [CrossRef]
35. Yeh, C.W.; Zhong, H.Q.; Ho, Y.F.; Tian, Z.H.; Yeh, K.W. The diurnal emission of floral scent in Oncidium hybrid orchid is controlled by CIRCADIAN CLOCK ASSOCIATED 1 (CCA1) through the direct regulation on terpene synthase. *BMC Plant Biol.* **2022**, *22*, 472. [CrossRef]
36. Huang, M.; Ma, C.; Yu, R.; Mu, L.; Hou, J.; Yu, Y.; Fan, Y. Concurrent changes in methyl jasmonate emission and the expression of its biosynthesis-related genes in Cymbidium ensifolium flowers. *Physiol. Plant.* **2015**, *153*, 503–512. [CrossRef]
37. Ma, D.; Xiao, W.F.; Li, Z.; Zhang, J.W.; Chen, H.M.; Lv, F.B. Research Progress on Floral Scent Components of Orchidaceae. *Chin. Agric. Sci. Bull.* **2023**, *39*, 52–60.
38. Wang, P.Y.; Zheng, S.Z.; Ye, W.; Yan, P.P.; Zhang, C.L.; Lin, Y.L. GC-MS analysis of volatile components of Oncidium hybridum. *Subtrop. Agric. Res.* **2020**, *16*, 5.

39. Nagegowda, D.A. Plant volatile terpenoid metabolism: Biosynthetic genes, transcriptional regulation and subcellular compartmentation. *FEBS Lett.* **2010**, *584*, 2965–2973. [CrossRef]
40. Yu, Y.Y. The Research on Circadian Rhythmic and Light-induced Emission of Methyl Jasmonate in Cymbidium ensifolium. Master's Thesis, South China Agricultural University, Guangzhou, China, 2017.
41. Kolosova, N.; Gorenstein, N.; Kish, C.M.; Dudareva, N. Regulation of Circadian Methyl Benzoate Emission in Diurnally and Nocturnally Emitting Plants. *Plant Cell* **2001**, *13*, 2333–2347. [CrossRef]
42. Chen, Y.Q.; F, N.Y.; Ye, X.X.; Luo, Y.H.; Zhong, H.Q.; Huang, M.L.; Fan, R.H. Analysis of fragrance formation in Oncidium based on transcriptome sequencing. *J. Nucl. Agric. Sci.* **2022**, *36*, 578–588.
43. Srivastava, D.; Shamim, M.; Kumar, M.; Mishra, A.; Maurya, R.; Sharma, D.; Pandey, P.; Singh, K.N. Role of circadian rhythm in plant system: An update from development to stress response. *Environ. Exp. Bot.* **2019**, *162*, 256–271. [CrossRef]
44. Moore-Ede, M. Sleeping as the world turns. *Nat. Hist.* **1982**, *91*, 28–36.
45. Mcclung, C.R. The Genetics of Plant Clocks. *Adv. Genet.* **2011**, *74*, 105–139.
46. Nagel, D.H.; Kay, S.A. Complexity in the Wiring and Regulation of Plant Circadian Networks. *Curr. Biol.* **2012**, *22*, R648–R657. [CrossRef]
47. Fankhauser, C.; Staiger, D. Photoreceptors in Arabidopsis thaliana: Light perception, signal transduction and entrainment of the endogenous clock. *Planta* **2002**, *216*, 1–16. [CrossRef]
48. Hong, G.J.; Xue, X.Y.; Mao, Y.B.; Wang, L.J.; Chen, X.Y. Arabidopsis MYC2 interacts with DELLA proteins in regulating sesquiterpene synthase gene expression. *Plant Cell* **2012**, *24*, 2635–2648. [CrossRef]
49. Grabherr, M.G.; Haas, B.J.; Yassour, M.; Levin, J.Z.; Thompson, D.A.; Amit, I.; Adiconis, X.; Fan, L.; Raychowdhury, R.; Zeng, Q. Full-length transcriptome assembly from RNA-Seq data without a reference genome. *Nat. Biotechnol.* **2011**, *29*, 644–652. [CrossRef]
50. Davidson, N.M.; Oshlack, A. Corset: Enabling differential gene expression analysis for de novo assembled transcriptomes. *Genome. Biol.* **2014**, *15*, 410. [CrossRef]
51. Buchfink, B.; Xie, C.; Huson, D.H. Fast and sensitive protein alignment using DIAMOND. *Nat Methods* **2015**, *12*, 59–60. [CrossRef]
52. Langmead, B.; Salzberg, S.L. Fast gapped-read alignment with Bowtie 2. *Nat. Methods* **2012**, *9*, 357–359. [CrossRef] [PubMed]
53. Li, B.; Dewey, C.N. RSEM: Accurate transcript quantification from RNA-Seq data with or without a reference genome. *BMC Bioinform.* **2011**, *12*, 323. [CrossRef] [PubMed]
54. Trapnell, C.; Williams, B.A.; Pertea, G.; Mortazavi, A.; Kwan, G.; Van Baren, M.J.; Salzberg, S.L.; Wold, B.J.; Pachter, L. Transcript assembly and quantification by RNA-Seq reveals unannotated transcripts and isoform switching during cell differentiation. *Nat. Biotechnol.* **2010**, *28*, 511–515. [CrossRef] [PubMed]
55. Ashburner, M.; Ball, C.A.; Blake, J.A.; Botstein, D.; Cherry, J.M. Gene ontology: Tool for the unification of biology. The Gene Ontology Consortium. *Nat. Genet.* **2000**, *25*, 25–29. [CrossRef]
56. Young, M.D.; Wakefield, M.J.; Smyth, G.K.; Oshlack, A. Gene ontology analysis for RNA-seq: Accounting for selection bias. *Genome Biol.* **2010**, *11*, R14. [CrossRef]
57. Kanehisa, M.; Goto, S.; Kawashima, S.; Okuno, Y.; Hattori, M. The KEGG resource for deciphering the genome. *Nucleic Acids Res.* **2004**, *32* (Suppl. S1), D277–D280. [CrossRef]
58. Chen, C.; Chen, H.; Zhang, Y.; Thomas, H.R.; Frank, M.H.; He, Y.; Xia, R. TBtools: An Integrative Toolkit Developed for Interactive Analyses of Big Biological Data. *Mol. Plant* **2020**, *13*, 1194–1202. [CrossRef]
59. Langfelder, P.; Horvath, S. WGCNA: An R package for weighted correlation network analysis. *BMC Bioinform.* **2008**, *9*, 559. [CrossRef]

plants

MDPI

Article

Transcriptomic Analysis Reveals the Mechanism of Color Formation in the Peel of an Evergreen Pomegranate Cultivar 'Danruo No.1' During Fruit Development

Xiaowen Wang [1,2,†], Chengkun Yang [1,2,†], Wencan Zhu [1,2], Zhongrui Weng [1,2], Feili Li [1,2], Yuanwen Teng [3], Kaibing Zhou [1,2], Minjie Qian [1,2,*] and Qin Deng [1,2,*]

1 School of Breeding and Multiplication (Sanya Institute of Breeding and Multiplication), Hainan University, Sanya 572025, China; xiaowenwang@hainanu.edu.cn (X.W.); hndxyck@hainanu.edu.cn (C.Y.); wencanzhu@hainanu.edu.cn (W.Z.); zhongrui.weng@hainanu.edu.cn (Z.W.); feili.li@hainanu.edu.cn (F.L.); zkb@hainanu.edu.cn (K.Z.)

2 Key Laboratory of Quality Regulation of Tropical Horticultural Crop in Hainan Province, School of Tropical Agriculture and Forestry, Hainan University, Haikou 570228, China

3 Hainan Institute of Zhejiang University, Sanya 572000, China; ywteng@zju.edu.cn

* Correspondence: minjie.qian@hainanu.edu.cn (M.Q.); dengqin_e@yeah.net (Q.D.)

† These authors contributed equally to this work.

Abstract

Pomegranate (*Punica granatum* L.) is an ancient fruit crop that has been cultivated worldwide and is known for its attractive appearance and functional metabolites. Fruit color is an important index of fruit quality, but the color formation pattern in the peel of evergreen pomegranate and the relevant molecular mechanism is still unknown. In this study, the contents of pigments including anthocyanins, carotenoids, and chlorophyll in the peel of 'Danruo No. 1' pomegranate fruit during three developmental stages were measured, and RNA-seq was conducted to screen key genes regulating fruit color formation. The results show that pomegranate fruit turned from green to red during development, with a dramatic increase in a^* value, indicating redness and anthocyanins concentration, and a decrease of chlorophyll content. Moreover, carotenoids exhibited a decrease–increase accumulation pattern. Through RNA-seq, totals of 30, 18, and 17 structural genes related to anthocyanin biosynthesis, carotenoid biosynthesis and chlorophyll metabolism were identified from differentially expressed genes (DEGs), respectively. Transcription factors (TFs) such as MYB, bHLH, WRKY and AP2/ERF were identified as key candidates regulating pigment metabolism by K-means analysis and weighted gene co-expression network analysis (WGCNA). The results provide an insight into the theory of peel color formation in evergreen pomegranate fruit.

Keywords: *Punica granatum*; pigment; anthocyanin; carotenoid; chlorophyll; RNA-Seq

1. Introduction

Pomegranate (*Punica granatum* L.) is an ancient deciduous fruit crop that originated from the area of Iran to the Himalayan Mountains in northern India, and is cultivated in many subtropical and temperate regions around the world [1]. It is a kind of excellent fruit tree that combines ecological and ornamental values with health care functions [2]. There are 30 evergreen pomegranate cultivars in Punjab India, and 'Vietnamese' is an evergreen cultivar from Vietnam [3]. 'Danruo No. 1' is an evergreen pomegranate cultivar bred in 2019, and is the only commercial cultivar cultivated in the tropical region of China. Fruit color is a crucial index of fruit appearance quality, which can serve as an indicator for assessing food quality and grade [4], thereby influencing consumers' preferences and the commercial value of the products. For example, consumers exhibit a pronounced preference for apples [5], peaches [6], and Chinese cherries [7] with uniform coloration and a predilection for red hues. The pigmentation of fruit primarily derives from the concentration and proportional distribution of three key pigments: anthocyanins, carotenoids, and chlorophyll [8].

Anthocyanins are water-soluble pigments belonging to flavonoids class, generating red, blue and purple colors in the flowers and fruits [9]. Anthocyanins are synthesized through the phenylpropanoid and flavonoid pathways, and the involved enzymes include phenylalanine ammonia-lyase (PAL), cinnamate 4-hydroxylase (C4H), 4-coumarate-CoA ligase (4CL), chalcone synthase (CHS), chalcone isomerase (CHI), flavanone 3-hydroxylase (F3H), flavonoid 3′-hydroxylase (F3′H), flavonoid 3′,5′-hydroxylase (F3′5′H), dihydroflavonol 4-reductase (DFR), anthocyanidin synthase (ANS), and UDP-glucose: flavonoid 3-O-glucosyltransferase (UFGT) [9]. Genes encoding these proteins are called structural genes. Anthocyanins are then transferred to vacuoles for storage, and the enzyme involved in this process is probably glutathione S-transferase (GST) [10]. The expression of structural genes is regulated by transcription factors (TFs), and MYB, bHLH and WD40 are the predominant TFs that can form a complex and participate in the regulation of anthocyanins biosynthesis [9].

Carotenoids constitute a significant class of natural pigments, and plant tissues or organs accumulating carotenoids exhibit orange, yellow or red colors. The carotenoids biosynthesis pathway has been clarified in diverse plant species [11]. Carotenoids are predominantly synthesized via the plastid-localized methylerythritol 4-phosphate (MEP) pathway in plants, which generates the precursor of carotenoids, i.e., geranylgeranyl diphosphate (GGPP) [11]. GGPP is then converted to the first colorless carotenoid 15-cis-phytoene catalyzed by phytoene synthase (PSY). PSY serves as a crucial rate-limiting enzyme of carotenoid biosynthesis [12, 13]. Phytoene is subsequently desaturated and isomerized to produce red-colored lycopene catalyzed by phytoene desaturase (PDS), ζ-carotene isomerase (ZISO), ζ-carotene desaturase (ZDS), and carotenoid isomerase (CRTISO). Lycopene is cyclized by lycopene ε-cyclase (LCYE) and/or lycopene β-cyclase (LCYB) to produce orange-colored α-carotene and β-carotene, respectively. The hydroxylation of α- and β-carotene is catalyzed by two cytochrome P450-type hydroxylases (CYP97A and CYP97C) and two non-heme β-ring hydroxylases (BCH1 and BCH2) to form yellow-colored lutein and zeaxanthin, respectively. Zeaxanthin is catalyzed by zeaxanthin epoxidase (ZEP) to produce violaxanthin and reversed back by violaxanthin de-epoxidase (VDE), which is the called xanthophyll cycle. Violaxanthin is transformed into neoxanthin by neoxanthin synthase (NSY). Violaxanthin and neoxanthin are catalyzed by 9-cis-epoxycarotenoid dioxygenase (NCED) to yield abscisic acid (ABA).

The chlorophyll molecule mainly consists of a porphyrin ring containing a magnesium atom and a long carbon side chain. The chlorophyll metabolic pathway can be classified into three phases [14,15]. The first phase involves the synthesis of chlorophyll a from glutamate, and enzymes participating in this process include glutamyl-tRNA synthetase (GluRS), glutamyl-tRNA reductase (HEMA), glutamate-1-semialdehyde 2,1-aminotransferase (GSA), 5-aminolevulinate dehydrogenase (HEMB), porphobilinogen deaminase (HEMC), uroporphyrinogen III synthase (HEMD), uroporphyrinogen III decarboxylase (HEME), coproporphyrinogen III oxidase (HEMF), protoporphyrinogen oxidase (HEMG), Mg-chelatase (CHLH/D/I), Mg-protoporphyrin IX methyltransferase (CHLM), Mg-protoporphyrin IX monomethylester cyclase (CRD), protochlorophyllide oxidoreductase (POR), divinyl chlorophyllide a 8-vinyl-reductase (DVR), and chlorophyll synthase (CHLG/P). The second phase encompasses the interconversion of chlorophyll a and chlorophyll b, named the chlorophyll cycle, and geranylgeranyl-diphosphate reductase (CAO), chlorophyllide a oxygenase (NYC1/NOL), and chlorophyll b reductase (HCAR) are the key enzymes involved in this process. The third phase includes the degradation of chlorophyll a, which includes hydroxymethyl chlorophyll a reductase (CLH), magnesium dechelatase (MCS), pheophytin phephorbide hydrolase (PPH), pheophorbide a oxygenase (PaO), and red chlorophyll catabolite reductase (RCCR).

Fruit color is largely associated with fruit development. In pear, chlorophyll content is decreased during fruit development in both occidental and oriental pears, while anthocyanins content shows a decrease pattern in occidental pears and an increase–decrease pattern in oriental pears [16]. In pomegranate, based on RNA-seq, the expression patterns of anthocyanin biosynthetic genes in aril and fruit peels at different developmental stages have been analyzed in the deciduous cultivars 'Taishanhong' and 'Dabenzi' [17,18]. During fruit ripening, total chlorophylls and carotenoids contents in 'Taishanhong' pomegranate peel are decreased, while anthocyanin content shows a continuous increase pattern [19]. Anthocyanin content in the peel of 'Moshiliu' pomegranate shows two peaks during fruit development, which are detected at the early and late developmental stages, respectively, while 'Hongbaoshi' shows an additional peak at the middle of the developmental stage [20]. However, fruit coloration and the accumulation pattern of different pigments in the fruit peel of evergreen pomegranate cultivar during fruit development has not been assessed. According to previous studies, our hypothesis is that the chlorophyll content in evergreen pomegranate might also decrease during fruit development, while carotenoids and anthocyanin could show a unique pattern.

In this study, in order to comprehensively analyze the formation of fruit color in an evergreen pomegranate cultivar 'Danruo No. 1' and the relevant molecular mechanism, fruit peel at three developmental stages was sampled and the contents of anthocyanins, carotenoids, and chlorophyll were measured. RNA-seq was conducted to identify key structural genes and transcription factors related to color formation. This research will provide new insights into the peel color formation mechanism in evergreen pomegranate.

2. Results

2.1. Coloration and Pigments Contents in 'Danruo No.1' Pomegranate Peel During Fruit Development

The peel coloration of 'Danruo No.1' pomegranate during fruit development is shown in Figure 1A. The fruit peel was green at stage 1 (S1), but turned to red at stage 2 (S2) and stage 3 (S3). The L^* value representing lightness in peel showed an increase–decrease patter, with the peak detected at S2, reaching the value of 50.06 (Figure 1B). The a^* value in peel increased constantly during fruit development, reaching the highest value of 50.23 at S3 (Figure 1C). Peel b^* values at S2 and S3 were significantly higher than those at S1, but there was no significant difference between S2 and S3 (Figure 1D).

The contents of chlorophyll a, chlorophyll b and total chlorophyll in pomegranate peel peaked at S1, but showed very low levels at S2 and S3, and there was no significant difference between these two stages (Figure 1E–G, Table S1). The total anthocyanin content in pomegranate peel increased significantly with fruit development (Table S1). The total anthocyanin content of S3 was 3.39 times and 1.36 times higher than that of S1 and S2, respectively (Figure 1H). Carotenoid content decreased first and then increased, and the content at S3 was still significantly lower than at S1 (Figure 1I, Table S1).

Figure 1. *Cont.*

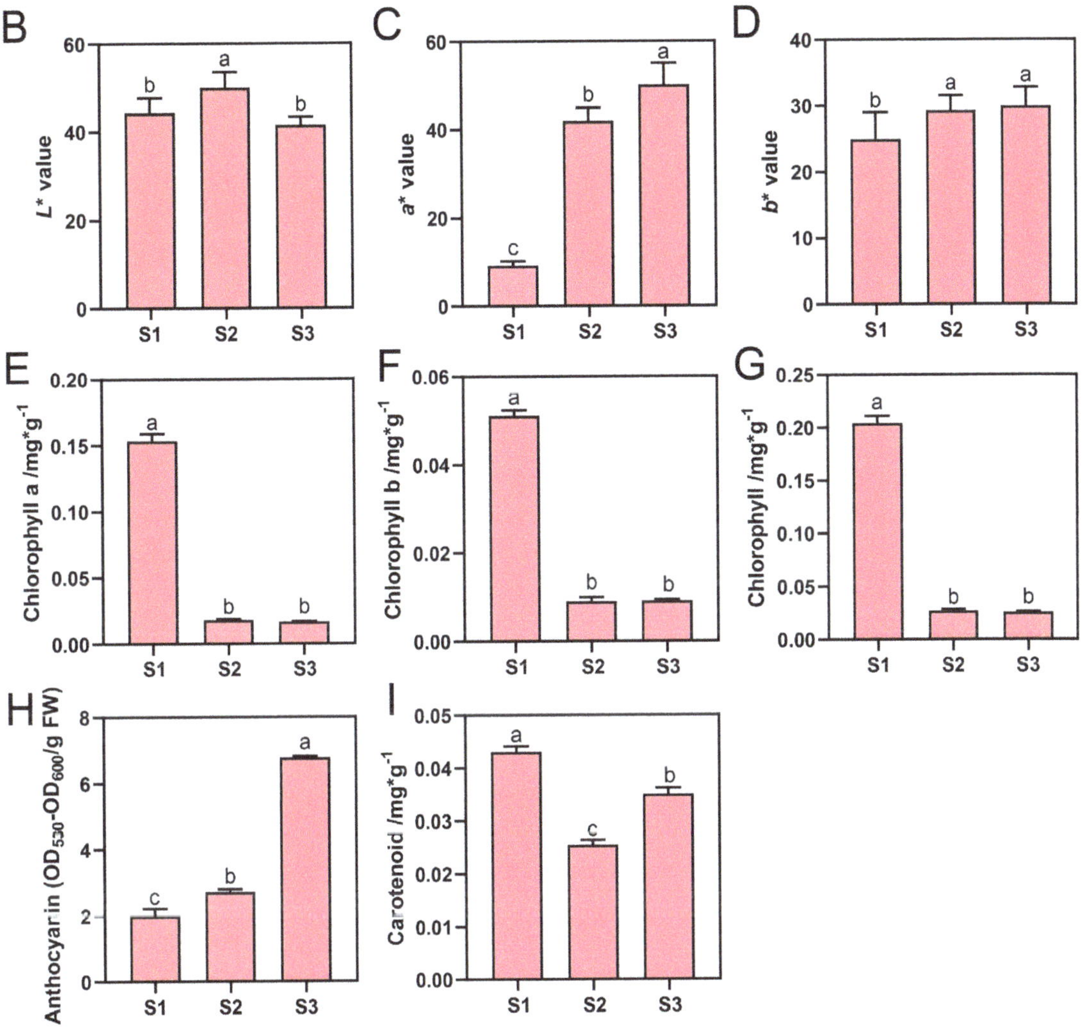

Figure 1. Coloration and pigment contents in 'Danruo No.1' pomegranate peel during fruit development. (**A**) Representative images of fruits at developmental stage 1 (S1), stage 2 (S2), and stage 3 (S3). (**B**) Fruit peel lightness (L^* value). (**C**) Fruit peel a^* value (higher value means redness and lower value means greenness). (**D**) Fruit peel b^* value (higher value means yellowness and lower value means blueness). (**E**) Chlorophyll a content. (**F**) Chlorophyll b content. (**G**) Total chlorophyll content. (**H**) Anthocyanin content. (**I**) Carotenoid content. Each value represents the mean ± standard deviation of three biological replicates. Different letters above the bars indicate significant differences ($p < 0.05$) according to one-way analysis of variance (ANOVA) followed by Tukey test.

2.2. RNA-seq Data Overview

A total of 42.34–57.38 million raw reads were obtained by the double-end sequencing of nine samples, resulting in a total of 41.70–55.07 million clean reads and a 6.25–8.26 G clean base (Table 1). The overall error rates were not more than 0.03%, the quality scores of 20 (Q20) values were around 98%, and the Q30 values were higher than 93.8% (Table 1). The GC contents ranged from 49.81% to 50.94% (Table 1). Principal component analysis (PCA) showed that PC1 and PC2 accounted for 49.82% and 23.52% of the total variation, respectively (Figure S1). Samples from three biological replicates of the same developmental stage clustered together, while samples from different developmental stages showed clear separation (Figure S1).

Table 1. Summary statistics of RNA-seq data quality for 9 samples.

Sample	Raw Reads	Clean Reads	Clean Base (G)	Error Rate (%)	Q20 (%)	Q30 (%)	GC Content (%)
S1-1	52,424,972	51,682,184	7.75	0.03	97.95	94.01	49.81
S1-2	51,211,690	50,438,940	7.57	0.03	97.89	93.9	49.87
S1-3	53,622,346	52,811,144	7.92	0.03	97.86	93.82	49.9
S2-1	57,381,414	55,067,352	8.26	0.03	97.99	94.41	50.65
S2-2	49,502,618	48,759,030	7.31	0.03	97.86	93.81	49.95
S2-3	53,041,172	50,933,314	7.64	0.03	97.94	94.33	50.94
S3-1	45,436,078	44,890,314	6.73	0.03	97.96	94.02	50
S3-2	54,455,452	52,331,886	7.85	0.02	98.06	94.52	50.57
S3-3	42,335,952	41,696,868	6.25	0.03	98.01	94.12	49.91

2.3. Differentially Expressed Genes (DEGs) Identification, and Gene Ontology (GO) and Kyoto Encyclopedia of Genes and Genomes (KEGG) Analyses

According to the DEGs multiple screening (log2 | FoldChange | > 1) and significant false discovery rate (FDR) < 0.05, totals of 2098, 2542, and 1340 up-regulated DEGs were detected in the comparison groups of S2 vs. S1, S3 vs. S1, and S3 vs. S2, respectively, and the corresponding numbers of downregulated DEGs were 3688, 3970, and 1265, respectively (Figure 2A–C).

The results of GO analyses of DEGs are shown in Figure S2. The DEGs were mainly annotated into cellular component, molecular function and biological process categories, which can be further divided into 40 subgroups. For DEGs from S2 vs. S1, the highest numbers of genes were annotated to secondary metabolism (159) in biological process, cytoskeleton (145) in cellular component, and tetrapyrrole binding (156) in molecular function, respectively. For S3 vs. S2, most DEGs were annotated to secondary metabolism (88) in biological process, plastid thylakoid (109) and chloroplast thylakoid (109) in cellular component, and oxidoreductase activity (95) and tetrapyrrole binding (95) in molecular function, respectively. For S3 vs. S1, most DEGs were annotated to secondary metabolism (180) in biological process, plastid thylakoid (186) and chloroplast thylakoid (186) in cellular component, and oxidoreductase activity (171) in molecular function, respectively.

KEGG analysis showed that most DEGs were annotated to metabolic pathway and the biosynthesis of secondary metabolites (Figure 2D–F). Regarding the pathways relevant to fruit coloration, flavonoid, carotenoid and phenylpropanoid biosynthesis pathways enriched DEGs from S2 vs. S1, S3 vs. S1, and S3 vs. S2 (Figure 2D–F). Besides this, DEGs from S3 vs. S2 were also enriched in flavone and flavonol biosynthesis, anthocyanin biosynthesis and porphyrin metabolism (Figure 2F).

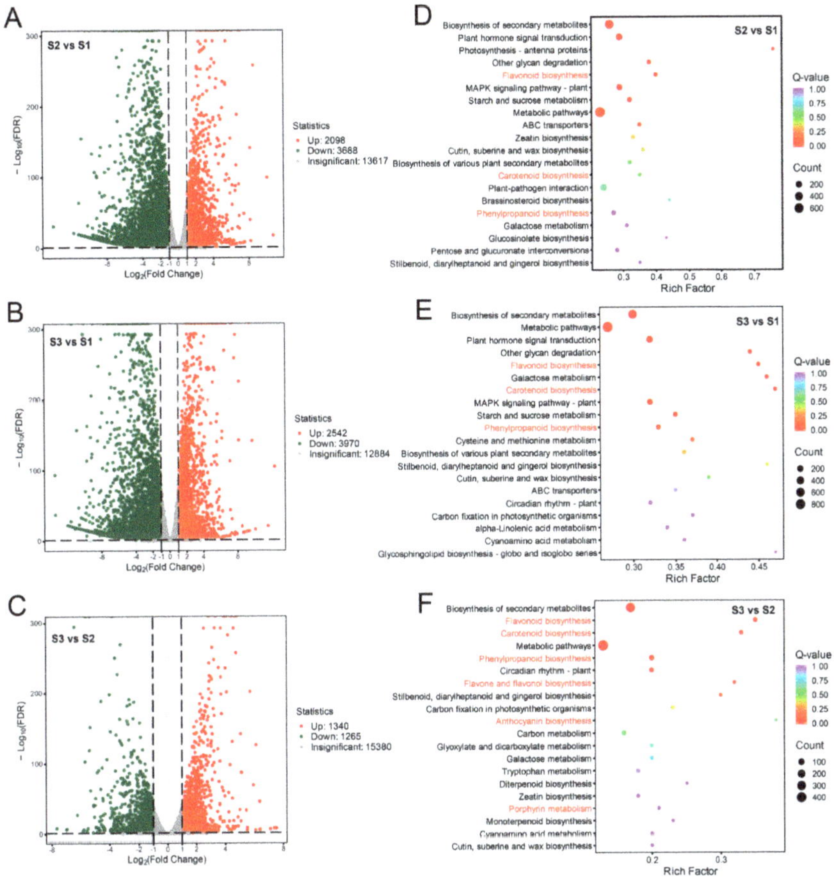

Figure 2. Differentially expressed genes (DEGs) identification and KEGG analysis. Volcano plots of DEGs from S2 vs. S1 (**A**), S3 vs. S1 (**B**), and S3 vs. S2 (**C**). Horizontal coordinates indicate the fold change of gene expression between different groups, and vertical coordinates indicate the significance level of gene expression difference in the two groups. Red dots indicate upregulated genes, green dots indicate downregulated genes, and grey dots indicate insignificant genes. Top 20 metabolic pathways analyzed by KEGG enrichment for DEGs from S2 vs. S1 (**D**), S3 vs. S1 (**E**), and S3 vs. S2 (**F**). The pathways associated with pigments metabolism are highlighted in red color.

2.4. DEGs Associated with Anthocyanin Biosynthesis in Pomegranate Peel During Fruit Development

A total of 30 DEGs involved in the anthocyanin biosynthetic pathway were identified (Table S2). Among them, genes including one *PAL*, one *C4H* (LOC116195737), one *F3′5′H*, one *ANS* (LOC116192123), one *FLS* (LOC116203029), one *ANR* (LOC116208810) and two *UFGTs* (LOC116209612 and LOC116201615) showed the highest expression levels at S1, with very low expression levels at S2 and S3 (Figure 3, Table S1). The expressions of one *DFR* (LOC116214145) and one *UFGT* (LOC116209568) showed a decrease–increase pattern, i.e., a high expression level was detected at S1 and S3 (Figure 3, Table S1). Most structural genes including one *C4H* (LOC116199390), two *4CLs*, two *CHSs*, two *CHIs* (LOC116196656 and LOC116187009), two *F3Hs* (LOC116201161 and LOC116211316), two *FLSs* (LOC116198870 and LOC116197460), one *DFR* (LOC116215495), four *ANSs* (LOC116213234, LOC116194289, LOC116197734, and LOC116205393), one *ANR* (LOC116193963), and two *UFGTs* (LOC116214735 and LOC116201378) showed a relatively higher expression level at S2 or S3, or both S2 and S3, when anthocyanin was highly accumulated (Figure 3, Table S1).

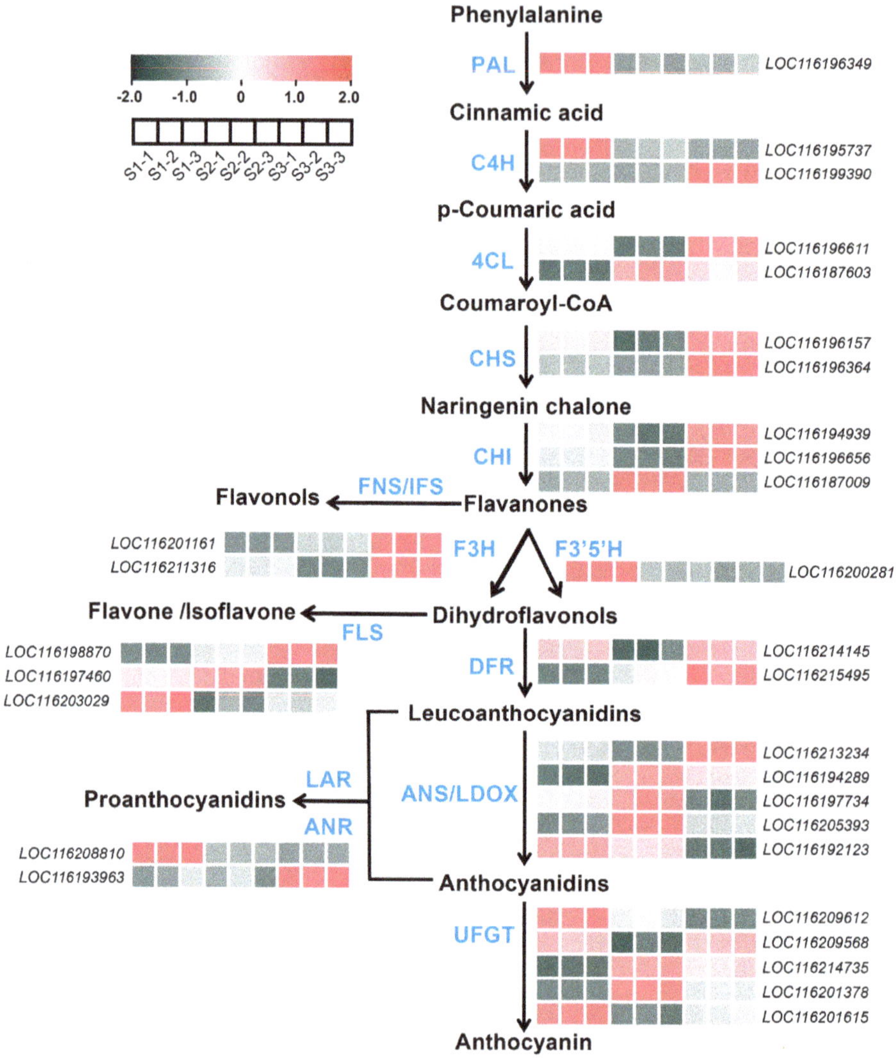

Figure 3. Expression patterns of the DEGs involved in anthocyanins synthesis in pomegranate peel at developmental stage 1 (S1), stage 2 (S2), and stage 3 (S3). The color scale from green to red represents the fragments per kilobase of transcript per million of fragments mapped (FPKM) values, from low to high.

2.5. DEGs Associated with Carotenoids Biosynthesis in Pomegranate Peel During Fruit Development

A total of 18 DEGs involved in the carotenoids biosynthetic pathway were identified (Table S2). One *ZEP* (LOC116194404) was highly expressed at S1 and S2 (Figure 4, Table S1). The expressions of one *PDS*, one *CRTISO* (LOC116203393), one *LCYB* (LOC116200932), one *VDE*, two *ZEPs* (LOC116207564 and LOC116210436), one *CCD* (LOC116187017) and one *NCED* (LOC116204963) showed the highest level at S3 (Figure 4, Table S1). For one *CYP97A*, one *CHYB*, one *ZEP* (LOC116196177), and one *NCED* (LOC116189437), relatively higher expression levels were detected only at S2 (Figure 4). In addition, the expressions of genes including one *CRTISO* (LOC116207720), one *LCYB* (LOC116201477), and one *NCED* (LOC116214029) were upregulated both at S2 and S3 (Figure 4, Table S1).

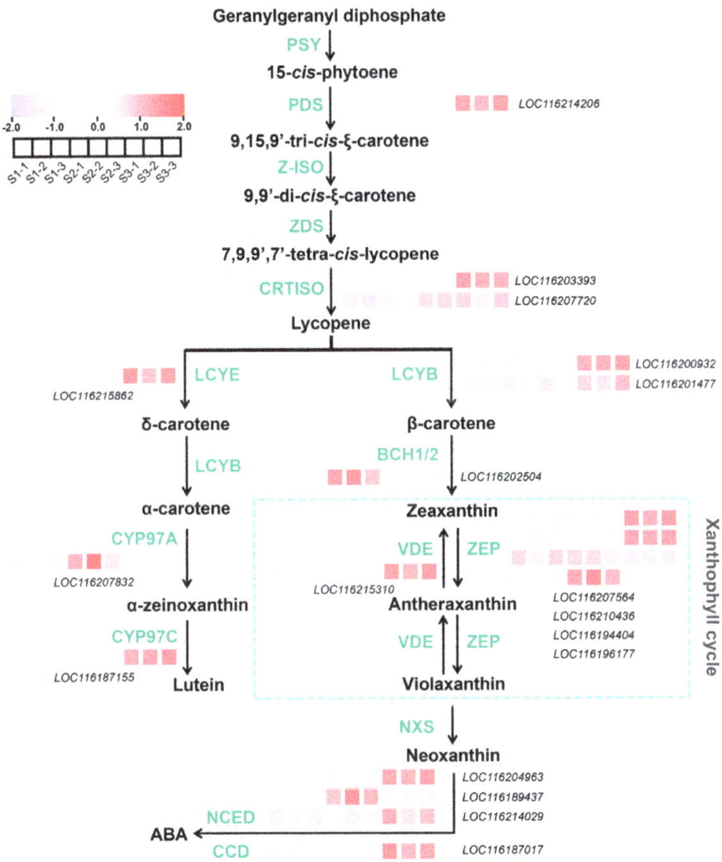

Figure 4. Expression pattern of the DEGs involved in carotenoids synthesis in pomegranate peel at developmental stage 1 (S1), stage 2 (S2), and stage 3 (S3). The color scale from blue to red represents the fragments per kilobase of transcript per million of fragments mapped (FPKM) values from low to high.

2.6. DEGs Associated with Chlorophyll Biosynthesis and Degradation in Pomegranate Peel During Fruit Development

A total of 17 DEGs involved in chlorophyll biosynthesis and degradation were identified (Tables S1 and S2). Among them, *POR* was the only gene that was highly expressed at S1, and it decreased to a very low level at S2 and S3 (Figure 5, Table S1). Higher expression levels of one *CHLM* (LOC116202463) and one *HEMD* (LOC116192834) were detected at S1 and S3, and S2 and S3, respectively (Figure 5, Table S1). All the rest of the chlorophyll biosynthetic genes, including one *HEME*, one *CHLH*, one *CHLM* (LOC116195758), two *DVRs*, and one *CAO*, and degradative genes including one *CLH*, one *NYC1*, one *NOL*, one *MCS*, and three *PAOs*, were only highly expressed at S3 (Figure 5, Table S1).

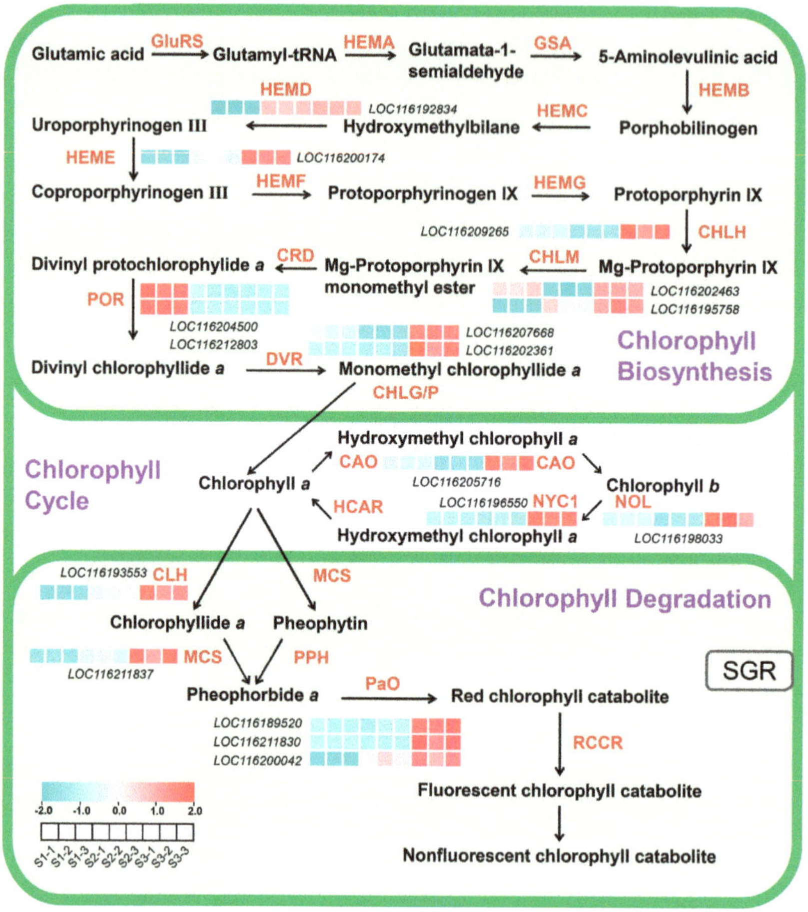

Figure 5. Expression pattern of the DEGs involved in chlorophyll biosynthesis and degradation in pomegranate peel at developmental stage 1 (S1), stage 2 (S2), and stage 3 (S3). The color scale from green to red represents the fragments per kilobase of transcript per million of fragments mapped (FPKM) values from low to high.

2.7. Screening Pigments Metabolism-Related TFs Based on K-Means and Weighted Gene Co-Expression Network Analysis (WGCNA)

We followed a two-step approach to screening regulatory genes that potentially participate in pigment metabolism. Firstly, K-means clustering analysis was performed for all DEGs (Figure 6A). Based on K-means analysis, genes in clusters 6/8, clusters 1/4, and clusters 9/7 were regarded as candidate genes positively/negatively involved in the accumulation of anthocyanins, chlorophyll, and carotenoids, respectively (Figure 6A). Secondly, WGCNA was conducted to analyze the correlation between DEGs and pigments concentration. All the DEGs were clustered into six modules, and the blue and red modules exhibited strong positive and negative correlations with anthocyanins content, respectively (Figure 6B). Genes in the turquoise module showed an extremely positive correlation coefficient with chlorophyll content, and these were also positively correlated to carotenoids content, while genes in the brown module displayed a strong negative correlation with these two substances (Figure 6B). In addition, genes from the yellow module were also negatively correlated with carotenoids content (Figure 6B).

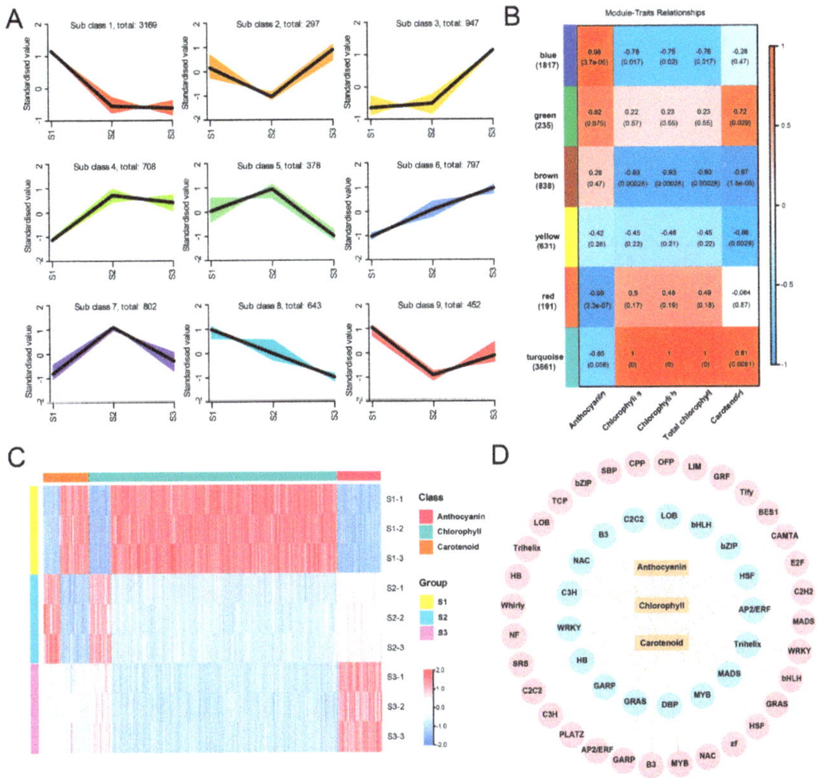

Figure 6. Identification of transcription factors (TFs) regulating pigments metabolism in pomegranate peel during fruit development. (**A**) K-means analysis of DEGs identified from transcriptome sequencing. The expression profiles of genes in each cluster are represented in different colors, and the average expression levels of all genes in developmental stage 1 (S1), S2, and S3 are represented in black. (**B**) Weighted gene co-expression network analysis (WGCNA) of DEGs identified from transcriptome sequencing. Module-trait correlations and corresponding *p*-values in parentheses. The left panel shows the six modules with gene numbers. The color scale on the right shows the module-trait correlations from −1 (blue) to 1 (red). 'Anthocyanin', 'Chlorophyll a', 'Chlorophyll b', 'Total chlorophyll' and 'Carotenoid' represent the changes in corresponding substances' concentrations. (**C**) Heatmap presenting the expression patterns of regulatory genes regulating pomegranate peel pigments metabolism during fruit development. (**D**) Correlation network between TFs' expression and pigments' contents; pink and blue circles represent positive and negative correlations, respectively. Purple, orange, and green lines representing the relation between TFs and anthocyanin, carotenoid, and chlorophyll, respectively.

By integrating the results from the K-means and WGCNA analyses, 45/1, 30/16, and 231/22 regulatory genes (encoding TFs) were identified as key candidate genes that positively/negatively regulate the accumulation of anthocyanins, carotenoid, and chlorophyll, respectively (Table S3). The expressions of these genes are shown in Figure 6C. For anthocyanin, the positive regulators were from 21 TF families including MYB, WRKY, bHLH, GRAS, etc., and an AP2/ERF TF was the only identified negative regulator (Figure 6D). For carotenoid, the positive regulators were from 12 TF families mainly including MYB, C2C2, AP2/ERF, bZIP, and LOB, and the negative regulators were from 10 families, mainly including HB, MYB, GARP, and HSF (Figure 6D). Furthermore, 29 and 13 TF families were regarded as positive and negative regulators for chlorophyll accumulation, respectively, which both contain MYB, C2H2, AP2/ERF, LOB, TCP, and Trihelix (Figure 6D).

2.8. Validation of DEGs by Quantitative Polymerase Chain Reaction (q-PCR)

Seven DEGs were chosen and analyzed by q-PCR to confirm the accuracy of RNA-seq data. The results show that gene expression in pomegranate peel during fruit development by q-PCR was consistent with the RNA-seq results (Figure 7, Table S1).

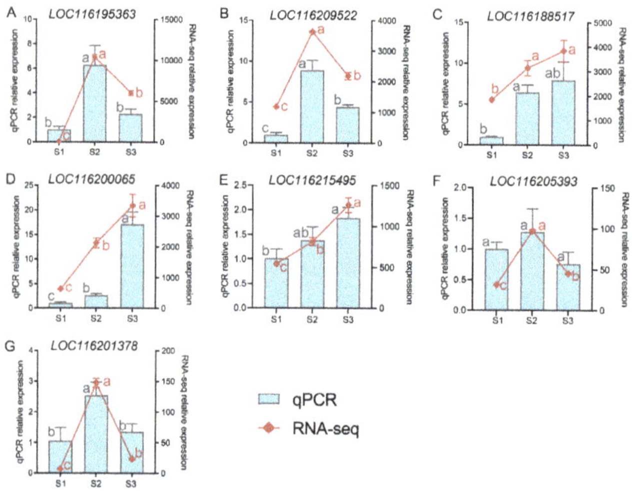

Figure 7. The expressions of seven genes in pomegranate peel at developmental stage 1 (S1), S2, and S3 from transcriptome data were examined by quantitative polymerase chain reaction (q-PCR). The expression levels obtained by RNA-seq and q-PCR are shown with a line chart and histogram, respectively. Data are presented as the mean ± standard deviation of three biological replicates. Different letters above the bars indicate significant differences ($p < 0.05$) according to one-way analysis of variance (ANOVA) followed by Tukey test. Data analyzed by qPCR (marked with gray letters) or RNA-seq (marked with red letters) were tested separately.

3. Discussion

Fruit color is an important index for judging fruit quality and maturity, in which chlorophyll, carotenoid and anthocyanin determine the color of plant tissue. In this study, with the growth and development of the fruit, anthocyanins content was significantly upregulated, carotenoids content was decreased first and then increased slightly, and chlorophyll content continuously declined, leading to the fruit peel color consequently turning from green to red (Figure 1E–I, Table S1). The trend of anthocyanin content in 'Danruo No. 1' pomegranate peel is similar to that in 'Taishanhong' pomegranate, mulberry and mango [19,21,22], but different from that in 'Moshiliu' and 'Hongbaoshi' pomegranate, pear and grape, in which the peak of anthocyanin content is observed at the early or middle developmental stages [16,20,23]. The trend of carotenoid content in 'Danruo No. 1' pomegranate peel is different from that of 'Taishanhong' pomegranate, in which carotenoid content is continuously decreased during fruit development [19]. A decrease in chlorophyll content during fruit development was also seen in 'Taishanhong' pomegranate [19], pear [16], and kiwi fruit [24]. All these results indicate that chlorophyll degradation during fruit development is quite conserved in fruits, while the anthocyanin and carotenoid accumulation patterns are diverse among different plant species, even among different cultivars of the same species.

Anthocyanin biosynthetic genes including *CHS*, *CHI*, *F3H*, *DFR*, *ANS*, and *UFGT* showed a relatively higher expression level at late developmental stages (Figure 3), which is associated with the anthocyanin content (Figure 1H). *CHI* and *CHS* are considered two crucial enzyme genes in the flavonoid biosynthesis pathway, closely related to fruit coloration, and their expression patterns increase with fruit development [25]. In pansy, the expression of *CHS* is closely associated with anthocyanin production [26]. It has been reported that *PgCHS*, *PgF3H*, and *PgDFR* isolated from pomegranates have higher expression levels in red fruits, and their transcriptional levels increase synchronously with flavonoid accumulation [27]. LDOX and UFGT catalyze the final two steps in the anthocyanin synthesis pathway. In the study, both *LDOX* and *UFGT* exhibited high levels of expression during the fruit color-changing period (Figure 3), which was consistent with the corresponding expression patterns observed in apple and jujube fruit [28,29]. In Arabidopsis, the ANS-deficient mutants are sensitive to high light due to impaired anthocyanin biosynthesis [30]. In grapes, *UFGT* is found to be expressed only in the skins of colored grape varieties after the color-changing period, but not in white varieties or pulp, with upregulated expression during the color-changing period [31]. The expression level of *UFGT* in wild pomegranates is in concurrence with the accumulation of anthocyanins observed in the leaves, flowers, green, and red fruits of the pomegranate plant [32]. *AcUFGT3* serves as a crucial gene that oversees the biosynthesis and accumulation of anthocyanins in red meat kiwifruit [33]. All these results suggest that the upregulation of anthocyanin biosynthetic genes is necessary for anthocyanin accumulation and red coloration.

Carotenoids content showed a decrease–increase pattern (Figure 1I). Pomegranate fruits undergo cell division and expansion periods from S1 to S2, probably resulting in a dilution effect and a decreased trend of carotenoids content from S1 to S2. Carotenoids biosynthesis is usually associated with fruit ripening, causing an increased trend of carotenoids content from S2 to S3. Most carotenoid biosynthetic genes, as well as the degradation genes, showed the expression peak at S2 or S3 (Figure 4), indicating that the relatively lower degradation rate at S1 and higher biosynthesis rate at S3 might be responsible for the higher carotenoids contents at S1 and S3, respectively. In celery, the expression of *AgCCD4* is negatively correlated with carotenoid content [34]. *PpCCD4* shows different expression patterns in different varieties of peaches, and its expression level is significantly higher in white-flesh peaches than in yellow-flesh peaches [35]. In mango, the expression of carotenoid biosynthesis-related genes such as *PSY*, *ZDS*, *CHYB* and *ZEP* is positively correlated with total carotenoid content [36,37]. In addition, the overexpression of the MADS transcription factor CsMADS6 in citrus healing tissues induces the expression of *LCYb1*, *PSY*, *PDS*, and *CCD1*, which increases carotenoid contents [38]. In tobacco, silencing the *PDS* gene resulted in a decrease in total carotenoid content [39]. The overexpression of *IbLCYB2* and *IbCCD4* genes increases and decreases carotenoid content in sweet potato, respectively [40,41]. Studies have shown that the downregulation of the *ZEP* gene in potato not only increases the accumulation of zeaxanthin in tubers, but also increases the total carotenoid level by 5.7-fold compared to the control group [42].

Chlorophyll content showed a peak at S1, but very low levels at S2 and S3 (Figure 1E–G), and most genes involved in chlorophyll biosynthesis and degradation, as well as cycling, showed the highest expression at S3, with *POR* showing the highest expression level at S1 (Figure 5), indicating that a high expression level of *POR* and lowly expressed degradative genes might contribute to high chlorophyll content at S1, and highly expressed degradative genes are responsible for the low chlorophyll content at S3. In Arabidopsis, different *POR* members have divergent roles in regulating chlorophyll biosynthesis, with *PORB* and *PORC* participating in low-light- and high-light-induced chlorophyll accumulation, respectively [43]. In the rice mutant (*pgl10*), *OsPORB* is downregulated, and Chl a and

Chl b contents are significantly reduced [44]. Our research has revealed that the expression pattern of chlorophyll catabolic genes exhibits a pronounced activity during the S3 phase (Figure 5), paralleling the trend observed in bananas [45]. Among genes involved chlorophyll degradation, all three *PaO* genes showed the highest expression level (Figure 5), indicating the key role of *PaO* in regulating chlorophyll degradation. PaO is a critical enzyme catalyzing the open-ring cleavage of the chlorophyll porphyrin to form a linear tetrapyrrole derivative. Studies on the induction of Arabidopsis leaf senescence in vitro under dark conditions showed that the expression, protein level and enzyme activity of *PaO* are consistently upregulated. They have noted that *PaO* predominantly regulates the degradation of chlorophyll at the transcriptional level, highlighting its significant role in chlorophyll catabolism during leaf senescence [46]. Increases in *PaO* activity or the expression of related genes are detected during fruit ripening in broccoli [47] and litchi [48], indicating a close relationship between enhanced chlorophyll catabolism and *PaO*. Similarly, in 'Zaosu' pear fruit, *PaO* has been identified as the key gene responsible for chlorophyll loss and chloroplast degradation in the fruit peel [49]. Therefore, we propose that increasing the activity of *PaO* genes during 'Danruo No. 1' pomegranate ripening could be a potential approach to maintaining good fruit appeal.

In this study, we analyzed the transcription factors at different developmental stages of 'Danruo No.1' pomegranate, and obtained 21 positive regulatory transcription factor families that may regulate anthocyanin accumulation, including MYB, WRKY, bHLH, GRAS, etc., and AP2/ERF is the only defined negative regulator (Figure 6D). MYB has been reported in several plant species, including Arabidopsis [50], tomato [51], apple [52] and tropical fruit [53]. The MBW complex composed of R2R3-MYB TFs, bHLH TFs and WD40 plays a key role in regulating the expression of structural genes in anthocyanin and flavonoid pathways, although specific members of each transcription factor family may differ in different species [54]. It has previously been reported that R2R3-MYB can activate the promoter of proanthocyanidin synthesis genes and regulate anthocyanin accumulation in peach flowers [55]. The activity of pomegranate regulatory gene *PgWD40* is confirmed in Arabidopsis *ttg1-9* mutant and proves to be dependent on MYB function [56]. In addition, some transcription factors such as WRKY [57], MYB (sweet cherry [58], grape [59], and kiwifruit [60]) and NAC [61] are also involved in the regulation of flavonoid metabolism. The transcriptional regulation mechanism of carotenoid biosynthesis has been well studied in different species. NAC22 [62] and AP2/ERF [63] can regulate the accumulation of carotenoids, thereby inducing fruit coloring. In our study, by constructing a correlation network, we obtained 12 classes of transcription factors (MYB, C2C2, AP2/ERF, bZIP, LOB, etc.) that positively regulate carotenoid components and 10 classes of negatively regulated transcription factors, mainly HB, MYB, GARP, HSF, etc. (Figure 6D). In addition, 29 TFs positively correlated with chlorophyll and 13 TFs negatively correlated with chlorophyll were obtained, mainly including MYB, C2H2, AP2/ERF, LOB, TCP, Trihelix and so on (Figure 6D). It has been reported that NAC016 directly binds to the *SGR1* promoter and regulates chlorophyll degradation [64]. Recently, the role of MYB in activating key chlorophyll biosynthesis genes to regulate chlorophyll accumulation and chloroplast development has also been confirmed in kiwifruit and tomato [65,66]. In addition, MYB [67], PIF [68] and NYC [69] transcription factors have also been shown to be involved in the regulation of chlorophyll metabolism.

4. Materials and Methods

4.1. Plant Materials

'Danruo No.1' pomegranate fruits were collected from a commercial pomegranate orchard located in Dongfang City, Hainan Province, China. Nine mature trees similar in size and number of fruit were chosen, with three trees per biological replicate. At 45 (Stage 1,

S1), 120 (Stage 2, S2) and 160 (Stage 3, S3) days after full bloom (DAFB), three fruits per tree, reaching a total of nine fruits per biological replicate, were harvested. After measuring the fruit peel color index using a portable colorimeter (LS170, Shenzhen Linshang Technology Co., Ltd., Shenzhen, China), fruit peel was collected, frozen in liquid nitrogen, and then stored at −80 °C for later use.

4.2. Measurement of Anthocyanin, Chlorophyll and Carotenoid Contents

Anthocyanin content was determined as described previously, with slight modifications [70]. Here, 0.2 g pomegranate peel samples were ground in liquid nitrogen, added to 1.2 mL of 1% HCl–methanol extraction solution (methanol:HCL = 99:1, *v*/*v*), and vortexed and extracted overnight at 4 °C in the dark. After centrifuging at 12,000 rpm for 10 min at 4 °C, the supernatant was used for anthocyanin detection. The absorbance was measured at 530 nm and 600 nm using a spectrophotometer (UV759, APL Instrument Co., Ltd., Shanghai, China). Anthocyanin content was expressed as the difference between absorbance values at 530 nm and 600 nm.

Chlorophyll and carotenoid contents were determined as described previously, with slight modifications [71]. Here, 0.2 g pomegranate peel samples were ground, added to 3 mL 95% ethanol, vortex-mixed and extracted for 3–5 min. After centrifuging at 12,000 rpm for 10 min, the supernatant was diluted 2 times and the absorbance values of chlorophyll a, chlorophyll b and carotenoids were observed at 665, 649 and 470 nm using a spectrophotometer (UV759, APL Instrument Co., Ltd., Shanghai, China), respectively.

4.3. RNA Extraction, Library Construction and RNA-Seq

Total RNA was extracted using the RNAprep Pure Plant Kit (Polysaccharides & Polyphenolics-rich) (Tiangen, Beijing, China) according to the manufacturer's protocol. RNA quality and concentration were analyzed by 1.0% agarose gel and a NanoDrop Lite Spectrophotometer (Thermo Scientific, Waltham, MA, USA), respectively.

RNA-Seq libraries were constructed by Wuhan Metware Biotechnology (Wuhan, China). The first strand of complementary DNA (cDNA) was synthesized from the fragmented mRNA template and random oligonucleotide primers. Then, deoxynucleotide triphosphates (dNTPs) were used as raw materials to synthesize the second strand of cDNA, and the double-stranded cDNA fragment was purified and recovered. The sequencing adaptor was ligated by end repair and A tail. The PCR products were purified by AMPure XP beads and the library was constructed. The library was diluted to 1.5 ng/uL by Qubit 2.0 Fluorometer (Thermo Scientific, Waltham, MA, USA), and the quality of the constructed library was checked by an Agilent 2100 bioanalyzer (Agilent, Santa Clara, CA, USA). When the insert size reached the expected value, the effective concentration of the library was accurately quantified by qRT-PCR to ensure the quality of the library. Each library was sequenced by the Illumina HiSeq2000 platform after the library inspection was qualified (Illumina Inc., San Diego, CA, USA). Low-quality data of the raw reads were removed by Fastp software 0.23.2 version (https://github.com/OpenGene/fastp, accessed on 16 May 2024), and TopHat [72] (http://tophat.cbcb.umd.edu, accessed on 16 May 2024) was then used to map clean reads to the pomegranate genome (NCBI accession number: PRJNA324150) [73]. Cufflinks (https://cole-trapnell-lab.github.io/cufflinks/, accessed on 17 May 2024) was used to assemble the transcripts [72], and gene expression was calculated and exhibited as fragments per kilobase of transcript per million of fragments mapped (FPKM). Differential expression analysis between two groups was performed using the DESeq2 v1.22.1 [74], and genes with $|\log2 \text{ fold change}| > 1$ and a significant p-value < 0.05 were considered as differentially expressed genes (DEGs). The raw data of the RNA-seq were submitted to NCBI with the following ID number: PRJNA1110799.

The interaction between co-expression modules was assessed using the Weighted Gene Co-Expression Network Analysis (WGCNA) algorithm. The concentrations of anthocyanins, chlorophyll and carotenoid in pomegranate peel during different developmental stages and all the expressed genes detected by RNA-seq (25,609 genes) were used for WGCNA. Highly correlated gene modules were identified using WGCNA in Metware Cloud (https://cloud.metware.cn/, accessed on 20 May 2024), and the results were visualized. The modules were built by the automatic network construction function 'blockwise'. The mergeCutHeight and minModuleSize were set to 0.25 and 50, respectively. The characteristic gene values were calculated for each module and used for testing the association with each sample or trait. Using the module characteristic genes to estimate the module–trait relationship, the samples were classified according to the corresponding traits, and the module with a |correlation coefficient| ≥ 0.8 and p-value ≤ 0.01 was selected for further analysis. K-means clustering was performed for all differential genes in Metware Cloud (https://cloud.metware.cn/, accessed on 21 May 2024). In total, 8193 genes with suitable FPKM values were used to construct K-means clustering. The K-means clustering was obtained using the default parameters. Kyoto Encyclopedia of Genes and Genomes (KEGG, https://www.genome.jp/kegg/, accessed on 22 May 2024) [75] and Gene Ontology (GO, https://www.geneontology.org/, accessed on 22 May 2024) [76] databases were selected for gene function annotation. All heatmaps were generated using TBtools, and the z-score normalization function was used to standardize the data between rows [77].

4.4. cDNA Synthesis and q-PCR

First-strand cDNAs were synthesized from 1 µg total RNA, and reverse transcription was performed using the HiScript III 1st Strand cDNA Synthesis Kit (+gDNA wiper) kit (vazyme, R312-01, Nanjing, China). For q-PCR assays, the cDNA templates were diluted 20 times to provide appropriate template concentration. The primers were designed using Primer3web version 4.1.0 (https://primer3.ut.ee/, accessed on 28 May 2024) and had optimum melting temperatures (Tm) and GC contents. Subsequently, the newly designed primers were checked for specificity using Tbtools, ensuring their suitability for further experimental applications. The designed gene-specific primers are listed in Supplementary Table S4. q-PCR was performed in a real-time fluorescent quantitative PCR instrument (qTOWER3 G, Jena, Germany). The quantitative program was set up as described previously [78]. The data were normalized using the cycle threshold (Ct) value corresponding to the pomegranate *actin* gene to minimize variation in the level of cDNA templates. Relative expression was calculated for each target gene as described by Livak and Schmittgen [79]. Analysis was performed in three biological replicates.

4.5. Data Analysis

SPSS 27.0 (SPSS Inc., Chicago, IL, USA) was used to analyze all data via a one-way Analysis of Variance (ANOVA), and the mean values were separated by Tukey test for multiple comparisons. The probability value of <0.05 was considered statistically significant. Cytoscape software (version 3.10.2) was used to visualize the network relationship.

5. Conclusions

During the development of 'Danruo No.1' pomegranate fruit, the color of fruit peel turns from green to red, with constant increases in the a^* value (representing redness) and anthocyanin concentration. In addition, the chlorophyll and carotenoid contents showed a decreasing pattern, and a decrease–increase pattern, respectively. Through RNA-seq, 8193 DEGs were identified, and KEGG enrichment analysis showed that flavonoid biosynthesis, carotenoid biosynthesis and phenylpropanoid biosynthesis were the most enriched pathways. Totals of 30, 18, and 17 structural genes related to anthocyanin biosynthesis, carotenoid biosynthesis and chlorophyll metabolism were screened from DEGs, respectively. Combined with WGCNA and K-means analyses, 45/1, 30/16, and 231/22 regulatory genes (encoding TFs) were identified as key candidate genes, which positively/negatively regulate the accumulation of anthocyanins, carotenoid, and chlorophyll, respectively, including MYB, WRKY, bHLH, AP2/ERF, bZIP, etc. The results of this study can provide a reference for the molecular mechanism of fruit peel color formation in evergreen pomegranate.

Supplementary Materials: The following supporting information can be downloaded at: https://www.mdpi.com/article/10.3390/plants13202903/s1, Figure S1: Principal component analysis of 9 samples; Figure S2: Go functional annotation of DEGs; Table S1: Results of one-way ANOVA analysis for the data; Table S2: Expression patterns of anthocyanin biosynthesis pathway, carotenoid biosynthesis pathway, and chlorophyll biosynthesis- and degradation-related differential genes; Table S3: Expression patterns of anthocyanin biosynthesis pathway and carotenoid biosynthesis pathway, and chlorophyll biosynthesis- and degradation-related differential transcription factor genes; Table S4: qRT-PCR primers of genes in pomegranate.

Author Contributions: X.W.: writing—review and editing, writing—original draft, methodology, formal analysis, data curation, conceptualization. C.Y.: writing—review and editing, writing—original draft, methodology, formal analysis, data curation, conceptualization. W.Z.: writing—review and editing, writing—original draft, methodology, data curation. Z.W.: writing—review and editing, writing—original draft, methodology, data curation. F.L.: writing—review and editing, writing—original draft, methodology, data curation. Y.T.: writing—review and editing, writing—original draft, funding acquisition. K.Z.: writing—review and editing, writing—original draft, conceptualization. M.Q.: writing—review and editing, writing—original draft, supervision, methodology, funding acquisition, data curation, conceptualization. Q.D.: writing—review and editing, writing—original draft, supervision, methodology, data curation, conceptualization. All authors have read and agreed to the published version of the manuscript.

Funding: This study was supported by the National Key Research and Development Plan of China (grant number: 2023YFD2300801), National Natural Science Foundation of China (grant numbers: 32360736 and 32160678), the Major Science and Technology Plan of Hainan Province (grant number: ZDKJ2021014), and Hainan Province Science and Technology Special Fund (grant number: ZDYF2023XDNY071).

Data Availability Statement: The data are contained within the manuscript and Supplementary Materials. The raw data of the RNA-seq were submitted to NCBI with the following ID number: PRJNA1110799.

Conflicts of Interest: The authors declare no conflicts of interest.

References

1. Da Silva, J.A.T.; Rana, T.S.; Narzary, D.; Verma, N.; Meshram, D.T.; Ranade, S.A. Pomegranate biology and biotechnology: A review. *Sci. Hortic.* **2013**, *160*, 85–107. [CrossRef]
2. Giménez-Bastida, J.A.; Ávila-Gálvez, M.Á.; Espín, J.C.; González-Sarrías, A. Evidence for health properties of pomegranate juices and extracts beyond nutrition: A critical systematic review of human studies. *Trends Food Sci. Technol.* **2021**, *114*, 410–423. [CrossRef]
3. Holland, D.; Hatib, K.; Bar-Ya'akov, I. Pomegranate: Botany, horticulture, breeding. *Hortic. Rev.* **2009**, *35*, 127–191. [CrossRef]
4. Cömert, E.D.; Mogol, B.A.; Gökmen, V. Relationship between color and antioxidant capacity of fruits and vegetables. *Curr. Res. Food Sci.* **2020**, *2*, 1–10. [CrossRef]
5. Ding, R.; Che, X.; Shen, Z.; Zhang, Y. Metabolome and transcriptome profiling provide insights into green apple peel reveals light- and UV-B-responsive pathway in anthocyanins accumulation. *BMC Plant Biol.* **2021**, *21*, 351. [CrossRef]
6. Campbell, B.L.; Nelson, R.G.; Ebel, R.C.; Dozier, W.A.; Adrian, J.L.; Hockema, B.R. Fruit quality characteristics that affect consumer preferences for satsuma mandarins. *HortScience* **2004**, *39*, 1664–1669. [CrossRef]

7. Wang, Y.; Tu, H.; Zhang, J.; Wang, H.; Liu, Z.; Zhou, J.; He, W.; Lin, Y.; Zhang, Y.; Li, M.; et al. Identifying potential anthocyanin biosynthesis regulator in Chinese cherry by comprehensive genome-wide characterization of the *R2R3-MYB* transcription factor gene family. *BMC Genom.* **2024**, *25*, 784. [CrossRef]
8. Abbott, J.A. Quality measurement of fruits and vegetables. *Postharvest Biol. Technol.* **1999**, *15*, 207–225. [CrossRef]
9. Jaakola, L. New insights into the regulation of anthocyanin biosynthesis in fruits. *Trends Plant Sci.* **2013**, *18*, 477–483. [CrossRef]
10. Yang, H.; Zhang, X.; Wu, R.; Tang, X.; Yang, Y.; Fan, X.; Gong, H.; Grierson, D.; Yin, X.; Li, J.; et al. Integrated metabolomic and transcriptomic analyses provide comprehensive new insights into the mechanism of chitosan delay of kiwifruit postharvest ripening. *Postharvest Biol. Technol.* **2024**, *210*, 112746. [CrossRef]
11. Sun, T.; Rao, S.; Zhou, X.; Li, L. Plant carotenoids: Recent advances and future perspectives. *Mol. Hortic.* **2022**, *2*, 3. [CrossRef]
12. Cazzonelli, C.I.; Pogson, B.J. Source to sink: Regulation of carotenoid biosynthesis in plants. *Trends Plant Sci.* **2010**, *15*, 266–274. [CrossRef]
13. Sun, T.; Yuan, H.; Cao, H.; Yazdani, M.; Tadmor, Y.; Li, L. Carotenoid Metabolism in Plants: The Role of Plastids. *Mol. Plant* **2018**, *11*, 58–74. [CrossRef]
14. Tanaka, A.; Tanaka, R. Chlorophyll metabolism. *Curr. Opin. Plant Biol.* **2006**, *9*, 248–255. [CrossRef]
15. Yu, Q.; Shen, Y.; Wang, Q.; Wang, X.; Fan, L.; Wang, Y.; Zhang, S.; Liu, Z.; Zhang, M. Light deficiency and waterlogging affect chlorophyll metabolism and photosynthesis in *Magnolia sinostellata*. *Trees* **2019**, *33*, 11–22. [CrossRef]
16. Yang, Y.-n.; Yao, G.-f.; Zheng, D.; Zhang, S.-l.; Wang, C.; Zhang, M.-y.; Wu, J. Expression differences of anthocyanin biosynthesis genes reveal regulation patterns for red pear coloration. *Plant Cell Rep.* **2015**, *34*, 189–198. [CrossRef]
17. Qin, G.; Xu, C.; Ming, R.; Tang, H.; Guyot, R.; Kramer, E.M.; Hu, Y.; Yi, X.; Qi, Y.; Xu, X.; et al. The pomegranate (*Punica granatum* L.) genome and the genomics of punicalagin biosynthesis. *Plant J.* **2017**, *91*, 1108–1128. [CrossRef]
18. Yuan, Z.; Fang, Y.; Zhang, T.; Fei, Z.; Han, F.; Liu, C.; Liu, M.; Xiao, W.; Zhang, W.; Wu, S.; et al. The pomegranate (*Punica granatum* L.) genome provides insights into fruit quality and ovule developmental biology. *Plant Biotechnol. J.* **2018**, *16*, 1363–1374. [CrossRef]
19. Zhao, X.Q.; Yuan, Z.H.; Yin, Y.L.; Feng, L.J. Patterns of Pigment Changes in Pomegranate (*Punica granatum* L.) Peel during Fruit Ripening. *Acta Hortic.* **2015**, *1089*, 83–89. [CrossRef]
20. Zhao, X.Q.; Yuan, Z.H.; Fang, Y.M.; Yin, Y.L.; Feng, L.J. Characterization and evaluation of major anthocyanins in pomegranate (*Punica granatum* L.) peel of different cultivars and their development phases. *Eur. Food Res. Technol.* **2013**, *236*, 109–117. [CrossRef]
21. Mo, R.; Zhang, N.; Li, J.; Jin, Q.; Zhu, Z.; Dong, Z.; Li, Y.; Zhang, C.; Yu, C. Transcriptomic Analysis Provides Insights into Anthocyanin Accumulation in Mulberry Fruits. *Horticulturae* **2022**, *8*, 920. [CrossRef]
22. Shi, B.; Wu, H.; Zheng, B.; Qian, M.; Gao, A.; Zhou, K. Analysis of Light-Independent Anthocyanin Accumulation in Mango (*Mangifera indica* L.). *Horticulturae* **2021**, *7*, 423. [CrossRef]
23. Qin, L.; Xie, H.; Xiang, N.; Wang, M.; Han, S.; Pan, M.; Guo, X.; Zhang, W. Dynamic Changes in Anthocyanin Accumulation and Cellular Antioxidant Activities in Two Varieties of Grape Berries during Fruit Maturation under Different Climates. *Molecules* **2022**, *27*, 384. [CrossRef]
24. Zhang, J.-Y.; Pan, D.-L.; Jia, Z.-H.; Wang, T.; Wang, G.; Guo, Z.-R. Chlorophyll, carotenoid and vitamin C metabolism regulation in *Actinidia chinensis* 'Hongyang' outer pericarp during fruit development. *PLoS ONE* **2018**, *13*, e0194835. [CrossRef]
25. Singh, K.; Kumar, A.; Kajal, M.; Singh, B. Characterization and expression analysis of chalcone synthase and chalcone isomerase genes in *Phyllanthus emblica* (L.). *J. Plant Biochem. Biotechnol.* **2019**, *28*, 105–113. [CrossRef]
26. Cui, Z.; Gu, J.; Li, J.; Zhao, A.; Fu, Y.; Wang, T.; Li, T.; Li, X.; Sheng, Y.; Zhao, Y.; et al. Tyrosine promotes anthocyanin biosynthesis in pansy (*Viola* × *wittrockiana*) by inducing ABA synthesis. *Trop. Plants* **2022**, *1*, 9. [CrossRef]
27. Kaur, R.; Aslam, L.; Kapoor, N.; Mahajan, R. Identification and comparative expression analysis of chalcone synthase, flavanone 3-hydroxylase and dihydroflavonol 4-reductase genes in wild pomegranate (*Punica granatum* L.) organs. *Braz. J. Bot.* **2020**, *43*, 883–896. [CrossRef]
28. Fei, X.; Wei, Y.; Qi, Y.; Luo, Y.; Hu, H.; Wei, A. Integrated LC-MS/MS and Transcriptome Sequencing Analysis Reveals the Mechanism of Color Formation during Prickly Ash Fruit Ripening. *Front. Nutr.* **2022**, *9*, 847823. [CrossRef]
29. Zhang, Q.; Wang, L.; Liu, Z.; Zhao, Z.; Zhao, J.; Wang, Z.; Zhou, G.; Liu, P.; Liu, M. Transcriptome and metabolome profiling unveil the mechanisms of *Ziziphus jujuba* Mill. peel coloration. *Food Chem.* **2020**, *312*, 125903. [CrossRef]
30. Zheng, X.-T.; Chen, Y.-L.; Zhang, X.-H.; Cai, M.-L.; Yu, Z.-C.; Peng, C.-L. ANS-deficient *Arabidopsis* is sensitive to high light due to impaired anthocyanin photoprotection. *Funct. Plant Biol.* **2019**, *46*, 756–765. [CrossRef]
31. Kobayashi, S.; Ishimaru, M.; Ding, C.; Yakushiji, H.; Goto, N. Comparison of UDP-glucose: Flavonoid 3-O-glucosyltransferase (UFGT) gene sequences between white grapes (*Vitis vinifera*) and their sports with red skin. *Plant Sci.* **2001**, *160*, 543–550. [CrossRef]
32. Kaur, R.; Kapoor, N.; Aslam, L.; Mahajan, R. Molecular characterization of PgUFGT gene and R2R3-PgMYB transcription factor involved in flavonoid biosynthesis in four tissues of wild pomegranate (*Punica granatum* L.). *J. Genet.* **2019**, *98*, 94. [CrossRef]

33. Liu, Y.; Liu, J.; Qi, Y.; Zhang, A.; Liu, Z.; Ren, X. Identification and characterization of AcUFGT6b, a xylosyltransferase involved in anthocyanin modification in red-fleshed kiwifruit (*Actinidia chinensis*). *Plant Cell Tissue Organ Cult.* **2019**, *138*, 257–271. [CrossRef]
34. Wang, H.; Yin, L.; Liu, J.; Jia, L.; Ding, X.; Shen, D.; Feng, K.; Xu, Z.; Xiong, A. The carotenoid cleavage dioxygenases gene AgCCD4 regulates the pigmentation of celery tissues with different colors. *Sci. Agric. Sin.* **2021**, *54*, 3279–3294. [CrossRef]
35. Fan, J.; Wu, J.; Li, Y.; Bie, H.; Wang, J.; Guo, J.; Cao, K.; Wang, L. Analysis of the Expression and Promoter Activity of the Key Gene PpCCD4 Involved in Carotenoid Synthesis in Peach. *J. Fruit Sci.* **2020**, *37*, 1271–1280. [CrossRef]
36. Dang, Z.; Zhu, M.; Chen, H.; Zhang, Y.; Gao, A.; Ma, W.; Chen, Y.; Wei, Y.; Zhang, H. MiMYB10 transcription factor regulates biosynthesis and accumulation of carotenoid involved genes in mango fruit. *Int. J. Biol. Macromol.* **2023**, *253*, 127665. [CrossRef]
37. Ma, X.; Luo, X.; Wei, Y.; Bai, T.; Shi, J.; Zheng, B.; Xu, W.; Li, L.; Wang, S.; Zhang, J.; et al. Chromosome-Scale Genome and Comparative Transcriptomic Analysis Reveal Transcriptional Regulators of β-Carotene Biosynthesis in Mango. *Front. Plant Sci.* **2021**, *12*, 749108. [CrossRef]
38. Lu, S.; Zhang, Y.; Zhu, K.; Yang, W.; Ye, J.; Chai, L.; Xu, Q.; Deng, X. The citrus transcription factor CsMADS6 modulates carotenoid metabolism by directly regulating carotenogenic genes. *Plant Physiol.* **2018**, *176*, 2657–2676. [CrossRef]
39. Wang, M.; Wang, G.; Ji, J.; Wang, J. The effect of *pds* gene silencing on chloroplast pigment composition, thylakoid membrane structure and photosynthesis efficiency in tobacco plants. *Plant Sci.* **2009**, *177*, 222–226. [CrossRef]
40. Kang, C.; Zhai, H.; Xue, L.; Zhao, N.; He, S.; Liu, Q. A lycopene β-cyclase gene, *IbLCYB2*, enhances carotenoid contents and abiotic stress tolerance in transgenic sweetpotato. *Plant Sci.* **2018**, *272*, 243–254. [CrossRef]
41. Zhang, J.; He, L.; Dong, J.; Zhao, C.; Wang, Y.; Tang, R.; Wang, W.; Ji, Z.; Cao, Q.; Xie, H.E. Integrated metabolic and transcriptional analysis reveals the role of carotenoid cleavage dioxygenase 4 (IbCCD4) in carotenoid accumulation in sweetpotato tuberous roots. *Biotechnol. Biofuels Bioprod.* **2023**, *16*, 45. [CrossRef]
42. Römer, S.; Lübeck, J.; Kauder, F.; Steiger, S.; Adomat, C.; Sandmann, G. Genetic engineering of a zeaxanthin-rich potato by antisense inactivation and co-suppression of carotenoid epoxidation. *Metab. Eng.* **2002**, *4*, 263–272. [CrossRef]
43. Masuda, T.; Fusada, N.; Oosawa, N.; Takamatsu, K.i.; Yamamoto, Y.Y.; Ohto, M.; Nakamura, K.; Goto, K.; Shibata, D.; Shirano, Y. Functional analysis of isoforms of NADPH: Protochlorophyllide oxidoreductase (POR), PORB and PORC, in *Arabidopsis thaliana*. *Plant Cell Physiol.* **2003**, *44*, 963–974. [CrossRef]
44. Yang, Y.-L.; Xu, J.; Rao, Y.-C.; Zeng, Y.-J.; Liu, H.-J.; Zheng, T.-T.; Zhang, G.-H.; Hu, J.; Guo, L.-B.; Qian, Q. Cloning and functional analysis of pale-green leaf (*PGL10*) in rice (*Oryza sativa* L.). *Plant Growth Regul.* **2016**, *78*, 69–77. [CrossRef]
45. Huang, H.; He, W. Application of exogenous cytokinin regulates cytokinin oxidase and antioxidant activity to maintain chlorophyll pigment during ripening of banana fruit. *Food Biosci.* **2023**, *55*, 102998. [CrossRef]
46. Pružinská, A.; Tanner, G.; Anders, I.; Roca, M.; Hörtensteiner, S. Chlorophyll breakdown: Pheophorbide a oxygenase is a Rieske-type iron–sulfur protein, encoded by the accelerated cell death 1 gene. *Proc. Natl. Acad. Sci. USA* **2003**, *100*, 15259–15264. [CrossRef]
47. Gomez-Lobato, M.E.; Civello, P.M.; Martínez, G.A. Effects of ethylene, cytokinin and physical treatments on *BoPaO* gene expression of harvested broccoli. *J. Sci. Food Agric.* **2012**, *92*, 151–158. [CrossRef]
48. Zou, S.-C.; Zhuo, M.-G.; Abbas, F.; Hu, G.-B.; Wang, H.-C.; Huang, X.-M. Transcription factor LcNAC002 coregulates chlorophyll degradation and anthocyanin biosynthesis in litchi. *Plant Physiol.* **2023**, *192*, 1913–1927. [CrossRef]
49. Jin, Y.; Li, C.; Zhang, S.; Liu, J.; Wang, M.; Guo, Y.; Xu, H.; Ge, Y. Sucrose, cell wall, and polyamine metabolisms involve in preserving postharvest quality of 'Zaosu' pear fruit by L-glutamate treatment. *Plant Physiol. Biochem.* **2024**, *208*, 108455. [CrossRef]
50. Onkokesung, N.; Reichelt, M.; van Doorn, A.; Schuurink, R.C.; van Loon, J.J.; Dicke, M. Modulation of flavonoid metabolites in *Arabidopsis thaliana* through overexpression of the *MYB75* transcription factor: Role of kaempferol-3, 7-dirhamnoside in resistance to the specialist insect herbivore *Pieris brassicae*. *J. Exp. Bot.* **2014**, *65*, 2203–2217. [CrossRef]
51. Jian, W.; Cao, H.; Yuan, S.; Liu, Y.; Lu, J.; Lu, W.; Li, N.; Wang, J.; Zou, J.; Tang, N. SlMYB75, an MYB-type transcription factor, promotes anthocyanin accumulation and enhances volatile aroma production in tomato fruits. *Hortic. Res.* **2019**, *6*, 22. [CrossRef]
52. Wang, S.; Zhang, Z.; Li, L.X.; Wang, H.B.; Zhou, H.; Chen, X.S.; Feng, S.Q. Apple MdMYB306-like inhibits anthocyanin synthesis by directly interacting with MdMYB17 and MdbHLH33. *Plant J.* **2022**, *110*, 1021–1034. [CrossRef]
53. Zhang, Y.; Xu, Y.; Huang, D.; Xing, W.; Wu, B.; Wei, Q.; Xu, Y.; Zhan, R.; Ma, F.; Song, S. Research progress on the MYB transcription factors in tropical fruit. *Trop. Plants* **2022**, *1*, 5. [CrossRef]
54. Ma, D.; Constabel, C.P. MYB repressors as regulators of phenylpropanoid metabolism in plants. *Trends Plant Sci.* **2019**, *24*, 275–289. [CrossRef]
55. Zhou, H.; Peng, Q.; Zhao, J.; Owiti, A.; Ren, F.; Liao, L.; Wang, L.; Deng, X.; Jiang, Q.; Han, Y. Multiple R2R3-MYB transcription factors involved in the regulation of anthocyanin accumulation in peach flower. *Front. Plant Sci.* **2016**, *7*, 1557. [CrossRef]
56. Ben-Simhon, Z.; Judeinstein, S.; Nadler-Hassar, T.; Trainin, T.; Bar-Ya'akov, I.; Borochov-Neori, H.; Holland, D. A pomegranate (*Punica granatum* L.) WD40-repeat gene is a functional homologue of *Arabidopsis TTG1* and is involved in the regulation of anthocyanin biosynthesis during pomegranate fruit development. *Planta* **2011**, *234*, 865–881. [CrossRef]

57. An, J.P.; Zhang, X.W.; You, C.X.; Bi, S.Q.; Wang, X.F.; Hao, Y.J. MdWRKY 40 promotes wounding-induced anthocyanin biosynthesis in association with Md MYB 1 and undergoes Md BT 2-mediated degradation. *New Phytol.* **2019**, *224*, 380–395. [CrossRef]
58. Jin, W.; Wang, H.; Li, M.; Wang, J.; Yang, Y.; Zhang, X.; Yan, G.; Zhang, H.; Liu, J.; Zhang, K. The R2R3 MYB transcription factor *PavMYB10.1* involves in anthocyanin biosynthesis and determines fruit skin colour in sweet cherry (*Prunus avium* L.). *Plant Biotechnol. J.* **2016**, *14*, 2120–2133. [CrossRef]
59. Fournier-Level, A.; Lacombe, T.; Le Cunff, L.; Boursiquot, J.-M.; This, P. Evolution of the VvMybA gene family, the major determinant of berry colour in cultivated grapevine (*Vitis vinifera* L.). *Heredity* **2010**, *104*, 351–362. [CrossRef]
60. Wang, W.Q.; Moss, S.M.; Zeng, L.; Espley, R.V.; Wang, T.; Lin-Wang, K.; Fu, B.L.; Schwinn, K.E.; Allan, A.C.; Yin, X.R. The red flesh of kiwifruit is differentially controlled by specific activation–repression systems. *New Phytol.* **2022**, *235*, 630–645. [CrossRef]
61. Sun, Q.; Jiang, S.; Zhang, T.; Xu, H.; Fang, H.; Zhang, J.; Su, M.; Wang, Y.; Zhang, Z.; Wang, N. Apple NAC transcription factor MdNAC52 regulates biosynthesis of anthocyanin and proanthocyanidin through MdMYB9 and MdMYB11. *Plant Sci.* **2019**, *289*, 110286. [CrossRef]
62. Gong, J.; Zeng, Y.; Meng, Q.; Guan, Y.; Li, C.; Yang, H.; Zhang, Y.; Ampomah-Dwamena, C.; Liu, P.; Chen, C. Red light-induced kumquat fruit coloration is attributable to increased carotenoid metabolism regulated by FcrNAC22. *J. Exp. Bot.* **2021**, *72*, 6274–6290. [CrossRef]
63. Dang, Q.; Sha, H.; Nie, J.; Wang, Y.; Yuan, Y.; Jia, D. An apple (*Malus domestica*) AP2/ERF transcription factor modulates carotenoid accumulation. *Hortic. Res.* **2021**, *8*, 223. [CrossRef]
64. Sakuraba, Y.; Kim, Y.-S.; Han, S.-H.; Lee, B.-D.; Paek, N.-C. The Arabidopsis transcription factor NAC016 promotes drought stress responses by repressing AREB1 transcription through a trifurcate feed-forward regulatory loop involving NAP. *Plant Cell* **2015**, *27*, 1771–1787. [CrossRef]
65. Ampomah-Dwamena, C.; Thrimawithana, A.H.; Dejnoprat, S.; Lewis, D.; Espley, R.V.; Allan, A.C. A kiwifruit (*Actinidia deliciosa*) R2R3-MYB transcription factor modulates chlorophyll and carotenoid accumulation. *New Phytol.* **2019**, *221*, 309–325. [CrossRef]
66. Wu, M.; Xu, X.; Hu, X.; Liu, Y.; Cao, H.; Chan, H.; Gong, Z.; Yuan, Y.; Luo, Y.; Feng, B.; et al. SlMYB72 Regulates the Metabolism of Chlorophylls, Carotenoids, and Flavonoids in Tomato Fruit. *Plant Physiol.* **2020**, *183*, 854–868. [CrossRef]
67. Wei, W.; Yang, Y.-Y.; Lakshmanan, P.; Kuang, J.-F.; Lu, W.-J.; Pang, X.-Q.; Chen, J.-Y.; Shan, W. Proteasomal degradation of MaMYB60 mediated by the E3 ligase MaBAH1 causes high temperature-induced repression of chlorophyll catabolism and green ripening in banana. *Plant Cell* **2023**, *35*, 1408–1428. [CrossRef]
68. Sakuraba, Y.; Jeong, J.; Kang, M.-Y.; Kim, J.; Paek, N.-C.; Choi, G. Phytochrome-interacting transcription factors PIF4 and PIF5 induce leaf senescence in *Arabidopsis*. *Nat. Commun.* **2014**, *5*, 4636. [CrossRef]
69. Tak, H.; Negi, S.; Gupta, A.; Ganapathi, T. A stress associated NAC transcription factor MpSNAC67 from banana (*Musa* x *paradisiaca*) is involved in regulation of chlorophyll catabolic pathway. *Plant Physiol. Biochem.* **2018**, *132*, 61–71. [CrossRef]
70. Zhang, D.; Quantick, P.C. Effects of chitosan coating on enzymatic browning and decay during postharvest storage of litchi (*Litchi chinensis* Sonn.) fruit. *Postharvest Biol. Technol.* **1997**, *12*, 195–202. [CrossRef]
71. Lichtenthaler, H.K.; Wellburn, A.R. Determinations of total carotenoids and chlorophylls a and b of leaf extracts in different solvents. *Analysis* **1983**, *11*, 591–592. [CrossRef]
72. Trapnell, C.; Roberts, A.; Goff, L.; Pertea, G.; Kim, D.; Kelley, D.R.; Pimentel, H.; Salzberg, S.L.; Rinn, J.L.; Pachter, L. Differential gene and transcript expression analysis of RNA-seq experiments with TopHat and Cufflinks. *Nat. Protoc.* **2012**, *7*, 562–578. [CrossRef]
73. Luo, X.; Li, H.; Wu, Z.; Yao, W.; Zhao, P.; Cao, D.; Yu, H.; Li, K.; Poudel, K.; Zhao, D. The pomegranate (*Punica granatum* L.) draft genome dissects genetic divergence between soft-and hard-seeded cultivars. *Plant Biotechnol. J.* **2020**, *18*, 955–968. [CrossRef]
74. Love, M.I.; Huber, W.; Anders, S. Moderated estimation of fold change and dispersion for RNA-seq data with DESeq2. *Genome Biol.* **2014**, *15*, 550. [CrossRef]
75. Kanehisa, M.; Goto, S.; Sato, Y.; Furumichi, M.; Tanabe, M. KEGG for integration and interpretation of large-scale molecular data sets. *Nucleic Acids Res.* **2012**, *40*, D109–D114. [CrossRef]
76. Ashburner, M.; Ball, C.A.; Blake, J.A.; Botstein, D.; Butler, H.; Cherry, J.M.; Davis, A.P.; Dolinski, K.; Dwight, S.S.; Eppig, J.T. Gene ontology: Tool for the unification of biology. *Nat. Genet.* **2000**, *25*, 25–29. [CrossRef]
77. Chen, C.; Wu, Y.; Li, J.; Wang, X.; Zeng, Z.; Xu, J.; Liu, Y.; Feng, J.; Chen, H.; He, Y.; et al. TBtools-II: A "one for all, all for one" bioinformatics platform for biological big-data mining. *Mol. Plant* **2023**, *16*, 1733–1742. [CrossRef]
78. Shi, B.; Wu, H.; Zhu, W.; Zheng, B.; Wang, S.; Zhou, K.; Qian, M. Genome-Wide Identification and Expression Analysis of *WRKY* Genes during Anthocyanin Biosynthesis in the Mango (*Mangifera indica* L.). *Agriculture* **2022**, *12*, 821. [CrossRef]
79. Livak, K.J.; Schmittgen, T.D. Analysis of relative gene expression data using real-time quantitative PCR and the $2^{-\Delta\Delta CT}$ method. *Methods* **2001**, *25*, 402–408. [CrossRef]

plants

MDPI

Article

Transcriptomic and Metabolomic Analysis Reveals the Potential Roles of Polyphenols and Flavonoids in Response to Sunburn Stress in Chinese Olive (*Canarium album*)

Yu Long [1,†], Chaogui Shen [1,†], Ruilian Lai [1], Meihua Zhang [2], Qilin Tian [3], Xiaoxia Wei [1,*] and Rujian Wu [1,*]

1 Fruit Research Institute, Fujian Academy of Agricultural Sciences, Fuzhou 350013, China; llyy103699@163.com (Y.L.); shenchaogui@faas.cn (C.S.); lairuilian@faas.cn (R.L.)

2 Minhou Meteorological Bureau of Fujian Province, Minhou 350100, China; fjsmhxqxj@163.com

3 Provincial Key Laboratory of Landscape Plants with Fujian and Taiwan Characteristics, Minnan Normal University, Zhangzhou 363000, China; tianql70@163.com

* Correspondence: weixiaoxia@faas.cn (X.W.); wurujian@faas.cn (R.W.); Tel.: +86-13705973257 (X.W.); +86-13706963696 (R.W.)

† These authors contributed equally to this work.

Abstract: Sunburn stress is one of the main environmental stress factors that seriously affects the fruit development and quality of Chinese olive, a tropical and subtropical fruit in south China. Therefore, the understanding of the changes in physiological, biochemical, metabolic, and gene expression in response to sunburn stress is of great significance for the industry and breeding of Chinese olive. In this study, the different stress degrees of Chinese olive fruits, including serious sunburn injury (SSI), mild sunburn injury (MSI), and ordinary (control check, CK) samples, were used to identify the physiological and biochemical changes and explore the differentially expressed genes (DEGs) and differentially accumulated metabolites (DAMs) by using transcriptomics and metabolomics. Compared with CK, the phenotypes, antioxidant capacity, and antioxidant-related enzyme activities of sunburn stress samples changed significantly. Based on DEG-based KEGG metabolic pathway analysis of transcriptomics, the polyphenol and flavonoid-related pathways, including phenylpropanoid biosynthesis, sesquiterpenoid, and triterpenoid biosynthesis, monoterpene biosynthesis, carotenoid biosynthesis, isoflavonoid biosynthesis, flavonoid biosynthesis, were enriched under sunburn stress of Chinese olive. Meanwhile, 33 differentially accumulated polyphenols and 99 differentially accumulated flavonoids were identified using metabolomics. According to the integration of transcriptome and metabolome, 15 and 8 DEGs were predicted to regulate polyphenol and flavonoid biosynthesis in Chinese olive, including 4-coumarate-CoA ligase (*4CL*), cinnamoyl-CoA reductase (*CCR*), cinnamoyl-alcohol dehydrogenase (*CAD*), chalcone synthase (*CHS*), flavanone-3-hydroxylase (*F3H*), dihydroflavonol 4-reductase (*DFR*), and anthocyanidin synthase (*ANS*). Additionally, the content of total polyphenols and flavonoids was found to be significantly increased in MSI and SSI samples compared with CK. Our research suggested that the sunburn stress probably activates the transcription of the structural genes involved in polyphenol and flavonoid biosynthesis in Chinese olive fruits to affect the antioxidant capacity and increase the accumulation of polyphenols and flavonoids, thereby responding to this abiotic stress.

Keywords: *Canarium album*; sunburn injury; polyphenol; flavonoid; metabolome; transcriptome

1. Introduction

Environmental stress is a common abiotic stress that affects plant growth and development. Under natural conditions, when the environmental temperature and light intensity reach a critical threshold of plant tolerance for a long time, plants undergo irreversible damage, resulting in high-temperature stress or sunburn injury. Fruit is an important economic crop, and it is often cultivated in open fields, which are always distributed around the tree crown, and the growth and development need to be across the hot summers [1]. Long periods of high-temperature sunburn lead to shrinking or wilting of branches and leaves, causing varying degrees of burns on the surface of the fruit and even fruit drop. Protecting fruits from sunburn has become a challenge for many fruit industries.

Sunburn is a type of light damage caused by excessive heat and light (visible or ultraviolet) [2]. It has negative effects on fruit maturity, appearance, color, photosynthesis, energy metabolism, stress response and defense, cell/cell membrane/tissue structure, and so on. The injuries caused by sunburn may be exacerbated by other stress factors, such as water deprivation [3]. The increase in the levels of reactive oxygen species (ROS) is the main reason for high-temperature sunburn injury [2]. Plants have evolved effective regulatory systems, such as by increasing the biosynthesis of photoprotective compounds, heat shock protein, and nonenzymatic antioxidant metabolites, to sense and respond to high-temperature or solar radiation stress to avoid oxidative stress caused by excessive ROS accumulation [2,4]. This phenomenon conversely activates antioxidant defense mechanisms while acting in concert with antioxidant enzymes in response to excessive ROS production to improve the adaptability of plants to abiotic or biotic stresses [2,5]. For example, sunburn affects metabolic pathways, such as heat shock response, stress response and defense, energy metabolism, photosynthesis, and protein biosynthesis, during the peel color change of loquat fruit [6]. The content of chloroplasts and anthocyanin in sunburned apples decreases [7], while the content of total flavonols, dihydrochalcones, total hydroxycinnamic acid, and peroxidase activity significantly increases [8]. Meanwhile, the content of antioxidants and polyphenols in apples increases after long-term strong light irradiation to adapt to the strong light environment [9]. With the increase in sunburn injury, photosynthetic pigments are destroyed, and polyphenol content is significantly increased to protect grapes from sunburn injury [10]. Sunburn injury leads to a decrease in the chlorophyll content, an increase in the flavonoid and carotenoid contents, and a higher ability to scavenge free radicals in the peel of satsuma mandarin fruits [11]. The content of total phenols and total flavonoids increases in sunburned pomegranate fruits, enhancing the total antioxidant capacity and thereby reducing and repairing membrane damage [12]. The sunburn injury promotes the accumulation of polyphenols and enhances the activity of antioxidant enzymes to increase the anti-oxidation ability of peels in pear [13].

The response of plants to sunburn stress depends on multiple signal transduction, metabolite biosynthesis, and gene transcription [14–16]. Metabolomics, based on high-throughput data, effectively identifies the dynamic composition of endogenous metabolites in organisms. Transcriptomics, on the other hand, can effectively monitor all gene transcripts in biological samples. Both of them are widely used in the study of plant stress response mechanisms. The transcriptomic and metabolomic studies revealed that sunburn caused significant changes in the content of 129 metabolites and the transcription of 447 genes in pomegranate. *NAC5*, *MYB93*, and *MYB111* transcription factors may be involved in phenylpropanoid and flavonoid biosynthesis in response to sunburn stress by regulating chalcone isomerase (*CHI*), flavonoid 3′, 5′-hydroxylase (*F3′5′H*), and chalcone synthase (*CHS*) [12]. Flavonoid biosynthesis and heat shock protein–heat shock factor pathway of *Sorbus pohuashanensis* played an important role in resisting sunburn stress [17]. The pathway involved in ROS clearance in apples was potentially crucial in response to sunburn stress [18]. These studies revealed that different plants showed different response modes to sunburn stress.

Chinese olive (*Canarium album* Raeusch.) is a traditional Chinese medicine and characteristic fruit and is widely distributed in southern China [19]. As a typical tropical and subtropical crop, the Chinese olive plant has a strong high-temperature tolerance. However, its fruits are often distributed outside the tree crown. During the fruit development and expansion period from July to August each year, they are prone to extremely high temperatures and strong sunlight, leading to severe diseases in fruits and physiological fruit drop. Hence, sunburn stress has become one of the essential environmental factors restricting the Chinese olive industry. This experiment identified the physiological and biochemical changes in Chinese olive fruits after sunburn stress and explored differentially accumulated metabolites (DAMs) and differentially expressed genes (DEGs) during sunburn stress occurrence through the integration of metabolomic and transcriptomic analyses. A metabolic regulatory network for the response of Chinese olive plants to sunburn stress was preliminarily constructed, laying a theoretical foundation for the study of genetic mechanisms of defense of Chinese olive plants and the selection of high-resistance varieties.

2. Results

2.1. Changes in Phenotypic and Antioxidant Capacity

The changes in the phenotype, color, and firmness of Chinese olive fruits under sunburn stress are shown in Figure 1. The results indicated that the sunburn caused varying degrees of burned patches on the surface of Chinese olive fruits (Figure 1A). In terms of fruit color (Figure 1B), the a^* value of the mild sunburn injury (MSI) sample significantly increased, whereas the h^* value significantly decreased. However, the a^* value of the serious sunburn injury (SSI) sample significantly increased, whereas the L^*, b^*, c^*, and h^* values all significantly decreased compared with those of the control check (CK) sample. Significant differences in chromatic aberration parameters were found between MSI and SSI samples, except for the a^* value. The fruit firmness of the MSI sample did not show significant change (Figure 1C), while that of the SSI sample significantly improved compared with that of the CK sample. The fruit firmness of the SSI sample also considerably increased compared with that of the MSI sample. In conclusion, MSI led to changes in the color of the fruit, whereas SSI affected both the color and firmness of the fruit.

Figure 1. The changes of phenotypes, antioxidant capacity, and antioxidant-related enzyme activities in response to sunburn stress of Chinese olive. (**A**) The phenotype changes of samples. (**B**) The color changes of samples are represented by chromatic aberration L^*, a^*, b^*, c^*, and h^* values. (**C**) The firmness changes of samples. (**D**) The T-AOC changes of samples (μmol·mL^{-1} FW). (**E**) The content of H_2O_2 in samples (μmol·g^{-1} FW). (**F**) The content of MDA in samples (nmol·g^{-1} FW). (**G**) The content of proline in samples (μg·g^{-1} FW). (**H**) The activity changes of CAT, SOD, POD, APX, and PPO of samples (U·g^{-1} FW). CK, MSI, and SSI indicate the samples of control check, mild sunburn injury, and serious sunburn injury, respectively. The upper- and lowercase letters on the curves represent significant differences at p value < 0.01 and <0.05 levels.

Further analysis found that the total antioxidant capacity (T-AOC) changed under sunburn stress. Compared with the CK sample, T-AOC was significantly upregulated in the MSI sample and significantly decreased in the SSI sample (Figure 1D). The hydrogen peroxide (H_2O_2) content significantly decreased in the MSI sample; however, there was no change in the SSI sample (Figure 1E). Moreover, the contents of malondialdehyde (MDA) and proline both significantly increased in MSI and SSI samples compared with the CK sample (Figure 1F,G). Compared with the CK sample, the activities of several antioxidant-related enzymes were significantly changed (Figure 1H); the catalase (CAT), superoxide dismutase (SOD), and ascorbate peroxidase (APX) activities increased in the MSI sample and then decreased in the SSI sample, while that of polyphenol oxidase (PPO) and peroxidase (POD) continuously increased.

2.2. Transcriptomic Analysis

2.2.1. Transcriptome Data Evaluation and DEGs Identification

A total of 72.16 GB of clean data were obtained from three biological duplicate samples of T-CK, T-MSI, and T-SSI using Illumina Hiseq 2000 (Biomarker Technologies, Beijing, China), and 43,160 unigenes were annotated. Meanwhile, the clean data range for each sample was from 7.36 to 8.88 GB, with the GC base content ranging from 43.74% to 44.06% and Q30 exceeding 96%. The data quality and reliability of the constructed library were found to be high and hence used for subsequent analysis.

A total of 3259 DEGs (Supplementary Materials Table S1) were selected based on the criteria of the corrected p value < 0.01 and $|\log2^{\text{fold change (FC)}}| \geq 1.0$. Among these, 2110, 2340, and 65 DEGs were identified between the comparison groups of T-CK vs. T-MSI, T-CK vs. T-SSI, and T-MSI vs. T-SSI, respectively. Among these, 1216, 34, and 8 DEGs were found to have differences between T-CK vs. T-MSI and T-CK vs. T-SSI, between T-CK vs. T-SSI and T-MSI vs. T-SSI, and between T-CK vs. T-MSI and T-MSI vs. T-SSI, respectively (Supplementary Materials Figure S1A). The cluster analysis results indicated that the variation patterns of DEGs in MSI and SSI samples were similar (Supplementary Materials Figure S1B), indicating that most DEGs had consistent responses to mild and serious sunburn injuries.

2.2.2. GO Pathway Enrichment Analysis Based on DEGs

Based on Gene Ontology (GO) enrichment analysis, the DEGs between samples were divided into cellular component (CC), molecular function (MF), and biological process (BP). Among these, 1635 DEGs between T-CK and T-MSI were enriched into 18 BPs, 15 CCs, and 12 MFs related to GO pathways (Supplementary Materials Table S2). Also, 1805 DEGs between T-CK and T-SSI were annotated into 18 BPs, 17 CCs, and 13 MFs related to GO pathways (Supplementary Materials Table S3). Additionally, DGEs between T-MSI and T-SSI samples were enriched into 24 GO pathways (12 BPs, 8 CCs, and 4 MFs) (Supplementary Materials Table S4). Among these, 44, 24, and 24 GO pathways co-enriched between T-CK vs. T-MSI and T-CK vs. T-SSI, T-CK vs. T-MSI and T-MSI vs. T-SSI, and T-CK vs. T-SSI and T-MSI vs. T-SSI, respectively. These results indicated that the GO pathways enriched in the SSI sample were similar to those in the MSI sample. Further analysis found that there were 20 unique co-enriched GO pathways between T-CK vs. T-MSI and T-CK vs. T-SSI without in T-MSI vs. T-SSI, including BPs such as reproduction, reproductive process, growth, rhythmic process, and detoxification, CCs such as cell junction, membrane-enclosed lumen, other organism, other organism part, extracellular region part, symplast, and supramolecular complex, MFs such as transcription factor activity, protein binding, signal transducer activity, structural molecule activity, electron carrier activity, antioxidant activity, nutrient reservoir activity, molecular transducer activity, molecular function regulator. Among these, it was worth noting that some DEGs were enriched in the antioxidant activity GO pathway, which was similar to the determination of physiological indicators related to antioxidant activity. Moreover, the main GO pathways enriched in T-CK vs. T-MSI, T-CK vs. T-SSI, and T-MSI vs. T-SSI (including more than 10%

of DEGs) were completely consistent, mainly including BPs such as metabolic processes, cellular processes, single organism processes, biological regulation, and response to stimuli, CCs such as cell, cell part, membrane, organelle, and membrane part, as well as the binding and catalytic activities of MFs. This indicated that these 12 GO pathways played potentially important roles in the whole response process of Chinese olive fruits to sunburn stress.

2.2.3. Enrichment Analysis of KEGG Metabolic Pathways Based on DEGs

The DEGs in comparisons of T-CK vs. T-MSI, T-CK vs. T-SSI, and T-MSI vs. T-SSI were enriched into 122, 126, and 31 Kyoto Encyclopedia of Genes and Genomes (KEGG) metabolic pathways, respectively. Among these, 120, 30, and 31 KEGG metabolic pathways co-enriched between T-CK vs. T-MSI and T-CK vs. T-SSI, T-CK vs. T-MSI and T-MSI vs. T-SSI, and T-CK vs. T-SSI and T-MSI vs. T-SSI, respectively. These results indicated that the KEGG metabolic pathway enriched after mild and serious sunburn stresses in Chinese olive fruits was similar. Further analysis of the top 20 enriched metabolic pathways revealed that T-CK vs. T-MSI and T-CK vs. T-SSI had 14 pathways in common (Figure 2A,B), including phenylpropanoid biosynthesis, plant hormone signal transduction, MAPK signaling pathway–plant, sesquiterpenoid and triterpenoid biosynthesis, monoterpenoid biosynthesis, tryptophan metabolism, cutin, suberine, and wax biosynthesis, galactose metabolism, isoflavonoid biosynthesis, flavonoid biosynthesis, pentose and glucuronate interconversions, starch and sucrose metabolism, carotenoid biosynthesis, and brassinosteroid biosynthesis. Moreover, the pathways of galactose metabolism and isoflavonoid biosynthesis were also found in the top 20 enriched KEGG metabolic pathways in comparison of T-MSI vs. T-SSI (Figure 2C). Interestingly, several metabolic pathways related to polyphenols and flavonoids, such as phenylpropanoid biosynthesis, sesquiterpenoid and triterpenoid biosynthesis, monoterpene biosynthesis, isoflavonoid biosynthesis, flavonoid biosynthesis, carotenoid biosynthesis, were found to be enriched under sunburn stress of Chinese olive, indicating that polyphenols and flavonoids might be potentially regulatory metabolites for sunburn injury of Chinese olive fruits.

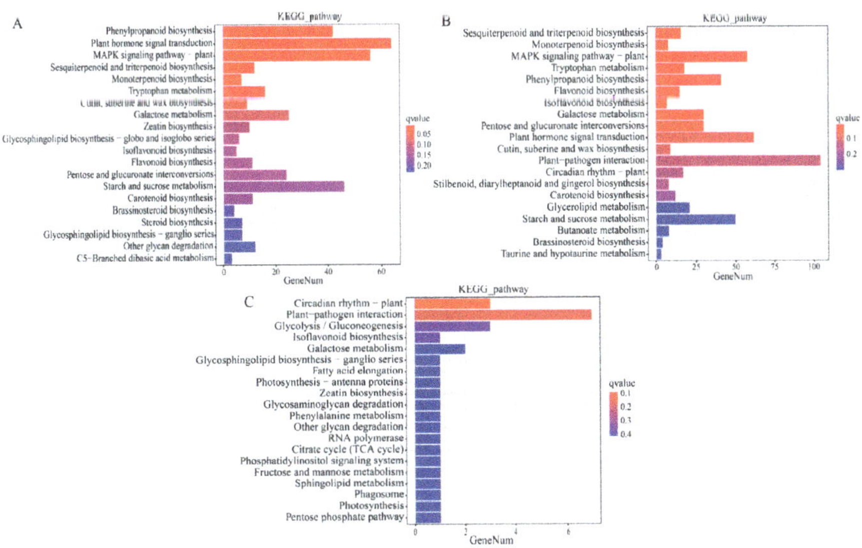

Figure 2. DEG-based enrichment analysis of KEGG metabolic pathway. (**A–C**) indicate the comparisons of T-CK vs. T-MSI, T-CK vs. T-SSI, and T-MSI vs. T-SSI. The red and blue colors indicate high and low q values, respectively.

2.3. Metabolomic Analysis

2.3.1. Metabolome Data Assessment and Metabolite Identification

Based on widely targeted metabolomics, 3380 metabolites (Supplementary Materials Table S5) were identified from M-CK, M-MSI, and M-SSI samples. Principal component analysis (PCA) showed that the biological duplicate samples within the group could aggregate well, indicating good-quality metabolomic data. Further, the first and second principal components reflected 36.89% and 19.03% of the data features, respectively (Supplementary Materials Figure S2A). The correlation analysis revealed that the square of the Spearman correlation coefficient r between biological repeated samples within the group was greater than 0.9, further indicating good biological repeatability (Supplementary Materials Figure S2B). The positive correlation between M-MSI and M-SSI was strong, indicating that the metabolic components after mild and serious sunburn stresses were similar compared with those for CK samples. These results indicated that metabolome data have high reliability and can be used for further analysis.

A total of 800 DAMs (Supplementary Materials Table S6) were identified using variable importance in projection (VIP) $\geq$ 1.0, $|\log2^{FC}| \geq 1.0$, and p value < 0.05 as the screening criteria. DAM cluster analysis revealed that the metabolite composition of M-MSI and M-SSI was relatively similar (Figure 3A), which was consistent with the transcriptome sequencing results. Among these, 532, 501, and 180 DAMs were identified in comparisons of M-CK vs. M-MSI, M-CK vs. M-SSI, and M-MSI vs. M-SSI, respectively. Further analysis found that 282, 81, and 66 DAMs shared differences both in M-CK vs. M-MSI and M-CK vs. M-SSI, M-CK vs. M-MSI and M-MSI vs. M-SSI, and M-CK vs. M-SSI and M-MSI vs. M-SSI, while 16 DAMs were differentially accumulated in all comparisons (Figure 3B).

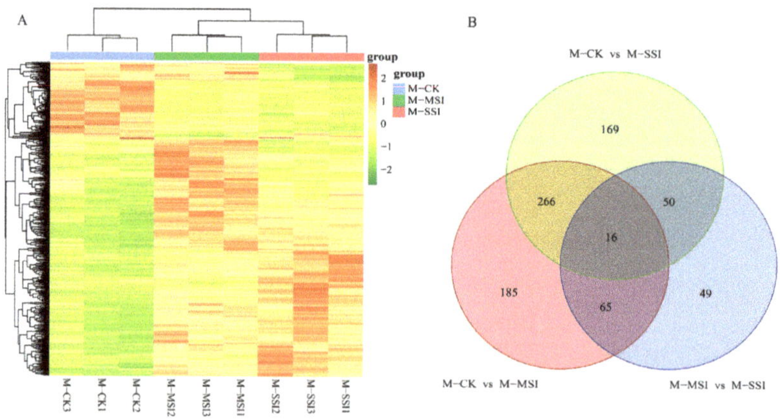

Figure 3. DAMs identification. (**A**,**B**) indicate the cluster analysis and Venn diagram. M-CK, M-MSI, and M-SSI indicate the metabolomes of control check, mild sunburn injury, and serious sunburn injury, respectively.

2.3.2. KEGG Pathway Analysis of DAMs

Based on DAMs, there were 97, 86, and 60 KEGG enrichment pathways in M-CK vs. M-MSI, M-CK vs. M-SSI, and M-MSI vs. M-SSI, respectively. According to the order of p value of the KEGG pathway (top 20), it was worth noting that some flavonoid biosynthesis-related pathways were significantly enriched or enriched in M-CK vs. M-MSI, M-CK vs. M-SSI, and M-MSI vs. M-SSI, such as isoflavonoid biosynthesis, flavone and flavonol biosynthesis, flavonoid biosynthesis, and anthocyanin biosynthesis (Supplementary Materials Figure S3). These results were consistent with the KEGG enrichment analysis of DEGs, which showed that the flavonoid-related metabolic pathways might take part in response to sunburn stress.

2.4. Integration of Transcriptome and Metabolome

2.4.1. Changes of Polyphenol Related Metabolites and Genes

The transcriptomic analysis results revealed that polyphenols and flavonoids were important metabolites of Chinese olive fruits that potentially responded to sunburn stress. Based on metabolomic data, a total of 33 differentially accumulated polyphenols (DAPs) were found in sunburn stress samples of Chinese olive, including 31 up-accumulated components, and involving cinnamaldehydes, cinnamic acids, coumarins, isocoumarins, diarylheptanoids, phenylpropanoic acids, tannins, and their derivatives (Supplementary Materials Table S6). Further, 9 DAPs in metabolic pathway of phenylpropanoid biosynthesis were found (Figure 4), of which one DAP was down-accumulated (4-hydroxycinnamic alcohol 4-*D*-glucoside) and eight DAPs were up-accumulated, including cinnamic acid, anethole, caffeoyl quinic acid, caffeoyl-CoA, caffeoyl aldehyde, *p*-coumaraldehyde, feruloyl-CoA, and coniferin. Meanwhile, 15 DEGs of this metabolic pathway were identified from transcriptomic data, including seven of 4-coumarate-CoA ligases (*4CLs*), three of caffeic acid 3-*O*-methyltransferases (*COMTs*), two of cinnamoyl-alcohol dehydrogenase (*CADs*), one of cinnamoyl-CoA reductase (*CCR*), phenylalanine ammonia lyase (*PAL*), and coniferyl-alcohol glucosyltransferase (*CAGT*). It was worth noting that the expression levels of several genes, such as *4CL* (BMK_Unigene_064108, BMK_Unigene_102104), *CCR* (BMK_Unigene_055431), and *CAD* (BMK_Unigene_068171), were changed more than two times in different sunburn stress samples, indicating that they might play important roles in response to this stress of Chinese olive.

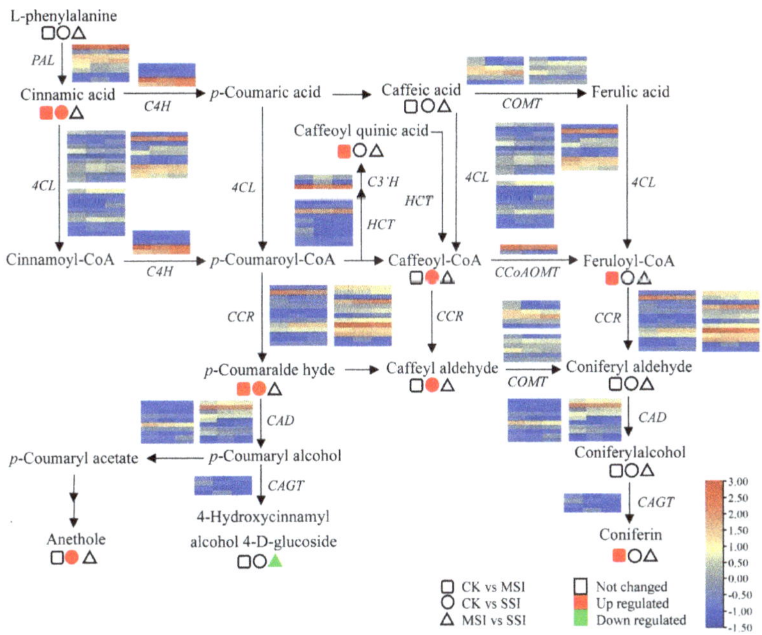

Figure 4. Changes of DEGs and DAMs in metabolic pathway of phenylpropanoid biosynthesis of Chinese olive under sunburn stress. The pathway of phenylpropanoid biosynthesis referring to the KEGG database (https://www.kegg.jp/kegg-bin/show_pathway?map00940) (accessed on 28 August 2023) [20]. Orange and blue colors indicate high and low expression levels of DEGs, while red and green colors represent the up- and down-accumulation of DAMs, respectively. No color indicates no significant change. Squares, circles, and triangles represent the comparisons of CK vs. MSI, CK vs. SSI, and MSI vs. SSI, respectively.

2.4.2. Changes of Flavonoid-Related Metabolites and Genes

A total of 99 differentially accumulated flavonoids (DAFs) in different sunburn stress samples were identified by metabolomics, including 88 up-accumulated components, which were involved in flavones, flavonols, isoflavonoids, isoflavans, chalcones, dihydrochalcones, flavans, flavanols and leucoanthocyanidins, flavanones, and anthocyanidins (Supplementary Materials Table S6). Among them, 16 DAFs were differentially accumulated in the metabolic pathway of flavonoid biosynthesis (Figure 5), of which 6 DAFs were down-accumulated in comparison to M-MSI vs. M-SSI, and 16 DAFs were significantly up-accumulated in comparison of M-CK vs. M-MSI or M-CK vs. M-SSI. Based on transcriptomic data, a total of eight structural genes of this pathway were identified as DEGs, including two of *CHSs* and one of *CHI*, flavanone 7-*O*-glucoside 2″-*O*-beta-L-rhamnosyltransferase (*FGR*), flavanone-3-hydroxylase (*F3H*), dihydroflavonol 4-reductase (*DFR*), anthocyanidin synthase (*ANS*), and flavonol synthase (*FLS*). Notably, the expression levels of several genes, such as *CHSs* (BMK_Unigene_068778 and BMK_Unigene_104115), *F3H* (BMK_Unigene_099272), *DFR* (BMK_Unigene_072482), and *ANS* (BMK_Unigene_068890), significantly increased during sunburn stress, indicating that they potentially regulated the response of Chinese olive to this stress by mediating flavonoid biosynthesis.

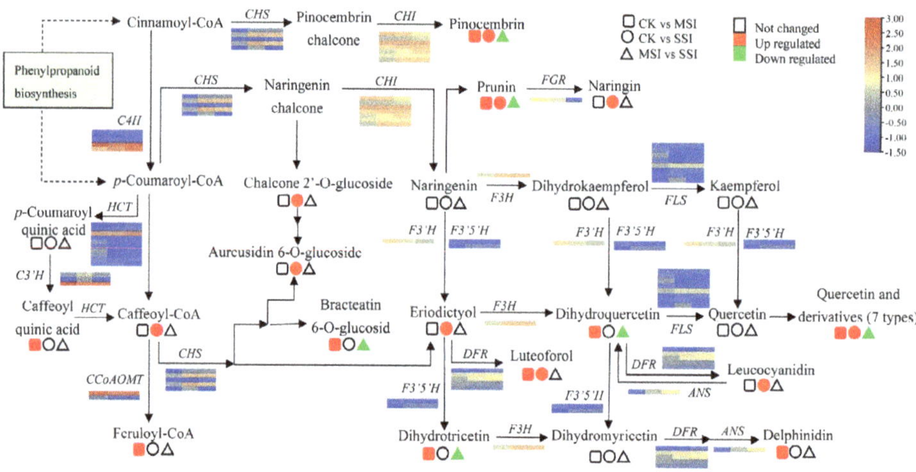

Figure 5. Changes in DEGs and DAMs in the metabolic pathway of flavonoid biosynthesis of Chinese olive under sunburn stress. The pathway of flavonoid biosynthesis referring to the KEGG database (https://www.kegg.jp/kegg-bin/show_pathway?map00941) (accessed on 28 August 2023) [20]. Orange and blue colors indicate high and low expression levels of DEGs, while red and green colors represent the up- and down-accumulation of DAMs, respectively. No color indicates no significant change. Squares, circles, and triangles represent the comparisons of CK vs. MSI, CK vs. SSI, and MSI vs. SSI, respectively.

2.4.3. Changes of Transcription Factors Related to Regulating Phenylpropanoid and Flavonoid Biosynthesis

The accumulation of polyphenols and flavonoids in plants is not only regulated by structural genes directly encoding enzymes related to phenylpropanoid and flavonoid synthesis but is also affected by transcription factors, including *MYB* and *bHLH*. Related studies indicated that *MYB* and *bHLH* can act as transcriptional activators or inhibitors, individually or jointly regulating the expression levels of structural genes related to phenylpropanoid and flavonoid biosynthesis pathways, thereby increasing or inhibiting the accumulation of polyphenols and flavonoids in plants, and mediating the regulation of plant tolerance to stress [21–23]. Based on transcriptomic data, a total of 32 differentially expressed *MYBs* after sunburn stress were identified (Supplementary Materials Table S1), of which 16 *MYBs* had higher expression levels in CK, MSI, and SSI samples, while the other 16 *MYBs* had lower expression abundances. What was noteworthiness was that the BMK_Unigene_013290 continuously down-expressed, while the BMK_Unigene_063026, BMK_Unigene_063155, and BMK_Unigene_064813 continuously up-expressed, which were potential negative and positive regulation of *MYB* transcription factors. Moreover, the expressions of 19 *bHLHs* were found to be significantly changed during this process. The overall expression abundances of seven *bHLHs* were higher, while the expression abundance of 12 *bHLHs* was lower. Especially, the expression levels of three *bHLHs* (BMK_Unigene_023520, BMK_Unigene_022454, BMK_Unigene_024971) were significantly decreased after sunburn stress, which potentially negatively regulated the response of Chinese olive to this stress.

2.4.4. Changes of Antioxidant Capacity-Related Genes

DEG-based GO pathway enriched analysis found that the antioxidant activity pathway was one of the specific co-enriched GO pathways in T-CK vs. T-MSI and T-CK vs. T-SSI. Further analysis indicated that a total of 13 differentially expressed genes related to antioxidant capacity were identified (Supplementary Materials Table S1), including one *CAT*, three *APX*, and nine *POD* genes. Among these, one *CAT* gene (BMK_Unigene_025872) was significantly down-expressed in both T-CK vs. T-MSI and T-CK vs. T-SSI. Three *APXs* were significantly up-expressed, and the fold change of BMK_Unigene_096922 was the highest in both T-CK vs. T-MSI and T-CK vs. T-SSI. Among the nine *POD* genes, there were four identical genes in both T-CK vs. T-MSI and T-CK vs. T-SSI, including three significantly up-expressed genes and one significantly down-expressed gene. It was worth noting that BMK_Unigene_099223 and BMK_Unigene_018627 were the two *POD* genes with the highest and lowest fold change, respectively. These genes might respond to sunburn stress by regulating the activity of corresponding enzymes in Chinese olive, including one *CAT* (BMK_Unigene_025872), one *APX* (BMK_Unigene_096922), and two *PODs* (BMK_Unigene_099223, BMK_Unigene_018627).

2.5. Changes in the Content of Total Polyphenols and Total Flavonoids

Metabolomic and transcriptomic data revealed that polyphenols and flavonoids potentially responded to sunburn stress in Chinese olive. Compared with the CK sample, the content of total polyphenols and total flavonoids (Figure 6) significantly increased by 1.65 and 2.00 times in the MSI sample and by 1.57 and 1.45 times in the SSI sample, respectively. And no significant difference in the content of total polyphenols was found between the MSI and SSI samples, while the content of total flavonoids significantly decreased in the SSI sample than those in the MSI sample. The results suggested that sunburn stress resulted in the accumulation of polyphenols and flavonoids in Chinese olive.

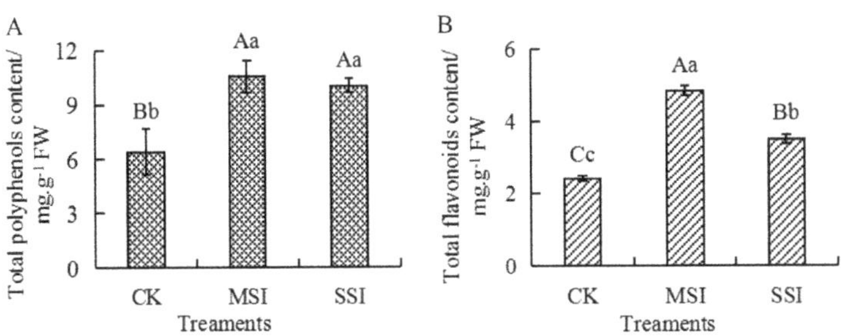

Figure 6. Changes in the content of total polyphenols (**A**) and total flavonoids (**B**) of Chinese olive fruits after sunburn stress. Upper- and lowercase letters on the curves represent significant differences at p value < 0.01 and <0.05 levels.

3. Discussion

3.1. Serious Sunburn Led to Significant Changes in Phenotype of Chinese Olive

Fruit color and firmness are essential phenotypic indicators for evaluating fruits. Studies on citrus [11], apple [7,8], and grape [24] have shown that sunburn affects the accumulation of pigments in the skin or flesh of fruit. In this study, the colors of Chinese olive after different types of sunburn stress significantly changed. These results were consistent with previous findings. Kim et al. [11] found that the firmness of high-temperature sunburn of satsuma mandarin fruits significantly increased, which might be related to the shrinkage, roughness, and cracking of the fruit surface under anatomical observation, ultimately leading to necrosis and browning caused by pathogen invasion. Torres et al. [25] revealed that photooxidative stress increased the firmness of apples, which was involved in the induction of lignin accumulation of stressed fruit, with the involvement of stress phytohormones such as ethylene. The changes in fruit color, phenotypic browning, or cell necrosis and hardening caused by high-temperature sunburn were closely related to oxidative stress caused by excessive ROS production [9,26]. In our present study, the mild sunburn stress did not cause obvious changes in the firmness of Chinese olive, while the serious sunburn injury significantly increased the firmness of the fruits. These phenomena might be related to the necrosis and pathological changes caused by serious sunburn.

3.2. Sunburn Stress Up-Accumulated the Polyphenol and Flavonoid in Chinese Olive to Involve in the Response to This Abiotic Stress

Sunburn is a common abiotic stress during growth and development of plants. Continuous high temperature and strong light cause metabolic disorders related to heat stress, leading to excessive formation of ROS and disruption of plant cell homeostasis [27]. Plants biosynthesize photoprotective compounds, heat shock proteins, and nonenzymatic antioxidants (polyphenols and flavonoids), enhance antioxidant enzyme activity, and improve plant oxidative stress capacity to alleviate the oxidative damage caused by excessive ROS to plants [2,5,28]. Polyphenols and flavonoids are important secondary metabolites in plants, which have been proven to have good antioxidant activity and are essential active substances to improve abiotic stress tolerance and adaptability [29–32]. Previous studies showed that sunburn promoted the accumulation of polyphenol and flavonoid compounds in pomegranate [12], apple [8,9,33], grape [10], satsuma mandarin [11], sea buckthorn [34], and strawberry [35]. Moreover, the polyphenol and flavonoid components have been proven to regulate plant antioxidant capacity. For example, the exogenous proanthocyanidin treatment inhibited the senescence of harvested banana fruit by enhancing antioxidant response and in vivo proanthocyanidin content [36]. Exogenous chlorogenic acid alleviated oxidative stress in apple leaves by enhancing antioxidant capacity [37]. Exogenous cinnamic acid regulated antioxidant enzyme activity and reduced lipid peroxidation in drought-stressed cucumber leaves [38]. In this study, the content of total polyphenols and total flavonoids significantly increased, and most of the DAPs and DAFs up-accumulated after sunburn stress in Chinese olive, suggesting that the polyphenol and flavonoids potentially regulated the response to this stress in Chinese olive. However, compared to the mild sunburn stress sample, the content of polyphenols and flavonoids in the serious sunburn stress sample decreased to a certain extent. Generally, when plants are exposed to stress, as the stress intensifies, the first-level response may not be sufficient, and plants may activate the second-level response [39]. It could be predicted that, after mild sunburn injury, the cells responded to the stress immediately and biosynthesized and accumulated the stress resistance-related metabolites, such as polyphenols, flavonoids, and antioxidant enzymes. However, the injury exceeded the threshold that the cells could withstand, resulting in significant cell or tissue necrosis.

3.3. The Potential Regulatory Network of Polyphenol and Flavonoid in Response to Sunburn Stress in Chinese Olive

Previous reports have shown that in pomegranates, sunburn stress led to significant changes in the expression levels of *CHI*, *F3′5′H*, and *CHS* genes [12]. Transcriptome DEG-based KEGG analysis revealed that phenylpropanoid biosynthesis was involved in the response of Chinese olive to sunburn stress. Moreover, 15 genes in this pathway were significantly differentially expressed during stress occurrence, including *4CL*, *COMT*, *PAL*, *CCR*, *CAD*, *CAGT*, and so forth. Among these, two *4CLs* (BMK_Unigene_064108, BMK_Unigene_102104), one *CCR* (BMK_Unigene_055431), and one *CAD* (BMK_Unigene_068171) were key genes potentially regulating polyphenol biosynthesis. Moreover, eight potential regulatory genes of flavonoid biosynthesis were identified, including *CHS*, *CHI*, *FGR*, *F3H*, *DFR*, *ANS*, and *FLS*, and the *CHSs* (BMK_Unigene_068778, BMK_Unigene_104115), *F3H* (BMK_Unigene_099272), *DFR* (BMK_Unigene_072482), and *ANS* (BMK_Unigene_068890) were screened as candidate genes for regulating flavonoid accumulation and sunburn stress response.

H_2O_2 is important for ROS; biotic and abiotic stresses have been revealed to induce the production and accumulation of H_2O_2 in plant cells [40]. MDA is the main product of membrane lipid peroxidation, and its content can be used to assess cell membrane damage and plant tolerance [41]. The accumulation of proline also reflects the injury of plants under stress [42], and the accumulated proline could function to clear ·OH [43]. Studies on citrus [44], apple [9], and pomegranate [12] have shown that sunburn stress led to a significant increase in O_2- and MDA content. As sunburn damage occurred, they up-accumulated and destroyed the cell membrane system. This study found that after sunburn stress, the MDA accumulation significantly increased, indicating damage to the cell membrane system in Chinese olive. During this process, the content of T-AOC and proline significantly increased, while the content of H_2O_2 was decreased, indicating that the accumulation of proline helped to eliminate excessive ROS and played an important role in antioxidants. Meanwhile, the antioxidant-related enzymes were activated, and multiple antioxidant enzymes were jointly involved in regulating the process [9]. In this study, MSI induced significant upregulation of activities of CAT, SOD, APX, POD, and PPO, indicating these enzymes are involved in sunburn stress in Chinese olive. Moreover, the expression levels of antioxidant regulatory genes have also been significantly increased, including *CAT*, *APX*, and *POD*. It could be seen that after sunburn stress, the antioxidant capacity in Chinese olive was enhanced by upregulating the regulatory gene expression to increase the activities of antioxidant-related enzymes.

In summary, flavonoids and polyphenols are important secondary metabolites in Chinese olive. Sunburn stress might activate the transcription of structural genes, such as *4CL*, *CCR*, *CAD*, *CHS*, *F3H*, *DFR*, and *ANS*, to mediate the biosynthesis and accumulation of polyphenols and flavonoids, thereby responding to sunburn stress (Figure 7).

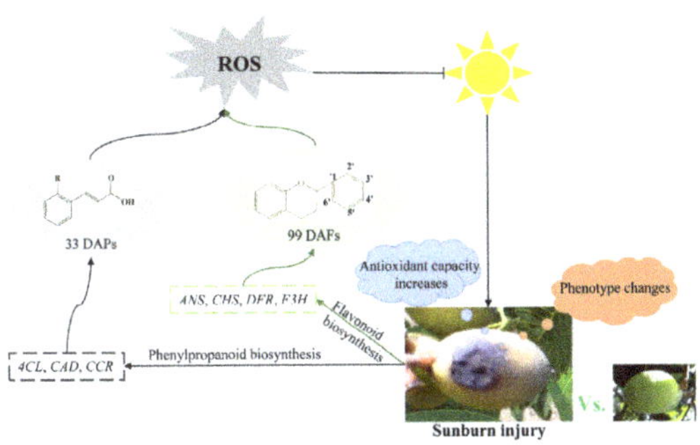

Figure 7. Potential regulation network of polyphenols and flavonoids in response to sunburn stress in Chinese olive.

4. Materials and Methods

4.1. Materials

The samples used in this study were the fruits of the *Canarium album* cv. *lingfeng*, a fresh edible cultivar selected by the Fruit Research Institute of Fujian Academy of Agricultural Sciences, and the samples were collected on July 29, 2022, from the experimental base located in Zhenjing Farm, Baisha, Minhou (26.21° N, 118.99° E, A119.2 m). Data from the orchard meteorological observation station revealed that the daily precipitation of the orchard on the sampling day was 0.0 mm. The daily average temperature was recorded at 31.4 °C, with a daily maximum of 36.9 °C and a daily minimum of 27.1 °C. The sunshine duration on that day was 9.8 h. The daily average temperature of the last 20 days before sampling was 31.6 °C, with a maximum temperature of 41.3 °C (average 38.0 °C). The daily maximum temperature of 18 days exceeded 35.0 °C, and the longest sunshine hours reached 12.7 h, of which 19 days had almost no precipitation (Supplementary Materials Table S7). The sunburn injury in Chinese olive fruits caused by sustained abnormal high temperature and strong sunlight is shown in Figure 1A. Normal fruit without sunburn injury was used as a control check (CK) sample, fruit with symptom and sunburn injury area of less than one-fourth of the whole fruit area were used as the mild sunburn injury (MSI) sample, and fruit with sunburn injury area of greater than one-fourth of the whole fruit area or those beginning to experience fruit tissue necrosis were used as the serious sunburn injury (SSI) sample. Five trees (numbered Ca1–Ca5) were randomly selected from the orchard; the grafted age of each tree was more than 9 years, and 10 fruits with different degrees of sunburn injury were collected from the same sunny side of each tree. The collected samples were taken to the laboratory, and the flesh was quickly cut. The samples were wrapped in tin foil, quick-frozen with liquid nitrogen, and placed in an ultra-low temperature freezer at −80 °C for later use.

4.2. Methods

4.2.1. Detection of Physiological and Biochemical Indicators

The color and firmness of the pericarp were determined using an NH310 chroma meter (3nh Technology Co., Ltd., Shenzhen, China) and a TMS-Pilot texture analyzer (Food Technology Corporation, Sterling, VA, USA). The content of total polyphenols was determined by a modified Folin–Ciocalteu method with gallic acid as a control [45]. Rutin was used as a control, and spectrophotometry was used to determine total flavonoids [46]. The contents of proline, T-AOC, H_2O_2, MDA, SOD, POD, APX, CAT, and PPO of samples were detected using the method described by ninhydrin colorimetric method [47], ferric reducing ability of plasma method [48], ammonium molybdate method [49], thiobarbituric acid method [50], nitro-blue tetrazolium method [50], guaiacol method [50], ascorbic acid colorimetric method [51], pyrocatechol method [47]. Each sample was measured for five biological replicates.

4.2.2. Transcriptome Sequencing

The CK, MSI, and SSI samples used for transcriptome sequencing were named T-CK, T-MSI, and T-SSI for library construction and sequencing, respectively, and three biological replicates were set up. First, the E.Z.N.A Plant RNA Kit (OMEGA Biotek Inc., Norcross, GA, USA) was used to extract the total RNA of the samples, agarose gel electrophoresis and TU-1810 spectrophotometer (Puxi General Instrument Co., Ltd., Beijing, China) were used to detect its integrity and purity, and an Agilent 2100 biological analyzer (Agilent Technologies, Santa Clara, CA, USA) was used to evaluate the RNA quality. After passing the inspection, 1 μg total RNA was taken, and sequencing libraries were constructed using the NEBNext Ultra RNA Library Prep Kit for Illumina (New England Biolabs, Ipswich, MAS, USA). Briefly, mRNA was purified from total RNA using poly-T oligo-attached magnetic beads. Fragmentation was carried out using divalent cations at elevated temperatures in NEBNext first-strand synthesis reaction buffer (5×). The first-strand cDNA was synthesized using a random hexamer primer and M-MuLV reverse transcriptase. The second-strand cDNA synthesis was subsequently performed using DNA polymerase I and RNase H. The library fragments were purified with an AMPure XP system (Beckman Coulter, Brea, CA, USA). Then, 3 μL USER enzyme (New England Biolabs, Ipswich, MAS, USA) was used with cDNA at 37 °C for 15 min and then at 95 °C for 5 min for polymerase chain reaction to construct the library. The library quality was assessed using the Agilent Bioanalyzer 2100 (Agilent Technologies, Santa Clara, CA, USA)and sequenced using the Illumina Hiseq 2000 platform (Biomarker Technologies, Beijing, China).

The *Arabidopsis thaliana* genome was the reference for transcriptome assembly. The transcripts were annotated based on NCBI non-redundant protein sequences (NR), Protein family (Pfam), Clusters of Orthologous Groups of proteins (KOG/COG/eggNOG), Swiss-Prot, Kyoto Encyclopedia of Genes and Genomes (KEGG), and Gene Ontology (GO). Differential expression analysis of two groups was performed using the DESeq R package (v 1.30.1). The expression abundance of the two groups was expressed by fragments per kilobase of transcript per million mapped reads. The corrected p value < 0.01 and fold change (FC) $\geq$ 2.0 or $\leq$0.5 (i.e., $|\log2^{FC}| \geq 1.0$) were used as the criteria to screen DEGs, and GO and KEGG pathway enrichment analyses were performed based on DEGs.

4.2.3. Metabolomic Assay

The CK, MSI, and SSI samples used for metabolomic detection were named M-CK, M-MSI, and M-SSI, respectively. The quality control (QC) samples were mixed in equal amounts, and three biological replicates were set up for nontargeted metabolomic detection. Then, 50 mg of the sample was used, and 1 mL of extraction solution containing internal standard (methanol–acetonitrile–water = 2:2:1, the internal standard concentration of 20 mg·L^{-1}) was added for extraction. A liquid chromatography–mass spectrometry system consisting of Waters UPLC Acquisition I-Class PLUS (Waters, Milford, MAS, USA) and a high-resolution mass spectrometer (UPLC Xevo G2-XS QTOF, Waters, Milford, MAS, USA) was used for metabolomic detection. The used column was Acquity UPLC HSS T3 (1.8 μm 2.1 mm × 100 mm Waters, Milford, MAS, USA). Both in positive and negative ion modes, mobile phase A and mobile phase B were composed of 0.1% formic acid aqueous solution and 0.1% formic acid acetonitrile, respectively, with injection volumes of 1 μL. Liquid chromatography mobile-phase conditions: flow rate, 400 μL·min^{-1}; 0–0.25 min, 98% A, 2% B; 0.25–10.00 min, 98% A, 2% B; 10.00–13.00 min, 2% A, 98% B; 13.00–13.10 min, 98% A, 2% B, and 13.10–15.00 min, 98% A, 2% B. In each data acquisition cycle, dual-channel data acquisition was performed on both low- and high-collision energies. The scanning frequency was 0.2 s, and the electron spray ionization (ESI) source parameters were set according to the method described by Wang et al. [52].

The MassLynx (v 4.2) was used to import the collected raw data into the Progenesis QI (v 2.0) software for peak extraction, peak alignment, and so on, based on the Progenesis QI software online METLIN database, public database, and Biomark's database for metabolite identification. Meanwhile, the theoretical fragment was identified, the precursor ion mass deviation was 100 ppm, and the fragment ion mass deviation was within 50 ppm [52]. The metabolites were quantitatively detected. KEGG, Human Metabolome Database (HMDB), and Lipid Metabolites and Pathways Strategy (LIPID MAPS) public databases were used for metabolite function and classification annotation. After normalizing the quantitative values of metabolites, principal component analysis (PCA) and Spearman's correlation analysis were used to judge the repeatability of the samples within the group and the quantity control samples. Orthogonal projections to latent structures–discriminant analysis was performed. FC $\geq$ 2.0 or $\leq$ 0.5 (i.e., $|\log2^{FC}| \geq 1.0$), variable importance in projection (VIP) $\geq$ 1.0, and significance p value < 0.05 were set to screen DAMs.

4.2.4. Statistical Analysis

Analyses of physiological and biochemical data were conducted using Microsoft Excel (v 2013) and IBM SPSS Statistics (v 19.0) software. Results were expressed as mean ± SD. p values of $\leq$0.01 and 0.05 were considered a statistically significant difference between treatments by the LSD method, which were represented in upper- and lowercase letters, respectively.

5. Conclusions

Chinese olive fruits underwent significant changes in color and increases in fruit firmness, antioxidant capacity, and antioxidant enzyme activities after sunburn injury. Based on transcriptomic identification, a total of 3259 DEGs were obtained, including 15 phenylpropanoid and 8 flavonoid biosynthesis regulatory structural genes and 7 key transcription factors. Based on metabolomic identification, a total of 800 DAMs were obtained, including 33 DAPs and 99 DAFs. DEG- and DAM-based KEGG revealed that sunburn stress significantly activated the polyphenol and flavonoid-related biosynthetic pathways in Chinese olive fruits, and the content of total polyphenols and total flavonoids significantly increased after the stress. These results indicated that sunburn stress might activate the transcription of structural genes, such as *4CL*, *CCR*, *CAD*, *CHS*, *F3H*, *DFR*, and *ANS*, so as to increase the accumulation of the polyphenols and flavonoids to enhance the tolerance of Chinese olive fruits to sunburn stress.

Supplementary Materials: The following supporting information can be downloaded at https://www.mdpi.com/article/10.3390/plants13172369/s1, Figure S1: DEGs identification. (A) and (B) indicate the Venn diagram and cluster analysis. T-CK, T-MSI, and T-SSI indicate the transcriptomes of control check, mild sunburn injury, and serious sunburn injury, respectively; Figure S2: Metabolome data assessment of Chinese olive fruits. (A) and (B) indicate the PCA analysis and Spearman's correlation coefficient analysis. M-CK, M-MSI, and M-SSI indicate the metabolomes of control check, mild sunburn injury, and serious sunburn injury, respectively; Figure S3: DAM-based enrichment analysis of KEGG pathway. (A), (B), and (C) indicate the comparisons of M-CK vs. M-MSI, M-CK vs. M-SSI, and M-MSI vs. M-SSI.; Table S1: DEG FPKMs; Table S2: GO classification based on DEGs in comparison of T-CK vs. T-MSI; Table S3: GO classification based on DEGs in comparison of T-CK vs. T-SSI; Table S4: GO classification based on DEGs in comparison of T-MSI vs. T-SSI; Table S5: All metabolites in different samples; Table S6: DAMs identification; Table S7: The 20 days' meteorological conditions before sampling.

Author Contributions: Conceptualization, R.L. and Y.L.; methodology, Y.L. and C.S.; software, Y.L.; validation, Y.L.; formal analysis, Y.L., C.S., R.L. and Q.T.; resources, M.Z.; data curation, Y.L.; writing—original draft preparation, Y.L. and C.S.; writing—review and editing, R.L., R.W. and X.W.; funding acquisition, R.W., R.L., and Q.T. All authors have read and agreed to the published version of the manuscript.

Funding: This work was supported by the Basic Sci-tech Project of Provincial Public Welfare Scientific Research Institution of Fujian (No. 2022R1028005, and No. 2020R10280010), the Natural Science Foundation of Fujian Province (No. 2021J05199, and No. 2023J01196), the Collaborative Innovation Project from the People's Government of Fujian Province and Chinese Academy of Agricultural Sciences (No. XTCXGC2021019-GSS01), the Project of Resource Protection of Species and Varieties (Tropical Crops) of Ministry of Agriculture and Rural Affairs of China (No. 18230171).

Data Availability Statement: The data supporting reported results can be found in Supplementary Materials. The raw sequence data reported in this paper have been deposited in the Genome Sequence Archive [53] in the National Genomics Data Center, China National Center for Bioinformation [54] (GSA: CRA012427) that are publicly accessible at https://ngdc.cncb.ac.cn/gsa (accessed on 29 August 2023).

Conflicts of Interest: The authors declare no conflicts of interest.

References

1. Jin, J.; Yang, L.; Fan, D.Y.; Feng, B.B.; Keram, I.; Hao, Q. Advances in the studies on high temperature tolerance of fruit trees. *Mol. Plant Breed.* **2019**, *17*, 1019–1027. [CrossRef]
2. Munne-Bosch, S.; Vincent, C. Physiological mechanisms underlying fruit sunburn. *Crit. Rev. Plant Sci.* **2019**, *38*, 140–157. [CrossRef]
3. Müller, K.; Keller, M.; Stoll, M.; Friedel, M. Wind speed, sun exposure and water status alter sunburn susceptibility of grape berries. *Front. Plant Sci.* **2023**, *14*, 1145274. [CrossRef] [PubMed]

4. Wang, F.; Xu, Z.; Fan, X.; Zhou, Q.; Cao, J.; Ji, G.; Jing, S.; Feng, B.; Wang, T. Transcriptome analysis reveals complex molecular mechanisms underlying UV tolerance of wheat (*Triticum aestivum* L.). *J. Agric. Food Chem.* **2019**, *67*, 563–577. [CrossRef] [PubMed]
5. Gupta, D.K.; Palma, J.M.; Corpas, F.J. *Antioxidants and Antioxidant Enzymes in Higher Plants*; Springer: Cham, Switzerland, 2018. [CrossRef]
6. Jiang, J.M.; Lin, Y.X.; Chen, Y.Y.; Deng, C.J.; Gong, H.W.; Xu, Q.Z.; Zheng, S.Q.; Chen, W. Proteomics approach reveals mechanism underlying susceptibility of loquat fruit to sunburn during color changing period. *Food Chem.* **2015**, *176*, 388–395. [CrossRef]
7. Xu, H.; Watanabe, Y.; Ediger, D.; Yang, X.; Iritani, D. Characteristics of sunburn browning fruit and rootstock-dependent damage-free yield of *Ambrosia*™ apple after sustained summer heat events. *Plants* **2022**, *11*, 1201. [CrossRef]
8. Zupan, A.; Mikulic-Petkovsek, M.; Slatnar, A.; Stampar, F.; Veberic, R. Individual phenolic response and peroxidase activity in peel of differently sun-exposed apples in the period favorable for sunburn occurrence. *J. Plant Physiol.* **2014**, *171*, 1706–1712. [CrossRef]
9. Feng, Y.; Li, S.; Jia, R.; Yang, J.; Su, Q.; Zhao, Z. Physiological characteristics of sunburn peel after apple debagged. *Molecules* **2022**, *27*, 3775. [CrossRef]
10. Gambetta, J.M.; Romat, V.; Schmidtke, L.M.; Holzapfel, B.P. Secondary metabolites coordinately protect grapes from excessive light and sunburn damage during development. *Biomolecules* **2021**, *12*, 42. [CrossRef]
11. Kim, M.; Park, Y.; Yun, S.K.; Kim, S.S.; Joa, J.; Moon, Y.E.; Do, G.R. The anatomical differences and physiological responses of sunburned satsuma mandarin (*Citrus unshiu* Marc.) fruits. *Plants* **2022**, *11*, 1801. [CrossRef]
12. Liu, C.; Su, Y.; Li, J.; Jia, B.; Cao, Z.; Qin, G. Physiological adjustment of pomegranate pericarp responding to sunburn and its underlying molecular mechanisms. *BMC Plant Biol.* **2022**, *22*, 169. [CrossRef] [PubMed]
13. Spera, N.; Vita, L.I.; Civello, P.M.; Colavita, G.M. Antioxident response and quality of sunburn Beurré D'Anjou pears (*Pyrus communis* L.). *Plant Physiol. Biochem.* **2023**, *198*, 107703. [CrossRef]
14. Guihur, A.; Rebeaud, M.E.; Goloubinoff, P. How do plants feel the heat and survive? *Trends Biochem. Sci.* **2022**, *47*, 824–838. [CrossRef]
15. Rosas, I.D.; Deis, L.; Baldo, Y.; Cavagnaro, J.B.; Cavagnaro, P.F. High temperature alters anthocyanin concentration and composition in grape berries of Malbec, Merlot, and Pinot Noir in a cultivar-dependent manner. *Plants* **2022**, *11*, 926. [CrossRef]
16. Zhu, J.K. Abiotic stress signaling and responses in plants. *Cell* **2016**, *167*, 313–324. [CrossRef]
17. Zhao, D.; Zhang, Y.; Lu, Y.; Fan, L.; Zhang, Z.; Zheng, J.; Chai, M. Genome sequence and transcriptome of *Sorbus pohuashanensis* provide insights into population evolution and leaf sunburn response. *J. Genet. Genom.* **2022**, *49*, 547–558. [CrossRef] [PubMed]
18. Waite, J.M.; Kelly, E.A.; Zhang, H.; Hargarten, H.; Waliullah, S.; Altman, N.S.; de Pamphilis, C.W.; Honaas, L.; Kalcsits, L. Transcriptomic approach to uncover dynamic events in the development of mid-season sunburn in apple fruit. *G3 Genes Genomes Genet.* **2023**, *13*, jkad120. [CrossRef] [PubMed]
19. He, C. *Canarium album* (Lour.) Raeusch. (Qingguo, Chinese Olive). In *Dietary Chinese Herbs*; Liu, Y., Wang, Z., Zhang, J., Eds.; Springer: Vienna, Austria, 2015; pp. 307–313. [CrossRef]
20. Kanehisa, M.; Furumichi, M.; Sato, Y.; Kawashima, M.; Ishiguro-Watanabe, M. KEGG for taxonomy-based analysis of pathways and genomes. *Nucleic Acids Res.* **2023**, *51*, D587–D592. [CrossRef]
21. Su, L.; Lv, A.; Wen, W.; Fan, N.; Li, J.; Gao, L.; Zhou, P.; An, Y. *MsMYB741* is involved in alfalfa resistance to aluminum stress by regulating flavonoid biosynthesis. *Plant J.* **2022**, *112*, 756–771. [CrossRef] [PubMed]
22. Cannavò, S.; Bertoldi, A.; Valeri, M.C.; Damiani, F.; Reale, L.; Brilli, F.; Paolocci, F. Impact of high light intensity and low temperature on the growth and phenylpropanoid profile of *Azolla filiculoides*. *Int. J. Mol. Sci.* **2023**, *24*, 8554. [CrossRef] [PubMed]
23. Wang, Y.; Jiang, W.; Li, C.; Wang, Z.; Lu, C.; Cheng, J.; Wei, S.; Yang, J.; Yang, Q. Integrated transcriptomic and metabolomic analyses elucidate the mechanism of flavonoid biosynthesis in the regulation of mulberry seed germination under salt stress. *BMC Plant Biol.* **2024**, *24*, 132. [CrossRef]
24. Gambetta, J.M.; Holzapfel, B.P.; Stoll, M.; Friedel, M. Sunburn in grapes: A review. *Front. Plant Sci.* **2021**, *11*, 604691. [CrossRef] [PubMed]
25. Torres, C.A.; Azocar, C.; Ramos, P.; Pérez-Díaz, R.; Sepulveda, G.; Moya-León, M.A. Photooxidative stress activates a complex multigenic response integrating the phenylpropanoid pathway and ethylene, leading to lignin accumulation in apple (*Malus domestica* Borkh.) fruit. *Hortic. Res.* **2020**, *7*, 22. [CrossRef] [PubMed]
26. Rustioni, L.; Fracassetti, D.; Prinsi, B.; Geuna, F.; Ancelotti, A.; Fauda, V.; Tirelli, A.; Espen, L.; Failla, O. Oxidations in white grape (*Vitis vinifera* L.) skins: Comparison between ripening process and photooxidative sunburn symptoms. *Plant Physiol. Biochem.* **2020**, *150*, 270–278. [CrossRef] [PubMed]
27. Chetty, K.A. The Effects of High Temperature Stress on the Enzymatic Antioxidant System in *Zea mays*. Ph.D. Thesis, University of the Western Cape, Cape Town, South Africa, 2017. Available online: https://etd.uwc.ac.za/ (accessed on 7 August 2018).
28. Sharma, A.; Shahzad, B.; Rehman, A.; Bhardwaj, R.; Landi, M.; Zheng, B. Response of phenylpropanoid pathway and the role of polyphenols in plants under abiotic stress. *Molecules* **2019**, *24*, 2452. [CrossRef]
29. Nabavi, S.; Šamec, D.; Tomczyk, M.; Milella, L.; Russo, D.; Habtemariam, S.; Suntar, I.; Rastrelli, L.; Daglia, M.; Xiao, J.; et al. Flavonoid biosynthetic pathways in plants: Versatile targets for metabolic engineering. *Biotechnol. Adv.* **2020**, *38*, 107316. [CrossRef]
30. Agati, G.; Azzarello, E.; Pollastri, S.; Tattini, M. Flavonoids as antioxidants in plants: Location and functional significance. *Plant Sci.* **2012**, *196*, 67–76. [CrossRef]

31. Vrhovsek, U.; Rigo, A.; Tonon, D.; Mattivi, F. Quantitation of polyphenols in different apple varieties. *J. Agric. Food Chem.* **2004**, *52*, 6532–6538. [CrossRef] [PubMed]
32. Cheynier, V.; Comte, G.; Davies, K.M.; Lattanzio, V.; Martens, S. Plant phenolics: Recent advances on their biosynthesis, genetics, and ecophysiology. *Plant Physiol. Biochem.* **2013**, *72*, 1–20. [CrossRef]
33. Mertoğlu, K.; Akkurt, E.; Evrenosoğlu, Y.; Çolak, A.; Esatbeyoglu, T. Horticultural characteristics of summer apple cultivars from Turkey. *Plants* **2022**, *11*, 771. [CrossRef]
34. Liu, R.; Yang, Y.; Zhao, M.; Wang, Y.; Meng, X.; Yan, T.; Ho, C. Effect of heat-treating methods on components, instrumental evaluation of color and taste, and antioxidant properties of sea buckthorn pulp flavonoids. *J. Food Sci.* **2022**, *87*, 5442–5454. [CrossRef] [PubMed]
35. Balasooriya, H.N.; Dassanayake, K.B.; Seneweera, S.; Ajlouni, S. Impact of elevated carbon dioxide and temperature on strawberry polyphenols. *J. Sci. Food Agric.* **2019**, *99*, 4659–4669. [CrossRef]
36. Chen, J.; Li, F.F.; Li, Y.X.; Wang, Y.S.; Wang, C.Z.; Yuan, D.B.; Jiang, Y.M. Exogenous procyanidin treatment delays senescence of harvested banana fruit by enhancing antioxidant responses and in vivo procyanidin content. *Postharvest Biol. Technol.* **2019**, *158*, 110999. [CrossRef]
37. Mei, Y.M.; Sun, H.L.; Du, G.D.; Wang, X.Q.; Lyu, D.G. Exogenous chlorogenic acid alleviates oxidative stress in apple leaves by enhancing antioxidant capacity. *Sci. Hortic.* **2020**, *274*, 109676. [CrossRef]
38. Sun, W.J.; Nie, Y.X.; Gao, Y.; Dai, A.H.; Bai, J.G. Exogenous cinnamic acid regulates antioxidant enzyme activity and reduces lipid peroxidation in drought-stressed cucumber leave. *Acta Physiol. Plant.* **2012**, *34*, 641–655. [CrossRef]
39. Leyva-Pérez, M.L.; Valverde-Corredor, A.; Valderrama, R.; Jiménez-Ruiz, J.; Muñoz-Merida, A.; Trelles, O.; Barroso, J.B.; Mercado-Blanco, J.; Luque, F. Early and delayed long-term transcriptional changes and short-term transient responses during cold acclimation in olive leaves. *DNA Res.* **2015**, *22*, 1–11. [CrossRef]
40. Song, X.G.; She, X.P. The generation and the role of hydrogen peroxide in plant. *J. Lianyungang Teach. Coll.* **2010**, *27*, 99–103. [CrossRef]
41. Fan, S.L. A Study on the Response Mechanism of Physiological and Biochemical Indicators of *Dahlia* to Water Stress. Master's Thesis, Shandong Agricultural University, Shandong, China, 2011. [CrossRef]
42. Zhu, H.; Zu, Y.G.; Wang, W.J.; Yan, Y.Q. The effect of proline on plant growth under adverse stress conditions. *J. Northeast For. Univ.* **2009**, *37*, 86–89. [CrossRef]
43. Jiang, M.Y.; Guo, S.C. Accumulation of proline in rice seedlings under ·OH stress and its antioxidant effects. *Sci. Bull.* **1997**, *42*, 646–649. [CrossRef]
44. Wan, J.F.; Li, J.; Yang, W.H.; Zeng, H.; Zhang, H.Z.; Chen, J.Z. ROS metabolism of citrus fruits in response to high-temperature-intense-light stress. *Fujian J. Agric. Sci.* **2019**, *34*, 920–924. [CrossRef]
45. Singleton, V.L.; Orthofer, R.; Lamuela-Raventós, R.M. Analysis of total phenols and other oxidation substrates and antioxidants by means of Folin-Ciocalteu reagent. *Methods Enzymol.* **1999**, *299*, 152–178. [CrossRef]
46. Song, Y.Q.; Wang, Y.P.; Mao, Y.J. Determination of total flavonoid in *Asparagus officinalis* Linn by spectrophotometry. *Chem. Anal. Meterage* **2005**, *14*, 52–53. [CrossRef]
47. Wang, X.K. *Principles and Techniques of Plant Physiological and Biochemical Experiments*; Higher Education Press: Beijing, China, 2006; pp. 169–173,278–279.
48. Benzie, I.F.; Strain, J.J. The ferric reducing ability of plasma (FRAP) as a measure of "antioxidant power": The FRAP assay. *Anal. Biochem.* **1996**, *239*, 70–76. [CrossRef]
49. Peng, J.; Wang, D.Y.; Xu, C.M.; Chen, L.P.; Deng, F.; Zhang, X.F. Ammonium molybdate method for detecting the activities of rice catalase. *Chin. Agric. Sci. Bull.* **2009**, *25*, 61–64.
50. Li, H.S. *Principles and Techniques of Plant Physiological Biochemical Experiment*; Higher Education Press: Beijing, China, 2000; pp. 164–169,246–248,260.
51. Ahn, T.; Paliyath, G.; Murr, D.P. Antioxidant enzyme activities in apple varieties and resistance to superficial scald development. *Food Res. Int.* **2007**, *40*, 1012–1019. [CrossRef]
52. Wang, J.; Zhang, T.; Shen, X.; Liu, J.; Xue, F. Serum metabolomics for early diagnosis of esophageal squamous cell carcinoma by UHPLC-QTOF/MS. *Metabolomics* **2016**, *12*, 116. [CrossRef]
53. Chen, T.; Chen, X.; Zhang, S.; Zhu, J.; Tang, B.; Wang, A.; Dong, L.; Zhang, Z.; Yu, C.; Sun, Y.; et al. The Genome Sequence Archive family: Toward explosive data growth and diverse data types. *Genom. Proteom. Bioinf.* **2021**, *19*, 578–583. [CrossRef]
54. Members and Partners, C.N. Database resources of the National Genomics Data Center, China National Center for Bioinformation in 2022. *Nucleic Acids Res.* **2022**, *50*, D27–D38. [CrossRef]

plants

MDPI

Article

Characterization of bZIP Transcription Factors in Transcriptome of *Chrysanthemum mongolicum* and Roles of *CmbZIP9* in Drought Stress Resistance

Xuan Wang [1], Yuan Meng [1], Shaowei Zhang [1], Zihan Wang [1], Kaimei Zhang [2], Tingting Gao [1] and Yueping Ma [1,*]

[1] College of Life and Health Sciences, Northeastern University, Shenyang 110169, China; wangxuan99000@163.com (X.W.); 2201455@stu.neu.edu.cn (Y.M.); 2201483@stu.neu.edu.cn (S.Z.); 2301526@stu.neu.edu.cn (Z.W.); gaotingting@mail.neu.edu.cn (T.G.)

[2] Co-Innovation Center for Sustainable Forestry in Southern China, College of Life Sciences, Nanjing Forestry University, Nanjing 210037, China; kaimeizhang@njfu.edu.cn

* Correspondence: yuepingma@mail.neu.edu.cn; Tel.: +86-024-83656098

Abstract: bZIP transcription factors play important roles in regulating plant development and stress responses. Although bZIPs have been identified in many plant species, there is little information on the bZIPs in *Chrysanthemum*. In this study, bZIP TFs were identified from the leaf transcriptome of *C. mongolicum*, a plant naturally tolerant to drought. A total of 28 full-length bZIP family members were identified from the leaf transcriptome of *C. mongolicum* and were divided into five subfamilies based on their phylogenetic relationships with the bZIPs from Arabidopsis. Ten conserved motifs were detected among the bZIP proteins of *C. mongolicum*. Subcellular localization assays revealed that most of the CmbZIPs were predicted to be localized in the nucleus. A novel bZIP gene, designated as *CmbZIP9*, was cloned based on a sequence of the data of the *C. mongolicum* transcriptome and was overexpressed in tobacco. The results indicated that the overexpression of *CmbZIP9* reduced the malondialdehyde (MDA) content and increased the peroxidase (POD) and superoxide dismutase (SOD) activities as well as the expression levels of stress-related genes under drought stress, thus enhancing the drought tolerance of transgenic tobacco lines. These results provide a theoretical basis for further exploring the functions of the bZIP family genes and lay a foundation for stress resistance improvement in chrysanthemums in the future.

Keywords: chrysanthemum; bZIP; transcriptome; drought tolerance; transgenic tobacco

1. Introduction

The basic leucine zipper (bZIP) family is one of the largest families of transcription factors in plants and plays important roles in plant development and responses to various biotic and abiotic stresses. The bZIP protein is also a well-known abscisic acid (ABA)-induced DNA-binding protein family that is essential for the activation of downstream gene expression during abiotic stress [1–5]. This family contains a highly conserved basic domain and a relatively divergent leucine chain that bind to a core sequence (ACGT) in the promoter of the target genes, such as G-box, C-box, and A-box [4,6]. To date, many bZIP TFs have been identified in plants. For example, 869 bZIPs with nine subgroups were identified from12 species of Rosaceae [7]; 75 bZIPs with 13 subfamilies were detected in *Arabidopsis thaliana* [8]; and 62 bZIP members with 12 subgroups were found in *Andrographis paniculatan* [9]. Several bZIP members are involved in plant tolerance to salt stress, drought, etc. [10–13].

Chrysanthemums is a famous traditional flower in China with high ornamental, medical, and economic value. However, the production and commercialization of chrysanthemum have been hindered by water shortages [14,15]. Therefore, cultivating new chrysanthemum varieties with high drought resistance is a key goal for chrysanthemum breeders. Compared to conventional breeding approaches, genetic transformation techniques are more powerful and efficient. Although many bZIP TFs have been identified in different plants, few bZIPs have been characterized in *Chrysanthemum*. *Chrysanthemum mongolicum* belongs to *Chrysanthemum* (Asteraceae), mainly distributed in Nei Mongol, China, and is a remarkable drought-tolerant plant that grows on rocky mountain slopes at an altitude of 1500–2500 m [16–18]. Its outstanding drought-resistant gene resources are an excellent choice for molecular breeding for chrysanthemum stress resistance. In this study, bZIP family-member genes were identified from the leaf transcriptome of *C. mongolicum*. The protein properties, phylogenies, and motif composition were assessed. In addition, a candidate *bZIP* gene, *CmbZIP9*, was isolated from *C. mongolicum* using RT-PCR and genomic PCR. The roles of *CmbZIP* were determined through heterologous expression in *Nicotiana tabacum* L. under drought conditions. This study provides useful information on exploring the functions of the bZIP family genes in *Chrysanthemum* species and lays a foundation for stress resistance improvement in chrysanthemums in the future.

2. Results

2.1. Phylogenetic Analysis of bZIP Family Members in C. mongolicum

In total, 28 full-length bZIP members were identified from the transcriptome of *C. mongolicum* (Table S1). The proteins of the bZIP family members in *C. mongolicum* range from 140 aa to 1822 aa in length. Most CmbZIPs were predicted to be localized in the nucleus (Table S1). The bZIP family members in *C. mongolicum* were divided into five subgroups by referencing the 13 subclasses of AtbZIPs in the phylogenetic tree. The homologous proteins in families B, D, F, H, J, K, and E of *A. thaliana* were not found in the obtained transcriptome data from *C. mongolicum* (Figure 1). The largest subgroup, S, contained 10 CmbZIP members, followed by group A (8). Only one member was homologous to family G of the AtbZIPs. In addition, the homologs of the family members of groups D and E in Arabidopsis were found in other Chrysanthemum bZIPs, suggesting that some bZIPs are expressed in other organs and developmental stages. This result suggested that the members of bZIP proteins in different subgroups might be involved in various biological functions in *C. mongolicum*.

2.2. Conserved Motif Analysis of CmbZIPs

A total of 10 protein-conserved motifs with 15~50 amino acid residues were identified in the CmbZIPs with low E-values ($<10^{-10}$) using the MEME tool (Table S2). At least one protein-conserved motif was found in all of the identified protein sequences (Figure 2). The number and type of conserved motifs in each of the five subclades were similar, but each subgroup had special characteristics. Motif 1, a conserved basic bZIP domain, was present in all of the members of the protein sequences, which confirmed the correctness of our identifications. Groups S and C had similar motif compositions, with most of the sequences containing motifs 1 and 2. We found that the protein sequences in group S had two different motif compositions: one consisted of motifs 1 and 9, and the other contained motifs 1, 2, and 8. In addition, groups A and I contained a significantly higher number of motifs than the other groups. Motif 7 predominantly existed in group I, while motifs 3 and 4 appeared only in group A. The specific motifs in each subgroup suggested their specific biological functions within the subgroup and the functional diversity of bZIP family genes.

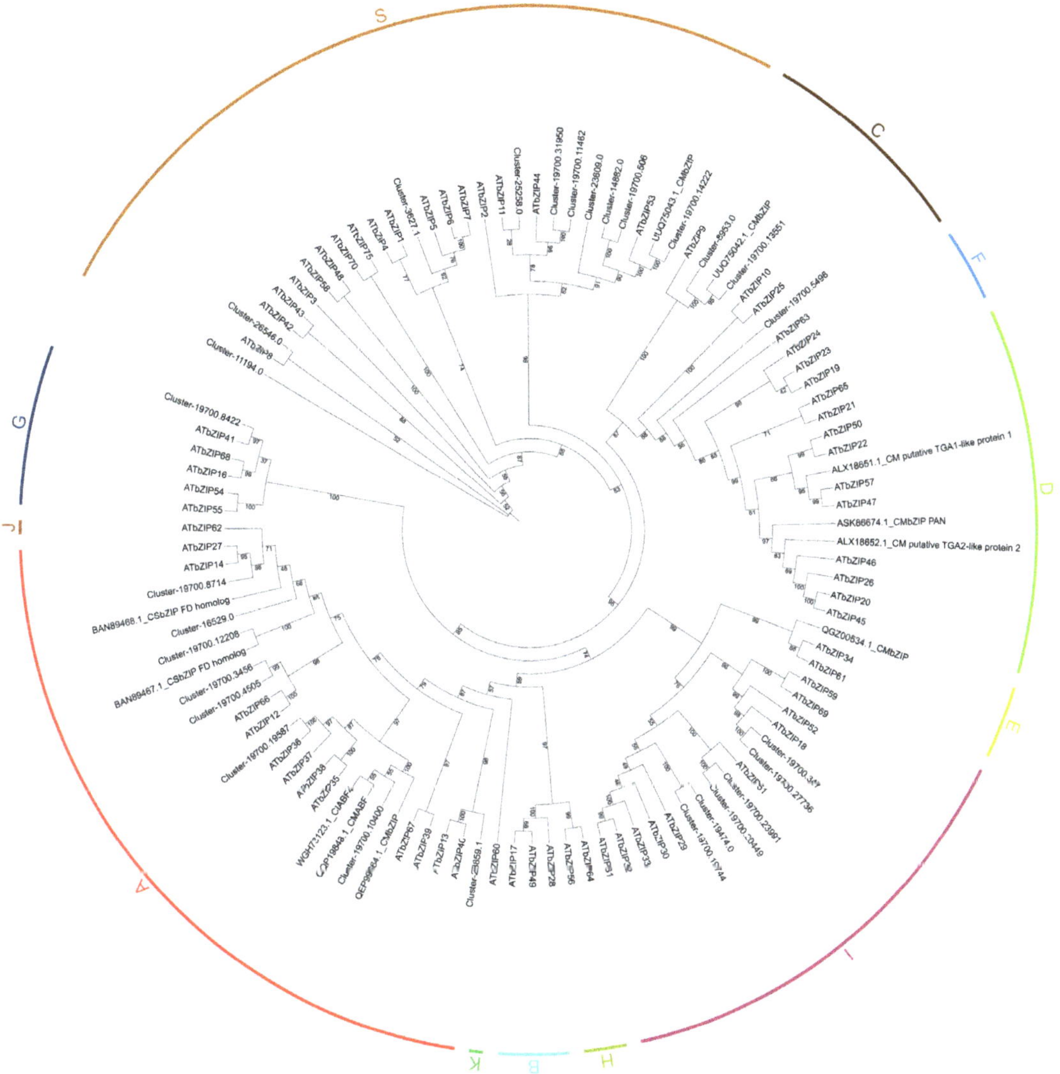

Figure 1. Phylogenetic analysis of bZIP proteins in *C. mongolicum*. The CmbZIPs are shown as clusters in the leaf transcriptome of *C. mongolicum*. Branch support was assessed with 1000 bootstrap replicates. The support values are provided at each node. The letters with different colors outside the circle indicate the various subgroups. AtbZIPs represent the bZIPs from *A. thaliana*. The 11 *Chrysanthemum* bZIPs are displayed with their sequence IDs and names.

2.3. Gene Isolation and Sequence Analysis

The phylogenetic analysis showed that the sequence of cluster-19700.13551 in the members of subgroup C was more closely related to that of AtbZIP9 (Figure 1), a protein that responds to drought stress. As such, we selected it to study its function in transgenic tobacco under drought stress. The full-length coding sequence of the *bZIP9* homolog, *CmbZIP9*, was successfully isolated from the leaves of *C. mongolicum* using RT-PCR and genomic PCR with a pair of primers (Table S3). It was 768 bp in length, encoding 255 amino acids (Figure S1A). The putative peptide sequence of CmbZIP9 contains a typical basic region leucine zipper (BRLZ) domain, which is conserved in bZIP family proteins (GenBank accession number OR831111, named *CmbZIP9*). *CmbZIP9* contains five exons and four introns (GenBank accession number OR842904, Figure S1B). The comparison of the putative amino acid sequences of CmbZIP9 with those of other bZIP homologs revealed that the sequence identity ranged from 57.8% to 94.12%, and the highest identity of 94.12% was shared with AabZIP from *Artemisia annua* (Figure S2A). The phylogenetic analysis showed that CmbZIP9 clusters with the bZIP9 homolog of *A. annua* and *Tanacetum cinerariifolium* into a clade with an 87% bootstrap support value, which is consistent with the biological evolutionary patterns (Figure S2B).

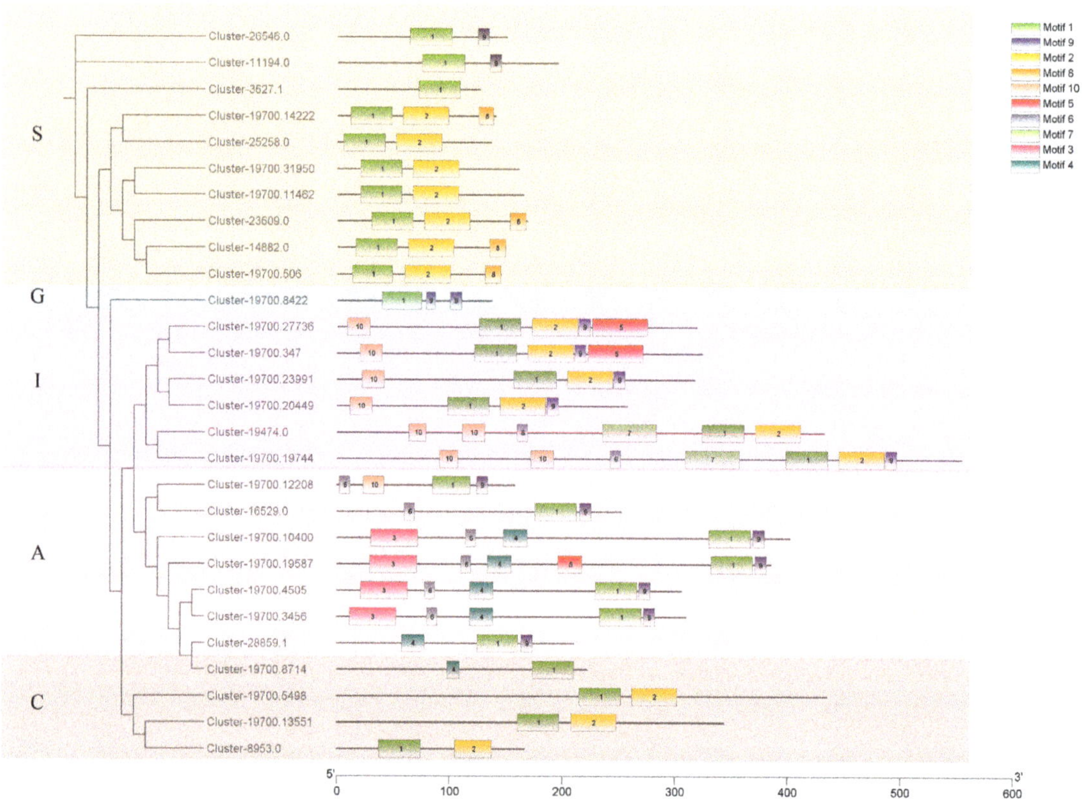

Figure 2. Conserved motif analysis of bZIP family in *C. mongolicum*. Conservative motifs are indicated with different colored boxes. The black line represents nonconserved sequences.

2.4. Heterologous Expression of CmbZIP9 in Tobacco

After rooting on MS medium containing kanamycin and rifampicin, more than 65 transgenic tobacco lines were obtained. Genomic PCR and qRT-PCR were performed to confirm the *CmbZIP9* transgenic lines (Figure 3). Lines 2, 4, and 6, which had high expression levels of *CmbZIP9*, were selected for further functional studies.

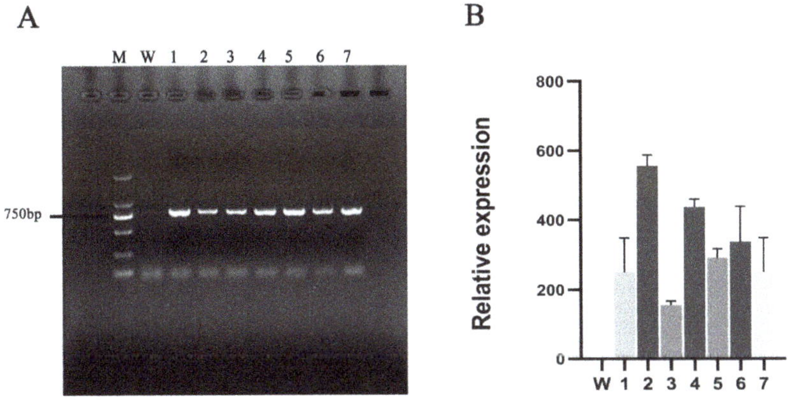

Figure 3. Identification of transgenic tobacco plants. (**A**) Genomic PCR for the identification of the transgenic plants. M: DL2000 marker, W: wild-type tobacco, lines 1–7: transgenic tobacco lines; uncropped gels are shown in Supplementary Figure S3. (**B**) Transcript levels of *CmbZIP9* in wild-type and transgenic tobacco plants. Data are shown as mean ± SD (n = 3).

2.5. *CmbZIP9 Improved the Tolerance of Plants to Drought Stress*

Four-week-old transgenic tobacco lines and wild-type seedlings were used to observe the phenotypes under drought stress. There were no obvious morphological differences between transgenic and wild-type plants under normal conditions. The WT tobacco wilted earlier than the transgenic tobacco, whereas the transgenic lines showed less wilting and grew well after 7 days of drought treatment (Figure 4A). Under drought stress, the relative water contents (RWCs) of the transgenic tobacco plants and wild-type plants were all reduced, but the transgenic tobacco plants had a significantly higher RWC than the wild-type plants (Figure 4B).

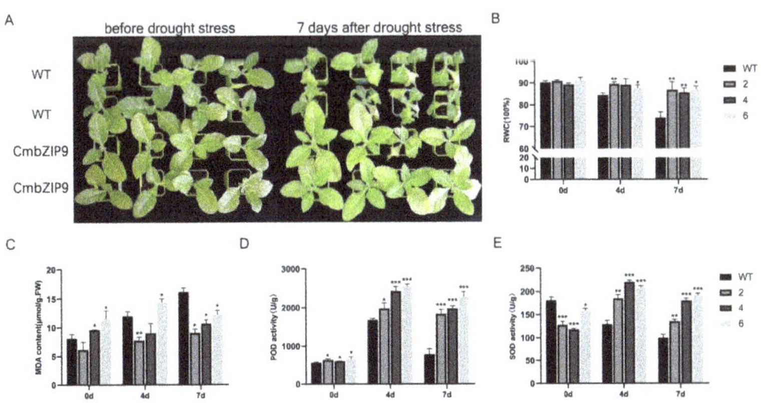

Figure 4. The constitutive expression of *CmbZIP9* in tobacco enhanced drought tolerance. (**A**) Morphology of wild-type and CmbZIP9 transgenic plants before and after 7 days of drought stress treatment. (**B**) Relative water content (RWC) before and after drought treatment for 4 and 7 d. (**C**) MDA content before and after drought treatment for 4 and 7 d. (**D**) Peroxidase (POD) activity before and after drought treatment for 4 and 7 d. (**E**) Superoxide dismutase (SOD) activity before and after drought treatment for 4 and 7 d. The numbers 2, 4, and 6 indicate T1 generations of three different transgenic lines. Asterisks represent values that are significantly different from those of the wild type (* $p < 0.05$, ** $p < 0.01$, *** $p < 0.001$).

The malondialdehyde (MDA) content gradually increased in all plants with increasing duration of drought stress, but a significantly higher content was observed in the wild-type plants than in the transgenic plants. After 7 days of exposure to drought stress, the MDA content of the wild-type plants was 1.65, 1.38 and 1.35 times higher than that in transgenic plants 2, 4 and 6, respectively (Figure 4C). Therefore, the overexpression of *CmbZIP9* reduced the degree of lipid peroxidation.

The peroxidase (POD) activity of the transgenic tobacco plants was slightly higher than that of the wild-type plants under normal conditions, after drought treatment for 4 days, and after drought treatment for 7 days. After 4 days of drought treatment, the POD contents of the transgenic tobacco plants 2, 4, and 6 was 1.18, 1.45, and 1.54 times higher than that in the wild-type plants, respectively. After 7 days of drought treatment, the POD content of transgenic plants 2, 4, and 6 was 2.33, 2.51 and 2.90 times higher than that of the wild-type plants, respectively (Figure 4D).

Under normal conditions, the activity of superoxide dismutase (SOD) in the leaves of the transgenic plants was slightly lower than that in the wild-type plants. After drought treatment, the SOD activity of the transgenic plants increased obviously. After 4 days of drought treatment, the SOD content of the transgenic plants 2, 4, and 6 was 1.44, 1.71 and 1.61 times higher than that of the wild-type plants, respectively. After 7 days of drought treatment, the SOD content of the transgenic plants was 1.35, 1.81, and 1.91 times higher than that of the wild-type plants, respectively (Figure 4E). These results indicated that the overexpression of *CmbZIP9* in tobacco increased the tolerance of tobacco to drought stress.

2.6. *CmbZIP9 Overexpression in Transgenic Tobacco Enhanced the Expression of Drought-Responsive Genes*

To further explore the possible molecular mechanism by which *CmbZIP9* improves drought tolerance, the transcript levels of four abiotic-stress-related genes, including ABA-responsive element binding factor (*AREB*), ethylene responsive factor (*ERF*), S-adenosylmethionine decarboxylase (*SAMDC*), and arginine decarboxylase (*ADC*) were investigated. The results showed that the expressions of the genes related to ROS detoxification (*NtSOD*) and stress (*NtERF*, *NtAREB*, *NtADC*, and *NtSAMDC*) in the transgenic lines were significantly higher than those in the WT plants after exposure to drought stress for 7 days (Figure 5). Therefore, *CmbZIP9* might regulate the expression of stress-related genes, thus improving the tolerance of transgenic tobacco to drought stress.

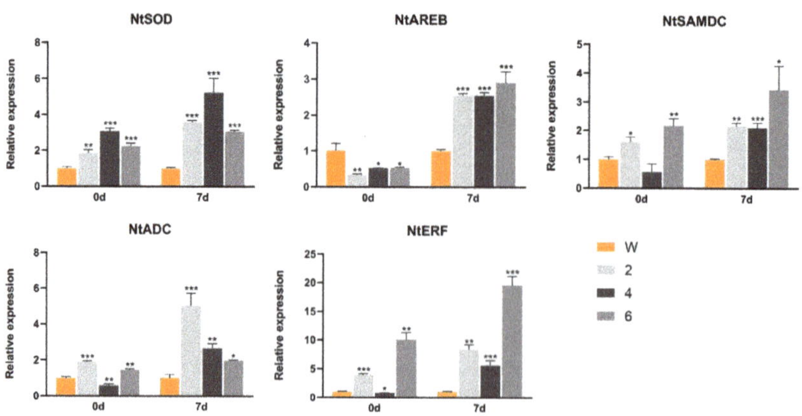

Figure 5. Expression levels of stress-related genes in wild-type and transgenic tobacco lines before and after 7 days of drought treatment. W: wild-type tobacco; 2, 4, and 6: T1 generations in three different transgenic lines. Data are presented as mean ± SD of three independent experiments. Asterisks represent values that are significantly different from those of the wild type (* $p < 0.05$, ** $p < 0.01$, *** $p < 0.001$).

3. Discussion

bZIP TFs have been shown to play important roles in plant development and the responses to various biotic and abiotic stress [15,19,20]. The bZIP domain is crucial to the functioning of the bZIP TF family members [21]. In this study, 28 CmbZIPs were identified in *C. mongolicum*, which were classified into five subgroups. The conserved motif of the basic region was present in all the members of the CmbZIPs, which indicated the evolutionary conservation of bZIP gene function in these plants. In addition, the CmbZIP proteins of the same subgroup shared the same conserved motif types and similar numbers of them, but the number of conserved motifs varied considerably between subclades, consistent with the bZIP findings in other plants [22,23]. The diversity of the conserved motifs outside of the bZIP domain in different bZIP members suggests that the bZIP genes might have diverse functions.

The bZIP structure varies among species, and even among different bZIP family members of the same species. For example, a maximum of 12 introns have been detected in Arabidopsis, cucumber, and rice; 11 bZIPs were found in mung bean and barley, and so on [24–27]. In our study, four introns were found within the ORF of *CmbZIP9*. The diverse structures of *bZIPs* suggest that different splicing states of exons and introns may have been of great significance in the evolution of *bZIP* genes in plants. The positions and phases of the introns in the basic and hinge regions of the bZIP domain are highly conserved, indicating multiple functions of this gene family in plant development and biotic/abiotic stress responses.

Usually, an appropriate balance between ROS production and scavenging is maintained in plants under normal conditions. However, large amounts of ROS are produced when plants are exposed to stress [28–30]. Excessive ROS results in plant metabolism disorders and oxidative damage, which inhibit normal plant growth and development and may even lead to plant death in severe cases [31–33]. In order to effectively reduce the accumulation of ROS, plants have evolved complex and precise antioxidant systems to neutralize ROS and prevent the oxidative damage of cells during stress conditions [34]. The MDA content is an important indicator of lipid peroxidation, which indirectly reflects the degree of plant membrane system damage and stress resistance [35,36]. The MDA content sharply increases in plants under abiotic stress [37]. In our study, the MDA content of *CmbZIP9* transgenic tobacco was significantly lower than that of the wild-type tobacco under drought stress, indicating lower cell membrane system damage and lipid peroxidation levels in *CmbZIP9* transgenic plants under drought stress than in wild-type plants.

SOD and POD are important components of the antioxidant enzyme defense system that can effectively remove excess ROS from plants [37–39]. SOD scavenges ROS by converting superoxide into H_2O_2 [40], whereas POD can consume ROS [41]. As a result, higher SOD and POD activities contribute to plant resistance to abiotic stresses [42,43]. The activities of SOD and POD have been found to be higher in many species after exposure to drought stress [44,45]. Overexpression of the grapevine transcription factor *VvABF2* in *Arabidopsis thaliana* led to considerably higher SOD and POD activities in transgenic plants than in wild-type plants under drought treatment and enhanced the drought tolerance of *Arabidopsis* [46]. The bZIP transcription factor *PtrABF*, isolated from *Poncirus trifoliata*, also increased the SOD and POD activities of transgenic tobacco lines compared with those of wild type after drought treatment [47]. In our study, the POD and SOD activities in the leaves of *CmbZIP9* transgenic tobacco were significantly increased under drought stress and these plants showed more stress resistance than wild-type tobacco (Figure 5). In addition, the transcripts of the *SOD* gene and some stress-related genes (*AREB*, *SAMDC*, *ERF*, and *ADC*) significantly increased in the transgenic lines under drought stress in this study. Previous studies have reported that these genes are involved in the response to abiotic stress and show higher expression levels under environmental stress [48–54]. Taking these results together, we found that the overexpression of *CmbZIP9* led to increases in POD and SOD activities and in the expression levels of genes related to environment stress under drought stress, which might enhance the drought tolerance of plants. These results demonstrated that *CmbZIP9* plays important roles in drought stress resistance.

4. Materials and Methods

4.1. Plant Materials

The *Chrysanthemum mongolicum* plants used in this study were collected from Baotou, Inner Mongolia, China, and transplanted into the nursery garden of Northeastern University, China. A voucher specimen was collected and stored in the herbarium of Northeastern University, with voucher number ZL-20170817-02. Plant leaves were collected, immediately frozen in liquid nitrogen, and stored at −80 °C. Total RNA was extracted using a Plant RNA kit (Omega Bio-Tek, Norcross, GA, USA) and treated with DNase I (Omega Bio-Tek, USA) to remove DNA contamination. Genomic DNA was extracted from fresh leaves using a DNA kit (DNeasy Plant Mini Kit, (Qiagen, Santa Clarita, CA, USA) and treated with RNase to remove the RNA.

4.2. CmbZIP Genes Identification, Motif Composition Analyses

The bZIP family members were detected from the transcriptome of *C. mongolicum* [26] using a hidden Markov model (HMM) (http://pfam.xfam.org/, accessed on 23 October 2023). The open reading frames (ORFs) were confirmed using NCBI's ORFfinder (ORFfinder Archives—NCBI Insights). The number of amino acids, isoelectric point (pI), and molecular weight (Mw) of the completely coded CmbZIP members were determined using the ExPASy website (https://www.expasy.org/, accessed on 2 November 2023) [55]. Thw WoLF PSORT website (https://wolfpsort.hgc.jp/, accessed on 10 November 2023) [56] was used to predict the localization of CmbZIPs. Online MEME (v 5.0) software was used to confirm the conserved motifs of the bZIPs (http://meme-suite.org/tools/meme, accessed on 15 November 2023) with a motif threshold of 10 for each sequence.

4.3. Phylogenetic Analysis

Seventy-one bZIP proteins of Arabidopsis thaliana were obtained from TAIR (https://www.arabidopsis.org/index.jsp, accessed on 1 December 2023). Eleven chrysanthemum bZIP protein sequences were obtained from NCBI (https://www.ncbi.nlm.nih.gov/, accessed on 5 December 2023). The putative bZIP protein sequences of *C. mongolicum* and AtbZIPs were aligned using MEGA [57] and manually checked. IQ-TREE was employed to construct a maximum likelihood (ML) phylogenetic tree with 1000 bootstrap replicates [58]. TBtools was used to visualize the phylogenetic tree and motif compositions [59].

4.4. Isolation of CmbZIP9 and Sequence Analysis

The quality of the total RNA was measured according to the method described by Zhang et al. [60]. First-strand cDNA was synthesized from a library of leaves using a Revert Aid First Strand cDNA Synthesis kit (Thermo Fisher Scientific Inc., Waltham, MA, USA), according to the manufacturer's instructions. The complete *CmbZIP9* was amplified via RT-PCR and genomic PCR with primers M551ZIP F and M551ZIP R (Table S1) which were designed according to cluster-19700.13551 in the transcriptome data of *C. mongolicum* [18] with Pfu DNA polymerase (Takara Tokyo, Japan). Purified PCR products were linked to the Kan-EZ-TA-TOPO vector and transformed into *Escherichia coli* competent cells according to the method previously described by Hu et al. [61]. Transformed colonies were identified by PCR with gene-specific primers; restriction digestion and sequenced by Zhongmeitaihe Company (Beijing, China). DNAMAN software (version 6.0) was used to align multiple amino acid sequences, and a neighbor-joining phylogenetic tree was constructed using MEGA 6 software with 1000 bootstrap replicates using Kimura two-parameter distances and the pairwise deletion of gaps. The intron–exon structures of *CmbZIP9* were analyzed using TBtools [59] and checked manually.

4.5. Transformation of *CmbZIP9* in Tobacco

The coding sequence of *CmbZIP9* was inserted into the pBI121 vector to generate the 35S:: CmbZIP9 expression vector, which was transformed into *Nicotiana tabacum* using the leaf disk method, mediated by *Agrobacterium tumefaciens* strain EHA105 following the method described by Hu et al. [61]. A series of Murashige and Skoog (MS) media with kanamycin and rifampicin were used to culture the leaf disks. Plant transformation was confirmed using genomic PCR, with specific primers used for initial *CmbZIP9* amplification. The qPCR was performed on ABI QuantStudio 5 (Applied Biosystems, Foster City, CA, USA) using an SYBR Green PCR master mix kit (Thermol Fisher Scientific, Waltham, MA, USA) following the manufacturer's instructions to explore the transcript levels of the transgenic tobacco lines. *NtACTIN* gene was used as an endogenous control [62].

4.6. Drought Stress Treatment

Tobacco seedlings (4–5 weeks old) of the transgenic and wild-type plants were cultured as described by Li et al. [62]. Plants with 8–10 leaves were used for drought stress treatment. After complete watering, the plants were cultured at 25 °C day/18 °C night with 70% relative humidity and were not watered again. A minimum of three replicates were performed for each experiment. The leaves of treated and control plants were harvested after drought stress treatment for 0, 4, and 7 days, immediately frozen in liquid nitrogen, and stored at −80 °C until use.

4.7. Measurement of Relative Water Content (RWC)

For relative water content (RWC) measurement, we followed the method described by Yaghoubian et al. [63]. with minor modifications to the method. Briefly, the fresh weight of each leaf was determined, then the leaf was immersed immediately in distilled water at 4 °C for 24 h to obtain the saturation weight (SW) and then placed in an oven at 80 °C for 24 h to measure the dry weight (DW).

4.8. Determination of MDA Content

MDA is a biomarker of lipid oxidation in plant membrane systems and indirectly reflects the level of damage to the plant membrane system and stress tolerance. The MDA content was measured using the method described by Li et al. [62] with minor modifications. Briefly, approximately 0.1 g of leaf tissue was homogenized with 5 mL of trichloroacetic acid (5%, w/v) and then centrifuged at 3000 rpm for 10 min. Next, 0.067% (w/v) thiobarbituric acid (2 mL) was added to 2 mL of the supernatant to be mixed, boiled for 30 min, and cooled on ice immediately. Then, the solution was centrifuged at 3000 rpm for 10 min. The absorbance of the supernatant was measured at 450, 532, and 600 nm to calculate the MDA content.

4.9. Measurement of the SOD and POD Activities

POD and SOD are antioxidant enzymes that play crucial roles in plant responses to various abiotic stresses. SOD and POD activities were determined using an ELISA reader (BioTek Instruments, Winooski, VT, USA), according to the manufacturer's instructions. About 0.3 g of leaves was homogenized with 5 mL of 0.1 mol/L Tris-HCl buffer (pH 8.5) to extract the crude enzyme solution after centrifuging at 4000 rpm for 5 min. The supernatant was used to measure the SOD content at 550 nm, and POD activity was assessed following the method described by Pan et al. [64].

4.10. Quantification Analysis of Stress-Related Genes

The transcript levels of antioxidative gene *NtSOD* and stress-related genes, including *NtERF*, *NtSAMD*, *AREB*, and *NtADC*, were examined in the wild-type and transgenic plants under drought stress conditions. The NtActin gene was used as a reference. The qRT-PCR reaction was performed in 20 μL volumes containing 10 μL of SYBR Green Master Mix (Thermo Fisher Scientific, USA), 1.0 μL of both forward and reverse primers to a final concentration of 200 μM, 2 μL of a template diluted 1: 5 of the cDNA, and 2.5 μL of PCR-grade ddH_2O to a final volume of 20 μL. The qPCR was performed on an ABI QuantStudio 5 real-time PCR system (Applied Biosystems, Foster City, CA, USA) with the following procedure: 55 °C for 2 min, 94°C for 5 min, and 40 cycles consisting of denaturation 94 °C for 15 s, 60 °C for 1 min. Each experiment was independently repeated in triplicate. Relative transcript levels were calculated using the $2^{-\Delta\Delta Ct}$ method [65]. The primers used for qRT-PCR are listed in Table S3.

4.11. Statistical Analyses

Three independent samples were used for each analysis. Statistical analyses were performed using GraphPad Prism 8.0 (San Diego, CA, USA). Multiple range tests were used to detect significant differences between the mean values, and statistical significance was defined as $p < 0.05$, $p < 0.01$, or $p < 0.001$.

5. Conclusions

In this study, 28 CmbZIP family genes were identified in the transcriptome of *C. mongolicum*, and 10 motifs were detected among these CmbZIP proteins. A candidate gene, *CmbZIP9*, was isolated. *CmbZIP9* contains five exons and four introns. The overexpression of *CmbZIP9* increased the drought tolerance of transgenic tobacco lines. The functions of the other *CmbZIPs* we identified in *C. mongolicum* and the additional functions of *CmbZIP9* need to be further explored.

Supplementary Materials: The following supporting information can be downloaded at: https://www.mdpi.com/article/10.3390/plants13152064/s1.

Author Contributions: Y.M. (Yueping Ma) collected the plants and designed the experiments; X.W., Y.M. (Yuan Meng), Z.W., S.Z. and T.G. performed the experiments and analyzed the data; X.W wrote the first draft; K.Z. and Y.M. (Yueping Ma) revised the manuscript. All authors have read and agreed to the published version of the manuscript.

Funding: This project was supported by the National Natural Science Foundation of China (No. 31872710).

Data Availability Statement: The data that support the findings of this study are openly available from NCBI GenBank at https://www.ncbi.nlm.nih.gov/. The accession numbers can be found in the article.

Conflicts of Interest: The manuscript was approved by all authors for publication and no conflicts of interest exist regarding the submission of this manuscript. The authors declare that this research was performed without any commercial or financial relationships that could be interpreted as a potential conflict of interest.

References

1. Wang, H.; Wang, H.; Shao, H.; Tang, X. Recent advances in utilizing transcription factors to improve plant abiotic stress tolerance by transgenic technology. *Front. Plant Sci.* **2016**, *7*, 67. [CrossRef]
2. Ma, H.; Liu, C.; Li, Z.; Ran, Q.; Xie, G.; Wang, B.; Fang, S.; Chu, J.; Zhang, J. ZmbZIP4 contributes to stress resistance in maize by regulating ABA synthesis and root development. *Plant Physiol.* **2018**, *178*, 753–770. [CrossRef] [PubMed]
3. Yao, L.; Hao, X.; Cao, H.; Ding, C.; Yang, Y.; Wang, L.; Wang, X. ABA-dependent bZIP transcription factor, CsbZIP18, from Camellia sinensis negatively regulates freezing tolerance in Arabidopsis. *Plant Cell Rep.* **2020**, *39*, 553–565. [CrossRef] [PubMed]
4. Joo, H.; Baek, W.; Lim, C.W.; Lee, S.C. Post-translational modifications of bZIP transcription factors in abscisic acid signaling and drought responses. *Curr. Genom.* **2021**, *22*, 4–15. [CrossRef]
5. Liang, Y.; Xia, J.; Jiang, Y.; Bao, Y.; Chen, H.; Wang, D.; Zhang, D.; Yu, J.; Cang, J. Genome-wide identification and analysis of bZIP gene family and resistance of TaABI5 (TabZIP96) under freezing stress in wheat (Triticum aestivum). *Int. J. Mol. Sci.* **2022**, *2*, 2351. [CrossRef]
6. An, J.; Yao, J.; Xu, R.; You, C.; Wang, X.; Hao, Y. Apple bZIP transcription factor MdbZIP44 regulates abscisic acid-promoted anthocyanin accumulation. *Plant Cell Environ.* **2018**, *41*, 2678–2692. [CrossRef] [PubMed]
7. Zhu, L.; Zhang, M.J.; Yang, X.Y.; Zi, Y.Q.; Yin, T.; Li, X.L.; Wen, K.; Zhao, K.; Wan, J.Q.; Zhang, H.Y.; et al. Genome-wide identification of bZIP transcription factors in 12 Rosaceae species and modeling of novel mechanisms of EjbZIPs response to salt stress. *Plant Genome* **2024**, e20468. [CrossRef] [PubMed]

8. Dröge-Laser, W.; Snoek, B.L.; Snel, B.; Weiste, C. The Arabidopsis bZIP transcription factor family-an update. *Curr. Opin. Plant Biol.* **2018**, *45*, 36–49. [CrossRef] [PubMed]
9. Guan, R.; Xu, S.; Lu, Z.; Su, L.; Zhang, L.; Sun, W.; Zhang, Y.; Jiang, C.; Liu, Z.; Duan, L.; et al. Genomic characterization of bZIP transcription factors related to andrographolide biosynthesis in *Andrographis paniculata*. *Int. J. Biol. Macromol.* **2022**, *223*, 1619–1631. [CrossRef]
10. Qu, D.; Wu, F.; Zhao, X.; Zhu, D.; Gu, L.; Yang, L.; Zhao, W.; Sun, Y.; Yang, J.; Tian, W.; et al. A bZIP transcription factor VabZIP12 from blueberry induced by dark septate endocyte improving the salt tolerance of transgenic Arabidopsis. *Plant Sci.* **2022**, *315*, 111135. [CrossRef]
11. Cao, L.; Lu, X.; Zhang, P.; Wang, G.; Wei, L.; Wang, T. Systematic analysis of differentially expressed maize ZmbZIP genes between drought and rewatering transcriptome reveals bZIP family members involved in abiotic stress responses. *Int. J. Mol. Sci.* **2019**, *20*, 4103. [CrossRef] [PubMed]
12. Joo, J.; Lee, Y.H.; Song, S.I. OsbZIP42 is a positive regulator of ABA signaling and confers drought tolerance to rice. *Planta* **2019**, *249*, 1521–1533. [CrossRef]
13. Wang, W.; Qiu, X.; Yang, Y.; Kim, H.S.; Jia, X.; Yu, H.; Kwak, S.S. Sweetpotato bZIP transcription factor IbABF4 confers tolerance to multiple abiotic stresses. *Front. Plant Sci.* **2019**, *10*, 630. [CrossRef] [PubMed]
14. Sahithi, B.M.; Razi, K.; Al Murad, M.; Vinothkumar, A.; Jagadeesan, S.; Benjamin, L.K.; Jeong, B.R.; Muneer, S. Comparative physiological and proteomic analysis deciphering tolerance and homeostatic signaling pathways in chrysanthemum under drought stress. *Physiol. Plant.* **2021**, *172*, 289–303. [CrossRef] [PubMed]
15. Wang, T.; Wei, Q.; Wang, Z.; Liu, W.; Zhao, X.; Ma, C.; Gao, J.; Xu, Y.; Hong, B. CmNF-YB8 affects drought resistance in chrysanthemum by altering stomatal status and leaf cuticle thickness. *J. Integr. Plant Biol.* **2022**, *64*, 741–755. [CrossRef]
16. Zhao, H.; Liu, Z.; Hu, X.; Yin, J.; Li, W.; Rao, G.; Zhang, X.; Huang, C.; Anderson, N.; Zhang, Q.; et al. Chrysanthemum genetic resources and related genera of Chrysanthemum collected in China. *Genet. Resour. Crop Evol.* **2009**, *56*, 937–946. [CrossRef]
17. Zhao, H.; Chen, F.; Chen, S.; Wu, G.; Guo, W. Molecular phylogeny of *Chrysanthemum*, *Ajania* and its allies (Anthemideae, Asteraceae) as inferred from nuclear ribosomal ITS and chloroplast trnL-F IGS sequences. *Plant Syst. Evol.* **2010**, *284*, 153–169. [CrossRef]
18. Ma, Y.; Meng, Y.; Lin, J. Comparative Transcriptome analyses reveal different regulatory mechanisms in ecological adaptation between *Chrysanthemum vestitum* and *Chrysanthemum mongolicum*. *Horticulturae* **2023**, *9*, 868. [CrossRef]
19. Yang, Y.; Kong, Q.; Tee, W.T.; Li, Y.; Low, P.M.; Patra, B.; Guo, L.; Yuan, L.; Ma, W. Transcription factor bZIP52 modulates Arabidopsis seed oil biosynthesis through interaction with WRINKLED1. *Plant Physiol.* **2023**, *192*, 2628–2639. [CrossRef] [PubMed]
20. Yue, L.; Pei, X.; Kong, F.; Zhao, L.; Lin, X. Divergence of functions and expression patterns of soybean bZIP transcription factors. *Front. Plant Sci.* **2023**, *14*, 1150363. [CrossRef]
21. de Souza, C.R.B.; Serrão, C.P.; Barros, N.L.F.; dos Reis, S.P.; Marques, D.N. Plant bZIP proteins: Potential use in agriculture—A Review. *Curr. Protein Pept. Sci.* **2023**, *25*, 107–119. [CrossRef] [PubMed]
22. Gu, C.H.; Hong, S.D.; Wang, J.; Shang, L.X.; Zhang, G.Z.; Zhao, Y.; Ma, Q.Q.; Ma, D.D. Identification and expression analysis of the bZIP and WRKY gene families during anthocyanins biosynthesis in *Lagerstroemia indica* L. *Hortic. Environ. Biotechnol.* **2024**, *65*, 169–180. [CrossRef]
23. Ye, F.T.; Zhu, G.; Wu, S.F.; Du, Y.Y.; Pan, X.F.; Wu, Y.C.; Qian, Z.Y.; Li, Z.W.; Lin, W.X.; Fan, K. Conserved and divergent evolution of the bZIP transcription factor in five diploid *Gossypium* species. *Planta* **2023**, *257*, 26. [CrossRef] [PubMed]
24. Wang, Z.; Cheng, K.; Wan, L.; Yan, L.; Jiang, H.; Liu, S.; Lei, Y.; Liao, B. Genome-wide analysis of the basic leucine zipper (bZIP) transcription factor gene family in six legume genomes. *BMC Genom.* **2015**, *16*, 1053. [CrossRef] [PubMed]
25. Jakoby, M.; Weisshaar, B.; Dröge-Laser, W.; Vicente-Carbajosa, J.; Tiedemann, J.; Kroj, T.; Parcy, F.; bZIP Research Group. bZIP transcription factors in Arabidopsis. *Trends Plant Sci.* **2002**, *7*, 106–111. [CrossRef] [PubMed]
26. Nijhawan, A.; Jain, M.; Tyagi, A.K.; Khurana, J.P. Genomic survey and gene expression analysis of the basic leucine zipper transcription factor family in rice. *Plant Physiol.* **2008**, *146*, 333–350. [CrossRef]
27. Pourabed, E.; Ghane, G.F.; Soleymani, M.P.; Razavi, S.M.; Shobbar, Z.S. Basic leucine zipper family in barley: Genome-wide characterization of members and expression analysis. *Molecular Biotech.* **2015**, *57*, 12–26. [CrossRef] [PubMed]
28. Zhang, H.; Zhu, J.; Gong, Z.; Zhu, J.K. Abiotic stress responses in plants. *Nat. Rev. Genet.* **2022**, *23*, 104–119. [CrossRef]
29. Miller, G.; Suzuki, N.; Ciftciyilmaz, S.; Mittler, R. Reactive oxygen species homeostasis and signalling during drought and salinity stresses. *Plant Cell Environ.* **2010**, *33*, 453–467. [CrossRef]
30. He, W.D.; Gao, J.; Dou, T.X.; Shao, X.H.; Bi, F.C.; Sheng, O.; Deng, G.M.; Li, C.Y.; Hu, C.H.; Liu, J.H.; et al. Early cold-induced peroxidases and aquaporins are associated with high cold tolerance in Dajiao (*Musa* spp. 'Dajiao'). Front. *Plant Sci.* **2018**, *9*, 282.
31. Li, L.; Yi, H. Differential expression of Arabidopsis defense-related genes in response to sulfur dioxide. *Chemosphere* **2012**, *87*, 718–724. [CrossRef] [PubMed]
32. Li, L.; Yi, H.; Liu, X.; Qi, H. Sulfur dioxide enhance drought tolerance of wheat seedlings through H2S signaling. *Ecotoxicol. Environ. Saf.* **2020**, *207*, 111248. [CrossRef]
33. Zhou, Q.; Li, Y.; Wang, X.; Yan, C.; Ma, C.; Liu, J.; Dong, S. Effects of different drought degrees on physiological characteristics and endogenous hormones of soybean. *Plants* **2022**, *11*, 2282. [CrossRef]

34. Mittler, R.; Zandalinas, S.I.; Fichman, Y.; Breusegem, F.V. Reactive oxygen species signalling in plant stress responses. *Nat. Rev. Mol. Cell Biol.* **2022**, *23*, 663–679. [CrossRef] [PubMed]
35. Luan, Y.; Cui, J.; Li, J.; Jiang, N.; Liu, P.; Meng, J. Effective enhancement of resistance to *Phytophthora infestans* by overexpression of miR172a and b in Solanum lycopersicum. *Planta* **2018**, *247*, 127–138. [CrossRef] [PubMed]
36. Cheng, S.; Tang, D.; Miller, W.B.; Shi, Y. Evaluation of salinity tolerance in honeysuckle (*Lonicera japonica*) using growth ion accumulation, lipid peroxidation, and non-enzymatic and enzymatic antioxidants system criteria. *J. Hortic. Sci. Biotechnol.* **2018**, *93*, 185–195. [CrossRef]
37. ElSayed, A.I.; El-Hamahmy, M.A.M.; Rafudeen, M.S.; Mohamed, A.H.; Omar, A.A. The impact of drought stress on antioxidant responses and accumulation of flavonolignans in Milk thistle (*Silybum marianum* (L.) Gaertn). *Plants* **2019**, *8*, 611. [CrossRef]
38. Khan, R.; Ma, X.; Zhang, J.; Wu, X.; Iqbal, A.; Wu, Y.; Zhou, L.; Wang, S. Circular drought-hardening confers drought tolerance via modulation of the antioxidant defense system, osmoregulation, and gene expression in tobacco. *Physiol. Plant.* **2021**, *172*, 1073–1088. [CrossRef] [PubMed]
39. Wang, H.; Zhang, Y.; Norris, A.; Jiang, C.Z. S1-bZIP transcription factors play important roles in the regulation of fruit quality and stress response. *Front. Plant Sci.* **2022**, *12*, 802802. [CrossRef]
40. Willekens, H.; Chamnongpol, S.; Davey, M.; Schraudner, M.; Langebartels, C.; Van Montagu, M.; Inze, D.; Van Camp, W. Catalase is a sink for H_2O_2 and is indispensable for stress defence in C3 plants. *EMBO J.* **1997**, *16*, 4806–4816. [CrossRef]
41. Kaushik, D.; Aryadeep, R. Reactive oxygen species (ROS) and response of antioxidants as ROS-scavengers during environmental stress in plants. *Front. Environ. Sci.* **2014**, *2*, 53.
42. Santos, A.D.A.; Silveira, J.A.G.D.; Bonifacio, A.; Rodrigues, A.C.; Figueiredo, M.D.V.B. Antioxidant response of cowpea co-inoculated with plant growth-promoting bacteria under salt stress. *Braz. J. Microbiol.* **2018**, *49*, 513–521. [CrossRef] [PubMed]
43. Zhou, Y.; Hu, L.; Ye, S.; Jiang, L.; Liu, S. Molecular cloning and functional characterization of a Cu/Zn superoxide dismutase gene (CsCSD1) from Cucumis sativus. *Plant Cell Tissue Organ* **2018**, *135*, 309–319. [CrossRef]
44. Liu, Y.; He, Z.; Xie, Y.; Su, L.; Zhang, R.; Wang, H.; Li, C.; Long, S. Drought resistance mechanisms of *Phedimus aizoon* L. *Sci. Rep.* **2021**, *11*, 13600. [CrossRef]
45. Song, S.; Qu, Z.; Zhou, X.; Wang, X.; Dong, S. Effects of weak and strong drought conditions on physiological stability of flowering soybean. *Plants* **2022**, *11*, 2708. [CrossRef] [PubMed]
46. Liu, J.; Chu, J.; Ma, C.; Jiang, Y.; Ma, Y.; Xiong, J.; Cheng, Z.M. Overexpression of an ABA-dependent grapevine bZIP transcription factor, VvABF2, enhances osmotic stress in Arabidopsis. *Plant Cell Rep.* **2019**, *38*, 587–596. [CrossRef]
47. Huang, X.S.; Liu, J.H.; Chen, X.J. Overexpression of PtrABF gene, a bZIP transcription factor isolated from *Poncirus trifoliata*, enhances dehydration and drought tolerance in tobacco via scavenging ROS and modulating expression of stress-responsive genes. *BMC Plant Biol.* **2010**, *10*, 230. [CrossRef]
48. Chen, N.; Qin, J.; Tong, S.; Wang, W.; Jiang, Y. One AP2/ERF Transcription Factor Positively Regulates Pi Uptake and Drought Tolerance in Poplar. *Int. J. Mol. Sci.* **2022**, *23*, 5241. [CrossRef]
49. Jia, T.; Hou, J.; Iqbal, M.Z.; Zhang, Y.; Cheng, B.; Feng, H.; Li, Z.; Liu, L.; Zhou, J.; Feng, G.; et al. Overexpression of the white clover TrSAMDC1 gene enhanced salt and drought resistance in Arabidopsis thaliana. *Plant Physiol. Biochem.* **2021**, *165*, 147–160. [CrossRef]
50. Ran, J.; Shang, C.; Mei, L.; Li, S.; Tian, T.; Qiao, G. Overexpression of CpADC from Chinese cherry (*Cerasus pseudocerasus* Lindl. 'Manaohong') promotes the ability of response to drought in *Arabidopsis thaliana*. *Int. J. Mol. Sci.* **2022**, *23*, 14943. [CrossRef]
51. Singh, D.; Laxmi, A. Transcriptional regulation of drought response: A tortuous network of transcriptional factors. *Front. Plant Sci.* **2015**, *6*, 895. [CrossRef]
52. Charfeddine, S.; Saıdi, M.N.; Charfeddine, M.; Gargouri-Bouzid, R. Genome-wide identification and expression profiling of the late embryogenesis abundant genes in potato with emphasis on dehydrins. *Mol. Biol. Rep.* **2015**, *42*, 1163–1174. [CrossRef]
53. Liu, J.H.; Nada, K.; Honda, C.; Kitashiba, H.; Wen, X.P.; Pang, X.M.; Moriguchi, T. Polyamine biosynthesis of apple callus under salt stress: Importance of the arginine decarboxylase pathway in stress response. *J. Exp. Bot.* **2006**, *57*, 2589–2599. [CrossRef] [PubMed]
54. Mo, H.J.; Sun, Y.X.; Zhu, X.L.; Wang, X.F.; Zhang, Y.; Yang, J.; Yan, G.J.; Ma, Z.Y. Cotton S-adenosylmethionine decarboxylase-mediated spermine biosynthesis is required for salicylic aci -and leucine-correlated signaling in the defense response to Verticillium dahliae. *Planta* **2016**, *243*, 1023–1039. [CrossRef] [PubMed]
55. Duvaud, S.; Gabella, C.; Lisacek, F.; Stockinger, H.; Ioannidis, V.; Durinx, C. Expasy, the Swiss Bioinformatics Resource Portal, as designed by its users. *Nucleic Acids Res.* **2021**, *49*, W216–W227. [CrossRef] [PubMed]
56. Horton, P.; Park, K.J.; Obayashi, T.; Fujita, N.; Harada, H.; Adams-Collier, C.J.; Nakai, K. WoLF PSORT: Protein localization predictor. *Nucleic Acids Res.* **2007**, *35*, W585–W587. [CrossRef]
57. Tamura, K.; Stecher, G.; Kumar, S. MEGA11: Molecular Evolutionary Genetics Analysis version 11. *Mol. Biol. Evol.* **2021**, *38*, 3022–3027. [CrossRef] [PubMed]
58. Nguyen, L.T.; Schmidt, H.A.; von Haeseler, A.; Minh, B.Q. IQ-TREE: A fast and effective stochastic algorithm for estimating maximum-likelihood phylogenies. *Mol. Biol. Evol.* **2015**, *32*, 268–274. [CrossRef] [PubMed]
59. Chen, C.; Wu, Y.; Li, J.; Wang, X.; Zeng, Z.; Xu, J.; Liu, Y.; Feng, J.; Chen, H.; He, Y.; et al. TBtools-II: A "one for all, all for one" bioinformatics platform for biological big-data mining. *Mol. Plant.* **2023**, *16*, 1733–1742. [CrossRef] [PubMed]

60. Zhang, W.; Xu, H.; Duan, X.; Hu, J.; Li, J.; Zhao, L.; Ma, Y. Characterizing the leaf transcriptome of *Chrysanthemum rhombifolium* (Ling et C. Shih), a drought resistant, endemic plant from China. *Front. Genet.* **2021**, *12*, 625985. [CrossRef]
61. Hu, J.; Jin, Q.; Ma, Y. AfLFY, a *LEAFY* homolog in *Argyranthemum frutescens*, controls flowering time and leaf development. *Sci. Rep.* **2020**, *10*, 1616. [CrossRef] [PubMed]
62. Li, J.; Xu, H.; Li, X.; Wang, L.; Wang, X.; Liu, Y.; Ma, Y. Molecular cloning and functional analysis of a *Chrysanthemum vestitum GME* homolog that enhances drought tolerance in transgenic tobacco. *Sci. Rep.* **2022**, *12*, 13551. [CrossRef] [PubMed]
63. Yaghoubian, I.; Modarres-Sanavy, S.A.M.; Smith, D.L. Plant growth promoting microorganisms (PGPM) as an eco-friendly option to mitigate water deficit in soybean (*Glycine max* L.): Growth, physio-biochemical properties and oil content. *Plant Physiol. Biochem.* **2022**, *191*, 55–66. [CrossRef] [PubMed]
64. Pan, Y.; Wu, L.; Yu, Z. Effect of salt and drought stress on antioxidant enzymes activities and SOD isoenzymes of liquorice (Glycyrrhiza uralensis Fisch). *Plant Growth Regul.* **2006**, *49*, 157–165. [CrossRef]
65. Kenneth, J.; Livak, T.D. Analysis of relative gene expression data using real-time quantitative PCR and the $2^{-\Delta\Delta CT}$ method. *Methods* **2001**, *25*, 402–408.

plants

Article

Genome-Wide Identification of the CYP716 Gene Family in *Platycodon grandiflorus* (Jacq.) A. DC. and Its Role in the Regulation of Triterpenoid Saponin Biosynthesis

Wuhua Zhang [1,2,†], Javed Iqbal [1,2,†], Zhihui Hou [1,2], Yingdong Fan [1,2], Jie Dong [1,2], Chengzhi Liu [1,2], Tao Yang [1,2], Daidi Che [1,2], Jinzhu Zhang [1,2,*] and Dawei Xin [3,*]

1 College of Horticulture and Landscape Architecture, Northeast Agricultural University, Harbin 150030, China; 15346549367@163.com (W.Z.); iqbal468@aup.edu.pk (J.I.); 18634445324@163.com (Z.H.); fanying_dong@163.com (Y.F.); jiedong1521@163.com (J.D.); liuchengzhi1103@163.com (C.L.); yangtao@neau.edu.cn (T.Y.); daidiche@neau.edu.cn (D.C.)
2 Key Laboratory of Cold Region Landscape Plants and Applications, Harbin 150030, China
3 Key Laboratory of Soybean Biology in Chinese Ministry of Education, College of Agriculture, Northeast Agricultural University, Harbin 150030, China
* Correspondence: jinzhuzhang@neau.edu.cn (J.Z.); xdawei@163.com (D.X.)
† These authors contributed equally to this work.

Abstract: The main type of saponins occurring in the root of *Platycodon grandiflorus* (Jacq.) A. DC. are oleanolic acid glycosides. The *CYP716* gene family plays a major role in catalyzing the conversion of β-amyrin into oleanolic acid. However, studies on the *CYP716* genes in *P. grandiflorus* are limited, and its evolutionary history remains poorly understood. In this study, 22 *PgCYP716* genes were identified, distributed among seven subfamilies. *Cis*-acting elements of the PgCYP716 promoters were mainly involved in plant hormone regulation and responses to abiotic stresses. *PgCYP716A264*, *PgCYP716A391*, *PgCYP716A291*, and *PgCYP716BWv3* genes were upregulated in the root and during saponin accumulation, as shown by RNA-seq analysis, suggesting that these four genes play an important role in saponin synthesis. The results of subcellular localization indicated that these four genes encoded membrane proteins. Furthermore, the catalytic activity of these four genes was proved in the yeast, which catalyzed the conversion of β-amyrin into oleanolic acid. We found that the content of β-amyrin, platycodin D, platycoside E, platycodin D3, and total saponins increased significantly when either of the four genes was over expressed in the transgenic hair root. In addition, the expression of *PgSS*, *PgGPPS2*, *PgHMGS*, and *PgSE* was also upregulated while these four genes were overexpressed. These data support that these four PgCYP716 enzymes oxidize β-amyrin to produce oleanolic acid, ultimately promoting saponin accumulation by activating the expression of upstream pathway genes. Our results enhanced the understanding of the functional variation among the *PgCYP716* gene family involved in triterpenoid biosynthesis and provided a theoretical foundation for improving saponin content and enriching the saponin biosynthetic pathway in *P. grandiflorus*.

Keywords: *P. grandiflorus*; triterpenoid saponins; gene family analysis; CYP716; oleanolic acid

1. Introduction

Platycodon grandiflorus (Jacq.) A. DC. is a perennial herbaceous plant of the Campanulaceae family. It is an economically important plant with ornamental, medicinal, and edible values [1]. It is a traditional herb in Chinese, Korean, and Japanese medicine. Its roots are rich in various bioactive compounds such as saponins, polysaccharides, and flavonoids. Among them, saponins, the primary bioactive components of *P. grandiflorus*, exhibit antitumor and hepatoprotective properties [2–5]. In recent years, the unique role of traditional Chinese medicines, including *P. grandiflorus*, in treating COVID-19 has gradually led to their increased demand and status [6].

Triterpenoid saponins are glycosides consisting of sapogenins combined with sugars. Most triterpenoids discovered so far are tetracyclic and pentacyclic triterpenoids, with the most common structure being the dammarane-type and oleanane-type [7]. Tetracyclic triterpenes are widely distributed in nature; ginsenosides and gibberellin are tetracyclic triterpenoids [8]. Pentacyclic saponins are more commonly observed in Chinese herbs such as *Glycyrrhiza uralensis*, *Bupleurum chinense*, and *Polygala tenuifolia* [9]. The saponins are mainly oleanane-type pentacyclic triterpenoids in *P. grandiflorus*, of which the main sapogenins is oleanolic acid. Plants mainly derive triterpenoids from the cytosolic mevalonic acid pathway (MVA), with a smaller contribution from the plastidial methylerythritol phosphate pathway (MEP) [10]. Farnesyl pyrophosphate (FPP) is derived from the precursor isopentenyl diphosphate (IPP) through a series of enzymatic reactions. With the assistance of key enzymes such as squalene epoxidase (SE) and farnesyl-diphosphate synthase (FPPS), Farnesyl-PP is transformed into 2,3-oxidosqualene. Subsequently, 2,3-oxidosqualene is cyclized by 2,3 oxidosqualene cyclase (OSC) to yield β-amyrin [11–13]. Further, β-amyrin is modified by cytochrome P450 to form oleanolic acid [14].

Cytochrome P450 is a class of heme-based B-type cytochrome superfamily protein that extensively participates in the biosynthesis of specialized metabolites (e.g., terpenes, plant hormones, and alkaloids) in plants [15]. The main enzymes involved in catalyzing the synthesis of oleanolic acid from β-amyrin are from the CYP716 family of the cytochrome P450 superfamily [16]. MtCYP716A12 is the first identified enzyme involved in the synthesis of pentacyclic triterpene saponins, which catalyzes the C-28 oxidation of β-amyrin into oleanolic acid [17]. With the development of sequencing technology, numerous CYP716 enzymes involved in catalyzing oleanolic acid formation have been identified and validated [18–20]. In *Kalopanax septemlobus* and *Chenopodium quinoa*, KsCYP716A94, CqCYP716A78, and CqCYP716A79 catalyze the C-28 oxidation of β-amyrin into oleanolic acid [21,22]. Meanwhile, five CYP716 enzymes (including PgCYP716A140, PgCYP716A141, PgCYP716D58, PgCYP716S7, and PgCYP716S6) were identified and validated in *P. grandiflorus* that oxidize β-amyrin at the C-28 position to produce oleanolic acid [18,23]. In addition to oxidizing β-amyrin at the C-28 position to produce oleanolic acid, some CYP716s catalyze triterpene oxidation at different positions, including C-6, C-2α, and C-12 [24–26]. For example, PgCYP716A47 catalyzes the C-12 hydroxylation of dammarenediol-II to form protopanaxadiol [27]. Subsequently, under the catalysis of CYP716A52v2, protopanaxadiol is oxidized to produce the downstream product, protopanaxatriol [28].

P. grandiflorus is an important traditional Chinese medicinal plant with a wide cultivation area [29]. Oleanolic acid, a pentacyclic oleanane-type triterpene, is the main active ingredient in *P. grandiflorus* [30]. The CPY716 family plays a major role in catalyzing the oxidation of β-amyrin into oleanolic acid. However, a systematic understanding of the *CYP716* gene family and its evolutionary history is lacking. In this study, a total of 22 *PgCYP716* genes were identified, distributed among seven subfamilies. Analysis of the *cis*-acting elements of *PgCYP716* genes revealed that they were mainly involved in plant hormone regulation and responses to abiotic stresses. The *PgCYP716A264*, *PgCYP716A391*, *PgCYP716A291*, and *PgCYP716BWv3* genes were cloned for subsequent experiments. The results of subcellular localization, yeast heterologous expression, and transgenic hairy root transformation indicated that all four enzymes were membrane proteins and oxidized β-amyrin to produce oleanolic acid. This study enhanced the understanding of the saponin synthesis pathway, expanding the gene pool for saponin biosynthesis and accelerating molecular breeding. Moreover, this study provided genetic resources for the cultivation of high-efficiency and high-quality varieties of *P. grandiflorus*.

2. Results

2.1. Identification of PgCYP716 Gene Family

Following the screening of the *P. grandiflorus* genome, 22 PgCYP716 genes were identified and their amino acid sequences were subsequently forwarded to the CYP450 Family Committee for nomenclature. The P450 family contained the complete P450 helix K region sequence FXXXR and the heme binding motif FXXGXXXCXG/A [31]. Multiple sequence alignment results indicated that all 22 genes contained the complete P450 helix K region sequence and the heme binding motif (Figure 1A). These 22 genes belonged to seven subfamilies, with the CYP716A subfamily containing 11 genes and the CYP716S subfamily containing six genes. The remaining subfamilies were CYP716BV, CYP716BW, CYP716D, CYP716U, and CYP716BWv, each containing only one gene. The length of the coding sequence of the 22 *PgCYP716* genes ranged from 288 to 1656 bp, with *PgCYP716BW2* being the longest, encoding 552 amino acids with a MW of 62,701 Da. PgCYP716A was the shortest, encoding 96 amino acids with a MW of 10,995.45 Da. The pI of the proteins within the PgCYP716s exhibited a wide range, with PgCYP716S5 having the lowest pI at 5.88, while PgCYP716A140 displayed the highest pI at 9.17. This variation in pI values among different members of the PgCYP716s highlighted the diverse biochemical properties and potential functional roles of these proteins (Supplementary Table S1).

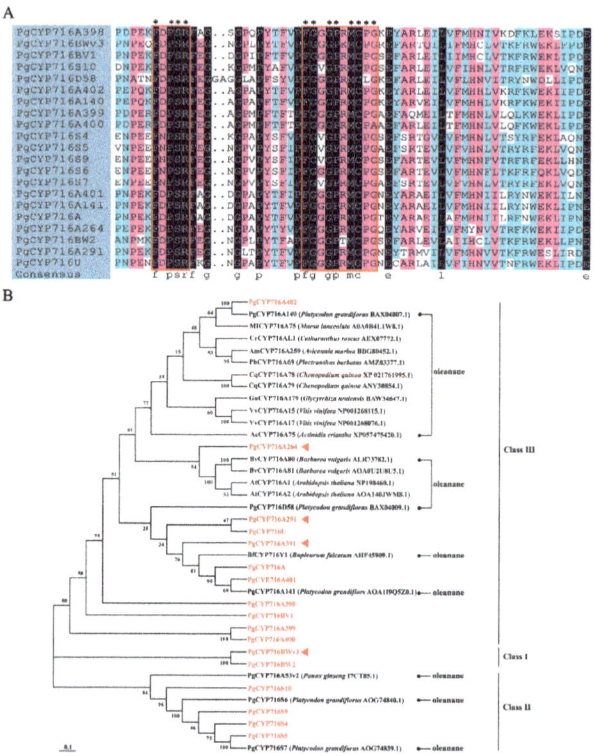

Figure 1. Identification of *PgCYP716s* in *P. grandiflorus*. (**A**) Multiple sequence alignment of the CYP450 domain of the PgCYP716 gene family. The asterisk denoted identical amino acids. (**B**) Phylogenetic tree of CYP716 proteins. The evolutionary tree was constructed based on the complete amino acid sequences of CYP716 proteins via MEGA 11 with the Maximum likelihood method. Bootstrap = 1000. CYP716s identified in this research are highlighted in red, while the filled arrowheads signify those CYP716s that have been functionally characterized in this study.

2.2. Evolutionary Analysis of the PgCYP716 Proteins in Different Species

In order to enhance our comprehension of the evolutionary connections within the PgCYP716 gene family, we performed a multiple sequence alignment of the 22 identified PgCYP716 amino acid sequences from *P. grandiflorus*. Further, 16 PgCYP716 amino acid sequences from 11 other plants, which oxidized β-amyrin to produce oleanolic acid, were used to construct an evolutionary tree. CYP716 proteins could be categorized into three major classes. Class I mainly included the PgCYP71BW and PgCYP716BWv subfamilies, and Class II mainly included the CYP716S subfamily from *P. grandiflorus* and PgCYP716A53v2 from *P. ginseng*. This suggested that the PgCYP716S subfamily may have the same function as the PgCYP716A53v2 of *Panax ginseng* in catalyzing the conversion of protopanaxatriol into protopanaxadiol. The remaining PgCYP716A, PgCYP716D, PgCYP716U, and PgCYP716BV subfamilies were grouped together with the CYP716A family of other species. This indicated that these four subfamilies, along with the majority of the CYP716A family, had the function of synthesizing oleanolic acid (Figure 1B). This analysis provided valuable insights into the evolutionary relationships among the CYP716 gene family in *P. grandiflorus*, establishing a basis for future explorations into the functional properties of this gene family.

2.3. Analysis of Conserved Motif and Structure of the PgCYP716 Genes

Plant genes often contain a high frequency of interruptions in the form of introns or exons, offering valuable insights into the evolutionary relationships within gene families. Additionally, the presence of diverse motif compositions can contribute to the functional versatility of genes [32,33]. This study integrated phylogenetic trees, gene structure diagrams, and motif analysis to evaluate the evolutionary relationships and gene structures of the PgCYP716 family. The PgCYP716A had only one motif (motif1). PgCYP716A291 and PgCYP716U had four motifs (motif1, motif2, motif3, and motif6). PgCYP716S4, PgCYP716S5, and PgCYP716S7 in the PgCYP716S subfamily contained five motifs (motif1, motif2, motif3, motif4, and motif6), suggesting that they were evolutionarily more closely related and may have similar functions. PgCYP716A391 and PgCYP716A398 had six and nine motifs, respectively. All the remaining proteins contained 10 motifs (Figure 2B). The gene structure analysis showed that the PgCYP716 family genes had varying numbers of exons, from two to six. Specifically, four genes in this family were found to be devoid of introns, containing only a single exon. (Figure 2C). Meanwhile, among the 10 motifs, some variation in the length of different motifs was observed, suggesting some differences in the functions (Figure 2D).

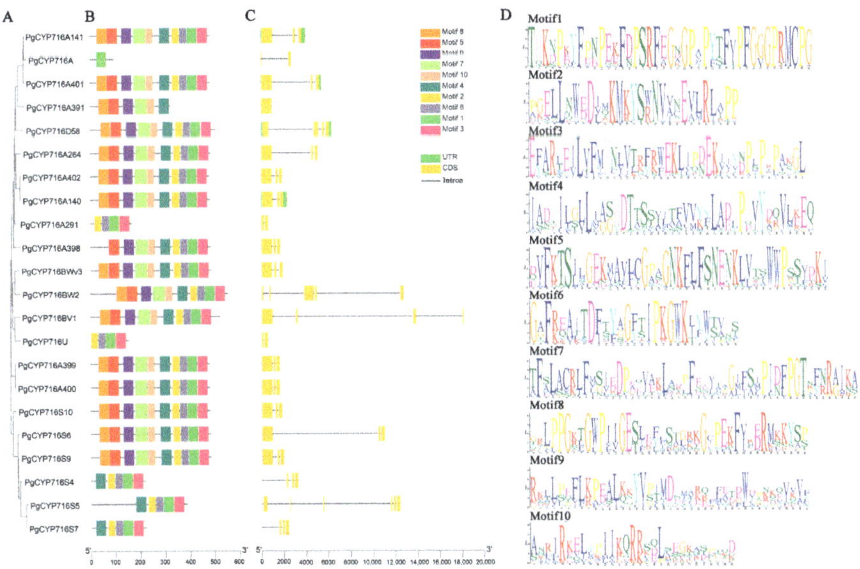

Figure 2. Phylogenetic tree and gene structure analysis of 22 *PgCYP716* genes in *P. grandiflorus*. (**A**) The construction of a phylogenetic tree for the *PgCYP716* gene family. (**B**) The distribution of 10 conserved domains in the PgCYP716. A total of 10 motifs were identified, and each color represents one motif. (**C**) Gene structure analysis of *PgCYP716s*. (**D**) Conserved structural motifs of PgCYP716 proteins. Different colors represented different amino acids.

2.4. Analysis of Chromosomal Localization and Duplication of the PgCYP716 Genes

By utilizing the genome annotation data of *P. grandiflorus*, our study delved into the precise mapping of *PgCYP716* genes across the chromosomes. The 22 *PgCYP716* genes were mainly distributed on five chromosomes in an uneven manner, with no genes present on chromosomes 3, 4, 6, and 7. Chromosome 8 had the maximum 10 genes, followed by five genes on chromosome 1, three genes each on chromosomes 5 and 9, and one gene on chromosome 2 (Figure 3A).

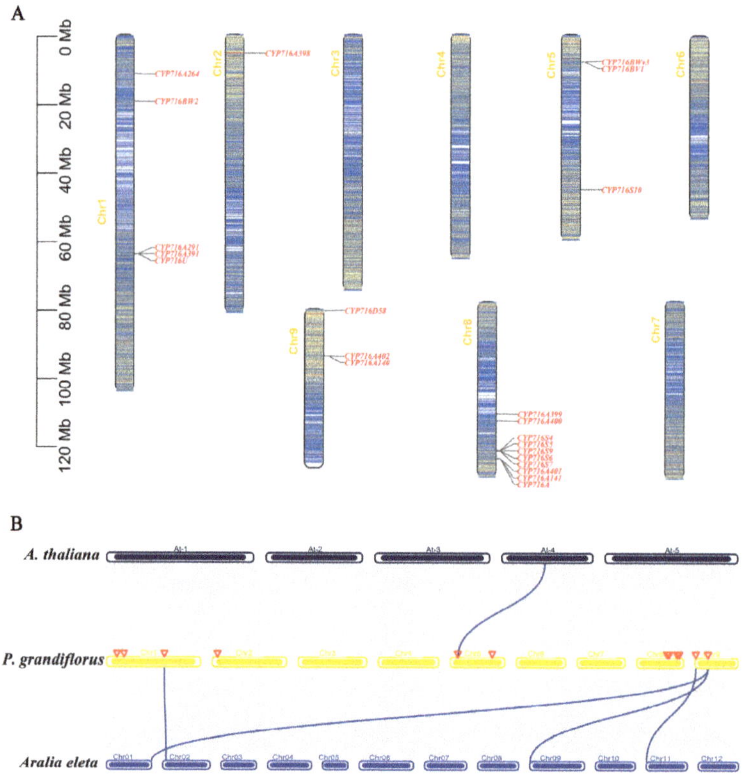

Figure 3. Chromosome localization and collinearity analysis of *PgCYP716* gene family in *P. grandiflorus*. (**A**) Chromosomal distribution of 22 *PgCYP716* genes in *P. grandiflorus*. (**B**) The collinearity analysis of the CYP716 gene family in *P. grandiflorus*, *A. eleta*, and *A. thaliana*. The red triangle represented the chromosomal position of *PgCYP716s*.

Gene duplication plays a crucial role in evolutionary dynamics by providing a fundamental mechanism for the expansion and diversification of gene families. This process entails the replication of individual genes within the genome of an organism, resulting in an augmentation of the gene pool through the generation of additional gene copies. The phenomenon of gene duplication underpins the enhancement of genetic diversity, the emergence of novel functions, and the evolution of complex biological systems across various species. The expansion of gene families through duplication events not only contributes to genetic diversity but also provides the raw material for evolutionary innovations and adaptations. To evaluate the evolutionary relationship of the CYP716 genes in various species, collinearity analysis was performed between *P. grandiflorus* and four plants (*P. pseudoginseng*, *P. ginseng*, *A. eleta*, and *A. thaliana*). No collinearity was observed between the CYP716 genes in *P. grandiflorus* and those in the traditional Chinese medicines *P. pseudoginseng* and *P. ginseng*. However, collinearity was observed with those in *A. eleta* and *A. thaliana* (Figure 3B). This indicated that the CYP716 genes in *P. grandiflorus* exhibited relatively distant relationships with *P. pseudoginseng* and *P. ginseng* and closer relationships with model plant *A. thaliana* and a woody plant *A. eleta*.

2.5. Analysis of Cis-Acting Element in the Promoter of PgCYP716s

Cis-acting elements were mainly involved in the regulation of gene expression. In order to delve deeper into the response of *PgCYP716* genes to challenging conditions, we utilized PlantCARE to identify cis-acting elements from the 2.0 kb 5′ upstream regions of *PgCYP716s*. The *cis*-acting elements present in the PgCYP716 promoters were predominantly associated with the regulation of plant hormones and responses to external environmental stresses (Figure 4). In terms of abiotic stresses, the cis-acting element of the PgCYP716 promoters was observed to be mainly involved in the response to cold stress and drought stress. Additionally, five types of hormone-regulation-related elements were found, including those related to gibberellin, methyl jasmonate, auxin, salicylic acid, and abscisic acid-responsiveness. These findings provide additional evidence that *PgCYP716s* plays a crucial role in the response of *P. grandiflorus* to hormonal signals and environmental stressors.

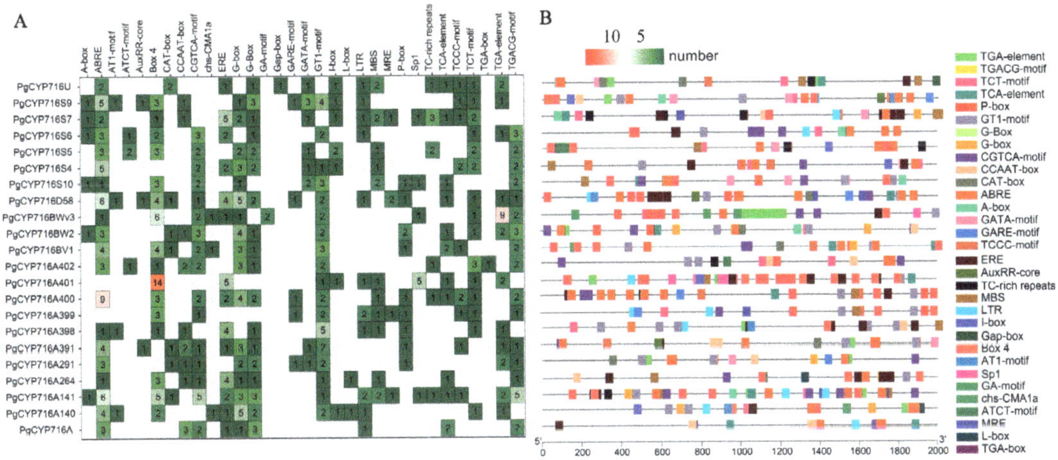

Figure 4. Analysis of cis-acting element in the promoter of *PgCYP716s*. (**A**) The number of different cis-acting elements in the *PgCYP716* gene promoter. (**B**) The positional distribution of different cis-acting elements in the *PgCYP716* gene promoter.

2.6. RNA-seq Analysis of PgCYP716s Gene Expression

The content of specialized metabolites changed with the growth cycle of the plant. In order to explore the expression patterns of the *PgCYP716* genes during different periods of saponin accumulation, the expression profiling of *PgCYP716* genes was further investigated based on the transcriptome data from our laboratory. Through analyzing the transcriptome data at different stages of saponin accumulation, the 22 *PgCYP716* genes were roughly divided into two major categories based on similar expressions. One category, represented by genes such as *PgCYP716A140* and *PgCYP716BWv3*, was mainly upregulated in the early stages of saponin accumulation. The other category mainly upregulated in the middle and late stages of saponin accumulation and included genes such as *PgCYP716A391* and *PgCYP716A291* (Figure 5A). This indicated that the *PgCYP716* genes may participate in saponin synthesis at all stages of saponin accumulation.

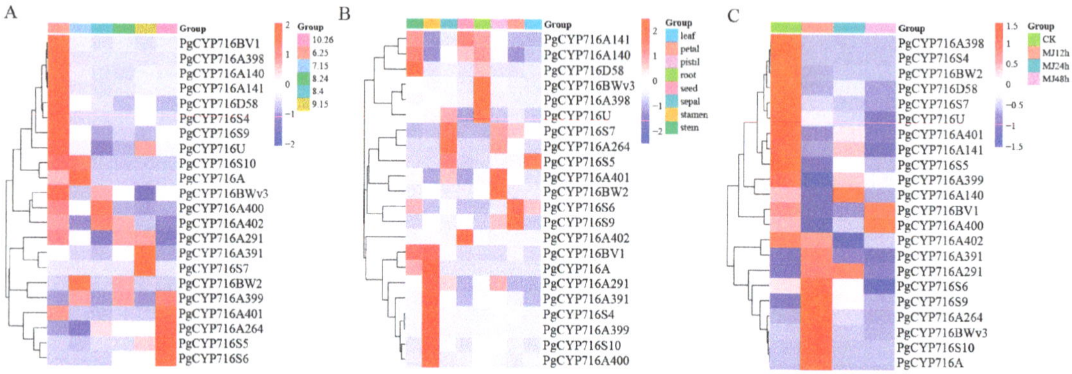

Figure 5. The expression of *PgCYP716s* genes based on RNA-seq. (**A**) The expression of *PgCYP716s* at different levels of saponin accumulation. (**B**) The changes in the expression of *PgCYP716s* in different tissues. (**C**) The changes in the expression of *PgCYP716s* under MeJA treatment.

To evaluate the transcript level of the *PgCYP716* genes in various tissues, the transcriptome data of eight tissues (petal, root, pistil, leaf, flower, seed, stamen, and stem) were downloaded from the SRA database. The results indicated that the 22 genes were expressed in all tissues and could be categorized into three groups based on the expression pattern. The expression of 22 *PgCYP716* genes was the lowest in the roots compared with other tissues. The genes in the first category were mainly upregulated in the root and stem, including *PgCYP716A141* and *PgCYP716BWv3*. The genes in the second category had high degree of expression in the seed, petal, and leaf and included *PgCYP716S6* and *PgCYP716A401*. The genes in the third category possessed a high level of expression in the stamen and included *PgCYP716S10* (Figure 5B). These findings indicated that the *PgCYP716* genes exhibit tissue-specific expression in *P. grandiflorus*.

Analysis of cis-acting elements of the *PgCYP716s* promoter revealed that most of the *PgCYP716* promoters contained a large number of MeJA cis-responsive elements (Figure 5C). To further clarify whether the *PgCYP716* gene responds to MeJA, the transcriptome data downloaded from SRA database were analyzed. The result indicated that most genes could be induced by MeJA, with *PgCYP716A264*, *PgCYP716BWv3*, *PgCYP716A291*, and *PgCYP716A391* getting significantly upregulated at 12–24 h. The expression of *PgCYP716A402* and *PgCYP716BW2* increased at 0 and 12 h and decreased at 24 and 48 h. A class of genes, represented by *PgCYP716A400*, exhibited upregulation mainly at 48 h after induction by MeJA (Figure 5C). These results indicated that the *PgCYP716* genes could respond to MeJA in *P. grandiflorus*.

2.7. Cloning and Expression Analysis of the PgCYP716 Genes

RNA sequencing analysis and the sequence similarity of PgCYP716s revealed that the *PgCYP716A264*, *PgCYP716A391*, *PgCYP716A291*, and *PgCYP716BWv3* genes may play an important role in saponin synthesis. *PgCYP716A264*, *PgCYP716A391*, *PgCYP716A291*, and *PgCYP716BWv3* genes were cloned for further research. The subcellular localization experiments indicated that these genes encoded membrane proteins (Figure 6A). Tissue-specific expression data indicated that these genes were expressed at low levels in the seeds compared to other tissues. *PgCYP716A264* and *PgCYP716BWv3* were mainly significantly upregulated in the roots. *PgCYP716A391* was predominantly expressed in the stem, followed by the leaf and stem. The upregulation of *PgCYP716A291* was the highest in the pistil, followed by the stamens and petal (Figure 6B). qRT-PCR analysis of the four genes under MeJA induction revealed that the transcript level of the *PgCYP716BWv3* gene was notably inhibited after MeJA induction. The expression level of *PgCYP716A264* and

PgCYP716A291 reached the highest at 12 h, and *PgCYP716A391* reached the highest at 24 and 48 h (Figure 6C).

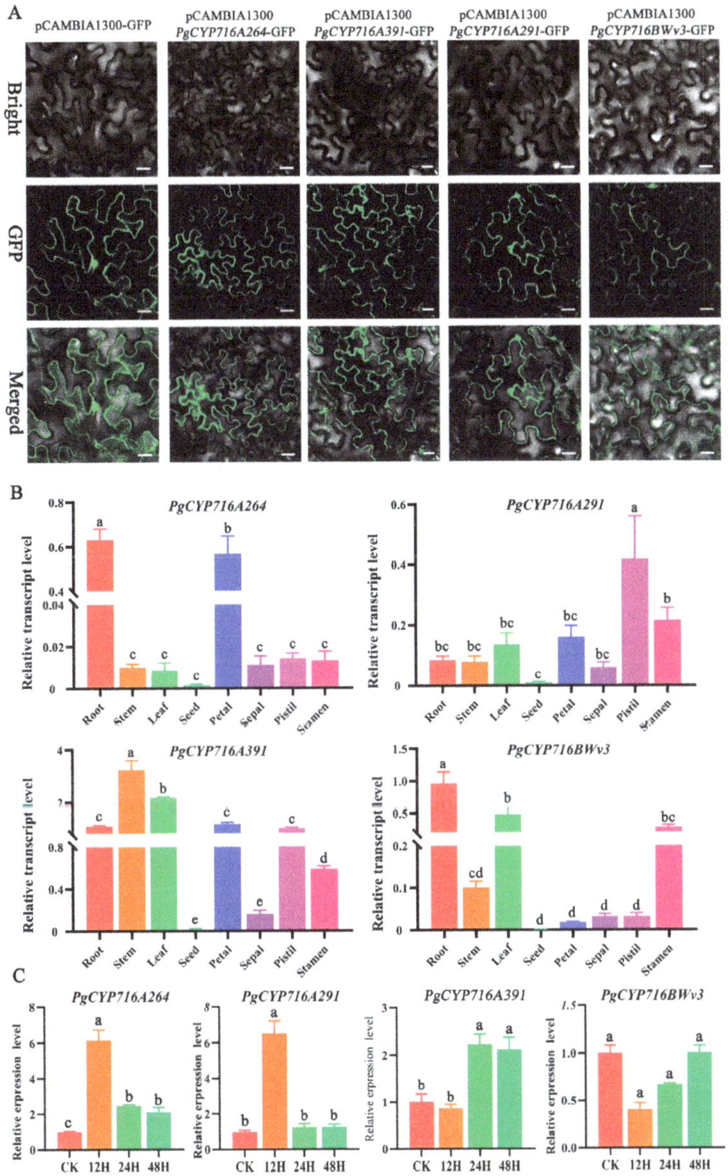

Figure 6. The subcellular localization and expression analysis of *PgCYP716A264*, *PgCYP716A391*, *PgCYP716A291*, and *PgCYP716BWv3*. (**A**) *PgCYP716A264*, *PgCYP716A391*, *PgCYP716A291*, and *Pg-CYP716BWv3* genes encoded membrane proteins. The top, middle, and bottom represent bright, GFP, and merged fields, respectively. (**B**) Expression levels of the four genes in eight tissues. (**C**) Expression levels of the four genes under MeJA induction. Different letters indicated significant differences ($p < 0.05$). Bar = 20 μm.

2.8. The Functional Verification of PgCYP716 Genes in Yeast

To verify whether these four PgCYP716 enzymes catalyze the production of oleanolic acid from β-amyrin, the yeast expression vectors of the pESC-*PgbAS1*-URA and pESC-*PgCYP716s*-TRP were constructed and transformed into the WAT11 yeast strain (Figure 7B). The metabolites isolated from yeast culture exhibited a color reaction in the vanillin-perchloric acid assay, indicating the presence of oleanolic acid (Figure 7A). Simultaneously, the metabolites isolated from yeast culture of the four PgCYP716s were concurrently analyzed using HPLC, revealing the presence of oleanolic acid (Figure 7C). These data indicated that all four PgCYP716 enzymes catalyzed the conversion of β-amyrin into oleanolic acid (Figure 7D).

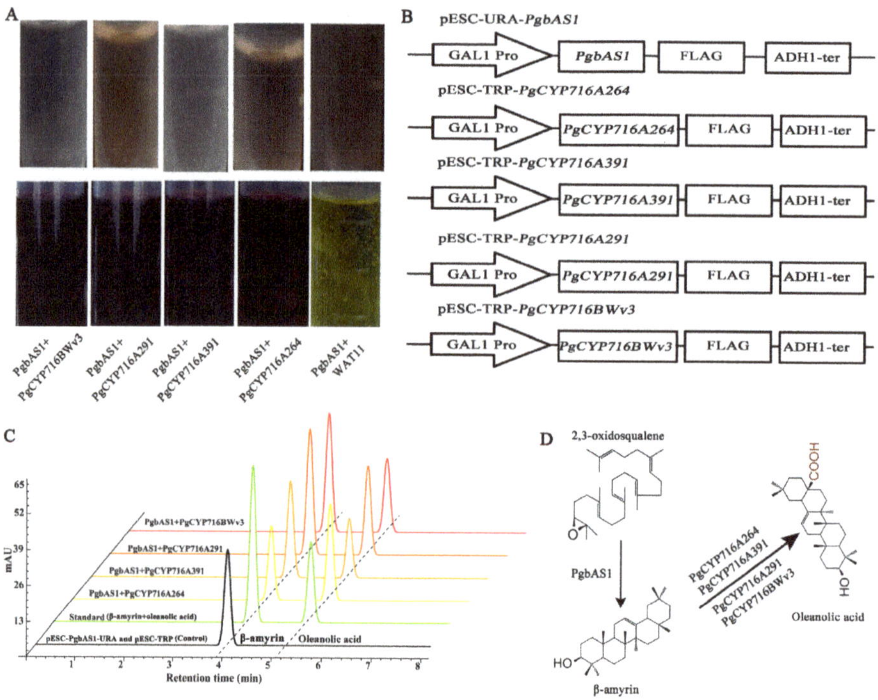

Figure 7. In vivo enzymatic activity assay in yeast. (**A**) Detection of the yeast products using the vanillin-perchloric acid method indicated that the four enzymes could catalyze the production of oleanolic acid from β-amyrin. The top figure shows yeast-induced products, and the bottom figure shows oleanolic acid detected using the vanillin-perchloric acid method. (**B**) Schematic representation of the construction of PgbAS1 and the yeast heterologous expression vectors with four PgCYP716s. (**C**) HPLC analysis of yeast cultures indicated that PgCYP716A264, PgCYP716A391, PgCYP716A291, and PgCYP716BWv3 could produce oleanolic acid. (**D**) A summary of the biochemical reactions catalyzed by four PgCYP716 enzymes.

2.9. PgCYP716s Overexpression Elevated the Saponin Content

Yeast heterologous expression analysis revealed that all four enzymes catalyzed the conversion of β-amyrin into oleanolic acid. To further confirm the function of these enzymes in *P. grandiflorus* ontogeny, their functions were validated using the hairy root transformation established in our laboratory [34]. The transgenic hairy roots of four *PgCYP716* genes were generated via *Agrobacterium*-mediated transformation (Figure 8A; Supplementary Figure S1). The metabolite content of the transgenic hairy roots was evaluated. The four genes could increase the content of β-amyrin, total saponin, platycoside E, platycodin D3, and platycodin

D (Figure 8B–E). Oleanolic acid content in the transgenic hairy roots with overexpressed *PgCYP716A264*, *PgCYP716A391*, and *PgCYP716A291* exhibited an increase, whereas that in the transgenic hairy roots with overexpressed *PgCYP716BWv3* exhibited a decreasing trend (Figure 8E). The macromolecules, namely, soluble sugars and starch, were the key upstream substrates for the synthesis of saponins. In the four transgenic hairy roots, the content of soluble sugars and starch exhibited a decreasing trend compared with the control group. This indicated that in the process of producing downstream specialized metabolites, the upstream macromolecules such as soluble sugars and starch underwent degradation and transformation, leading to reduced content (Figure 8B–E). qRT-PCR was performed to detect the expression levels of functional genes in the saponin synthesis pathway (*PgSS*, *PgGPPS2*, *PgMCS*, *PgHMGS*, and *PgSE*). The results showed that the overexpression of *PgCYP716A264*, *PgCYP716A391*, and *PgCYP716A291* and *PgCYP716BWv3* genes could promote the expression of saponin synthesis pathway genes (Figure 8F–I). These data revealed that four PgCYP716 enzymes participated in catalyzing the production of oleanolic acid from β-amyrin, ultimately contributing to the accumulation of saponin content.

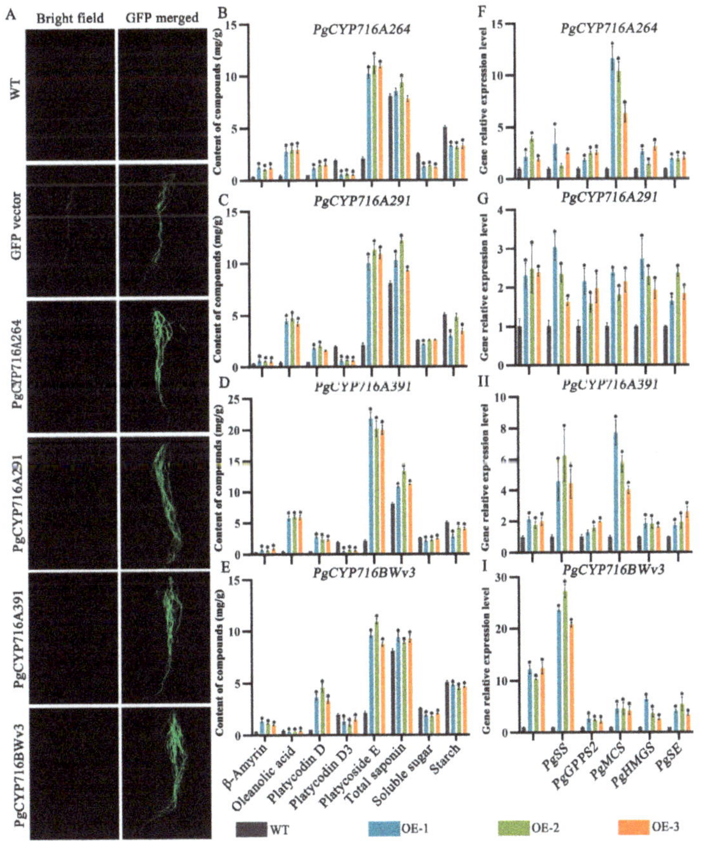

Figure 8. *PgCYP716A264*, *PgCYP716A391*, *PgCYP716A291*, and *PgCYP716BWv3* genes were functionally validated in the hairy roots of *P. grandiflorus*. (**A**) GFP fluorescence was detected in the hairy roots overexpressing *PgCYP716A264*, *PgCYP716A391*, *PgCYP716A291*, and *PgCYP716BWv3* genes. (**B–E**) Content of platycoside E, platycodin D3, β-amyrin, oleanolic acid, platycodin D, total saponin, soluble sugar, and starch in the WT and transgenic hairy roots. (**F–I**) qRT-PCR analysis of the expression of saponin synthesis pathway genes in overexpressed hairy roots. Asterisks indicated significant differences ($p < 0.05$).

3. Discussion

CYP716A is the major subfamily that exhibits catalytic activity in the conversion of β-amyrin to oleanolic acid [35]. Ever since CYP716A12 was first described as a triterpenoid-oxidizing enzyme [17], the CYP716 family has been systematically identified and named in plants such as *Aralia elata* and has been reported to play complex roles in different plants [36]. *P. grandiflorus* contains more than 70 saponin monomers, which mainly belong to oleanane-type saponins [29]. However, systematic identification and nomenclature for the CYP716 family has not been performed. A total of 22 *PgCYP716* genes were identified and classified into seven subfamilies in this study. Multiple sequence alignment results indicated that all 22 genes contained the complete P450 helix K region sequence and heme binding motif (Figure 1A). Through the analysis of the evolutionary tree, the CYP716U, CYP716D, and CYP716BV subfamilies were clustered with the CYP716A subfamily, which catalyzes the synthesis of oleanolic acid from β-amyrin (Figure 1B). It was hypothesized that these three gene families also have the same catalytic function. In addition, the *PgCYP716s* subfamily was clustered with PgCYP716A53v2 from *P. ginseng* (Figure 1B). The function of PgCYP716A53v2 is to catalyze the formation of protopanaxatriol from protopanaxadiol. Therefore, the PgCYP716S subfamily in *P. grandiflorus* may be involved in catalyzing the formation of protopanaxatriol from protopanaxadiol (Figure 1B) [28]. *P. ginseng, P. grandiflorus*, and *P. quiquefolium* are traditional Chinese medicinal plants, with saponins as their main active ingredients [37]. No collinearity was observed between the *CYP716* genes in *P. grandiflorus* and those in the traditional Chinese medicines *P. pseudoginseng* and *P. ginseng*; however, collinearity was observed in the model plant A. thaliana and a woody plant *A. eleta* (Figure 3B), presumably due to the different types of saponin compounds.

Various abiotic stresses such as drought, temperature fluctuations, UV radiation, and heavy metals are common challenges in plant growth environments. These environmental factors enhance plant resistance to external stresses by regulating the levels of phytohormones (e.g., methyl jasmonate (MeJA), abscisic acid (ABA), and salicylic acid (SA)) in plants. Phytohormones, as important signaling molecules, play a key role in regulating the synthesis of bioactive compounds in medicinal plants and are also one of the important mechanisms by which external biotic and abiotic stresses promote the accumulation of secondary metabolites in medicinal plants [38]. Analysis of the *cis*-acting elements of the promoter of the *PgCYP716* genes revealed a large number of abiotic stress responses, including cold and drought stress, and hormone response elements, such as MeJA and SA (Figure 4). The analysis of the expression of *PgCYP716s* under MeJA treatment indicated that MeJA could induce the expression of *PgCYP716s* (Figure 5C; Figure 6C). In addition to abiotic stresses, transcription factors play an important role in regulating the biosynthesis of active ingredients in medicinal plants [39]. In *Artemisia annua*, the activation of *AaMYB15*, an R2R3-MYB transcription factor, was triggered by darkness and MeJA treatment. Moreover, *AaMYB15* suppresses the biosynthesis of artemisinin by directly interacting with the promoter of *AaORA* [40]. The promoters of *PgCYP716s* contained a large number of MYB and MYC transcription factor binding sites, suggesting that MYB and MYC transcription factors may affect the accumulation of saponins by regulating the transcription level of *PgCYP716s* (Figure 4).

Five *PgCYP716* genes were identified and functionally verified in *P. grandiflorus*; they catalyze the synthesis of oleanolic acid from β-amyrin (Figure 1B). For more comprehensive understanding of the functions of the remaining genes in this family, we cloned the remaining 17 genes of this family, successfully isolating 11 genes. We speculated that the genomic heterozygosity of *P. grandiflorus* is too high or repetitive sequences may be folded during the assembly process, leading to assembly errors and affecting the gene cloning [35,41,42]. Based on the sequence similarity of PgCYP716s and the gene expression changes under MeJA treatment and different saponin accumulation periods, the function of *PgCYP716A264*, *PgCYP716A391*, *PgCYP716A291*, and *PgCYP716BWv3* genes were further studied. Similar to the majority of CYP450 proteins, subcellular localization analysis revealed that the four PgCYP716 genes in *P. grandiflorus* encoded membrane proteins (Figure 6A) [36,43]. Yeast heterologous expression revealed that PgCYP716A264, PgCYP716A391, PgCYP716A291, and PgCYP716BWv3 oxidize β-amyrin to produce oleanolic acid (Figure 7). This indicated that the CYP716BWv subfamily may also play an important role in catalyzing the production of oleanolic acid from β-amyrin [44]. Through hairy root transformation, transgenic and positive hairy roots of four genes were successfully obtained (Figure 8A; Supplementary Figure S1). Overexpression in hairy roots showed a notable enhancement in β-amyrin and total saponin levels in comparison to the wild-type (WT). Gene expression analysis further confirmed that the genes involved in the saponin biosynthesis pathway were up-regulated in the overexpressed hairy roots (Figure 8B–I). However, overexpression of the *PgCYP716BWv3* gene resulted in decreased oleanolic acid content (Figure 8E). This suggested that PgCYP716BWv3 has a continuous catalytic function, catalyzing the formation of other downstream substances after producing oleanolic acid and leading to its reduced content [23]. Platycodin D, platycodin D3, and platycoside E are the main saponin monomers in the *P. grandiflorus* root [1]. Though the contents of platycodin D and platycoside E increased in overexpressing hairy roots compared with WT, the content of platycodin D3 decreased (Figure 8B–E). This indicated that enzymes involved in later modifications (e.g., PgCYP450 and PgUGT) may contribute to the differences in the content of different saponin monomers.

4. Materials and Methods

4.1. Identification of the PgCYP716 Genes

The genome data of *P. grandiflorus*, CYP450 protein sequence of *A. thaliana* and annotation file of CYP450 domain (PF00067) were downloaded from the NGDC (https://ngdc.cncb.ac.cn/gwh, accessed on 15 May 2024), TAIR database (https://www.arabidopsis.org/, accessed on 15 May 2024), and Pfam database (http://pfam-legacy.xfam.org/, accessed on 15 May 2024), respectively [33,41]. Initially, the BioEdit was used to screen the genome of *P. grandiflorus* through BLASP search, using the AtCYP450 protein sequence as the query with a threshold of *E*-value $\leq 10^{-5}$ to acquire the initial set of candidate genes [45]. Subsequently, the HMM (PF00067) was applied to build the HMM profile against the *P. grandiflorus* protein dataset (*E*-value $\leq 10^{-10}$ and bits $\geq$ 85) to obtain a secondary set of candidate genes. After merging and removing duplicate sequences from both gene sets, the resulting amino acid sequences of the CYP450 family of *P. grandiflorus* were forwarded to the Cytochrome P450 Homepage (https://drnelson.uthsc.edu/, accessed on 5 May 2024) for nomenclature. Finally, the number of *PgCYP716* genes was determined. Additionally, the features of the PgCYP716 sequences were determined using the ExPasy website (https://web.expasy.org/protparam/, accessed on 15 May 2024).

4.2. Phylogenetic Analysis and Classification

The MUSCLE software (v 3.8.31) was employed to align the amino acid sequences from *P. grandiflorus* (Pg), *Vitis vinifera* (Vv), *Maesa lanceolata* (Ml), *Chenopodium quinoa* (Cq), *Barbarea vulgaris* (Bv), *Glycyrrhiza uralensis* (Gu), *Catharanthus roseus* (Cr), and *Bupleurum falcatum* (Bf). Subsequently, the phylogenetic tree was constructed via MEGA 11 (https://www.megasoftware.net/, accessed on 15 May 2024) with the Maximum likelihood method [46]. To enhance the stability of the tree, the Bootstrap parameter was set to 1000. Subsequently, the ggtree package was used to visualize the phylogenetic tree [47].

4.3. Analysis of the Gene Structure and Conserved Protein Motifs of PgCYP716 Genes

The MEME online tool (http://meme-suite.org/meme/tools/meme, accessed on 5 May 2024) was utilized to predict the motif positional distribution and quantity within the PgCYP716 protein sequences. The analysis was conducted with specified parameters, setting a maximum of 10 deviations and an optimal motif width ranging from six to 200 amino acid residues. TBtools (v 2.088) was used to analyze the intron–exon structure of the *PgCYP716* genes. Finally, the evolutionary tree, gene structure, and conserved domains of CYP716 were visualized using TBtools [48].

4.4. Analysis of Cis-Acting Elements in the Promotors of PgCYP716 Genes

To further explore the regulatory elements of the PgCYP716 gene family, the sequence 2000 bp upstream of the translational start codon of *PgCYP716* genes was extracted using the Seqkit (v0.11.0), bedtools (v2.30.0), and samtools packages (v1.6) as described in previous studies [49–51]. Subsequently, PlantCARE (http://www.plantcare.co.uk/, accessed on 15 May 2024) was used to predict cis-acting elements. Finally, cis-acting element positions and numbers were visualized using TBtools, tidyverse, ggplot2, and RColorBrewer packages [48,52].

4.5. Chromosome Localization and Gene Collinearity

The chromosome position of the *PgCYP716* genes were extracted based on the annotation file of the *P. grandiflorus* genome. The genomic data of *A. thaliana*, *P. notoginseng*, *P. ginseng*, and *Aralia elata* were downloaded from the Tair database (https://www.arabidopsis.org/, accessed on 15 May 2024) and the TCM Plant Genome Database (http://cbcb.cdutcm.edu.cn/TCMPG/, accessed on 15 May 2024). TBtools was used for collinearity analysis and visualization [48].

4.6. Expression Pattern of PgCYP716s Based on RNA-seq

The transcriptome data of eight different tissues (SRR8712510–SRR8712517) and the root treated with MeJA (SRR8712518–SRR8712529) were downloaded from the SRA database. HISAT2 (v2.2.1) and featureCounts (v2.0.1) were used to analyze the transcriptome data and obtain gene expression matrices [53,54]. Additionally, based on the existing transcriptome data of root saponin accumulation from June to October in our laboratory, the expression level of the *PgCYP716* genes in different periods of saponin accumulation was analyzed. Finally, heatmap package (v1.0.12) was used to construct the diagrams [55].

4.7. Treatment of P. grandiflorus Roots with MeJA

In this experiment, 30-day-old seedlings of *P. grandiflorus* were cultivated on 1/2 MS medium containing 30 g/L sucrose and 5 g/L agar (pH 5.8). The seedlings were further treated with 100 μM MeJA dissolved in dimethyl sulfoxide within a growth environment of 16 h light followed by 8 h of darkness, maintained at a temperature of 25 °C. The samples treated with MeJA were collected at 0, 12, 24, and 48 h.

4.8. Cloning of Full-Length of PgCYP716s and PgbAS1

Total RNA from *P. grandiflorus* was extracted using Trizol method and the RNA was converted into cDNA using reverse transcription kit (Takara, Beijing, China). The sequences of four *PgCYP716* genes and the *PgbAS1* sequence (Accession number: KY412556.1) were obtained from the genome database of *P. grandiflorus* and the NCBI database, respectively. The open reading frames of *PgbAS1* and *PgCYP716s* were amplified from the *P. grandiflorus* cDNA via PCR using KeyPo DNA polymerase (Vazyme Biotech Co., Ltd., Nanjing, China). Additionally, the DNA sequences were inserted into the pMD19-T vector (Takara, Beijing, China) for cloning, followed by sequencing to verify their accuracy and integrity. Primers used to amplify both *PgCYP716s* and *PgbAS1* genes were designed using Primer premier 5. The primers used in this step are all listed in Supplementary Table S1.

4.9. Quantitative Real-Time PCR

Plant RNA was acquired from eight different tissues (roots, stems, leaves, petals, stamens, pistils, petals, and seeds) and the root tissue treated with 100 μmol MeJA for 0, 12, 24, and 48 h. qRT-PCR was conducted using the BIO-Rad CFX96 instrument. The *18S rRNA* gene was used as the internal reference gene in this study [1]. All primers are listed in Supplementary Table S2.

4.10. Yeast Heterologous Expression

The full-length 2,3-oxidosqualene cyclase *PgbAS1* gene was inserted into pESC-URA vector, and the full-length of *PgCYP716* genes were inserted into the pESC-TRP vector. The engineered yeast strain Saccharomyces cerevisiae WAT11, which has been modified to incorporate the P450 reductase from *A. thaliana*, serves as an optimal heterologous host for the expression of P450 enzymes. This strain is equipped with the mevalonate (MVA) metabolic pathway, enabling it to synthesize the triterpenoid precursor 2,3-oxidosqualene [56]. To validate the function of the *PgCYP716s*, the pESC-*PgbAS1*-URA vector was used to transform WAT11, after which the cyclization of 2, 3-oxidosqualene was catalyzed to generate β-amyrin [57]. Further, the pESC-*PgCYP716s*-TRP vector was used to transform the WAT11 containing pESC-*PgbAS1*-URA as described previously [10]. pESC-*PgbAS1*-URA and pESC-TRP (the empty vector) were also co-transformed into WAT11 as the control. The transformed yeast was inoculated in 5 mL SD/-TRP-URA liquid medium and grown till the OD_{600} reached 1.5. Further, the yeast cells were harvested by centrifugation at 1000 rpm for 1 min, washed two times using sterile water, resuspended in 100 mL SC/-TRP-URA liquid medium, which added 2 g galactose, and cultured for 3 days in the dark at 28 °C to induce metabolite biosynthesis in yeast. Further, we use a mixture of acetone and ethyl acetate (1:1 v/v) to extract yeast metabolites as described previously [56]. The liquid was evaporated using a low-temperature concentration method, and the remaining residue was dissolved in 1.5 mL methanol for the next step of analysis. The metabolites isolated from yeast culture were detected using the vanillin-perchloric acid method and high-performance liquid chromatography (HPLC) [58,59]

4.11. Subcellular Localization

The full coding sequences of *PgCYP716* genes lacking a stop codon were inserted into the pCAMBIA1300-GFP vector at the *Bam*HI and *Kpn*I sites. Subsequently, the sequences were aligned with the genome sequences, and the correctly aligned sequences of pCAMBIA1300-*PgCYP716s*-GFP vector were used to transform *Agrobacterium tumefaciens* GV3101 by referring to the instructions from GV3101 Chemically Competent Cell (Weidibio, Shanghai, China). Cell suspensions with pCAMBIA1300-*PgCYP716s*-GFP vectors were transiently injected into 4-week-old tobacco leaves. After 2 days of incubation in the dark at 23 °C, the subcellular localization of the PgCYP716s was visualized and photographed using a laser-scanning confocal microscope at 488 nm (FV3000, Olympus, Tokyo, Japan).

4.12. Functional Verification of PgCYP716 Genes Using Hairy Root Transformation

The *PgCYP716* genes were integrated into the pCAMBIA1300-sGFP vector for overexpression and then introduced into the *A. tumefaciens* strain K599. After selecting single colonies, they were cultured in 100 mL LB medium, which contained 50 mg/L kanamycin and 200 μmol/L acetosyringone (AS), until reaching an OD_{600} of about 0.8. Subsequently, the cells were collected by centrifugation at 4000 rpm for 20 min, resuspended in infiltration solution (200 μmol/L AS, 10 mmol/L 2-Morpholinoethanesulphonic acid (MES), and 10 mmol/L $MgCl_2$). This cell suspension was used to infect 30-day-old seedlings without roots using the vacuum infiltration method as described in a previous study [34,60,61]. Further, the seedlings were cultivated in sterilized soil. Positive hairy roots were identified after 60 days and measured for the target gene expression and saponin content.

4.13. Statistical Analysis

All experiments were conducted with three biological replicates. Results are reported as mean ± standard error (SE). GraphPad Prism 8.0 (San Diego, CA, USA), Origin (Pro) (OriginLab Corporation, Northampton, MA, USA) and R were used to plot graphs. Asterisks denoted statistical significance using Student's t-test or one-way analysis of variance (ANOVA). Different letters indicate significant differences ($p < 0.05$).

5. Conclusions

A total of 22 *PgCYP716* genes from seven subfamilies were identified in this study. The PgCYP716 gene family was comprehensively analyzed through phylogenetic, collinearity, cis-acting element, and gene expression analyses in different tissues under MeJA treatment and at various developmental stages. Four genes, namely, *PgCYP716A264*, *PgCYP716A391*, *PgCYP716A291*, and *PgCYP716BWv3*, were finally cloned. Subcellular localization, yeast heterologous expression, and overexpression in hairy roots revealed that all four PgCYP716 enzymes were membrane proteins. All of them could catalyze the formation of oleanolic acid from β-amyrin, thereby increasing the content of saponin. This study laid the foundation for further understanding of the role of *PgCYP716* genes in saponin synthesis, providing suitable candidate genes for breeding varieties of *P. grandiflorus* that produce high levels of saponin.

Supplementary Materials: The following supporting information can be downloaded at: https://www.mdpi.com/article/10.3390/plants13141946/s1, Figure S1: Detection of transgenic hairy roots using agarose gel electrophoresis; Table S1: Prediction of physicochemical properties of PgCYP716 family proteins; Table S2: Primer design of qRT-PCR and vector construction.

Author Contributions: Investigation, W.Z. and J.I.; Writing-original draft, W.Z.; Data curation, W.Z., Y.F., C.L. and J.I.; Software, W.Z., J.I. and Z.H.; Visualization, T.Y., J.D. and D.C.; Conceptualization, J.Z. and D.C.; Methodology, J.Z.; Supervision, D.X. All authors have read and agreed to the published version of the manuscript.

Funding: This research was supported by the National Natural Science Foundation of China (No. 31971700) and Joint Guiding Project of Natural Science Foundation of Heilongjiang Province (No. LH2020C014).

Data Availability Statement: The original contributions presented in the study are included in the article/Supplementary Materials, further inquiries can be directed to the corresponding author/s.

Acknowledgments: We appreciate all the people who have collaborated on this project.

Conflicts of Interest: The authors declare no conflicts of interest.

References

1. Su, X.; Liu, Y.; Han, L.; Wang, Z.; Cao, M.; Wu, L.; Jiang, W.; Meng, F.; Guo, X.; Yu, N.; et al. A Candidate Gene Identified in Converting Platycoside E to Platycodin D from *Platycodon grandiflorus* by Transcriptome and Main Metabolites Analysis. *Sci. Rep.* **2021**, *11*, 9810. [CrossRef] [PubMed]
2. Wu, J.; Yang, G.; Zhu, W.; Wen, W.; Zhang, F.; Yuan, J.; An, L. Anti-Atherosclerotic Activity of Platycodin D Derived from Roots of *Platycodon grandiflorum* in Human Endothelial Cells. *Biol. Pharm. Bull.* **2012**, *35*, 1216–1221. [CrossRef] [PubMed]
3. Kwon, J.; Lee, H.; Kim, N.; Lee, J.-H.; Woo, M.H.; Kim, J.; Kim, Y.S.; Lee, D. Effect of Processing Method on Platycodin D Content in *Platycodon grandiflorum* Roots. *Arch. Pharm. Res.* **2017**, *40*, 1087–1093. [CrossRef] [PubMed]
4. Ahn, K.S.; Noh, E.J.; Zhao, H.L.; Jung, S.H.; Kang, S.S.; Kim, Y.S. Inhibition of Inducible Nitric Oxide Synthase and Cyclooxygenase II by *Platycodon grandiflorum* Saponins via Suppression of Nuclear Factor-kappaB Activation in RAW 264.7 Cells. *Life Sci.* **2005**, *76*, 2315–2328. [CrossRef] [PubMed]
5. Lee, E.J.; Kang, M.; Kim, Y.S. Platycodin D Inhibits Lipogenesis through AMPKα-PPARγ2 in 3T3-L1 Cells and Modulates Fat Accumulation in Obese Mice. *Planta Med.* **2012**, *78*, 1536–1542. [CrossRef]
6. Kim, T.Y.; Jeon, S.; Jang, Y.; Gotina, L.; Won, J.; Ju, Y.H.; Kim, S.; Jang, M.W.; Won, W.; Park, M.G.; et al. Platycodin D, a Natural Component of *Platycodon grandiflorum*, Prevents Both Lysosome- and TMPRSS2-Driven SARS-CoV-2 Infection by Hindering Membrane Fusion. *Exp. Mol. Med.* **2021**, *53*, 956–972. [CrossRef] [PubMed]
7. Zhao, Y.-J.; Li, C. Biosynthesis of Plant Triterpenoid Saponins in Microbial Cell Factories. *J. Agric. Food Chem.* **2018**, *66*, 12155–12165. [CrossRef] [PubMed]
8. Kim, Y.-J.; Lee, O.R.; Oh, J.Y.; Jang, M.-G.; Yang, D.-C. Functional Analysis of 3-Hydroxy-3-Methylglutaryl Coenzyme a Reductase Encoding Genes in Triterpene Saponin-Producing Ginseng. *Plant Physiol.* **2014**, *165*, 373–387. [CrossRef] [PubMed]
9. Deng, B.; Zhang, P.; Ge, F.; Liu, D.-Q.; Chen, C.-Y. Enhancement of Triterpenoid Saponins Biosynthesis in Panax Notoginseng Cells by Co-Overexpressions of 3-Hydroxy-3-Methylglutaryl CoA Reductase and Squalene Synthase Genes. *Biochem. Eng. J.* **2017**, *122*, 38–46. [CrossRef]
10. Wang, Y.; Zhang, H.; Ri, H.C.; An, Z.; Wang, X.; Zhou, J.-N.; Zheng, D.; Wu, H.; Wang, P.; Yang, J.; et al. Deletion and Tandem Duplications of Biosynthetic Genes Drive the Diversity of Triterpenoids in Aralia Elata. *Nat. Commun.* **2022**, *13*, 2224. [CrossRef]

11. Abe, I. Enzymatic Synthesis of Cyclic Triterpenes. *Nat. Prod. Rep.* **2007**, *24*, 1311–1331. [CrossRef] [PubMed]
12. Hoshino, T. β-Amyrin Biosynthesis: Catalytic Mechanism and Substrate Recognition. *Org. Biomol. Chem.* **2017**, *15*, 2869–2891. [CrossRef] [PubMed]
13. Rodríguez-Concepción, M.; Campos, N.; María Lois, L.; Maldonado, C.; Hoeffler, J.F.; Grosdemange-Billiard, C.; Rohmer, M.; Boronat, A. Genetic Evidence of Branching in the Isoprenoid Pathway for the Production of Isopentenyl Diphosphate and Dimethylallyl Diphosphate in *Escherichia coli*. *FEBS Lett.* **2000**, *473*, 328–332. [CrossRef]
14. Moses, T.; Pollier, J.; Faizal, A.; Apers, S.; Pieters, L.; Thevelein, J.M.; Geelen, D.; Goossens, A. Unraveling the Triterpenoid Saponin Biosynthesis of the African Shrub *Maesa lanceolata*. *Mol. Plant* **2015**, *8*, 122–135. [CrossRef] [PubMed]
15. Finnigan, J.D.; Young, C.; Cook, D.J.; Charnock, S.J.; Black, G.W. Cytochromes P450 (P450s): A Review of the Class System with a Focus on Prokaryotic P450s. *Adv. Protein Chem. Struct. Biol.* **2020**, *122*, 289–320. [CrossRef] [PubMed]
16. Yasumoto, S.; Seki, H.; Shimizu, Y.; Fukushima, E.O.; Muranaka, T. Functional Characterization of CYP716 Family P450 Enzymes in Triterpenoid Biosynthesis in Tomato. *Front. Plant Sci.* **2017**, *8*, 21. [CrossRef] [PubMed]
17. Carelli, M.; Biazzi, E.; Panara, F.; Tava, A.; Scaramelli, L.; Porceddu, A.; Graham, N.; Odoardi, M.; Piano, E.; Arcioni, S.; et al. Medicago Truncatula CYP716A12 Is a Multifunctional Oxidase Involved in the Biosynthesis of Hemolytic Saponins. *Plant Cell* **2011**, *23*, 3070–3081. [CrossRef] [PubMed]
18. Miettinen, K.; Pollier, J.; Buyst, D.; Arendt, P.; Csuk, R.; Sommerwerk, S.; Moses, T.; Mertens, J.; Sonawane, P.D.; Pauwels, L.; et al. The Ancient CYP716 Family Is a Major Contributor to the Diversification of Eudicot Triterpenoid Biosynthesis. *Nat. Commun.* **2017**, *8*, 14153. [CrossRef]
19. Yu, G.; Lam, T.T.-Y.; Zhu, H.; Guan, Y. Two Methods for Mapping and Visualizing Associated Data on Phylogeny Using Ggtree. *Mol. Biol. Evol.* **2018**, *35*, 3041–3043. [CrossRef]
20. Tamura, K.; Seki, H.; Suzuki, H.; Kojoma, M.; Saito, K.; Muranaka, T. CYP716A179 Functions as a Triterpene C-28 Oxidase in Tissue-Cultured Stolons of Glycyrrhiza Uralensis. *Plant Cell Rep.* **2017**, *36*, 437–445. [CrossRef]
21. Fiallos-Jurado, J.; Pollier, J.; Moses, T.; Arendt, P.; Barriga-Medina, N.; Morillo, E.; Arahana, V.; de Lourdes Torres, M.; Goossens, A.; Leon-Reyes, A. Saponin Determination, Expression Analysis and Functional Characterization of Saponin Biosynthetic Genes in *Chenopodium quinoa* Leaves. *Plant Sci.* **2016**, *250*, 188–197. [CrossRef] [PubMed]
22. Han, J.Y.; Chun, J.-H.; Oh, S.A.; Park, S.-B.; Hwang, H.-S.; Lee, H.; Choi, Y.E. Transcriptomic Analysis of Kalopanax Septemlobus and Characterization of KsBAS, CYP716A94 and CYP72A397 Genes Involved in Hederagenin Saponin Biosynthesis. *Plant Cell Physiol.* **2018**, *59*, 319–330. [CrossRef] [PubMed]
23. Tamura, K.; Teranishi, Y.; Ueda, S.; Suzuki, H.; Kawano, N.; Yoshimatsu, K.; Saito, K.; Kawahara, N.; Muranaka, T.; Seki, H. Cytochrome P450 Monooxygenase CYP716A141 Is a Unique Β-Amyrin C-16b Oxidase Involved in Triterpenoid Saponin Biosynthesis in *Platycodon grandiflorus*. *Plant Cell Physiol.* **2017**, *58*, 1119. [CrossRef] [PubMed]
24. Boutanaev, A.M.; Moses, T.; Zi, J.; Nelson, D.R.; Mugford, S.T.; Peters, R.J.; Osbourn, A. Investigation of Terpene Diversification across Multiple Sequenced Plant Genomes. *Proc. Natl. Acad. Sci. USA* **2015**, *112*, E81-88. [CrossRef] [PubMed]
25. Khakimov, B.; Kuzina, V.; Erthmann, P.Ø.; Fukushima, E.O.; Augustin, J.M.; Olsen, C.E.; Scholtalbers, J.; Volpin, H.; Andersen, S.B.; Hauser, T.P.; et al. Identification and Genome Organization of Saponin Pathway Genes from a Wild Crucifer, and Their Use for Transient Production of Saponins in *Nicotiana benthamiana*. *Plant J.* **2015**, *84*, 478–490. [CrossRef] [PubMed]
26. Yasumoto, S.; Fukushima, E.O.; Seki, H.; Muranaka, T. Novel Triterpene Oxidizing Activity of Arabidopsis Thaliana CYP716A Subfamily Enzymes. *FEBS Lett.* **2016**, *590*, 533–540. [CrossRef] [PubMed]
27. Han, J.-Y.; Kim, H.-J.; Kwon, Y.-S.; Choi, Y.-E. The Cyt P450 Enzyme CYP716A47 Catalyzes the Formation of Protopanaxadiol from Dammarenediol-II during Ginsenoside Biosynthesis in Panax Ginseng. *Plant Cell Physiol.* **2011**, *52*, 2062–2073. [CrossRef] [PubMed]
28. Han, J.-Y.; Hwang, H.-S.; Choi, S.-W.; Kim, H.-J.; Choi, Y.-E. Cytochrome P450 CYP716A53v2 Catalyzes the Formation of Protopanaxatriol from Protopanaxadiol during Ginsenoside Biosynthesis in Panax Ginseng. *Plant Cell Physiol.* **2012**, *53*, 1535–1545. [CrossRef] [PubMed]
29. Zhang, S.; Chai, X.; Hou, G.; Zhao, F.; Meng, Q. *Platycodon grandiflorum* (Jacq.) A. DC.: A Review of Phytochemistry, Pharmacology, Toxicology and Traditional Use. *Phytomedicine* **2022**, *106*, 154422. [CrossRef]
30. Yang, Y.-H.; Dai, S.-Y.; Deng, F.-H.; Peng, L.-H.; Li, C.; Pei, Y.-H. Recent Advances in Medicinal Chemistry of Oleanolic Acid Derivatives. *Phytochemistry* **2022**, *203*, 113397. [CrossRef]
31. Chen, Y.; Klinkhamer, P.G.L.; Memelink, J.; Vrieling, K. Diversity and Evolution of Cytochrome P450s of Jacobaea Vulgaris and Jacobaea Aquatica. *BMC Plant Biol.* **2020**, *20*, 342. [CrossRef] [PubMed]
32. Malik, W.A.; Wang, X.; Wang, X.; Shu, N.; Cui, R.; Chen, X.; Wang, D.; Lu, X.; Yin, Z.; Wang, J.; et al. Genome-Wide Expression Analysis Suggests Glutaredoxin Genes Response to Various Stresses in Cotton. *Int. J. Biol. Macromol.* **2020**, *153*, 470–491. [CrossRef] [PubMed]
33. Shen, C.; Li, X. Genome-Wide Analysis of the P450 Gene Family in Tea Plant (*Camellia sinensis*) Reveals Functional Diversity in Abiotic Stress. *BMC Genom.* **2023**, *24*, 535. [CrossRef] [PubMed]

34. Zhang, W.; Zhang, J.; Fan, Y.; Dong, J.; Gao, P.; Jiang, W.; Yang, T.; Che, D. RNA Sequencing Analysis Reveals PgbHLH28 as the Key Regulator in Response to Methyl Jasmonate-Induced Saponin Accumulation in *Platycodon grandiflorus*. *Hortic. Res.* **2024**, *11*, uhae058. [CrossRef] [PubMed]
35. Kim, J.; Kang, S.-H.; Park, S.-G.; Yang, T.-J.; Lee, Y.; Kim, O.T.; Chung, O.; Lee, J.; Choi, J.-P.; Kwon, S.-J.; et al. Whole-Genome, Transcriptome, and Methylome Analyses Provide Insights into the Evolution of Platycoside Biosynthesis in *Platycodon grandiflorus*, a Medicinal Plant. *Hortic. Res.* **2020**, *7*, 112. [CrossRef] [PubMed]
36. Cheng, Y.; Liu, H.; Tong, X.; Liu, Z.; Zhang, X.; Li, D.; Jiang, X.; Yu, X. Identification and Analysis of CYP450 and UGT Supergene Family Members from the Transcriptome of *Aralia elata* (Miq.) Seem Reveal Candidate Genes for Triterpenoid Saponin Biosynthesis. *BMC Plant Biol.* **2020**, *20*, 214. [CrossRef] [PubMed]
37. Shin, B.-K.; Kwon, S.W.; Park, J.H. Chemical Diversity of Ginseng Saponins from *Panax ginseng*. *J. Ginseng Res.* **2015**, *39*, 287–298. [CrossRef] [PubMed]
38. Zheng, H.; Fu, X.; Shao, J.; Tang, Y.; Yu, M.; Li, L.; Huang, L.; Tang, K. Transcriptional Regulatory Network of High-Value Active Ingredients in Medicinal Plants. *Trends Plant Sci.* **2023**, *28*, 429–446. [CrossRef]
39. Wang, M.; Qiu, X.; Pan, X.; Li, C. Transcriptional Factor-Mediated Regulation of Active Component Biosynthesis in Medicinal Plants. *Curr. Pharm. Biotechnol.* **2021**, *22*, 848–866. [CrossRef]
40. Wu, Z.; Li, L.; Liu, H.; Yan, X.; Ma, Y.; Li, Y.; Chen, T.; Wang, C.; Xie, L.; Hao, X.; et al. AaMYB15, an R2R3-MYB TF in Artemisia Annua, Acts as a Negative Regulator of Artemisinin Biosynthesis. *Plant Sci.* **2021**, *308*, 110920. [CrossRef]
41. Jia, Y.; Chen, S.; Chen, W.; Zhang, P.; Su, Z.; Zhang, L.; Xu, M.; Guo, L. A Chromosome-Level Reference Genome of Chinese Balloon Flower (*Platycodon grandiflorus*). *Front. Genet.* **2022**, *13*, 869784. [CrossRef] [PubMed]
42. Ranallo-Benavidez, T.R.; Jaron, K.S.; Schatz, M.C. GenomeScope 2.0 and Smudgeplot for Reference-Free Profiling of Polyploid Genomes. *Nat. Commun.* **2020**, *11*, 1432. [CrossRef] [PubMed]
43. Williams, P.A.; Cosme, J.; Sridhar, V.; Johnson, E.F.; McRee, D.E. Mammalian Microsomal Cytochrome P450 Monooxygenase: Structural Adaptations for Membrane Binding and Functional Diversity. *Mol. Cell* **2000**, *5*, 121–131. [CrossRef] [PubMed]
44. Han, J.-Y.; Kim, M.-J.; Ban, Y.-W.; Hwang, H.-S.; Choi, Y.-E. The Involvement of β-Amyrin 28-Oxidase (CYP716A52v2) in Oleanane-Type Ginsenoside Biosynthesis in Panax Ginseng. *Plant Cell Physiol.* **2013**, *54*, 2034–2046. [CrossRef] [PubMed]
45. Hall, T.A. BioEdit: A User-Friendly Biological Sequence Alignment Editor and Analysis Program for Windows 95/98/NT. *Nucleic Acids Symp. Ser.* **1999**, *41*, 95–98. [CrossRef]
46. Kumar, S.; Stecher, G.; Li, M.; Knyaz, C.; Tamura, K. MEGA X: Molecular Evolutionary Genetics Analysis across Computing Platforms. *Mol. Biol. Evol.* **2018**, *35*, 1547–1549. [CrossRef] [PubMed]
47. Yu, G. Using Ggtree to Visualize Data on Tree-Like Structures. *Curr. Protoc. Bioinform.* **2020**, *69*, e96. [CrossRef] [PubMed]
48. Chen, C.; Chen, H.; Zhang, Y.; Thomas, H.R.; Frank, M.H.; He, Y.; Xia, R. TBtools: An Integrative Toolkit Developed for Interactive Analyses of Big Biological Data. *Mol. Plant* **2020**, *13*, 1194–1202. [CrossRef] [PubMed]
49. Shen, W.; Le, S.; Li, Y.; Hu, F. SeqKit: A Cross-Platform and Ultrafast Toolkit for FASTA/Q File Manipulation. *PLoS ONE* **2016**, *11*, e0163962. [CrossRef]
50. Quinlan, A.R.; Hall, I.M. BEDTools: A Flexible Suite of Utilities for Comparing Genomic Features. *Bioinformatics* **2010**, *26*, 841–842. [CrossRef]
51. Li, H.; Handsaker, B.; Wysoker, A.; Fennell, T.; Ruan, J.; Homer, N.; Marth, G.; Abecasis, G.; Durbin, R. 1000 Genome Project Data Processing Subgroup The Sequence Alignment/Map Format and SAMtools. *Bioinformatics* **2009**, *25*, 2078–2079. [CrossRef] [PubMed]
52. Wickham, H.; Averick, M.; Bryan, J.; Chang, W.; McGowan, L.D.; François, R.; Grolemund, G.; Hayes, A.; Henry, L.; Hester, J.; et al. Welcome to the Tidyverse. *J. Open Source Softw.* **2019**, *4*, 1686. [CrossRef]
53. Liao, Y.; Smyth, G.K.; Shi, W. featureCounts: An Efficient General Purpose Program for Assigning Sequence Reads to Genomic Features. *Bioinformatics* **2014**, *30*, 923–930. [CrossRef] [PubMed]
54. Kim, D.; Paggi, J.M.; Park, C.; Bennett, C.; Salzberg, S.L. Graph-Based Genome Alignment and Genotyping with HISAT2 and HISAT-Genotype. *Nat. Biotechnol.* **2019**, *37*, 907–915. [CrossRef] [PubMed]
55. Ding, W.; Goldberg, D.; Zhou, W. PyComplexHeatmap: A Python Package to Visualize Multimodal Genomics Data. *Imeta* **2023**, *2*, e115. [CrossRef] [PubMed]
56. Zhan, C.; Ahmed, S.; Hu, S.; Dong, S.; Cai, Q.; Yang, T.; Wang, X.; Li, X.; Hu, X. Cytochrome P450 CYP716A254 Catalyzes the Formation of Oleanolic Acid from β-Amyrin during Oleanane-Type Triterpenoid Saponins Biosynthesis in *Anemone flaccida*. *Biochem. Biophys. Res. Commun.* **2018**, *495*, 1271–1277. [CrossRef] [PubMed]
57. Um, Y.; Jin, M.L.; Lee, D.Y.; Kim, C.K.; Hong, C.P.; Lee, Y.; Kim, O.T. Functional Characterization of the β-Amyrin Synthase Gene Involved in Platycoside Biosynthesis in *Platycodon grandiflorum*. *Hortic. Environ. Biotechnol.* **2017**, *58*, 613–619. [CrossRef]
58. Yu, H.; Liu, M.; Yin, M.; Shan, T.; Peng, H.; Wang, J.; Chang, X.; Peng, D.; Zha, L.; Gui, S. Transcriptome Analysis Identifies Putative Genes Involved in Triterpenoid Biosynthesis in *Platycodon grandiflorus*. *Planta* **2021**, *254*, 34. [CrossRef]
59. Misra, R.C.; Sharma, S.; Sandeep; Garg, A.; Chanotiya, C.S.; Ghosh, S. Two CYP716A Subfamily Cytochrome P450 Monooxygenases of Sweet Basil Play Similar but Nonredundant Roles in Ursane- and Oleanane-Type Pentacyclic Triterpene Biosynthesis. *New Phytol.* **2017**, *214*, 706–720. [CrossRef]

60. Cao, X.; Xie, H.; Song, M.; Lu, J.; Ma, P.; Huang, B.; Wang, M.; Tian, Y.; Chen, F.; Peng, J.; et al. Cut-Dip-Budding Delivery System Enables Genetic Modifications in Plants without Tissue Culture. *Innovation* **2023**, *4*, 100345. [CrossRef]
61. Liu, G.; Yuan, Y.; Jiang, H.; Bao, Y.; Ning, G.; Zhao, L.; Zhou, X.; Zhou, H.; Gao, J.; Ma, N. *Agrobacterium tumefaciens*-Mediated Transformation of Modern Rose (*Rosa hybrida*) Using Leaf-Derived Embryogenic Callus. *Hortic. Plant J.* **2021**, *7*, 359–366. [CrossRef]

plants

MDPI

Article

Selection of Stable Reference Genes for QRT-PCR in Tree Peony 'Doulv' and Functional Analysis of *PsCUC3*

Shuang Zhou, Chao Ma, Wenbin Zhou, Shuangcheng Gao, Dianyun Hou, Lili Guo and Guoan Shi *

College of Agriculture, Henan University of Science and Technology, Luoyang 471023, China; zhoushuang2001@163.com (S.Z.); machao840508@163.com (C.M.); zwl26088@163.com (W.Z.); gsczml@163.com (S.G.); dianyun518@163.com (D.H.); guolili0928@126.com (L.G.)
* Correspondence: gashi1963@163.com; Tel.: +86-138-3843-2986

Abstract: (1) Background: Tree peonies display extensive cultivar diversity due to widespread hybridization, resulting in a complex genetic architecture. This complexity complicates the selection of universal reference genes across different cultivars for qRT-PCR analyses. *Paeonia suffruticosa* 'Doulv', notable for its unique green blooms in China, exhibits chlorosis post-flowering and features petaloid stamens and pistils. (2) Methods: Based on published literature and RNA-seq data from 'Doulv', nine candidate reference genes—*ACT* (*Actin*), *TUB* (*β-Tubulin*), *UBC* (*Ubiquitin Conjugating Enzyme*), *UBQ* (*Ubiquitin*), *UPL* (*Ubiquitin Protein Ligase*), *PP2A* (*Protein Phosphatase 2A*), *PP2C* (*Protein Phosphatase 2C*), *MBF1A* (*Multiprotein Bridging Factor 1A*), and *GAPDH* (*Glyceraldehyde-3-Phosphate Dehydrogenase*)—were selected. Their expression stability was assessed across various tissues and developmental stages of 'Doulv' flowers using qRT-PCR, with evaluations conducted via GeNorm_v3.5, NormFinder_v20, and BestKeeper_v1.0. Gene cloning and expression analyses of *PsCUC3*, including its subcellular localization, were performed. (3) Results: *GAPDH* and *ACT* were identified as the most stable reference genes in petaloid stamens across various developmental stages of 'Doulv', whereas *UBC* and *MBF1A* were optimal across different tissues. Notably, specific conserved amino acids in PsCUC3 from 'Doulv' diverged from those in NAM/CUC3 proteins of other species, impacting its protein structure. *PsCUC3* expression analysis revealed no correlation with chlorophyll content in petaloid stamens but an association with petaloid organ development. Furthermore, PsCUC3 was predominantly localized in the nucleus. (4) Conclusions: This study comprehensively evaluated suitable reference genes using GeNorm_v3.5, NormFinder_v20, and BestKeeper_v1.0 software, establishing a robust qRT-PCR detection system for 'Doulv' peony. These results provide a solid experimental foundation for further research on 'Doulv' peony. Building on this experimental foundation, the functional analysis of the *PsCUC3* gene was conducted. The findings suggest a potential association between the *PsCUC3* gene and floral morphology alterations in 'Doulv', identifying *PsCUC3* as crucial for understanding the molecular mechanisms influencing floral structure in tree peonies.

Keywords: 'Doulv' peony; reference gene; *PsCUC3*; functional analysis; floral type mutation

1. Introduction

Tree peonies, scientifically known as *Paeonia suffruticosa* Andrews, are indigenous to China and boast a heritage extending over 1400 years. Renowned for their large, exquisite blooms, these peonies are celebrated as the 'king of flowers', symbolizing prosperity and joy in Chinese culture. China is home to over 1000 cultivars of tree peonies [1]. The flower colors of tree peonies can be divided into nine types: red, pink, yellow, purple, green, white, black, blue, and multi-color. Most research on tree peonies has focused on red, pink, yellow, purple, and multi-color blooms [2–9] and petal blotch [10,11], which are bright and conspicuous, with no prior studies reported on green-colored peonies. There are more than ten cultivars of green flower peonies in China, but all their blooms transition to white or pink upon opening, significantly diminishing their ornamental appeal. The molecular mechanisms underlying flower chlorosis in green flower peonies remain unclear.

Globally, there are more than 2000 cultivars of tree peonies, each resulting from extensive hybridization during cultivar development, leading to a complex genetic makeup. Lv et al. mapped four publicly available RNA-seq datasets to the draft genome of the tree peony cultivar "Luo shen xiao chun", and the mapping ratios were very low, ranging from 0.04% to 68.41% [12], indicating high genetic diversity among tree peony cultivars and the lack of universality among reference genes for qRT-PCR analysis. Zhou et al. analyzed the expression of 12 candidate reference genes across various tissues and five developmental stages of tree peony in six distinct colors using qPCR. Their findings identified *PP2A*, *UPL*, and *UBQ* as the most suitable reference genes across all tested samples. *Helicase*, *TUA* (*alpha-tubulin*), and *EIF5A* (*eukaryotic translation initiation factor 5A*) also exhibited notable expression stability across different samples [13]. In the cultivars 'Feng Dan' and 'Xi Shi', *GAPDH* and *UBC* were highlighted as optimal reference genes, whereas *EF-1α* (*elongation factor 1α*) and *UBC* showed the greatest stability in 'Que Hao' [14]. In *Paeonia ostii*, *UBQ* and *Tip41* (*tip41-like protein*) were the most stable reference genes across various tissues and seed developmental stages [15]. The genes *ERVTP* (*endoplasmic reticulum vesicle transporter protein*) and *PP2CFP* (*protein phosphatase 2C family protein*) were identified as the best for petals during different stages of floral development in *Paeonia ostii* cultivars 'Feng Dan', 'Xi Shi', and 'Que Hao'. *RPS9* (*ribosomal protein S9*) and *ARFA1C* (*ADP-ribosylation factor A1C*) maintained high expression stability across flowers of different colors in 'Feng Dan'. Similarly, *AMPDS* (*AMP-dependent synthetase family protein*) and *PUF1639* (*protein of unknown function 1639*) showed stable expression across various tissues, with *RPS9* and *PUF1639* being the most stable in seeds at different developmental stages. *PUF1639*, *MBF1A*, *PP2CFP*, and *RPS9* outperformed traditional reference genes in tree peony [16]. In the Itoh peony 'Bartzella', *UBQ* and *TIP41* were the best reference genes for petals at different developmental stages, whereas *β-TUB*, *UBQ*, and *ACT* were the most stable in various organs [17]. It is necessary to screen stable reference genes when conducting gene expression analysis in previously unreported peony varieties.

Paeonia suffruticosa 'Doulv' is the most famous tree peony cultivar in China, particularly notable for its unique green flowers, which are considered precious and rare in China. Initially green during the early flowering stages, the blooms of 'Doulv' transition to white upon opening. The presence of petaloid stamens and pistils gives the 'Doulv' peony a crown-like or globular floral morphology, making it an excellent candidate for exploring the molecular mechanisms involved in flower type development in tree peonies. The selection of appropriate reference genes was a necessary experimental foundation for further study of the molecular mechanisms underlying flower chlorosis and flower type formation in 'Doulv' peony. In this study, stable reference genes were comprehensively evaluated using GeNorm_v3.5, NormFinder_v20, and BestKeeper_v1.0 software, and a robust qRT-PCR detection system for 'Doulv' peony was successfully established. Based on this experimental foundation, functional analysis of the *PsCUC3* gene was carried out. *CUP-SHAPED COTYLEDONs* (*CUCs*) are members of the NAC transcription factor family, pivotal in pattern formation [18], organ separation [19], organ boundary formation [20], and floral development [21,22]. Comprehensive methodologies including gene cloning, expression analysis, and subcellular localization of *PsCUC3* were executed. These procedures have established a foundational platform for subsequent inquiries into the molecular mechanisms influencing floral morphology in 'Doulv' peony.

2. Results

2.1. Flower Characteristics of 'Doulv' Peony

The floral morphology of 'Doulv' peony is characterized by a crown-like or globular shape. Externally, the flowers consist of intact petals, while internally, they feature petaloid stamens and pistils. During Stage III, the blooms of 'Doulv' exhibit a distinct green hue, transitioning to white by Stage V (Figure 1a). Throughout the blooming period, the sepals, petals, and petaloid pistils maintain their green coloration. In contrast, the petaloid stamens, which constitute a significant portion of the 'Doulv' floral structure, exhibit chlorosis following the flower's opening (Figure 1b).

Figure 1. Floral characteristics of 'Doulv' peony. (**a**) Developmental stages of 'Doulv' flowers. Stage I: sepals crack open, exposing the outer petals. Stage II: inner petals become visible as the buds fluff up. Stage III: outer petals fully expand, inner petals loosen, though the innermost petals remain closed. Stage IV: inner petals expand fully, revealing the carpels. Stage V: all inner petals are fully expanded, with carpels completely exposed. Stage VI: outer petals start to dehydrate and undergo senescence. (**b**) Tissue differentiation in 'Doulv' flower at Stage III and Stage V. Stage III features fully expanded outer petals and loosened inner petals, while innermost petals are still closed, maintaining overall green coloration. By Stage V, all inner petals are fully expanded, and carpels are entirely exposed, with a shift from green to white in color.

2.2. Gene-Specific PCR Amplification Efficiency Analysis

RT-PCR amplification results revealed specificity in nine candidate reference genes, with the exception of *UPL*, which displayed no band formation. The specific PCR products for the remaining eight genes presented as single bands at expected lengths (Figure 2), qualifying them for further analysis. Additionally, the melting curves for each primer set exhibited a single peak, as shown in Figure 3. The amplification efficiency and correlation coefficients (R^2) were determined from the standard curve slope. The R^2 values for all detected candidate reference genes ranged from 0.9911 to 0.9979. However, the amplification efficiency of *PP2A* was 165.38%, exceeding the optimal threshold of 105%, rendering it unsuitable for qRT-PCR. The amplification efficiencies of the other seven genes were between 90.63% and 102.94%, with amplicon sizes ranging from 140 to 233 bp. These seven genes were deemed appropriate for qRT-PCR analyses and were utilized to assess the stability of reference genes (Table 1).

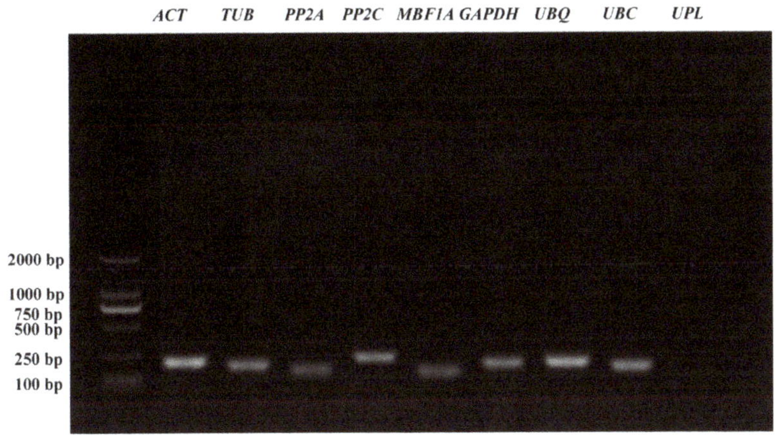

Figure 2. The agarose gel electrophoresis map of candidate reference genes in 'Doulv'.

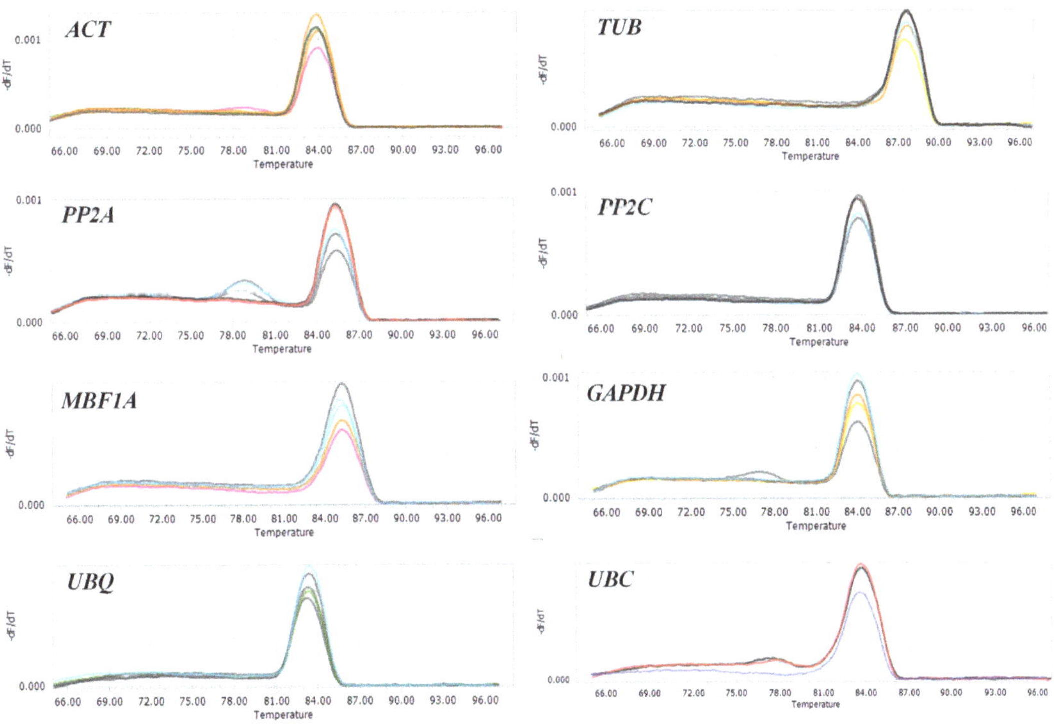

Figure 3. Melting curve of candidate reference genes.

Table 1. Primers used for qRT-PCR.

Reference Gene	Primer Sequence (5′-3′)	Product Length	Tm/°C	Efficiency	R^2
ACT	F: GAGAGATTCCGTTGCCCTGA R: TAGGTGCAAGAGCCGTGATT	211	60	99.88%	0.9979
TUB	F: CATTTCTTCATGGTTGGGTTCG R: CTTTGTGCTCATCTTTCCTCGG	183	62	101.57%	0.9928
UPL	F: CTTTATCACCTCCATTGTTTCGC R: ACACTATTTGTTTCCACCCGTCT	207	60	No band	
PP2A	F: TGTTTGGATGTTCTCAAGGCAG R: CAGAATTATCCGCTCCTCGTC	147	60	165.38%	0.9911
PP2C	F: GTTGGTGGCGTTCTTGCTGTTT R: CCTCGTGCGTAGGCTTCTTGTAT	233	63	96.76%	0.9949
MBF1A	F: GCGTTGACGACTTTCTCATCCTTC R: CGCACATCCTTGACAAACCCTA	140	63	97.04%	0.9976
GAPDH	F: TGTTCACTCCATCACTGCTAC R: ACATCCACAGTAGGAACACGA	189	55	98.70%	0.9968
UBQ	F: AAGACGCTGACTGGCAAGACA R: GCAAGACGAGATGAAGGGTAGACT	199	61	102.94%	0.9973
UBC	F: ACCTCCCGAAACTCTCTATGACG R: GGGTCTTCACCAGGAGGATGTAG	171	56	90.63%	0.9962

2.3. Expression Stability Analysis of Candidate Reference Genes

The expression stability of candidate reference genes was analyzed using GeNorm_v3.5, NormFinder_v20, and BestKeeper_v1.0 software, as per Zhang et al. [23]. A geomean of ranking values, derived from M values (GeNorm), S values (NormFinder), and SD (BestKeeper), was utilized to comprehensively assess the expression stability of these genes. Additionally, paired difference analysis ($Vn/n + 1$) conducted by GeNorm determined the optimal number of internal control genes. As demonstrated in Figure 4a, for the petaloid stamens at different developmental stages of 'Doulv', a V2/3 value less than 0.15 indicated that two reference genes were sufficient for qRT-PCR. *GAPDH* and *ACT* were identified as the most stable reference genes for these stages (Table 2). Similarly, in different tissues of the 'Doulv' flower, a V2/3 value less than 0.15 (Figure 4b) suggested the necessity of two reference genes. *UBC* and *MBF1A* were chosen as the optimal internal reference gene combination for these tissues (Table 3).

Table 2. The comprehensive analysis of stability of candidate reference genes in petaloid stamens at different developmental stages of 'Doulv'.

Reference Gene	GeNorm M Value	NormFinder S Value	BestKeeper SD	Geomean of Ranking Values	Ranking
GAPDH	0.567	0.132	0.96	0.416	1
ACT	0.661	0.205	0.98	0.510	2
UBC	0.638	0.271	0.91	0.540	3
UBQ	0.682	0.308	0.98	0.590	4
MBF1A	0.824	0.487	0.79	0.682	5
PP2C	0.734	0.323	1.39	0.691	SD > 1
TUB	1.102	0.723	1.55	1.073	SD > 1

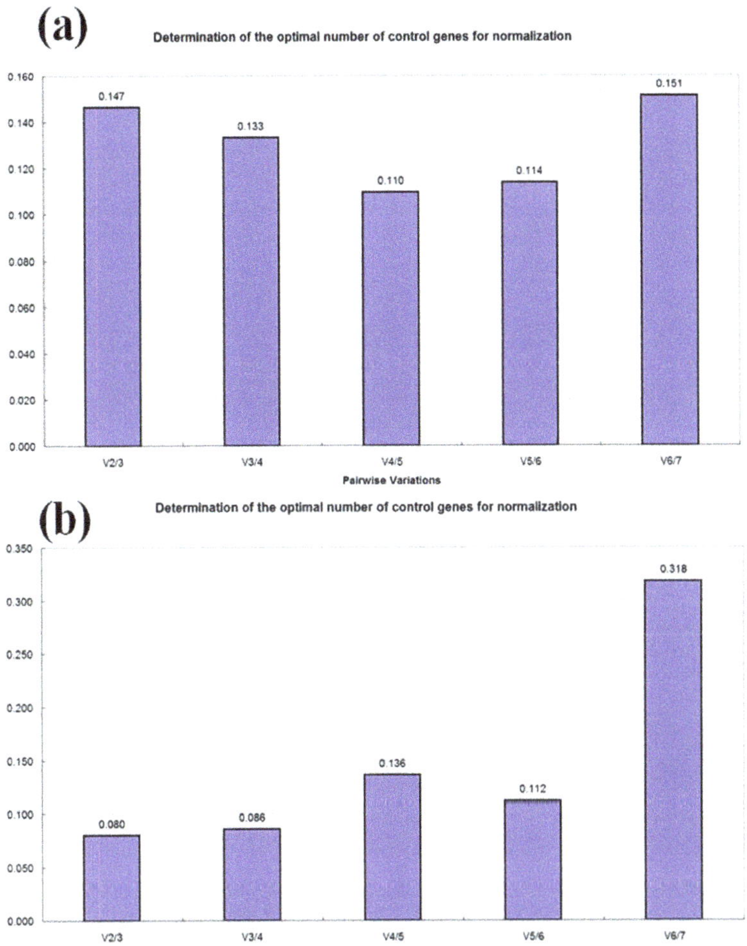

Figure 4. Paired difference analysis (V$n/n + 1$) of candidate reference genes by GeNorm. (**a**) Pairwise variation value (V value) of reference genes in petaloid stamens at different developmental stages of 'Doulv'; (**b**) V value of reference genes in different tissues of 'Doulv' flower.

Table 3. The comprehensive analysis of stability of candidate reference genes in different tissues of 'Doulv' flower.

Reference Gene	GeNorm M Value	NormFinder S Value	BestKeeper SD	Geomean of Ranking Values	Ranking
UBC	0.73	0.058	0.28	0.228	1
MBF1A	0.723	0.132	0.33	0.316	2
PP2C	0.788	0.356	0.24	0.407	r = 0.034
UBQ	0.931	0.533	0.14	0.411	r = −0.315
ACT	0.898	0.162	0.84	0.496	3
GAPDH	0.868	0.244	0.69	0.527	4
TUB	2.248	1.538	1.74	1.819	M value > 1.5, SD > 1

2.4. Cloning and Bioinformatics Analysis of *PsCUC3* from 'Doulv' Peony

The full-length cDNA of *PsCUC3* spans 1253 bp (Figure S1), encompassing an open reading frame (ORF) of 1125 bp that encodes an NAC protein consisting of 374 amino acids (Figure 5). The *PsCUC3* gene has been deposited in GenBank under the accession number OP745631.

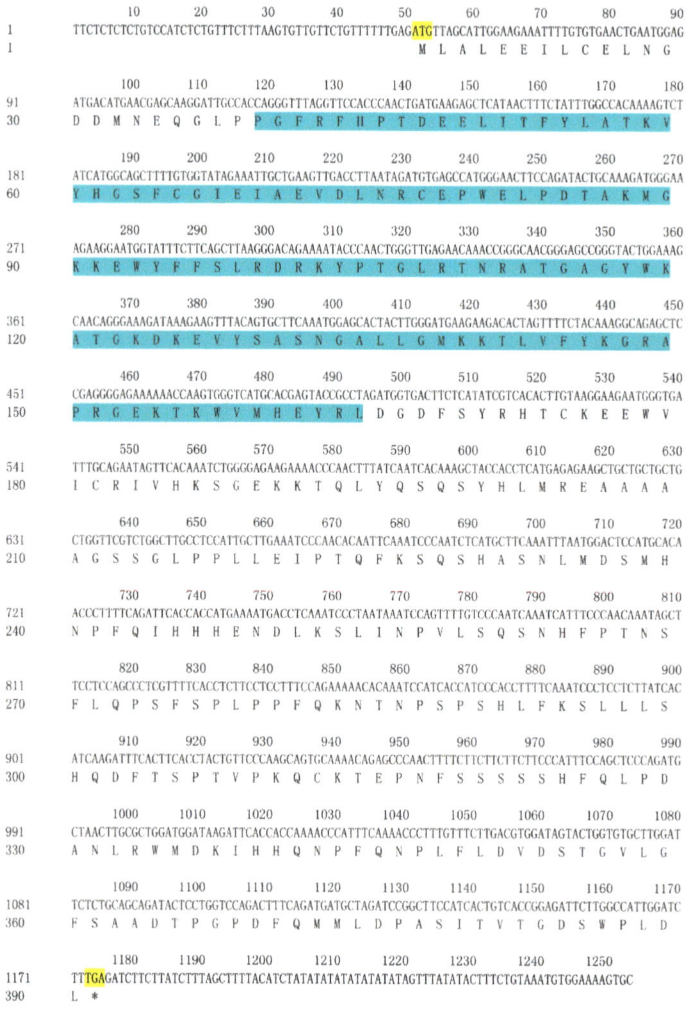

Figure 5. Nucleotide sequence and deduced amino acid sequence of the *PsCUC3* from 'Doulv' peony. Yellow: initiation codon and termination codon, respectively. Blue: NAC domain (23-148 aa). *: biosynthesis termination of PsCUC3 protein.

The molecular weight of the PsCUC3 protein is calculated at 42.38 kDa, with a theoretical isoelectric point (pI) of 6.38. The stability of the PsCUC3 protein is suggested by an instability index of 39.83, classifying it as stable. The protein's aliphatic index stands at 65.75, indicating robust thermostability, and the grand average of hydropathicity (GRAVY) at −0.560 confirms its hydrophilic nature. The secondary structure of PsCUC3 comprises 24.33% alpha helices, 16.04% extended strands, 5.61% beta turns, and 54.01% random coils. Predictive analyses suggest that the PsCUC3 protein localizes primarily within the nucleus.

Multiple sequence alignments of NAM/CUC3 proteins from various species—including *Paeonia suffruticosa* 'Doulv' (OP745631), *Cephalotus follicularis* (GAV65838.1), *Theobroma cacao* (XP_017981595.1), *Durio zibethinus* (XP_022769445.1), *Herrania umbratica* (XP_021284753.1), *Populus deltoides* (KAH8514964.1), *Corchorus capsularis* (OMO64504.1), *Hevea brasiliensis* (XP_021665943.1), *Nyssa sinensis* (KAA8537376.1), *Vitis riparia* (XP_034675316.1), *Gossypium hirsutum* (NP_001313889.1), and *Arabidopsis thaliana* (NM_106292.3)—revealed high sequence identity, particularly in the N-terminal regions, which were highly conserved across all examined NAM/CUC3 proteins (Figure 6).

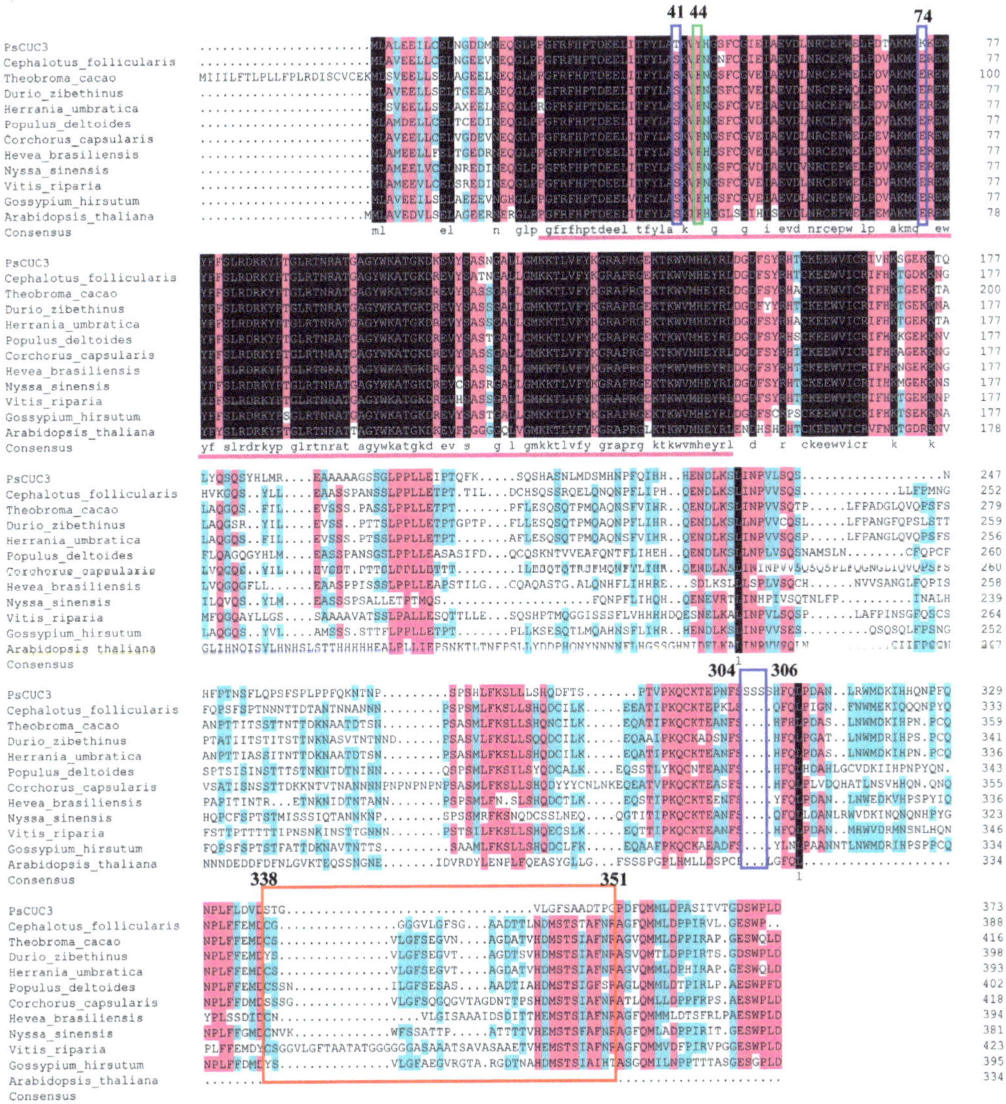

Figure 6. Alignment of the amino acid sequences of PsCUC3 with NAM/CUC3 proteins from 11 different plant species, including *Arabidopsis thaliana*. The NAC domain is highlighted in rose red. Boxes denote positions where conserved amino acids in PsCUC3 from 'Doulv' peony differ from those in other species.

The N-terminal region of the PsCUC3 protein, especially the NAC domain, demonstrated significant conservation. However, the C-terminal region of PsCUC3 exhibited high variability. Notably, some conserved amino acids in NAM/CUC3 proteins from other species underwent substitutions in PsCUC3 from 'Doulv' peony. Specifically, amino acids at positions 41, 44, and 74 changed from serine (S), phenylalanine (F), and glutamate (E) to threonine (T), tyrosine (Y), and lysine (K), respectively. Additionally, three serine residues (S) were inserted between positions 304 and 306 in PsCUC3. The amino acid sequence from positions 338 to 351 in the C-terminal region of PsCUC3 from 'Doulv' also showed clear differences compared to other species (Figure 6).

Multiple sequence alignments revealed that PsCUC3 possesses the highest amino acid sequence similarity (65.83%) to the NAM domain-containing protein from *Cephalotus follicularis* (GAV65838.1). SOPMA was employed to predict the protein's secondary structure. According to the analysis, alterations in conserved amino acids at positions 41, 74, and 304-306 in PsCUC3 did not affect its secondary structure. However, the modification at residue 44, where phenylalanine (F) in CfNAM of *Cephalotus follicularis* was replaced by tyrosine (Y) in PsCUC3, transitioned the structure from a random coil (c) to an alpha helix (h). Additionally, the amino acid sequence from residues 338-351 in the C-terminal region of PsCUC3 differed markedly from those of other NAM/CUC proteins, resulting in a significant reduction in alpha helix content (Figure 7).

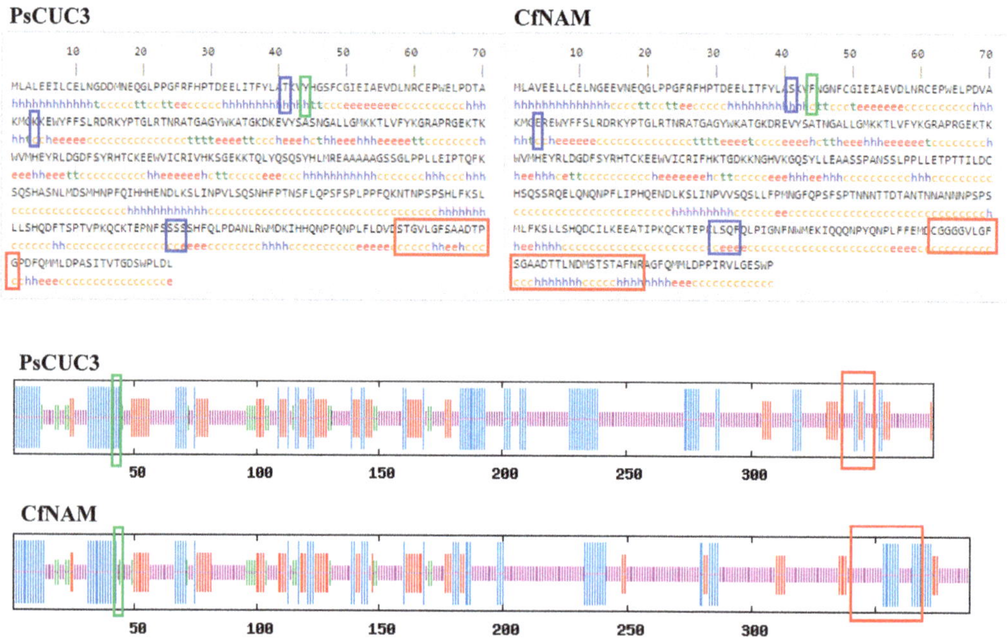

Figure 7. Protein secondary structures of PsCUC3 from 'Doulv' peony and CfNAM from *Cephalotus follicularis* were predicted using SOPMA. Blue boxes highlight regions where the secondary structures of PsCUC3 and CfNAM are identical. Green and red boxes denote regions where the secondary structures of PsCUC3 have altered. h: alpha helix; e: extended strand; t: beta turn; c: random coil.

2.5. Protein–Protein Interaction Analysis of PsCUC3

The phylogenetic analysis of PsCUC3 and Arabidopsis NAC proteins revealed that PsCUC3 is homologous to AtNAC31 (CUC3, AT1G76420) (Figure S2). STRING_v12.0 was utilized to predict the protein–protein interactions of CUC3 using the Arabidopsis protein database. As depicted in Figure 8, abCUC3 potentially interacts with ten proteins categorized into three functional groups: The first group is proteins associated with boundary specification, meristem initiation and maintenance, and organ patterning, including MYB105, MYB117, STM, and KNAT6. Notably, MYB117 also plays a role in floral organ development and ovule outgrowth initiation. The second group is proteins involved in structural formation and metabolism, such as AT5G52650 (structural constituent of ribosome), AT4G37810 (regulates epidermal patterning and stomatal patterning), ACX4 (acyl-coenzyme A oxidase 4, crucial for the initial step of peroxisomal fatty acid beta-oxidation in post-germinative growth of oilseed species), and KNAT7 (implicated in secondary cell wall biosynthesis and xylem development). The third group is proteins related to hormone regulation, including CKX5 (cytokinin oxidase/dehydrogenase, catalyzing cytokinin degradation) and GA2OX2 (Gibberellin 2-beta-dioxygenase 2, regulating endogenous gibberellin levels).

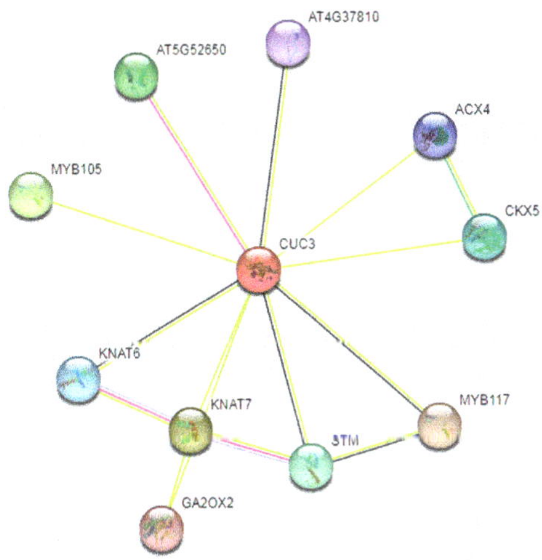

Figure 8. Interaction network analysis of CUC3 protein in *Arabidopsis thaliana*.

2.6. Chlorophyll Contents in the Petaloid Stamens of 'Doulv' and Expression Patterns of PsCUC3 Gene

The blooms of 'Doulv' peony exhibited green hues at Stage III, transitioning to white by Stage V (Figure 1). The primary floral components, the petaloid stamens, showed an increase in chlorophyll content, peaking at Stage III before significantly diminishing by Stages V and VI (Figure 9a). This chlorophyll profile aligns with the phenotypic changes observed in 'Doulv' peony. *GAPDH* and *ACT* were used as internal controls during various developmental stages in the petaloid stamens. Within these stamens, *PsCUC3* expression was highest at Stage I and declined progressively as the flowers opened (Figure 9a). These findings suggest no direct correlation between *PsCUC3* expression levels and changes in chlorophyll content in the petaloid stamens of 'Doulv'. In different floral tissues, *UBC* and *MBF1A* acted as internal controls. The expression levels of *PsCUC3* were relatively uniform across the petals, petaloid stamens, and petaloid pistils, all displaying high expression levels, whereas its expression was markedly reduced in the sepals (Figure 9b).

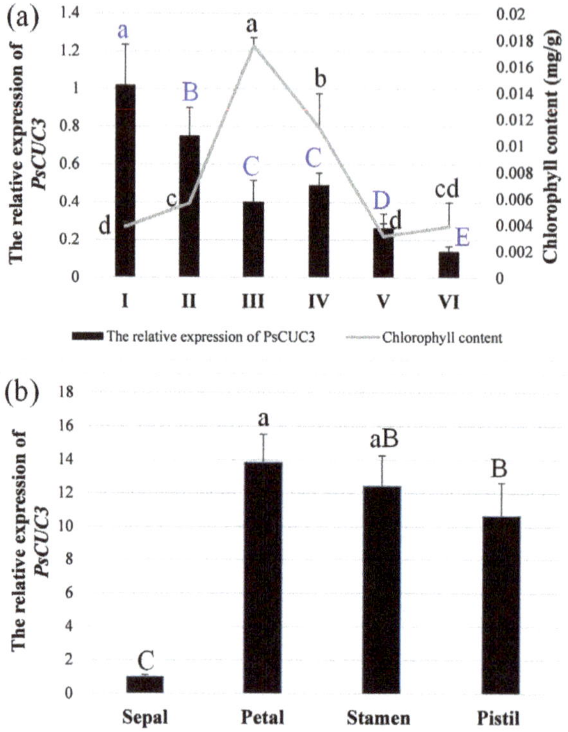

Figure 9. Chlorophyll contents and expression analysis of *PsCUC3*. (**a**) Chlorophyll contents and relative expression levels of *PsCUC3* in petaloid stamens at different developmental stages of 'Doulv'. (**b**) Relative expression levels of *PsCUC3* in different tissues of 'Doulv' flower. Different letters indicate significant differences at $p < 0.05$ (lower case letters) or $p < 0.01$ (capital letters), respectively (based on the one-way ANOVA test).

2.7. Subcellular Localization of PsCUC3 Protein

The PsCUC3 protein was anticipated to localize within the nucleus and potentially function in transcriptional activation and DNA binding. To validate these predictions, the subcellular localization of PsCUC3 was examined in tobacco (*Nicotiana benthamiana*) leaves through transient transformation using a super-promoter-driven GFP-PsCUC3 fusion protein. Additionally, mCherry was employed as a nuclear localization marker. Laser confocal microscopy (LSCM) revealed that GFP fluorescence signals from GFP-PsCUC3 were specifically localized in the nucleus. Similarly, the nuclear localization of mCherry, which shifted from red to yellow when merged with GFP signals, confirmed the nuclear presence. In contrast, control experiments using the empty super1300-GFP vector showed GFP localization in both the nucleus and the cytomembrane (Figure 10). These observations are consistent with predictions made by Plant-mPLoc, supporting the role of PsCUC3 as a transcription factor involved in regulating floral development in 'Doulv' peony.

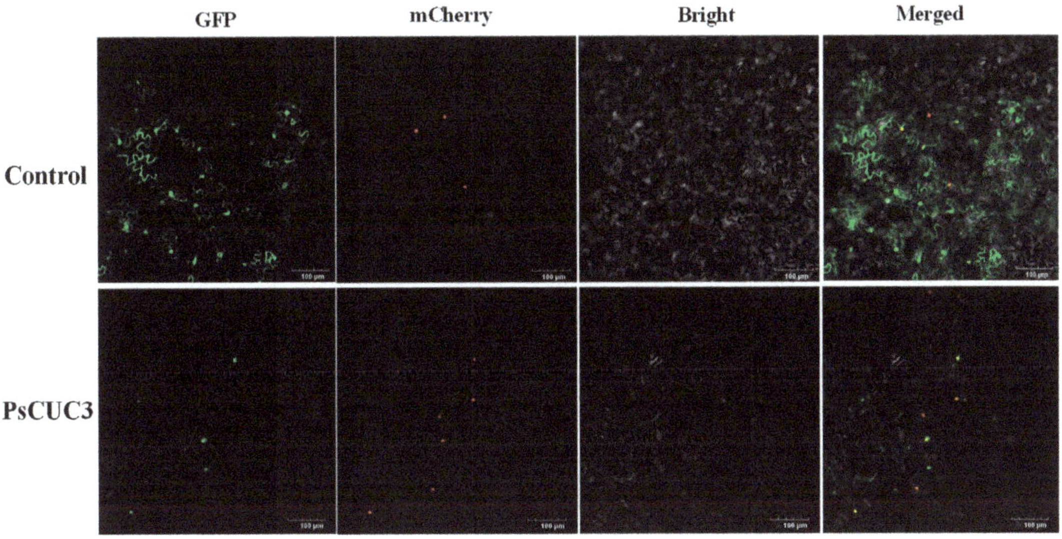

Figure 10. Subcellular localization analysis of PsCUC3 protein.

3. Discussion

Tree peonies exhibit a broad array of varieties, each resulting from extensive hybridization during cultivar development, leading to a complex and somewhat obscure genetic background. Notably, chromosome-level genome assembly for *Paeonia ostii* and a draft genome for the *Paeonia suffruticosa* cultivar "Luo shen xiao chun" have been reported [12,24], highlighting the substantial genetic diversity across different cultivars of *Paeonia suffruticosa*. Prior studies focusing on reference genes in tree peonies have demonstrated the lack of universality among existing reference genes for qRT-PCR analysis across different peony varieties [13–17]. These findings underscore the absence of a universal reference gene suitable for all peony varieties, emphasizing the need to screen for appropriate reference genes when conducting gene expression analysis in previously unreported peony varieties.

Green flower peonies are considered precious and rare and are very popular in China. There are more than ten cultivars of green flower peonies in China, but none of them can stay green during flower opening; all their blooms transition to white or pink upon opening, diminishing their ornamental appeal. The molecular mechanism underlying flower chlorosis remains unclear.

'Doulv' peony is the most famous and representative cultivar of green flower peony in China. In this context, our study analyzed nine candidate reference genes in 'Doulv' peony. The results identified *GAPDH* and *ACT* as the most stable reference genes in the petaloid stamens across various developmental stages of 'Doulv', while *UBC* and *MBF1A* proved to be the most stable in different tissues of the 'Doulv' flower. The selection of appropriate reference genes was a necessary experimental foundation for the further study of the molecular mechanisms underlying flower chlorosis and flower type formation in 'Doulv' peony, providing technical support for prolonging the coloring time of 'Doulv' in practice, as well as improving its ornamental quality, which has high economic value and practical significance. Based on this experimental foundation, gene functional analysis could be performed for 'Doulv' peony.

NAC transcription factors constitute a plant-specific family initially identified in Petunia [18]. *Arabidopsis thaliana* harbors 105 genes within this family [25,26], while *Oryza sativa* contains 140 NAC or NAC-like genes [27]. NAC proteins are distinguished by their highly conserved N-terminal domains, comprising approximately 150 amino acids capable of binding both DNA and other proteins [28]. The highly variable C-terminal regions of these proteins play a critical role in transcriptional regulation [29,30]. NAC proteins are integral to a multitude of plant biological processes, including responses to abiotic stress [31,32], lignin biosynthesis [33,34], fruit ripening regulation [35,36], leaf senescence [37–39], pattern formation [18], organ separation [19], organ boundary formation [20], and flower development [21,22]. *CUC* genes, members of the NAC family, are named due to the cup-like appearance of the fused cotyledons they produce [18,19]. *Arabidopsis thaliana* expresses three such genes, which are implicated in embryonic, boundary, and shoot meristem formation, alongside leaf and flower development [18–22,40,41]. In Arabidopsis, while most single *cuc* mutants appear phenotypically normal, a minority exhibit heart-shaped cotyledons; however, double and triple mutants predominantly display cup-shaped cotyledons [19,20]. *CUC1*, *CUC2*, and *CUC3* demonstrate partial functional redundancy [40]. The expression of *CUC2* and *CUC3*, mediated by Arabidopsis histidine kinase 2 (AHK2) and AHK3, may regulate flower development through cytokinin enhancement [21]. Overexpression of *CUC3* in Arabidopsis results in compact flower arrangements and stunted inflorescences [21]. In Petunia, diminished expression of the *CUC3* ortholog *NH16* through virus-induced gene silencing (VIGS) leads to increased fusion between petals and stamens, as well as stamens and carpels, frequently resulting in the formation of additional petals [22].

The *PsCUC3* gene was selected for investigation based on previous studies suggesting its involvement in the floral development of 'Doulv' peony. This study involved gene cloning and expression analysis of *PsCUC3*. The results from qRT-PCR demonstrated no correlation between the expression levels of *PsCUC3* and changes in chlorophyll content within the petaloid stamens of 'Doulv'. The expression levels of *PsCUC3* across petaloid tissues (petals, petaloid stamens, and petaloid pistils) displayed no significant variation, whereas a marked reduction was observed in the sepals. These findings suggest that the *PsCUC3* gene may play a role in petaloid tissue formation in 'Doulv' peony.

Multiple sequence alignments revealed that PsCUC3 is a homologous protein to the NAM/CUC3 proteins from other species. In Arabidopsis and Petunia, CUC3 is involved in boundary formation [18,19] and flower development [21,22]. Subcellular localization analysis showed that the PsCUC3 protein is localized in the nucleus. A mutation at amino acid position 44 in PsCUC3 led to alterations in the protein's secondary structure; this residue is located within the NAC domain, which is capable of binding DNA and regulating gene expression. Changes at residues 338-351 in the C-terminal region of PsCUC3 resulted in a significant reduction in alpha helices, a feature of the transcriptional activation domain that may be involved in protein–protein interactions and transcriptional activation. Protein–protein interaction analysis suggested that CUC3 could interact with MYB105, MYB117, STM, and KNAT6, influencing boundary formation and organ patterning. Additionally, interactions with CKX2 and GA2OX2 indicate that CUC3 may also regulate growth and development by modulating hormone levels. PsCUC3 might act as a transcription factor, interacting with other proteins to regulate flower development in 'Doulv' peony.

4. Materials and Methods

4.1. Plant Materials

Five-year-old 'Doulv' peonies were cultivated at Shenzhou Peony Garden. Plants displaying uniform developmental progress were selected and labeled prior to blooming. The developmental stages of 'Doulv' peony flowers were categorized into six distinct phases based on the methodology described by Wang et al. [42]. Stage I involved the cracking of sepals, revealing the outer petals. In Stage II, the inner petals became visible, and the buds appeared fluffy. By Stage III, the outer petals had fully expanded, and the inner petals loosened, though the innermost petals remained closed. Stage IV saw the full expansion of the inner petals and the exposure of the carpels. During Stage V, the inner petals were completely expanded, with the carpels fully exposed. Finally, in Stage VI, the outer petals started to dehydrate and undergo senescence. Tissue samples, including petaloid stamens from Stages I through VI, as well as sepals, petals, petaloid stamens, and petaloid pistils from Stage III, were collected for further analysis.

4.2. RNA Extraction and Reverse Transcription

RNA extraction from 'Doulv' peony was performed using the MiniBEST Plant RNA Extraction Kit (TaKaRa Bio, Beijing, China). The integrity of the extracted RNA was verified via 1% (*w*/*v*) agarose gel electrophoresis, ensuring only samples with clear and distinct 28S/18S rRNA bands were selected for further analysis. Subsequent quantification was conducted using a Multiskan Go microplate spectrophotometer (Thermo Scientific, Waltham, MA, USA), with A_{260}/A_{280} absorbance ratios ranging from 1.8 to 2.0, indicative of high-quality RNA. For reverse transcription, 1000 ng of RNA from each sample was utilized as a template in conjunction with a PrimeScript™ RT reagent Kit with gDNA Eraser (TaKaRa Bio, Beijing, China). The resulting cDNA samples were then stored at −20 °C for future use.

4.3. Selection of Candidate Reference Genes and Primer Design

Nine candidate reference genes were selected based on RNA-seq data from 'Doulv' and previous studies in other varieties of tree peony [13–17]. The genes selected for this study include *Actin* (*ACT*), *β-Tubulin* (*TUB*), *ubiquitin conjugating enzyme* (*UBC*), *ubiquitin* (*UBQ*), *ubiquitin protein ligase* (*UPL*), *protein phosphatase 2A* (*PP2A*), *protein phosphatase 2C* (*PP2C*), *multiprotein bridging factor 1A* (*MBF1A*) and *glyceraldehyde-3-phosphate dehydrogenase* (*GAPDH*). Primer design was carried out using Primer Premier 5.0. The primer sequences for *PP2C* were adopted from the work of Liu et al. [16]. Details of all primer sequences are provided in Table 1.

4.4. PCR Amplification and Amplification Efficiency for QRT-PCR

A mixed cDNA sample consisting of equal volumes from sepals, petals, petaloid stamens, and petaloid pistils at Stage III was utilized for PCR amplification. The reaction was set up in a final volume of 20 µL, containing 10 µL of 2 × Hieff®PCR Master Mix (With Dye) (Yeasen, Shanghai, China), 7.0 µL of ddH_2O, 1.0 µL of the diluted mixed cDNA, 1.0 µL of forward primer (10 µM), and 1.0 µL of reverse primer (10 µM). The PCR cycling conditions were as follows: an initial denaturation at 94 °C for 5 min, followed by 35 cycles of 94 °C for 30 s, 60 °C for 30 s, and 72 °C for 20 s, with a final elongation step at 72 °C for 10 min. The PCR products were then visualized through 1.5% agarose gel electrophoresis.

For qRT-PCR analysis, the mixed cDNA was diluted in a five-fold gradient, and a standard curve was prepared with dilution factors of 1, 5, 5^2, 5^3, 5^4, 5^5. The qRT-PCR was performed on a LightCycler® 96 system (Roche, Basel, BC, Switzerland). Each 10 µL reaction volume included 5.0 µL of TB Green™ Premix Ex Taq™ II (Tli RNaseH Plus) (TaKaRa), 3.0 µL of ddH_2O, 1.0 µL of diluted mixed cDNA, 0.5 µL of forward primer (10 µM), and 0.5 µL of reverse primer (10 µM). The reaction conditions were set to 95 °C for 60 s, followed by 40 cycles of 95 °C for 10 s and 60 °C for 30 s. A melting curve analysis was conducted at the end of the cycles by heating from 95 °C for 10 s and 65 °C for 60 s to 97 °C at a gradual rate. Each sample was run in triplicate for technical accuracy, and ddH_2O was included as a negative control. The amplification efficiency (E) and correlation coefficient (R^2) were calculated based on the standard curve slope.

4.5. QRT-PCR Analysis

For qRT-PCR analysis, cDNAs from petaloid stamens of 'Doulv' peony at developmental Stages I through VI, along with sepals, petals, petaloid stamens, and petaloid pistils at Stage III, were utilized. The analysis was performed using a LightCycler® 96 system (Roche). The reaction mixture and PCR cycling parameters followed those previously described in Section 4.4. Each sample was tested in triplicate to ensure technical reliability, and ddH_2O was included as a negative control in parallel. Expression stability (M value, GeNorm_3.5), stability value (S value, NormFinder_v20), and standard deviation (SD, BestKeeper_v1.0) were used to evaluate the stability of the candidate genes. BestKeeper_v1.0 was used to calculate the coefficient of correlation (r), standard deviation (SD), and coefficient of variation (CV). If the SD value was greater than 1, the reference gene was unstable.

If the SD value was less than 1, the stable reference gene had a larger r value, as well as smaller SD and CV values. A geomean of ranking values, derived from M values, S values, and SD, was utilized to comprehensively assess the expression stability of these genes. Additionally, paired difference analysis (V value) conducted by GeNorm_v3.5 determined the optimal number of internal control genes. The default V value is 0.15. If the $Vn/n + 1$ value is less than 0.15, the optimal number of reference genes is n.

4.6. Gene Cloning and Bioinformatics Analysis of PsCUC3 in 'Doulv' Peony

The *PsCUC3* gene was selected for investigation based on its potential involvement in the floral development of 'Doulv' peony. Primer design was conducted using Primer Premier 5.0, with the sequences being 5′ TTCTCTCTCTGTCCATCTCTG 3′ for the forward primer and 5′ GCACTTTTCCACATTTACAGA 3′ for the reverse primer. The PCR reaction was carried out in a 50 μL mixture containing 25 μL of 2 × Phanta® Flash Master Mix (Dye Plus) (Vazyme, Nanjing, China), 19.0 μL of ddH_2O, 2.0 μL of diluted mixed cDNA, 2.0 μL of forward primer (10 μM), and 2.0 μL of reverse primer (10 μM). The PCR cycling conditions were set as follows: an initial denaturation at 98 °C for 30 s, followed by 35 cycles of 98 °C for 10 s, 52 °C for 5 s, and 72 °C for 15 s, with a final elongation step at 72 °C for 1 min. The resulting PCR product was resolved by 1.5% gel electrophoresis and subsequently extracted using a Gel Extraction Kit (CWBIO, Taizhou, China). For cloning, the *TaKaRa Ex Taq®* (TaKaRa) and dNTPs (TaKaRa) were used to add an adenine to the 3′ end of the *PsCUC3* PCR product. The A-tailed product was then cloned into the pMD™ 19-T Vector (TaKaRa), and the ligation product was transformed into *E. coli* DH5α competent cells (TaKaRa). Positive clones were identified through PCR screening and subsequently sent for sequencing at Shanghai Sangon Biological Engineering Technology & Services, Shanghai, China.

An ORF search was performed according to the NCBI ORF Finder (https://www.ncbi.nlm.nih.gov/orffinder/, accessed on 6 October 2022), and conserved domain analysis was carried out using NCBI Conserved Domains Search (https://www.ncbi.nlm.nih.gov/Structure/cdd/wrpsb.cgi?, accessed on 6 October 2022). The properties of PsCUC3 were analyzed on ExPASy (https://web.expasy.org/protparam/, accessed on 6 October 2022). The protein secondary structure of PsCUC3 was predicted by SOPMA (https://npsa-prabi.ibcp.fr/cgi-bin/npsa_automat.pl?page=npsa_sopma.html, accessed on 6 October 2022). Subcellular localization of PsCUC3 was analyzed on Plant-mPLoc (http://www.csbio.sjtu.edu.cn/bioinf/plant-multi/, accessed on 6 October 2022). A homology search was performed according to NCBI-BLAST (https://blast.ncbi.nlm.nih.gov/Blast.cgi, accessed on 6 October 2022), and DNAMAN6.0 software was used for homology comparison. The annotated Arabidopsis NAC proteins were downloaded from PlantTFDB (https://planttfdb.gao-lab.org/, accessed on 6 October 2022), and the phylogenetic tree of PsCUC3 and Arabidopsis NAC proteins was inferred using MEGA-X with the neighbor-joining method. Bootstrap values were calculated for 1000 iterations. The protein–protein interaction network was analyzed by STRING_v12.0 (https://cn.string-db.org/, accessed on 6 October 2022).

4.7. Chlorophyll Content Determination of Petaloid Stamens in 'Doulv' Peony

The petaloid stamens of 'Doulv' peony at developmental Stages I through VI were harvested for chlorophyll content analysis. Fresh samples were pulverized in liquid nitrogen and subsequently extracted with 5 mL of 95% ethanol for 48 h under dark conditions. Following the extraction, samples were centrifuged at 10,000 rpm for 10 min, and the absorbance of the supernatants was measured at 665 nm and 649 nm wavelengths. Chlorophyll (Chl) concentrations were calculated using the following equation: C_{Chl} (mg/L) = $18.16A_{649} + 6.63A_{665}$. The total chlorophyll content was then calculated using the following formula: Chl content (mg/g) = C_{Chl} (mg/L) × V (L)/FW (g), where C_{Chl} represents the chlorophyll concentration, V is the volume of the extract, and FW is the fresh weight of the samples. Quantitative analysis was performed on three individual plants. Data analysis, including statistical evaluations, was conducted using Microsoft

Excel. Differences in chlorophyll content across the different stages were assessed using a one-way ANOVA test.

4.8. Gene Expression Analysis

The expression pattern of the *PsCUC3* gene was investigated using qRT-PCR, conducted on a LightCycler 96 system (Roche, Germany). Primer sequences for qRT-PCR were designed using Primer Premier 5.0, with the forward primer being 5′ TGCTGGTTCGTCTGGCTTG 3′ and the reverse primer 5′ AGGTGAAAACGAGGGCTGGAG 3′. *GAPDH* and *ACT* served as internal controls for assessing expression in the petaloid stamens at different developmental stages of the 'Doulv' flowers. For other tissues of the 'Doulv' flower, *UBC* and *MBF1A* were utilized as internal controls. The reaction mixture composition and PCR cycle parameters followed those previously described in Section 4.4. Relative expression levels of *PsCUC3* were calculated using the $2^{-\Delta\Delta Ct}$ method as described by Schmittgen and Livak [43]. The expression levels in petaloid stamens at Stage I and in sepals at Stage III were used as reference controls. Analysis was performed on three individual plants, with each reaction run in triplicate. Statistical analyses were conducted using Microsoft Excel, employing a one-way ANOVA test to evaluate the data.

4.9. Subcellular Localization Analysis of the PsCUC3 Protein

The full-length coding region of *PsCUC3* was successfully cloned using specific primers designed to include restriction enzyme sites: the forward primer was CGG<u>GGTACC</u>CCGTTCTCTCTCTGTCCATCTCTG (*Kpn* I site underlined), and the reverse primer was CCG<u>GAATTC</u>CGGGCACTTTTCCACATTTACAGA (*EcoR* I site underlined). This *PsCUC3* fragment was subsequently inserted into the super1300-GFP vector using a restriction-enzyme-mediated cloning method. This construct, driven by a super-promoter, resulted in the super-GFP-PsCUC3 construct. For control purposes, the empty super1300-GFP vector was employed, and p2300-35S-*H2B-mcherry* was utilized to confirm nuclear localization. These plasmids were transformed into *Agrobacterium tumefaciens* GV3101 (Biomed). For transient expression, an *Agrobacterium* strain containing the super-GFP-PsCUC3 construct and another containing p2300-35S-*H2B-mcherry*, both at an optical density (OD_{600}) of 0.3, were mixed in equal volumes. A similar procedure was followed for the strain containing the empty super1300-GFP vector. The *Agrobacterium* mixtures were incubated under dark conditions for 3 h and then injected into tobacco (*Nicotiana benthamiana*) leaves. Post infiltration, the transformed tobacco plants were kept under dark conditions for 48–72 h. Subsequently, the leaves were examined using an FV3000 laser confocal microscope (LSCM, Olympus, Tokyo, Japan) to visualize and capture the GFP and mCherry fluorescence signals, providing insights into the subcellular localization and expression dynamics of the PsCUC3 protein within plant cells.

5. Conclusions

In this study, nine candidate reference genes were selected based on published literature and transcriptome sequencing data. Their expression stability was comprehensively evaluated using GeNorm_v3.5, NormFinder_v20, and BestKeeper_v1.0 software. *GAPDH* and *ACT* were identified as the most stable reference genes in the petaloid stamens across various developmental stages of 'Doulv', while *UBC* and *MBF1A* were determined to be the optimal reference genes in different tissues of 'Doulv' flowers. A robust qRT-PCR detection system for 'Doulv' peony was successfully established, enhancing future gene function studies in this cultivar. Gene cloning of *PsCUC3* was conducted, revealing that some conserved amino acids in the PsCUC3 protein from 'Doulv' peony differed from those in NAM/CUC3 proteins from other species, resulting in alterations to its protein secondary structure. QRT-PCR results indicated that *PsCUC3* expression did not correlate with changes in chlorophyll content in the petaloid stamens at different developmental stages of 'Doulv'. However, its expression was associated with the formation of petaloid organs in 'Doulv' flowers. Subcellular localization analysis demonstrated that the PsCUC3 protein was localized in the nucleus, suggesting its potential role as a transcription factor.

In summary, the findings suggest that the *PsCUC3* gene may be involved in the floral type mutation of 'Doulv' peony. Further research is needed to elucidate the molecular and physiological mechanisms underlying this association.

Supplementary Materials: The following supporting information can be downloaded at https://www.mdpi.com/article/10.3390/plants13131741/s1: Figure S1: PCR product of *PsCUC3*; Figure S2: Phylogenetic analysis of PsCUC3 and Arabidopsis NAC proteins.

Author Contributions: Conceptualization, G.S.; data curation, S.Z. and W.Z.; funding acquisition, C.M. and S.G.; methodology, L.G.; supervision, D.H.; writing—original draft, S.Z.; writing—review and editing, G.S. All authors discussed the results and commented on the manuscript. All authors have read and agreed to the published version of the manuscript.

Funding: This study was funded by Natural Science Foundation of Henan Province (242300421319), Scientific and Technological Project in Henan Province (232102110175), National Key R & D Projects of China (2018YFD1000400), the Science and Technology Innovation Talents in Universities of Henan Province (22HASTIT036), and the Training Program for University Young Key Teachers in Henan Province (2021GGJS050).

Data Availability Statement: Data are contained within the article and Supplementary Materials.

Conflicts of Interest: The authors declare no conflicts of interest.

References

1. Zhou, S.L.; Zou, X.H.; Zhou, Z.Q.; Liu, J.; Xu, C.; Yu, J.; Wang, Q.; Zhang, D.M.; Wang, X.Q.; Ge, S.; et al. Multiple species of wild tree peonies gave rise to the 'king of flowers', *Paeonia suffruticosa* Andrews. *Proc. R. Soc. B* **2014**, *281*, 20141687. [CrossRef] [PubMed]
2. Zhang, L.L.; Yan, L.; Zhang, C.; Kong, X.; Zheng, Y.Q.; Dong, L. Glucose supply induces PsMYB2-mediated anthocyanin accumulation in *Paeonia suffruticosa* 'Tai Yang' cut flower. *Front. Plant Sci.* **2022**, *13*, 874526. [CrossRef] [PubMed]
3. Qi, Y.; Zhou, L.; Han, L.L.; Zou, H.Z.; Miao, K.; Wang, Y. *PsbHLH1*, a novel transcription factor involved in regulating anthocyanin biosynthesis in tree peony (*Paeonia suffruticosa*). *Plant Physiol. Biochem.* **2020**, *154*, 396–408. [CrossRef] [PubMed]
4. Zhang, Y.Z.; Zhang, Y.; Duan, X.J.; Liu, X.; Yuan, S.Q.; Han, J.M.; Cheng, Y.W. Anthocyanins in tree peony (*Paeonia suffruticosa*) and their relationship with flower color. *Hortic. Sci. Technol.* **2020**, *38*, 776–784. [CrossRef]
5. Zhang, Y.Z.; Yu, C.Y.; Lu, X.; Cheng, Y.W. Transcriptome profiling of flower development reveals key genes mediating yellow formation in tree peony. *Phyton-Int. J. Exp. Bot.* **2022**, *91*, 2505–2518. [CrossRef]
6. Zhang, X.P.; Jia, J.S.; Zhao, M.Y.; Li, C.; Han, X.; Xu, Z.D. Identification and characterization of microRNAs involved in double-color formation in *Paeonia suffruticosa 'Shima Nishiki'* by high-throughput sequencing. *Hortic. Environ. Biotechnol.* **2022**, *63*, 125–135. [CrossRef]
7. Zhang, X.P.; Xu, Z.D.; Yu, X.Y.; Zhao, L.Y.; Zhao, M.Y.; Han, X.; Qi, S. Identification of two novel R2R3-MYB transcription factors, *PsMYB114L* and *PsMYB12L*, related to anthocyanin biosynthesis in *Paeonia suffruticosa*. *Int. J. Mol. Sci.* **2019**, *20*, 1055. [CrossRef] [PubMed]
8. Zhao, Y.Z.; Cheng, Y.W.; Xu, S.Z.; Ma, H.P.; Han, J.M.; Zhang, Y. Tree peony variegated flowers show a small insertion in the *F3'H* gene of the acyanic flower parts. *BMC Plant Biol.* **2020**, *20*, 211.
9. Zhang, Y.Z.; Xu, S.Z.; Ma, H.P.; Duan, X.J.; Gao, S.X.; Zhou, X.J.; Cheng, Y.W. The R2R3-MYB gene *PsMYB58* positively regulates anthocyanin biosynthesis in tree peony flowers. *Plant Physiol. Biochem.* **2021**, *164*, 279–288. [CrossRef]
10. Luan, Y.T.; Tang, Y.H.; Wang, X.; Xu, C.; Tao, J.; Zhao, D.Q. Tree peony R2R3-MYB transcription factor PsMYB30 promotes petal blotch formation by activating the transcription of the anthocyanin synthase gene. *Plant Cell Physiol.* **2022**, *63*, 1101–1116. [CrossRef]
11. Shi, Q.Q.; Yuan, M.; Wang, S.; Luo, X.N.; Luo, S.; Fu, Y.Q.; Li, X.; Zhang, Y.L.; Li, L. *PrMYB5* activates anthocyanin biosynthetic *PrDFR* to promote the distinct pigmentation pattern in the petal of *Paeonia rockii*. *Front. Plant Sci.* **2022**, *13*, 955590. [CrossRef]
12. Lv, S.Z.; Cheng, S.; Wang, Z.Y.; Li, S.M.; Jin, X.; Lan, L.; Yang, B.; Yu, K.; Ni, X.M.; Li, N.; et al. Draft genome of the famous ornamental plant *Paeonia suffruticosa*. *Ecol. Evol.* **2020**, *10*, 4518–4530. [CrossRef]
13. Zhou, L.; Shi, Q.Q.; Wang, Y.; Li, K.; Zheng, B.Q.; Miao, K. Evaluation of candidate reference genes for quantitative gene expression studies in tree peony. *J. Am. Soc. Hortic. Sci.* **2016**, *141*, 99–111. [CrossRef]
14. Li, J.; Han, J.G.; Hu, Y.H.; Yang, J. Selection of reference genes for quantitative real-time PCR during flower development in tree peony (*Paeonia suffruticosa* Andr.). *Front. Plant Sci.* **2016**, *7*, 516. [CrossRef] [PubMed]

15. Li, C.Q.; Hu, L.Z.; Wang, X.Q.; Liu, H.Z.; Tian, H.H.; Wang, J.S. Selection of reliable reference genes for gene expression analysis in seeds at different developmental stages and across various tissues in *Paeonia ostii*. *Mol. Biol. Rep.* **2019**, *46*, 6003–6011. [CrossRef]
16. Liu, H.F.; Gao, L.X.; Hu, Y.H. Reference genes discovery and selection for quantitative real-time PCR in tree peony seed and petal tissue of different development stages. *J. Agric. Biotechnol.* **2015**, *23*, 1639–1648. (In Chinese)
17. Shang, S.S.; Fan, L.T.; Zhou, S.; Xu, M.Q.; Gao, S.C.; Shi, G.A. Screening of reference genes and expression analysis of petal senescence associated genes in Itoh peony 'Bartzella' cut flowers. *Plant Physiol. J.* **2023**, *59*, 153–164. (In Chinese)
18. Souer, E.; Houwelingen, A.V.; Kloos, D.; Mol, J.; Koes, R. The no apical meristem gene of petunia is required for pattern formation in embryos and flowers and is expressed at meristem and primordia boundaries. *Cell* **1996**, *85*, 159–170. [CrossRef]
19. Aida, M.; Ishida, T.; Fukaki, H.; Fujisawa, H.; Tasaka, M. Genes involved in organ separation in Arabidopsis: An analysis of the cup-shaped cotyledon mutant. *Plant Cell* **1997**, *9*, 841–857. [CrossRef]
20. Hibara, K.; Karim, M.R.; Takada, S.; Taoka, K.; Furutani, M.; Aida, M.; Tasaka, M. *Arabidopsis CUP-SHAPED COTYLEDON3* regulates postembryonic shoot meristem and organ boundary formation. *Plant Cell* **2006**, *18*, 2946–2957. [CrossRef]
21. Li, X.G.; Su, Y.H.; Zhao, X.Y.; Li, W.; Gao, X.Q.; Zhang, X.S. Cytokinin overproduction-caused alteration of flower development is partially mediated by *CUC2* and *CUC3* in Arabidopsis. *Gene* **2010**, *450*, 109–120. [CrossRef] [PubMed]
22. Zhong, J.S.; Powell, S.; Preston, J.C. Organ boundary NAC-domain transcription factors are implicated in the evolution of petal fusion. *Plant Biol.* **2016**, *18*, 893–902. [CrossRef] [PubMed]
23. Zhang, W.C.; Xu, J.; Wang, Q.; Li, J.; Li, Y.D.; Dong, M.; Sun, H.Y. Transcriptome-based identification of the optimal reference genes for quantitative real-time polymerase chain reaction analyses of lingonberry fruits throughout the growth cycle. *Plants* **2023**, *12*, 4180. [CrossRef] [PubMed]
24. Yuan, J.H.; Jiang, S.J.; Jian, J.B.; Liu, M.Y.; Yue, Z.; Xu, J.B.; Li, J.; Xu, C.Y.; Lin, L.H.; Jing, Y.; et al. Genomic basis of the giga-chromosomes and giga-genome of tree peony *Paeonia ostii*. *Nat. Commun.* **2022**, *13*, 7328. [CrossRef] [PubMed]
25. Riechmann, J.L.; Ratcliffe, O.J. A genomic perspective on plant transcription factors. *Curr. Opin. Plant Biol.* **2000**, *3*, 423–434. [CrossRef] [PubMed]
26. Ooka, H.; Satoh, K.; Doi, K.; Nagata, T.; Otomo, Y.; Murakami, K.; Matsubara, K.; Osato, N.; Kawai, J.; Carninci, P.; et al. Comprehensive analysis of NAC family genes in *Oryza sativa* and *Arabidopsis thaliana*. *DNA Res.* **2003**, *10*, 239–247. [CrossRef] [PubMed]
27. Fang, Y.J.; You, J.; Xie, K.B.; Xie, W.B.; Xiong, L.Z. Systematic sequence analysis and identification of tissue-specific or stress-responsive genes of NAC transcription factor family in rice. *Mol. Genet. Genom.* **2008**, *280*, 547–563. [CrossRef] [PubMed]
28. Ernst, H.A.; Olsen, A.N.; Skriver, K.; Larsen, S.; Leggio, L.L. Structure of the conserved domain of ANAC, a member of the NAC family of transcription factors. *EMBO Rep.* **2004**, *5*, 297–303. [CrossRef]
29. Kim, S.G.; Kim, S.Y.; Park, C.M. A membrane-associated NAC transcription factor regulates salt-responsive flowering via *FLOWERING LOCUS T* in *Arabidopsis*. *Planta* **2007**, *226*, 647–654. [CrossRef]
30. Kou, X.H.; Wang, S.; Wu, M.S.; Guo, R.Z.; Xue, Z.H.; Meng, N.; Tao, X.M.; Chen, M.M.; Zhang, Y.F. Molecular characterization and expression analysis of NAC family transcription factors in tomato. *Plant Mol. Biol. Rep.* **2014**, *32*, 501–516. [CrossRef]
31. Saimi, G.; Wang, Z.Y.; Liusui, Y.H.; Guo, Y.J.; Huang, G.Q.; Zhao, H.X.; Zhang, J.B. The functions of an NAC transcription factor, GhNAC2-A06, in cotton response to drought stress. *Plants* **2023**, *12*, 3755. [CrossRef] [PubMed]
32. Liu, J.H.; Wang, H.F.; Su, M.X.; Li, Q.Q.; Xu, H.L.; Song, J.Q.; Li, C.P.; Li, Q.L. A transcription factor *Sl*NAC4 gene of *Suaeda liaotungensis* enhances salt and drought tolerance through regulating ABA synthesis. *Plants* **2023**, *12*, 2951. [CrossRef] [PubMed]
33. Chen, L.J.; Wu, F.; Zhang, J.Y. NAC and MYB families and lignin biosynthesis-related members identification and expression analysis in *Melilotus albus*. *Plants* **2021**, *10*, 303. [CrossRef] [PubMed]
34. Li, X.T.; Wang, N.Y.; She, W.Q.; Guo, Z.X.; Pan, H.L.; Yu, Y.; Ye, J.W.; Pan, D.M.; Pan, T.F. Identification and functional analysis of the *CgNAC043* gene involved in lignin synthesis from *Citrus grandis* "San Hong". *Plants* **2022**, *11*, 403. [CrossRef] [PubMed]
35. Kou, X.H.; Zhou, J.Q.; Wu, C.E.; Yang, S.; Liu, Y.F.; Chai, L.P.; Xue, Z.H. The interplay between ABA/ethylene and NAC TFs in tomato fruit ripening: A review. *Plant Mol. Biol.* **2021**, *106*, 223–238. [CrossRef] [PubMed]
36. Liu, G.S.; Li, H.L.; Grierson, D.; Fu, D.Q. NAC transcription factor family regulation of fruit ripening and quality: A review. *Cells* **2022**, *11*, 525. [CrossRef] [PubMed]
37. Kim, H.J.; Nam, H.G.; Lim, P.O. Regulatory network of NAC transcription factors in leaf senescence. *Curr. Opin. Plant Biol.* **2016**, *33*, 48–56. [CrossRef] [PubMed]
38. Wang, H.L.; Zhang, Y.; Wang, T.; Yang, Q.; Yang, Y.L.; Li, Z.; Li, B.S.; Wen, X.; Li, W.Y.; Yin, W.L.; et al. An alternative splicing variant of PtRD26 delays leaf senescence by regulating multiple NAC transcription factors in *Populus*. *Plant Cell* **2021**, *33*, 1594–1614. [CrossRef] [PubMed]
39. Nagahage, I.S.P.; Matsuda, K.; Miyashita, K.; Fujiwara, S.; Mannapperuma, C.; Yamada, T.; Sakamoto, S.; Ishikawa, T.; Nagano, M.; Ohtani, M.; et al. NAC domain transcription factors VNI2 and ATAF2 form protein complexes and regulate leaf senescence. *Plant Direct* **2023**, *7*, e529. [CrossRef] [PubMed]
40. Vroemen, C.W.; Mordhorst, A.P.; Albrecht, C.; Kwaaitaal, M.A.C.J.; de Vries, S.C. The *CUP-SHAPED COTYLEDON3* gene is required for boundary and shoot meristem formation in Arabidopsis. *Plant Cell* **2003**, *15*, 1563–1577. [CrossRef]
41. Hasson, A.; Plessis, A.; Blein, T.; Adroher, B.; Grigg, S.; Tsiantis, M.; Boudaoud, A.; Damerval, C.; Laufs, P. Evolution and diverse roles of the *CUP-SHAPED COTYLEDON* genes in *Arabidopsis* leaf development. *Plant Cell* **2011**, *23*, 54–68. [CrossRef] [PubMed]

42. Wang, R.H.; Liu, Y.L.; Li, J.R. Studies on the blossom physiology in the different development stage of peony and Chinese peony flower. *Acta Hortic. Sin.* **2005**, *32*, 861–865. (In Chinese)
43. Schmittgen, T.D.; Livak, K.J. Analyzing real-time PCR data by the comparative C_T method. *Nat. Protoc.* **2008**, *3*, 1101–1108. [CrossRef]

plants

MDPI

Article

Regulatory Mechanism of Proanthocyanidins in Grape Peels Using vvi-miR828a and Its Target Gene *VvMYBPA1*

Lingqi Yue [1,†], Jingjing He [1,†], Tian Gan [1], Songtao Jiu [2], Muhammad Khalil-Ur-Rehman [3], Kunyu Liu [1], Miao Bai [1], Guoshun Yang [1] and Yanshuai Xu [1,*]

[1] College of Horticulture, Hunan Agricultural University, Changsha 410128, China; 18737690195@163.com (L.Y.); baimiao@hunau.edu.cn (M.B.); guoshunyang@aliyun.com (G.Y.)
[2] Department of Plant Science, School of Agriculture and Biology, Shanghai Jiao Tong University, Shanghai 200240, China; jiusongtao@sjtu.edu.cn
[3] Horticultural Sciences, The Islamia University of Bahawalpur, Bahawalpur 62300, Pakistan; muhammad.khalil@iub.edu.pk
* Correspondence: yx56@hunau.edu.cn
† These authors contributed equally to this work.

Abstract: Anthocyanins and proanthocyanidins are considered to be essential secondary metabolites in grapes and are used to regulate metabolic processes, while miRNAs are involved in their synthesis of anthocyanins and proanthocyanidins to regulate metabolic processes. The present research work was carried out to investigate the underlying regulatory mechanism of target genes in the grape cultivars 'Italia' and 'Benitaka'. miRNA and transnscriptomic sequencing technology were employed to characterize both the profiles of miRNAs and the transcripts of grape peels at 10 and 11 weeks post flowering (10 wpf and 11 wpf). The results revealed that the expression level of vvi-miR828a in 'Italia' at 10 and 11 wpf was significantly higher than that in 'Benitaka'. miRNA-seq analysis predicted *MYBPA1* to be the target gene of vvi-miR828a. In transcriptome analysis, the expression level of the *VvMYBPA1* gene in 'Benitaka' was significantly higher than that in 'Italia'; in addition, the TPM values (expression levels) of *VvMYBPA1* and miR828a also showed an evident negative correlation. The determination of the proanthocyanidin (PA) content in 'Italia' and 'Benitaka' peels at 11 wpf demonstrated that the PA content of 'Benitaka' was significantly higher than that of 'Italia'. The outcomes of RT-qRCR analysis exhibited that the expression levels of the *VdPAL*, *VdCHS*, *VdCHI*, *VdDFR*, *VdMYB5b*, *VdANR*, and *VdMYBPA1* genes related anthocyanin and proanthocyanidin pathways were reduced, while the expression levels of all of the above genes were increased after the transient expression of the *VvMYBPA1* vector into grape leaves. The results of the transient overexpression experiment of vvi-miR828a before the veraison period of strawberry fruits showed that vvi-miR828a can significantly slow down the coloration of strawberries. The vvi-miR828a negatively regulates the accumulation of proanthocyanidins in grape fruits by inhibiting the expression of *VvMYBPA1*.

Keywords: grape; miRNA-seq; vvi-miR828a; *VvMYBPA1*; PA

1. Introduction

Anthocyanins not only contribute to fruit coloring but also offer health benefits by providing protection against diseases such as cancers, diabetes, and neurological disorders.

Anthocyanins and proanthocyanidins (PAs) belong to the class of flavonoids, which are divided into several subgroups including flavones, flavanols, flavanones, catechins, anthocyanins, and chalcones. Anthocyanins not only offer health nutritional benefits but are also associated with color development in fruits as colors, while proanthocyanidins provide astringency and bitterness in the fruits and prevent the picking of early or unripe fruits to avoid commercial yield loss [1–4]. Anthocyanins and proanthocyanidins are considered as essential metabolites and important components of processed agricultural products due to their nutritional value. These metabolites also provide protection from heart disease, cancer, and many other diseases due to their high medicinal importance [5,6].

The anthocyanins and PAs are regulated by similar metabolic pathways in grapes and synthesized from the amino acid phenylalanine through the phenylpropanoid metabolism pathway [6,7]. The flavonoids and stilbenes are generally associated with the phenylalanine pathway [8]. The biosynthesis of flavonoid compounds is initiated by one p-coumarol coenzyme A and three propylene glycol coenzyme A molecules, while p-coumarol coenzyme A is synthesized from phenolic acids [9]. Many other enzymes are involved in the phenylalanine pathway, including chalcone isomerase (CHI), flavonoid 3-hydroxylase (F3H), flavanol synthase (FLS), and dihydro flavanol reductase (DFR). Anthocyanin synthase (ANS)/glucose transferase (UFGT) and *MYBPA1* regulate the synthesis of proanthocyanidins by activating the expression of these genes or enzymes, which further promote the formation of naringenin, dihydro flavanols, flavanols, and anthocyanins [10–12]. Several metabolic changes occur during the plant growth and development period. The fruit development cycle in grapes is a double S-curve; the first cycle starts from the growth and development period, from flowering and fruit setting to veraison, while second cycle starts from veraison to ripening [13]. The biosynthesis of proanthocyanidins usually starts after flowering and fruit setting at 3~9 weeks post grape fruit anthesis, and the expression of *MYBPA1, ANR*, and *LAR* genes promotes the synthesis of proanthocyanidins that gradually increase in the peel and seeds. Proanthocyanidin accumulation initiates after fruit set and attains its peak during the veraison period. The anthocyanins begin to accumulate significantly after the veraison period, and the expression of *MYBA1* gradually increases until fruit ripening [14,15].

Previous studies have revealed the importance of enzymes and structural genes in the flavonoid biosynthetic pathway and the regulatory role of some key genes/transcription factors in grapes. Among these transcription factors (TFs), unique ones are MYB, bHLH, and WD40. They form a transcription complex protein and are involved in regulating the synthesis of anthocyanins and proanthocyanidins. The MYB transcription factor is considered important among these three TFs [16,17]. The MYB gene/transcription factors have been shown to be involved in the control of flavonoid synthesis; among these TFs, *VvMYB5a* and *VvMYB5b* are involved in the general flavonoid metabolism pathway [13,18], *VvMYBPA1* is involved in the PA synthesis pathway [19], and *VvMYBA1* and *VvMYBA2* are involved in the anthocyanin synthesis or biosynthesis pathway [20–23].

Recent literature has elucidated the role of microRNA (miRNAs) in the regulation the flavonoid biosynthetic pathway [24]. Plant miRNA is a type of short non-coding RNA (21–24 nucleotides) that regulates the post-transcriptional expression of genes [25]. It plays a regulatory role in plant growth and development, as well as instress response, and secondary metabolism [26]. miRNA is typically initiated by its own promoter and transcribed by RNA polymerase II. Initially, it produces the primary transcript pri-miRNA; which is then, recognized and cleaved by DCL1 (RNase III endoribonuclease DICER-like 1). his process forms precursor-miRNA (pre-miRNA), characterized by a stem-loop structure and specific folding free energy. The pre-miRNA is subsequently cleaved again by DCL1 to form a miRNA–miRNA* double-stranded complex. Subsequently, HEN1 (S-adenosylmethionine-dependent methyltransferase HEN1) methyltransferase methylates the 3′ terminal base of the double strand to protect it from degradation. Methylated miRNA double strands are exported from the nucleus to the cytoplasm by the plant export protein HASTY. In the cytoplasm, one of the mature miRNA strands is degraded, and the other strand enters the RNA-induced gene silencing complex RISC and combines with the Argonaute protein to form the RISC silencing complex [27–29]. miRNA mainly cleaves or inhibits the target gene by complementary pairing with the sequence of the target mRNA, which means that when the sequence of the miRNA and the target mRNA are completely complementary, the miRNA cleaves the target mRNA, leading to its degradation. When the sequences of miRNA and target mRNA are not perfectly complementary, the translation of target gene mRNA is inhibited [30,31]. In Arabidopsis, apple, and tomato, the *SPL*-miR156 module (*SPL* gene) regulates anthocyanin synthesis and is highly conserved in plants. *SPL* affects anthocyanin biosynthesis by destroying the MBW complex [32,33]. *SPL*-miR156 has also been found to regulate the synthesis of anthocyanins in grapes, pomegranates, blueberries, and other fruit trees [34–36]. Many other miRNAs, such as miR157a, miR159, miR162, miR167a, miR172, miR395, miR396, miR396a, miR828, miR858a, etc., were also found to regulate the structural genes *4CL, CHI, F3H, LDOX,* and *UFGT,* as well as WD40, bHLH, and MYB transcription factors in the flavonoid biosynthetic pathway, thereby affecting the synthesis of secondary metabolites such as anthocyanins, proanthocyanidins, and flavanols [15,24]. The proanthocyanidins in persimmons are formed through the activation of laccase enzyme polymers of flavan-3-ols stored in the vacuoles. The laccase gene *DkLAC2* is involved in proanthocyanidin biosynthesis and is regulated by miR397. After overexpression of DkmiR397, the laccase gene *DkLAC2* is inhibited in persimmon, and the synthesis of proanthocyanidins is reduced [37].

In grapes, vvi-miR828 and vvi-miR858 inhibit anthocyanin synthesis by negatively regulating *VvMYB114* [38,39]. However, there is barely any research on the regulation of proanthocyanins by miRNAs in grapes. In this study, the fruit peels of 'Italia' and 'Benitaka' bud sport cultivar before the veraison period (10 weeks post flowering) and at the veraison period (11 weeks post flowering) were used as test materials by conducting small RNA and transcriptome sequencing to explore candidate miRNAs that regulate proanthocyanidins and predict their target genes. The effect of vvi-miR828a on anthocyanin synthesis was studied by detecting the PA content in the peels of 'Italia' and 'Benitaka' and transiently expressing candidate miRNA (vvi-miR828a) in strawberry fruits during the veraison period. By transiently expressing vvi-miR828a and *VvMYBPA1* in spine grape (*Vitis davidii*) leaves and comparing the expression of key genes in the anthocyanin and proanthocyanidin biosynthetic pathways with RT-qRCR, we observed the negative regulation of vvi-miR828a on its target genes and its inhibitory effect on proanthocyanidin synthesis. The outcomes of our study will ultimately be helpful in understanding the regulatory role of vvi-miR828a in PA biosynthesis as well as in proposing future research to comprehend the mechanism involved in PA biosynthesis.

2. Results

2.1. Sequencing Depth and Quality Control of Small RNA

The number of raw reads obtained from each sample in the 10 wpf and 11 wpf periods of 'Italia' and 'Benitaka' grape cultivars was 144,895,190 (It10), 113,765,862 (It11), 154,764,914 (Be10), and 181,890,944 (Be11). After filtering the data and removing low-quality fragments (reads), including those with incorrect adapters and those with plot Ns, the clean reads were 144,890,936 (It10), 113,750,206 (It11), 154,760,108 (Be10), and 181,878,014 (Be11). Fragments with a length of <18 nt were removed, and each sample obtained fragments ranging from 58.4% to 70.6% for comparison with the reference genome.

The total length of small RNAs measured via comparison was 18~25 nt. Among the four samples, the most abundantly expressed small RNA was 21 nt in length followed by small RNA of 24 nt in length (Figure 1). Using the miRBase and Rfam databases, we compared the fragments that aligned with the reference genome in order to analyze small RNAs and to screen out known non-coding RNAs (snoRNA, snRNA, tRNA, and RNAs). The composition of all non-coding RNAs was similar among the four samples, with only minor differences occurring within each sample (Table 1). Among the components of all non-coding RNAs, known miRNAs account for the largest average proportion of read fragments, reaching an average of 21.62% (Table 1), followed by rRNA (21.44%); the proportions of other RNAs are the newly predicted 5.01% for miRNA, 0.43% for tRNA, 0.15% for snoRNA, and 0.015% for snRNA.

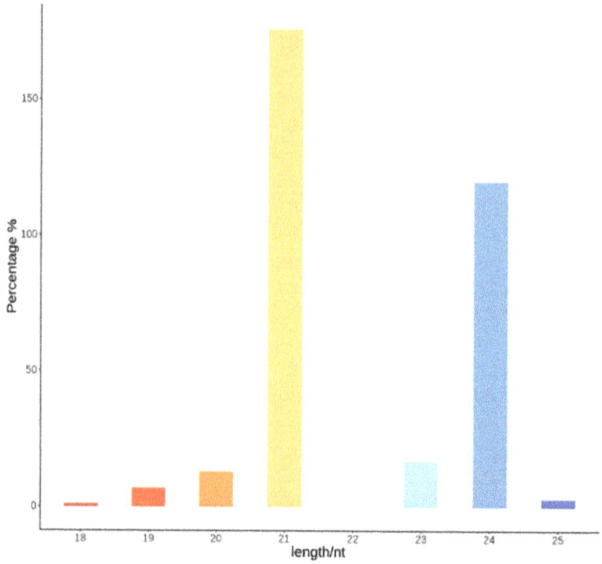

Figure 1. Expression proportions of small RNAs of different lengths in four samples.

Table 1. Small RNA sequencing data of four samples.

Type	It10	It11	Be10	Be11
Known miRNA	1,461,607 (21.93%)	1,349,557 (22.46%)	1,686,465 (21.27%)	1,441,903 (20.83%)
Novel miRNA	326,182 (4.89%)	315,922 (5.26%)	362,619 (4.57%)	365,986 (5.29%)
rRNA	1,880,625 (28.21%)	1,140,941 (18.99%)	1,635,203 (20.62%)	1,242,239 (17.94%)
tRNA	46,149 (0.69%)	17,279 (0.29%)	28,357 (0.36%)	25,122 (0.36%)
snoRNA	8354 (0.13%)	6624 (0.11%)	18,816 (0.24%)	9239 (0.13%)
snRNA	527 (0.01%)	660 (0.01%)	1341 (0.02%)	1151 (0.02%)
Repbase	3117 (0.05%)	3035 (0.05%)	3919 (0.05%)	2856 (0.04%)
Exon	680,058 (10.2%)	535,576 (8.92%)	722,475 (9.11%)	720,531 (10.41%)
Intron	403,736 (6.06%)	423,030 (7.04%)	533,595 (6.73%)	557,451 (8.05%)
Unknown	1,856,008 (27.84%)	2,214,860 (36.87%)	2,937,031 (37.04%)	2,556,046 (36.92%)
Total	6,666,363	6,007,484	7,929,821	6,922,524

Note: Novel miRNA, newly predicted miRNA; rRNA, ribosomal RNA; tRNA, transfer RNA; snoRNA, small nucleolar RNA; snRNA, small nuclear RNA; unknown, unannotated fragment.

2.2. Identification of Known and New miRNAs

Through sequence alignment analysis, a total of 306 miRNAs were identified in the It10 sample (including 127 known miRNAs and 179 new miRNAs); a total of 310 miRNAs were identified in the It11sample (including 119 known miRNAs and 191 new miRNAs); a total of 330 miRNAs were identified in the Be10 sample (126 known miRNAs and 204 new miRNAs); and 332 miRNAs were identified in the Be11 sample (including 135 known miRNAs and 197 new miRNAs). In total, 267 miRNAs were detected between the two samples of 'Italia', while 287 miRNAs were detected in the two samples of 'Benitaka'; in the comparative analysis of 'Italia' and 'Benitaka', 248 miRNAs (Figure 2A) were detected. There were 265 detected miRNAs in the four samples, and the following eight were specifically expressed in 'Italia': vvi-miR171c, vvi-miR171d, vvi-miR171j, vvi-miR156e, vvi-miR399e, 13_23166, 14_28417, and Un_40751. A total of 48 genes were specifically expressed in the 'Benitaka' cultivar (Figure 2B). The expression levels of all specifically expressed miRNAs were relatively low.

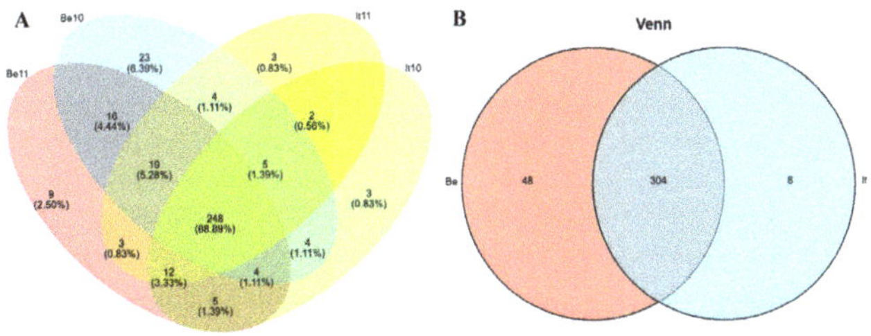

Figure 2. Venn diagram of the number and proportion of miRNA expressions in each sample. (**A**) The expression number and proportion of all identified miRNAs in each sample. (**B**) The number of co-expressed and specifically expressed miRNAs in the 'Italia' and 'Benitaka' cultivar.

The identified known miRNAs that were identified are divided into 31 conserved miRNA families, most of which are found in a variety of plants; in addition to the conserved miRNA families, sequencing also detected some known but less conserved miRNA families including the following: MIR3623, MIR3624, MIR3625, MIR3626, MIR3627, MIR3628, MIR3629, MIR3632, MIR3633, MIR3634, MIR3635, MIR3636, MIR3637, MIR3638, MIR3639, and MIR3640. Among the known conserved miRNA families, MIR169_2 had the most members, with 12, followed by the MIR156, MIR166, and MIR159 families, with 7–8 members; other families have one to six members. Members of the same family of non-coding miRNAs show different expression profiles; for example, the expression of most miRNAs of the MIR169 family in 'Benitaka' decreases with fruit development, while the expression of miR169t increases with fruit development.

2.3. Expression Trends of All Obtained miRNAs

The relative expression of miRNA was measured in TPM. A notable variation was observed in the expression pattern of each miRNA in the different samples. Some miRNAs had a high expression in all samples, while some had the opposite expression. A total of 27 miRNAs had a TPM value greater than 500. In order to gain a deeper understanding about the expression of miRNA in the different samples, the k-means clustering method was used to conduct a sub-clustering analysis of all miRNAs through the transformation value of Log10 (TPM + 1); all miRNAs were divided according to their expression trend. The miRNAs were divided into eight gene clusters (Supplementary Figure S1). Among the eight gene clusters, three showed consistent expression trends in the four samples of 'Italia' and 'Benitaka', while the other five gene clusters showed different expression trends in the two cultivars. Overall, diverse expression trends of miRNAs among the four samples were noticed.

The results of the miRNA gene cluster analysis showed that the number of members of the above eight gene clusters were 20, 8, 20, 22, 8, 20, 15, and 11, respectively. The eighth gene cluster contains 11 miRNAs; during the fruit development period of 'Italia' grapes, the expression levels of these 11 miRNAs increased significantly, while in the 'Benitaka' cultivar, the expression levels did not change significantly. The second and fifth gene clusters each contain eight miRNAs; their expression in the 'Benitaka' cultivar shows a gradual decreasing trend with fruit development, while the 'Italia' cultivar shows the opposite trend. The fourth and sixth gene clusters contain 22 and 20 miRNAs, respectively, and their expression levels increased significantly with fruit development in the 'Benitaka' cultivar, while the opposite trend was shown in the 'Italia' cultivar. The first and the third gene clusters comprise 20 miRNAs, and the expression trends in the two cultivars were consistent. The seventh gene cluster contains 15 miRNAs in the 'Benitaka' cultivar, and these miRNAs increased significantly with fruit development. Their expression was not obvious before the veraison period but increased significantly after it. In contrast, in the 'Italia' variety, the expression trends of these miRNAs showed an insignificant decrease with fruit development. In particular, miRNAs such as vv-miR171c, vv-miR171d, vv-miR171j, and vv-miR399e were not expressed in the 'Italia' cultivar but were significantly reduced with fruit development in the 'Benitaka' cultivar.

The predicted target genes were detected in small numbers as specifically expressed miRNAs in both the 'Italia' and 'Benitaka' cultivars; the target genes of these miRNAs were annotated as *SPL* (scale promoter binding protein), laccase, auxin response factor, E3 ubiquitin ligase, WRKY transcription factor, ultraviolet receptor UVR8, cytochrome P450, UDP-glycosyltransferase, copper transporter, ethylene response factor, and bZIP transcription factor, respectively.

2.4. *Differential Expression of miRNA in Two Cultivars*

In total, 161 differentially expressed miRNAs were present in the Be10 and Be11 samples, including 72 up-regulated and 89 down-regulated miRNAs. On the other hand, 90 differentially expressed miRNAs were observed between the It10 and It11 samples (20 were up-regulated and 70 were down-regulated). Additionally, there were 181 differentially expressed miRNAs between the It10 and Be10 samples, including 119 up-regulated and 62 down-regulated miRNAs. There were also 105 differentially expressed miRNAs between the It11 and Be11 samples (63 were up-regulated and 42 were down-regulated) (Figure 3A). Among the differentially expressed miRNAs in the four samples, the number of up-regulated miRNAs in 'Benitaka' was greater than that in 'Italia', in which most of the differentially expressed miRNAs showed a down-regulated expression. In addition, some miRNAs were also found to be specifically expressed in 'Italia' and 'Benitaka' at the 11 wpf stage (Figure 3B).

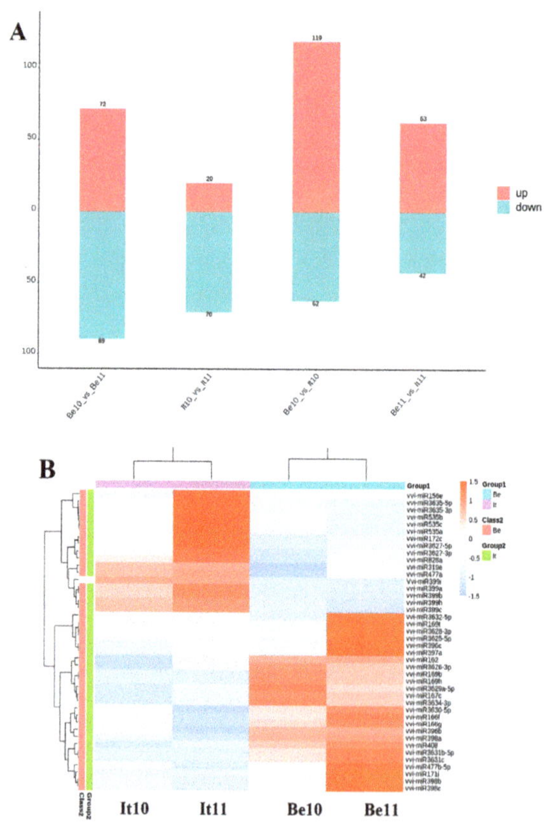

Figure 3. Number and cluster analysis of up-regulated and down-regulated differentially expressed miRNAs. (**A**) Pairwise comparison of differentially expressed miRNAs between different samples.

(**B**) Expression clustering heat map of differentially expressed miRNAs in four samples; the red color represents a high miRNA expression, while the blue color represents a low miRNA expression.

2.5. Target Gene Prediction of miRNAs

A total of 4356 target genes were predicted; out of these genes, 73 were differentially expressed miRNAs($|log_2FC| > 1$). vvi-miR160a had the lowest number of target genes (49), while vvi-miR397a had the highest number (182). The results depicted that the same miRNA is involved in multiple biological functions during plant growth and development.

All predicted target genes were subjected to GO enrichment and KEGG analysis; a total of 4356 target genes were annotated into three functional categories: biological processes, cellular components, and molecular functions. The three functional categories have 10, 3, and 7 GO items, respectively. The majority of genes were observed in metabolic processes, protein binding, and cellular component processes. In the KEGG pathway analysis, target genes were enriched in 19 pathways, among which, genes related to metal ion fixation, metal ion transport, and phenylalanine metabolism were highly enriched (Supplementary Figure S2).

The results of cluster analysis showed that among the differentially expressed miRNAs of the samples It11 and Be11, the expression of vvi-miR828a, vvi-miR156e, and vvi-miR399a gradually increased with fruit development in 'Italia'; in 'Benitaka', its expression gradually decreased during fruit development. The expression of vvi-miR408, vvi-miR2111-5p, vvi-miR479, vvi-miR160c, and vvi-miR477b-3p showed the opposite trend (Supplementary Figure S3). For target gene prediction, it was found that the target genes of these miRNAs were annotated as ethylene response factors, laccases, ABC transcription factors, sucrose transporters, glutathione transferases, bHLH transcription factors, WRKY transcription factors, and MYB transcription factors.

The target gene enrichment analysis indicated that vvi-miR828a targeted the most MYB transcription factors; therefore, it was selected for further analysis. Among all the target genes of vvi-miR828a, it was observed that the level of the *VvMYBPA1* gene was relatively and significantly different in expression levels between the two cultivars (Table 2).

Table 2. TPM values and functional annotations of target genes predicted by vvi-miR828a.

miRNA and Target Gene Name	Be10 (TPM)	It10 (TPM)	Be11 (TPM)	It11 (TPM)	Annotation from NCBI
vvi-miR828a	0	3.2054	1.2463	5.4922	
VIT_08s0007g04830	0	0.31	0	0	MYB-related protein
VIT_15s0046g00170	204.97	52.37	46.43	23.55	MYBPA1 protein
VIT_04s0008g01470	0.66	3.55	24.74	8.61	Probable WRKY transcription factor 50
VIT_09s0070g00240	87.97	19.2	22.24	10.81	Cinnamoyl-CoA reductase 1
VIT_09s0002g00800	0	0.21	0	0.11	Zinc finger CCCH domain-containing protein
VIT_15s0048g02120	0.61	3.63	0.77	1.61	Transcriptional activator MYB
VIT_04s0008g00660	9.17	32.02	17.65	41.37	Protein early flowering 3
VIT_14s0006g01280	0.76	0	0	0	MYBA7-R2R3 MYB transcription factor
VIT_06s0004g00570	9.66	12.69	18.48	13.44	*MYB5b*
VIT_08s0007g07230	27.14	19.46	28.73	24.12	MYBCS1-MYB transcription factor
VIT_17s0000g06190	175.15	105.52	133.45	87.57	MYB30-R2R3 MYB30 transcription factor

Note: Red font represents the expression of vvi-miR828a and the most obviously regulated target gene/protein MYBPA1.

Our previous laboratory report studied the transcript expression levels in the peels of 'Italia' and 'Benitaka' grapes at 10 and 11 wpf [40]; it was found that there was an obvious negative correlation between them (Figure 4A). In four sample tissues, the expression level of vvi-miR828a was detected, and the results were consistent with the RNA-seq results. The expression level of vvi-miR828a was not detected in the Be10 sample; in the Be11, It10, and It11 samples, its expression was detected. The qRT-PCR results were consistent with this. Among all tissue samples, vvi-miR828a expression was highest in the peel of the 'Italia' cultivar at 11 weeks post flowering (Figure 4B). After measuring the PA content in the peel of It11 and Be11 at 11wpf, the results indicated a higher PA content in 'Benitaka' (0.151 mg/mL) compared to 'Italia' (0.017 mg/mL). The PA contents significantly differed in the peels of the two cultivars at 11 wpf (Figure 4C).

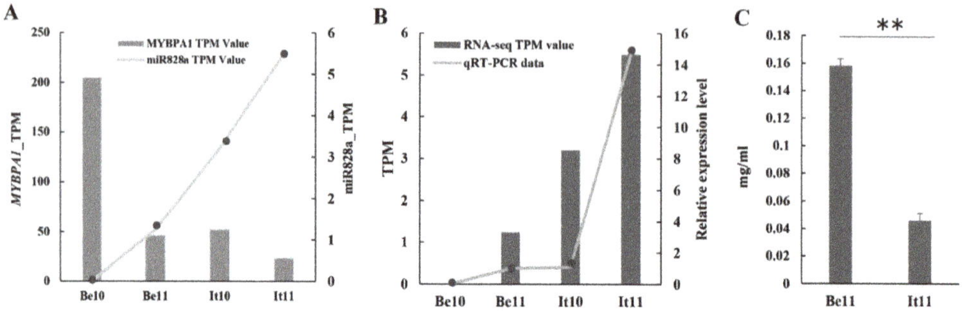

Figure 4. MYBPA1 gene expression, qRT-PCR verification, and proanthocyanidin content. (**A**) VvMYBPA1 and miR828a transcriptome sequencing TPM values. (**B**) qRT-PCR detection of vvi-miR828a expression to verify the accuracy of miRNA sequencing data. (**C**) Proanthocyanidin (PA) content in the peel of 'Italia' (It11) and 'Benitaka' (Be11) cultivars at 11 weeks post flowering; ** indicates $p < 0.01$.

2.6. Verification of Dual-Luciferase Assay between vvi-miR828a and Target Gene VvMYBPA1

vvi-MIR828a was constructed into the pSAK277 vector as an effector; the target gene *VvMYBPA1* and its mutation target sequence were constructed into the 3′ end of the pGreenII 0800-LUC vector as a reporter (Figure 5A,B). OX-miR828a (OX, overexpressed) and the target gene *VvMYBPA1* were mixed and injected into tobacco leaves. The results of the dual-luciferase experiments showed that after injection, the expression of the LUC reporter gene was significantly inhibited. miR828a targeted *VvMYBPA1*, which was consistent with the results of target gene prediction (Figure 5C), indicating that miR828a targets and negatively regulates *VvMYBPA1*.

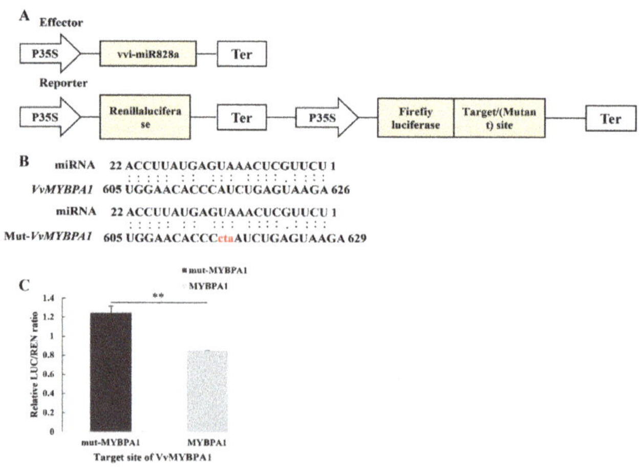

Figure 5. Verification of vvi-miR828a and target gene *VvMYBPA1* using a dual-luciferase assay. (**A**) Schematic diagram of dual-luciferase vector. (**B**) Recognition sites of vvi-miR828a, target gene *VvMYBPA1*, and its mutant sequence mut-*VvMYBPA1*; the red "cta" base indicates the mutant bases. (**C**) Luc/Ren ratio (*MYBPA1* indicates the Luc/Ren ratio of the vvi-miR828a target sequence; mut-*MYBPA1* indicates the Luc/Ren ratio of the vvi-miR828a target sequence mutation; ** indicates $p < 0.01$).

2.7. Effects of Transient Expression of miR828a and VvMYBPA1 on the Expression of Genes Related to Anthocyanin and Proanthocyanidin Synthesis

After the transient overexpression of vvi-miR828a, the expression levels of the *VdPAL, VdCHS, VdCHI, VdDFR, VdMYB5b, VdANR,* and *VdMYBPA1* genes were all significantly reduced. On the other hand, after the transient overexpression of *VvMYBPA1*, the expression of these genes significantly increased; this result was significantly different from the results of the transient overexpression of vvi-miR828a (Figure 6A–G). It is speculated that vvi-miR828a inhibits the expression of genes related to anthocyanin and proanthocyanidin biosynthesis. A bacterial solution containing the vectors of miR828a and *VvMYBPA1* was injected into the leaves of *Vitis davidii*, and the contents of proanthocyanidins were detected in the leaves 5 days later. The results showed that miR828a significantly inhibited the synthesis of proanthocyanidins, while *VvMYBPA1* significantly increased the synthesis of proanthocyanidins (Figure 6H).

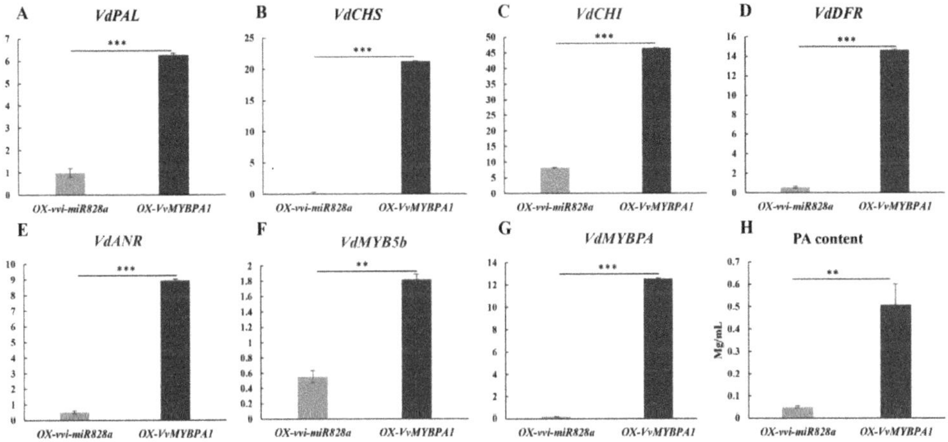

Figure 6. Expression levels of genes related to anthocyanin and proanthocyanidin synthesis after the transient expression of miR828a and *VvMYBPA1*. (**A–G**) These represent the expression levels of the *VdPAL, VdCHS, VdCHI, VdDFR, VdMYB5b, VdANR,* and *VdMYBPA1* genes, respectively, after transiently expressing pre-miR828a and *VvMYBPA1* in grape leaves for 3 days. The gray bar graph (OX-vvi-miR828a) represents the expression levels of anthocyanin and proanthocyanidin synthesis-related genes after the injection of bacterial fluid containing the vvi-miR828a vector; the black bar graph part (*OX-VvMYBPA1*) represents the expression level of related genes after the injection of bacterial fluid containing the *VvMYBPA1* vector. (**H**) PA content after the transient expression of pre-miR828a and *VvMYBPA1* in *Vitis davidii* leaves; ** indicates $p < 0.01$, indicating significant difference, *** indicates $p < 0.001$, indicating highly significant difference.

2.8. Transient Expression to Verify the Biological Function of vvi-miR828a

After continuous observation of strawberry fruits that transiently overexpressed vvi-miR828a but did not change color, the results indicated that the coloration of strawberry fruits was inhibited after the transient overexpression of vvi-miR828a, which slowed down strawberry fruit coloration compared to the control (Figure 7A). This result indicates that vvi-miR828a may affect the accumulation of anthocyanins in fruits. A bacterial solution containing vectors of miR828a and *VvMYBPA1* was injected into the strawberry fruits, respectively. The content of proanthocyanidins in strawberry fruits after the overexpression of miR828a and *VvMYBPA1* was detected 5 days post inoculation. The results showed that miR828a significantly inhibited the synthesis of proanthocyanidins, *VvMYBPA1* significantly increased the synthesis of proanthocyanidins (Figure 7B), and miR828a inhibited the biosynthesis of proanthocyanidins.

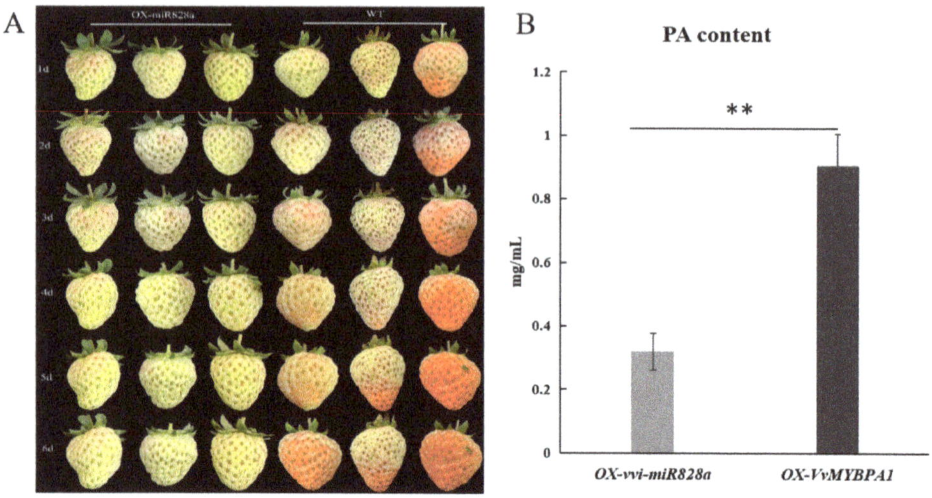

Figure 7. Transient overexpression of vvi-miR828a in strawberry fruits at the veraison stage. (**A**) Phenotypes were observed 1~6 days after the transient overexpression of vvi-miR828a. The three columns on the left represent strawberry fruits 1~6 days after the injection of a bacterial solution that contains the vvi-miR828a vector. The three columns on the right represent strawberry fruits 1~6 days after the injection of a bacterial solution that contains an empty vector. (**B**) Proanthocyanidin content in strawberry fruits after the transient overexpression of vvi-miR828a and *VvMYBPA1*; ** indicates $p < 0.01$, indicating significant difference.

3. Discussion

In grapes, anthocyanins and PAs are regulated by identical metabolic pathways; many structural genes and transcription factors in these pathways have been analyzed. *VvMYBA1* and *VvMYBA2* are considered as key genes, which specifically regulate anthocyanin biosynthesis in grape berries; *VvMYBPA1* and *VvMYBPA2* mainly regulate the biosynthesis of proanthocyanidins in grapes [41]. *VvMYB5a* and *VvMYB5b* are also involved in the synthesis of grape anthocyanins and proanthocyanidins [42]. miRNA is a type of endogenous non-coding small RNA that negatively regulates gene expression post-transcriptionally and participates in plant signal transduction, hormone response, secondary metabolism, and other processes.

Some studies on miR828 and miR858 functioning and characterization have been reported in plants; their main biological function is to participate in the regulation of the synthesis and accumulation of plant anthocyanins and proanthocyanidins. In tomatoes, miR828 negatively regulates the accumulation of anthocyanins; in fruits the expression of anthocyanins and related enzyme genes in the synthesis pathway decreased in the T1 generation of miR828 transgenic tomatoes [43]. In grapes under root-limited cultivation, it was observed that the anthocyanin content in the fruits was higher. Previous sequencing reports revealed that the expression of miR828 was decreased under root-limited cultivation conditions (treatment). The over expression of miR828 in Arabidopsis showed light-colored leaf types, consistent with the phenotype of grape berries in controls, suggesting that miR828 acts as a repressor of anthocyanin biosynthesis [38]. A recent report stated that miR828 inhibits the intensity of petal color during the growth and development of pink strawberries [44]. Our results indicated that miR828 may be involved in the regulation of PA biosynthesis (Figures 4 and 6). miRNA has also been found to be involved in regulating the biosynthesis of PA in apples, persimmons, cotton [45], and leguminous plants [46]. Among them, miR858 negatively regulates PA accumulation in apple peel by targeting *MdMYB9/11/12* [47]. An miRNA (miR858b) that negatively regulates proanthocyanidin synthesis was identified in persimmons; it can target *MYB19* and *MYB20* by transiently expressed miR858b in persimmon fruits, and the results showed that it inhibited the PA content in persimmons. The results of the transient expression of *MYB19* and *MYB20* showed that these genes increased the PA content in persimmons [48].

The results of miRNA-seq found that the expression level of vvi-miR828a gradually increased with berry development in 'Italia' grapes, while the expression level demonstrated the opposite trend in the 'Benitaka' grape berries; there were significant differences in the grape berries of the two cultivars before and after the veraison stage. The target gene prediction results demonstrated that most of the target genes of vvi-miR828a are MYB transcription factors, including *MYBPA1*, *MYB5b*, etc. Among them, *MYBPA1* has the highest expression level [40]. These transcription factors are involved in regulating the synthesis of flavonoids (including anthocyanins, proanthocyanidins, flavonols, flavones, etc.). Therefore, determining whether or not vvi-miR828a regulates *MYBPA1* became the main focus of this study.

Variations in the expression of *VvMYBPA1* in 'Italia' and 'Benitaka' cultivars at 10 and 11 wpf were noticed, and there was a significant negative correlation in the expression of vvi-miR828a at the 10 and 11wpf of the two cultivars. The expression of *VvMYBPA1* can regulate early fruit development and PA accumulation in seeds. The transient expression of *VvMYBPA1* in grape leaves revealed that the PA content in grape leaves increased significantly. *VvMYBPA1* can activate the expression of *VvLAR*, *VvANR*, and several flavonoid pathway genes, but cannot activate *VvUFGT*. *VvMYBPA1* may specifically regulate grape PA biosynthesis [19,41]. PA content in the peel of the 'Italia' and 'Benitaka' cultivars was detected at 11 wpf; the results indicated that the PA content in the peel of the 'Benitaka' variety was significantly higher than that of the 'Italia' variety. A vector bacterial solution containing miR828a and the target gene *VvMYBPA1* was mixed at a ratio of 9:1 and was injected into tobacco leaves. The results showed that, compared with tobacco leaves injected with 'miR828a + the target gene *VvMYBPA1* CDS mutant sequence', the reporter gene LUC/REN value of the 'miR828a + normal *VvMYBPA1* CDS sequence (non-mutated sequence)' was significantly reduced, which is consistent with the predicted results of miR828a targeting the negative regulation of *VvMYBPA1* (Figure 5C). It is speculated that miR828a may be involved in the regulation of grape proanthocyanidin biosynthesis.

The experimental results of the transient expression of vvi-miR828a during the veraison stage of strawberry fruits showed that vvi-miR828a significantly slowed down the coloration of strawberry fruits (Figure 7), indicating that vvi-miR828a may also negatively regulate anthocyanin biosynthesis. The results of the transient expression of vvi-miR828a and *VvMYBPA1* experiments, as well as qRT-RCR experiments in grape leaves, showed that the expression of the *VdPAL*, *VdCHS*, *VdCHI*, *VdDFR*, *VdMYB5b*, *VdANR*, and *VdMYBPA1* genes related to the biosynthesis of anthocyanins and proanthocyanidins was significantly reduced; contrarily, after the transient expression of *VvMYBPA1*, the expression of these genes increased significantly, which was significantly different from the transient overexpression of vvi-miR828a (Figure 6). vvi-miR828a inhibits the expression of anthocyanin and proanthocyanidin synthesis genes, which may affect the accumulation of anthocyanins and proanthocyanidins in grape berries. Therefore, vvi-miR828a may have a bidirectional effect in regulating grape anthocyanins and proanthocyanidins. In the early stages of grape fruit development, the expression of vvi-miR828a was very low, and its regulatory effect was not significant; after fruit development, its expression was increased.

One gene, *MYB5b*, which was involved in anthocyanin synthesis, was also exhibited in the prediction results of miR828a target genes in this study. In grape berries, *VvMYB5b* can activate genes for anthocyanin biosynthesis (except *VvUFGT*, which cannot be activated), promoting the accumulation of anthocyanins. At the same time, *VvMYB5b* can also activate the expression of the *VvLAR* and *VvANR* genes that play an important role in the regulatory mechanism of PA biosynthesis [13]. *MYB5b* activates several genes of the flavonoid biosynthetic pathway during the ripening of grape berries. Changes in *MYB5b* expression also affect flavonoid biosynthesis during this developmental stage. The *VvMYB5b* gene was transgenically overexpressed in petunias, and it was found that *VvMYB5b* induced the accumulation of anthocyanins. *MYBPA1* and *MYB5b* can activate genes in the flavonoid biosynthetic pathway [19], thereby affecting anthocyanin and PA content. In grapes, in addition to targeting and regulating *VvMYBPA1*, miR828a may also target *MYB5b*; therefore, the relationship between miR828a and *VvMYB5b* and whether *VvMYB5b* is involved in the regulation of PA synthesis still needs to be studied.

4. Conclusions

The prediction results of miR828a and the target genes found that *VvMYBPA1* is the target gene of vvi-miR828a. During the veraison period (11 wpf), the PA content of 'Benitaka' peels was significantly higher than that of 'Italia' peels. The transient expression of miR828a during the veraison period of strawberry fruits can significantly slow down their coloration. The results of the transient expression of miR828a and *VvMYBPA1* in grape leaves and qRT-RCR testing showed that miR828a inhibits the expression of PA synthesis-related genes. The above results illustrate that vvi-miR828a can negatively regulate the synthesis of proanthocyanidins in grape berries by inhibiting the expression of the *VvMYBPA1* gene.

5. Materials and Methods

5.1. Plant Materials

The 'Italia' and 'Benitaka' grapes were obtained from the national grape germplasm grapeyard (Zhengzhou Fruit Research Institute, Chinese Academy of Agricultural Sciences, Zhengzhou, China). Grape peels were collected before the veraison period (10 weeks after flowering, 10 wpf) and after the veraison period (11 weeks after flowering, 11 wpf), and were named It10, It11, Be10, and Be11, respectively. Two vines and 10 clusters from each cultivar were selected for the current study. A total of 5~10 berries (clean and healthy with uniform size) were collected from the upper, middle, and lower clusters, then quickly and immediately put into liquid nitrogen and transferred to the laboratory to store them in a −80 °C refrigerator for the extraction of total RNA and subsequent experiments. The peels of 'Italia' and 'Benitaka' berries collected during the 10 wpf and 11 wpf periods (Figure 8) were used as small RNA sequencing (seq) materials to generate 4 independent small RNA-seq libraries.

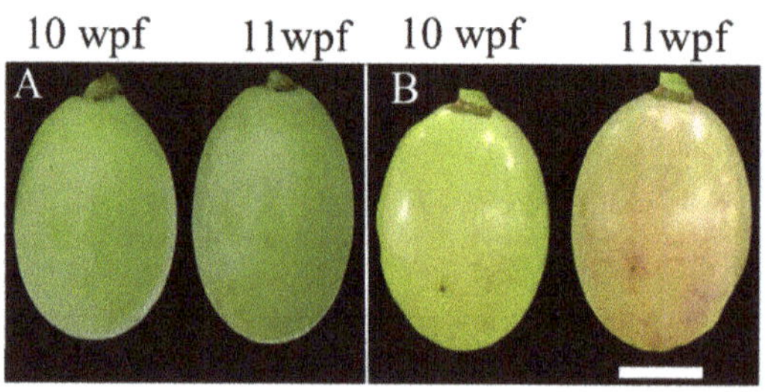

Figure 8. Fruits of 'Italia' and 'Benitaka' before and after the veraison period. (**A**) representative single fruits of 'Italia' at 10 weeks (10 wpf) and 11 weeks (11 wpf) post flowering. (**B**) representative single fruits of 'Benitaka' at 10 weeks and 11 weeks post flowering; scale bar length is equal to 6 mm.

5.2. Total RNA Extraction

Total RNA was extracted from sample tissues using a kit (Solebao Biotechnology Co., Ltd., Shanghai, China); a Nanodrop2000 (Thermo Fisher Scientific, Waltham, MA, USA) was used to detect the concentration and purity of the extracted RNA; agarose gel electrophoresis was used to detect RNA integrity; and Agilent5300 was used to determine the RQN value. Requirements for single library construction: the total amount of RNA was $\geq$ 1 ug, the concentration of RNA was $\geq$50 ng/μL, RQN was >7, and OD260/280 was between 1.8 and 2.2.

5.3. RNA Library Preparation

Isolate and purify the above-mentioned mRNA, use T4 RNA ligase to add adapter sequence 1 to the 3′ end, and then use T4 RNA ligase to add adapter sequence 2 to the 5′ end. Use the reverse transcription kit (TruSeqTM RNA sample prep Kit) to reverse transcribe the mRNA by using random primers; the RNA which were connected with the adapter are used as a template to reverse synthesize one-strand cDNA and then perform second-strand synthesis to form a stable double-stranded structure to construct a cDNA library. The sequencing primers are used to perform PCR amplification (11–12 cycles) to enrich the library concentration. According to the different characteristics of the length distribution of miRNA, target fragments are recovered by cutting the PAGE gel (6% Novex TBE PAGE gel, 1.0 mm, 10 well). After single-strand circularization of library DNA and the removal of uncirculated sequences and purification, DNA nanoballs (DNB) are prepared and sequenced using the Illumina HiSeq 4000 (300 cycles) system (Majorbio Biomedical Technology Co., Ltd. Company, Shanghai, China).

5.4. RNA Sequencing and Comparison

The sequencing image signals were converted into text signals through CASAVA base calling and stored in fasta format as raw data. The data of each sample was separated according to the index sequence for subsequent analysis.

Fastx_toolkit (Version 0.0.14) software (http://hannonlab.cshl.edu/fastx_toolkit/ (accessed on 5 April 2022) was used for quality control. To obtain clean reads, the sequence data were filtered and removed adapter sequences containing poly-N sequence low-quality reads and sequences with reads length < 18 nt. HISAT2 (Version 2.1.0) (https://daehwankimlab.github.io/hisat2/ (accessed on 5 April 2022) software was utilized to compare clean reads with the grape 12× reference genome (https://plants.ensembl.org/Vitis_vinifera/Info/Index?db=core (accessed on 5 April 2022) and to perform sequence comparison. StringTie (Version 1.3.3b) (https://ccb.jhu.edu/software/stringtie/index.shtml?t=example (accessed on 5 April 2022) software was used to assemble transcripts from the clean read data for each sample after quality control. After the completion of reference genome annotation, the read alignment rate was generally higher than 65% (total mapped reads). By searching in the Rfam11.0 database, the filtered sequences were annotated. The Repeat, rRNA, tRNA, snRNA, and snoRNA were first determined, and then the remaining sequences (sequences with a length of 18–25 nt) were used to identify known miRNAs and predict new miRNAs (unknown miRNAs).

5.5. Identification of Known miRNAs and Prediction of New miRNAs

The reads of the reference genome with the miRBase and Rfam databases were compared to obtain known miRNA and ncRNA annotation information, respectively. The miRNA identification and prediction process includes three parts: identification of known miRNAs, prediction of new miRNAs, and sRNA statistics. (1) Identification of known miRNAs: compared the mapped reads to the reference genome with the miRNA precursor and mature sequences in the miRBase database, counted the detail of the matched miRNAs in each sample, and predicted its secondary structure. (2) Prediction of new miRNAs: miReap software (Version 3) (http://mireap.sourceforge.net/ (accessed on 11 April 2022) was used to predict new miRNAs from reads that do not have annotation information. sRNA (that was unable to align) was aligned with Rfam and miRBase to the reference the genome and intercepted its near sequences. Rfam and miRBase software was used to predict the secondary structure of pre-miRNAs as well. Based on the prediction results, dicer digestion site information, energy values, and other features were used to filter and identify new miRNA. (3) sRNA statistics: the Rfam database (http://rfam.xfam.org/ (accessed on 20 April 2022) was used to annotate reads of known miRNAs that were not aligned—as well as filtered ribosomal RNA (rRNA), transfer RNA (tRNA), nuclear small RNA (snRNA), small nucleolar RNA (snoRNA), and other ncRNAs and repetitive sequences—counting the types and numbers of these sequences to obtain unannotated reads containing potential miRNAs.

5.6. Differential Expression Analysis of miRNAs

After gene expression analysis, the read count of transcripts was carried out; the expression level of each transcript was calculated based on the average number of transcripts per million, corresponding to a specific gene or transcript in the sample (TPM) method. RSEM (http://deweylab.biostat.wisc.edu/rsem/ (accessed on 22 April 2022) was used to quantify gene abundance. DEGseq was utilized to perform differential gene expression analysis, and the screening parameters were $|\log_2 FC| > 1$, Q value (p-adjust value) ≤ 0.05. The Majorbio cloud platform (Majorbio Biomedical Technology Co., Ltd., Shanghai, China) was used to get TPM values and DEG miRNAs (differentially expressed miRNAs) data.

5.7. Target Gene Prediction and Enrichment Analysis of miRNA

The online software (Version 2, 2017 release)PsRNATarget (https://www.zhaolab.org/psRNATarget/analysis?function=1 (accessed on 18 June 2022) was used to predict target genes of differentially expressed miRNAs; the maximum expected value was set to 5. Enrichment analysis of predicted target genes was performed through Gene Ontology (GO) and Kyoto Encyclopedia of Genes and Genomes (KEGG) analysis. The software Goatools (Version 0.6.5) and the Fisher method were used to test p value (p-adjust value) < 0.05.

5.8. Dual Luciferase Reporter Gene Assay to Verify Target Genes

A dual-luciferase reporter gene transient expression system was used to study the relationship between miRNA and target genes. Using vvi-miR828a as the effector and the corresponding target gene *VvMYBPA1* as the reporter, we explored the regulatory effect of vvi-miR828a on its target gene. Using 'Italia' grape cDNA as a template, the target sequence of vvi-miR828a and its flanking sequences (150 bp fragments upstream and downstream of the target sequence) were amplified and cloned into the intermediate vector pEASY®-Blunt. At the same time, the target sequence was artificially changed, and three base CTAs were added to the 10~11th bases of the target sequence. The sequence was synthesized by Sangon Biotech; it was then cloned into the intermediate vector pEASY®-Blunt. After verification by DNA sequencing, the plasmid was extracted and cloned into the pGreenII 0800-LUC vector (at the 3′ end of the LUC gene) that had been double digested with EcoRI and HindIII to complete the construction of the vector. The *VvMYBPA1* CDS sequence was subsequently cloned using the 'Italia' grape cDNA template. The *VvMYBPA1* CDS sequence was constructed into the pBI-121 vector through a homologous recombination method. It was then transformed into *E. coli* competent Trans5α, and positive clones were selected after overnight culture on a resistant medium. After bacterial liquid PCR and sequencing verification, the positive bacterial liquid was saved. Finally, a plasmid extraction kit (TransGen Biotech Co., Ltd., Beijing, China) was used to extract the plasmid from the positive bacterial fluid, and the fusion plasmid was transformed into an Agrobacterium GV3101 (pSoup) strain by the freeze–thaw method for further experimentation.

The cultured Agrobacterium single colony was activated in a LB liquid medium containing 50 ng/μL Kan/Spec and 25 ng/μL Rif. After collecting the bacterial liquid, a centrifugation step was carried out, the supernatant was discarded, and the bacteria dried. The bacteria were resuspended in the preprepared infection solution and incubated for about 4 h under dark conditions. During the induction period, a mixture of effectors and reporters (9:1) was prepared and injected back into the selected tobacco leaves. After the injection, the tobacco seedlings were cultured in dark conditions for 12 h and then transferred to the conventional culture conditions to continue culturing for 2 d. After the culture, the corresponding dual-luciferase activity was measured using a GloMax® Navigator System (Promega, Madison, WI, USA). After taking the corresponding number of tobacco leaf samples, they were ground into a powder form and treated with a lysis stock solution. The solution was centrifuged at 6000 rpm for 5~10 min, and the supernatant was used for dual-luciferase activity measurement. Four biological replicates were performed for each combined sample and protected from light during the determination. The supernatant was transferred to the enzyme plate and immediately the luciferase reaction solution was added to the LUC value. Stop & Glo reaction solution was added to the same mixture to observe the REN value. The LUC/REN ratio was calculated, and SPSS v17.0 software was used for statistical analysis. A T-test was used to determine the significant difference ($p < 0.05$).

5.9. Construction of Transient Overexpression Vector

The method of cloning the vvi-miR828a precursor stem-loop sequence using the cDNA of 'Italia' grapes as a template was as follows: the PCR reaction system included 1 μL of each forward and reverse primer, 1 μL of cDNA template, 22 μL of ddH_2O, 25 μL of high-fidelity enzyme, and 50 μL for the total system. The PCR program was set as follows: pre-denature at 98 °C for 1 min, then perform 30 cycles of the following steps: denaturation at 98 °C for 30 s, annealing at 60 °C for 15 s, extension at 72 °C for 30 s, and finally store at 4 °C. The purified target fragment pre-miR828a was then ligated into the intermediate vector pEASY®-Blunt (TransGen Biotech Biotechnology Co., Ltd., Beijing, China) vector using T4 DNA ligase. The ligated products were transformed into *E. coli* to confirm positive clones. Positive clones were inoculated into LB liquid medium containing Kan, and after culturing for 12 h at 200 rpm and 37 °C, the pEASY-pre-miR828a plasmid was extracted using the EasyPure plasmid extraction kit (TransGen Biotech Biotechnology Co., Ltd., Beijing, China). Using the extracted plasmid as a template, the seamless cloning kit (Beyotime Biotechnology Co., Ltd., Shanghai, China) was used to clone it into the pBI-121 vector, which was driven by the CaMV35S strong promoter to generate the recombinant overexpression vector pBI-121-OX-miR828a. Finally, OX-miR828a was used as a template to transform it into Agrobacterium competent GV3101 for later use. The cDNA of 'Italia' grapes was used to clone the CDS sequence of *VvMYBPA1*, and the subsequent method is the same as 1.8.

5.10. Transient Overexpression Test

After selecting a single colony from the constructed pBI-121-OX-miR828a and pBI-121-*VvMYBPA1* vectors, it was inoculated in 20 mL LB liquid culture medium containing 50 ng/mL Kan and 25 ng/mL Rif and incubated at 28 °C at 200 rpm to inactivate the strains overnight. When the OD600 of the bacterial solution reached approximately 0.8~1.0, the bacterial solution was centrifuged at 6000 rpm for 5~10 min to collect the bacterial cells. Subsequently, the collected bacteria were re-suspended in a suspension containing 200 mM acetosyringone, 200 mM magnesium chloride, and 200 mM MES (pH 5.7), and the optical density was adjusted to 1.0 and placed in the dark at room temperature for 4 h. The 2nd to 3rd spine grape leaves were selected below the growth. Three leaves from each group were selected for treatment. The vacuum infiltration method was utilized to infect the grape leaves. The leaves were removed after 10 min of infection and the leaf surface was gently wiped with filter paper to remove excess infection fluid. The petiole was covered with soaked absorbent cotton and the leaf placed in the LED light incubator. Samples were collected on the 4th day post treatment for subsequent analysis. The strawberry cultivar 'Benihoppe' (*Fragaria* × *ananassa*) was grown in a glass greenhouse and used in the transient overexpression experiment. A quantity of 1 mL extract from the prepared bacterial liquid was injected in the selected strawberry fruits before the veraison stage. After injection, fruit phenotype changes were continuously observed every day for 7 days.

5.11. Real-Time Fluorescence Quantitative PCR (qRT-PCR) Verification

Verification of miRNA828a expression by qRT-PCR: extract from grape peels was used to perform reverse transcription of miR828a using the stem-loop primer method as described by Shi [49]; after the reverse transcription, a qRT-PCR test was performed using SYBR Green Premix Pro Taq HS qPCR Kit on a fluorescence quantitative PCR instrument

(Hangzhou Bioer Technology Co., Ltd., Hangzhou, China). qRT-PCR conditions were as follows: pre-denaturation at 95 °C for 1 min; denaturation at 95 °C for 15 s, extension at 60 °C for 1 min, 40 cycles. U6 (primer: F: 3′-CCGATAAAATTGGAACGATACAGAG-5′; R: 3′TCGATTTGTGCGTGT-CATCCT-5′) was used as housekeeping miRNA. All qRT-PCR reactions were performed with 3 technical replicates. SPSS Statistics 20 software was used to analyze the significant differences in expression levels between different samples. qRT-PCR verification of genes related to anthocyanin and proanthocyanidin synthesis in transient overexpression and control RNA samples was performed. The expression levels of genes related to anthocyanin and proanthocyanidin synthesis *VdPAL, VdCHS, VdCHI, VdDFR, VdMYB5b, VdANR*, and *VdMYBPA1* in the RNA of transient overexpression and control samples were analyzed using qRT-PCR, and *VdActin* was used as the housekeeping gene. The PCR reaction procedure refers to the instructions of the SYBR Green Premix Pro Taq HS qPCR Kit: pre-denaturation at 94 °C for 30 s; denaturation at 94 °C for 5 s, annealing at 60 °C for 15 s, extension at 72 °C for 10 s, 40 cycles. The relative expression of each gene was calculated using the $2^{-\Delta\Delta Ct}$ method.

Supplementary Materials: The following supporting information can be downloaded at: https://www.mdpi.com/article/10.3390/plants13121688/s1, Figure S1: Clustering trend analysis of miRNA subclusters. Figure S2: Prediction of miRNA Target Genes. Figure S3: Expression Clustering Heatmap of miRNAs. Table S1: All known differentially expressed miRNAs.

Author Contributions: Conceptualization, T.G., S.J., K.L. and G.Y.; Methodology, M.B.; Data curation, J.H.; Writing—review & editing, L.Y.; Supervision, M.K.-U.-R. and Y.X. All authors have read and agreed to the published version of the manuscript.

Funding: This work was supported by grants from the National Natural Science Foundation of China Youth Fund (No. 32002015). Natural Science Foundation Youth Fund of Hunan Provincial (2021JJ40252).

Data Availability Statement: The raw sequencing data has been uploaded to NCBI; The login number is: PRJNA1121963.

Conflicts of Interest: The authors declare no conflict of interest.

References

1. Liu, Z.; Wang, H.; Zhang, J.; Chen, Q.; He, W.; Zhang, Y.; Luo, Y.; Tang, H.; Wang, Y.; Wang, X. Comparative metabolomics profiling highlights unique color variation and bitter taste formation of Chinese cherry fruits. *Food Chem.* **2024**, *439*, 138072. [CrossRef]
2. Vidal, S.; Francis, L.; Noble, A.; Kwiatkowski, M.; Cheynier, V.; Waters, E. Taste and mouth-feel properties of different types of tannin-like polyphenolic compounds and anthocyanins in wine. *Anal. Chim. Acta* **2004**, *513*, 57–65. [CrossRef]
3. Forkner, R.E.; Marquis, R.J.; Lill, J.T. Feeny revisited: Condensed tannins as anti-herbivore defences in leaf-chewing herbivore communities of Quercus. *Ecol. Entomol.* **2004**, *29*, 174–187. [CrossRef]
4. Wrangham, R.W.; Conklin-Brittain, N.L.; Hunt, K.D. Dietary response of chimpanzees and cercopithecines to seasonal variation in fruit abundance. I. Antifeedants. *Int. J. Primatol.* **1998**, *19*, 949–970. [CrossRef]
5. García-Lafuente, A.; Guillamón, E.; Villares, A.; Rostagno, M.A.; Martínez, J.A. Flavonoids as anti-inflammatory agents: Implications in cancer and cardiovascular disease. *Inflamm. Res.* **2009**, *58*, 537–552. [CrossRef]
6. Xia, H.; Zhang, X.; Shen, Y.; Guo, Y.; Wang, T.; Wang, J.; Lin, L.; Deng, H.; Deng, Q.; Xu, K.; et al. Comparative analysis of flavonoids in white and red table grape cultivars during ripening by widely targeted metabolome and transcript levels. *J. Food Sci.* **2022**, *87*, 1650–1661. [CrossRef] [PubMed]
7. Czemmel, S.; Heppel, S.C.; Bogs, J. R2R3 MYB transcription factors: Key regulators of the flavonoid biosynthetic pathway in grapevine. *Protoplasma* **2012**, *249* (Suppl. S2), 109–118. [CrossRef] [PubMed]
8. Du, H.; Zhang, L.; Liu, L.; Tang, X.F.; Yang, W.J.; Wu, Y.M.; Huang, Y.B.; Tang, Y.X. Biochemical and molecular characterization of plant MYB transcription factor family. *Biochemistry* **2009**, *74*, 1–11. [CrossRef] [PubMed]
9. Li, X.; Hong, Y.; Jackson, A.; Guo, F. Dynamic regulation of small RNAs in anthocyanin accumulation during blueberry fruit maturation. *Sci. Rep.* **2021**, *11*, 15080. [CrossRef]
10. Gu, K.D.; Wang, C.K.; Hu, D.G.; Hao, Y.J. How do anthocyanins paint our horticultural products? *Sci. Hortic.* **2019**, *249*, 257–262. [CrossRef]
11. Naing, A.H.; Kim, C.K. Roles of R2R3-MYB transcription factors in transcriptional regulation of anthocyanin biosynthesis in horticultural plants. *Plant Mol. Biol.* **2018**, *98*, 1–18. [CrossRef] [PubMed]

12. Wang, N.; Liu, W.; Zhang, T.; Jiang, S.; Xu, H.; Wang, Y.; Zhang, Z.; Wang, C.; Chen, X. Transcriptomic analysis of red-fleshed apples reveals the novel role of MdWRKY11 in flavonoid and anthocyanin biosynthesis. *J. Agric. Food Chem.* **2018**, *66*, 7076–7086. [CrossRef] [PubMed]
13. Deluc, L.; Bogs, J.; Walker, A.R.; Ferrier, T.; Decendit, A.; Merillon, J.M.; Robinson, S.P.; Barrieu, F. The transcription factor VvMYB5b contributes to the regulation of anthocyanin and proanthocyanidin biosynthesis in developing grape berries. *Plant Physiol.* **2008**, *147*, 2041–2053. [CrossRef] [PubMed]
14. Hichri, I.; Barrieu, F.; Bogs, J.; Kappel, C.; Delrot, S.; Lauvergeat, V. Recent advances in the transcriptional regulation of the flavonoid biosynthetic pathway. *J. Exp. Bot.* **2011**, *62*, 2465–2483. [CrossRef] [PubMed]
15. Wang, C.; Wang, L.; Ye, J.; Xu, F. Fruit quality of Vitis vinifera: How plant metabolites are affected by genetic, environmental, and agronomic factors. *Sci. Hortic.* **2022**, *305*, 111404. [CrossRef]
16. Zhao, M.; Li, J.; Zhu, L.; Chang, P.; Li, L.; Zhang, L. Identification and characterization of MYB-bHLH-WD40 regulatory complex members controlling anthocyanidin biosynthesis in blueberry fruits development. *Genes* **2019**, *10*, 496. [CrossRef] [PubMed]
17. Schaart, J.G.; Dubos, C.; Romero De La Fuente, I.; van Houwelingen, A.M.; de Vos, R.C.; Jonker, H.H.; Xu, W.; Routaboul, J.M.; Lepiniec, L.; Bovy, A.G. Identification and characterization of MYB-b HLH-WD 40 regulatory complexes controlling proanthocyanidin biosynthesis in strawberry (F ragaria × ananassa) fruits. *New Phytol.* **2013**, *197*, 454–467. [CrossRef] [PubMed]
18. Deluc, L.; Barrieu, F.; Marchive, C.; Lauvergeat, V.; Decendit, A.; Richard, T.; Carde, J.P.; Mérillon, J.M.; Hamdi, S. Characterization of a grapevine R2R3-MYB transcription factor that regulates the phenylpropanoid pathway. *Plant Physiol.* **2006**, *140*, 499–511. [CrossRef]
19. Bogs, J.; Jaffé, F.W.; Takos, A.M.; Walker, A.R.; Robinson, S.P. The grapevine transcription factor VvMYBPA1 regulates proanthocyanidin synthesis during fruit development. *Plant Physiol.* **2007**, *143*, 1347–1361. [CrossRef]
20. Kobayashi, S.; Ishimaru, M.; Hiraoka, K.; Honda, C. Myb-related genes of the Kyoho grape (*Vitis labruscana*) regulate anthocyanin biosynthesis. *Planta* **2002**, *215*, 924–933.
21. Kobayashi, S.; Goto-Yamamoto, N.; Hirochika, H. Retrotransposon-induced mutations in grape skin color. *Science* **2004**, *304*, 982. [CrossRef] [PubMed]
22. Walker, A.R.; Lee, E.; Bogs, J.; McDavid, D.A.; Thomas, M.R.; Robinson, S.P. White grapes arose through the mutation of two similar and adjacent regulatory genes. *Plant J.* **2007**, *49*, 772–785. [CrossRef]
23. Azuma, A.; Kobayashi, S.; Mitani, N.; Shiraishi, M.; Yamada, M.; Ueno, T.; Kono, A.; Yakushiji, H.; Koshita, Y. Genomic and genetic analysis of Myb-related genes that regulate anthocyanin biosynthesis in grape berry skin. *Theor. Appl. Genet.* **2008**, *117*, 1009–1019. [CrossRef] [PubMed]
24. Yang, K.; Han, H.; Li, Y.; Ye, J.; Xu, F. Significance of miRNA in enhancement of flavonoid biosynthesis. *Plant Biol.* **2022**, *24*, 217–226. [CrossRef] [PubMed]
25. Huang, X.; Xia, R.; Liu, Y. microRNA-mediated regulation in fruit quality. *Curr. Opin. Food Sci.* **2022**, *46*, 100837. [CrossRef]
26. Zhang, C.J.; Mo, B.X.; Chen, X.M.; Cui, J. Advances on the molecular action mechanisms of plant miRNA. *Biotechnol. Bull.* **2020**, *36*, 1.
27. Carthew, R.W.; Sontheimer, E.J. Origins and mechanisms of miRNAs and siRNAs. *Cell* **2009**, *136*, 642–655. [CrossRef] [PubMed]
28. Voinnet, O. Origin, biogenesis, and activity of plant microRNAs. *Cell* **2009**, *136*, 669–687. [CrossRef]
29. Wang, J.; Mei, J.; Ren, G. Plant microRNAs: Biogenesis, homeostasis, and degradation. *Front. Plant Sci.* **2019**, *10*, 433008. [CrossRef]
30. Wahid, F.; Shehzad, A.; Khan, T.; Kim, Y.Y. MicroRNAs: Synthesis, mechanism, function, and recent clinical trials. *Biochim. Biophys. Acta (BBA)-Mol. Cell Res.* **2010**, *1803*, 1231–1243. [CrossRef]
31. Bartel, D.P. MicroRNAs: Genomics, biogenesis, mechanism, and function. *Cell* **2004**, *116*, 281–297. [CrossRef]
32. Gou, J.Y.; Felippes, F.F.; Liu, C.J.; Weigel, D.; Wang, J.W. Negative regulation of anthocyanin biosynthesis in Arabidopsis by a miR156-targeted SPL transcription factor. *Plant Cell* **2011**, *23*, 1512–1522. [CrossRef] [PubMed]
33. Ramsay, N.A.; Glover, B.J. MYB–bHLH–WD40 protein complex and the evolution of cellular diversity. *Trends Plant Sci.* **2005**, *10*, 63–70. [CrossRef]
34. Cui, M.; Wang, C.; Zhang, W.; Pervaiz, T.; Haider, M.S.; Tang, W.; Fang, J. Characterization of Vv-miR156: Vv-SPL pairs involved in the modulation of grape berry development and ripening. *Mol. Genet. Genom.* **2018**, *293*, 1333–1354. [CrossRef]
35. Li, X.; Hou, Y.; Xie, X.; Li, H.; Li, X.; Zhu, Y.; Zhai, L.; Zhang, C.; Bian, S. A blueberry MIR156a–SPL12 module coordinates the accumulation of chlorophylls and anthocyanins during fruit ripening. *J. Exp. Bot.* **2020**, *71*, 5976–5989. [CrossRef] [PubMed]
36. Saminathan, T.; Bodunrin, A.; Singh, N.V.; Devarajan, R.; Nimmakayala, P.; Jeff, M.; Aradhya, M.; Reddy, U.K. Genome-wide identification of microRNAs in pomegranate (*Punica granatum* L.) by high-throughput sequencing. *BMC Plant Biol.* **2016**, *16*, 122.
37. Zaman, F.; Zhang, M.; Liu, Y.; Wang, Z.; Xu, L.; Guo, D.; Luo, Z.; Zhang, Q. DkmiR397 regulates proanthocyanidin biosynthesis via negative modulating DkLAC2 in Chinese PCNA persimmon. *Int. J. Mol. Sci.* **2022**, *23*, 3200. [CrossRef] [PubMed]
38. Chen, Q.; Deng, B.; Gao, J.; Zhao, Z.; Chen, Z.; Song, S.; Wang, L.; Zhao, L.; Xu, W.; Zhang, C.; et al. Comparative analysis of miRNA abundance revealed the function of Vvi-miR828 in fruit coloring in root restriction cultivation grapevine (*Vitis vinifera* L.). *Int. J. Mol. Sci.* **2019**, *20*, 4058. [CrossRef]
39. Tirumalai, V.; Swetha, C.; Nair, A.; Pandit, A.; Shivaprasad, P.V. miR828 and miR858 regulate VvMYB114 to promote anthocyanin and flavonol accumulation in grapes. *J. Exp. Bot.* **2019**, *70*, 4775–4792. [CrossRef]

40. Wen, W.; Fang, H.; Yue, L.; Khalil-Ur-Rehman, M.; Huang, Y.; Du, Z.; Yang, G.; Xu, Y. RNA-Seq Based Transcriptomic Analysis of Bud Sport Skin Color in Grape Berries. *Horticulturae* **2023**, *9*, 260. [CrossRef]
41. Zhang, X.; Ma, W.; Guan, X.; Wang, F.; Fan, Z.; Gao, S.; Yao, Y. VvMYB14 participates in melatonin-induced proanthocyanidin biosynthesis by upregulating expression of VvMYBPA1 and VvMYBPA2 in grape seeds. *Hortic. Res.* **2023**, *10*, uhac274. [CrossRef] [PubMed]
42. He, F.; Mu, L.; Yan, G.L.; Liang, N.N.; Pan, Q.H.; Wang, J.; Reeves, M.J.; Duan, C.Q. Biosynthesis of anthocyanins and their regulation in colored grapes. *Molecules* **2010**, *15*, 9057–9091. [CrossRef] [PubMed]
43. Xiao-Yun, J.I.; Zhi-Qin, Y.U.; Gui-Liang, T.A.; Lei-Hao, J.I.; Li, Z.H.; Li-Heng, H.E.; Run-Zhi, L.I. Cloning of Arabidopsis At-pri-miR828 gene and its genetic transformation into tomato. *Acta Hortic. Sin.* **2013**, *40*, 2419.
44. Yue, J.; Liu, Z.; Zhao, C.; Zhao, J.; Zheng, Y.; Zhang, H.; Tan, C.; Zhang, Z.; Xue, L.; Lei, J. Comparative transcriptome analysis uncovers the regulatory roles of MicroRNAs involved in petal color change of pink-flowered strawberry. *Front. Plant Sci.* **2022**, *13*, 854508. [CrossRef] [PubMed]
45. Nayak, S.P.; Prasad, P.; Singh, V.; Tripathi, A.M.; Bag, S.K.; Mohanty, C.S. Role of miRNAs in the regulation of proanthocyanidin biosynthesis in the legume Psophocarpus tetragonolobus (L.) DC. *Plant Growth Regul.* **2024**, *102*, 23–38. [CrossRef]
46. Mei, J.; Niu, Q.; Xu, K.; Huang, Y.; Bai, S.; Zhu, J.; Li, H.; Miao, M.; Tong, F.; Yu, D.; et al. GhmiR858 Inhibits the Accumulation of Proanthocyanidins by Targeting GhTT2L in Cotton (*Gossypium hirsutum*). *J. Agric. Food Chem.* **2023**, *71*, 15341–15351. [CrossRef]
47. Zhang, B.; Yang, H.J.; Qu, D.; Zhu, Z.Z.; Yang, Y.Z.; Zhao, Z.Y. The MdBBX22–miR858–MdMYB9/11/12 module regulates proanthocyanidin biosynthesis in apple peel. *Plant Biotechnol. J.* **2022**, *20*, 1683–1700. [CrossRef]
48. Yang, S.; Zhang, M.; Xu, L.; Luo, Z.; Zhang, Q. MiR858b inhibits proanthocyanidin accumulation by the repression of DkMYB19 and DkMYB20 in persimmon. *Front. Plant Sci.* **2020**, *11*, 576378. [CrossRef]
49. Shi, F.; Xu, H.; Liu, C.; Tan, C.; Ren, J.; Ye, X.; Feng, H.; Liu, Z. Whole-transcriptome sequencing reveals a vernalization-related ceRNA regulatory network in chinese cabbage (*Brassica campestris* L. ssp. pekinensis). *BMC Genom.* **2021**, *22*, 819.

plants MDPI

Article

Genome-Wide Investigation of Class III Peroxidase Genes in *Brassica napus* Reveals Their Responsiveness to Abiotic Stresses

Obaid Ullah Shah [1], Latif Ullah Khan [1], Sana Basharat [2], Lingling Zhou [1], Muhammad Ikram [1], Jiantao Peng [1], Wasi Ullah Khan [1], Pingwu Liu [1,*] and Muhammad Waseem [1,*]

1 School of Breeding and Multiplication (Sanya Institute of Breeding and Multiplication), College of Tropical Agriculture and Forestry, Hainan University, Sanya 572025, China; obaidus890@gmail.com (O.U.S.); latif.hainu3103@gmail.com (L.U.K.); m15237352151@163.com (L.Z.); drikram@gzhu.edu.cn (M.I.); jiantaopeng@hainanu.edu.cn (J.P.); wasibiotechnologist@gmail.com (W.U.K.)

2 Department of Botany, University of Agriculture Faisalabad, Faisalabad 38000, Pakistan; sanabasharat@gmail.com

* Correspondence: hnulpw@hainanu.edu.cn (P.L.); m.waseem.botanist@gmail.com (M.W.)

Abstract: *Brassica napus* (*B. napus*) is susceptible to multiple abiotic stresses that can affect plant growth and development, ultimately reducing crop yields. In the past, many genes that provide tolerance to abiotic stresses have been identified and characterized. Peroxidase (POD) proteins, members of the oxidoreductase enzyme family, play a critical role in protecting plants against abiotic stresses. This study demonstrated a comprehensive investigation of the *POD* gene family in *B. napus*. As a result, a total of 109 *POD* genes were identified across the 19 chromosomes and classified into five distinct subgroups. Further, 44 duplicate events were identified; of these, two gene pairs were tandem and 42 were segmental. Synteny analysis revealed that segmental duplication was more prominent than tandem duplication among *POD* genes. Expression pattern analysis based on the RNA-seq data of *B. napus* indicated that *BnPOD* genes were expressed differently in various tissues; most of them were expressed in roots rather than in other tissues. To validate these findings, we performed RT-qPCR analysis on ten genes; these genes showed various expression levels under abiotic stresses. Our findings not only furnish valuable insights into the evolutionary dynamics of the *BnPOD* gene family but also serve as a foundation for subsequent investigations into the functional roles of *POD* genes in *B. napus*.

Keywords: peroxidases; expression pattern; abiotic stress; rapeseed

1. Introduction

Peroxidases are a large family of ubiquitous isozymes that play an important role in plant growth, development, and defense mechanisms [1] and have been categorized into hemoglobin peroxidases and non-hemoglobin peroxidases [2,3]. Non-hemoglobin peroxidases include (i) class I peroxidases, which are mainly involved in oxidative stress tolerance [4]; (ii) class II peroxidases, which are found in fungi and play an essential role in lignin metabolism [5]; and (iii) class III peroxidases, which are a plant-specific ubiquitous multigenic family and play pivotal roles in plant growth, development, and stress responses [6,7]. POD proteins have highly conserved amino acid residues essential for glycosylation and cysteine–histidine catalytic reactions [7] and are involved in different physiological and developmental processes, including cell wall metabolism, lignification, auxin metabolism, and tolerance and defense against biotic and abiotic stresses [8–10].

Previously, many functional studies have identified the role of PODs in the tolerance of crop plants to different environmental stresses [11–13]. For example, the overexpression of AtPrx64 is involved in aluminum tolerance in *Nicotiana tabacum* [11], whereas the overexpression of the cold-inducible peroxidase gene (RCI3) in Arabidopsis shows increases in dehydration and salt tolerances [12]. Similarly, the CrPrx and CrPrx1 genes from

Catharanthus roseus are involved in germination and exhibit better cold tolerance in tobacco [13], while a mutant of TaPRX-2A in Triticum aestivum is involved in salinity tolerance [14]. In RNA-seq and microarrays, the expression profiles of several peroxidase (PRX) genes showed differential expressions under different stresses in wheat and maize, indicating the role of peroxidase in response to abiotic stresses [15,16]. These findings suggested that plant PODs are anticipated to be involved in abiotic stresses in multiple crop plants.

In recent years, genome-scale bioinformatics analyses of multigenic families, like peroxidases, have been greatly helpful in understanding the physiological roles and characteristics of these gene families. The *POD* gene family has been reported in numerous plant species, including *Arabidopsis thaliana* [17], *Manihot esculenta* [18], *Pyrus pyrifolia* [19], *Zea mays* [16], *Oryza sativa* [7], *Citrus sinensis* [20], *Vitis vinifera* [21], *Glycine max* [22], and *Capsicum annuum* L. [23]. Genetic evidence has indicated that *POD* serves as a family of enzymes responsive to abiotic stresses in diverse plant species [24].

Brassica napus is the second most abundant oilseed crop worldwide, contributing an estimated 12% of the world's vegetable oil production [25], and this production is influenced by environmental stresses [26]. For example, seed survival and germination were reduced owing to heat stress or high temperatures, while low growth and development were reportedly caused by drought stress, which ultimately led to lower crop production. In previous studies, the POD gene family was involved in stress tolerance in different crops, including rice, maize, sorghum, and soybeans. Therefore, the present study selected the POD gene family for genome-wide identification in *B. napus*, using multi-omics analysis for the characterization, classification, structural diversification, genomic distribution, and identification of POD genes in response to multiple abiotic stresses.

2. Results

2.1. Characterization of BnPOD Gene Family

A total of 109 genes were identified in the *B. napus* genome, using Arabidopsis POD proteins as query sequences after removing redundant sequences. All the identified POD proteins were renamed based on their physical position in the *B. napus* genome ZA11 viz. *BnAPOD01_BnCPOD109*. The predicted protein length and MW of the *POD* genes in *B. napus* varies from 279 aa/31.8118 kDa to 568 aa/62.91441 kDa. The theoretical pI ranged from 4.4 (*BnCPOD57*) to 10.11 (*BnAPOD13*), while GRAVY ranged from −0.355 (*BnAPOD18*) to 0.244 (*BnCPOD59*). Moreover, the in silico subcellular localization of all the *BnPOD*-predicted proteins revealed a wide range of cellular compartmentalization, including extracellular spaces, chloroplasts, and mitochondria (Table S1). Detailed information about the predicted protein length, pI, MW, chromosomal location, and other related information is shown in Table S1.

2.2. Phylogeny of POD Genes in B. napus

To determine the homology and similarity among the BnPOD proteins and with those in Arabidopsis, a phylogenetic tree was generated with a bootstrap set at 1000 replicates. The phylogenetic analyses showed that all the *PODs* were clustered into five phylogenetic subclades, namely from A to E (Figure 1), based on the gene structure. The large subclades, A and B, had 62 (35 *BnPODs* and 27 *AtPODs*) and 38 (26 *BnPODs* and 12 *AtPODs*) *PODs*, respectively, whereas 37 and 28 *PODs* were found in subclades E and D, respectively. Similarly, 21 *PODs*, including 7 *BnPODs* and 14 *AtPODs*, were found in subclade C (Figure 1). These findings suggested the diversification of POD proteins in *B. napus*.

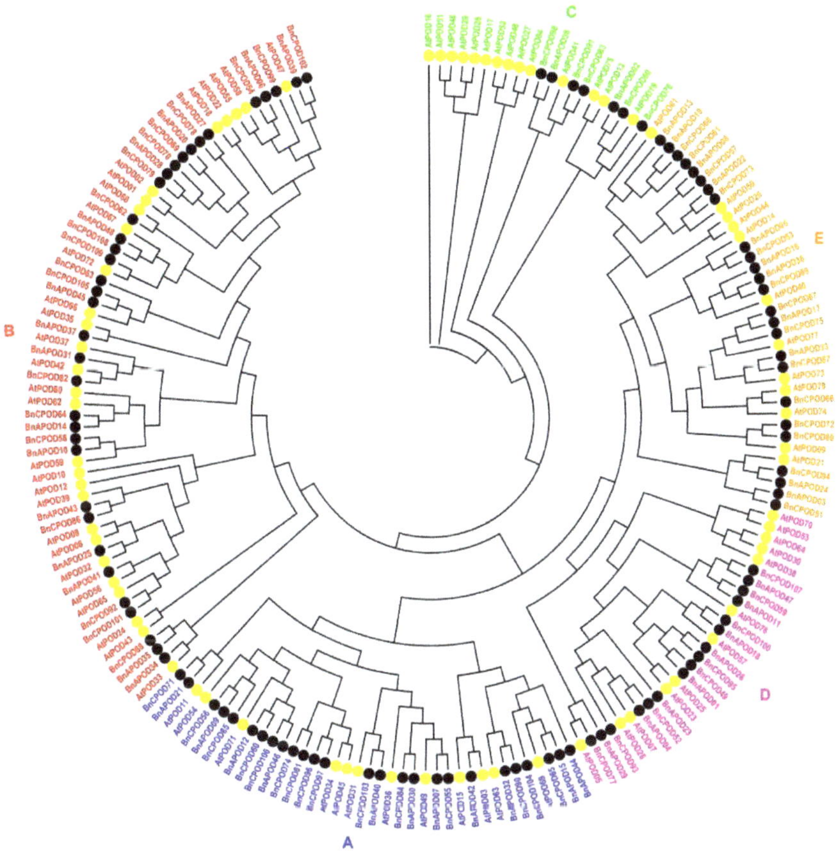

Figure 1. The phylogenetic tree of *B. napus* and *A. thaliana* POD proteins. The maximum likelihood phylogenetic tree was constructed using MEGA 11 with 1000 bootstrap replicates. The phylogenetic tree was clustered into 5 subclades (A–E). A distinct color represents each subclade.

2.3. Gene Structural Analysis of BnPOD Proteins

For this purpose, we used all the identified BnPOD genes for the detection of the exon–intron organization and conserved protein domains (Figure 2d). The exon–intron analysis revealed that the number of exons in *BnPODs* ranged from 2 to 7 among all the phylogenetic clades. For example, *BnCPOD81* had seven exons in clade A, while *BnAPOD14*, *BnAPOD15*, *BnCPOD57*, and *BnCPOD65* had three exons (Figure 2d). Similarly, *BnAPOD33* and *BnCPOD80* had four exons in clades B and E, whereas *BnAPOD38*, *BnCPOD83*, *BnCPOD91*, and *BnCPOD98* had two exons, but *BnCPOD50* had six exons in clade C. Generally, *POD* genes in the same subgroup show similar exon–intron features, providing evidence of their phylogenetic relationship (Figure 2d).

Furthermore, motif analysis was performed using the MEME database in accordance with the phylogenetic relationship. As a result, *BnPODs* that clustered in the same clade shared common motif compositions, indicating functional similarity among the members of the same *BnPOD* subclade. The InterProScan annotation of the MEME motif analyses revealed that eight motifs (motifs 1–4, 6, 7, 9, and 10) were noted as POD protein motifs, which are characteristic features of the peroxidase gene family. At least six motifs (1–4 and 7–10) were annotated as Heme peroxidases and were found in 95% of the sequences (Table 1 and Figure 2b). The *cis*-regulatory elements are involved in gene expression regulation.

We used 2 kb long promoter sequences of all the *BnPODs* to scan the different types of regulatory elements, using the PlantCARE database.

All the identified promoters were classified into two categories: (i) hormone-responsive and (ii) stress-responsive elements. Among the hormone-responsive elements, we identified ABRE (abscisic-acid-responsive element), the CGTCA/TGACG motif (methyl-jasmonate-responsive element), the TGA element (auxin-responsive element), and the TCA element (salicylic-acid-responsive element) (Figure 3 and Table S2). Similarly, among the stress-related responsive elements were LTR (low-temperature-responsive element), G-box (pathogen-inducible element/positive regulator of senescence), the WUN motif (wound-responsive element), MBS (MYB binding site in drought inducibility), TC-rich repeats (defense and stress responses), and STRE (stress-responsive element) (Figure 3 and Table S2). These findings indicate the diverse roles of class III peroxidase genes based on their gene structure.

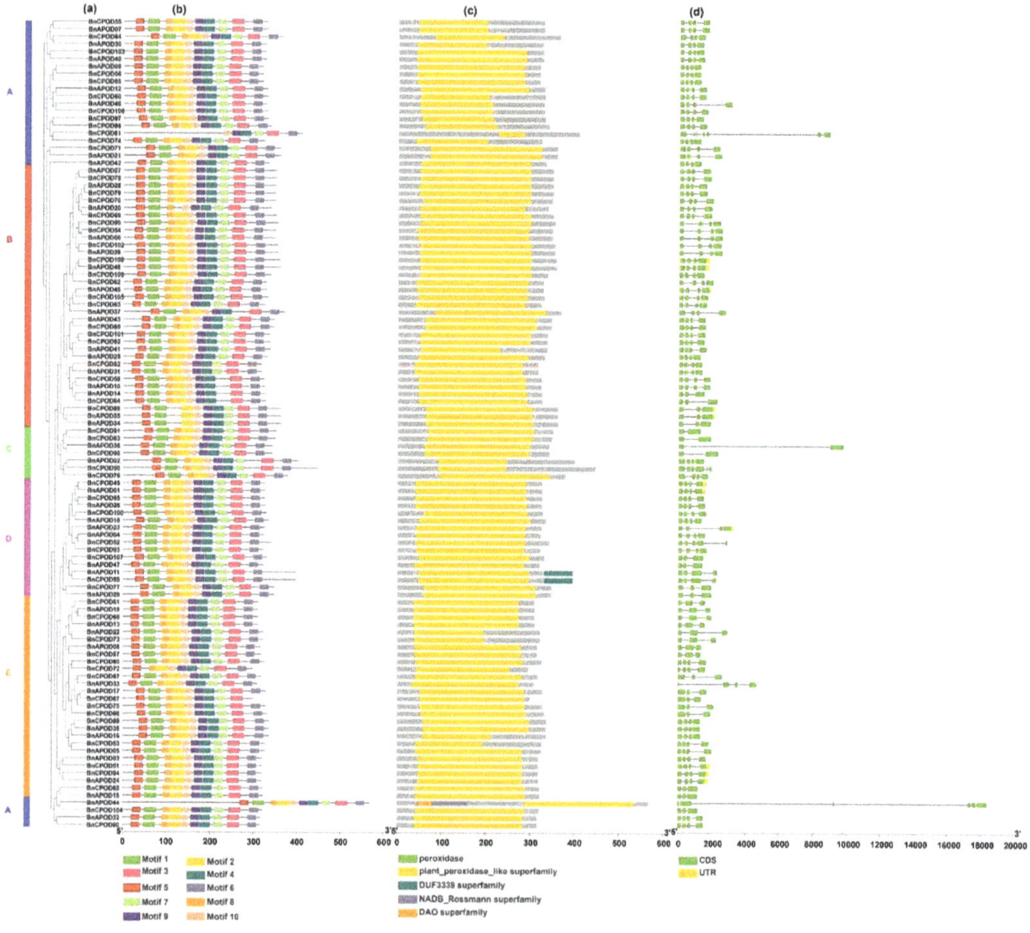

Figure 2. Gene structure of *POD* genes in *B. napus*. (**a**) The phylogenetic tree shows all the *BnPOD* genes in the five subclades. (**b**) Conserved motif analysis conducted using MEME Suite. A total of 10 motifs were predicted. (**c**) The domain organization of *BnPODs*. (**d**) Exon–intron organization of *BnPODs*. The A, B, C, D and E represent each subclades.

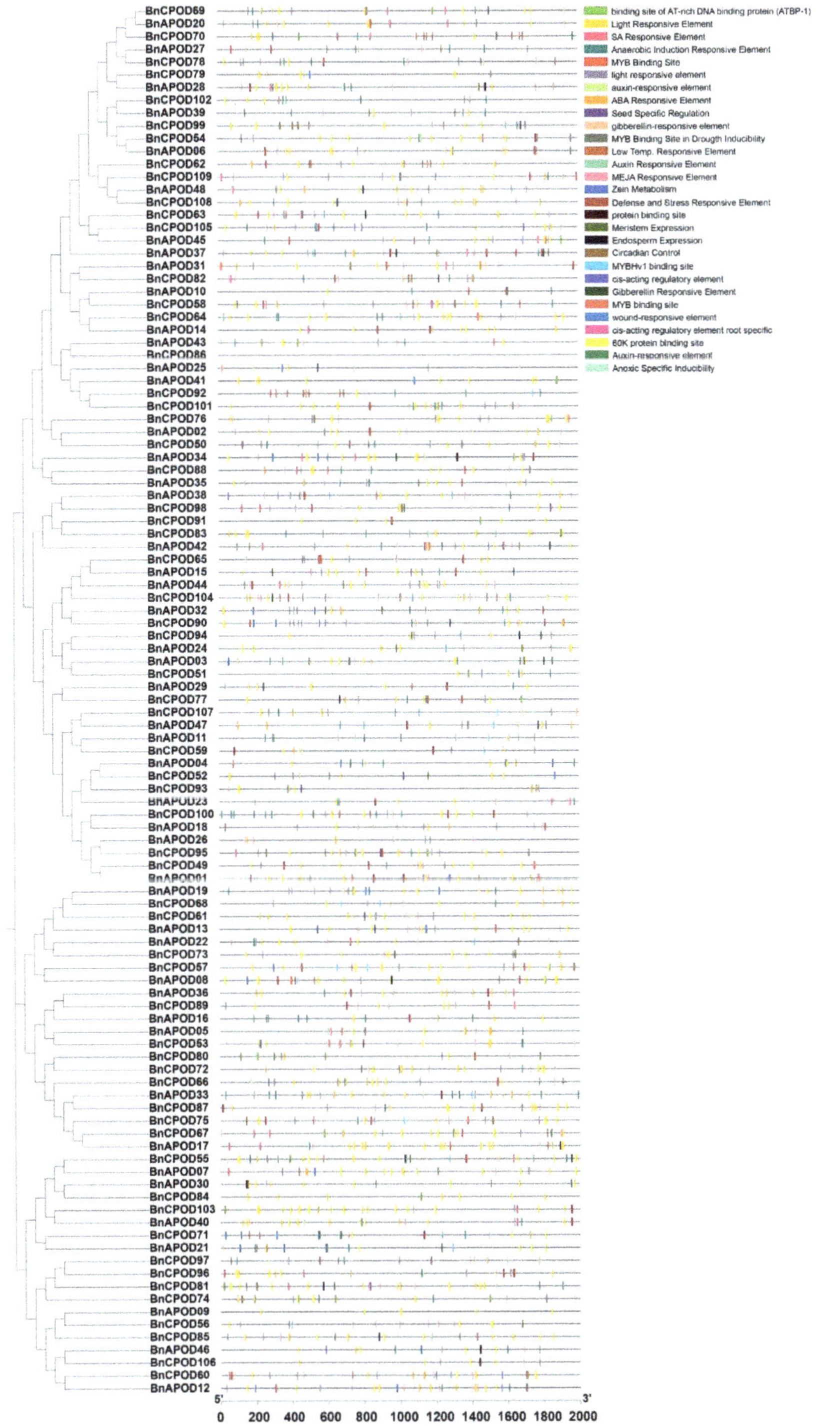

Figure 3. The analysis of the *BnPOD* promoter regions. The 2 kb sequences of the *BnPOD* gene-promoter regions were extracted from and analyzed using the PlantCARE database.

Table 1. Sequences of ten predicted MEME motifs with InterProScan annotations.

Motif	Sequence	InterProScan Annotation
Motif-1	DPRJAAALLRLHFHDCFVNGCDASVLLDS	Heme peroxidases
Motif-2	AACPGVVSCADILALAARDSVVLAGGPSW	Heme peroxidases
Motif-3	DPGTPNTFDNSYFKNLRQGKGLLQSDQAL	Heme peroxidases
Motif-4	KDLVALSGAHTIGFAHCGSFTBRL	Heme peroxidases
Motif-5	AQLSPGFYDKSCPNAESIVRN	-
Motif-6	FFRAFAKAMVKMGNIGVLTGSQ	Heme peroxidases
Motif-7	GDPDPTLBPTYAAQLRKKCPR	Heme peroxidases
Motif-8	SLRGFEVIDDIKAALE	-
Motif-9	PSPFDNVSQLITKFAAKGLNV	Heme peroxidases
Motif-10	VPLGRRDGRVSNASE	Heme peroxidases

2.4. Chromosomal Distribution and Duplication Events of POD Genes

The *BnPOD* genes were unevenly distributed in all the chromosomes in the *B. napus* genome. For example, chromosomes C02 and C09 had 10 BnPODs; A01, A02, C01, and C03 had 9 BnPODs each; A10/C07 had 6 BnPODs; A07/C05 had 5 BnPODs; A09, C04, C06, and C08 had 4 BnPODs; A04, A05, and A08 had 2 BnPODs; and A06 had only 1 BnPOD (Table S1). Moreover, the expansion of a gene family occurs owing to the duplication events arising at the whole genome, and we identified duplicate gene pairs among the BnPODs based on sequence similarity. As a result, 44 duplicate events corresponding to 109 BnaPODs were identified with >80% sequence similarity. Gene pairs were selected as tandemly duplicated if the sequence was found within a 100 kb window of the duplicated genomic region. According to these criteria, the *BnAPOD35–BnAPOD34* and *BnCPOD97–BnCPOD96* gene pairs were identified as tandemly duplicated (Table S3 and Figure 4). The remaining 43 BnPODs, including *BnCPOD55–BnAPOD07*, *BnCPOD84–BnAPOD30*, *BnCPOD100–BnAPOD18*, *BnAPOD04–BnCPOD52*, and *BnCPOD107–BnAPOD47*, were identified as segmentally duplicated (Table S3). C09, A01, A02, A03, A10, C01, C02, and C03 had undergone the highest number of segmental duplication events, indicating their key contribution to the expansion of the *BnPOD* family in *B. napus* (Figure 4).

The Ka, Ks, and Ka/Ks values are key indicators of selection pressures. In general, a Ka/Ks ratio of >1 refers to positive selection; a Ka/Ks ratio of <1 refers to purifying selection, and a Ka/Ks ratio equal to 1 refers to neutral selection. The Ka/Ks ratio varies between 0.05 and 0.65, indicating that all the *BnPOD* gene pairs exhibited negative or purifying selection (Ka/Ks < 1) (Table S3). These findings demonstrated that the purifying or negative selection was involved in maintaining the conservation of the *BnPOD* gene structure during the domestication or evolution process. Further, we estimated the divergence time between duplication events, and the results showed that the estimated divergence time ranged from 0.86 to 32.67 million years ago (MYA), with an average of ~6.15 MYA (Table S3). The syntenic relationships between different chromosome segments of different species provide a source of information about the origin of gene family members. For this purpose, we performed synteny analysis between *POD* genes from the *B. napus*, *B. rapa*, *B. oleracea*, and *Arabidopsis* genomes (Table S4 and Figure 5). Our results revealed that 55 *BnPODs* were placed in 106 colinear blocks with *B. oleracea*, 49 *BnPODs* in 102 colinear blocks in *B. rapa*, and 94 *BnPODs* in 127 colinear blocks in the *Arabidopsis* genome (Table S4 and Figure 5).

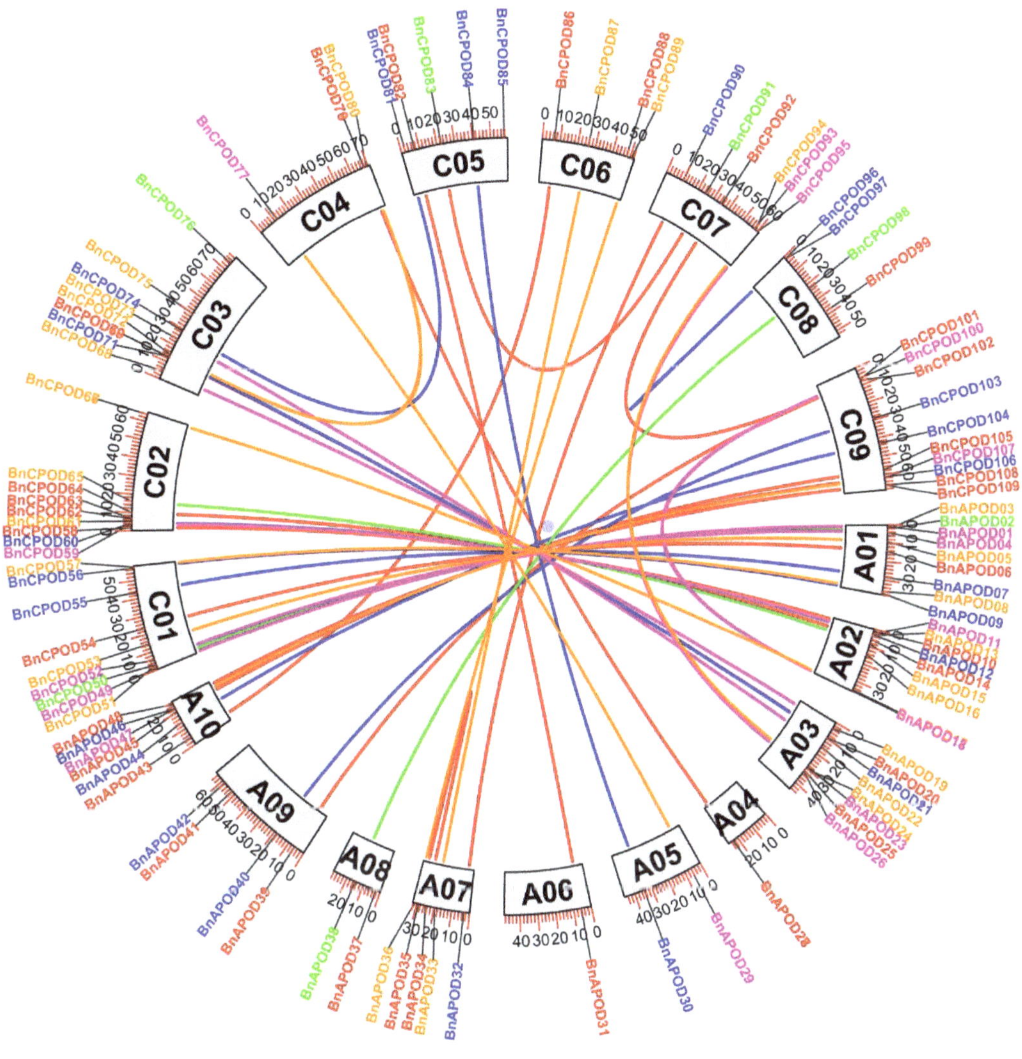

Figure 4. Circos plot of *POD* gene duplication in *B. napus*. The different colors represent the genes found in different (A–E subclades) subgroups, and the lines in the middle show segmental and tandem duplications between different chromosomes.

2.5. miRNA-Mediated BnPOD Regulation

In recent years, various studies have unveiled the importance of miRNA-mediated gene expression regulation in response to biotic and abiotic stresses. In this study, we identified two putative miRNAs targeting three *BnPOD* genes. The results showed that members of the Bn-miR167 family targeted *BnCPOD66* and members of the Bn-miR395 family targeted *BnAPOD44* and *BnCPOD104* (Table S5 and Figure 6).

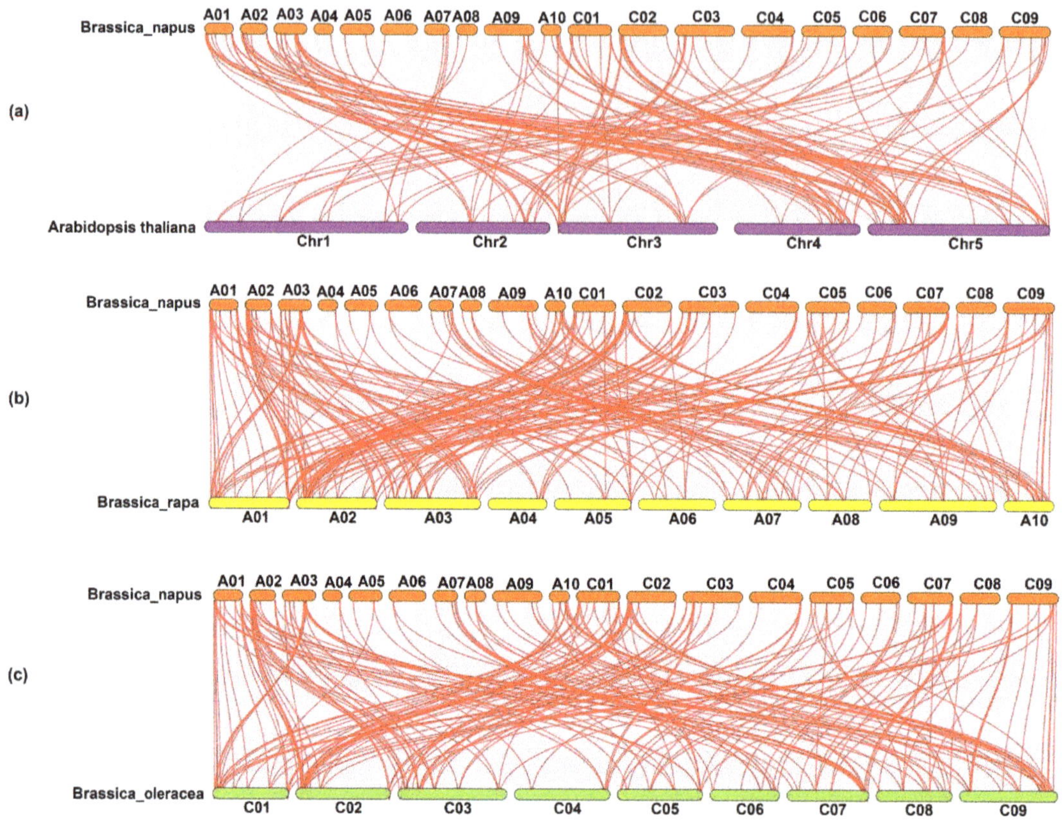

Figure 5. Dual synteny plots between (**a**) *B. napus* and *A. thaliana*, (**b**) *B. napus* and *B. rapa*, and (**c**) *B. napus* and *B. oleracea* genomes, with orthologous *POD* genes shown with red connecting lines.

2.6. Expression Profiling of BnPOD Genes in Various Tissues

The expression patterns of the *BnPODs* genes in different tissues, including the leaves, stems, roots, and seeds at 14 DAF (days after flowering), 28 DAF, 42 DAF, and 56 DAF and the siliquae at 10 DAF, 20 DAF, 30 DAF, 40 DAF, 50 DAF, and 60 DAF, were investigated. The analysis of the RNA-seq revealed spatial expression patterns of *BnPODs* in different tissues. For instance, most genes were expressed in roots compared with other tissues. The expressions of *BnCPOD50*, *BnAPOD17*, and *BnCPOD87* were confined to leaves, but *BnCPOD62*, *BnAPOD06*, *BnCPOD51*, *BnAPOD31*, and *BnCPOD71* were expressed only in 60 DAF siliquae. The expressions of *BnAPOD18*, *BnAPOD11*, *BnCPOD59*, *BnAPOD26*, *BnCPOD95*, *BnAPOD47*, and *BnCPOD107* were higher in seeds at 14 DAF, while *BnCPOD76*, *BnAPOD16*, and *BnCPOD87* were expressed in 56 DAF seeds (Figure 7). This temporal expression pattern indicated that the dynamic roles of the candidate *BnPODs* might play an important role in the growth and developmental processes of *B. napus*.

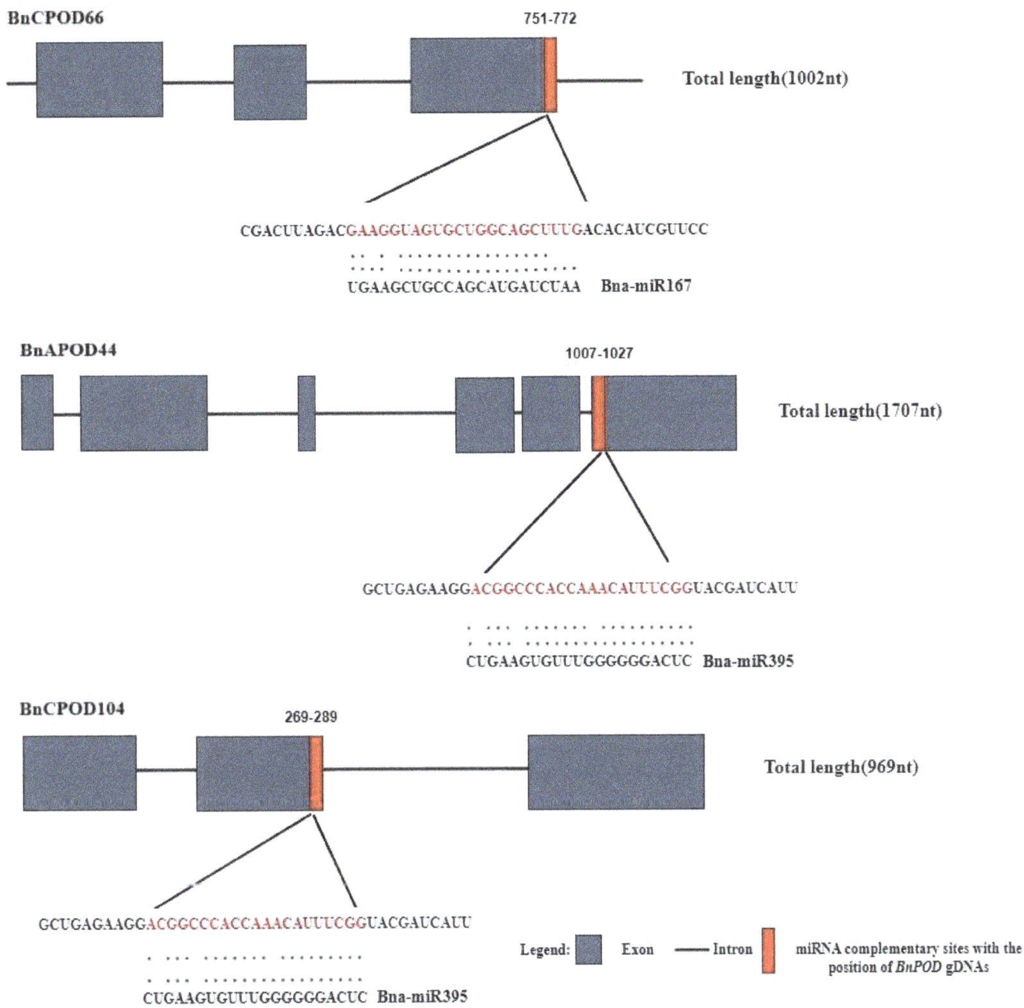

Figure 6. A network illustration of the regulatory associations among the putative miRNAs and *BnPOD* genes. The red color indicates the miRNA complimentary site with the position of *BnPODs* gDNAs.

2.7. Expression Patterns of BnPODs during Abiotic Stresses

We selected POD genes according to the RNA-seq expression profile; for example, *BnAPOD24*, *BnAPOD34*, and *BnCPOD99* were highly expressed in all the tissues; *BnCPOD62* and *BnCPOD84* were expressed in partial tissues, such as roots, leaves, stems, and siliquae; and *BnAPOD08* was highly expressed in leaf tissues. To investigate the expressions of the above *BnPODs* in *B. napus* during abiotic stresses (salinity, drought, heat, cold, and cadmium chloride), these tissues were analyzed at 0 h, 2 h, 4 h, and 6 h. With RT-qPCR, the expressions of ten selected *BnPODs* were examined under the applied abiotic stresses. The expression patterns of all the *BnPODs* were significantly changed under the applied stresses. For instance, under drought stress, all the *BnPODs* were initially downregulated at 2 h and 4 h but significantly upregulated at 6 h except for *BnCPOD62* and *BnCPOD84*. Under cold stress, only *BnAPOD34* was upregulated at 6 h, but *BnAPOD24* and *BnCPOD99* were upregulated at 4 h. Although *BnCPOD87* and *BnCPOD106* were upregulated at 2 h, they were downregulated in other time intervals. Likewise, under heat stress, the expressions of *BnAPOD08*, *BnAPOD24*, and *BnCPOD87* were upregulated at 6 h. However, it is interesting that all the selected genes were downregulated under the salt and heavy-metal stresses (Figure 8). Our results indicate that PODs might play a pivotal role in tolerance to applied abiotic stresses.

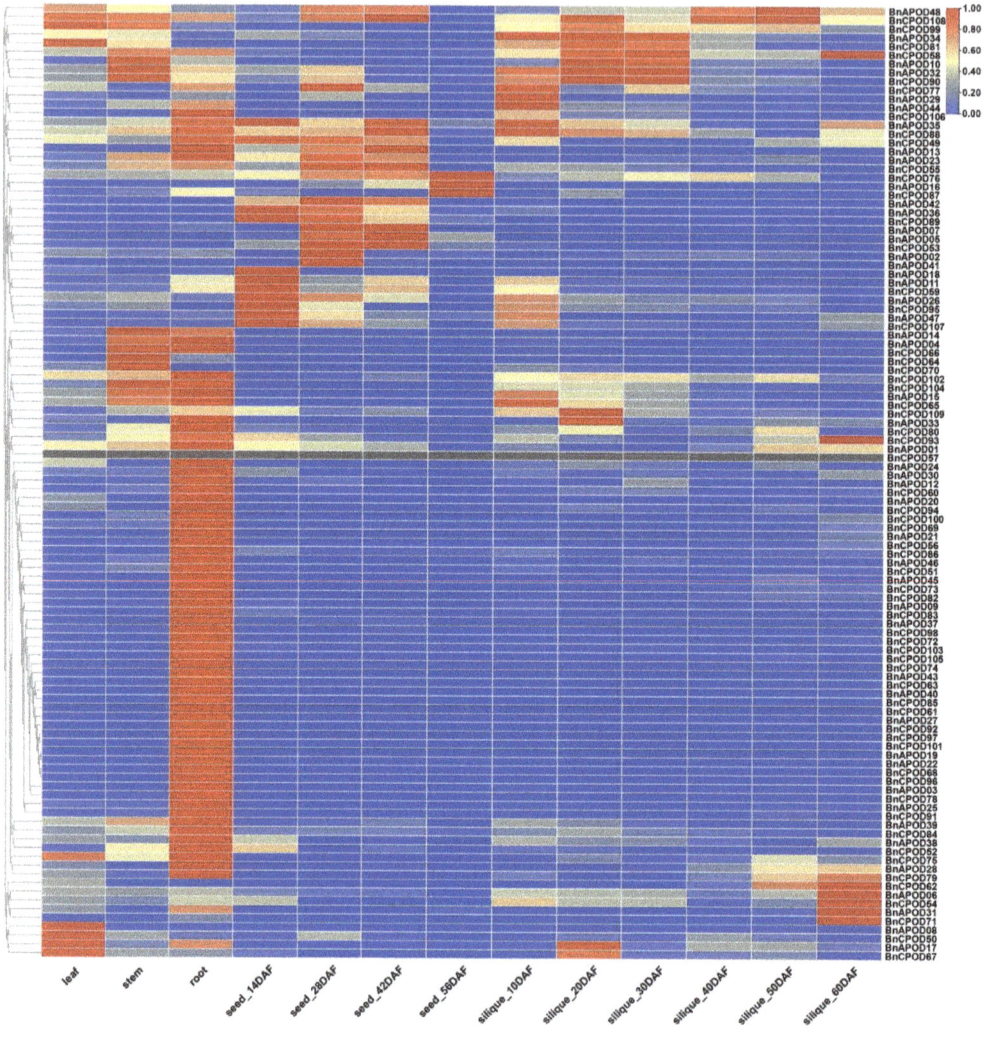

Figure 7. Tissue/organ-specific expression patterns of *BnPOD* genes in *B. napus* according to in silico RNA-seq data. DAF; days after flowering.

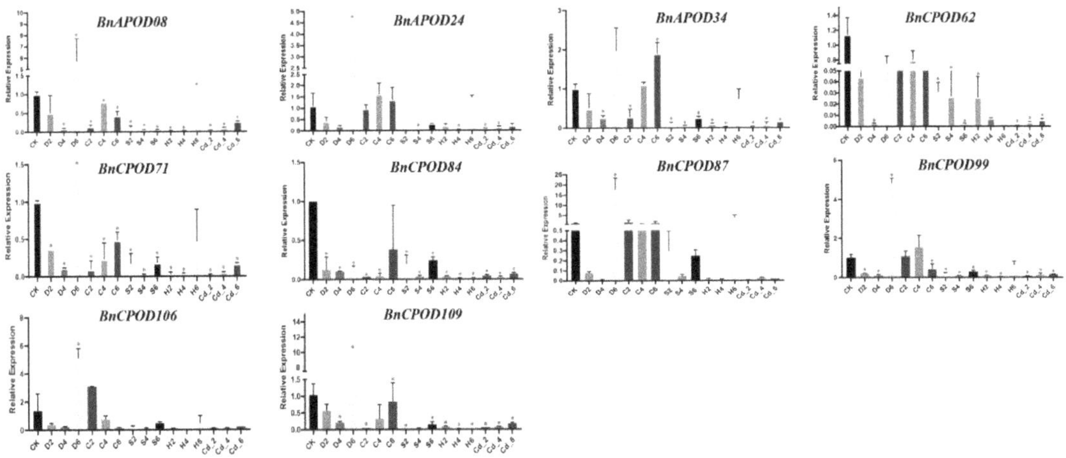

Figure 8. Expression patterns of selected *BnPODs* subjected to different abiotic stresses. The data are presented with ±standard errors. Statistically significant differences are denoted by asterisks * $p \leq 0.05$. CK, control; D, drought; C, cold; S, salt; H, heat; Cd, cadmium; numbers 2, 4, and 6 indicate time intervals of 2 h, 4 h, and 6 h, respectively. The samples at 0 h were used as a control (CK).

3. Discussion

Brassica napus is an allotetraploid crop and is susceptible to environmental stresses, including cold, heat, salt, drought, and heavy metals [27], which negatively influence the growth and final production [28]. Class III peroxidases (PODs) play critical roles in regulating plant physiology, particularly responses to abiotic stresses [24]. Consequently, the systematic genome-wide characterization of the *POD* genes in different crops is essential to clarify their biological roles in plants, particularly in growth and defense responses. The *POD* gene family has been extensively studied in different plant species, such as *A. thaliana* [17], *Saccharum officinarum* [29], *O. sativa* [7], *N. tabacum* [30], *Z. mays* [16], *Betula pendula* [31], *P. bretschneideri* [19], and *M. esculenta* [18], while the comprehensive identification of *PODs* in *B. napus* is still lacking.

In this study, we identified a total of 109 *POD* genes in *B. napus*, which is more than those reported in *A. thaliana* (73) [17] and *M. esculenta* (91) [18] but fewer than those in *O. sativa* (138) [7] and *G. max* (124) [22]. This higher number of *POD* genes in *B. napus* might be owing to its allopolyploid nature. We first analyzed the physical locations of the *BnPODs* in *B. napus* throughout the A and C genomes. We found that all the *BnPODs* are unevenly distributed in all the chromosomes, which is consistent with their chromosomal distributions in *A. thaliana* [17], *S. officinarum* [29], *O. sativa* [7], *N. tabacum* [30], and *Z. mays* [16]. Gene duplication, such as segmental duplication, tandem duplication, and whole-genome duplication, is responsible for the evolution and expansion of the gene family [18]. Next, 44 paralogous *POD* genes were identified in *B. napus*, indicating that segmental duplication contributed to the *BnPOD* expansion. Accumulated evidence has demonstrated that duplication events have been important for gene expansion in the *POD* family. The results of our study are consistent with *POD* expansion in *O. sativa* [7], *P. pyrifolia* [19], *G. max* [22], and *M. esculenta* [18].

Various *cis*-regulatory elements have been identified and can regulate the expression of stress-related genes [32]. We found multiple *cis*-acting elements in the upstream regions of the *BnPOD* family, including hormone-responsive and stress-responsive elements. Similar findings have also been reported for *V. vinifera POD* genes [21]. Antioxidant genes, such as *SOD*, *CAT*, and *APX*, regulate the underlying stress-tolerant mechanisms [33–35]. For instance, *PODs* significantly play an important role in mitigating different abiotic stresses [11,12,36]. In this study, we selected a few *BnPOD* genes for various abiotic stresses, which are located on ChrA02, A06, C02, C03, C04, C05, and C09. The phylogeny analysis of the selected genes revealed that *BnAPOD08* and *BnCPOD87* belong to subclade E; *BnAPOD34*, *BnCPOD62*, and *BnCPOD109* belong to subclade B; and *BnCPOD71*, *BnCPOD84*, *BnCPOD99*, and *BnCPOD106* belong to subclade A (Figure 2).

Notably, *BnPOD* genes showed large variations in their expression profile patterns, revealing their roles in different stress responses and defense responses in *B. napus*. Meanwhile, individual *POD* genes are usually sensitive to one specific external stress, and few genes are widely responsive to various biotic and abiotic stresses. For example, all the selected *BnPOD* genes are responsive to all the applied stresses, including cold, drought, salt, and heavy metals. Previous studies have suggested that the overexpression of *POD* genes, such as *AtPrx22*, *AtPrx39*, and *AtPrx69*, in *A. thaliana* is involved in cold tolerance [37]; also, *POD* family genes play a role in stresses in *Populus* and *G. max* [22,38] under drought, heat [39], salinity, and cadmium stresses. In summary, the expression patterns of the *BnPODs* under various abiotic stresses are complex and diverse and may be related to the functional diversity of the *POD* gene family members. These results provide useful insights into the potential capabilities of *BnPODs* under various abiotic stresses.

4. Materials and Methods

4.1. Mining of BnPOD Family in B. napus

POD protein sequences of *A. thaliana* were retrieved from the TAIR database (https://www.arabidopsis.org/, accessed on 2 August 2023) [40] and used as a query sequence in the *B. napus* genome, using the BlastP tool (https://blast.ncbi.nlm.nih.gov/Blast.cgi, accessed on 2 August 2023). After the removal of duplications, the non-redundant protein sequences were subjected to SMART (http://smart.emblheidelberg.de/, accessed on 2 August 2023) and NCBI conserved-domain searches (https://www.ncbi.nlm.nih.gov/Structure/cdd/wrpsb.cgi, accessed on 4 August 2023). The final POD proteins were renamed according to their physical positions in the *B. napus* genome. Furthermore, the physicochemical properties of the BnPOD proteins, including the protein length (aa), theoretical isoelectric point (pI), molecular weight (kilodaltons; kDa), and grand average of hydropathicity (GRAVY) were computed using ProtParam (http://web.expasy.org/protparam/, accessed on 6 August 2023). WoLF PSORT (https://wolfpsort.hgc.jp/, accessed on 9 August 2023) was used to predict the in silico subcellular localizations of all the deduced POD proteins [41].

4.2. Phylogeny of B. napus POD Proteins

To draw the phylogenetic tree of the POD proteins, the protein sequences of *B. napus* and *A. thaliana* were first aligned using MEGA software (Version 11) [42], and the maximum likelihood tree was then generated using iq-tree software (http://www.iqtree.org, accessed on 14 August 2023). The bootstrap values were set at 1000 replicates. The generated phylogeny was further modified using ITOL (https://itol.embl.de/, accessed on 15 August 2023) [43].

4.3. Gene Structural Analysis of B. napus POD Proteins

The intron and exon organizations within the *BnPOD* genes were predicted using TBtools (version 2.003) [44]. For the identification of the conserved motifs within the BnPODs, we used MEME Suite (https://meme-suite.org, accessed on 19 August 2023) [45], with the number of motifs set at 10.

For the identification of the *cis*-regulatory elements within the promoters of the *BnPOD* genes, a 2 kb sequence upstream from the start codons was submitted to the PlantCARE website (http://bioinformatics.psb.ugent.be/webtools/plantcare/html/, accessed on 25 August 2023) [46].

4.4. Gene Duplication and Evolutionary Analysis of BnPOD

We analyzed the duplication events of the BnaPOD gene family using MCScanX (https://github.com/wyp1125/MCScanX, accessed on 26 August 2023), and visualized them in Circos, as described by Waseem et al. [47]. Furthermore, the Ka and Ks substitutions between the gene pairs were also calculated as described by Aslam et al. [48]. The divergence time (T, MYA; millions of years ago) was calculated as follows: $T = Ks/2R$ (where $R = 1.5 \times 10^{-8}$ and is synonymous with the number of substitutions per site per year) [49]. Moreover, the syntenic relationship among *B. napus*, *B. rapa*, *B. oleracea*, and *Arabidopsis thaliana* was assessed using MCScanX and visualized in TBtools.

4.5. Prediction of Putative miRNAs Targeting BnPOD Genes

To predict the potential targets of *BnPOD* genes by miRNAs, the psRNATarget database (http://plantgrn.noble.org/psRNATarget/, accessed on 27 August 2023) [50] was employed with default parameters, including the number of expectations being set at 3. The interaction network between the predicted miRNAs and their corresponding *BnPOD* genes was visualized in Cytoscape (version 3.10.1) [51,52].

4.6. Expression Analysis of BnPOD Genes in Different Tissues

The expression data of the *BnPOD* genes were retrieved from the BnPIR database (http://cbi.hzau.edu.cn/bnapus/index.php, accessed on 3 September 2023) [53]. We evaluated the temporal and spatial expression data of the 109 *BnPOD* genes using RNA-seq data for various *B. napus* tissues, including the roots, stems, leaves, seeds, and siliquae [34], as well as at different developmental stages.

4.7. Plant Materials and Abiotic Stresses

We conducted a controlled laboratory experiment to investigate the *BnPOD* gene responses under various abiotic stresses. Plants of *B. napus* var. 383-5 were grown in a hydroponic solution, using the full Hoagland solution (2.5 mM $Ca(NO_3)_2$, 2 mM KCl, 1 mM $MgSO_4$, 0.5 mM $NH_4H_2PO_4$, 0.5 mM $CaCl_2$, 25 μM H_3BO_3, 5 μM $MnSO_4$, 0.8 μM $ZnSO_4$, 0.3 μM $CuSO_4$, 0.1 μM $(NH_4)_6Mo_7O_{24}$, and 100 μM EDTA, 2NaFe), with continuous aeration [54]. Other experimental conditions, including a 14 h light and 10 h dark cycle, were maintained. Subsequently, the plants were exposed to various abiotic stresses, including NaCl (150 mM), $CdCl_2$ (cadmium chloride, 2.5 mM), cold stress at 8 °C, heat stress at 42 °C, and drought stress. For the drought stress, the plants were subjected to a drought stress with a 15% PEG-6000 solution. Each stress treatment was carried out with three independent biological replicates. The samples were collected at 0 h, 2 h, 4 h, and 6 h. All the samples were immediately frozen in liquid nitrogen and stored at −80 °C for the subsequent analysis.

4.8. RNA Isolation and Real-Time Quantitative PCR Expression Analysis

The total RNA was extracted from the collected samples, using an RNAprep Pure Plant Kit (Lot no. Q5510, Tiangen, Beijing, China) following the manufacturer's protocol. The concentration of the extracted RNA samples was measured using a NanoDrop (2000 C spectrophotometer from Thermo Fisher Scientific, Waltham, MA, USA). DNase I treatment was used to remove the genomic DNA [55]. The first complementary strand of DNA (cDNA) was synthesized using HiScript 1st strand cDNA synthesis kit (V22.1, Vazyme Biotech Co., Ltd., Sanya, China). The cDNA of samples was assessed by RT-qPCR using ChamQTM SYBR RT-qPCR Master Mix (Vazyme Biotech Co., Ltd., Sanya, China). The expression level of the *Actin* was used as internal control. The RT-qPCR was performed with three independent biological replicates. The relative expression level of *BnPODs* was calculated using the $2^{-\Delta\Delta CT}$ method [56]. The primers used in this study are listed in Table S6.

5. Conclusions

This research provides the first comprehensive genome-wide analysis of the *POD* family genes in *B. napus*. We identified 109 *POD* family genes from the BnPIR database of *B. napus*. Then, we studied their basic classification, gene structure, motif distribution and duplication events. Additionally, RT-qPCR analysis revealed that the selected *BnPOD* genes were significantly expressed under abiotic stresses, suggesting that these genes can be potential candidates for *B. napus* crop improvement. In the future, attempts to increase the stress tolerance of *B. napus* will require knowledge of the specific peroxidases that are up-regulated by stresses. In summary, our discoveries offer valuable perspectives into the functional dimensions of *POD* family genes, which open the way for upcoming investigations in economically significant *B. napus* crops.

Supplementary Materials: The following supporting information can be downloaded at https://www.mdpi.com/article/10.3390/plants13070942/s1: Table S1: Characteristics of PODs in *Brassica napus;* Table S2: Identification of different *cis*-regulatory elements in promoters of *BnaPODs;* Table S3: Parameters related to gene duplication and evolution of *BnPODs* in *B. napus;* Table S4: Collinearity analysis among *Brassica napus*, *Brassica rapa*, *Brassica oleracea*, and *Arabidopsis thaliana;* Table S5: *BnaPOD*–miRNA prediction using psRNATarget server; Table S6: List of all the primers used in this study.

Author Contributions: O.U.S. conceived the study design and analyzed the results. O.U.S. and L.U.K. performed all the bioinformatics analyses. M.I. and M.W. reviewed the data. O.U.S., S.B. and M.W. drafted the manuscript. W.U.K., J.P. and L.Z. managed the multiple stresses and seedling protection, and M.W., M.I. and P.L. revised the manuscript. P.L. and M.W. supervised and provided funding for the whole research work. All authors have read and agreed to the published version of the manuscript.

Funding: This work was supported by grants from the China NSFC Research Fund for International Young Scientists (grant number: 32250410291), Key Research Program of Hainan Province (grant number: ZDYF2022XDNY185), and Special Project for the Academician Team Innovation Center of Hainan Province (grant number: YSPTZX202206).

Data Availability Statement: The authors confirm that the data supporting the findings of this study are available within the article and its Supplementary Materials.

Acknowledgments: The authors thank all the editors and reviewers for their valuable and insightful comments on this manuscript.

Conflicts of Interest: The authors have no conflicts of interest to declare, either financial or non-financial.

References

1. Hiraga, S.; Sasaki, K.; Ito, H.; Ohashi, Y.; Matsui, H. A Large Family of Class III Plant Peroxidases. *Plant Cell Physiol.* **2001**, *42*, 462–468. [CrossRef]
2. Hiraga, S.; Ichinose, C.; Onogi, T.; Niki, H.; Yamazoe, M. Bidirectional Migration of SeqA-Bound Hemimethylated DNA Clusters and Pairing of oriC Copies in *Escherichia coli*: Migration of SeqA-DNA Complex and oriC. *Genes Cells* **2000**, *5*, 327–341. [CrossRef] [PubMed]
3. Koua, D.; Cerutti, L.; Falquet, L.; Sigrist, C.J.A.; Theiler, G.; Hulo, N.; Dunand, C. PeroxiBase: A Database with New Tools for Peroxidase Family Classification. *Nucleic Acids Res.* **2009**, *37*, D261–D266. [CrossRef]
4. Zámocký, M.; Furtmüller, P.G.; Obinger, C. Evolution of Structure and Function of Class I Peroxidases. *Arch. Biochem. Biophys.* **2010**, *500*, 45–57. [CrossRef] [PubMed]
5. Piontek, K.; Smith, A.T.; Blodig, W. Lignin Peroxidase Structure and Function. *Biochem. Soc. Trans.* **2001**, *29*, 111–116. [CrossRef]
6. Mathé, C.; Barre, A.; Jourda, C.; Dunand, C. Evolution and Expression of Class III Peroxidases. *Arch. Biochem. Biophys.* **2010**, *500*, 58–65. [CrossRef] [PubMed]
7. Passardi, F.; Longet, D.; Penel, C.; Dunand, C. The Class III Peroxidase Multigenic Family in Rice and Its Evolution in Land Plants☆,☆☆. *Phytochemistry* **2004**, *65*, 1879–1893. [CrossRef] [PubMed]
8. Herrero, J.; Fernández-Pérez, F.; Yebra, T.; Novo-Uzal, E.; Pomar, F.; Pedreño, M.Á.; Cuello, J.; Guéra, A.; Esteban-Carrasco, A.; Zapata, J.M. Bioinformatic and Functional Characterization of the Basic Peroxidase 72 from Arabidopsis Thaliana Involved in Lignin Biosynthesis. *Planta* **2013**, *237*, 1599–1612. [CrossRef]
9. Pandey, V.P.; Awasthi, M.; Singh, S.; Tiwari, S.; Dwivedi, U.N. A Comprehensive Review on Function and Application of Plant Peroxidases. *Biochem. Anal. Biochem.* **2017**, *6*, 308. [CrossRef]

10. Passardi, F.; Tognolli, M.; De Meyer, M.; Penel, C.; Dunand, C. Two Cell Wall Associated Peroxidases from Arabidopsis Influence Root Elongation. *Planta* **2006**, *223*, 965–974. [CrossRef]
11. Wu, Y.; Yang, Z.; How, J.; Xu, H.; Chen, L.; Li, K. Overexpression of a Peroxidase Gene (AtPrx64) of Arabidopsis Thaliana in Tobacco Improves Plant's Tolerance to Aluminum Stress. *Plant Mol. Biol.* **2017**, *95*, 157–168. [CrossRef] [PubMed]
12. Llorente, F.; López-Cobollo, R.M.; Catalá, R.; Martínez-Zapater, J.M.; Salinas, J. A Novel Cold-inducible Gene from *Arabidopsis*, *RCI3*, Encodes a Peroxidase That Constitutes a Component for Stress Tolerance. *Plant J.* **2002**, *32*, 13–24. [CrossRef] [PubMed]
13. Jaggi, M.; Kumar, S.; Sinha, A.K. Overexpression of an Apoplastic Peroxidase Gene CrPrx in Transgenic Hairy Root Lines of Catharanthus Roseus. *Appl. Microbiol. Biotechnol.* **2011**, *90*, 1005–1016. [CrossRef]
14. Su, P.; Yan, J.; Li, W.; Wang, L.; Zhao, J.; Ma, X.; Li, A.; Wang, H.; Kong, L. A Member of Wheat Class III Peroxidase Gene Family, TaPRX-2A, Enhanced the Tolerance of Salt Stress. *BMC Plant Biol.* **2020**, *20*, 392. [CrossRef]
15. Sečenji, M.; Lendvai, Á.; Miskolczi, P.; Kocsy, G.; Gallé, Á.; Szűcs, A.; Hoffmann, B.; Sárvári, É.; Schweizer, P.; Stein, N.; et al. Differences in Root Functions during Long-Term Drought Adaptation: Comparison of Active Gene Sets of Two Wheat Genotypes: Adaptation of Wheat Root Genes to Drought. *Plant Biol.* **2010**, *12*, 871–882. [CrossRef]
16. Wang, Y.; Wang, Q.; Zhao, Y.; Han, G.; Zhu, S. Systematic Analysis of Maize Class III Peroxidase Gene Family Reveals a Conserved Subfamily Involved in Abiotic Stress Response. *Gene* **2015**, *566*, 95–108. [CrossRef] [PubMed]
17. Tognolli, M.; Penel, C.; Greppin, H.; Simon, P. Analysis and Expression of the Class III Peroxidase Large Gene Family in Arabidopsis Thaliana. *Gene* **2002**, *288*, 129–138. [CrossRef] [PubMed]
18. Wu, C.; Ding, X.; Ding, Z.; Tie, W.; Yan, Y.; Wang, Y.; Yang, H.; Hu, W. The Class III Peroxidase (POD) Gene Family in Cassava: Identification, Phylogeny, Duplication, and Expression. *Int. J. Mol. Sci.* **2019**, *20*, 2730. [CrossRef]
19. Cao, Y.; Han, Y.; Meng, D.; Li, D.; Jin, Q.; Lin, Y.; Cai, Y. Structural, Evolutionary, and Functional Analysis of the Class III Peroxidase Gene Family in Chinese Pear (*Pyrus bretschneideri*). *Front. Plant Sci.* **2016**, *7*, 1874. [CrossRef]
20. Li, Q.; Dou, W.; Qi, J.; Qin, X.; Chen, S.; He, Y. Genome-wide Analysis of the CIII Peroxidase Family in Sweet Orange (Citrus Sinensis) and Expression Profiles Induced by Xanthomonas Citri Subsp. Citri and Hormones. *J. Genet.* **2020**, *99*, 10. [CrossRef]
21. Xiao, H.; Wang, C.; Khan, N.; Chen, M.; Fu, W.; Guan, L.; Leng, X. Genome-Wide Identification of the Class III POD Gene Family and Their Expression Profiling in Grapevine (*Vitis vinifera* L). *BMC Genom.* **2020**, *21*, 444. [CrossRef] [PubMed]
22. Aleem, M.; Riaz, A.; Raza, Q.; Aleem, M.; Aslam, M.; Kong, K.; Atif, R.M.; Kashif, M.; Bhat, J.A.; Zhao, T. Genome-Wide Characterization and Functional Analysis of Class III Peroxidase Gene Family in Soybean Reveal Regulatory Roles of GsPOD40 in Drought Tolerance. *Genomics* **2022**, *114*, 45–60. [CrossRef] [PubMed]
23. González-Gordo, S.; Muñoz-Vargas, M.A.; Palma, J.M.; Corpas, F.J. Class III Peroxidases (POD) in Pepper (*Capsicum annuum* L.): Genome-Wide Identification and Regulation during Nitric Oxide (NO)-Influenced Fruit Ripening. *Antioxidants* **2023**, *12*, 1013. [CrossRef] [PubMed]
24. Almagro, L.; Gómez Ros, L.V.; Belchi-Navarro, S.; Bru, R.; Ros Barceló, A.; Pedreño, M.A. Class III Peroxidases in Plant Defence Reactions. *J. Exp. Bot.* **2009**, *60*, 377–390. [CrossRef] [PubMed]
25. Ayadi, J.; Debouba, M.; Rahmani, R.; Bouajila, J. Brassica Genus Seeds: A Review on Phytochemical Screening and Pharmacological Properties. *Molecules* **2022**, *27*, 6008. [CrossRef]
26. Lohani, N.; Jain, D.; Singh, M.B.; Bhalla, P.L. Engineering Multiple Abiotic Stress Tolerance in Canola, *Brassica napus*. *Front. Plant Sci.* **2020**, *11*, 3. [CrossRef]
27. Song, J.-M.; Guan, Z.; Hu, J.; Guo, C.; Yang, Z.; Wang, S.; Liu, D.; Wang, B.; Lu, S.; Zhou, R.; et al. Eight High-Quality Genomes Reveal Pan-Genome Architecture and Ecotype Differentiation of *Brassica napus*. *Nat. Plants* **2020**, *6*, 34–45. [CrossRef]
28. Mehmood, S.S.; Lu, G.; Luo, D.; Hussain, M.A.; Raza, A.; Zafar, Z.; Zhang, X.; Cheng, Y.; Zou, X.; Lv, Y. Integrated Analysis of Transcriptomics and Proteomics Provides Insights into the Molecular Regulation of Cold Response in *Brassica napus*. *Environ. Exp. Bot.* **2021**, *187*, 104480. [CrossRef]
29. Shang, H.; Fang, L.; Qin, L.; Jiang, H.; Duan, Z.; Zhang, H.; Yang, Z.; Cheng, G.; Bao, Y.; Xu, J.; et al. Genome-Wide Identification of the Class III Peroxidase Gene Family of Sugarcane and Its Expression Profiles under Stresses. *Front. Plant Sci.* **2023**, *14*, 1101665. [CrossRef]
30. Cheng, L.; Ma, L.; Meng, L.; Shang, H.; Cao, P.; Jin, J. Genome-Wide Identification and Analysis of the Class III Peroxidase Gene Family in Tobacco (Nicotiana Tabacum). *Front. Genet.* **2022**, *13*, 916867. [CrossRef]
31. Cai, K.; Liu, H.; Chen, S.; Liu, Y.; Zhao, X.; Chen, S. Genome-Wide Identification and Analysis of Class III Peroxidases in Betula Pendula. *BMC Genom.* **2021**, *22*, 314. [CrossRef]
32. Ijaz, U. Plant Cis-Regulatory Elements: Methods of Identification and Applications. *Asian J. Agric. Biol.* **2020**, *8*, 207–222. [CrossRef]
33. Khan, W.U.; Khan, L.U.; Chen, D.; Chen, F. Comparative Analyses of Superoxide Dismutase (SOD) Gene Family and Expression Profiling under Multiple Abiotic Stresses in Water Lilies. *Horticulturae* **2023**, *9*, 781. [CrossRef]
34. Raza, A.; Su, W.; Gao, A.; Mehmood, S.S.; Hussain, M.A.; Nie, W.; Lv, Y.; Zou, X.; Zhang, X. Catalase (CAT) Gene Family in Rapeseed (*Brassica napus* L.): Genome-Wide Analysis, Identification, and Expression Pattern in Response to Multiple Hormones and Abiotic Stress Conditions. *Int. J. Mol. Sci.* **2021**, *22*, 4281. [CrossRef]
35. Zaman, Q.U.; Hussain, M.A.; Khan, L.U.; Hui, L.; Khan, D.; Khokhar, A.A.; Cui, J.; Raza, A.; Wang, H.-F. Genome-Wide Identification and Expression Profiling of APX Gene Family under Multifactorial Stress Combinations and Melatonin-Mediated Tolerance in Pitaya. *Sci. Hortic.* **2023**, *321*, 112312. [CrossRef]

36. Kumar, S.; Jaggi, M.; Sinha, A.K. Ectopic Overexpression of Vacuolar and Apoplastic Catharanthus Roseus Peroxidases Confers Differential Tolerance to Salt and Dehydration Stress in Transgenic Tobacco. *Protoplasma* **2012**, *249*, 423–432. [CrossRef]
37. Kim, B.H.; Kim, S.Y.; Nam, K.H. Genes Encoding Plant-Specific Class III Peroxidases Are Responsible for Increased Cold Tolerance of the Brassinosteroid-Insensitive 1 Mutant. *Mol. Cells* **2012**, *34*, 539–548. [CrossRef] [PubMed]
38. Ren, L.-L.; Liu, Y.-J.; Liu, H.-J.; Qian, T.-T.; Qi, L.-W.; Wang, X.-R.; Zeng, Q.-Y. Subcellular Relocalization and Positive Selection Play Key Roles in the Retention of Duplicate Genes of *Populus* Class III Peroxidase Family. *Plant Cell* **2014**, *26*, 2404–2419. [CrossRef]
39. Li, X.; Ma, W.; Jiang, Y. Expression Patterns of Heat Shock Protein Genes and Antioxidase Genes in *Apis cerana cerana* (Hymenoptera: Apidae) under Heat Stress. *J. Entomol. Sci.* **2023**, *58*, 95–103. [CrossRef]
40. Rhee, S.Y. The Arabidopsis Information Resource (TAIR): A Model Organism Database Providing a Centralized, Curated Gateway to Arabidopsis Biology, Research Materials and Community. *Nucleic Acids Res.* **2003**, *31*, 224–228. [CrossRef]
41. Horton, P.; Park, K.-J.; Obayashi, T.; Fujita, N.; Harada, H.; Adams-Collier, C.J.; Nakai, K. WoLF PSORT: Protein Localization Predictor. *Nucleic Acids Res.* **2007**, *35*, W585–W587. [CrossRef] [PubMed]
42. Tamura, K.; Stecher, G.; Kumar, S. MEGA11: Molecular Evolutionary Genetics Analysis Version 11. *Mol. Biol. Evol.* **2021**, *38*, 3022–3027. [CrossRef] [PubMed]
43. Letunic, I.; Bork, P. Interactive Tree Of Life (iTOL) v4: Recent Updates and New Developments. *Nucleic Acids Res.* **2019**, *47*, W256–W259. [CrossRef]
44. Chen, C.; Chen, H.; Zhang, Y.; Thomas, H.R.; Frank, M.H.; He, Y.; Xia, R. TBtools: An Integrative Toolkit Developed for Interactive Analyses of Big Biological Data. *Mol. Plant* **2020**, *13*, 1194–1202. [CrossRef]
45. Bailey, T.L.; Boden, M.; Buske, F.A.; Frith, M.; Grant, C.E.; Clementi, L.; Ren, J.; Li, W.W.; Noble, W.S. MEME SUITE: Tools for Motif Discovery and Searching. *Nucleic Acids Res.* **2009**, *37*, W202–W208. [CrossRef]
46. Lescot, M. PlantCARE, a Database of Plant Cis-Acting Regulatory Elements and a Portal to Tools for in Silico Analysis of Promoter Sequences. *Nucleic Acids Res.* **2002**, *30*, 325–327. [CrossRef]
47. Waseem, M.; Aslam, M.M.; Shaheen, I. The DUF221 Domain-Containing (DDP) Genes Identification and Expression Analysis in Tomato under Abiotic and Phytohormone Stress. *GM Crops Food* **2021**, *12*, 586–599. [CrossRef] [PubMed]
48. Aslam, M.M.; Fritschi, F.B.; Di, Z.; Wang, G.; Li, H.; Lam, H.; Waseem, M.; Weifeng, X.; Zhang, J. Overexpres-sion of LaGRAS Enhances Phosphorus Acquisition via Increased Root Growth of Phosphorus-deficient White Lupin. *Physiol. Plant.* **2023**, *175*, e13962. [CrossRef]
49. Koch, M.A.; Haubold, B.; Mitchell-Olds, T. Comparative Evolutionary Analysis of Chalcone Synthase and Al-cohol Dehydrogenase Loci in Arabidopsis, Arabis, and Related Genera (Brassicaceae). *Mol. Biol. Evol.* **2000**, *17*, 1483–1498. [CrossRef]
50. Dai, X.; Zhao, P.X. psRNATarget: A Plant Small RNA Target Analysis Server. *Nucleic Acids Res.* **2011**, *39*, W155–W159. [CrossRef]
51. Franz, M.; Lopes, C.T.; Huck, G.; Dong, Y.; Sumer, O.; Bader, G.D. Cytoscape.Js: A Graph Theory Library for Visualisation and Analysis. *Bioinformatics* **2016**, *32*, 309–311. [CrossRef] [PubMed]
52. Shannon, P.; Markiel, A.; Ozier, O.; Baliga, N.S.; Wang, J.T.; Ramage, D.; Amin, N.; Schwikowski, B.; Ideker, T. Cytoscape: A Software Environment for Integrated Models of Biomolecular Interaction Networks. *Genome Res.* **2003**, *13*, 2498–2504. [CrossRef] [PubMed]
53. Song, J.; Liu, D.; Xie, W.; Yang, Z.; Guo, L.; Liu, K.; Yang, Q.; Chen, L. BnPIR: *Brassica napus* Pan-genome Information Resource for 1689 Accessions. *Plant Biotechnol. J.* **2021**, *19*, 412–414. [CrossRef] [PubMed]
54. Hammer, P.A.; Tibbitts, T.W.; Langhans, R.W.; McFarlane, J.C. Base-Line Growth Studies of 'Grand Rapids' Lettuce in Controlled Environments1. *J. Am. Soc. Hort. Sci.* **1978**, *103*, 649–655. [CrossRef]
55. Huang, Z.; Fasco, M.J.; Kaminsky, L.S. Optimization of DNase I Removal of Contaminating DNA from RNA for Use in Quantitative RNA-PCR. *BioTechniques* **1996**, *20*, 1012–1020. [CrossRef] [PubMed]
56. Livak, K.J.; Schmittgen, T.D. Analysis of Relative Gene Expression Data Using Real-Time Quantitative PCR and the $2^{-\Delta\Delta CT}$ Method. *Methods* **2001**, *25*, 402–408. [CrossRef]

plants

MDPI

Article

Analysis of the UDP-Glucosyltransferase (*UGT*) Gene Family and Its Functional Involvement in Drought and Salt Stress Tolerance in *Phoebe bournei*

Hengfeng Guan [1,†], Yanzi Zhang [2,†], Jingshu Li [1], Zhening Zhu [1], Jiarui Chang [1], Almas Bakari [1], Shipin Chen [1], Kehui Zheng [3,*] and Shijiang Cao [1,*]

[1] College of Forestry, Fujian Agriculture and Forestry University, Fuzhou 350002, China; g1290570831@163.com (H.G.); chuchu7613@163.com (J.L.); zheningzhu@163.com (Z.Z.); cherichang@163.com (J.C.); almasbakari154@gmail.com (A.B.); fjcsp@126.com (S.C.)

[2] Center for Plant Metabolomics, Haixia Institute of Science and Technology, Fujian Agriculture and Forestry University, Fuzhou 350002, China; zhangyanzi163@163.com

[3] College of Computer and Information Sciences, Fujian Agriculture and Forestry University, Fuzhou 350002, China

* Correspondence: zhkehui@fafu.edu.cn (K.Z.); 000q813040@fafu.edu.cn (S.C.)

† These authors contributed equally to this work.

Abstract: Uridine diphosphate glycosyltransferases (UDP-GTs, UGTs), which are regulated by *UGT* genes, play a crucial role in glycosylation. In vivo, the activity of *UGT* genes can affect the availability of metabolites and the rate at which they can be eliminated from the body. *UGT* genes can exert their regulatory effects through mechanisms such as post-transcriptional modification, substrate subtype specificity, and drug interactions. *Phoebe bournei* is an economically significant tree species that is endemic to southern China. Despite extensive studies on the *UGT* gene family in various species, a comprehensive investigation of the *UGT* family in *P. bournei* has not been reported. Therefore, we conducted a systematic analysis to identify 156 *UGT* genes within the entire *P. bournei* genome, all of which contained the PSPG box. The *PbUGT* family consists of 14 subfamilies, consistent with *Arabidopsis thaliana*. We observed varying expression levels of *PbUGT* genes across different tissues in *P. bournei*, with the following average expression hierarchy: leaf > stem xylem > stem bark > root xylem > root bark. Covariance analysis revealed stronger covariance between *P. bournei* and closely related species. In addition, we stressed the seedlings with 10% NaCl and 10% PEG-6000. The *PbUGT* genes exhibited differential expression under drought and salt stresses, with specific expression patterns observed under each stress condition. Our findings shed light on the transcriptional response of *PbUGT* factors to drought and salt stresses, thereby establishing a foundation for future investigations into the role of *PbUGT* transcription factors.

Keywords: bioinformatics analysis; *Phoebe bournei*; UDP-glycosyltransferase; expression analysis

1. Introduction

Glycosylation, a widespread compound modification pathway in plant secondary metabolism, is a crucial transformation process in organisms. It serves as the final stage for many biosynthetic products, altering the activity, stability, and solubility of receptors [1]. Glycosylation plays a vital role in plant hormone balance, detoxification processes, and secondary metabolite production. Glycosyltransferases (GTs) are responsible for plant glycosylation and constitute a highly diverse multi-gene family present in various organisms. These enzymes catalyze the specific transfer of glycosyl groups from donors to receptors, forming α- and β-glycosidic bonds and generating oligosaccharides, polysaccharides, complex sugars (such as glycoproteins and glycolipids), and glycoside compounds [2]. The most prevalent glycosyl donor in plants is UDP-glucose, while receptors include glycolipids, proteins, and nucleic acids.

The carbohydrate active enzyme database (CAZy) classifies glycosyltransferases (GTs) from different species into 114 gene families based on amino acid sequence similarity, substrate specificity, catalytic mechanism, signal membrane, and conformation of the glycosidic chain. The GT1 branch represents the largest division, indicating its close functional association with plants. Within the GT1 branch, there exists a subclass known as UDP-glucosyltransferases (UGTs) due to their utilization of UDPG as the glycosyl donor [3]. One prominent characteristic of the plant *UGT* family is the presence of a highly conserved segment, known as the PSPG box, comprising 44 amino acids located at the protein's C-terminus [4]. This conserved motif facilitates the recognition of and interaction with uridine diphosphate glycosyl donors. Conversely, the N-terminal sequence, which is considerably variable and less conserved, governs the recognition and binding of various receptors, contributing to the vast substrate diversity and complex structures of plant *UGT* genes, such as their involvement in the glycosylation of diverse compounds including hormones, terpenoids, and sterols [5].

The *UGT* gene family has been extensively identified and classified in various plants, including *Arabidopsis thaliana* [6], *Linum usitatissimum* [7], *Triticum aestivum* [8], *Zea mays* [9], *Gossypium hirsutum* [10], and *Prunus persica* [11]. *Arabidopsis thaliana*, a model plant, possesses 122 *UGT* genes, among which 8 are pseudogenes. These genes are categorized into 14 distinct groups (A–N) based on their amino acid sequences [6]. Subsequent studies have revealed the presence of O and P groups in higher plants [12]. *Oryza sativa* has a total of 609 *UGT* genes, divided into 40 groups along with an unknown class [13]. *Citrus grandis* exhibits 145 *UGT* genes distributed among 16 groups [14], while *Zea mays* contains 147 *UGT* genes classified into 17 groups, with 18 genes shared with *Arabidopsis thaliana* and 2 genes shared with *Oryza sativa* [9]. The functional roles of the *UGT* gene family have been validated in various plants, including hormone balance regulation, localization control, secondary metabolite synthesis, detoxification, and defense responses [15].

Phoebe bournei, a species of the Lauraceae family, is characterized by a straight trunk and limited branches. It serves as an excellent timber species and is endemic to China [16]. Noteworthy for its unique fragrance, *P. bournei* holds significant value as a timber tree. Its wood exhibits strong resistance to corrosion, making it resistant to cracking and bending during the drying process. Consequently, it possesses high economic worth and finds applications in shipbuilding and furniture production [17]. However, due to excessive logging driven by its economic value and the adverse effects of environmental factors such as soil fertility degradation, *P. bournei* has experienced a decline in its wild population and is now listed as a secondary nationally protected plant.

While several gene families in other plants have been extensively studied, limited research has been conducted on the gene family of *P. bournei*. This dearth of studies can be attributed to the relatively recent sequencing of the *P. bournei* genome. The complete genome data of *P. bournei* was initially obtained by Liu Zhongjian and Chen Shiping from Fujian Agriculture and Forestry University in 2020 [18]. However, this study did not unveil the chromosomal-level information. Subsequently, in 2022, the genome of *P. bournei* was deciphered by the Tong Zaikang team from Zhejiang Agriculture and Forestry University, shedding light on the synthesis mechanism of *P. bournei* sesquiterpenes [19].

The gene families in *P. bournei* primarily include commonly found families like *WRKY* [20], *NF-Y* [21], and *PLR* [22], while the *UGT* gene family has received limited attention. Previous studies have primarily focused on whole-gene family analyses, with few investigations on stress responses. Furthermore, existing research primarily addresses nutrient deficiency stress, neglecting drought and salt stresses. Therefore, our study aims to elucidate the response mechanisms of *PbUGT* genes under drought and salt stress conditions.

Previously, we studied transcriptional analysis of the *GATA* gene family under abiotic stress in Phoebe bournei [23]. We know that under drought stress, the photosynthesis of *P. fortunei* is inhibited, the net photosynthetic rate, transpiration rate, and stomatal conductance of its leaves decrease, while the activities of oxidoreductase (POD and SOD) increase, and the ABA content in its roots increases. In terms of morphological characteristics, the root color of the seedlings becomes darker with longer drought treatment [24]. Moreover, in our previous experiments, we found that excessive stress time and excessive solution concentration may lead to the death of the seedlings. Therefore, we will use the method used in these previous experiments to stress *Phoebe bournei*, using 10% NaCl to apply salt stress to annual seedlings and 10% PEG-6000 to apply drought stress. Finally, qRT-PCR was used to detect the expression profile of the genes under drought and salt stress. This study provides a new idea for studying the evolution of *UGT* genes and helps to provide a better understanding of the abiotic stress response of *Phoebe bournei*.

2. Results

2.1. Identification and Physicochemical Analysis of PbUGT Gene Members

Through rigorous verification using various methods, a comprehensive set of 156 *PbUGT* genes was successfully identified within the entirety of the genome of *P. bournei*. These *UGT* genes were subsequently designated *PbUGT*1-156 based on their chromosomal distribution and are presented in Table 1. To gain further insights into the physicochemical properties of the *UGT* proteins, ProtParam, MEME, and other online tools were employed for analysis. The results revealed that the amino acid count of the 156 *PbUGT* proteins ranged from 132 aa (*PbUGT*66) to 1019 aa (*PbUGT*89). The molecular weight of the proteins varied between 14948.04 Da (*PbUGT*66) and 113398.82 Da (*PbUGT*89). The isoelectric points spanned a range from 4.62 to 9.80. Specifically, there were 136 acidic proteins (PI < 6.5), 12 neutral proteins (6.5 < PI < 7.5), and 8 basic proteins (PI > 7.5). Subcellular localization analysis of these *PbUGT* proteins indicated that 88 proteins were localized within the chloroplast, 43 were in the cytoplasm, and 15 were in the nucleus, while the remaining proteins were distributed across various organelles such as vacuoles, peroxisomes, and mitochondria. Furthermore, instability coefficient analysis unveiled that 46 proteins were classified as stable, whereas the remaining 110 proteins exhibited instability. Notably, the maximum instability coefficient reached a significant value of 73.59, suggesting that the majority of the *PbUGT* proteins were characterized as unstable acidic proteins.

Table 1. Physical and chemical properties of *PbUGT* family members.

Name	Average Molecular Weight	Theoretical pI	Asp + Glu	Arg + Lys	Instability Index	Aliphatic Index	GRAVY
Distribution range	14,948.04–113,398.82	4.62–9.8	17–127	11–110	32.1–72.59	71.5–102.3	−0.467–0.273
Average value	50,994.75	5.87	54	44	44.31	88.18	−0.121
Minimum value member	*PbUGT*66	*PbUGT*140	*PbUGT*154	*PbUGT*154	*PbUGT*144	*PbUGT*136	*PbUGT*66
Maximum value member	*PbUGT*89	*PbUGT*1	*PbUGT*89	*PbUGT*9	*PbUGT*122	*PbUGT*75	*PbUGT*154

2.2. Chromosome Localization Analysis of the PbUGT Gene Family

To determine the precise chromosomal positions of the *UGT* genes, we utilized MapChart 2.32 software to construct a visual representation of their distribution. Our analysis revealed that the 156 *UGT* genes were dispersed across all 12 chromosomes. Notably, chromosome 10 harbored the highest number of *PbUGT*s, with a total of 32 genes, followed by chromosome 2, with 28 *PbUGT*s. Furthermore, the distribution pattern of *UGT* gene members within different subfamilies of *P. bournei* displayed irregularities, characterized by the presence of gene clusters. These clusters varied in size, ranging from 2 to 13 genes, with the largest cluster residing on chromosome 10. We further examined the chromosomal distribution within different gene groups. Chromosome 2 contained 28 genes belonging to eight distinct groups. Among these, group A exhibited the greatest number of members in the *P. bournei UGT* family, totaling 36 genes. Specifically, thirteen members from this group formed a cluster on chromosome 7, while four members clustered on chromosomes 2 and 4. The remaining six members were dispersed across chromosomes 1, 2, 5, and other regions encompassing chromosomes 8 to 12 (Figure 1). This uneven distribution of *PbUGT* genes across chromosomes indicates the presence of genetic variations during the evolution of *P. bournei*.

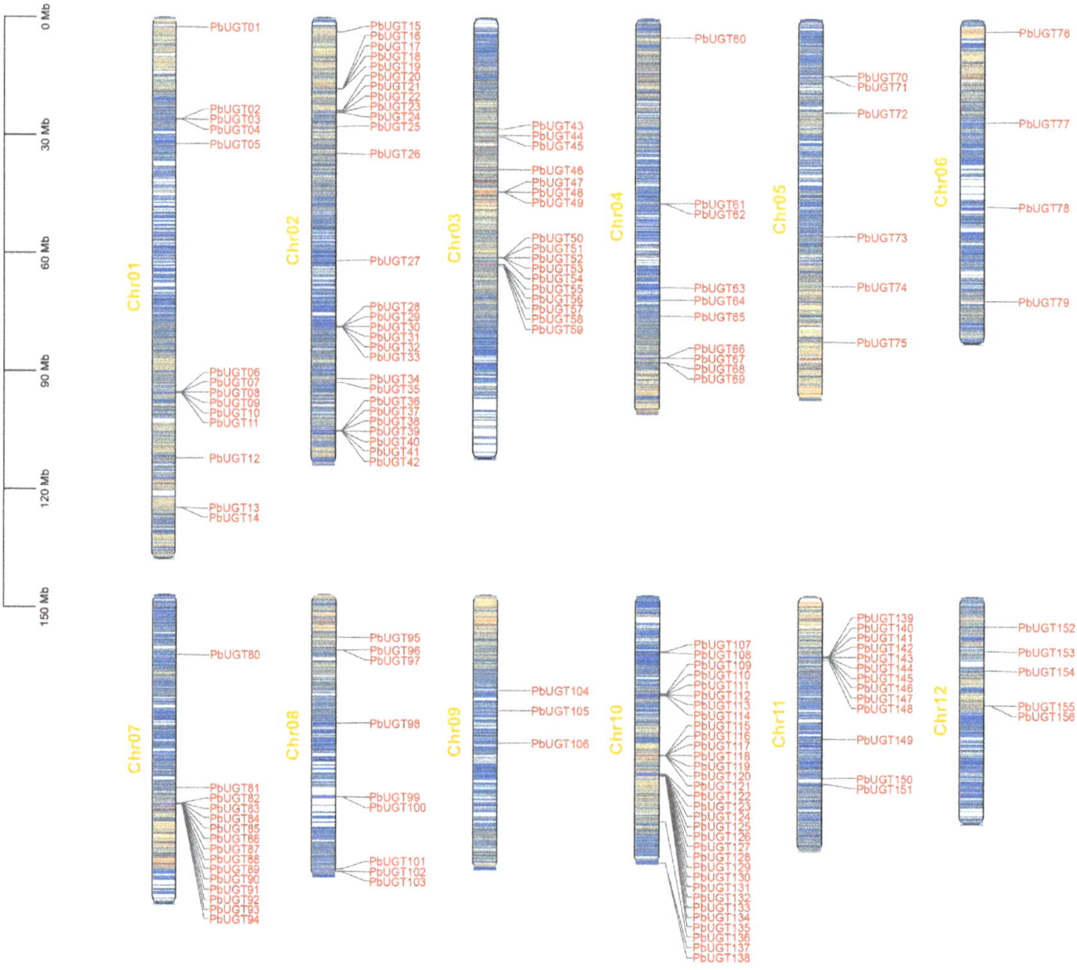

Figure 1. Chromosomal distribution and duplication of *PbUGT* genes.

2.3. *PbUGT Protein Conserved Motif Analysis*

To gain further insights into the conserved domain characteristics of the *UGT* family in *Phoebe bournei*, we employed the online tool MEME to identify and assign 10 distinct motifs, designated as motifs 1 to 10 (Figure 2). The motif analysis revealed that the number of motifs present in each *PbUGT* family member ranged from three to thirteen. Notably, 114 out of the 156 members exhibited a combination of motifs 4 to 9. This unique motif combination serves as a distinctive feature of the family, setting it apart from other gene families and implying its close association with specific functional and performance sequences. Additionally, a notable observation was that members belonging to the same evolutionary group exhibited a high degree of similarity in the distribution of conserved motifs. This suggests that members classified within the same group are likely to share similar functions due to the similarity in motif distribution. Domain analysis unveiled the presence of four conserved domains, namely glycosyltransferase, GTB-type superfamily, GT1-Gtf-like, and PLN02448 domains, in all 156 *UGT* members. This observation suggests that the composition of these three conserved motifs is intricately linked to the synthesis and regulation of glycosyltransferases.

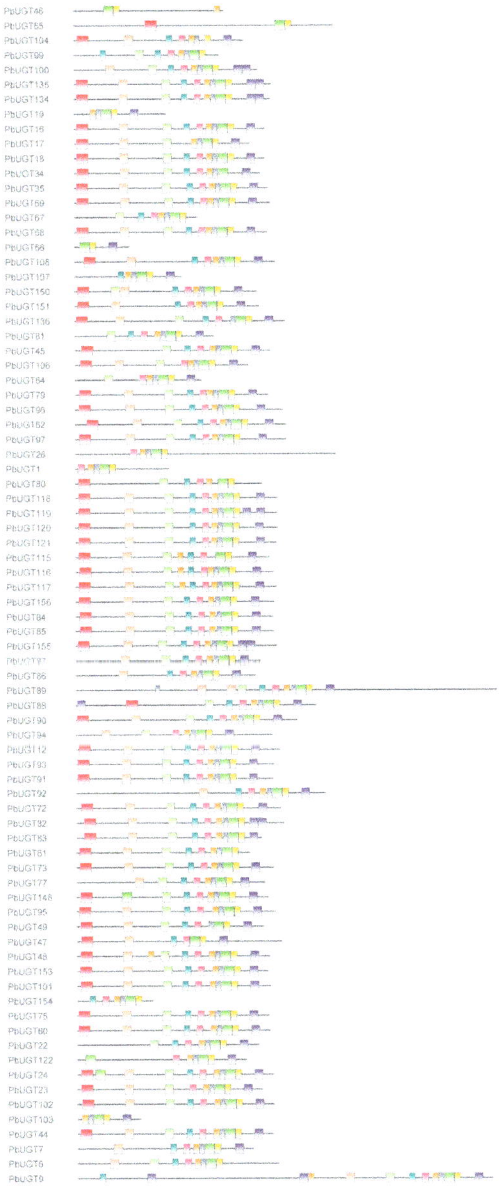

Figure 2. *Cont.*

Figure 2. Conserved motif of the *PbUGT* gene family. Different colors correspond to different types of motifs with the numbers 1–10.

2.4. Phylogenetic Analysis of the PbUGT Gene Family

Previous studies have classified *UGT* genes in *Arabidopsis* into 14 distinct groups, labeled A to N, with the identification of three additional groups, O, P, and Q, in higher plants. To elucidate the functional attributes of each *PbUGT*-family gene and determine its phylogenetic group, we obtained 122 sequences from *A. thaliana* via the *Arabidopsis* resource website. Subsequently, we constructed a phylogenetic tree by incorporating these sequences with the 156 identified *P. bournei* sequences. The phylogenetic tree revealed the presence of 14 *A. thaliana* groups within *P. bournei*, with an uneven distribution across these groups (Figure 3). Notably, the majority of *PbUGT* genes were found in groups A, G, E, and L, with the largest representation in group A, consisting of 36 members. Conversely, groups K, M, and N each contained only one member. The differences observed among various subfamilies may be linked to the evolutionary trajectory and functional diversity within *P. bournei*.

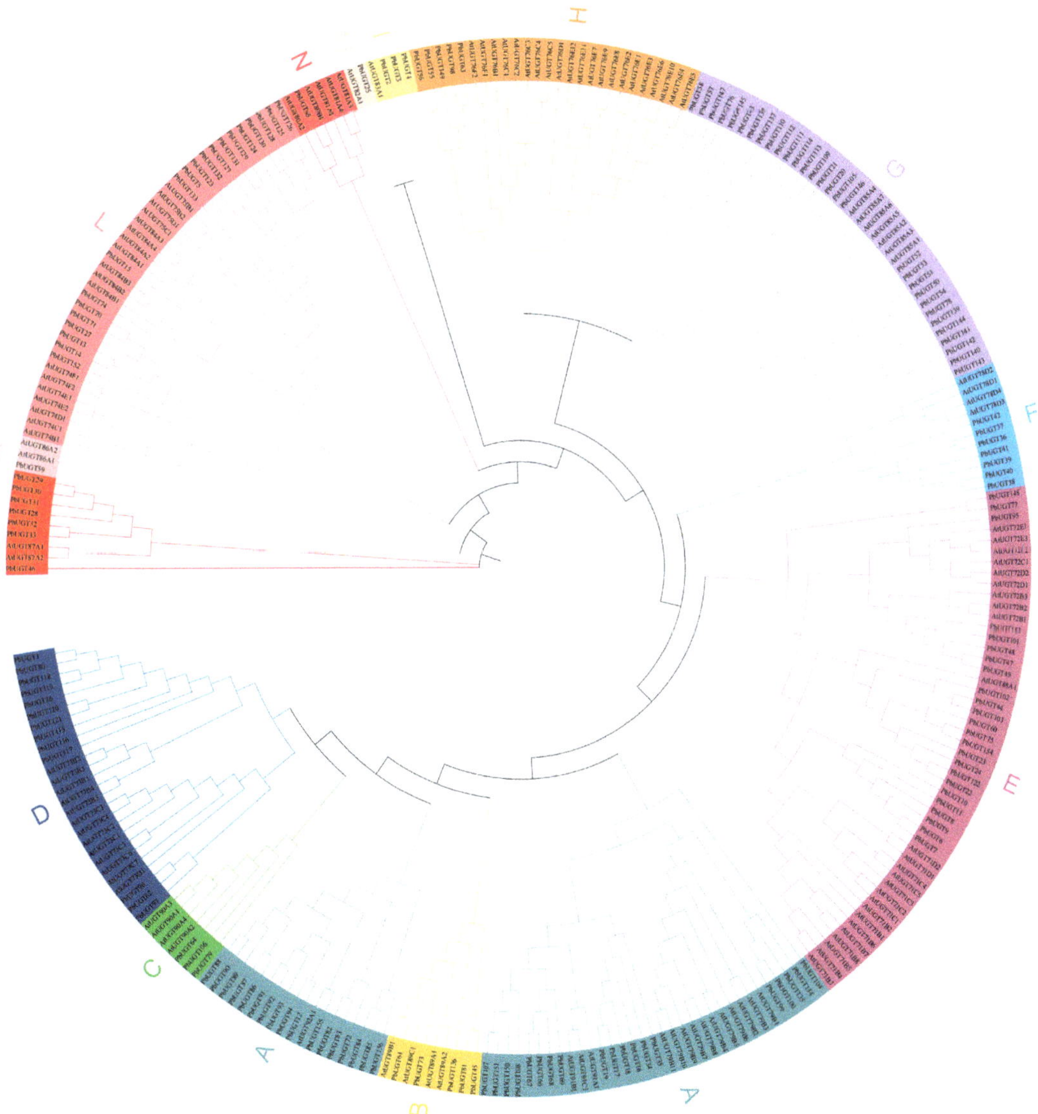

Figure 3. Phylogenetic tree of two plants' *UGT* proteins. The different colored arcs represent the UGT protein subfamilies. The tree was built using 156 PbUGTs from *Phoebe bournei*, 122 AtUGTs from *Arabidopsis thaliana*.

2.5. Organ and Tissue Expression Profile Analysis of the PbUGT Gene Family

By leveraging the original transcriptome sequencing data obtained from five distinct tissues in *P. bournei* (PRJNA628065), we constructed an expression heatmap to examine the tissue-specific expression patterns of the 156 *PbUGT* genes (Figure 4). The heatmap displayed expression levels ranging from low (blue) to high (dark red), highlighting considerable variations in the expression patterns across different tissues. Among the genes analyzed, *PbUGT*1, *PbUGT*21, *PbUGT*26, *PbUGT*27, *PbUGT*35, *PbUGT*55, *PbUGT*84, *PbUGT*104, *PbUGT*117, *PbUGT*121, *PbUGT*145, and *PbUGT*154 lacked corresponding expression data in the reported transcriptome, suggesting their limited involvement in the growth of *P. bournei*. In contrast, *PbUGT*101 and *PbUGT*136 exhibited significantly higher expression levels (FPKM > 15) in all tissues, indicating their crucial roles in the growth and development of *P. bournei*. On average, the expression levels ranked as follows: leaf > stem xylem > stem epidermis > root xylem > root epidermis. Among the fourteen subfamilies, the B subfamily displayed the highest average expression level, surpassing other subfamilies by a substantial margin. Conversely, the C, D, and K subfamilies exhibited notably low overall expression levels (FPKM < 1). The differential expression of *PbUGT* genes across various tissues signifies functional disparities among genes within distinct subfamilies.

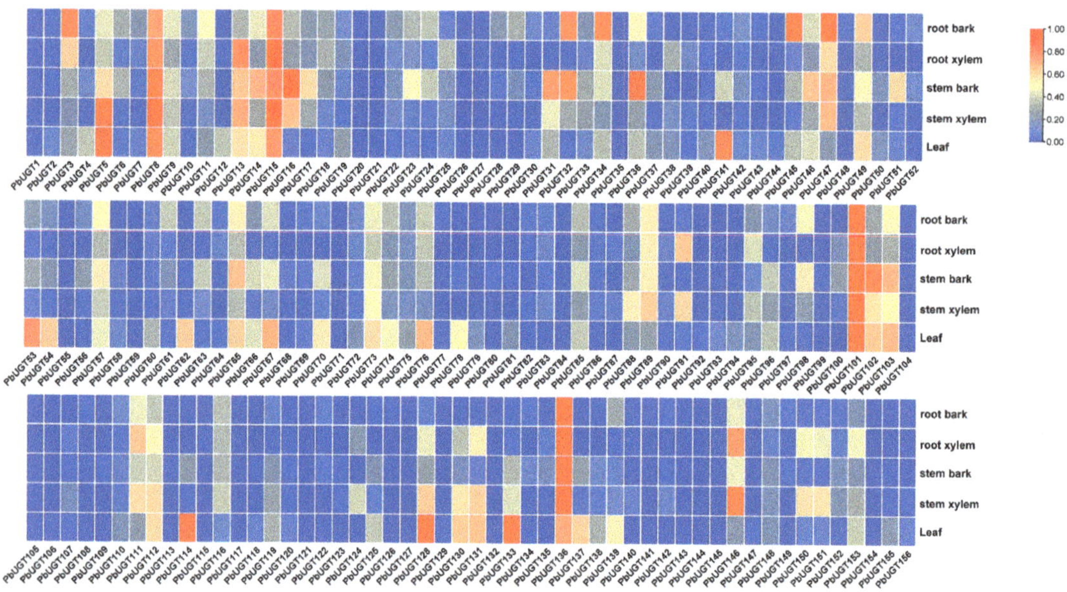

Figure 4. Expression heatmap of the *PbUGT* family in different organ and tissues. Differences in gene expression changes are shown by color, with warmer colors indicating higher expression.

2.6. Synteny Analysis of Homologous UGT Genes among Phoebe bournei, Arabidopsis thaliana, Populus trichocarpa, and Cinnamomum camphora

To investigate the collinearity relationships within the *UGT* gene family among *Phoebe bournei*, *Arabidopsis thaliana*, *Populus trichocarpa*, and *Cinnamomum camphora*, a comprehensive genome-wide collinearity analysis was conducted. The aim was to unveil the interconnections within the *UGT* family. As depicted in Figure 5, a total of 13 *PbUGT* genes on 8 chromosomes displayed collinearity with *UGT* genes from *A. thaliana*, while 39 *UGT* genes on 10 chromosomes exhibited collinearity with *UGT* genes from *P. trichocarpa*. Additionally, 60 *UGT* genes on all 12 chromosomes displayed collinearity with *UGT* genes from *Cinnamomum camphora*. The collinearity analysis revealed that species with closer genetic relationships exhibited stronger collinearity within the same gene family. This observation suggests a plausible association with species evolution, where these genes might have arisen earlier, with each gene family demonstrating homology prior to species differentiation. Over time, as the species diverged, these homologous gene relationships gradually diminished.

Figure 5. Synteny analysis of *UGT* genes among *Phoebe bournei*, *Arabidopsis thaliana*, *Populus trichocarpa*, and *Cinnamomum camphora*. The two rings in the middle represent the gene density of each chromosome, the gray lines represent the collinearity, and the red lines represent the highlighted collinearity of *UGT* genes.

2.7. Duplication Analysis of PbUGT Genes in Chromosomal Replication

To elucidate the expansion and evolutionary mechanisms underlying the *PbUGT* gene family, potential gene duplication events in the *P. bournei* genome were investigated. The entire *P. bournei* genome was compared using MCScanX software, leveraging amino acid sequence homology. The analysis revealed the presence of 14 tandem repeat gene clusters within the *UGT* gene family members (Figure 6), indicating that tandem repeats played a pivotal role in the gene family's expansion. Additionally, this study examined the ratio of nonsynonymous substitutions (Ka) to synonymous substitutions (Ks) between the duplicated genes. The results demonstrated that the ratio was significantly below one, implying a negative selection effect on the *PbUGT* gene during the evolutionary process, indicative of purifying selection.

2.8. Enrichment Analysis of Stress-, Hormone-, and Light-Responsive Cis-Elements in the Promoter Region of Phoebe bournei UGT Genes

Cis-acting elements serve as indicators of gene function and predictors of transcriptional regulation patterns. The promoter regions of *PbUGT* genes encompass numerous cis-acting elements with distinct functions, which can be categorized into the stress response, hormone response, and light response. In total, 7, 35, and 11 cis-elements associated with stress responsiveness, hormone responsiveness, and light responsiveness, respectively, were identified across all *PbUGT* promoter sequences (Figure 7). Among them, the G-box (a light-response cis-element) exhibited the highest frequency, occurring 366 times in the promoter sequences of 131 genes. It was followed by ABRE (a hormone-response cis-element), which appeared 326 times in the promoter sequences of 124 genes. Previous studies have indicated that cis-element responses in various plants are involved in diverse stresses and hormone regulation. Hence, *PbUGT* genes may play a role in stress, hormone, and light regulation during the growth and development of *P. bournei*. This finding provides valuable insights for subsequent stress treatments in this experimental analysis.

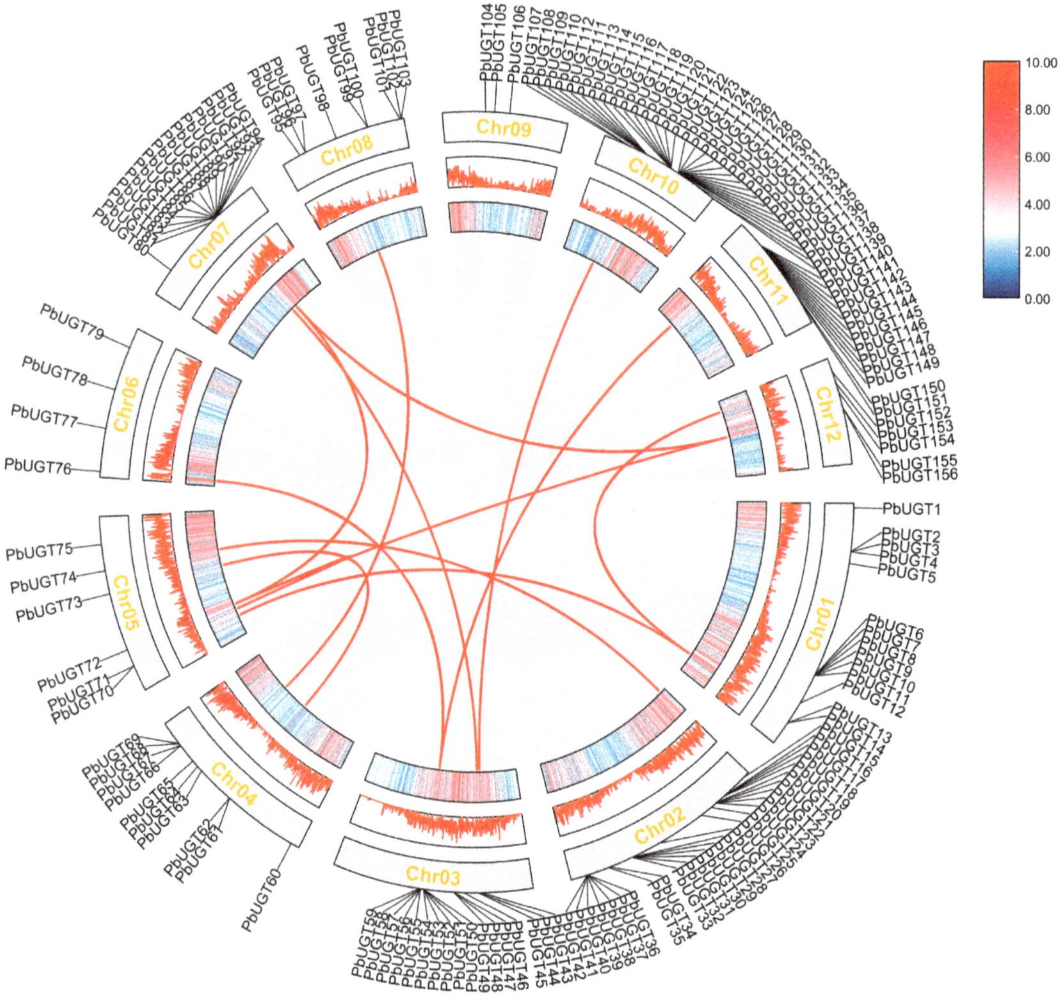

Figure 6. Distribution and collinearity analysis of *PbUGT* genes. The two rings in the middle represent the gene density of each chromosome, the gray lines represent the collinearity, and the red lines represent the highlighted collinearity of *UGT* genes.

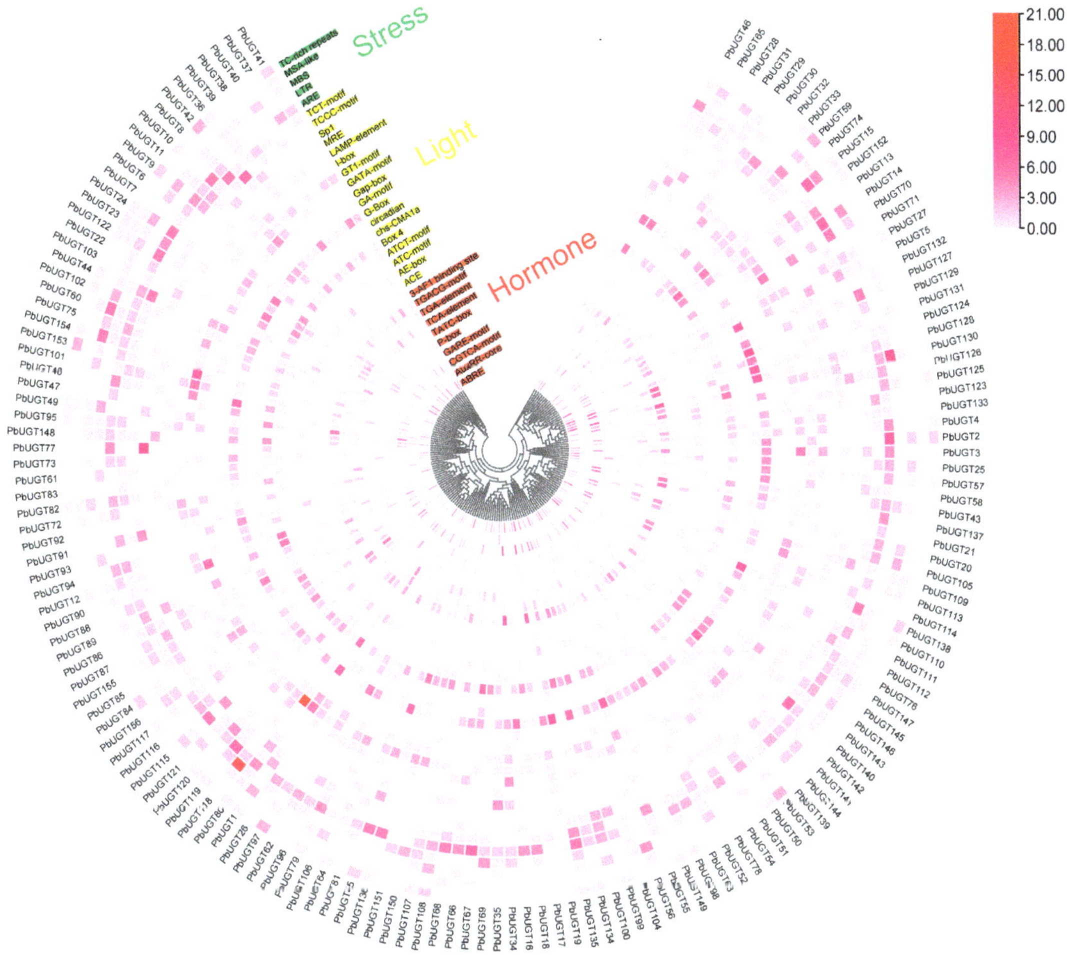

Figure 7. Enrichment analysis of stress-, hormone-, and light-responsive cis-elements in the promoter regions of the *PbUGT* gene family. Differences in the number of cis-elements are indicated by color, with darker colors indicating more cis-elements.

2.9. Screening and Verification of Stress Resistance Genes in PbUGT Genes

Based on the cis-acting elements, five genes—*PbUGT*14, *PbUGT*15, *PbUGT*56, *PbUGT*60, and *PbUGT*154—displaying the highest number of elements corresponding to adversity were identified. Subsequently, their transient expression levels were verified in response to various adversity stresses using RT-qPCR. The expression level analysis of *PbUGT* genes under 10% NaCl salt stress revealed that *PbUGT*14, *PbUGT*15, *PbUGT*56, *PbUGT*60, and *PbUGT*154 were all affected by salt stress (Figure 8A). Generally, their expression levels exhibited a "rise–fall–rise" trend over time. *PbUGT*14, *PbUGT*56, *PbUGT*60, and *PbUGT*154 reached their peak expression levels at 4 h, followed by an upward trend after a sudden decline. In contrast, the expression level of *PbUGT*15 continuously decreased with time.

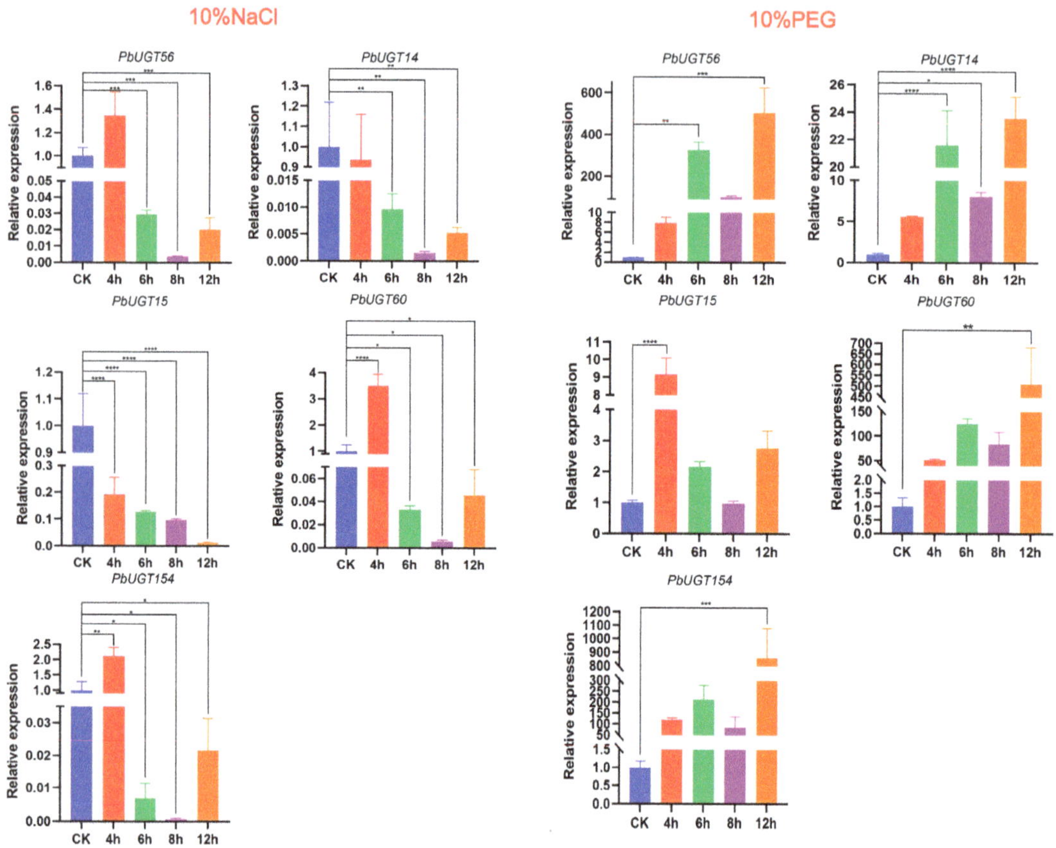

(A) **(B)**

Figure 8. Expression profile of *PbUGT* genes responding to salt and drought stresses tested using RT-qPCR. (**A**) The relative gene expression levels under salt (10% NaCl) treatments for the same periods (0, 4, 6, 8,12 h). Control seedlings were treated with distilled water. (**B**) The relative gene expression levels under drought (10% PEG) treatments for the same periods (0, 4, 6, 8,12 h). Control seedlings were treated with distilled water. (* $p < 0.05$, ** $p < 0.01$, *** $p < 0.0005$, **** $p < 0.0001$).

Under 10% PEG-6000 drought stress (Figure 8B), except for *PbUGT*15, *PbUGT*14, *PbUGT*56, *PbUGT*60, and *PbUGT*154 displayed peak expression levels at 6 h, followed by a sharp decrease and subsequent upward trend. *PbUGT*15, on the other hand, peaked at 4 h, exhibiting an overall "rise–decline–rise" pattern. The expression levels of the five genes demonstrated significant variations under 10% NaCl salt stress and 10% PEG-6000 drought stress, displaying both up-regulation and down-regulation. This indicates that these members of the gene family play crucial roles in the response of *P. bournei* to drought and salt stresses.

3. Discussion

In the present investigation, a total of 156 *PbUGT* genes were identified in the genome of *Phoebe bournei*. The number of *PbUGT* genes in *P. bournei* exceeded that of *Cucumis sativus L.* (94), *Dendrobium officinale* (44), and *pomegranate* (145), but was lower than *Arachis hypogaea L.* (267) and *Glycine max* (182). In comparison to other gene families, the *UGT* gene family demonstrated a higher abundance across all plant species. The initial stage of species differentiation showed a certain covariance, and the analysis of species covariance showed that *UGT* genes may have differentiated earlier in the evolutionary process, showing higher homology and closer linkage relationship with related plants. As species continue to differentiate, covariance gradually decreases.

Analysis of the physicochemical properties of *P. bournei* identified that 145 genes (93%) had an average molecular mass greater than 30 KDa, with an average of 50 KDa, and contained more than 100 amino acids. Regarding the proteins, 136 (87%) were found to be acidic, and 130 (83%) were hydrophilic. The instability coefficients of 110 proteins exceeded 40, indicating that most of the proteins regulated by *PbUGT* genes were both unstable and acidic. Subcellular localization analysis revealed an uneven distribution of *PbUGT* genes across organelles, with the majority located in the mitochondria (88), followed by 43 in the cytoplasm, and 15 in the nucleus, with the remaining in organelles such as vesicles and peroxisomes. The abundance of *PbUGT* genes in mitochondria may be related to the substantial energy requirements of glycosyltransferase, a process regulated by *UGT*. This finding provides a significant foundation and insight for further *UGT* gene research.

During natural growth, forest trees frequently encounter various abiotic stresses, including drought, low temperature, and salinity stresses, which impede their growth. These stresses can cause damage to plants, necessitating optimal water and fertilizer conditions for forest trees to achieve expected growth rates. However, this also results in greater nutrient consumption and longer recovery periods for the forest land. Genes play a vital role in enhancing the tolerance of forest trees to abiotic stresses [25]. Forest trees respond to stress signals by modulating the expression of various genes, which, in turn, regulate metabolic responses and control tree growth under stressful conditions.

The current experiment, in conjunction with previous studies and cis-element analysis, has demonstrated that *UGT* genes across species respond to diverse abiotic stresses. For instance, maize mutants lacking the *ZmUGT2* gene failed to complete flavonol glycosylation in vivo, resulting in inhibited germination rates and suppressed stem and root growth under abiotic stress conditions, particularly salt and drought stresses [26,27]. In tea trees, abiotic stress suppressed the gene regulation of *CsUGT91Q2* responsible for nerolidol glycol production, leading to impaired reactive oxygen species scavenging and reduced cold resistance, thereby increasing sensitivity to cold stress [28,29].

As transcription factors, *UGT* genes can activate the expression of stress-related genes, enhancing plant stress resistance and participating in the regulatory network governing adversity stress in forest trees. Phylogenetic analysis, transcriptome data, and RT-qPCR data confirmed that the five *PbUGT* genes can be rapidly induced and activated in response to abiotic stresses such as drought and salt stress, playing a significant role in the plant's response to these stresses.

Furthermore, transcriptome data analysis of *P. bournei* seedlings subjected to drought and salt stress revealed the involvement of all five *UGT* genes in abiotic stress responses. Their expression varied significantly with the duration of stress. The majority of genes exhibited up-regulated expression under drought and salt stress, displaying a "rising–declining–rising" expression pattern under both 10% NaCl salt stress and 10% PEG-6000 drought stress. Particularly, under the 10% NaCl salt stress, all five *UGT* genes exhibited a high expression pattern after only 4 h. These findings suggest that *PbUGT* genes play a positive regulatory role in the abiotic stress response. This provides new insights for *P. bournei* species selection and stress regulation studies, aiming to expand the planting area of *P. bournei* and maximize its growth potential under stressful conditions.

4. Materials and Methods

4.1. Identification, Chromosomal Localization, and Analysis of Protein Physicochemical Properties

The whole-genome data and annotation files of *Phoebe bournei* were acquired from the China National GeneBank DataBase (CNGBdb) with lookup number CNP0002030 (https://db.cngb.org/ (accessed on 1 June 2023)). The Pfam website was utilized to obtain the *UGT* structural domain (query code PF00201) for screening conserved sequences (E-value < 1×10^{-5}) in all *P. bournei* protein sequences using Hmmer 3.3.2 software (http://hmmer.org/ (accessed on 2 June 2023)) [30]. Furthermore, the CDD function of the NCBI website (https://www.ncbi.nlm.nih.gov/cdd (accessed on 3 June 2023)) and SMART online (https://smart.embl.de/ (accessed on 4 June 2023)) [31] were employed to analyze and integrate the obtained protein sequences. Erroneous sequences were manually removed, ensuring the presence of UDP-glycosyltransferase structural domains. Ultimately, 156 sequences containing the complete *UGT* structural domain were selected.

Based on the whole-genome data and annotation files of *P. bournei*, specific location information for *UGT* genes on each chromosome was obtained. Subsequently, Mapchart 2.32 software [32] was employed to map the location information of each *PbUGT* gene onto the 12 *P. bournei* chromosomes. The ExPASy online (https://web.expasy.org/protparam (accessed on 15 July 2023)) [33] was utilized to predict and calculate the molecular weight, number of amino acids, theoretical PI (isoelectric point), number of positive (negative) residues, instability coefficient, lipolysis coefficient, and total average hydrophilicity of each *UGT* protein in *P. bournei*. The subcellular localization of each member of the *P. bournei UGT* gene family was predicted using the WoLFPSORT website (https://wolfpsort.hgc.jp (accessed on 17 July 2023)) [34].

4.2. Gene Structure, Protein Conserved Motifs, and Exon–Intron Analysis

The annotation file of the *P. bournei* genome and the amino acid sequences of the 156 *UGT* genes were used to construct a phylogenetic evolutionary tree for *PbUGT* genes using the One Step Build an ML Tree function in TBtools v1.120 [35]. The MEME 5.5.2 online software (http://meme-suite.org/tools/meme (accessed on 19 July 2023)) [36] was employed to analyze the conserved motifs of *PbUGT* genes with a motif set of 10. The GSDS2.0 online tool (http://gsds.cbi.pku.edu.cn/ (accessed on 21 July 2023)) [37] was utilized to map the gene structure, including the UTRs, exons, and introns of each *UGT* gene. The Batch CD Search function of the NCBI website (https://www.ncbi.nlm.nih.gov/Structure/bwrpsb/bwrpsb.cgi (accessed on 23 July 2023)) was used to query the structural domains by submitting the gene sequences. The Gene Structure View function in TBtools [35] facilitated the visualization and mapping of phylogenetic relationships, conserved motif composition, structural domains, and exon–intron gene structures for each *PbUGT* gene.

4.3. Phylogenetic Analysis and Construction of the Phylogenetic Tree

Nucleotide and amino acid sequences of 122 *Arabidopsis thaliana UGT* genes were obtained from the *A. thaliana* glycosyltransferase family 1 information website (http://www.p450.kvl.dk/UGT.shtml (accessed on 27 July 2023)) [38–40] as a reference. Multi-protein sequence alignment was performed using MUSCLE [41] on the combined sequences of 156 *UGT* genes from *P. bournei*. Phylogenetic trees of related proteins were constructed using the maximum likelihood method in the MEGA11 program with a bootstrap value set to 1000 and default parameters [42]. The constructed phylogenetic trees were further organized and embellished using the iTol 6.8.2 online website (https://itol.embl.de/ (accessed on 30 July 2023)) [43].

4.4. Analysis of Tissue Expression Profiles

The sequence number PRJNA628065 of *P. bournei* was searched on the European Bioinformatics Centre repository EBI (https://www.ebi.ac.uk (accessed on 1 October 2023)). Transcriptome expression data of *P. bournei* in five different tissues were downloaded [19]. RNA-seq analysis was conducted on the data to investigate the differential expression of the 156 *PbUGT* genes. Tbtools v1.120 software was used to algebraically transform the FPKM values from the analyzed data into log2 values, which were then used to generate heat maps.

4.5. Covariance Analysis

To elucidate the covariance patterns among *PbUGT* family homologous genes in *Arabidopsis thaliana, Populus hirsutus,* and *Cinnamomum camphora*, a genome-wide covariance analysis was conducted. This analysis integrated the chromosome-level genome data of *P. bournei* with the genome-wide data of *A. thaliana* [44], *P. hirsutus* [45], and *C. camphora* [46]. The Synteny Visualization function in Tbtools was employed to map the covariance across species. The whole-genome data of *A. thaliana* and *P. trichocarpa* from the Ensembl Plants database (https://plants.ensembl.org/index.html (accessed on 3 October 2023)) and *C. camphora* genome-wide data from the National Genomics Data Center (https://ngdc.cncb.ac.cn/ (accessed on 3 October 2023)) (login number: GWH-BGXC00000000) were utilized.

4.6. Chromosomes Location, Gene Duplication, and Synteny Analysis

Circos [47] was used to identify each *PbUGT* gene on individual chromosomes. The results were then visualized using the Multiple Covariance Scanning Toolkit MCScan [48] with default settings. A comparison within the entire *P. bournei* genome was performed based on amino acid sequence homology. The duplication events in each *PbUGT* gene were analyzed, and the results were visualized using Circos and Dual Synteny Plot in Tbtools.

4.7. Enrichment Analysis of Stress-, Hormone-, and Light-Responsive Cis-Elements in the Promoter Region

The GTF/GFF3 Sequences Extract function of Tbtools was employed to obtain a 2000 bp segment upstream of the promoter region of each *PbUGT* gene based on the whole-genome sequencing sequence of *Phoebe bournei*. PlantCare (https://bioinformatics.psb.ugent.be/webtools/plantcare/html/ (accessed on 5 October 2023)) [49], an online software, was used for the prediction and screening of cis-acting elements associated with *PbUGT* genes. Subsequently, the data were categorized, organized, and analyzed, and the Heatmap function in Tbtools was utilized to classify and visualize the number of cis-acting elements for each stress-responsive *PbUGT* gene.

4.8. Material and Stress Treatments

One-year-old *P. bournei* seedlings from the Fujian Academy of Forestry were selected as the plant material and incubated in artificial climate chambers under different treatments. Before stress, the soil for planting the seedlings was prepared using peat soil, humus soil, sandy soil, perlite, etc., in a ratio of 5:2:2:1, and was cultivated in pots with a diameter of 150 mm. During stress treatment, the setting parameters of the artificial climate chambers were as follows: light cycle 12 h/d, temperature 25 °C, LED lights used for lighting, photosynthetically active radiation set to 1200 $\mu mol \cdot mol^{-1} \cdot s^{-1}$. The capacity of each pot was about 3.75 L. The seedlings with consistent growth potential were divided into two groups: the control group and the stress treatment group. The treatment group comprised thirty plants, while the control group consisted of three plants. For each biological replicate, two plants were included in the treatment group. Three sets of biological replicates were established for each time point. During stress treatment, we removed the *P. bournei* seedlings from the basin and cleaned their roots with pure water. The PEG treatment was transplanted into a beaker containing 10% PEG 6000 of ¼-strength Hogland solution. The salt treatment was transplanted into a beaker containing 10% NaCl of ¼-strength Hogland solution. The control group was maintained under normal moisture conditions. All stress treatments were conducted for 4, 6, 8, and 12 h. After the designated treatment periods, leaf samples from the *P. bournei* seedlings were collected, immediately frozen in liquid nitrogen, and stored for subsequent RNA extraction.

4.9. Real-Time Quantitative Polymerase Chain Reaction (RT-qPCR) Analysis of PbUGT Genes

RNA extraction was performed using the HiPure Plant RNA Mini Kit (Magenbio, Canton, China), and cDNA synthesis was carried out using the PrimeScript RT reagent Kit (Perfect Real-Time) kit (TaKaRa, Dalian, China). Based on the types of cis-acting elements present in the *PbUGT* gene family members, the genes with the highest number of elements associated with drought and salt stress responses (*PbUGT*14, *PbUGT*15, *PbUGT*56, *PbUGT*60, and *PbUGT*154) were selected for RT-qPCR analysis. The primer sequences for the quantitative RT-qPCR are shown in Table 2. Specific primers in the non-conserved regions of the target genes were designed using Primer 3.0 software and synthesized by Fuzhou Qingbaiwang Biotechnology Company. Real-time fluorescence quantitative analysis was performed using 1 μL of cDNA template, 10 μL of SYBR Premix Ex TaqTM II, 2 μL of specific primers, and 7 μL of ddH_2O. The reaction program consisted of 40 cycles: 95 °C for 30 s, 95 °C for 5 s, 60 °C for 30 s, 95 °C for 5 s, 60 °C for 60 s, and 50 °C for 30 s. The internal reference gene used was PbEF1α (GenBank no. KX682032) [50]. The expression of the target gene was calculated using the $2^{-\Delta\Delta Ct}$ method. GraphPad Prism 8.0 software was utilized for statistical analysis, performing *t*-tests and generating graphs. A significance level of $p < 0.05$ was used to determine statistical significance.

Table 2. Primer sequences of quantitative RT-qPCR.

Primer	Sequence (5′~3′)
PbUGT14(OF21457-F)	TGCATGGGGCCTCAAAGG
PbUGT14(OF21457-R)	GGGCGAGCACTTCCAACT
PbUGT15(OF04928-F)	GTTTCCGCAGTGGGGTGA
PbUGT15(OF04928-R)	CGGTCCCCCAACAACCTC
PbUGT56(OF23900-F)	GTTCGTGGGGTGGAGTGG
PbUGT56(OF23900-R)	GGGTGGGCCAGAACTTCC
PbUGT60(OF22991-F)	TCGCCGCATCGAATCCTC
PbUGT60(OF22991-R)	AGGCGAGGGATTCAGGGT
PbUGT154(OF17153-F)	GCGGGCAGAGGTTCTTGT
PbUGT154(OF17153-R)	TTGCGACGCCCATGAGTT
Internal reference gene (*EF1α-F*)	CATTCAAGTATGCGTGGGT
Internal reference gene (*EF1α-R*)	ACGGTGACCAGGAGCA

5. Conclusions

The *UGT* gene family holds considerable significance in plant growth and development. In the case of *Phoebe bournei*, a comprehensive set of 156 *PbUGT* genes were successfully identified. Evolutionary analysis facilitated the classification of *P. bournei UGT* genes into fourteen distinct subfamilies. Notably, genes within the same subfamily often exhibited similar exon–intron structures, motifs, and protein configurations. The expression patterns of *PbUGT* genes demonstrated noteworthy variations under salt and drought stresses. Our experimental data substantiated the significant differences exhibited by *UGT* genes in response to diverse stresses, thereby emphasizing their crucial role in stress adaptation. This study serves as a foundational exploration into the functional attributes of the *PbUGT* gene family, offering a theoretical framework for fellow researchers interested in investigating the response of *PbUGT* genes under abiotic stress conditions. In light of the mounting challenges posed by land salinization and drought, further research on enhancing *P. bournei* species could contribute to expanding the cultivation area dedicated to this species. The anti-stress mechanism of *UGT* gene needs to be further explored.

Author Contributions: S.C. (Shijiang Cao) and K.Z. conceived the study. H.G. and Y.Z. performed the experiments, data analysis, and manuscript writing. J.L. worked on the RT-qPCR tests. Z.Z. and J.C. performed the RNA isolation. A.B. helped in the experiments. H.G. and J.C. revised the manuscript. S.C. (Shipin Chen) supported the project with funding. All authors have read and agreed to the published version of the manuscript.

Funding: This research was jointly funded by the Fujian Agriculture and Forestry University Forestry Peak Discipline Construction Project (71201800739, to Shijiang Cao) and the Seventh Project of Forest Seeding Breaking in Fujian Province (LZKG-202208, to Shipin Chen).

Data Availability Statement: The whole-genome data and annotation files of *Phoebe bournei* were acquired from the China National GeneBank DataBase (CNGBdb) with lookup number CNP0002030 (https://db.cngb.org/ (accessed on 1 January 2023)).

Conflicts of Interest: The authors declare no conflicts of interest.

References

1. Vogt, T.; Jones, P. Glycosyltransferases in plant natural product synthesis: Characterization of a supergene family. *Trends Plant Sci.* **2000**, *5*, 380–386. [CrossRef] [PubMed]
2. Keegstra, K.; Raikhel, N. Plant glycosyltransferases. *Curr. Opin. Plant Biol.* **2001**, *4*, 219–224. [CrossRef]
3. Hughes, J.; Hughes, M.A. Multiple secondary plant product UDP-glucose glucosyltransferase genes expressed in cassava (Manihot esculenta Crantz) cotyledons. *DNA Seq.* **1994**, *5*, 41–49. [CrossRef] [PubMed]
4. Kanzaki, S.; Kamikawa, S.; Ichihi, A.; Tanaka, Y.; Shimizu, K.; Koeda, S. Isolation of UDP:flavonoid 3-O-glycosyltransferase (UFGT)-like Genes and Expression Analysis of Genes Associated with Anthocyanin Accumulation in Mango 'Irwin' skin. *Hortic. J.* **2019**, *88*, 435–443. [CrossRef]
5. Wang, B.; Jin, S.H.; Hu, H.Q.; Sun, Y.G.; Wang, Y.W.; Han, P.; Hou, B.K. UGT87A2, an Arabidopsis glycosyltransferase, regulates flowering time via *FLOWERING LOCUS C*. *New Phytol.* **2012**, *194*, 666–675. [CrossRef] [PubMed]
6. Paquette, S.; Møller, B.L.; Bak, S. On the origin of family 1 plant glycosyltransferases. *Phytochemistry* **2003**, *62*, 399–413. [CrossRef]
7. Barvkar, V.T.; Pardeshi, V.C.; Kale, S.M.; Kadoo, N.Y.; Gupta, V.S. Phylogenomic analysis of UDP glycosyltransferase 1 multigene family in *Linum usitatissimum* identified genes with varied expression patterns. *BMC Genom.* **2012**, *13*, 175. [CrossRef] [PubMed]
8. He, Y.; Ahmad, D.; Zhang, X.; Zhang, Y.; Wu, L.; Jiang, P.; Ma, H. Genome-wide analysis of family-1 UDP glycosyltransferases (UGT) and identification of UGT genes for FHB resistance in wheat (*Triticum aestivum* L.). *BMC Plant Biol.* **2018**, *18*, 67. [CrossRef]
9. Li, Y.; Li, P.; Wang, Y.; Dong, R.; Yu, H.; Hou, B. Genome-wide identification and phylogenetic analysis of Family-1 UDP glycosyltransferases in maize (*Zea mays*). *Planta* **2014**, *239*, 1265–1279. [CrossRef]
10. Xiao, X.; Lu, Q.; Liu, R.; Gong, J.; Gong, W.; Liu, A.; Ge, Q.; Li, J.; Shang, H.; Li, P.; et al. Genome-wide characterization of the UDP-glycosyltransferase gene family in upland cotton. *3 Biotech* **2019**, *9*, 453. [CrossRef]
11. Wu, B.; Gao, L.; Gao, J.; Xu, Y.; Liu, H.; Cao, X.; Zhang, B.; Chen, K. Genome-Wide Identification, Expression Patterns, and Functional Analysis of UDP Glycosyltransferase Family in Peach (*Prunus persica* L. Batsch). *Front. Plant Sci.* **2017**, *8*, 389. [CrossRef] [PubMed]
12. Caputi, L.; Malnoy, M.; Goremykin, V.; Nikiforova, S.; Martens, S. A genome-wide phylogenetic reconstruction of family 1 UDP-glycosyltransferases revealed the expansion of the family during the adaptation of plants to life on land. *Plant J. Cell Mol. Biol.* **2012**, *69*, 1030–1042. [CrossRef] [PubMed]
13. Cao, P.; Bartley, L.E.; Jung, K.; Ronald, P.C. Construction of a Rice Glycosyltransferase Phylogenomic Database and Identification of Rice-Diverged Glycosyltransferases. *Mol. Plant* **2008**, *1*, 858–877. [CrossRef] [PubMed]
14. Wu, B.; Liu, X.; Xu, K.; Zhang, B. Genome-wide characterization, evolution and expression profiling of UDP-glycosyltransferase family in pomelo (*Citrus grandis*) fruit. *BMC Plant Biol.* **2020**, *20*, 459. [CrossRef] [PubMed]
15. Poppenberger, B.; Berthiller, F.; Lucyshyn, D.; Sieberer, T.; Schuhmacher, R.; Krska, R.; Kuchler, K.; Glössl, J.; Luschnig, C.; Adam, G. Detoxification of the Fusarium mycotoxin deoxynivalenol by a UDP-glucosyltransferase from *Arabidopsis thaliana*. *J. Biol. Chem.* **2003**, *278*, 47905–47914. [CrossRef] [PubMed]
16. Li, S.; Li, X.; Li, J.; Huang, P.; Wei, F.; Cui, H.; van der Werff, H. Lauraceae. In *Flora of China*; Science Press: Beijing, China; Missouri Botanical Garden Press: St. Louis, MO, USA, 2008; Volume 7.
17. Wu, D.; Zhu, Z. Preliminary study on structure and spatial distribution pattern of *Phoebe bournei* in Luo Boyan nature reserve in Fujian province. *Sci. Silvae Sin.* **2003**, *39*, 23–30.
18. Chen, S.P.; Sun, W.H.; Xiong, Y.F.; Jiang, Y.T.; Liu, X.D.; Liao, X.Y.; Zhang, D.Y.; Jiang, S.Z.; Li, Y.; Liu, B.; et al. The Phoebe genome sheds light on the evolution of magnoliids. *Hortic. Res.* **2020**, *7*, 146. [CrossRef]
19. Han, X.; Zhang, J.; Han, S.; Chong, S.L.; Meng, G.; Song, M.; Wang, Y.; Zhou, S.; Liu, C.; Lou, L.; et al. The chromosome-scale genome of *Phoebe bournei* reveals contrasting fates of terpene synthase (TPS)-a and TPS-b subfamilies. *Plant Commun.* **2022**, *3*, 100410. [CrossRef]
20. Zhang, M.; Wu, M.J.; Tong, Z.K.; Han, X.; Zhang, J.H.; Cheng, L.J.; Zhou, S.C. Identification of the PbWRKY gene family and its expression analysis under deficiency of phosphorus in *Phoebe bournei*. *Sci. Silvae Sin.* **2022**, *58*, 133–147.

21. Han, S.; Han, X.; Li, Y.B.; Zhang, Y.T.; Zhang, J.H.; Tong, Z.K. Identification of NF-Y gene family and expression analysis in response to drought stress in *Phoebe bournei*. *J. Agric. Biotechnol.* **2022**, *30*, 1112–1127.
22. Wang, Q.Q.; Zhang, Y.T.; Zhang, J.H.; Liu, H.; Tong, Z.K. Identification of PLR gene family and expression of responsive hormones in *Phoebe bournei*. *J. Zhejiang AF Univ.* **2022**, *39*, 1173–1182.
23. Yin, Z.; Liao, W.; Li, J.; Pan, J.; Yang, S.; Chen, S.; Cao, S. Genome-Wide Identification of GATA Family Genes in *Phoebe bournei* and Their Transcriptional Analysis under Abiotic Stresses. *Int. J. Mol. Sci.* **2023**, *24*, 10342. [CrossRef] [PubMed]
24. Yu, J.; Yin, K.; Liu, Y.; Li, Y.; Zhang, J.; Han, X.; Tong, Z. Co-expression network analysis reveals PbTGA4 and PbAPRR2 as core transcription factors of drought response in an important timber species *Phoebe bournei*. *Front. Plant Sci.* **2024**, *14*, 1297235. [CrossRef] [PubMed]
25. Bressan, R.; Bohnert, H.; Zhu, J. Abiotic Stress Tolerance: From Gene Discovery in Model Organisms to Crop Improvement. *Mol. Plant* **2009**, *2*, 1–2. [CrossRef] [PubMed]
26. Li, Y.J.; Li, P.; Wang, T.; Zhang, F.J.; Huang, X.X.; Hou, B.K. The maize secondary metabolism glycosyltransferase UFGT2 modifies flavonols and contributes to plant acclimation to abiotic stresses. *Ann. Bot.* **2018**, *122*, 1203–1217. [CrossRef]
27. Sharif, R.; Su, L.; Chen, X.; Qi, X. Involvement of Auxin in Growth and Stress Response of Cucumber. *Veg. Res.* **2022**, *2*, 13. [CrossRef]
28. Zhao, M.; Zhang, N.; Gao, T.; Jin, J.; Jing, T.; Wang, J.; Wu, Y.; Wan, X.; Schwab, W.; Song, C. Sesquiterpene glucosylation mediated by glucosyltransferase UGT91Q2 is involved in the modulation of cold stress tolerance in tea plants. *New Phytol.* **2020**, *226*, 362–372. [CrossRef]
29. Sharif, R.; Su, L.; Chen, X.; Qi, X. Hormonal Interactions Underlying Parthenocarpic Fruit Formation in Horticultural Crops. *Hortic. Res.* **2022**, *9*, uhab024. [CrossRef]
30. Finn, R.D.; Clements, J.; Eddy, S.R. HMMER web server: Interactive sequence similarity searching. *Nucleic Acids Res.* **2011**, *39*, 29–37. [CrossRef]
31. Ivica, L.; Tobias, D.; Peer, B. SMART 7: Recent updates to the protein domain annotation resource. *Nucleic Acids Res.* **2012**, *40*, 302–305.
32. Voorrips, R. MapChart: Software for the graphical presentation of linkage maps and QTLs. *J. Hered.* **2002**, *93*, 77–78. [CrossRef]
33. Gasteiger, E.; Gattiker, A.; Hoogland, C.; Ivanyi, I.; Appel, R.D.; Bairoch, A. ExPASy: The proteomics server for in-depth protein knowledge and analysis. *Nucleic Acids Res.* **2003**, *31*, 3784–3788. [CrossRef] [PubMed]
34. Horton, P.; Park, K.J.; Obayashi, T.; Fujita, N.; Harada, H.; Adams-Collier, C.J.; Nakai, K. WoLF PSORT: Protein localization predictor. *Nucleic Acids Res.* **2007**, *35*, 585–587. [CrossRef] [PubMed]
35. Chen, C.; Chen, H.; Zhang, Y.; Thomas, H.R.; Frank, M.H.; He, Y.; Xia, R. TBtools: An Integrative Toolkit Developed for Interactive Analyses of Big Biological Data. *Mol. Plant* **2020**, *13*, 1194–1202. [CrossRef]
36. Bailey, T.L.; Boden, M.; Buske, F.A.; Frith, M.; Grant, C.E.; Clementi, L.; Ren, J.; Li, W.W.; Noble, W.S. MEME SUITE: Tools for motif discovery and searching. *Nucleic Acids Res.* **2009**, *37*, 202–208. [CrossRef] [PubMed]
37. Hu, B.; Jin, J.; Guo, A.Y.; Zhang, H.; Luo, J.; Gao, G. GSDS 2.0: An upgraded gene feature visualization server. *Bioinformatics* **2015**, *31*, 1296–1297. [CrossRef] [PubMed]
38. Mackenzie, P.I.; Owens, I.S.; Burchell, B.; Bock, K.W.; Bairoch, A.; Belanger, A.; Fournel-Gigleux, S.; Green, M.; Hum, D.W.; Iyanagi, T.; et al. The UDP glycosyltransferase gene superfamily: Recommended nomenclature update based on evolutionary divergence. *Pharmacogenetics* **1997**, *7*, 255–269. [CrossRef] [PubMed]
39. Li, Y.; Baldauf, S.; Lim, E.K.; Bowles, D.J. Phylogenetic analysis of the UDP-glycosyltransferase multigene family of *Arabidopsis thaliana*. *J. Biol. Chem.* **2001**, *276*, 4338–4343. [CrossRef] [PubMed]
40. Ross, J.; Li, Y.; Lim, E.K.; Bowles, D.J. Higher plant glycosyltransferases. *Genome Biol.* **2001**, *2*, reviews3004. [CrossRef]
41. Edgar, R.C. MUSCLE: Multiple sequence alignment with high accuracy and high throughput. *Nucleic Acids Res.* **2004**, *32*, 1792–1797. [CrossRef]
42. Tamura, K.; Stecher, G.; Kumar, S. MEGA11: Molecular Evolutionary Genetics Analysis version 11. *Mol. Biol. Evol.* **2021**, *38*, 2725–2729. [CrossRef] [PubMed]
43. Letunic, I.; Bork, P. Interactive Tree Of Life (iTOL) v5: An online tool for phylogenetic tree display and annotation. *Nucleic Acids Res.* **2021**, *49*, 293–296. [CrossRef] [PubMed]
44. Cheng, C.Y.; Krishnakumar, V.; Chan, A.P.; Thibaud-Nissen, F.; Schobel, S.; Town, C.D. Araport11: A complete reannotation of the *Arabidopsis thaliana* reference genome. *Plant J. Cell Mol. Biol.* **2017**, *89*, 789–804. [CrossRef] [PubMed]
45. Tuskan, G.A.; DiFazio, S.; Jansson, S.; Bohlmann, J.; Grigoriev, I.; Hellsten, U.; Putnam, N.; Ralph, S.; Rombauts, S.; Salamov, A. The Genome of Black Cottonwood, *Populus trichocarpa* (Torr. &Gray). *Science* **2006**, *313*, 1596–1604. [PubMed]
46. Sun, W.; Xiang, S.; Zhang, Q.; Xiao, L.; Zhang, D.; Zhang, P.; Chen, D.; Hao, Y.; Liu, D.; Ding, L. The camphor tree genome enhances the understanding of magnoliid evolution. *J. Genet. Genom.* **2022**, *49*, 249–253. [CrossRef]
47. Krzywinski, M.; Schein, J.; Birol, I.; Connors, J.; Gascoyne, R.; Horsman, D.; Jones, S.J.; Marra, M.A. Circos: An information aesthetic for comparative genomics. *Genome Res.* **2009**, *19*, 1639–1645. [CrossRef]
48. Wang, Y.; Tang, H.; DeBarry, J.D.; Tan, X.; Li, J.; Wang, X.; Lee, T.H.; Jin, H.; Marler, B.; Guo, H.; et al. *MCScanX*: A toolkit for detection and evolutionary analysis of gene synteny and collinearity. *Nucleic Acids Res.* **2012**, *40*, e49. [CrossRef]

49. Lescot, M.; Déhais, P.; Thijs, G.; Marchal, K.; Moreau, Y.; Van de Peer, Y.; Rouzé, P.; Rombauts, S. PlantCARE, a database of plant cis-acting regulatory elements and a portal to tools for in silico analysis of promoter sequences. *Nucleic Acids Res.* **2002**, *30*, 325–327. [CrossRef]
50. Zhang, J.; Zhu, Y.; Pan, Y.; Huang, H.; Li, C.; Li, G.; Tong, Z. Transcriptomic profiling and identification of candidate genes in two *Phoebe bournei* ecotypes with contrasting cold stress responses. *Trees* **2018**, *32*, 1315–1333. [CrossRef]

plants

Article

Genome-Wide Analysis of Aux/IAA Gene Family in *Artemisia argyi*: Identification, Phylogenetic Analysis, and Determination of Response to Various Phytohormones

Conglong Lian [1,2], Jinxu Lan [1,2], Rui Ma [1,2], Jingjing Li [1,2], Fei Zhang [1,2], Bao Zhang [1,2], Xiuyu Liu [1,2] and Suiqing Chen [1,2,*]

1 School of Pharmacy, Henan University of Chinese Medicine, 156 East Jin-shui Rd., Zhengzhou 450046, China; liancl00@163.com (C.L.); baozhang923@163.com (B.Z.); liuxiuyuzy@163.com (X.L.)
2 Henan Key Laboratory of Chinese Medicine Resources and Chemistry, 156 East Jin-shui Rd., Zhengzhou 450046, China
* Correspondence: chsq@hactcm.edu.cn

Abstract: *Artemisia argyi* is a traditional herbal medicine plant, and its folium artemisia argyi is widely in demand due to moxibustion applications globally. The Auxin/indole-3-acetic acid (Aux/IAA, or IAA) gene family has critical roles in the primary auxin-response process, with extensive involvement in plant development and stresses, controlling various essential traits of plants. However, the systematic investigation of the Aux/IAA gene family in *A. argyi* remains limited. In this study, a total of 61 Aux/IAA genes were comprehensively identified and characterized. Gene structural analysis indicated that 46 Aux/IAA proteins contain the four typical domains, and 15 Aux/IAA proteins belong to non-canonical IAA proteins. Collinear prediction and phylogenetic relationship analyses suggested that Aux/IAA proteins were grouped into 13 distinct categories, and most Aux/IAA genes might experience gene loss during the tandem duplication process. Promoter *cis*-element investigation indicated that Aux/IAA promoters contain a variety of plant hormone response and stress response *cis*-elements. Protein interaction prediction analysis demonstrated that AaIAA26/29/7/34 proteins are possibly core members of the Aux/IAA family interaction. Expression analysis in roots and leaves via RNA-seq data indicated that the expression of some AaIAAs exhibited tissue-specific expression patterns, and some AaIAAs were involved in the regulation of salt and saline-alkali stresses. In addition, RT-qPCR results indicated that AaIAA genes have differential responses to auxin, with complex response patterns in response to other hormones, indicating that Aux/IAA may play a role in connecting auxin and other hormone signaling pathways. Overall, these findings shed more light on AaIAA genes and offer critical foundational knowledge toward the elucidation of their function during plant growth, stress response, and hormone networking of Aux/IAA family genes in *A. argyi*.

Keywords: *Artemisia argyi*; Aux/IAA family; expression pattern; function prediction

1. Introduction

Artemisia argyi Lev. et Vant., a perennial herb with a strong aroma belonging to the genus *Artemisia* and the family Asteraceae, is widely distributed in Asian countries, such as China, Japan, and Korea (https://www.plantplus.cn/info/Artemisia%20argyi?t=foc; accessed on 5 December 2023). In China, *A. argyi* is widely distributed except in extremely arid or cold regions, and the main growth areas include Nanyang in Henan Province, Qichun in Hubei Province, and Anguo in Hebei Province. *A. argyi* is known as a medicinal plant. The dried leaf of *A. argyi* is a well-known traditional Chinese medicine (TCM) known as "folium artemisiae argyi" or "Chinese mugwort", which is commonly used in moxibustion for the treatment of hemostasis, epistaxis, hemorrhage, inflammation, abdominal cold pain, uterine cold infertility, itches, and menstruation-related ailments throughout Asia [1]. Modern pharmacological research uncovered that folium artemisiae argyi can be applied as an antibacterial or antiviral treatment because of its rich volatile oils [2,3]. In addition, *A. argyi* can be used in food, cosmetics, and daily necessities. This versatility of *A. argyi* leads to a massive market demand. Although the current production of *A. argyi* is considerable, and the output of fresh folium artemisiae argyi in China reached 192,000 tons in 2020, its production still cannot meet the market demand. Therefore, it is critical to study the growth and development of *A. argyi*.

Plant hormones are critical for the regulation of plant growth and development in various stages, especially auxin. Many of these rely on the auxin signaling pathway, which determines the fate of plants from birth to death and plays essential roles during plant life cycles, including embryogenesis, cell division and elongation, stress response, and secondary metabolite biosynthesis [4]. Auxin/indole-3-acetic acid (Aux/IAA, or IAA) proteins are important components of the auxin signaling pathway and are established as principal targets of the auxin transport inhibitor response 1 (TIR1) as well as the similar proteins AUXIN RECEPTOR F-BOX/AFBs [5,6]. Aux/IAA genes encode short-lived nuclear proteins that are degraded via the 26S proteasome and can be ubiquitinated by binding with TIR1/AFB receptors, causing ARF homodimers to activate the transcription of auxin-responsive genes under high auxin levels [7,8]. Aux/IAA proteins contain four highly conserved domains: I, II, III, and IV. Domains I and II and domains III and IV are situated at the N-terminus and C-terminus, respectively, producing distinct roles of repression, degradation, or interaction [9,10]. Domain I contains an amphiphilic motif LxLxLx that is associated with TOPLESS (TPL)/TPL-related (TPR) corepressors and is required for the transcriptional inhibitory function of Aux/IAA proteins [11]. Domain II contains the GWPPV motif, which acts as the target sequence of Aux/IAA protein ubiquitination for degradation [12]. Domains III and IV combine with the ARF and contribute to the homo- and hetero-dimerization between Aux/IAA and ARF proteins [13]. The diversity of conserved structures and functions indicates functional differences among different Aux/IAA members.

In the past years, the biological functions of Aux/IAA family genes have been revealed. In particular, Aux/IAA plays a pivotal role in monocots and dicots by affecting the growth and development of different tissues. In *Arabidopsis thaliana*, *Aux/IAA1* and *Aux/IAA8* genes have many roles in the regulation of the development of diverse organs, such as seed germination, root development, rosette leaves, shoot tropisms, and inflorescence [14,15]. In rice, *OsIAA1*, *OsIAA4*, *OsIAA11*, and *OsIAA12* are involved in plant height and leaf angle, tiller angle and gravity tropism, lateral roots formation, and leaf angle, respectively [16–18]. Moreover, *Aux/IAA* genes are involved in adversity stress responses. In *A. thaliana*, *AtIAA5/AtIAA6/AtIAA19* mediate drought response by regulating glucosinolate levels [19]. In rice, *OsIAA9* and *OsIAA20* genes are significantly upregulated under high salt stress [20]. The overexpression of *OsIAA6* triggered noticeable drought resistance enhancements [21]. *OsIAA20* plays a key role in plant responses to drought and salt stresses through the ABA signal transduction pathway [22]. In addition, Aux/IAA genes also impact secondary metabolism, such as the biosynthesis of volatile compounds, including α-humulene, α-terpinene, β-caryophyllene, β-phellandrene, and γ-element [23].

Recently, with the release of reference genomes for different species, Aux/IAA gene families are being identified and analyzed in an increasing number of plant species through bioinformatics approaches, including model plants *A. thaliana* (29 members) [24], *Populus* (35 members) [25], and maize (40 members) [26], as well as medicinal plants such as *Salvia miltiorrhiza* (23 members) [27], *Dendrobium officinale* (14 members) [28], and *Boehmeria nivea* (16 members) [29]. However, the identification and functional information of Aux/IAA genes in medicinal plants are also limited and uncharacterized, especially in the *A. argyi* genome. The recent publication of the whole genome data of *A. argyi* enabled a massive progression in the comprehension of genome organization, which allowed for the identification of Aux/IAA genes [30]. Examining the role of Aux/IAA genes in *A. argyi* could provide an improved understanding of the functional attributes of this gene family in *A. argyi* and stimulate further research in related medicinal plant organisms. In this study, we endeavored to uncover genome-wide characterization and organization of Aux/IAA genes, predicting the related protein domains, architecture, GO annotation, promoter *cis*-elements, phylogenetic relationships, protein interaction, and expression patterns in different tissues, and response to various hormones. These results contributed to the clarification of Aux/IAA protein functions. They provided valuable information for further studies in *A. argyi* and other medicinal plants to uncover the functional roles of Aux/IAA genes in growth and progression.

2. Results

2.1. Identification of the Aux/IAA Genes in A. argyi

Based on biosequence analysis using profile hidden Markov models (HMMER) and Basic Local Alignment Search Tool for proteins (BLASTP) selection, and after determining the presence of the Aux/IAA (PF02309) domain using the Pfam, a total of 61 Aux/IAA members were identified in the genome data of *A. argyi* (Table 1) [30]. Identified genes were named from *AaIAA1* to *AaIAA61* based on the AaIAA location on the chromosomes (Figure 1). The analysis of physicochemical properties demonstrated that the predicted AaIAA proteins varied in size from 155 (AaIAA20, AaIAA52, and AaIAA61) to 1089 amino acids (AaIAA7) with molecular masses and theoretical isoelectric points ranging from 17.8 kDa (AaIAA5) to 121.9 kDa (AaIAA7) and from 4.96 (AaIAA47) to 9.18 (AaIAA8), respectively. The protein instability coefficient of Aux/IAA proteins ranged from 28.42 (AaIAA49) to 70.7 (AaIAA29), and most are unstable except 13 proteins, namely AaIAA6/11/16/17/22/23/33/48/49/50/54/55/56 proteins, as their instability indexes were below 40. The aliphatic index varies from 57.47 (AaIAA18, AaIAA25, and AaIAA57) to 88.81 (AaIAA54). The average hydropathicity varies from −0.863 (AaIAA49) to −0.248 (AaIAA45), indicating that these proteins are hydrophilic. In addition, all proteins are situated in the nucleus, suggesting that they are likely to be transcription factors and play essential roles in the nucleus. The sizable variability in fundamental physical and chemical characteristics indicates that they may possess functions in varied microenvironments and have diverse functions in plant growth and development (Table 1).

Table 1. Physical and chemical characters of AaIAAs proteins in *A. argyi*.

Gene Name	Gene ID	Amino Acids Number (aa)	Molecular Weight (Da)	Isoelectric Point (pI)	Instability Index	Aliphatic Index	Grand Average of Hydropathicity	Subcellular Localization
AaIAA1	Aarg01G001560.1	255	28,351.05	8.37	50.14	58.9	−0.687	Nuclear
AaIAA2	Aarg01G004030.1	206	23,404.46	6.88	47.62	69.95	−0.702	Nuclear
AaIAA3	Aarg01G011540.1	191	21,670.62	5.62	60.92	63.25	−0.699	Nuclear
AaIAA4	Aarg02G020290.1	194	22,104.6	8.98	40.02	81.24	−0.424	Nuclear
AaIAA5	Aarg02G020300.1	158	17,806.49	5.64	40.72	65.95	−0.477	Nuclear
AaIAA6	Aarg02G020570.1	239	26,431.46	8.59	30.31	74.18	−0.562	Nuclear
AaIAA7	Aarg02G023990.1	1089	121,872.7	6.28	62.05	69.94	−0.642	Nuclear
AaIAA8	Aarg03G002350.1	241	26,448.04	9.18	42.86	75.56	−0.637	Nuclear
AaIAA9	Aarg03G012510.1	326	34,600.87	7.53	42.41	71.81	−0.495	Nuclear
AaIAA10	Aarg03G012760.1	328	34,784.98	7.55	42.21	69.88	−0.525	Nuclear
AaIAA11	Aarg04G026560.1	237	26,131.52	8.29	39.93	60.89	−0.691	Nuclear
AaIAA12	Aarg04G026570.1	175	19,816.62	5.39	64.26	71.83	−0.574	Nuclear
AaIAA13	Aarg05G010530.1	177	20,054.95	7.66	57.36	66.05	−0.685	Nuclear
AaIAA14	Aarg05G010900.1	177	20,024.97	8.39	55.97	67.12	−0.653	Nuclear
AaIAA15	Aarg05G011700.1	217	23,924.21	7	50.43	79.4	−0.464	Nuclear
AaIAA16	Aarg05G020910.1	265	29,066.46	8.78	37.29	80.11	−0.34	Nuclear
AaIAA17	Aarg06G011190.1	183	20,467.39	7.02	37.98	88.36	−0.431	Nuclear
AaIAA18	Aarg06G014790.1	190	21,153	8.3	45.9	57.47	−0.745	Nuclear
AaIAA19	Aarg06G021200.1	229	25,531.29	8.16	40.71	70.61	−0.512	Nuclear
AaIAA20	Aarg07G015030.1	155	18,391.91	5.46	58.75	80.45	−0.623	Nuclear
AaIAA21	Aarg07G020250.1	332	35,671.32	8.66	45.05	70.42	−0.48	Nuclear
AaIAA22	Aarg08G015630.1	207	22,925.83	5.97	32.7	81.4	−0.529	Nuclear
AaIAA23	Aarg08G015670.1	209	23,189.17	5.61	32.82	82.49	−0.489	Nuclear

Table 1. *Cont.*

Gene Name	Gene ID	Amino Acids Number (aa)	Molecular Weight (Da)	Isoelectric Point (pI)	Instability Index	Aliphatic Index	Grand Average of Hydropathicity	Subcellular Localization
AaIAA24	Aarg08G018190.1	313	34,615.93	6.4	57.62	68.79	−0.486	Nuclear
AaIAA25	Aarg09G003120.1	190	21,183.09	8.3	46.6	57.47	−0.731	Nuclear
AaIAA26	Aarg09G037610.1	654	72,827.47	6.04	55.49	71.68	−0.471	Nuclear
AaIAA27	Aarg10G000170.1	257	28,609.35	8.37	48.75	60.7	−0.654	Nuclear
AaIAA28	Aarg10G001790.1	254	28,301.06	8.87	48.75	60.67	−0.693	Nuclear
AaIAA29	Aarg10G002700.1	1044	117,163.7	6.56	70.7	68	−0.716	Nuclear
AaIAA30	Aarg10G004460.1	206	23,406.49	6.88	49.76	68.54	−0.712	Nuclear
AaIAA31	Aarg10G012770.1	179	20,275.95	5.97	62.44	63.69	−0.765	Nuclear
AaIAA32	Aarg10G013430.1	191	21,642.52	5.39	61.81	62.72	−0.702	Nuclear
AaIAA33	Aarg10G054010.1	238	26,322.34	8.59	29.37	72.02	−0.575	Nuclear
AaIAA34	Aarg10G056370.1	162	18,394.02	7.95	61.36	88.46	−0.577	Nuclear
AaIAA35	Aarg11G002340.1	241	26,451.1	9.11	41.04	75.98	0.611	Nuclear
AaIAA36	Aarg11G002950.1	263	28,731.7	8.13	40.33	76.65	−0.393	Nuclear
AaIAA37	Aarg11G003340.1	169	18,728.47	6.31	41.01	77.81	−0.439	Nuclear
AaIAA38	Aarg11G003350.1	212	23,901.29	7.53	48.43	78.07	−0.412	Nuclear
AaIAA39	Aarg11G014830.1	326	34,716.01	8.14	43	71.81	−0.52	Nuclear
AaIAA40	Aarg11G030150.1	235	26,733.46	6.24	44.76	76.26	−0.397	Nuclear
AaIAA41	Aarg11G030160.1	197	22,766.72	6.39	50.71	63.71	−0.724	Nuclear
AaIAA42	Aarg12G024700.1	256	27,610.54	6.45	49.77	62.03	−0.674	Nuclear
AaIAA43	Aarg12G025240.1	255	27,696.68	6.45	49.29	63.8	−0.677	Nuclear
AaIAA44	Aarg13G015770.1	177	20,088.97	7.66	56.27	63.84	−0.69	Nuclear
AaIAA45	Aarg13G016280.1	217	23,777.13	5.87	43.99	87.51	−0.248	Nuclear
AaIAA46	Aarg13G016670.1	188	20,803.73	5.59	48.35	87.5	−0.393	Nuclear
AaIAA47	Aarg13G025060.1	229	25,974.51	4.96	40.33	73.58	−0.758	Nuclear
AaIAA48	Aarg14G002120.1	221	24,696.54	6.92	29.79	59.05	−0.854	Nuclear
AaIAA49	Aarg14G003710.1	222	24,697.49	6.92	28.42	57.93	−0.863	Nuclear
AaIAA50	Aarg14G012600.1	190	21,158.13	6.6	39.98	85.63	−0.453	Nuclear
AaIAA51	Aarg14G023130.1	229	25,503.23	8.16	40.34	69.78	−0.522	Nuclear
AaIAA52	Aarg15G014380.1	155	18,391.91	5.46	58.75	80.45	−0.623	Nuclear
AaIAA53	Aarg15G018930.1	332	35,698.35	8.66	43.89	70.42	−0.486	Nuclear
AaIAA54	Aarg15G024900.1	193	22,277.12	5.22	37.4	88.81	−0.439	Nuclear
AaIAA55	Aarg16G015470.1	214	23,544.67	5.95	32.54	86.45	−0.323	Nuclear
AaIAA56	Aarg16G015550.1	214	23,557.75	5.95	32.54	88.74	−0.288	Nuclear
AaIAA57	Aarg17G001730.1	190	21,183.09	8.3	46.6	57.47	−0.731	Nuclear
AaIAA58	Aarg0G001160.1	237	26,068.55	8.6	42.49	64.18	−0.673	Nuclear
AaIAA59	Aarg0G013210.1	175	19,772.52	5.2	61.38	71.83	−0.57	Nuclear
AaIAA60	Aarg0G017000.1	326	34,716.01	8.14	43	71.81	−0.52	Nuclear
AaIAA61	Aarg0G047700.1	155	18,391.91	5.46	58.75	80.45	−0.623	Nuclear

2.2. Chromosomal Distribution, Collinear Prediction, and Sequence Similarity Analysis of AaIAA Genes

According to the study by Hongyu Chen et al., *A. argyi* can be assembled into 17 pseudochromosomes on the chromosome scale [30]. A total of 57 *AaIAA* genes in *A. argyi* were distributed on 17 chromosomes (Chr), with *AaIAA58*, *AaIAA59*, *AaIAA60*, and *AaIAA61* distributed in the scaffolds of GWHBRAE00001633, GWHBRAE00001586, GWHBRAE00001446, and GWHBRAE00002123, respectively (Figure 1). The distribution of *AaIAAs* genes varies: there was just a single *AaIAA57* gene on Chr17; Chr10 had the largest number with eight total genes; Chr11 had seven genes; and the remaining 15 chromosomes were distributed with two to four genes. In addition, there are tandem duplication phenomena in the distribution of *AaIAA* genes, including *AaIAA5* and *AaIAA6* genes on Chr2, *AaIAA22* and *AaIAA23* genes on Chr8, and *AaIAA55* and *AaIAA56* genes on Chr16 (Figure 1).

Gene divergence and duplication events are critical causes of evolutionary momentum and can also significantly promote the expansion of gene families and the diversification of protein functionalities [31]. In this study, 57 IAA family gene collinear events were used to reveal the expansion of the *A. argyi* IAA family genes during evolution (Figure 2A). A total of 45 *AaIAA* genes have 106 collinear events. Among them, the majority of the

A. argyi AaIAA genes had one (18 genes) or two (14 genes) collinear events, while 4, 2, 2, and 5 *AaIAA* genes had 3, 4, 5, and 6 collinear events, respectively (Figure 2A, Table S2). These findings implied that homologous genes may exert similar functions, indicating functional redundancy or functional diversity in the Aux/IAA family in *A. argyi*. Furthermore, a visualization of the sequence similarity of *AaIAA* genes in *A. argyi* was analyzed by Circoletto [32]. As depicted in Figure 2B, 11 genes with red ribbons indicate that they have the highest similarity homologous genes, including *AaIAA21* to *AaIAA53*, *AaIAA7* to *AaIAA29*, *AaIAA1/27* to *AaIAA28*, and *AaIAA9/10/39* to *AaIAA60*. A total of 47 genes with orange ribbons indicate they also have 75% < orange ≤ 99% similarly homologous genes.

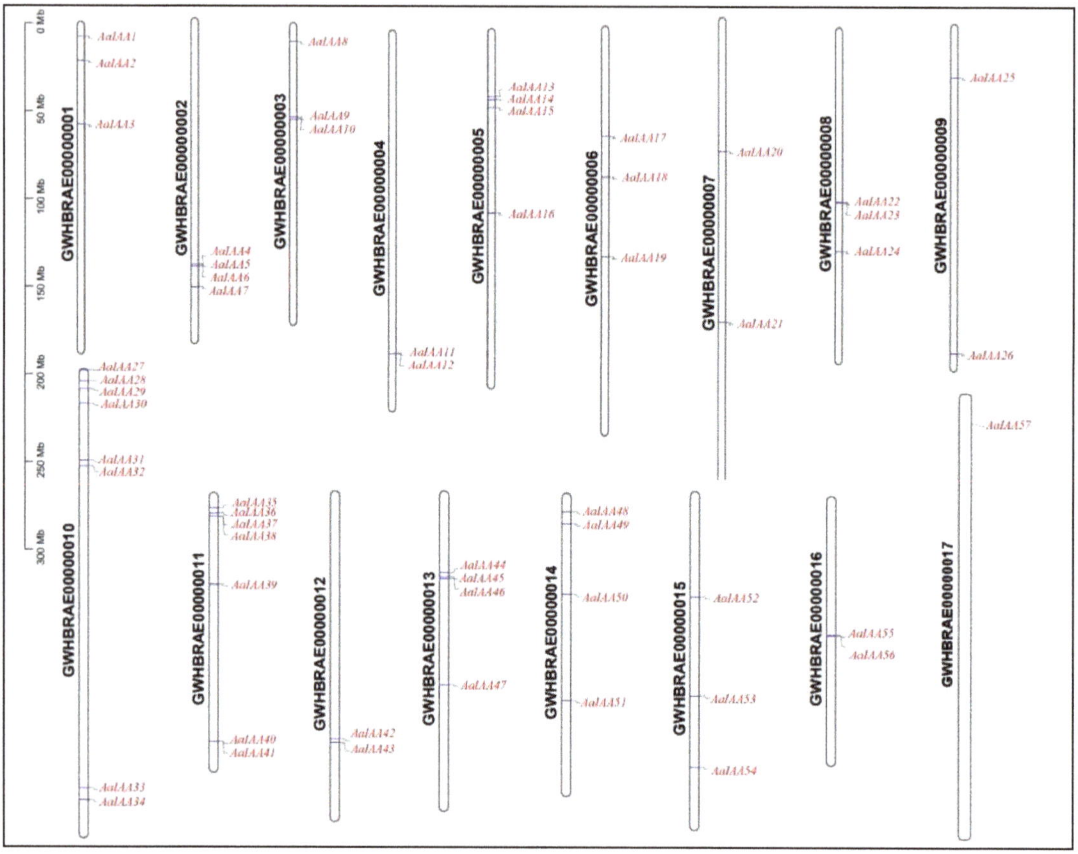

Figure 1. Chromosome distribution mapping of Aux/IAA gene family in *A. argyi*.

Additionally, we analyzed the collinear events of *AaIAA* genes within the *A. argyi* genome and between the *A. thaliana* genome (Figure 2C, Table S3). Compared to *A. argyi* and *A. thaliana*, 35 *AaIAA* genes in *A. argyi* had orthologs with 21 members of *A. thaliana*. Most of the *A. argyi AaIAA* genes had one (13 genes) or two (13 genes) orthologs in the *A. thaliana* genome, seven *A. argyi AaIAA* genes each had three orthologs in *A. thaliana*, and another two *A. argyi AaIAA* genes had four orthologs in *A. thaliana*. In contrast, there were one to seven orthologs when comparing *A. thaliana* to *A. argyi* (Figure 2C, Table S3). These findings implied that most *AaIAA* genes experienced gene loss in the tandem duplication process of the AaIAA gene family throughout *A. argyi*.

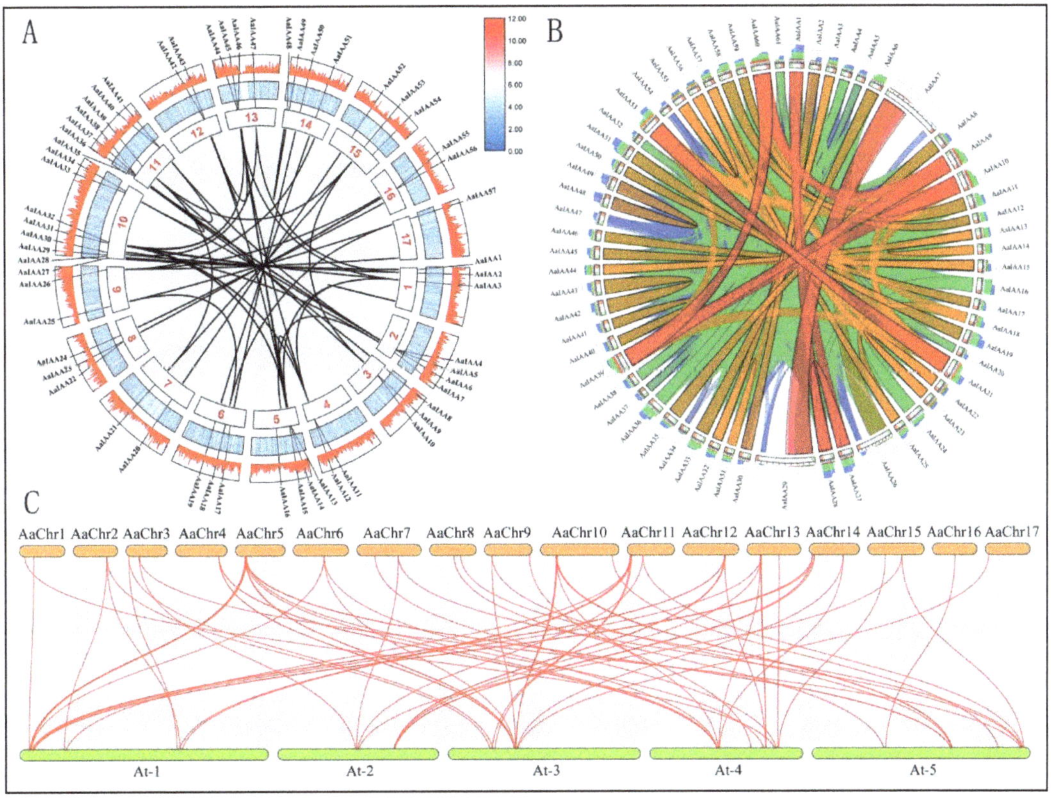

Figure 2. Collinear prediction and sequence similarity analysis of the Aux/IAA genes in *A. argyi*. (**A**): Mapping of collinear prediction of the AaIAA family genes in *A. argyi*. Black lines show AaIAA paralogs, red maths indicates chromosome numbers. (**B**): Visualising sequence similarity of the AaIAA genes in *A. argyi*. Different colored ribbons represent different levels of similarity, blue ≤ 50%, green ≤ 75%, orange ≤ 99%, red > 99%. (**C**): Collinear prediction of the Aux/IAA family genes between *A. thaliana* and *A. argyi*. Red lines show the collinear events.

2.3. Gene Structure, Motif Composition, and Promoter Cis-Element Analysis of AaIAA Genes

Gene structure schematics of exon–intron organizations of all identified AaIAA genes were generated using GSDS, which can provide more insight into the evolution of the Aux/IAA family in *A. argyi* (Figure 3). The number of introns in all AaIAA genes was between 1 and 14. The coding sequences of most (67.2%) of the AaIAA genes were disrupted by three (nine AaIAAs) or four (32 AaIAAs) introns, and the intron locations and situations are highly conserved (Figure 3C). Subsequent assessments suggested that the dispersal and situations of introns coincided with the evolutionary alignment of all AaIAA genes (Figure 3A).

The Multiple Em for Motif Elicitation (MEME) web server, including 20 motifs, was used to analyze the conserved motif distributions of AaIAA proteins. The proteins of AaIAAs vary from 3 to 13 motifs, with most of the proteins having motif 1, motif 2, motif 3, motif 5, and motif 6 (Figure 3B). Furthermore, a substantially similar genetic organization is identified in the same evolutionary group of AaIAA genes, such as AaIAA22, AaIAA23, AaIAA55, and AaIAA56, indicating that duplicated genes may exert the same function. Moreover, four typical conserved domains, including domain I (motif 6), domain II (motif 3), domain III (motif 5), and domain IV (motif 1), were mapped (Figure 3C). A total of 46 AaIAA proteins possess four typical domains, whereas a subset of the IAA proteins possesses reduced numbers of domains, including motif 6 (domain I), which is missing in AaIAA2/7/18/20/25/26/29/34/47/52/54/57/61; AaIAA7 and AaIAA29 only have domain IV; AaIAA20, AaIAA52, AaIAA54, and AaIAA61 contain domain III and domain IV; AaIAA26 and AaIAA34 contain domain II and domain IV; and AaIAA31 is missing domain IV (Figure 3B). Notably, proteins in the same subset exhibit comparable distributions.

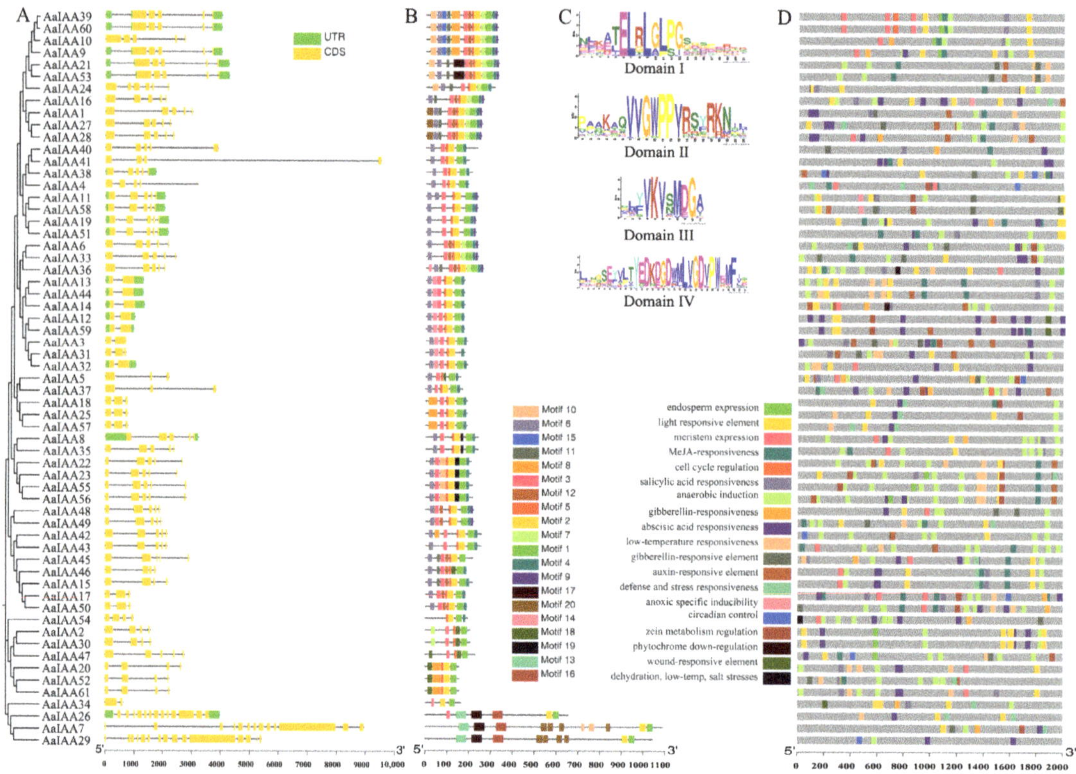

Figure 3. Gene structure, conserved motif, conserved domains, and promoter *cis*-element analysis of the Aux/IAA family genes in *A. argyi*. (**A**) Exon–intron organization of *A. argyi* Aux/IAA genes. (**B**) Conserved motif analysis of *A. argyi* Aux/IAA proteins. (**C**) The amino acid sequence of four typical conserved domains. (**D**) *Cis*-element analysis in the promoters of *A. argyi* Aux/IAA genes.

The 2000 bp regions of the *AaIAA* genes' promoter sequences were interrogated to find *cis*-elements to determine the expression level responses of *AaIAA* genes to adversity or hormonal stresses. The findings exhibited that AaIAA genes contain various plant hormone response and stress response elements, encompassing auxin, abscisic acid (ABA), salicylic acid (SA), methyl jasmonate (MeJA) and gibberellin-responsive and drought-inducible cis-elements (Figure 3D). Regarding the auxin-responsive elements, 38 auxin-responsive elements, including 30 *AaIAA* genes, were identified. In particular, four auxin-responsive elements were found in the promoter regions of *AaIAA3*, indicating that *AaIAA3* has a strong function in the auxin-responsive domain. There are 41, 31, 34, and 43 *AaIAAs* genes containing MeJA-responsive, salicylic acid, gibberellin-responsive, and abscisic acid-responsive elements, respectively. Furthermore, various classes of hormone-associated *cis*-elements were identified in the promoter sequences of many AaIAA genes, including *AaIAA1*, *AaIAA3*, *AaIAA5*, and *AaIAA9*. The presence of *cis*-regulatory sequences exhibits a possible role of the Aux/IAA genes in crosstalk regulation by various hormones. Additionally, all *AaIAA* genes contain light-responsive elements varying from 5 to 21, and some genes contain tissue specificity *cis*-elements, such as *cis*-regulatory elements involved in endosperm expression in *AaIAA2*, *AaIAA9*, *AaIAA10*, *AaIAA23*, *AaIAA60*, *AaIAA30*, *AaIAA34*, *AaIAA36*, *AaIAA39*, *AaIAA47*, *AaIAA55*, and *AaIAA56*; *AaIAA18*, *AaIAA19*, *AaIAA25*, *AaIAA26*, *AaIAA34*, *AaIAA38*, and *AaIAA57* have *cis*-regulatory elements involved in flavonoid biosynthetic gene regulation; and *AaIAA20*, *AaIAA27*, *AaIAA28*, *AaIAA33*, and *AaIAA61* have *cis*-elements involved in differentiation of palisade mesophyll cells (Figure 3D).

2.4. GO Annotation of AaIAA Genes

Gene Ontology (GO) annotation was performed for all 61 Aux/IAA genes to examine the biological influences they exert (Table S4). In biological processes, a maximum of 60 genes responded to the regulation of transcription, DNA-templated; followed by the auxin-activated signaling pathway (59 genes); response to auxin (56 genes); and transcription, DNA-templated (55 genes). In terms of cellular components, 59, 12, and 2 AaIAA genes were classified into the nucleus, intracellular, and integral components of the membrane, respectively. In addition, at the molecular level, 56 of 61 AaIAA unigenes were recognized as having DNA-binding transcription factor activity, while 48, 15, 11, and 7 AaIAA genes responded to protein binding, sequence-specific DNA binding, transcription regulatory region DNA binding, and DNA binding, respectively (Figure 4).

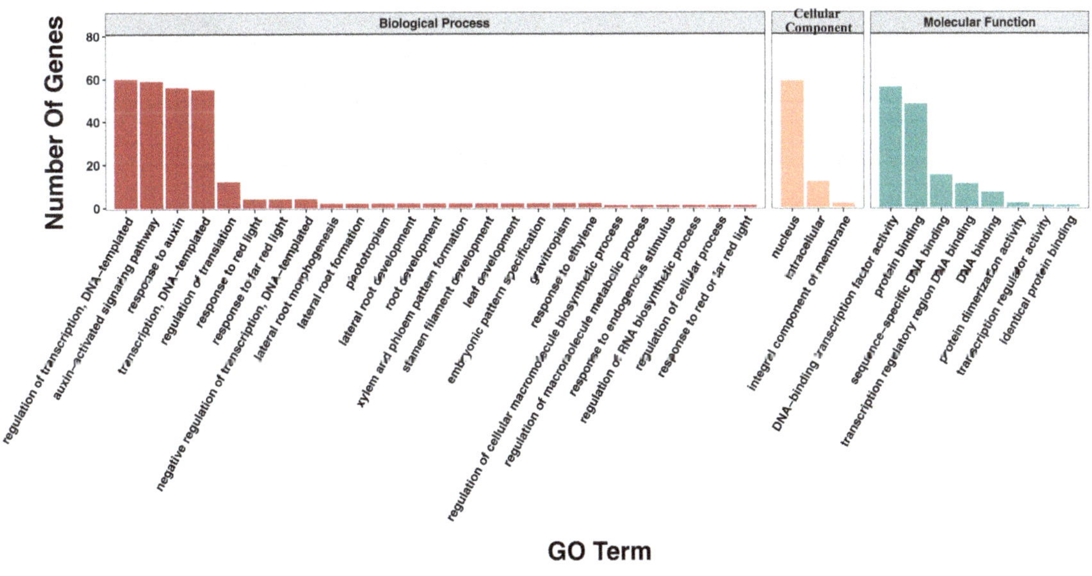

Figure 4. GO annotation of the Aux/IAA gene family in *A. argyi*.

2.5. Phylogenetic Relationship of Aux/IAA Genes in A. thaliana and A. argyi

To examine the relationship of Aux/IAAs proteins between *A. argyi* and *A. thaliana*, phylogenetic analysis was performed using the predicted complete amino acid sequences of Aux/IAAs from *A. argyi* and *A. thaliana*. Each Aux/IAA protein was grouped into 13 distinct classes named Group 1 to 13 (Figure 5). *A. argyi* proteins were distributed across all groups except Group 3. In *A. thaliana*, Group 1 was restricted, and other groups except Group 7, 8, and 12 were expanded in *A. argyi*, including three Aux/IAA genes from *A. thaliana* and nine genes from *A. argyi* in Group 4 and five genes from *A. thaliana* and fifteen genes from *A. argyi* in Group 9. Orthologous genes typically have similar biological functions. There exist orthologous genes between *A. argyi* and *A. thaliana*, clustered into the same branch, including AaIAA34 to AtIAA33 in Group 1 and AaIAA5 and AaIAA37 to AtIAA6 and AtIAA39 in Group 7 (Figure 5). In addition, 21 pairs of proteins paralogous to sister pairs of AaIAA proteins in the same species that are clustered to the same branch were found, similar to the clustering result in Figure 3A, suggesting similar biological functions or functional redundancy.

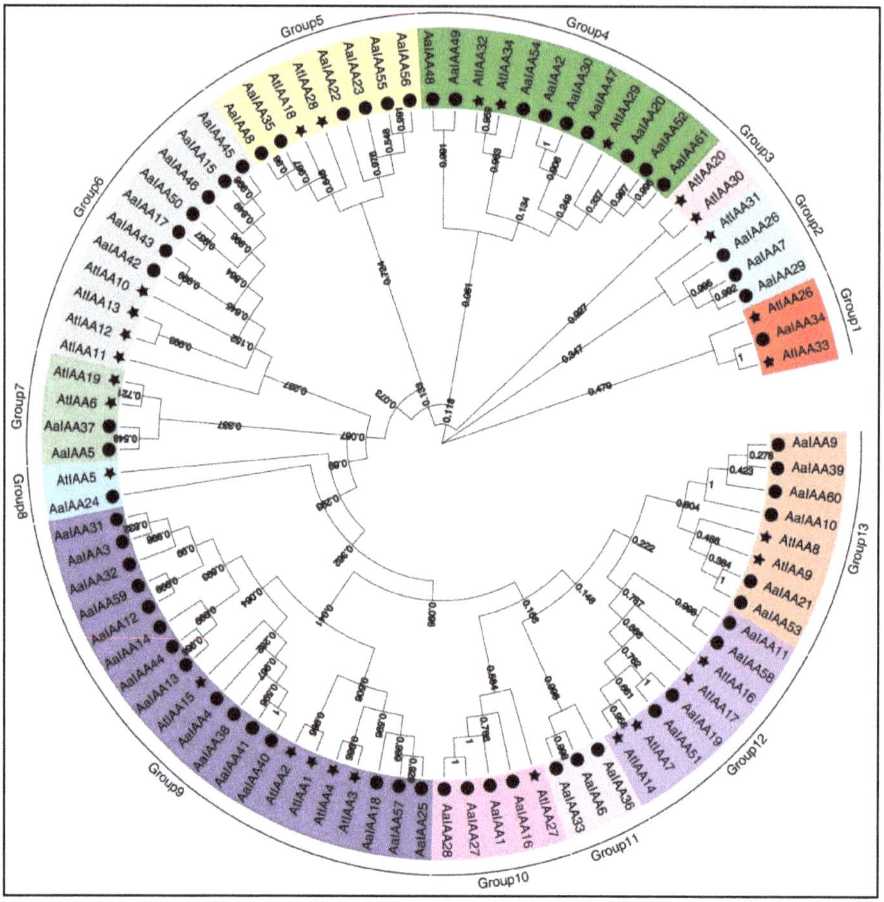

Figure 5. Phylogenetic investigation of Aux/IAA proteins in *A. argyi* and *A. thaliana*. The phylogenetic tree was developed using MEGA software (Version: 6.06) and the neighbor-joining method with 1000 bootstraps. Each Aux/IAA group (1 to 13) is indicated by a specific color. AaIAAs in *A. argyi* are indicated by a solid circle dot, and AtIAAs in *Arabidopsis* are indicated by a star.

2.6. Protein Interaction Prediction Analysis of Aux/IAA Proteins in A. argyi

To deeply comprehend the potential function of Aux/IAA proteins and the interaction between its family members in *A. argyi*, 61 AaIAA proteins were examined using STRING software (Version: 12.0) based on the homologous genes of *Artemisia annua* in the STRING database (Figure 6). There were 34 nodes and 145 edges, indicating that 34 proteins have 145 interactions, including 29 Aux/IAA proteins. Among them, AaIAA26 had the largest number of potentially interacting proteins and can potentially interact with 24 Aux/IAA members, followed by AaIAA29 which can potentially interact with 23 Aux/IAA members, and AaIAA32/36/38/4/41/44/47/50/51/54/57/59/61 proteins that had only one potential interacting Aux/IAA protein. These results suggest that AaIAA26/29/7/34 proteins are possibly core members of the IAA family interaction and potential functional collaborators between Aux/IAA proteins. Furthermore, for more protein interaction analysis, 26 and 25 Aux/IAA proteins can interact with the ARF proteins CTI12_AA579330 and CTI12_AA519960, respectively. AaIAA30, AaIAA47, and AaIAA61 proteins can potentially interact with Ubiquitin-related modifier 1 homolog proteins (CTI12_AA267880 and CTI12_AA498100), which operate as a sulfur carrier necessary for the 2-thiolation of mcm(5)S(2)U at the tRNA wobble positions of cytosolic tRNA(Lys), tRNA(Glu), and tRNA(Gln), acting as a sulfur donor in tRNA 2-thiolation reaction through thiocarboxylation (-COSH) at its C-terminus by MOCS3 [33]. Additionally, AaIAA7 proteins can potentially interact with the Mediator of RNA polymerase II transcription subunit 14 protein (CTI12_AA036660), which operates as a constituent of the Mediator complex, a coactivator acting to regulate the transcription of almost all RNA polymerase II-associated genes [34]. Therefore, this protein interaction network analysis offers evidence for further validation of the function of IAA proteins, supporting the idea that IAA proteins are involved in multiple biological processes through their interactions with target proteins.

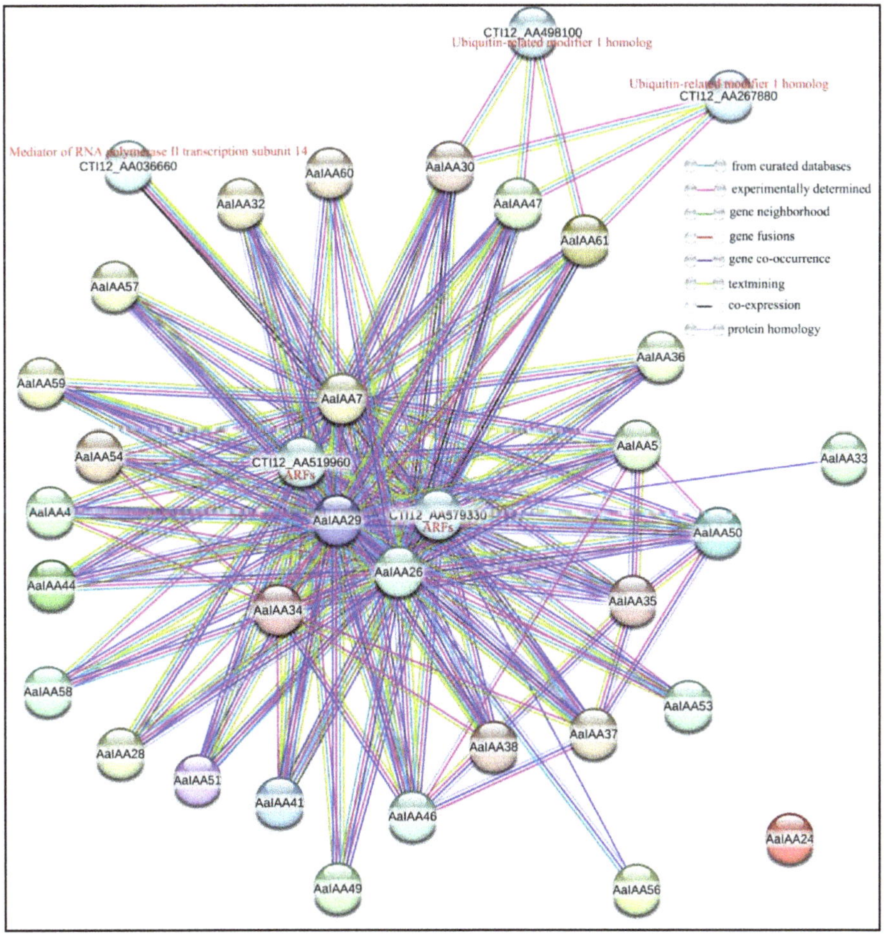

Figure 6. Prediction analysis of Aux/IAA protein interaction in *A. argyi*. Different line colors indicate the type of interaction evidence.

2.7. Expression Analysis of AaIAA Genes in Leaf and Root, and in Response to Salt and Saline-Alkali Stresse

To comprehend the spatial pattern expression and the putative function of *IAA* genes in *A. argyi*, we analyzed their expression profiles base on fragments per kilobase of exon model per million mapped fragments (FPKM) values in leaves and roots using available RNA-seq datasets (NCBI accession: PRJNA1054765) (Figure 7A, Table S5). Tissue expression analysis in roots and leaves through RNA-seq data exhibited that *AaIAA40*, *AaIAA41*, *AaIAA47*, *AaIAA50*, and *AaIAA54* genes were undetected in the leaf and root of RNA-seq data, indicating that these genes may be found in a distinct tissue-specific expression pattern in other tissues or organs, such as stems. *AaIAA2*, *AaIAA17*, *AaIAA20*, *AaIAA30*, *AaIAA34*, *AaIAA37*, *AaIAA52*, and *AaIAA61* genes were lowly expressed, or extremely lowly expressed, or had no expression in leaves or roots. The remaining 48 *AaIAA* genes were used for analysis, and a heatmap was constructed (Figure 7A). The results demonstrated that the expression levels of *AaIAA5*, *AaIAA11*, *AaIAA19*, *AaIAA51*, and *AaIAA58* genes were significantly higher in leaves than in roots. Most remaining members were significantly higher in roots than in leaves, including *AaIAA1/4/8/10/13/14/15/18/22/23/24/27/28/29/35/36/38/39/42/43/44/45/46/48/49/53/55/56/57* genes (Figure 7A). In addition, low abundance expression analysis exhibited some members with average FPKM < 1 with low expression levels in leaves [35], including *AaIAA4/14/22/38/45/55/56*. Only *AaIAA5* exhibited low expression levels in roots among the remaining 48 members. These results revealed that *AaIAA* genes exhibited tissue-specific expression patterns in *A. argyi*.

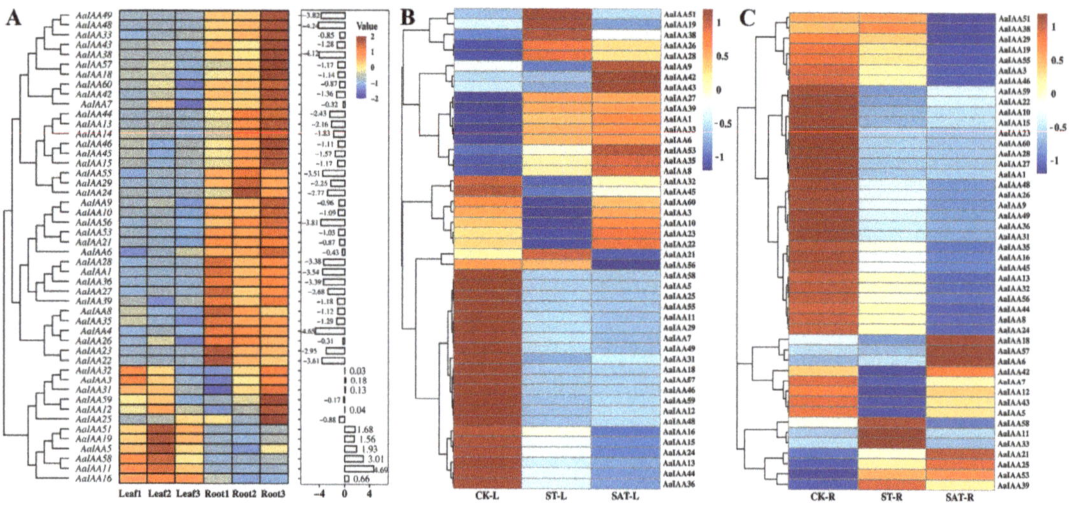

Figure 7. Expression profiles of AaIAA genes in the leaves and roots of *A. argyi* in response to salt and saline-alkali stresses based on RNA-seq data. (**A**) The expression profiles of *AaIAA* genes in the leaves and roots of *A. argyi*; FPKM data were normalized by Z-score; the bar chart shows the $\log_2$FC values of root vs. leaf. (**B**) Expression profiles of AaIAA genes in response to salt and saline–alkali stresses in leaves. (**C**) Expression profiles of AaIAA genes in response to salt and saline–alkali stresses in roots. The $\log_2$(FPKM values) data were normalized by Z-score. R: root, L: leaf, CK: control group, ST: salt treatment, SAT: saline–alkali treatment. Each set had three replicates.

Furthermore, to comprehend the expression patterns of *Aux/IAA* genes under salt and saline–alkali treatment using available RNA-seq data (NCBI accession: PRJNA1054765) in *A. argyi*, *Aux/IAA* gene expression in salt and saline–alkali stress RNA-seq data were analyzed (Table S6). Members with FPKM value = 0 in any set were excluded, and the remaining 46 *AaIAA* genes were used for analysis. The results showed that most

Aux/IAA members responded to salt and saline–alkali stresses (Figure 7B,C). In leaves, more AaIAA members exhibited reductions under salt and saline–alkali stresses. Among them, twelve genes were significantly responsive to salt stress, including nine genes (*AaIAA5/11/12/18/48/49/57/58/59*) downregulated and two genes (*AaIAA8* and *AaIAA35*) upregulated; sixteen genes significantly responded to saline–alkali stress, including twelve genes (*AaIAA5/11/12/13/15/16/18/24/58/59/44/48/49/57*) downregulated and two genes (*AaIAA8* and *AaIAA35*) upregulated (Figure 7B). In roots, most of the *AaIAA* genes exhibited a decrease under salt and saline–alkali stresses. Among them, three genes (*AaIAA1*, *AaIAA27*, and *AaIAA28*) significantly responded to salt stress in downregulated form, and six genes (*AaIAA1/24/26/27/28/29*) exhibited significant downregulation in response to saline–alkali stress (Figure 7C). Additionally, some *AaIAA* genes exhibited opposite expression patterns in response to salt and saline–alkali stresses, such as *AaIAA35*, *AaIAA8*, *AaIAA27*, *AaIAA1*, and *AaIAA28*, showing increases in leaves and decreases in roots (Figure 7B,C). Generally, most *Aux/IAA* genes exhibited similar expression patterns in response to salt stress and saline–alkali stress, and the expression of most *Aux/IAA* genes was decreased under salt and saline–alkali stresses, indicating that the growth of *A. argyi* was inhibited under salt and saline-alkali stresses.

2.8. Differential Expression Patterns of AaAux/IAA Genes in Response to Different Hormones Treatment

Auxin represents a critical growth modulator controlling plant growth, recruiting the assistance of various endogenous signals, encompassing various plant hormones like abscisic acid, gibberellic acid, jasmonates, and salicylic acid [36]. The Aux/IAA gene family plays a regulatory role in auxin regulating plant growth, and it is also involved in the response to other hormones. To assess the physiological functions of *AaIAA* genes and their reactions to auxin and different hormones, 12 genes, with FPKM > 10 and no sister pair genes, were chosen to examine their expression under indole-3-acetic acid (IAA), salicylic acid (SA), methyl jasmonate (MeJA), and abscisic acid (ABA) treatments, and the results of quantitative real time polymerase chain reaction (RT-qPCR) showing the expression of 12 *AaIAA* genes responses to the hormones of IAA, SA, MeJA, and ABA (Figure 8). Under IAA treatment, 11 members exhibited significantly increased expression, especially *AaIAA13* after 3 h and *AaIAA12/21/33/35* after 12 h. While *AaIAA19* exhibited decreased expression, especially after 12 h, *AaIAA7/9/26/36* also showed reduced expression after 3 h (Figure 8A). Under ABA treatment, *AaIAA21/26/33/35/36* exhibited increased expression; *AaIAA7/9/11/12/13/19/57* exhibited both upregulated and downregulated expression at different time points of ABA treatment, such as the expression of *AaIAA7* exhibiting significantly increased expression after 12 h, while it decreased after 3 h and 24 h. Among them, *AaIAA35* exhibited significant upregulation (up to 14.48-fold). The *AaIAA35* gene plays a core role in the response to ABA of the Aux/IAA family in *A. argyi* (Figure 8B). Under SA treatment, *AaIAA21/36* showed increased expression; *AaIAA26/33/35/57* showed decreased expression; and the remaining six *AaIAAs* genes showed both upregulated and downregulated expression at different time points of SA treatment, including the expression of *AaIAA7* showing significant decreases after 3 h and 12 h, while it increased after 24 h and 48 h. Among them, *AaIAA11* and *AaIAA12* were significantly up-regulated (up to 4.24-fold). The *AaIAA11* and *AaIAA12* genes play core roles in response to SA of the Aux/IAA family in *A. argyi* (Figure 8C). Under MeJA treatment, *AaIAA7/13/19* showed decreased expression; *AaIAA21/33/35/36* showed increased expression; *AaIAA9/11/12/26/57* showed both upregulated and downregulated expression at different time points of MeJA treatment, such as the expression of *AaIAA9* showing significant increases after 3 h and 12 h, while it was decreased after 24 h and 48 h. Among them, *AaIAA35* exhibited significant upregulation (up to 33.27-fold). The *AaIAA35* gene plays a core role in response to ABA of the Aux/IAA family in *A. argyi* (Figure 8D).

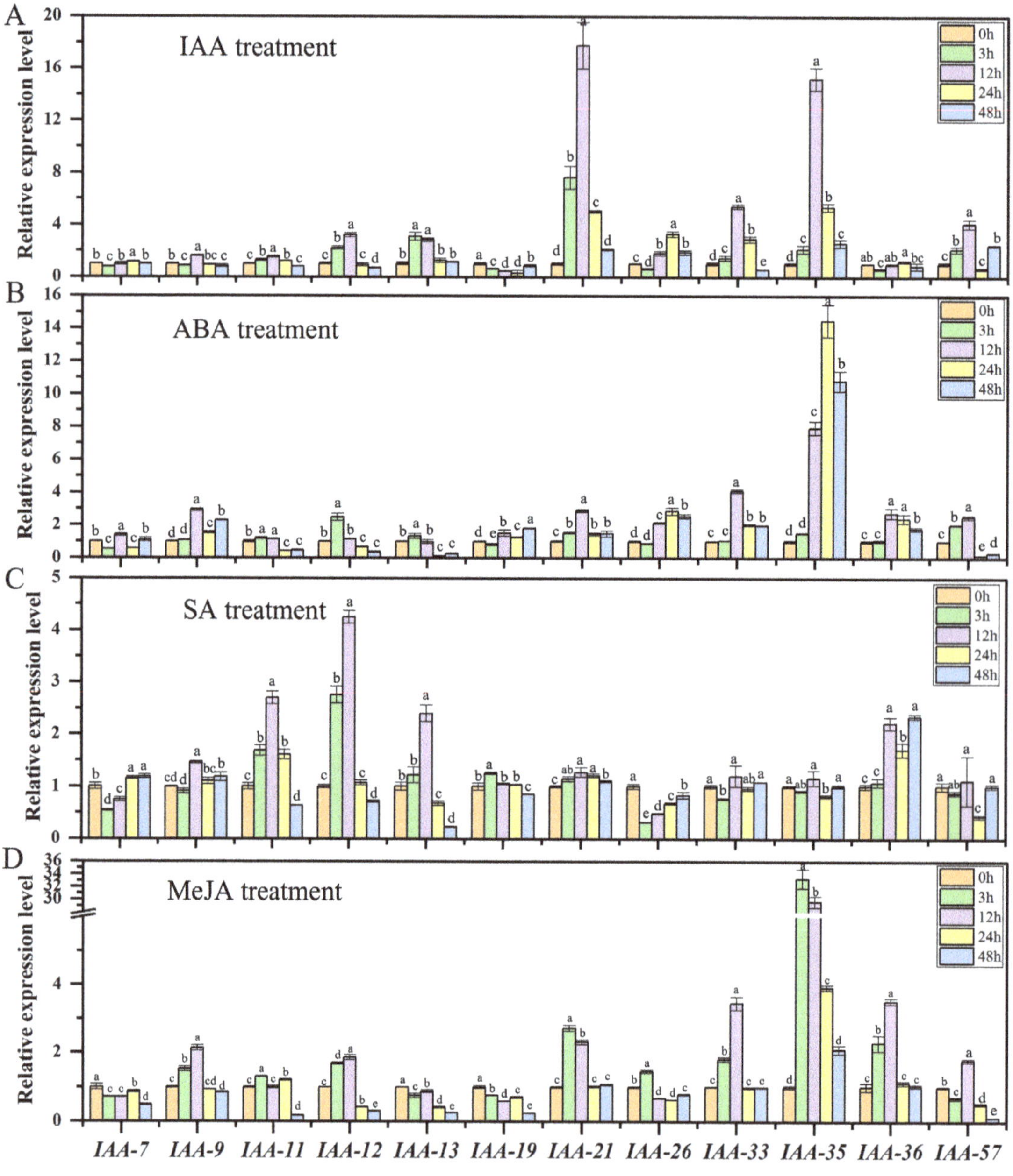

Figure 8. Expression profiles of AaIAA genes under various hormonal treatments. (**A**) IAA treatment. (**B**) ABA treatment. (**C**) SA treatment. (**D**) MeJA treatment. Error bars represent standard error of the mean; data shown are means ± SE. Letters "a,b,c. . ." indicate significant differences at $p < 0.05$.

3. Discussion

Recently, with the continuous examination of the genomes of medicinal plants, genome sequencing research has enabled an opportunity to extract gene classes through a genome-wide test. The analysis of the *A. argyi* genome provides important genetic resources for the mining and functional analysis of gene families. Auxin signaling acts as a critical signaling avenue throughout various plant biological processes, including development, organogenesis, and response to various environmental alterations [37]. Aux/IAA genes modulate auxin-triggered gene expression and varied aspects of plant growth through the Aux/IAA–ARF and Aux/IAA–TIR complexes [37,38]. Consequently, research into the function of the Aux/IAA gene family in *A. argyi* is beneficial for analyzing *A. argyi* growth and development, stress resistance, and other biological processes. In our study, the systematic characterization of the *A. argyi* Aux/IAA gene class members outlined offer novel insights into the possible roles of various Aux/IAA genes in controlling plant reactions to auxin and their potential functions.

3.1. Characterization and Structure of the AaIAA Family Genes in A. argyi

Many Aux/IAA genes that regulate auxin signal transduction and auxin degradation have been identified in various plants by the application of physiological, genetic, molecular, and biochemical methods, primarily through complete genomic sequencing. In this study, 61 AaIAA genes were identified, and the number of AaIAA members in *A. argyi* was much higher than in tomato (26 members), *A. thaliana* (29 members), rice (31 members), and maize (31 members), and similar in number to diploid *Brassica rapa* (55 members) [39], polyploid *Brassica oleracea* (60 members), and the paleopolyploid soybean (63 members) [40]. *A. argyi* is an allotetraploid with 34 chromosomes and 34 pseudochromosomes, separated into two haplotypes based on self-comparison. The longer and more complete chromosomes were assigned to haplotype group A, which contained pseudochromosomes 1 to 17 and was used for our genomic analysis [30]. Therefore, 61 AaAux/IAA genes identified in our study might represent half the members of *A. argyi*, similar to the same allotetraploid *Brassica napus* in Aux/IAA genes (119 members) [41]. In our study, compared to *A. argyi* and *A. thaliana*, gene divergence and duplication events showed that 35 Aux/IAA genes in *A. argyi* had orthologs with 21 members of *A. thaliana*. Most of the *A. argyi* Aux/IAA genes had one (13 genes) or two (13 genes) orthologs in the *A. thaliana* genome, seven *A. argyi* Aux/IAA genes each had three orthologs in *A. thaliana*, and another two *A. argyi* Aux/IAA genes had four orthologs in *A. thaliana*. Theoretically, there should be four ortholog genes for *A. thaliana* Aux/IAA genes in *A. argyi*, implying that most Aux/IAA genes experienced gene loss in the tandem duplication process of the Aux/IAA gene family in *A. argyi*. Moreover, the phylogenetic relationship of Aux/IAA genes analysis showed that Aux/IAA genes mostly exist as sister pairs of genes, suggesting there were gene duplication events, consistent with the whole gene duplication event that *A. argyi* underwent after divergence from *Artemisia annua* [30].

Previous studies have demonstrated that Aux/IAA genes are functionally different. The differentiation of the Aux/IAA genes in *A. thaliana* may depend on both the molecular structure of proteins and expression patterns. In our study, 46 members possess the four typical domains, I, II, III, and IV. A total of 15 non-canonical Aux/IAA proteins were found to lack single or multiple domains, contributing to their divergence. Among them, domain I contains conserved leucine residues and the LXLXLX motif. The LXLXLX motif plays a vital role in the repression of IAA proteins [42]. Mutation at any of these three Leu residues in this motif results in total loss of repression or strongly reduced repression in the case of the mutation in the third Leu in the motif [42]. In this study, AaIAA2/7/18/20/25/26/29/34/47/52/54/57/61 proteins do not contain domain I, implying that these proteins could not act as a repressor in auxin signaling and may function differently. Eight Aux/IAA proteins (AaIAA2/7/26/29/30/34/54) do not contain Domain II, consistent with evidence from the AtIAA20 lacking domain II, and cannot be rapidly degraded in the presence of basal or increased levels of auxin in *A. thaliana*, suggesting that these genes can repress auxin response gene expression [43]. Domains III and IV can bind to the ARF, and are responsible for homo- and hetero-dimerization between Aux/IAA and ARF proteins [38]. Five Aux/IAA proteins (AaIAA5/7/26/29/34) do not contain domain III, suggesting that these genes cannot bind ARF or function independently. In addition, most of the AaIAA proteins contain domains III–IV, which are important for binding to ARFs. AaIAA proteins usually directly bind to ARF proteins and regulate the expression of downstream auxin response genes [44].

3.2. Tissue-Specific Expression of Aux/IAA Genes in A. argyi

Tissue-specific and developmental stage-specific expression of Aux/IAA genes has been established across various species. In *Hedychium coronarium*, the tissue-preferential and stage-specific expressions exhibited several Aux/IAA genes expressed in flowers, suggesting their involvement in the biology of specific tissues and flower scent formation [45]. In *Triticum aestivum*, *TaIAA27* and *TaIAA31* exhibit tissue-specific expression in roots and seeds [46]. In *Brassica napus*, the RNA-seq data indicated that the expression of Aux/IAA members was highly variable among tissues, and 24 Aux/IAA genes had tissue(s)-specific expression patterns [41]. In our study, tissue expression analysis in roots and leaves in RNA-seq data demonstrated that the expression level of AaIAAs differed, such as *AaIAA11*, *AaIAA19*, *AaIAA20*, *AaIAA30*, *AaIAA51*, *AaIAA58*, and *AaIAA61* genes being significantly higher in leaves than in roots, and 25 remaining members exhibited significantly higher levels in roots than that in leaves. These results indicated that more AaIAA members act on the roots and regulate plant growth and development, similar to findings in *Fagopyrum tataricum*, which also exhibited significantly lower expression levels of most *FtAux/IAA* genes in the leaves than in root tissues [47]. Previous studies have shown that Aux/IAA genes are involved in root growth and development, and *OsIAA11* is expressed in the root tip, lateral root, middle column, and lateral root primordia [48]. In *A. thaliana*, *AtIAA33* plays a crucial role in maintaining apical stem cell activity [49]. In addition, *AaIAA40*, *AaIAA41*, *AaIAA47*, *AaIAA50*, and *AaIAA54* genes were not detected in leaf and root of RNA-seq data, indicating that these genes may illustrate a distinct tissue-specific expression pattern in other tissues in *A. argyi*, and tissue-specific expression of these members requires further analysis.

3.3. Adversity and Hormone Responses of Aux/IAA Genes in A. argyi

Gene expression patterns in *A. argyi* responses to salt and saline–alkali stresses and different hormone stimuli were identified using RNA-seq data or RT-qPCR analysis, delivering novel insights into the potential functions in governing plant growth and development by *AaIAA* genes. Aux/IAA is critical, acting as a class of primary auxin-responsive genes, rapidly induced by auxin. In *Fagopyrum tataricum*, all Aux/IAA genes exhibited an intricate response pathway under IAA treatment; among them, different genes exhibited differing trends upon treatment for extended durations, and the expression of some genes differed in different tissues [47]. In our study, 12 *AaIAA* genes also showed a complex pattern in response to IAA treatment by RT-qPCR. Although only two members of the selected 12 members contained auxin-responsive *cis*-elements in their promoters, the IAA family acted as a class of primary auxin-responsive genes, involved in auxin responses. Furthermore, the Aux/IAA family regulates different abiotic stresses. The level of glucosinolates (GLS) is regulated by *IAA5*, *IAA6*, and *IAA19* in the auxin Aux/IAA protein family, and *AtIAA5/6/19* deficiency causes decreased GLS levels and decreased drought tolerance, indicating that Aux/IAA proteins regulate drought tolerance in *Arabidopsis* by modulating glucosinolate levels [19]. In rice, some Aux/IAA genes were regulated by abiotic stresses such as drought, salt, and low temperature, such as *OsIAA9* and *OsIAA20* being significantly upregulated under high-salt conditions [20]. In our study, RNA-seq data analysis exhibited that 12 and 16 *AaIAA* genes were significantly responsive to salt stress and saline–alkali stress in leaves, respectively. Most *AaIAA* genes exhibited decreased expression under salt and saline–alkali stresses in roots, indicating that the growth of *A. argyi* was inhibited under salt and saline–alkali stresses. These findings indicated that the Aux/IAA family regulates the response to salt and saline–alkali stresses in *A. argyi*.

Additionally, auxin is a critical growth regulator that governs plant development, and it is recruited by other endogenous signals, including abscisic acid, brassinosteroids, cytokinins, ethylene, gibberellic acid, jasmonates, and salicylic acid [36]. As an essential gene family that responds to auxin signaling, it is speculated that other hormones regulate the Aux/IAA gene family. Prior studies have focused on the Aux/IAA family response to ethylene regulation. The downregulation of *SlIAA3* results in auxin- and ethylene-related developmental defects, supporting the hypothesis that *SlIAA3* is a molecular link between ethylene and auxin signaling in tomatoes [50]. The Aux/IAA gene family also mediates the interaction between auxin and the signaling of other hormones, such as abscisic acid, cytokinin, and salicylic acid. *IAA16* confers reduced responses to auxin and abscisic acid and impedes plant growth and fertility [51]. In cotton, the enhanced resistance in *GhIAA43*-silenced cotton plants is because of the activation of SA-related defenses, suggesting that *GhIAA43* can act as a molecular link between SA and auxin signaling [52]. In a study of tobacco mosaic virus-directed reprogramming of Aux/IAA proteins, transcriptional expression studies demonstrate a role for Aux/IAA-interacting proteins in the regulation of SA and JA host defense responses, as well as virus-specific movement factors [53]. In our study, 12 *AaIAA* genes can respond to ABA, SA, and MeJA treatments, but their expression levels varied considerably. For ABA responsiveness, 8 of 12 *AaIAA* selected genes except *AaIAA11/13/21/35* contained a *cis*-acting element involved in abscisic acid responsiveness. For MeJA responsiveness, 9 of 12 *AaIAA* selected genes except for *AaIAA11/26/57* contained *cis*-acting regulatory elements involved in MeJA responsiveness. Only five members, including AaIAA9/11/13/35/36, for SA responsiveness, contain cis-acting elements involved in salicylic acid responsiveness. RT-qPCR results demonstrated that some members could respond to related hormones. However, they did not contain corresponding regulatory *cis*-elements in their promoters, and there may be other regulatory modes responding to hormone regulation, suggesting that *AaIAA* genes may exert an essential role in regulating plant growth and development by auxin interactions with other hormones. In addition, we know that hormone concentrations influence the response patterns of different genes to hormones. For example, high concentrations cause non-specific effects and cannot be considered as only specific hormones. Thus, the next study should develop a dose-response curve to provide a more accurate regulatory model for AaIAA family members to respond to different hormone regulation. Overall, the functional characterization and expression analysis of Aux/IAA family genes assist in revealing the mechanisms underlying how auxin signaling is involved in plant growth, responses to environmental changes, and interactions with other hormones in a spatiotemporal-specific manner in *A. argyi*.

3.4. Possible Correlation between Aux/IAA Gene and Switching to Secondary Metabolism in A. argyi

Previous studies have shown that various stresses and hormones can regulate the accumulation of secondary metabolites, which in turn controls the quality of medicinal plants [54,55]. In *Glycyrrhiza uralensis*, exogenous ABA has the ability to stimulate the synthesis of Glycyrrhiza uralensis's active components, such as isoliquiritin, glycyrrhizin, liquiritigenin, and isoliquiritin [54]. In *Ruta graveolens*, salicylic acid can promote the accumulation of phenols and flavonoids in callus [55]. In cultured *Onosma paniculatum* cells, brassinolide together with IAA (indoleacetic acid) and BAP (6-benzylaminopurine) can promote the accumulation of shikonin at appropriate concentrations [56]. In apple, it has been reported that auxin has the ability to regulate anthocyanin biosynthesis via the Aux/IAA-ARF signaling pathway [57]. In *A. argyi*, previous studies have shown that exogenous hormone (MeJA, SA and ABA) treatments changed the eight secondary metabolites, including phenylpropanoids, flavonoids, terpenoids, alkaloids, and others [58]. In our study, the expression levels of 12 *AaIAA* genes changed to varying degrees in response to IAA, SA, MeJA, and ABA, and various hormones had an impact on the accumulation of secondary metabolites in *A. argyi*. it is speculated that MeJA, SA, IAA, and ABA may have regulated the accumulation of secondary metabolites in *A. argyi* through the Aux/IAA genes or Aux/IAA-related signaling pathway. In addition, Aux/IAA proteins, which are a crucial part of the auxin signaling system, are essential for controlling the growth process of plants [5]. Plant secondary metabolite accumulation is a dynamic process that is ever-changing during the growth phase of the plant [59]. It is suggested that AaIAA proteins may present as a potential mediator of plant development and secondary metabolite accumulation. Above all, more research is required to understand how the Aux/IAA protein in *A. argyi* regulates the accumulation of secondary metabolites.

4. Materials and Methods

4.1. Plant Material, Growth Conditions, and Hormone Treatment

The plant materials used in our study were obtained from the medicinal botanical garden of Henan University of Chinese Medicine, Zhengzhou, Henan Province, China. The root buds of *A. argyi* were used as explants for disinfected, and sterile tissue culture seedlings of *A. argyi* were obtained as described by Li et al. [60].

Aseptic tissue culture seedlings of *A. argyi* plants were grown in tissue culture bottles with ½ Murashige and Skoog (½MS) culture medium containing 0.7% agar and 3% sucrose [60]. The medium has a high concentration of inorganic salts and ions, also includes iodine and chloride, is a relatively stable ion balance solution, has a high nitrate content, and can meet the nutritional and physiological needs of most plant cells [61]. The seedlings were grown in sterile conditions under a 16 h/8 h light/dark photoperiod (120 μmol m^{-2} s^{-1}) at 24 to 26 °C. After 1 month of growth, the *A. argyi* seedlings were treated with different hormones. For the salicylic acid (SA), methyl jasmonate (MeJA), and abscisic acid (ABA) treatments, 200 μmol SA, 200 μmol MeJA, and 200 μmol ABA were evenly sprayed on the leaves of *A. argyi* after uncapping the bottle, then the bottle was capped after the spraying was finished. The control group was sprayed with sterile water. For the indole-3-acetic acid (IAA) treatment, two sets of *A. argyi* seedlings were incubated in liquid ½MS medium with and without 10 μM IAA for 0 h, 3 h, 12 h, 24 h, and 48 h. The leaves of all samples between three and seven nodes were harvested at 0 h, 3 h, 12 h, 24 h, and 48 h following treatment. All collected samples were frozen in liquid nitrogen immediately and preserved at −80 °C for subsequent RT-qPCR analysis.

For RNA-seq analysis, the 3-month-old *A. argyi* seedlings that were transplanted from tissue culture seedlings and grown in an illumination incubator with a relative humidity of 70% during a 16 h/8 h light/dark photoperiod (120 μmol m^{-2} s^{-1}) at 24 to 26 °C were selected for stress treatments. The uniformly developed seedlings of *A. argyi* were fully watered using a 200 mM total solution ($NaCl/Na_2SO_4$ = 1:1, pH = 7.0) for the salt stress treatment, 200 mM total ($NaCl/Na_2SO_4/NaHCO_3/Na_2CO_3$ = 1:1:1:1, pH = 9.0) for the saline–alkali stress treatment, and water treatment as a control. After three days of treatment, the leaves from four to seven nodes were removed, and as soon as the roots were thoroughly cleaned in water, samples were taken. Upon collection, every sample was promptly frozen in liquid nitrogen and stored at −80 °C for later use.

4.2. Identification of Aux/IAA Genes in A. argyi

A. argyi genome annotations were obtained from the National Genomics Data Center (NGDC, https://ngdc.cncb.ac.cn; accessed on 10 November 2022) with the BioProject accession number PRJCA010808 [30]. HMMER and BLASTP were applied to search for *A. argyi* Aux/IAA proteins using the published *Arabidopsis* Aux/IAA protein sequences as the initial protein query. A Hidden Markov Model of Pfam3 tools was used to examine the Aux/IAA (PF02309) domain for all the obtained protein sequences. Physiochemical parameters of each gene were determined using ExPASy (https://web.expasy.org/protparam/; accessed on 12 March 2023). Information about genomic sequences, cDNA sequences, protein sequences, and feature information was obtained from the GFF file available on the *A. argyi* genome project webpages [30].

4.3. Gene Structural, Motif Scanning and Phylogenetic Analysis of Aux/IAA Genes in A. argyi

All AaAux/IAA genes were mapped to the chromosomes from the physical positional information obtained from the *A. argyi* genomic GFF files downloaded from the NGDC database using Tbtools [62]. The exon–intron organization of the AaIAA genes was examined using the Gene Structure Display Server (GSDS). The IrTIFY protein conserved domains were analyzed by Pfam (https://pfam.xfam.org; accessed on 15 March 2023), and the MEME website (https://meme-suite.org/meme/; accessed on 15 March 2023) was used to predict protein conserved motifs with the motif value established at 10. Multiple sequence alignments were conducted on the AaIAA protein sequences using ClustalW with default parameters. A phylogenetic tree was constructed using the aligned AaIAA protein sequences using MEGA (Version: 6.06) [63] and selecting the neighbor-joining method, with 1000 bootstrap replicates. The same methods were applied to analyze the evolutionary relationships between *A. argyi* and *Arabidopsis*. TBtools was employed to visualize the conserved domain, motif, and IrTIFY evolutionary tree.

4.4. Cis-Elements Analysis in the Promoter Regions of Aux/IAA Genes

To characterize *cis*-elements in the promoter region of each Aux/IAA gene in *A. argyi*, 2000 bp of genomic sequence upstream of the translation start site was chosen from the available *A. argyi* genome sequence using TBtools, and analyzed with PlantCARE (http://bioinformatics.psb.ugent.be/webtools/plantcare/html/; accessed on 25 March 2023). A total of 21 *cis*-elements were selected for analysis, including hormone-related *cis*-elements and stress-response *cis*-elements. Different colors represent different *cis*-elements, and *cis*-elements distribution was visualized using TBtools [62].

4.5. Collinearity and Specific Duplication Events Analysis

Multiple collinear scanning toolkits (MCScanX) were employed to analyze gene duplication events with default parameters [64]. To uncover the synteny relationship of orthologous Aux/IAA genes between *A. argyi* and *Arabidopsis*, syntenic analysis maps were constructed using the Dual Synteny Plotter software (available online: https://github.com/CJ-Chen/TBtools; accessed on 15 June 2023) [61]. Furthermore, the visualization of sequence similarity of the AaIAA genes in *A. argyi* was analyzed by Circoletto [32].

4.6. GO Functional Identification

The Gene Ontology (GO) terms for functional annotation were analyzed using Blast2GO software (https://www.blast2go.com/; accessed on 6 April 2023) with default parameters. The annotations of GO terms were investigated using the Gene Ontology Consortium (http://geneontology.org/; accessed on 6 April 2023).

4.7. Transcriptome Analysis Based on RNA-Seq Data

To uncover the expression profile of Aux/IAA genes in various tissues in response to salt and saline–alkali stress in *A. argyi*, RNA-seq data from our study, including leaves and roots with or without salt and saline–alkali stress for three days, were used for analysis (NCBI accession: PRJNA1054765). Fragments per kilobase per million mapped reads (FPKM) were calculated using the Cufflinks tool (v2.2.1) to estimate gene expression levels. Heatmaps were constructed using the OmicStudio tools at https://www.omicstudio.cn/tool; accessed on 10 December 2023 [65] based on $\log_2$(FPKM values), data were normalized using Z-score, and each set underwent three replicates.

4.8. RNA Extraction and RT-qPCR Analysis

Total RNA from different samples of *A. argyi* was extracted using the EASYspin Plus Plant RNA Kit (AidLab, Beijing, China) following the manufacturer's instructions. A sample of 1 μg of quantified RNA samples was prepared for the synthesis of cDNA strands using the EasyScript All-in-One First-Strand cDNA Synthesis SuperMix for qPCR (One-Step gDNA Removal) (Transgen, Beijing, China) in accordance with the manufacturer's instructions. The RT-qPCR experimental steps were performed according to a previously published protocol [66]. RT-qPCR amplification was conducted using TransStart Top Green qPCR SuperMix (Transgen, Beijing, China) according to the manufacturer's instructions and was performed on the ABI QuantStudio 5 (Thermofisher Scientific, Waltham, MA, USA). Each RT-qPCR reaction contained 20 μL, including 10 μL 2× TransStart Top Green qPCR SuperMix, 1 μL cDNA template, 0.4 μL gene-specific primers (10 μM), 0.4 μL 50× Passive Reference Dye II, and 7.8 μL Nuclease-free water. The reaction conditions were 94 °C for 30 s, 45 cycles of 94 °C for 5 s, and 60 °C for 30 s. Each assay included three biological replicates and three technical replicates. *Actin* was used as an internal reference gene. The relative expression level was determined using the $2^{-\Delta\Delta Ct}$ method [67]. All primers are listed in Table S1.

5. Conclusions

In this study, a genome-wide analysis of the Aux/IAA gene family in *A. argyi* was conducted, and 61 Aux/IAA genes were identified in the *A. argyi* genome by bioinformatic analysis. Their physical and chemical characters, chromosome locations, gene structure, motif scanning, GO annotation, promoter *cis*-elements, phylogenetic relationships, and protein interaction were examined. Furthermore, Aux/IAA gene expression patterns in roots and leaves, salt, and saline-alkali stresses were examined by RNA-seq data, and the expression patterns in response to hormones, including IAA, ABA, SA, and MeJA, were also confirmed through RT-qPCR, exhibiting a complex response pattern, suggesting that Aux/IAA may have a role in connecting auxin to other hormones. Overall, our study analyzed the Aux/IAA family of *A. argyi* and its expression patterns. Our data will act as a significant resource for further investigations on functional genomics of the Aux/IAA genes and expand the possibilities for studying plant growth and development in medicinal plant breeding.

Supplementary Materials: The following supporting information can be downloaded at: https://www.mdpi.com/article/10.3390/plants13050564/s1, Table S1: All primers used in this paper; Table S2: Collinear prediction of AaIAA family genes in *A. argyi*; Table S3: Collinear prediction of AuxIAA family genes between *A. thaliana* and *A. argyi*; Table S4: GO annotation of AaIAA family genes in *A. argyi*; Table S5: RNA-seq data of AaIAA genes in leaves and roots; Table S6: RNA-seq data of AaIAA genes in response to salt and saline-alkali stress in leave and roots.

Author Contributions: Conceptualization, C.L. and S.C.; formal analysis, C.L., R.M., J.L. (Jingjing Li) and B.Z.; funding acquisition, R.M. and S.C.; investigation, J.L. (Jinxu Lan) and J.L. (Jingjing Li); resources, J.L. (Jinxu Lan) and F.Z.; supervision, C.L., F.Z. and S.C.; validation, J.L. (Jinxu Lan) and R.M.; writing—original draft, C.L. and X.L.; writing—review and editing, C.L. and S.C. All authors have read and agreed to the published version of the manuscript.

Funding: This research was funded by the Key Scientific Research Project of Higher Education Institutions in Henan Province (22A360012); Chinese herbal medicine industry technology system of Henan Province (No. 14 [2018]); National Natural Science Foundation of China (82204551); and Key project at central government level: The ability establishment of sustainable use for valuable Chinese medicine resources (2060302).

Data Availability Statement: Data are contained within the article and supplementary materials.

Conflicts of Interest: The authors declare no conflicts of interest.

References

1. Liu, Y.; He, Y.A.; Wang, F.; Xu, R.C.; Yang, M.; Ci, Z.M.; Wu, Z.F.; Zhang, D.K.; Lin, J.Z. From longevity grass to contemporary soft gold: Explore the chemical constituents, pharmacology, and toxicology of *Artemisia argyi* H.Lev. & vaniot essential oil. *J. Ethnopharmacol.* **2021**, *279*, 114404.
2. Wang, W.; Zhang, X.-K.; Wu, N.; Fu, Y.-J.; Zu, Y.-G. Antimicrobial activities of essential oil from *Artemisiae argyi* leaves. *J. For. Res.* **2006**, *17*, 332–334. [CrossRef]
3. Jiang, Z.; Guo, X.; Zhang, K.; Sekaran, G.; Cao, B.; Zhao, Q.; Zhang, S.; Kirby, G.M.; Zhang, X. The Essential Oils and Eucalyptol from *Artemisia vulgaris* L. Prevent Acetaminophen-Induced Liver Injury by Activating Nrf2–Keap1 and Enhancing APAP Clearance through Non-Toxic Metabolic Pathway. *Front. Pharmacol.* **2019**, *10*, 782. [CrossRef] [PubMed]
4. Anfang, M.; Shani, E. Transport mechanisms of plant hormones. *Curr. Opin. Plant Biol.* **2021**, *63*, 102055. [CrossRef] [PubMed]
5. Kepinski, S.; Leyser, O. The *Arabidopsis* F-box protein TIR1 is an auxin receptor. *Nature* **2005**, *435*, 446–451. [CrossRef] [PubMed]

6. Tan, X.; Calderon-Villalobos, L.I.A.; Sharon, M.; Zheng, C.X.; Robinson, C.V.; Estelle, M.; Zheng, N. Mechanism of auxin perception by the TIR1 ubiquitin ligase. *Nature* **2007**, *446*, 640–645. [CrossRef] [PubMed]
7. Dharmasiri, N.; Dharmasiri, S.; Weijers, D.; Lechner, E.; Yamada, M.; Hobbie, L.; Ehrismann, J.S.; Jürgens, G.; Estelle, M. Plant development is regulated by a family of auxin receptor F box proteins. *Dev. Cell* **2005**, *9*, 109–119. [CrossRef] [PubMed]
8. Meng, N.; Tan, X.; Caldeon-Villalobos, L.I.A.; Estelle, M. Mechanism of Auxin Perception by the SCF-TIR1 Ubiquitin Ligase. *FASEB J.* **2008**, *22*, 260.1.
9. Hagen, G.; Guilfoyle, T. Auxin-responsive gene expression: Genes, promoters and regulatory factors. *Plant Mol. Biol.* **2002**, *49*, 373–385. [CrossRef]
10. Abel, S.; Nguyen, M.D.; Theologis, A. ThePS-IAA4/5-like Family of Early Auxin-inducible mRNAs in *Arabidopsis thaliana*. *J. Mol. Biol.* **1995**, *251*, 533–549. [CrossRef]
11. Szemenyei, H.; Hannon, M.; Long, J.A. TOPLESS mediates auxin-dependent transcriptional repression during *Arabidopsis* embryogenesis. *Science* **2008**, *319*, 1384–1386. [CrossRef]
12. Gray, W.M.; Kepinski, S.; Rouse, D.; Leyser, O.; Estelle, M. Auxin regulates SCFTIR1-dependent degradation of AUX/IAA proteins. *Nature* **2001**, *414*, 271–276. [CrossRef] [PubMed]
13. Korasick, D.A.; Westfall, C.S.; Lee, S.G.; Nanao, M.H.; Dumas, R.; Hagen, G.; Guilfoyle, T.J.; Jez, J.M.; Strader, L.C. Molecular basis for AUXIN RESPONSE FACTOR protein interaction and the control of auxin response repression. *Proc. Natl. Acad. Sci. USA* **2014**, *111*, 5427–5432. [CrossRef] [PubMed]
14. Yang, X.Q.; Lee, S.; So, J.H.; Dharmasiri, S.; Dharmasiri, N.; Ge, L.; Jensen, C.; Hangarter, R.; Hobbie, L.; Estelle, M. The IAA1 protein is encoded by *AXR5* and is a substrate of SCF^{TIR1}. *Plant J.* **2004**, *40*, 772–782. [CrossRef] [PubMed]
15. Wang, J.; Yan, D.W.; Yuan, T.T.; Gao, X.; Lu, Y.T. A gain-of-function mutation in *IAA8* alters *Arabidopsis* floral organ development by change of jasmonic acid level. *Plant Mol. Biol.* **2013**, *82*, 71–83. [CrossRef] [PubMed]
16. Song, Y.L.; You, J.; Xiong, L.Z. Characterization of *OsIAA1* gene, a member of rice Aux/IAA family involved in auxin and brassinosteroid hormone responses and plant morphogenesis. *Plant Mol. Biol.* **2009**, *70*, 297–309. [CrossRef] [PubMed]
17. Song, Y.; Xu, Z.-F. Ectopic Overexpression of an AUXIN/INDOLE-3-ACETIC ACID (Aux/IAA) Gene *OsIAA4* in Rice Induces Morphological Changes and Reduces Responsiveness to Auxin. *Int. J. Mol. Sci.* **2013**, *14*, 13645–13656. [CrossRef]
18. Chen, S.-H.; Zhou, L.-J.; Xu, P.; Xue, H.-W. SPOC domain-containing protein Leaf inclination3 interacts with LIP1 to regulate rice leaf inclination through auxin signaling. *PLoS Genet.* **2018**, *14*, e1007829. [CrossRef]
19. Salehin, M.; Li, B.H.; Tang, M.; Katz, E.; Song, L.; Ecker, J.R.; Kliebenstein, D.J.; Estelle, M. Auxin-sensitive Aux/IAA proteins mediate drought tolerance in *Arabidopsis* by regulating glucosinolate levels. *Nat. Commun.* **2019**, *10*, 4021. [CrossRef] [PubMed]
20. Jain, M.; Khurana, J.P. Transcript profiling reveals diverse roles of auxin-responsive genes during reproductive development and abiotic stress in rice. *FEBS J.* **2009**, *276*, 3148–3162. [CrossRef]
21. Jung, H.; Lee, D.-K.; Choi, Y.D.; Kim, J.-K. *OsIAA6*, a member of the rice Aux/IAA gene family, is involved in drought tolerance and tiller outgrowth. *Plant Sci.* **2015**, *236*, 304–312. [CrossRef]
22. Zhang, A.; Yang, X.; Lu, J.; Song, F.; Sun, J.; Wang, C.; Lian, J.; Zhao, L.; Zhao, B. *OsIAA20*, an Aux/IAA protein, mediates abiotic stress tolerance in rice through an ABA pathway. *Plant Sci.* **2021**, *308*, 110903. [CrossRef]
23. Deng, W.; Yan, F.; Liu, M.; Wang, X.; Li, Z. Down-regulation of *SlIAA15* in tomato altered stem xylem development and production of volatile compounds in leaf exudates. *Plant Signal. Behav.* **2013**, *7*, 911–913. [CrossRef]
24. Overvoorde, P.J.; Okushima, Y.; Alonso, J.M.; Chan, A.; Chang, C.; Ecker, J.R.; Hughes, B.; Liu, A.; Onodera, C.; Quach, H.; et al. Functional Genomic Analysis of the AUXIN/INDOLE-3-ACETIC ACID Gene Family Members in *Arabidopsis thaliana*. *Plant Cell* **2005**, *17*, 3282–3300. [CrossRef]
25. Kalluri, U.C.; DiFazio, S.P.; Brunner, A.M.; Tuskan, G.A. Genome-wide analysis of Aux/IAA and ARF gene families in *Populus trichocarpa*. *BMC Plant Biol.* **2007**, *7*, 59. [CrossRef]
26. Jiang, L.; Li, Z.; Yu, X.; Liu, C. Bioinformatics analysis of Aux/IAA gene family in maize. *Agron. J.* **2021**, *113*, 932–942. [CrossRef]
27. Huang, B.; Qi, Y.X.; Huang, X.S.; Yang, P. Genome-wide identification and co-expression network analysis of Aux/IAA gene family in *Salvia miltiorrhiza*. *PeerJ* **2023**, *11*, e15212. [CrossRef]
28. Si, C.; Zeng, D.Q.; da Silva, J.A.T.; Qiu, S.X.; Duan, J.; Bai, S.; He, C.M. Genome-wide identification of Aux/IAA and ARF gene families reveal their potential roles in flower opening of *Dendrobium officinale*. *BMC Genom.* **2023**, *24*, 199. [CrossRef] [PubMed]
29. Huang, X.; Bao, Y.N.; Wang, B.; Liu, L.J.; Chen, J.; Dai, L.J.; Peng, D.X. Identification and expression of Aux/IAA, ARF, and LBD family transcription factors in *Boehmeria nivea*. *Biol. Plant.* **2016**, *60*, 244–250. [CrossRef]
30. Chen, H.; Guo, M.; Dong, S.; Wu, X.; Zhang, G.; He, L.; Jiao, Y.; Chen, S.; Li, L.; Luo, H. A chromosome-scale genome assembly of *Artemisia argyi* reveals unbiased subgenome evolution and key contributions of gene duplication to volatile terpenoid diversity. *Plant Commun.* **2023**, *4*, 100516. [CrossRef] [PubMed]
31. Bowers, J.E.; Chapman, B.A.; Rong, J.; Paterson, A.H. Unravelling angiosperm genome evolution by phylogenetic analysis of chromosomal duplication events. *Nature* **2003**, *422*, 433–438. [CrossRef]
32. Darzentas, N. Circoletto: Visualizing sequence similarity with Circos. *Bioinformatics* **2010**, *26*, 2620–2621. [CrossRef]
33. Wang, F.; Liu, M.; Qiu, R.; Ji, C. The dual role of ubiquitin-like protein Urm1 as a protein modifier and sulfur carrier. *Protein Cell* **2011**, *2*, 612–619. [CrossRef] [PubMed]
34. Richter, W.F.; Nayak, S.; Iwasa, J.; Taatjes, D.J. The Mediator complex as a master regulator of transcription by RNA polymerase II. *Nat. Rev. Mol. Cell Biol.* **2022**, *23*, 732–749. [CrossRef] [PubMed]

35. Brooks, M.J.; Rajasimha, H.K.; Roger, J.E.; Swaroop, A. Next-generation sequencing facilitates quantitative analysis of wild-type and *Nrl*$^{-/-}$ retinal transcriptomes. *Mol. Vis.* **2011**, *17*, 3034–3054. [PubMed]
36. Mazzoni-Putman, S.M.; Brumos, J.; Zhao, C.S.; Alonso, J.M.; Stepanova, A.N. Auxin Interactions with Other Hormones in Plant Development. *Cold Spring Harb. Perspect. Biol.* **2021**, *13*, a039990. [CrossRef] [PubMed]
37. Enders, T.A.; Strader, L.C. Auxin Activity: Past, Present, and Future. *Am. J. Bot.* **2015**, *102*, 180–196. [CrossRef] [PubMed]
38. Chandler, J.W. Auxin response factors. *Plant Cell Environ.* **2016**, *39*, 1014–1028. [CrossRef] [PubMed]
39. Paul, P.; Dhandapani, V.; Rameneni, J.J.; Li, X.; Sivanandhan, G.; Choi, S.R.; Pang, W.; Im, S.; Lim, Y.P. Genome-Wide Analysis and Characterization of Aux/IAA Family Genes in *Brassica rapa*. *PLoS ONE* **2016**, *11*, e0151522. [CrossRef] [PubMed]
40. Singh, V.K.; Jain, M. Genome-wide survey and comprehensive expression profiling of Aux/IAA gene family in chickpea and soybean. *Front. Plant Sci.* **2015**, *6*, 918. [CrossRef]
41. Li, H.T.; Wang, B.; Zhang, Q.H.; Wang, J.; King, G.J.; Liu, K.D. Genome-wide analysis of the auxin/indoleacetic acid (Aux/IAA) gene family in allotetraploid rapeseed (*Brassica napus* L.). *BMC Plant Biol.* **2017**, *17*, 204. [CrossRef]
42. Tiwari, S.B.; Hagen, G.; Guilfoyle, T.J. Aux/IAA proteins contain a potent transcriptional repression domain. *Plant Cell* **2004**, *16*, 533–543. [CrossRef] [PubMed]
43. Dreher, K.A.; Brown, J.; Saw, R.E.; Callis, J. The *Arabidopsis* Aux/IAA protein family has diversified in degradation and auxin responsiveness. *Plant Cell* **2006**, *18*, 699–714. [CrossRef]
44. Cho, H.; Ryu, H.; Rho, S.; Hill, K.; Smith, S.; Audenaert, D.; Park, J.; Han, S.; Beeckman, T.; Bennett, M.J.; et al. A secreted peptide acts on BIN2-mediated phosphorylation of ARFs to potentiate auxin response during lateral root development. *Nat. Cell Biol.* **2014**, *16*, 66–76. [CrossRef]
45. Ke, Y.G.; Abbas, F.; Zhou, Y.W.; Yu, R.C.; Yue, Y.C.; Li, X.Y.; Yu, Y.Y.; Fan, Y.P. Genome-Wide Analysis and Characterization of the Aux/IAA Family Genes Related to Floral Scent Formation in *Hedychium coronarium*. *Int. J. Mol. Sci.* **2019**, *20*, 3235. [CrossRef] [PubMed]
46. Qiao, L.Y.; Zhang, X.J.; Han, X.; Zhang, L.; Li, X.; Zhan, H.X.; Mat, J.; Luo, P.G.; Zhane, W.P.; Cui, L.; et al. A genome-wide analysis of the *auxin/indole-3-acetic acid* gene family in hexaploid bread wheat (*Triticum aestivum* L.). *Front. Plant Sci.* **2015**, *6*, 770. [CrossRef] [PubMed]
47. Yang, F.; Zhang, X.X.; Tian, R.F.; Zhu, L.W.; Liu, F.; Chen, Q.F.; Shi, X.J.; Huo, D.A. Genome-Wide Analysis of the Auxin/Indoleacetic Acid Gene Family and Response to Indole-3-Acetic Acid Stress in *Tartary Buckwheat* (*Fagopyrum tataricum*). *Int. J. Genom.* **2021**, *2021*, 3102399. [CrossRef]
48. Zhu, Z.X.; Liu, Y.; Liu, S.J.; Mao, C.Z.; Wu, Y.R.; Wu, P. A Gain-of-Function Mutation in *OsIAA11* Affects Lateral Root Development in Rice. *Mol. Plant* **2012**, *5*, 154–161. [CrossRef] [PubMed]
49. Lv, B.S.; Yu, Q.Q.; Liu, J.J.; Wen, X.J.; Yan, Z.W.; Hu, K.Q.; Li, H.B.; Kong, X.P.; Li, C.L.; Tian, H.Y.; et al. Non-canonical AUX/IAA protein IAA33 competes with canonical AUX/IAA repressor IAA5 to negatively regulate auxin signaling. *EMBO J.* **2020**, *39*, e101515. [CrossRef]
50. Salma, C.; Alain, L.; Claude, P.J.; Mondher, B. Tomato Aux/IAA3 and HOOKLESS are important actors of the interplay between auxin and ethylene during apical hook formation. *Plant Signal. Behav.* **2009**, *4*, 559–560. [CrossRef]
51. Rinaldi, M.A.; Liu, J.; Enders, T.A.; Bartel, B.; Strader, L.C. A gain-of-function mutation in *IAA16* confers reduced responses to auxin and abscisic acid and impedes plant growth and fertility. *Plant Mol. Biol.* **2012**, *79*, 359–373. [CrossRef]
52. Su, Y.X.; Wang, G.L.; Huang, Z.Y.; Hu, L.L.; Fu, T.; Wang, X.Y. Silencing *GhIAA43*, a member of cotton AUX/IAA genes, enhances wilt resistance via activation of salicylic acid-mediated defenses. *Plant Sci.* **2022**, *314*, 111126. [CrossRef]
53. Collum, T.D.; Padmanabhan, M.S.; Hsieh, Y.C.; Culver, J.N. Tobacco mosaic virus-directed reprogramming of auxin/indole acetic acid protein transcriptional responses enhances virus phloem loading. *Proc. Natl. Acad. Sci. USA* **2016**, *113*, E2740–E2749. [CrossRef] [PubMed]
54. Qiao, J.; Luo, Z.; Li, Y.; Ren, G.; Liu, C.; Ma, X. Effect of abscisic acid on accumulation of five active components in root of *Glycyrrhiza uralensis*. *Molecules* **2017**, *22*, 1982. [CrossRef] [PubMed]
55. Sharifi, Y.; Nematzadeh, G.A.; Ghasemi Omran, V.; Tavabe Ghavami, T.S.; Ebrahimzadeh, M.A. Effect of salicylic acid on phenols and flavonoids content in callus culture of Iranian sodab (*Ruta graveolens*): A threatened medicinal plant of north of Iran. *Tabari Biomed. Stud. Res. J.* **2019**, *1*, 32–36.
56. Yang, Y.; Huang, J.; Ding, J. Interaction between exogenous brassinolide, IAA and BAP in secondary metabolism of cultured *Onosma paniculatum* cells. *Plant Growth Regul.* **2003**, *39*, 253–261. [CrossRef]
57. Wang, Y.; Wang, N.; Xu, H.; Jiang, S.; Fang, H.; Su, M.; Zhang, Z.; Zhang, T.; Chen, X. Auxin regulates anthocyanin biosynthesis through the Aux/IAA–ARF signaling pathway in apple. *Hortic. Res.* **2018**, *5*, 59. [CrossRef] [PubMed]
58. Yang, L.; Yan, Y.; Zhao, B.; Xu, H.; Su, X.; Dong, C. Study on the Regulation of Exogenous Hormones on the Absorption of Elements and the Accumulation of Secondary Metabolites in the Medicinal Plant *Artemisia argyi* Leaves. *Metabolites* **2022**, *12*, 984. [CrossRef]
59. Li, Y.; Kong, D.; Fu, Y.; Sussman, M.R.; Wu, H. The effect of developmental and environmental factors on secondary metabolites in medicinal plants. *Plant Physiol. Biochem.* **2020**, *148*, 80–89. [CrossRef]
60. Li, J.; Wang, L.; Chen, S.; Yang, H.; Guo, K.; Ma, R.; Lian, C. Establishment of Plant Tissue Culture System for *Artemisia argyi*. *Mol. Plant Breed.* 2023. Available online: http://kns.cnki.net/kcms/detail/46.1068.S.20230517.0916.004.html (accessed on 17 May 2023). (In Chinese)

61. Pasternak, T.P.; Steinmacher, D. Plant Growth Regulation in Cell and Tissue Culture In Vitro. *Plants* **2024**, *13*, 327. [CrossRef]
62. Chen, C.; Chen, H.; Zhang, Y.; Thomas, H.R.; Frank, M.H.; He, Y.; Xia, R. TBtools: An Integrative Toolkit Developed for Interactive Analyses of Big Biological Data. *Mol. Plant* **2020**, *13*, 1194–1202. [CrossRef]
63. Tamura, K.; Stecher, G.; Peterson, D.; Filipski, A.; Kumar, S. MEGA6: Molecular Evolutionary Genetics Analysis Version 6.0. *Mol. Biol. Evol.* **2013**, *30*, 2725–2729. [CrossRef] [PubMed]
64. Wang, Y.; Tang, H.; DeBarry, J.D.; Tan, X.; Li, J.; Wang, X.; Lee, T.-H.; Jin, H.; Marler, B.; Guo, H.; et al. MCScanX: A toolkit for detection and evolutionary analysis of gene synteny and collinearity. *Nucleic Acids Res.* **2012**, *40*, e49. [CrossRef]
65. Lyu, F.; Han, F.; Ge, C.; Mao, W.; Chen, L.; Hu, H.; Chen, G.; Lang, Q.; Fang, C. OmicStudio: A composable bioinformatics cloud platform with real-time feedback that can generate high-quality graphs for publication. *iMeta* **2023**, *2*, e85. [CrossRef]
66. Lian, C.L.; Zhang, B.; Yang, J.F.; Lan, J.X.; Yang, H.; Guo, K.H.; Li, J.J.; Chen, S.Q. Validation of suitable reference genes by various algorithms for gene expression analysis in *Isodon rubescens* under different abiotic stresses. *Sci. Rep.* **2022**, *12*, 19599. [CrossRef] [PubMed]
67. Livak, K.J.; Schmittgen, T.D. Analysis of relative gene expression data using real-time quantitative PCR and the $2^{-\Delta\Delta CT}$ Method. *Methods* **2001**, *25*, 402–408. [CrossRef] [PubMed]

plants

MDPI

Article

Functional and Transcriptome Analysis Reveal Specific Roles of *Dimocarpus longan DlRan3A* and *DlRan3B* in Root Hair Development, Reproductive Growth, and Stress Tolerance

Qilin Tian [1,†], Xiying Xie [1,2], Ruilian Lai [1], Chunzhen Cheng [1], Zihao Zhang [1], Yukun Chen [1], Xu XuHan [1,3], Yuling Lin [1] and Zhongxiong Lai [1,*]

1 Institute of Horticultural Biotechnology, Fujian Agriculture and Forestry University, Fuzhou 350002, China; tianql70@163.com (Q.T.); xxiying@foxmail.com (X.X.); lairuilian@faas.cn (R.L.); ld0532cheng@126.com (C.C.); zhangzihao863@163.com (Z.Z.); cyk68@163.com (Y.C.); xxuhan@163.com (X.X.); buliang84@163.com (Y.L.)
2 School of Media and Design, Nantong Institute of Technology, Nantong 226019, China
3 Institut de la Recherche Interdisciplinaire de Toulouse, IRIT-ARI, 31300 Toulouse, France
* Correspondence: laizx01@163.com; Tel.: +86-591-8378-9484
† Current address: Provincial Key Laboratory of Landscape Plants with Fujian and Taiwan Characteristics, Minnan Normal University, Zhangzhou 363000, China.

Abstract: Ran GTPases play essential roles in plant growth and development. Our previous studies revealed the nuclear localization of DlRan3A and DlRan3B proteins and proposed their functional redundancy and distinction in *Dimocarpus longan* somatic embryogenesis, hormone, and abiotic stress responses. To further explore the possible roles of *DlRan3A* and *DlRan3B*, gene expression analysis by qPCR showed that their transcripts were both more abundant in the early embryo and pulp in longan. Heterologous expression of *DlRan3A* driven by its own previously cloned promoter led to stunted growth, increased root hair density, abnormal fruits, bigger seeds, and enhanced abiotic stress tolerance. Conversely, constitutive promoter *CaMV 35S* (*35S*)-driven expression of *DlRan3A*, *35S*, or *DlRan3B* promoter-controlled expression of *DlRan3B* did not induce the alterations in growth phenotype, while they rendered different hypersensitivities to abiotic stresses. Based on the transcriptome profiling of longan *Ran* overexpression in tobacco plants, we propose new mechanisms of the *Ran*-mediated regulation of genes associated with cell wall biosynthesis and expansion. Also, the transgenic plants expressing *DlRan3A* or *DlRan3B* genes controlled by *35S* or by their own promoter all exhibited altered mRNA levels of stress-related and transcription factor genes. Moreover, *DlRan3A* overexpressors were more tolerant to salinity, osmotic, and heat stresses, accompanied by upregulation of oxidation-related genes, possibly involving the Ran-RBOH-CIPK network. Analysis of a subset of selected genes from the *Ran* transcriptome identified possible cold stress-related roles of brassinosteroid (BR)-responsive genes. The marked presence of genes related to cell wall biosynthesis and expansion, hormone, and defense responses highlighted their close regulatory association with *Ran*.

Keywords: *Dimocarpus longan* Lour.; ras-related nuclear protein (Ran); promoter; root hair; reproduction; stress responses; transcriptome analysis

1. Introduction

The longan tree (*Dimocarpus longan* Lour., Sapindaceae), a tropical evergreen fruit tree in southern China, is of great edible value and exhibits medicinal properties. However, there are certain challenges such as lacking aborted-seed varieties and the influence of environmental stresses on vegetative growth, fruit yield, and quality [1]. Being a Sapindaceae family of fruit crops, thicker arils and abortive seeds are important quality traits. In recent years, therefore, there have been several reports on the mechanisms of longan fruit and seed development which are controlled by various regulators [2,3].

Ras-related nuclear protein (Ran) GTPases, also known as molecular switches, play a universal role in nucleocytoplasmic transport, specifically expressed during embryogenesis, and are essential for cell division in animal development [4,5]. Plant Ran GTPases, along with their cofactors and nucleoporins, are involved in diverse biological processes including plant growth and root development [6,7], regulation of hormone (such as auxin and abscisic acid, ABA) sensitivity [6,8,9], enhancement of disease resistance [10], and stress responses, such as those to drought, salinity, osmotic stress, and cold and oxidative stresses [7,8,11]. This indicates that plant *Ran* may have a remarkable impact on plant growth and, hence, its economic traits when overexpressed.

Previous studies by our research group have cloned *DlRan3A* and *DlRan3B*, two *Ran* genes from somatic embryos in *D. longan*, and revealed their involvement in longan somatic embryogenesis [12]. *DlRan3A* and *DlRan3B* promoters (hereafter, referred to as 'p*DlRan3A*' and 'p*DlRan3B*', respectively), 1256 bp and 1569 bp in length, were cloned from embryogenic callus (EC) [13,14]. The deletion experiments showed that p*DlRan3A* and p*DlRan3B* might be involved in transcriptional control of plant hormones and specific defense reactions [14,15]. However, the molecular mechanism of longan *Ran* and its relationship with plant growth (especially seed and fruit development) and responses to the external environment need further investigation.

In this study, expression profiling of *DlRan3A* and *DlRan3B* was analyzed in different longan tissues, during zygotic embryo development, and in the pulp of ripening fruits, to reveal the potential role of *Ran* in embryo and fruit development in woody fruit trees. Currently, the molecular basis of the *Ran* gene and its promoter for the regulation in plant embryo and fruit development is poorly understood, in part due to the fact that few studies have aimed to investigate global changes in the gene expression elicited by overexpression of the plant *Ran* gene. For selection and further application of *Ran* genes in molecular breeding using transgenic technology, we generated transgenic tobacco plants overexpressing *DlRan3A* and *DlRan3B*, using both *CaMV 35S* (*35S*) and their own promoters. Based on the analysis of plant phenotype and environmental stress response, transcriptome analysis of transgenic lines was subsequently performed to identify the possible downstream genes responsible for the specific plant growth and stress tolerance phenotypes among different transgenic lines. The selected genes shared or were distinct between transgenic lines, respectively, will be useful to examine the effect and underlying molecular mechanism of longan *Ran* overexpression, and for genetic engineering to enhance the stress tolerance of woody fruit trees, including longan.

2. Results

2.1. Longan DlRan3A and DRan3B Are Highly Expressed in Early Embryos and Pulp (Aril)

To explore the potential developmental role of *DlRan3A* and *DlRan3B*, we first investigated their expression patterns in *D. longan* Lour. cv. Honghezi in diverse tissues (Figure 1a). We found that, apart from the fundamental expressions in longan tissues, *DlRan3A* and *DlRan3B* were both predominantly expressed in the seeds and then in the pulp, but displayed the lowest levels in the anthers (Figure 1a). This indicates that *Ran* is indispensable to longan growth and development while *DlRan3A* and *DlRan3B* genes might play more significant roles in seed and pulp development.

To further characterize *DlRan3A* and *DlRan3B* expression patterns during longan seed (embryo) and pulp development, we next examined their expressions during 'Honghezi' zygotic embryo development and in the pulp of ripening fruits. In the early stages (S1 and S2), both genes showed high expression with a fluctuating trend that declined during the zygotic embryo development (Figure 1b). However, both gene expressions showed an increase in the pulp of ripening fruits (Figure 1c).

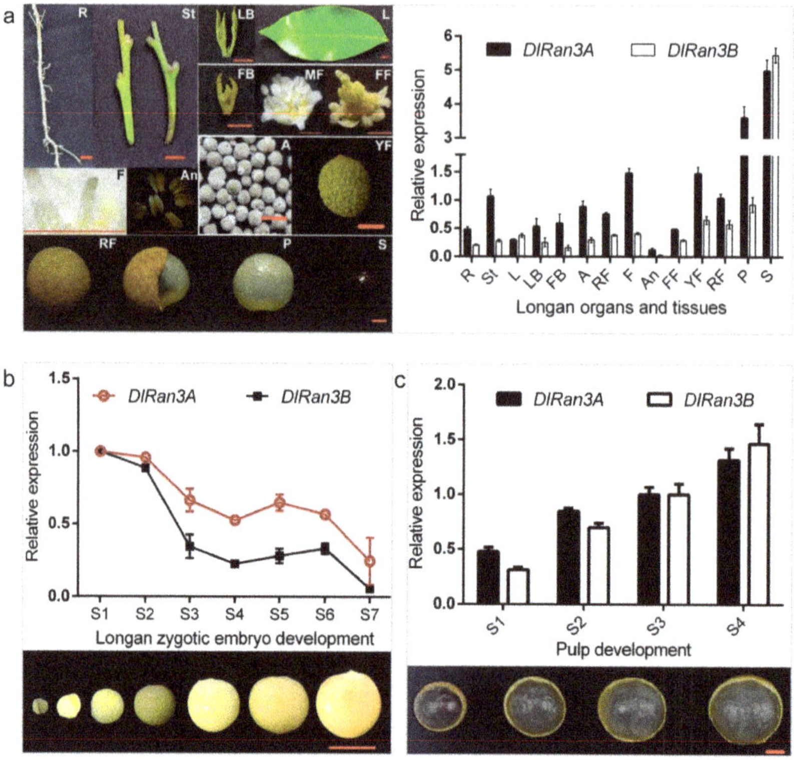

Figure 1. Relative expression levels of *DlRan3A* and *DlRan3B* in different tissues, developing zygotic embryos and pulp in longan. (**a**) Relative expression of *DlRan3A* and *DlRan3B* in different tissues (updated from Chen [16]). R: root; St: stem; L: leaf; LB: leaf bud; FB: floral bud; A: alabastrum; MF: male flower; F: filament; An: anther; FF: female flower; YF: young fruit; RF: ripe fruit; P: pulp; S: seed; bar = 5 mm. (**b**) Relative expression of *DlRan3A* and *DlRan3B* during zygotic embryo development. Longan zygotic embryos from stage S1 to S7 were collected from 16 June to 12 July in 2015, every four or every five days; bar = 10 mm. (**c**) Relative expression of *DlRan3A* and *DlRan3B* in different sizes of longan pulp. Longan pulp from stage S1 to S4 were collected in August 2015; bar = 10 mm.

2.2. Characterization of DlRan3A and DlRan3B Genes and the Use of Their Own Promoters by Heterologous Expression in Tobacco Plants

To further investigate the function of *DlRan3A* and *DlRan3B*, we cloned each gene into pCAMBIA1301 driven by *35S* or *Ran* promoters and transformed into *Agrobacterium tumefaciens* strain EHA105. The T2 line *35S*-driven expression of *DlRan3A* or *DlRan3B* (henceforth, referred to as '35S_A' or '35S_B', respectively) and those driven by the *Ran* promoter (hereinafter, referred to as 'PA_A' or 'PB_B', respectively) were employed for the following experiments.

At the same planting time, excluding PA_A, there were no significant differences in plant height and blade generation rate among the transgenic tobaccos (Figure S3). Notably, PA_A showed a dwarf phenotype with reduced height and thick stems, along with phenotypes of late flowering and reduced axillary buds (Figure 2a). After vertical cultivation for 7 d, 14 d, and 21 d, excluding PA_A, there were no significant differences in root growth among the transgenic tobaccos. Notably, PA_A exhibited a sturdier root system along with increased numbers of root hairs (Figures 2b and S4). Even after 50–60 d of growth, PA_A maintained distinct phenotypes in fruit and seed development, showing abnormal fruits, larger seeds, decreased seed setting rate, and increased seed weight (Figures 2c and S5). This indicates that p*DlRan3A*-driven expression of *DlRan3A* caused stunted growth and significantly affected the development of root, fruit, and seed. Moreover, the histochemical staining assay conducted to assess p*DlRan3A*- or p*DlRan3B*-driven expression of the *GUS* gene in transgenic tobaccos (hereafter, referred to as 'PA_GUS' or 'PB_GUS', respectively) revealed that both the promoters led to *GUS* expression in roots, leaves, flowers, fruits, and seeds. However, the *DlRan3A* promoter could drive greater *GUS* accumulation in petals (Figures S6 and S7).

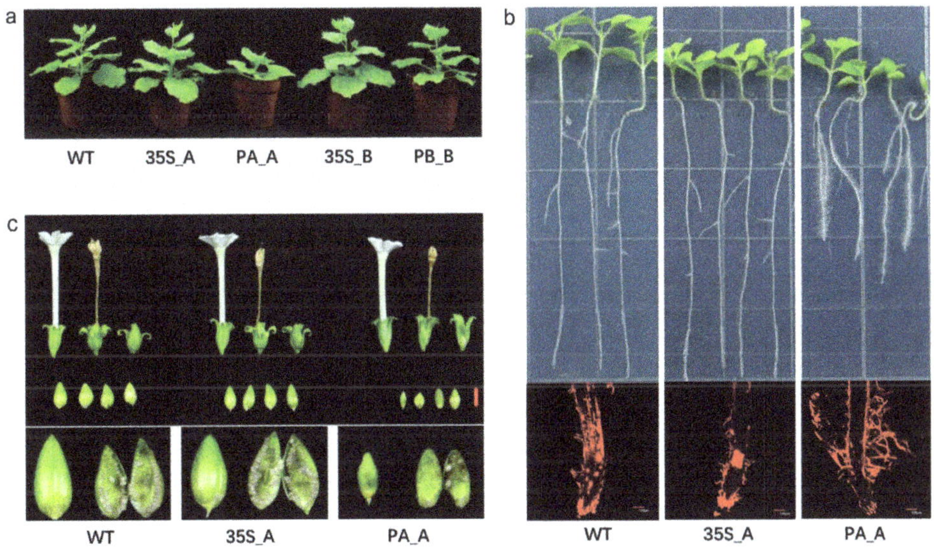

Figure 2. Phenotype of transgenic tobaccos. (**a**) The entire phenotypes of transgenic tobaccos. WT and T2 lines of P35S_A, PA_A, P35S_B, and PB_B (45 d) are displayed in the graph from left to right. (**b**) The phenotype of transgenic tobacco roots (21 d). WT and T2 lines of P35S_A and PA_A are displayed in the graph from left to right. The upper images illustrate the transgenic tobaccos germinated and vertically cultivated on MS medium for 21 d, and the lower images illustrate the 21-day-cultivated tobacco roots stained with propidium iodide, as observed under a confocal microscope (bar = 100 μm). (**c**) The phenotypes of flowers, fruits, and seeds of transgenic tobaccos. The WT and T2 lines of P35S_A and PA_A are displayed in the graph from left to right (bar = 5 mm).

2.3. Heterologous Expression in Tobaccos Reveals Shared and Distinct Roles of DlRan3A and DlRan3B in Various Stresses

Under 100 mM NaCl stress, 200 mM mannitol stress, 10 μM ABA, 35 °C heat stress, and 15 °C cold stress, tobaccos expressing *DlRan3A* (i.e., P35S_A and PA_A) exhibited stronger tolerance to salinity, osmotic, and heat stresses. In contrast, tobaccos expressing *DlRan3B* (i.e., P35S_B, and PB_B) showed relatively weaker tolerance against abiotic stress. Remarkably, P35S_B tobaccos were hypersensitive to salinity, osmotic, and heat stresses, while P35S_A tobaccos were hypersensitive to ABA. Additionally, tobaccos expressing *DlRan3A* or *DlRan3B* showed mild tolerance against cold stress (Figure 3).

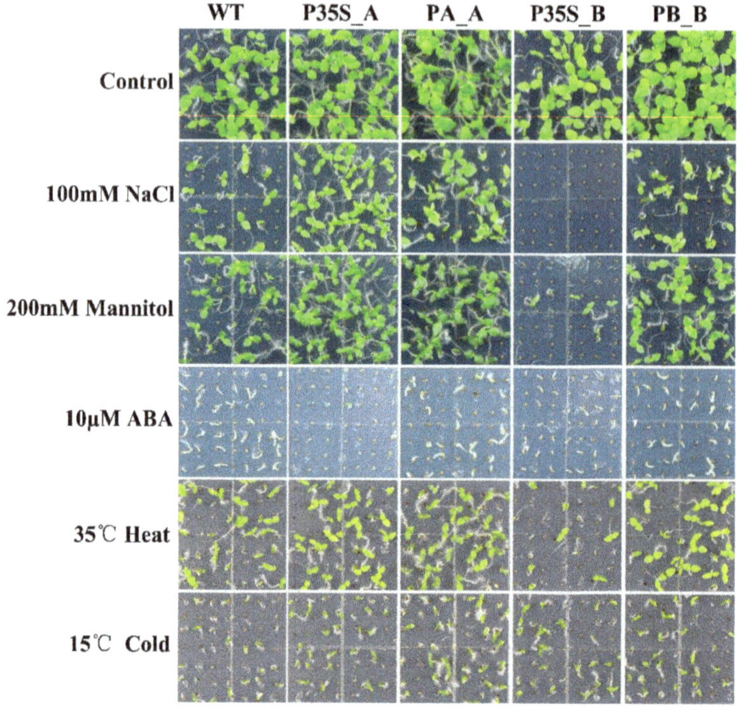

Figure 3. Abiotic stress tolerance of transgenic tobaccos. Graphics related to the WT and T2 lines of P35S_A, PA_A, P35S_B, and PB_B are displayed from left to right; a control group without any treatment and the treatment groups belonging to 100 mM NaCl, 200 mM mannitol, 10 μM ABA, 35 °C heat, and 15 °C cold stresses are displayed from top to bottom.

2.4. Transcriptome Profiling of Transgenic Tobaccos Revealed Alteration in Genes Involved in Biosynthetic Processes Related to the Cell Wall, Hormone Signaling, and Stress Responses

To understand the molecular functions of longan *Ran* GTPases, RNA-sequencing (RNA-Seq) profiling was conducted to estimate the effects on the entire transcriptome. Sequencing libraries were generated from WT, P35S_A, P35S_B, PA_A, and PB_B transgenic tobacco plants cultivated in Murashige and Skoog (MS) medium for 21 d. Reads alignment to the reference genome is shown in Table S5. The differentially expressed genes (DEGs) with higher transcript levels in transgenic lines than the corresponding control sample were denoted "up-regulated genes", whereas those with lower transcript levels were defined as "down-regulated genes". As shown in Table S6, the PB_B versus WT group (hereinafter, referred to as 'PB_B/WT') showed the largest number of DEGs, followed by PA_A versus WT groups (hereinafter, referred to as 'PA_A/WT'); the DEGs in the PA_A versus P35S_A groups or PB_B versus P35S_B groups (hereinafter, referred to as 'PA_A/P35S_A' or 'PB_B/P35S_B', respectively) were less than any transgenic lines versus the WT groups.

Notably, not many DEGs in P35S_A versus WT groups (hereinafter, referred to as '35S_A/WT') were shared with the P35S_B versus WT groups (hereinafter, referred to as '35S_B/WT') (Figure S8a and Table S7). Furthermore, functional annotation revealed that the shared DEGs are mainly involved in cell wall organization or biogenesis (such as cellulose synthase, CESA) and plant resistance (such as asparagine synthetase, AS, and heat shock protein, HSP). Importantly, PA_A/WT and 35S_A/WT shared a certain number of DEGs (Figure S8b and Table S8). These are also mainly involved in cell wall organization or biogenesis (such as cellulose synthase) and plant resistance, such as AS,

HSP, pleiotropic drug resistance protein (PDR), calmodulin-binding protein (CBP), peroxide (PER), and respiratory burst oxidase homolog (RBOH). Among the DEGs, most DEGs were up-regulated, except some like *CYP450* (cytochrome P450) and *HSP*. The shared DEGs between PB_B/WT and 35S_B/WT are listed in Figure S8c and Table S9, including calcium-binding protein, CESA, expansin, and genes involved in plant resistance (such as *AS*, *NAC*, *ERF*, *WRKY*, *bHLH*, *C3H*, *GRAS*, and *HSP*). Among these, most DEGs associated with plant resistance were down-regulated.

Additionally, we also analyzed the expression profiles of transcription factor genes, cell wall biosynthesis and expansin/extensin genes, hormone-related genes, and stress-related genes between non-transgenic and transgenic tobaccos. As shown in Figure S9, the heat map indicated that certain transcription factors (MYB, NAC, WRKY, ERF, GRAS, bHLH, etc.) were down-regulated in the transgenic lines, especially in the 35S_B and PB_B lines. Regarding hormone-related genes, auxin-related DEGs were up- or down-regulated; ethylene-related *ERFs* were mostly down-regulated, especially in the 35S_B and PB_B lines. Also, all of the brassinosteroid (BR)-related *EXORDIUMs* were up-regulated in the transgenic lines, along with down-regulated *MYC2*s (jasmonic acid-related) and up-regulated *ARR9* (cytokinin-related). Notably, most extensin genes were up-regulated in PA_A along with three extensin genes (Niben101Scf02042g00001, Niben101Scf02191g03003, and Niben101Scf03036g00006). The two *expansins* (Niben101Scf03913g01045 and Niben101Scf20887g00008) were significantly up-regulated or down-regulated in the PA_A/P35S_A comparisons, respectively (Figure S10). The expression profiles of stress-related genes are shown in Figures S11 and S12.

2.5. Functional Classification, and Kyoto Encyclopedia of Genes and Genomes (KEGG) Analyses of DEGs

Gene Ontology (GO) enrichment analysis was performed for global functional analysis of DEGs related to longan *Ran* expression. Compared to WT, PA_A showed the greatest number of significantly enriched pathways and related DEGs (Table S10).

In P35S_A (compared to WT), the oxidoreductase activity pathway acting on peroxide as an acceptor was significantly enriched for four DEGs: the peroxide gene, *RBOH*, and Novel00358, were up-regulated, while *CAT1* (catalase isozyme 1) was down-regulated. In PA_A (compared to WT), among the sixteen significantly enriched pathways, "cell wall organization or biogenesis" or related pathways were predominant along with six up-regulated DEGs (mainly extensin gene). Other pathways were mainly enriched for upregulated *CESAs*, *XET/XTH* (xyloglucan endotransglucosylase/hydrolase), and *UGE* (Bifunctional UDP-glucose 4-epimerase and UDP-xylose 4-epimerase), while *UGT73C3* (UDP-glycosyltransferase 73C3), *GT* (Anthocyanidin 3-O-glucosyltransferase), and *CIPK* (Calcineurin B-like-interacting serine/threonine-protein kinase 11) were down-regulated.

Further comparative analysis of PA_A/P35S_A or PB_B/P35S_B illuminated the biological function of longan *Ran* promoters. For PA_A/P35S_A or PB_B/P35S_B, all the listed cell wall biogenesis pathways were significantly enriched for the three extensin genes mentioned above. However, these were up-regulated in PA_A/P35S_A but down-regulated in PB_B/P35S_B.

Likewise, in PA_A/PB_B, the significantly enriched pathways were mainly related to cell wall and oxidative stress. Notably, the oxidative stress-related pathways were significantly enriched for four up-regulated peroxide genes (Niben101Scf03990g00010, Niben101Scf00416g01009, Niben101Scf02709g01005, and Niben101Scf02349g03001) and a *GPX* (glutathione peroxidase) gene (Niben101Scf06369g03006) (also see Figure S13).

Moreover, the KEGG biochemical pathways analysis of related DEGs was implemented. Compared to WT, the pathways related to alanine, aspartate, and glutamate metabolism were significantly enriched in P35S_A, P35S_B, or PA_A, lines, respectively (Table S11), suggesting the key role of longan *Ran* in these processes. Also, we found five significantly enriched pathways in PA_A (compared to PB_B). Most of these pathways were related to up-regulated DEGs, including five *GST* (glutathione S-transferase) (Niben101Scf00069g02013, Niben101Scf10316g03004,

Niben101Scf03147g10010, Niben101Scf03482g0101, and Niben101Scf11037g00006) and a *GPX* (mentioned above in the GO analysis of PA_A/PB_B) genes.

2.6. Verification qPCR Validation of Candidate DEGs

To validate the RNA-Seq results, 14 DEGs were selected for qPCR assay. The gene symbols, FPKM values, and relative expression levels are shown in Figure 4. Although the fold changes in transcript levels detected by RNA-Seq and qPCR were not perfectly matched, most of the candidate DEGs showed similar trends of change in gene expression as in the RNA-Seq results. In general, the expression patterns measured by the two methods were consistent.

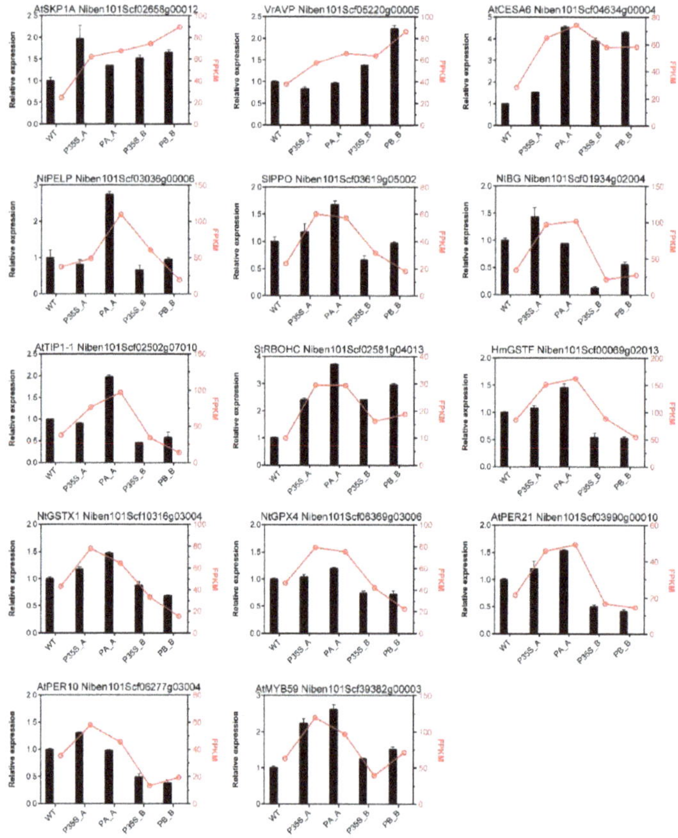

Figure 4. Validation of RNA-seq results by qPCR. The corresponding gene ID is mentioned at the top of each graph. The columns represent the relative expression levels measured by qPCR using *NbEF1a* as a reference gene, and the red lines portray the FPKM change from RNA-seq.

3. Discussion

3.1. *DlRan3A and DlRan3B Up-Regulate Cell Wall Biosynthesis and Expansion Genes to Regulate Early Embryo and Pulp Development in Longan*

Studies have proved that plant Ran is ubiquitously expressed in all tissues, such as in tomato [17], Arabidopsis [18], tall fescue [19], sugarcane [20], etc. In longan, *DlRan3A* and *DlRan3B* also exhibited a certain fundamental expression in almost the whole tree, suggesting their essential roles in development. *Ran* is more expressed in meristematic tissues, embryos, or roots, which provides a clue to its functional diversity in plants [18–20].

Arabidopsis RAN1 interaction with DEM1 (defective embryo and meristems 1) is required for organized cell divisions during embryonic and post-embryonic growth in tomato [21]. Moreover, it was shown to mediate seed development by affecting the onset of endosperm cellularization [22]. A previous study revealed that, akin to its homolog *AtRan3*, longan Ran might have similar roles in cell proliferation during somatic embryogenesis [12]. Notably, *DlRan3A* and *DlRan3B* indicated functional similarity and specificity in transcriptional regulation during longan somatic embryogenesis [14,15]. Here, based on expression profiling of *DlRan3A* and *DlRan3B* at different development stages of seeds, fruits, and other tissues, we found that both genes were highly expressed in seeds and pulp (aril), as well as in the young embryos of early-stage zygotic embryo development. Additionally, they may positively regulate fruit pulp swelling. Importantly, based on the previous similar findings of their roles in late embryogenesis (cotyledon embryos), *DlRan3A* and *DlRan3B* might regulate early embryo development, especially the formation of cotyledons.

Proteins involved in cell wall synthesis and expansion in plants are closely related to plant development including embryo and fruit development [23]. For instance, XET/XTH, an important enzyme for cell wall loosening, has a key role in plant somatic embryogenesis [24]. Additionally, the genes encoding for cellulose synthase-like protein AtCSLA7, hydroxyproline-rich glycoprotein RSH (root-shoot-hypocotyl-defective), and AtEXT3 (extensin 3), are essential for cell wall structure and normal embryo development in Arabidopsis [25,26]. The current study suggests that longan *Ran* expression led to up-regulation of numerous cell wall biosynthesis and expansin/extensin genes, indicating the role of *Ran* in cell wall development. Additionally, during the fruit expansion stage, the cell wall needs to turn loose. We found that *DlRan3A* and *DlRan3B* were highly expressed in longan pulp, and their expression levels increased with the expansion of longan pulp (aril), along with several up-regulated genes related to cell wall loosening or fruit ripening (encoding expansins, XETs, XTHs, etc.). This indicates the potential role of *Ran* in fruit development by enhancing the accumulation of cell wall-loosening proteins. The current study provides significant new insights for *Ran* function during embryo and pulp (aril) development. However, the potential mechanism(s) of *Ran*-mediated transcript accumulation of cell wall related genes during early embryo and pulp (aril) development needs to be further revealed.

3.2. pDlRan3A Driven Expression of DlRan3A Led to Stunted Plant Growth, Higher Root Hair Density, Abnormal Fruits, and Bigger Seeds, Potentially via Partial Regulation of Expansin- and Extensin-like Genes

Ran GTPase is closely related to both vegetative and reproductive growth in plants. Expression of wheat *TaRAN1* prolongs the life cycle, elevates mitotic index, decreases lateral roots, increases primordial tissue, and enhances auxin hypersensitivity [6]. Similarly, ectopic expression of *FaRan* in Arabidopsis increases axillary buds and reduces apical dominance, suggesting its potential role in the initiation of meristem and subsequent growth and development [19]. In this study, p*DlRan3A*-controlled expression of *DlRan3A* led to stunted plant growth, increased root hair density, abnormal fruits, and larger seeds, suggesting the important role of *Ran* in certain organs. It seems that an excessive or ectopic presence of *Ran*, beyond its optimal concentration, may significantly affect plant development.

The agronomic characteristic dwarfism, such as reduced height and thick stems, is often desirable to bolster lodging resistance [27]. The current results suggest that p*DlRan3A*-driven expression of *DlRan3A* can be an additional functional factor for stunted plant growth. Importantly, stunted or dwarfed plant growth has also been linked to transcriptional regulation of genes related to cell wall synthesis and expansion [28], which is consistent with our finding that two *expansin* genes, *AtEXLA2* Niben101Scf03913g01045 and *OsEXPA4* Niben101Scf20887g00008, were, respectively, up-regulated and down-regulated in the PA_A/P35S_A comparison. Despite the fact that the suppressed expression of the *expansin* gene is known to reduce plant growth while the overexpression does not necessarily promote it [28–30], our finding highlights the role of *DlRan3A* in the regulation of stunted plant growth and provides a novel clue for understanding the molecular mechanism of *Ran* in the activation or suppression of *expansin* genes.

Root hairs, offering the plant a competitive advantage in the acquisition of water and nutrients from the soil, can enhance plant tolerance to abiotic stress [31]. Though previous studies have revealed the functional involvement of small GTPases in root hair growth [32,33], detailed investigations are required to understand the underlying mechanism(s). In particular, the role of Ran GTPases has not yet been reported in root hair growth. Here, we found that three extensin-like genes were up-regulated in PA_A (compared to P35S_A), and down-regulated in PB_B (compared to P35S_B), while both types were significantly enriched in pathways of cell wall organization or biogenesis. Based on the phenotypes of transgenic plants, we propose that *DlRan3A* and *DlRan3B*, driven by their native promoters, might play different roles in cell wall organization or biogenesis in longan tissues, especially via regulation of extensin-like genes. Emerging evidence suggests that extensin- and extensin-like proteins are essential for cell wall self-assembly and, hence, root hair development, such as in tomato [34], barley [35], Arabidopsis [36], etc. Here, we showed that *DlRan3A* plays a very specific role in root hair development via up-regulation of extensin-like genes, but further studies are required to discover and validate the downstream effectors of *Ran* in root hair development.

In addition, one of the three extensin-like genes was annotated as pistil-specific extensin-like protein gene '*PELP*' (Niben101Scf03036g00006). Notably, extensins have been reported in reproductive development in some plants. For instance, the class III pistil-specific extensin-like proteins (PELPIIIs) of Nicotiana were associated with interspecific incompatibility, specific inhibition of pollen tube, and growth [37,38]. p*DlRan3A*-driven expression of *DlRan3A* led to abnormal fruits, larger seeds, a reduced seed setting rate, and increased seed weight, coinciding with increased and abnormal expression of *PELP* in PA_A. Thus, *DlRan3A* and *PELP* might be closely interrelated and involved in reproductive development, providing new perspectives on *Ran*-mediated transcriptional regulation of plant reproductive processes.

3.3. *DlRan3A and DlRan3B Function in Stress Tolerance by Regulating Different Stress-Responsive Genes*

A close association between *Ran* and plant responses to environmental stresses, such as salinity [39], osmotic stress [8], drought [7], low temperatures [9,40], aluminum toxicity [41], and oxidative stress [11], has been elucidated in numerous studies. However, the understanding of the underlying mechanisms is slowly improving, and there have been rare reports about the role of longan *Ran* in defense responses. In this study, using transcriptome analysis, we revealed numerous DEGs related to cell wall biosynthesis and stress response (*PER, aquaporin, AS, GST, HSP, DNAJ*, etc.), as well as stress-related transcription factor (MYB, NAC, WRKY, ERF, GRAS, bHLH, etc.) DEGs. The data strongly suggest the important roles of *Ran* in a complex stress regulatory network. We found that the tobacco overexpressing *DlRan3A* (P35S_A and PA_A) were more tolerant to abiotic stress (salinity, osmotic, and heat stress). Conversely, tobacco overexpressing *DlRan3B* (P35S_B and PB_B) exhibited relatively poor tolerance compared to the P35S_A and PA_A lines. In particular, compared to WT and *DlRan3B* overexpressing tobaccos, P35S_A tobaccos showed a little more sensitivity to ABA, which plays a critical role in various stress responses. More studies are needed to further elucidate the role of *DlRan3A* in relationship with ABA accumulation and stress (salinity, osmotic, and heat stress) tolerance.

Previous studies in rice suggesting that *RAN* overexpression improved cold tolerance are partially supported by our results. Despite *Ran's* associations with cell division, the hormone BR might also participate in *Ran*-regulated cold tolerance. Notably, many previous studies have shown BR-induced cold tolerance in plants, such as in Arabidopsis and *Brassica napus* [42], *Cucumis sativus* [43], *Chorispora bungeana* [44], *Elymus nutans* [45], *Medicago truncatula* [46], etc. Importantly, plants overexpressing longan *Ran* showed increased accumulation of *EXORDIUM* transcripts than the WT (Figure S7), and well-characterized

BR-responsive genes, such as *expansins*, *XETs*, *aquaporins*, and *ASs* [47–49], were significantly up-regulated in plants with increased longan *Ran* expression. The phenotypic observations, together with the transcriptome analysis of the BR-responsive genes, suggest a possible role of longan *Ran* in activated BR signaling and BR-mediated plant defense mechanisms, especially in cold tolerance.

Despite the similarities, the functional divergence between the two members of the longan *Ran* family in stress responses (salinity, osmotic, and heat) might be partially attributed to the inhibition of stress-regulated transcription factor (WRKY, ERF, GRAS, bHLH, C3H, C2H2, etc.) genes in P35S_B or PB_B. Moreover, the activation of oxidative stress-related genes (*PER*, *GST*, *GPX*, *PPO*, *RBOH*, etc.) can stimulate the antioxidant defense system in P35S_vA and PA_A, resulting in the differences between *DlRan3A* overexpressing tobaccos and *DlRan3B* overexpressing ones in terms of their stress responses (Figure S13). Plant RBOHs are the key enzymes that catalyze the generation of reactive oxygen species (ROS) during plant defense responses and are involved in the modulation of root growth [50]. Plant CBL (calcineurin B-like) and CIPK (CBL-interacting protein kinase) proteins, form one of the important Ca^{2+} decoding complexes to decipher Ca^{2+} signals elicited by environmental challenges [51]. Rac/Rop GTPases and CBL-CIPKs involve integration of calcium signaling into ROS regulation via direct interaction with RBOH [52]. Using GO analysis, we showed that *CIPK* was significantly down-regulated in PA_A, while the RBOH was up-regulated. Considering the fact that RBOH acts as a convergence point targeted by a complex regulatory network, these data endorse the notion that the resistant phenotypes and associated molecular changes in plants overexpressing *DlRan3A* are at least partly due to the Ran–RBOH–CIPK regulatory network. This new evidence suggests the involvement of *Ran* in plant defense responses via calcium signaling and ROS regulation.

4. Materials and Methods

4.1. Plant Materials

The different tissues of longan were collected from Fujian Agriculture and Forestry University in Fuzhou. All materials were mixed samples from at least six rootstock longan plants, collected and stored at 80 °C for further studies. *Nicotiana benthamiana* tobaccos were used for stable genetic transformation.

4.2. Gene Expression Analysis

Total RNA was extracted using the RNAprep Pure Plant Kit (TIANGEN Code, DP441, Beijing, China) or total plant RNA extraction kit (BioTeke Code, RP3312, Beijing, China) following the manufacturer's protocol. cDNAs were synthesized using the PrimeScript^{TM} Perfect Real-Time RT Reagent Kit (TaKaRa Code, RR037A (Dalian, China)). Quantitative real-time PCR analysis (qPCR) was performed to evaluate the transcript levels of the *DlRan3A* and *DlRan3B* genes in longan tissues, during zygotic embryo and pulp developments. Typical reactions were prepared using the SYBR Premix Ex Taq kit (Takara) and all the qPCR reactions were performed in triplicate. QPCR assays were implemented using the LightCycler 480 qPCR instrument (Roche Applied Science, Basel, Switzerland) and cycling conditions were chosen according to the manufacturer's protocol. The expression profiles of *DlRan3A* and *DlRan3B* in longan tissues were quantified using three pre-microRNAs (*pre-miR167f3p*, *pre-miR171f*, and *pre-miR394a*) as the reference genes; and the expression levels during zygotic embryo development were quantified using the $2^{-\Delta\Delta CT}$ method, with longan *Fe-SOD* as a reference gene. The expression levels of *DlRan3A* and *DlRan3B* in other longan samples were quantified using the internal standards as described previously [53]. Primer names and sequences are provided in Table S1.

4.3. Vector Construction

The coding sequences of *DlRan3A* (JQ775539) and *DlRan3B* (JQ279697) were PCR-amplified from longan cDNA using primers with *BamH* I/*Sal* I restriction sites at the 5′/3′ ends, respectively. To generate expression constructs pCAMBIA1301-*35S*-*DlRan3A* (hereinafter, referred to as '35S_A') and pCAMBIA1301-*35S-DlRan3B* (hereinafter, referred to as '35S_B'), the PCR-amplified products were in-fusion cloned into the corresponding sites of pCAMBIA1301SN vector (modified by Feng [54]) having a *35S* promoter and nos terminator, respectively. p*DlRan3A* and p*DlRan3B* were amplified from longan DNA, using primers with *EcoR* I/*Kpn* I restriction enzyme sites at the $5'/3'$ ends. The *35S* promoter was removed from the construct 35S_A and 35S_B by *EcoR* I and *Kpn* I digestion, and the amplified products were in-fusion cloned into the corresponding sites of *35S*-removed vectors, respectively, to generate expression constructs pCAMBIA1301-p*DlRan3A(1256bp)-DlRan3A* (hereinafter, referred to as 'PA_A') and pCAMBIA1301-p*DlRan3B (1569bp)-DlRan3B* (hereinafter, referred to as 'PB_B'). To construct pCAMBIA1301-p*DlRan3A(1256bp)-GUS* (hereinafter, referred to as 'PA_GUS') and pCAMBIA1301-p*DlRan3A(1256bp)-GUS* (hereinafter, referred to as 'PB_GUS'), p*DlRan3A* and p*DlRan3B* were amplified using primers with *Hind* III/*Nco* I restriction sites, and then in-fusion cloned into pCAMBIA1301 with removed *35S*. These expression constructs were transformed into *N. benthamiana* using *Agrobacterium* strain EHA105. Primer sequences for the isolation of the *DlRan3A* and *DlRan3B* genes, as well as their promoter sequences, are provided in Supplementary data Table S2. The design of these constructs is depicted in Supplementary data Figure S1.

4.4. Generation of T2 Transgenic Tobacco Plants

The *Agrobacterium*-mediated transformation of tobacco leaf segments and regeneration of T0 transgenic tobaccos were performed following the method by Feng [54]. To generate T1 tobacco plants, the T0 lines of P35S_A, PA_A, P35S_B, and PB_B plants were grown in a growth chamber until flowering, and then the self-pollination stage. The resulting generations were selected against hygromycin pressure (40 mg/L), PCR verified, and validated with GUS staining by using GUS histochemical assays kit (Real-Times, Beijing, China). Similarly, the final T2 lines, used for further analysis, were obtained through self-pollination of T1 plants and then subjected to similar screening. The single T2 line of P35S_A or PA_A was obtained from T1 lines exhibiting the highest expression levels of *DlRan3A*. Similarly, the single T2 line of P35S_B or PB_B was obtained from T1 lines exhibiting the highest expression levels of *DlRan3B* (Figure S2). The used primer sequences are provided in Table S3.

4.5. Phenotypic Analysis

The effects of *DlRan3A* or *DlRan3B* overexpression were analyzed by scoring a range of specific plant phenotypes: flowering time, plant height, blade generation rate, and the development of roots, flowers, fruits, and seeds. The data for phenotype analysis were acquired from a minimum of fifteen independent plants. The tobacco root tips were stained by propidium iodide as described previously and observed by laser scanning confocal microscopy (Olympus, Tokyo, Japan; FV1200) [40].

4.6. Analysis of Environmental Stress Response

Seeds of T1 lines of P35S_A, PA_A, P35S_B, and PB_B tobacco plants (T2 lines) were germinated on MS medium with 100 mM NaCl, 200 mM mannitol, or 10 μM ABA. All seedlings were cultivated for 10 d in a chamber of 25 °C. For the heat and cold treatments, the tobacco seeds germinated on MS medium were exposed to 35 or 15 °C, respectively, for 6 h and then cultivated in the same condition as mentioned above. The control transgenic and non-transgenic tobaccos groups were cultivated without any treatments.

4.7. RNA-Sequencing

A cDNA library was generated from a pool of equal quantities of total RNA from 21 d cultivated tobacco seedlings of non-transgenic WT and transgenic T2 lines, 35S_A, PA_A, P35S_B, and PB_B, respectively. Plant samples were sent to the Novogene company (Beijing, China) for RNA extraction, cDNA library preparation, sequencing, quality control, reads mapping to the reference genome, and the gene expression quantification. RNA-seq of the mRNA and libraries was performed on an Illumina® HiSeq™ 2000 platform according to Cui's method [55]. DEGs were screened according to their expression profiles meeting the following criteria: $|\log2|$ (transgenic lines/corresponding control) > 1 and the adjusted p value < 0.005. Expression profiles of the screened DEGs were visualized by heatmap using the TBtools software (v0.6673) [56]. The iTAK database was used to define the transcription factor family according to relative rules [57,58]. Hormone-related genes and stress-related genes were analyzed based on annotations in the TAIR (The Arabidopsis Information Resource) database. The GO and KEGG enrichment analyses of the DEGs were conducted following the method by Zhao [59]. The GO terms and KEGG pathways with corrected p-values < 0.05 were considered significantly enriched.

4.8. qPCR Validation

Total RNAs were isolated from transgenic tobaccos using the TriPure Isolation Reagent (Roche) following the manufacturer's protocol. cDNA synthesis and qPCR analysis were conducted as above described, using the $2^{-\Delta\Delta CT}$ method with *NbEF1a* as a reference gene [16]. Primer sequences are provided in Table S4.

5. Conclusions

In this study, we investigated the expression and function of longan Ran GTPase genes *DlRan3A* and *DlRan3B*. The results of the study indicated that *DlRan3A* and *DlRan3B* have important roles in longan early embryo and pulp development. Heterologous expression of *DlRan3A* and *DlRan3B* by employing *35S* or *Ran* promoters, combined with *Ran*-overexpression transcriptome of transgenic plants, showed that *DlRan3A* and *DlRan3B* regulate cell wall-related genes to affect plant growth. Particularly, p*DlRan3A*-driven expression of *DlRan3A* led to stunted plant growth, higher root hair density, abnormal fruits, and bigger seeds, potentially via partial regulation of the expansin- and extensin-like genes. Furthermore, *DlRan3A* and *DlRan3B* might function via shared molecular mechanisms in cold stress response but differed in conferring plants with salinity, osmotic, and heat stress tolerance or sensitivity, by up-regulated or down-regulated different stress- or hormone-related genes. Overall, the present study provided a widespread characterization on longan Ran GTPase genes *DlRan3A* and *DlRan3B*, highlighting the associated gene expression mechanisms incorporated by *Ran*.

Supplementary Materials: The following supporting information can be downloaded at: https://www.mdpi.com/article/10.3390/plants13040480/s1, Figure S1: Constructed vectors of *DlRan3A* and *DlRan3B*; Figure S2: Gene-relative expression in transgenic modified tobacco (T1 lines); Figure S3: Height (a) and leaf number (b) of transgenic tobacco; Figure S4: Phenotype of transgenic tobacco root (7 d and 14 d); Figure S5: Hundred-grain weight of transgenic tobacco; Figure S6: GUS staining of transgenic tobacco seedlings (T1 lines); Figure S7: GUS staining of flowers, fruits, and seeds of transgenic tobacco; Figure S8: Expression profiles of the shared DEGs in transgenic lines; Figure S9: Expression profiles of DEGs classified as genes encoding transcription factors (TFs) (a) and hormone-related genes (HRGs) (b); Figure S10: Expression profiles of DEGs involved in cell wall development; Figure S11: Expression profiles of stress-related DEGs; Figure S12: Mapman software (v3.6.0) visualization of stress-related genes; Figure S13: Expression profiles of the antioxidant-defense-related DEGs; Table S1: Sequences of primers used for qPCR analysis of *DlRan3A* and *DlRan3B* in longan; Table S2: Primer sequences used for *DlRan3A* and *DlRan3B* constructs for genetic transformation of *Nicotiana benthamiana*; Table S3: Sequences of primers used for qPCR detection in transgenic tobaccos; Table S4: Primer sequences used for qPCR verification; Table S5: Comparison of reads by transcriptome sequencing to reference genome; Table S6: Differentially expressed genes (DEGs) of the GM *N. benthamiana* lines; Table S7: Shared DEGs between the P35S_A and P35S_B lines; Table S8: Shared DEGs between the P35S_A and PA_A lines; Table S9: Shared DEGs between the P35S_B and PB_B lines; Table S10: Significant GO enrichment pathways of the genetically modified (GM) *N. benthamiana* lines; Table S11: Significant KEGG enrichment pathways of the GM *N. benthamiana* lines.

Author Contributions: Conceptualization, Z.L. and X.X. (Xuhan Xu); methodology, Q.T. and X.X. (Xiying Xie); validation, Q.T., X.X. (Xiying Xie) and R.L.; resources, Z.Z.; data curation, Y.C.; writing—original draft preparation, Q.T.; writing—review and editing, Y.L.; supervision, C.C.; funding acquisition, Z.L. and Y.L. All authors have read and agreed to the published version of the manuscript.

Funding: This research was funded by the National Natural Science Foundation of China (31572088, 31672127), the Special fund for scientific and technological innovation of Fujian Agricultural and Forestry University Applied Basic Research (CXZX2017189 and CXZX2017314), the Natural Science Foundation of Fujian Province (2020N0061 and 2021J05199), and the startup fund of Minnan Normal University (4206-L21715).

Data Availability Statement: The RNA sequencing data have been uploaded to the NCBI (National Center for Biotechnology Information) SRA (Sequence Read Archive) under accession number PRJNA607013.

Conflicts of Interest: The authors declare no conflicts of interest.

References

1. Lin, Y.; Min, J.; Lai, R.; Wu, Z.; Chen, Y.; Yu, L.; Cheng, C.; Jin, Y.; Tian, Q.; Liu, Q.; et al. Genome-wide sequencing of longan (*Dimocarpus longan* Lour.) provides insights into molecular basis of its polyphenol-rich characteristics. *Gigascience* **2017**, *6*, gix023. [CrossRef] [PubMed]
2. Jue, D.; Liu, L.; Sang, X.; Shu, B.; Wang, J.; Wang, Y.; Zhang, C.; Shi, S. SNP-based high-density genetic map construction and candidate gene identification for fruit quality traits of *Dimocarpus longan* Lour. *Sci. Hortic.* **2021**, *284*, 110086. [CrossRef]
3. Xu, X.; Zhang, C.; Xu, X.; Cai, R.; Guan, Q.; Chen, X.; Chen, Y.; Zhang, Z.; XuHan, X.; Lin, Y.; et al. Riboflavin mediates m6A modification targeted by miR408, promoting early somatic embryogenesis in longan. *Plant Physiol.* **2023**, *192*, 1799–1820. [CrossRef] [PubMed]
4. Salus, S.S.; Demeter, J.; Sazer, S. The Ran GTPase system in fission yeast affects microtubules and cytokinesis in cells that are competent for nucleocytoplasmic protein transport. *Mol. Cell. Biol.* **2002**, *22*, 8491–8505. [CrossRef] [PubMed]
5. Scrofani, J.; Sardon, T.; Meunier, S.; Vernos, I. Microtubule Nucleation in Mitosis by a RanGTP-Dependent Protein Complex. *Curr. Biol.* **2015**, *25*, 131–140. [CrossRef] [PubMed]
6. Wang, X.; Xu, Y.; Han, Y.; Bao, S.; Du, J.; Yuan, M.; Xu, Z.; Chong, K. Overexpression of RAN1 in rice and *Arabidopsis* alters primordial meristem, mitotic progress, and sensitivity to auxin. *Plant Physiol.* **2006**, *140*, 91–101. [CrossRef]
7. Malambane, G.; Madumane, K.; Sewelo, L.T.; Batlang, U. Drought stress tolerance mechanisms and their potential common indicators to salinity, insights from the wild watermelon (*Citrullus lanatus*): A review. *Front. Plant Sci.* **2022**, *13*, 1074395. [CrossRef]
8. Zang, A.; Xu, X.; Neill, S.; Cai, W. Overexpression of OsRAN2 in rice and *Arabidopsis* renders transgenic plants hypersensitive to salinity and osmotic stress. *J. Exp. Bot.* **2010**, *61*, 777–789. [CrossRef]
9. Chen, N.; Xu, Y.; Wang, X.; Du, C.; Du, J.; Yuan, M.; Xu, Z.; Chong, K. OsRAN2, essential for mitosis, enhances cold tolerance in rice by promoting export of intranuclear tubulin and maintaining cell division under cold stress. *Plant Cell Environ.* **2011**, *34*, 52–64. [CrossRef]
10. Xu, P.; Ma, W.; Liu, J.; Hu, J.; Cai, W. Overexpression of a small GTP-binding protein Ran1 in *Arabidopsis* leads to promoted elongation growth and enhanced disease resistance against *P. syringae* DC3000. *Plant J.* **2021**, *108*, 977–991. [CrossRef]
11. Yan, J.J.; Xie, B.; Zhang, L.; Li, S.J.; van Peer, A.F.; Wu, T.J.; Chen, B.Z.; Xie, B.G. Small GTPases and Stress Responses of *vvran1* in the Straw Mushroom *Volvariella volvacea*. *Int. J. Mol. Sci.* **2016**, *17*, 1527. [CrossRef]
12. Fang, Z.-Z.; Zhang, Y.-L.; Lai, C.C.; Lai, Z. Developmental regulation of Ran 3untranslated region during somatic embryogenesis in *Dimocarpus longan* Lour. *Sci. Hortic.* **2014**, *176*, 297–302. [CrossRef]
13. Tian, Q.; Lin, Y.; Lai, Z.; Wang, T. Cloning and Bioinformatic Analysis of the Promoters of DlRan3A and DlRan3B from Embryogenic Callus in *Dimocarpus longan*. *Chin. J. Trop. Crops* **2014**, *35*, 82–89.
14. Tian, Q.; Lin, Y.; Yang, M.; Zhang, D.; Lai, R.; Lai, Z. DlRan3A is involved in hormone, light, and abiotic stress responses in embryogenic callus of *Dimocarpus longan* Lour. *Gene* **2015**, *569*, 267–275. [CrossRef]
15. Tian, Q.; Lin, Y.; Zhang, D.; Lai, R.; Lai, Z. Ras-Related Nuclear Protein Ran3B Gene Is Involved in Hormone Responses in the Embryogenic Callus of *Dimocarpus longan* Lour. *Int. J. Mol. Sci.* **2016**, *17*, 873. [CrossRef]
16. Chen, Y.; Xu, X.; Chen, X.; Chen, Y.; Zhang, Z.; Xuhan, X.; Lin, Y.; Lai, Z. Seed-Specific Gene MOTHER of FT and TFL1(MFT) Involved in Embryogenesis, Hormones and Stress Responses in *Dimocarpus longan* Lour. *Int. J. Mol. Sci.* **2018**, *19*, 2403. [CrossRef] [PubMed]
17. Ach, R.A.; Gruissem, W. A small nuclear GTP-binding protein from tomato suppresses a Schizosaccharomyces pombe cell-cycle mutant. *Proc. Natl. Acad. Sci. USA* **1994**, *91*, 5863–5867. [CrossRef] [PubMed]
18. Haizel, T.; Merkle, T.; Pay, A.; Fejes, E.; Nagy, F. Characterization of proteins that interact with the GTP-bound form of the regulatory GTPase Ran in *Arabidopsis*. *Plant J.* **1997**, *11*, 93–103. [CrossRef]

19. Lü, S.H.; Fan, Y.L.; Jin, C.X. Overexpression of a Ran GTPase homologous gene, FaRan from tall fescue, in transgenic *Arabidopsis*. *Biol. Plant.* **2011**, *55*, 331–334. [CrossRef]
20. Huang, N.; Ling, H.; Zhang, X.; Mao, H.; Su, Y.; Su, W.; Liu, F.; Xu, L.; Chen, R.; Que, Y. A Small GTP-Binding Gene ScRan from Sugarcane is Involved in Responses to Various Hormone Stresses and *Sporisirium scitamineum* Challenge. *Sugar Technol.* **2018**, *20*, 669–680. [CrossRef]
21. Reynolds, J.P. Characterisation of the Defective Embryo and Meristems (DEM) Genes of Plants. Ph.D. Thesis, The University of Queensland, Townsville, QLD, Australia, 2013.
22. Liu, P.; Qi, M.; Wang, Y.; Chang, M.; Liu, C.; Sun, M.; Yang, W.; Ren, H. *Arabidopsis* RAN1 mediates seed development through its parental ratio by affecting the onset of endosperm cellularization. *Mol. Plant* **2014**, *7*, 1316–1328. [CrossRef] [PubMed]
23. Wolf, S.; Hématy, K.; Höfte, H. Growth control and cell wall signaling in plants. *Annu. Rev. Plant Biol.* **2012**, *63*, 381–407. [CrossRef] [PubMed]
24. Malinowski, R.; Filipecki, M.; Tagashira, N.; Wiśniewska, A.; Gaj, P.; Plader, W.; Malepszy, S. Xyloglucan endotransglucosylase/hydrolase genes in cucumber (*Cucumis sativus*)-differential expression during somatic embryogenesis. *Physiol. Plant.* **2004**, *120*, 678–685. [CrossRef] [PubMed]
25. Hall, Q.; Cannon, M.C. The cell wall hydroxyproline-rich glycoprotein RSH is essential for normal embryo development in *Arabidopsis*. *Plant Cell* **2002**, *14*, 1161–1172. [CrossRef] [PubMed]
26. Cannon, M.C.; Terneus, K.; Hall, Q.; Tan, L.; Wang, Y.; Wegenhart, B.L.; Chen, L.; Lamport, D.T.; Chen, Y.; Kieliszewski, M.J. Self-assembly of the plant cell wall requires an extensin scaffold. *Proc. Natl. Acad. Sci. USA* **2008**, *105*, 2226–2231. [CrossRef] [PubMed]
27. Liu, Q.; Wu, K.; Harberd, N.P.; Fu, X. Green Revolution DELLAs: From translational reinitiation to future sustainable agriculture. *Mol. Plant* **2021**, *14*, 547–549. [CrossRef] [PubMed]
28. Budot, B.O.; Encabo, J.R.; Ambita, I.D.; Atienza-Grande, G.A.; Satoh, K.; Kondoh, H.; Ulat, V.J.; Mauleon, R.; Kikuchi, S.; Choi, I.R. Suppression of cell wall-related genes associated with stunting of *Oryza glaberrima* infected with *Rice tungro spherical virus*. *Front. Microbiol.* **2014**, *5*, 26. [CrossRef] [PubMed]
29. Choi, D.; Lee, Y.; Cho, H.T.; Kende, H. Regulation of expansin gene expression affects growth and development in transgenic rice plants. *Plant Cell* **2003**, *15*, 1386–1398. [CrossRef]
30. Ilias, I.A.; Negishi, K.; Yasue, K.; Jomura, N.; Morohashi, K.; Baharum, S.N.; Goh, H.H. Transcriptome-wide effects of expansin gene manipulation in etiolated *Arabidopsis* seedling. *J. Plant Res.* **2019**, *132*, 159–172. [CrossRef]
31. Koevoets, I.T.; Venema, J.H.; Elzenga, J.T.; Testerink, C. Roots Withstanding their Environment: Exploiting Root System Architecture Responses to Abiotic Stress to Improve Crop Tolerance. *Front. Plant Sci.* **2016**, *7*, 1335. [CrossRef]
32. Lei, M.J.; Wang, Q.; Li, X.; Chen, A.; Luo, L.; Xie, Y.; Li, G.; Luo, D.; Mysore, K.S.; Wen, J.; et al. The small GTPase ROP10 of Medicago truncatula is required for both tip growth of root hairs and nod factor-induced root hair deformation. *Plant Cell* **2015**, *27*, 806–822. [CrossRef]
33. Kang, E.; Zheng, M.; Zhang, Y.; Yuan, M.; Yalovsky, S.; Zhu, L.; Fu, Y. The Microtubule-Associated Protein MAP18 Affects ROP2 GTPase Activity during Root Hair Growth. *Plant Physiol.* **2017**, *174*, 202–222. [CrossRef]
34. Bucher, M.; Brunner, S.; Zimmermann, P.; Zardi, G.I.; Amrhein, N.; Willmitzer, L.; Riesmeier, J.W. The expression of an extensin-like protein correlates with cellular tip growth in tomato. *Plant Physiol.* **2002**, *128*, 911–923. [CrossRef] [PubMed]
35. Kwasniewski, M.; Janiak, A.; Mueller-Roeber, B.; Szarejko, I. Global analysis of the root hair morphogenesis transcriptome reveals new candidate genes involved in root hair formation in barley. *J. Plant Physiol.* **2010**, *167*, 1076–1083. [CrossRef] [PubMed]
36. Velasquez, S.M.; Ricardi, M.M.; Dorosz, J.G.; Fernandez, P.V.; Nadra, A.D.; Pol-Fachin, L.; Egelund, J.; Gille, S.; Harholt, J.; Ciancia, M.; et al. O-glycosylated cell wall proteins are essential in root hair growth. *Science* **2011**, *332*, 1401–1403. [CrossRef] [PubMed]
37. Eberle, C.A.; Anderson, N.O.; Clasen, B.M.; Hegeman, A.D.; Smith, A.G. PELPIII: The class III pistil-specific extensin-like Nicotiana tabacum proteins are essential for interspecific incompatibility. *Plant J.* **2013**, *74*, 805–814. [CrossRef] [PubMed]
38. Alves, C.M.L.; Noyszewski, A.K.; Smith, A.G. Structure and function of class III pistil-specific extensin-like protein in interspecific reproductive barriers. *BMC Plant Biol.* **2019**, *19*, 118. [CrossRef] [PubMed]
39. Ralevski, A. The Interaction of Early Salt Stress-Induced 2 (ESI2) and the Ran G Protein in *Arabidopsis*. Master's Thesis, Concordia University, Montreal, QC, Canada, 2013.
40. Xu, P.; Cai, W. RAN1 is involved in plant cold resistance and development in rice (*Oryza sativa*). *J. Exp. Bot.* **2014**, *65*, 3277–3287. [CrossRef] [PubMed]
41. Karuppanapandian, T.; Rhee, S.-J.; Kim, E.-J.; Han, B.K.; Hoekenga, O.A.; Lee, G.P. Proteomic analysis of differentially expressed proteins in the roots of Columbia-0 and Landsberg erecta ecotypes of *Arabidopsis thaliana* in response to aluminum toxicity. *Can. J. Plant Sci.* **2012**, *92*, 1267–1282. [CrossRef]
42. Kagale, S.; Divi, U.K.; Krochko, J.E.; Keller, W.A.; Krishna, P. Brassinosteroid confers tolerance in *Arabidopsis thaliana* and *Brassica napus* to a range of abiotic stresses. *Planta* **2007**, *225*, 353–364. [CrossRef]
43. Xia, X.-J.; Wang, Y.-J.; Zhou, Y.-H.; Tao, Y.; Mao, W.-H.; Shi, K.; Asami, T.; Chen, Z.; Yu, J.-Q. Reactive Oxygen Species Are Involved in Brassinosteroid-Induced Stress Tolerance in Cucumber. *Plant Physiol.* **2009**, *150*, 801–814. [CrossRef]
44. Liu, Y.; Jiang, H.; Zhao, Z.; An, L. Abscisic acid is involved in brassinosteroids-induced chilling tolerance in the suspension cultured cells from *Chorispora bungeana*. *J. Plant Physiol.* **2011**, *168*, 853–862. [CrossRef]

45. Fu, J.; Sun, P.; Luo, Y.; Zhou, H.; Gao, J.; Zhao, D.; Pubu, Z.; Liu, J.; Hu, T. Brassinosteroids enhance cold tolerance in Elymus nutans via mediating redox homeostasis and proline biosynthesis. *Environ. Exp. Bot.* **2019**, *167*, 103831. [CrossRef]
46. Arfan, M.; Zhang, D.W.; Zou, L.J.; Luo, S.S.; Tan, W.R.; Zhu, T.; Lin, H.H. Hydrogen Peroxide and Nitric Oxide Crosstalk Mediates Brassinosteroids Induced Cold Stress Tolerance in *Medicago truncatula*. *Int. J. Mol. Sci.* **2019**, *20*, 144. [CrossRef] [PubMed]
47. Sánchez, M.A.; Mateos, I.; Labrador, E.; Dopico, B. Brassinolides and IAA induce the transcription of four alpha-expansin genes related to development in *Cicer arietinum*. *Plant Physiol. Biochem.* **2004**, *42*, 709–716. [CrossRef] [PubMed]
48. Lisso, J.; Steinhauser, D.; Altmann, T.; Kopka, J.; Müssig, C. Identification of brassinosteroid-related genes by means of transcript co-response analyses. *Nucleic Acids Res.* **2005**, *33*, 2685–2696. [CrossRef] [PubMed]
49. Divi, U.K.; Rahman, T.; Krishna, P. Gene expression and functional analyses in brassinosteroid-mediated stress tolerance. *Plant Biotechnol. J.* **2016**, *14*, 419–432. [CrossRef] [PubMed]
50. Orman-Ligeza, B.; Parizot, B.; de Rycke, R.; Fernandez, A.; Himschoot, E.; Van Breusegem, F.; Bennett, M.J.; Perilleux, C.; Beeckman, T.; Draye, X. RBOH-mediated ROS production facilitates lateral root emergence in *Arabidopsis*. *Development* **2016**, *143*, 3328–3339. [CrossRef] [PubMed]
51. Verma, P.; Sanyal, S.K.; Pandey, G.K. Ca^{2+}–CBL–CIPK: A modulator system for efficient nutrient acquisition. *Plant Cell Rep.* **2021**, *40*, 2111–2122. [CrossRef] [PubMed]
52. Qu, Y.; Yan, M.; Zhang, Q. Functional regulation of plant NADPH oxidase and its role in signaling. *Plant Signal. Behav.* **2017**, *12*, 1356970. [CrossRef] [PubMed]
53. Lin, Y.L.; Lai, Z.X. Reference gene selection for qPCR analysis during somatic embryogenesis in longan tree. *Plant Sci.* **2010**, *178*, 359–365. [CrossRef]
54. Feng, X. Genome-Wide Identification and Function Analysis of the Superoxide Dismutase Gene Family in *Musa* spp. Ph.D. Thesis, Fujian Agriculture and Forestry University, Fuzhou, China, 2016.
55. Cui, Y.; Zhu, Y.; Lin, Y.; Chen, L.; Feng, Q.; Wang, W.; Xiang, H. New insight into the mechanism underlying the silk gland biological process by knocking out fibroin heavy chain in the silkworm. *BMC Genome* **2018**, *19*, 215. [CrossRef] [PubMed]
56. Chen, C.; Chen, H.; Zhang, Y.; Thomas, H.R.; Frank, M.H.; He, Y.; Xia, R. TBtools: An Integrative Toolkit Developed for Interactive Analyses of Big Biological Data. *Mol. Plant* **2020**, *13*, 1194–1202. [CrossRef]
57. Pérez-Rodríguez, P.; Riaño-Pachón, D.M.; Corrêa, L.G.; Rensing, S.A.; Kersten, B.; Mueller-Roeber, B. PlnTFDB: Updated content and new features of the plant transcription factor database. *Nucleic Acids Res.* **2010**, *38*, D822–D827. [CrossRef]
58. Jin, J.; Zhang, H.; Kong, L.; Gao, G.; Luo, J. PlantTFDB 3.0: A portal for the functional and evolutionary study of plant transcription factors. *Nucleic Acids Res.* **2014**, *42*, D1182–D1187. [CrossRef]
59. Zhao, W.; Yang, X.; Yu, H.; Jiang, W.; Sun, N.; Liu, X.; Liu, X.; Zhang, X.; Wang, Y.; Gu, X. RNA-Seq-based transcriptome profiling of early nitrogen deficiency response in cucumber seedlings provides new insight into the putative nitrogen regulatory network. *Plant Cell Physiol.* **2015**, *56*, 455–467. [CrossRef] [PubMed]

plants

MDPI

Article

Bioinformatic Analysis of Codon Usage Bias of HSP20 Genes in Four Cruciferous Species

Huiyue Ji, Junnan Liu, Yineng Chen, Xinyi Yu, Chenlu Luo, Luxi Sang, Jiayu Zhou * and Hai Liao *

School of Life Science and Engineering, Southwest Jiaotong University, Chengdu 610031, China; jhywork123@163.com (H.J.); liujunnan2001623@icloud.com (J.L.); chenyineng1028@163.com (Y.C.); yuxinyi205@163.com (X.Y.); chenluuu525@163.com (C.L.); sangluxi8671@163.com (L.S.)
* Correspondence: spinezhou@home.swjtu.edu.cn (J.Z.); ddliaohai@home.swjtu.edu.cn (H.L.); Tel.: +86-28-87603202 (H.L.); Fax: +86-28-87603202 (H.L.)

Abstract: Heat shock protein 20 (HSP20) serves as a chaperone and plays roles in numerous biological processes, but the codon usage bias (CUB) of its genes has remained unexplored. This study identified 140 *HSP20* genes from four cruciferous species, *Arabidopsis thaliana*, *Brassica napus*, *Brassica rapa*, and *Camelina sativa*, that were identified from the Ensembl plants database, and we subsequently investigated their CUB. As a result, the base composition analysis revealed that the overall GC content of *HSP20* genes was below 50%. The overall GC content significantly correlated with the constituents at three codon positions, implying that both mutation pressure and natural selection might contribute to the CUB. The relatively high ENc values suggested that the CUB of the *HSP20* genes in four cruciferous species was relatively weak. Subsequently, ENc exhibited a negative correlation with gene expression levels. Analyses, including ENc-plot analysis, neutral analysis, and PR2 bias, revealed that natural selection mainly shaped the CUB patterns of *HSP20* genes in these species. In addition, a total of 12 optimal codons (ΔRSCU > 0.08 and RSCU > 1) were identified across the four species. A neighbor-joining phylogenetic analysis based on coding sequences (CDS) showed that the 140 *HSP20* genes were strictly and distinctly clustered into 12 subfamilies. Principal component analysis and cluster analysis based on relative synonymous codon usage (RSCU) values supported the fact that the CUB pattern was consistent with the genetic relationship at the gene level and (or) species levels. These results will not only enrich the *HSP20* gene resource but also advance our understanding of the CUB of *HSP20* genes, which may underlie the theoretical basis for exploration of their genetic and evolutionary pattern.

Keywords: Brassicaceae; codon usage bias; *HSP20* genes; mutation pressure; natural selection

1. Introduction

The genetic code is the coding information of mRNA and uses 61 codons to encode the 20 amino acids, in addition to three stop codons UAG, UGA, and UAA. Among the 20 amino acids, Met and Trp are encoded by the single codons AUG and UGG, respectively, while the remaining 18 amino acids are encoded by at least two codons, resulting in several synonymous codons. However, the choice of synonymous codons is not, at present, at uniform frequencies in the gene transcript to encode all amino acids, a phenomenon called codon usage bias (CUB) [1]. Prior researchers have identified a variety of factors contributing to the CUB patterns, including natural selection, mutation pressure, genomic structure, gene length, structure of anticodon binding domain in tRNA, and gene expressional levels [2,3]. To date, the usage frequency of synonymous codons has been revealed to be unequal across various species, between nuclei and organelle genes, and even within genes [4]. Notably, it was evident that genes in monocots and dicots exhibited a significant difference in synonymous codon usage [5] characterized by higher AT content in dicots and higher GC content in monocots, implying a connection between CUB and the evolutionary history of organisms [6]. Furthermore, CUB analysis has shed light on the evolutionary adaptation of virus to plant host [7] and the domestication of cultivated crops [8]. CUB was also a predictor of gene expression levels, as a high CUB was concluded to be associated with high gene expression levels [9,10]. For instance, CUB has been utilized in developing a live-cell monitoring system to evaluate the bioavailability of compounds in cyanobacteria, since the cells containing preferred-codon genes showed more robust growth over those containing rare-codon genes [11]. In addition, the high CUB is conducive to improving the stability of mRNA [12], aiding in protein function prediction [13], and enhancing the global translation rates [14]. An approach involving the substitution of rare codons with preferred synonymous codons, known as codon optimization, is a promising method for boosting target protein production, especially in heterologous hosts with a different CUB [14]. At present, CUB analyses based on plant genomes are highly available due to the extensive sequencing of nuclei and organellar genomes, leading to valid and credible results [4,15]. Accumulating advances based on the genomes revealed that several factors, in particular natural selection, played a significant role in CUB [10,16,17]. Conversely, the significance of CUB in gene families has not been well studied. A gene family, composed of genes derived from common ancestral genes, shows diverse functions in the regulation of plant growth and development, the participation of hormone signal transduction, and the response to biotic and abiotic stresses [18]. Therefore, it will be of great significance to investigate the CUB of gene families by which their evolutionary history and complex expressional patterns might be understood.

Among plant gene families, heat shock protein 20 (HSP20) is particularly attractive for its ubiquitous presence in both prokaryotes and eukaryotes [19]. HSP20 acts as a chaperone and is involved not only in plant morphogenesis and physiological growth but also in responses to various environmental stresses [20]. These proteins are relatively small, with a molecular weight spanning from 14 to 42 kDa. HSP20 is composed of three domains: an N-terminal domain that binds to the target protein, an intermediate α-crystallin domain that enhances the stability of the target protein, and a short C-terminal extension domain facilitating homologous oligomerization [21]. In accordance with its subcellular localization, sequence identity, and phylogenetic relationship, HSP20 can be divided into at least 17 subfamilies [22]. Subfamilies CI-CXII, MI, MII, P, ER, and Px are localized to the nucleus/cytoplasm, mitochondria, mitochondria, plastids, endoplasmic reticulum, and peroxisomes, respectively. To date, the reported HSP20 families from *Coix lacryma-jobi* [23], *Cucumis sativus* [22], *Glycine max* [24], and *Solanum lycopersicum* [25] have been reported to consist of 32, 30, 51, and 42 putative members, respectively. Phylogenetic analyses revealed that HSP20 in the same species was divided into different clades, implying that the duplication that led to these different clades occurred before the divergence of these species [24]. Selective pressure during the evolutionary process might be a driving factor influencing the expression abundance of HSP20 [26]. Furthermore, the expression patterns of various HSP20 members were diversified under different conditions, such as drought, salinity, and heat [25]. More importantly, heterologous expression of *LimHSP16.45* can enhance the cell viability of the latter under high temperature, high salt, and oxidative stress [27], while the overexpression of *AsHSP17* increases salt sensitivity during germination and post-germinative growth [28]. Given that HSP20 confers resistance to various environmental stresses, improving the expression efficiency of *HSP20* genes using codon optimization might be an important method to develop crops that are resistant to adverse environmental stresses. Overall, these traits made the *HSP20* gene family a model for CUB research.

The *Brassicaceae* family, encompassing a wide variety of vegetable and ornamental plants, is rich in genetic diversity, with significant potential for improving food crops. Specifically, *Brassica napus* and *Brassica juncea*, used as vegetable foods for millions of people, are rich in carotenoids, nutrients, and vitamins [29]. *Camelina sativa*, known for its strong tolerance to environmental stress and easy cultivation, presents many beneficial properties [30,31]. *Arabidopsis thaliana*, acting as a model plant, has been widely used in basic botanical research [32]. Currently, several efforts are focused on the development of *Brassicaceae* varieties that endure biotic and abiotic stresses but possess higher yields. In this study, *HSP20* genes in the nuclei genomes of *Arabidopsis thaliana*, *Brassica napus*, *Brassica rapa*, and *Camelina sativa* were wholly identified. Several CUB parameters, including ENc, RSCU, CAI, CBI, and Fop, were estimated to understand the CUB patterns of *HSP20* genes in different cruciferous species [33], which is different from previous studies. To our knowledge, this is the first study to explore the factors shaping the CUB of *HSP20* genes in various cruciferous species and might provide information for improving genetic breeding on the basis of *HSP20* genes.

2. Results

2.1. Identification and Nucleotide Composition of HSP20 Genes

A total of 140 nonredundant *Hsp20* genes were identified from four cruciferous species, comprising 27 genes from *A. thaliana*, 48 genes from *B. napus*, 36 genes from *B. rapa*, and 29 genes from *C. sativa* (Table S1). The CDS lengths of the *HSP20* genes ranged from 393 bp (*Csa19g056510*) to 1116 bp (*Csa20g006100*), with a maximum difference of 723 bp, and encoded proteins ranging from 130 to 371 amino acids in length. In addition, the predicted isoelectric point (pI) of HSP20 varied from 4.37 (*Csa19g056510*) to 9.78 (*CDY70225*), with 44.29% having a pI value greater than 7, indicating that more than half of the HSP20 proteins were neutral or acidic.

The A base is the richest base in the 140 *HSP20* genes, with an average content of 29.39%, followed by the G base, which averaged 26.60%. The average GC content of the 140 *HSP20* genes is 47.4% and ranged from 36.2% to 58.2%, with the lowest GC content contained in *CDY45944* and *Bra008274* and the highest in *Bra024931*. In terms of species, *HSP20* genes in *B. napus* have the highest GC content (48.1%), while those in *A. thaliana* have the lowest (46.5%).

2.2. Codon Usage Indices of HSP20 Genes

As shown in Figure 1, the average GC1, GC2, and GC3 contents of 140 *HSP20* genes from four species were 53.06%, 38.69% and 50.20%, respectively. The GC1 content across the four species varied from 42.31% to 60.24%, while the GC2 content of the four species varied from 24.61% to 48.28%. The average GC3s value of 48.5% indicated that the *HSP20* genes preferred an A or T base as the third base. Furthermore, the indices of A3s, T3s, G3s, and C3s indicated that *HSP20* genes tended to end with a T base, followed by the G, C, and A bases, with respective values of 35.2%, 33.1%, 29.3%, and 29.2%. The CAI values in 140 *HSP20* genes ranged from 0.154 (*CDX79100*) to 0.318 (*Bra024931*), with a mean value of 0.225. As the expression level of genes and preference often showed a positive correlation, the CAI value of *HSP20* genes indicated the relatively low expression levels of *HSP20* genes. Among the four species, the *HSP20* genes in *C. sativa* had the highest CAI values, followed by those of *A. thaliana*, *B. napus*, and *B. rapa*, implying that those in *C. sativa* had relatively high expression levels. The CBI values of global *HSP20* genes ranged from −0.177 (*CDY70473*) to 0.304 (*Csa11g019390*), with a mean CBI value of 0.033. Among the four species, *A. thaliana* had the lowest CBI value of 0.018, while *C. sativa* had the highest CBI value of 0.048. With respect to Fop, its ranging region was from 0.327 (*Csa05g059340*) to 0.591 (*Csa11g019390*), with an average of 0.437. In terms of species, *C. sativa* showed the highest Fop value (0.446), whereas *A. thaliana* showed the lowest (0.429). Additionally, the ENc values were calculated to be from 37.900 (*Bra018383*) to 61.000 (*Bra022078*, *Bra028253*, and *CDY00476*), with an average of 51.814. Regarding the four species, *HSP20* genes in *B. rapa* were considered to have the strongest CUB due to the lowest ENc value of 51.344, while those in *B. napus* showed the weakest CUB due to the highest ENc value of 52.310. The negative GRAVY values of *HSP20* genes in the four species confirmed their hydrophilic nature, suggesting their unlike location in membrane structure. The aromaticity (AROMO) values, representing the usage frequency of Phe, Tyr, and Trp residues, varied from 0.033 to

0.136, with a mean value of 0.069, confirming that the composition of aromatic amino acid residues varied little.

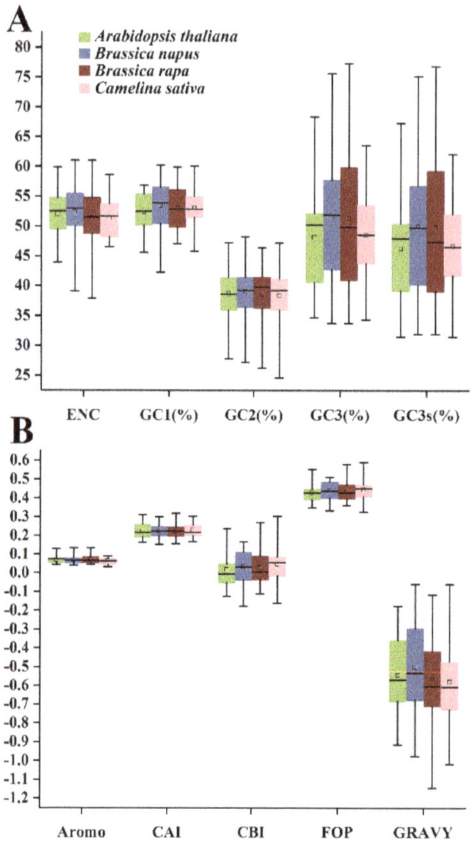

Figure 1. The codon usage indices of *HSP20* genes in four cruciferous species. (**A**) ENc, GC1, GC2, GC3, and GC3s indicators. (**B**) Aromo, CAI, CBI, FOP, and Gravy indicators.

2.3. ENc-GC3s Plot Analysis

To reflect the correlation between the ENc and GC3s values, a correlation curve was generated. According to Table S1, the ENc values of *HSP20* genes varied widely among genes, with an average of 51.814, indicating a generally low CUB of *HSP20* genes in the four studied species. Statistically, no *HSP20* gene had low ENc values (≤35) in the four species with a threshold value of strong bias, while 94 genes had high ENc values (50–61) with weak bias. Given that strong- and weak-bias genes were 0 and 67.14%, respectively, in the four species, the CUB of *HSP20* genes was determined to be relatively low.

As shown in Figure 2, most of the genes resided below the expected curve (blackline), except that a few genes of *B. napus* and *B. rapa* were situated above and (or) on the expected curve. The expected curve denotes the expected position of genes where the CUB was governed solely by mutation pressure, while the CUB of those genes distributed below the expected curve was mainly determined by natural selection. Consequently, natural selection was inferred as the primary influencer of the CUB pattern of *HSP20* genes in this study, a finding reflected in similar analyses of apple, *B. napus*, and citrus [4,34,35].

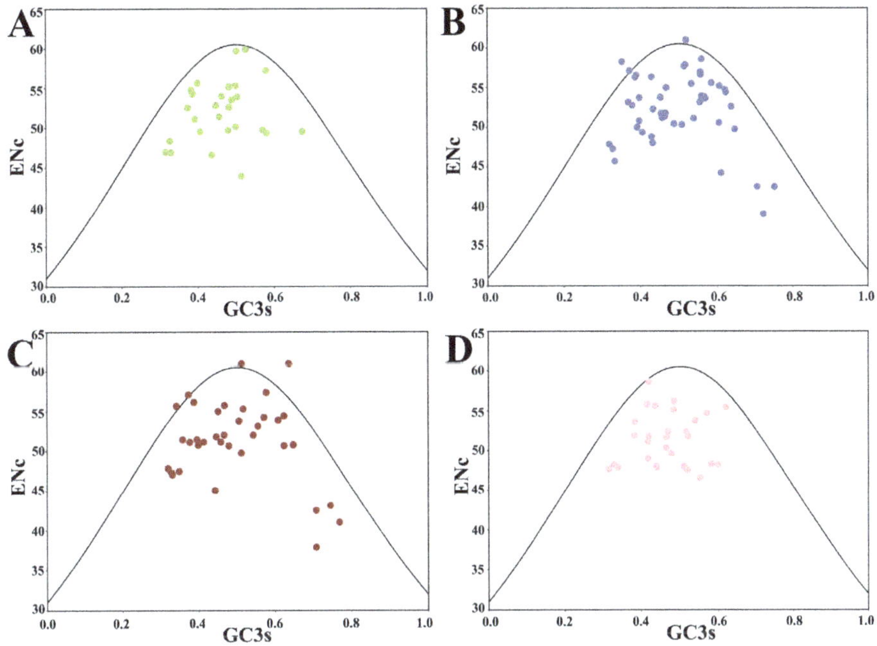

Figure 2. ENc-GC3s plot of *HSP20* genes in four species. The black line represents the expected curve when the codon usage bias is determined only by mutation pressure. (**A**) *A. thaliana*; (**B**) *B. napus*; (**C**) *B. rapa*; and (**D**) *C. sativa*.

2.4. PR2-Plot Analysis

Parity rule 2 (PR2) plots were conducted to ascertain any inequality between pyrimidines and purines at the third position (Figure 3). The plot, centered on 0.5, was divided into four quadrants. In case the mutation pressure alone affects codon bias, the occurrence of A/T or C/G at codon position three is equal, while natural selection pressure leads to the unequal usage of A/T or C/G at the third position [36]. As shown in Figure 3, the *HSP20* genes of the four cruciferous species were not evenly distributed in the four regions, indicating that strong bias played a role in the composition of the third codon position. Over half of the genes were located in the lower region (A3/(A3 + T3) < 0.5), revealing the preponderance of T over A. Similarly, more than half of the genes were located in the right region (G3/(G3 + C3) > 0.5), demonstrating that the occurrence of G in the third position was more frequent than that of C. Interestingly, a few dots in the upper left quadrant suggested a distant codon usage preference in *HSP20* genes.

2.5. Neutrality Plot

The relationships between GC12 and GC3 in the four species were separately examined using a neutrality plot (Figure 4). The content of GC12 ranged from 36.86% to 52.64%, while that of GC3 ranged from 33.74% to 77.33%. A weak positive correlation between GC12 and GC3 was observed in the four species, with R values ranging from 0.197 to 0.282, suggesting that CUB was affected by both natural selection and mutation pressure. Additionally, the contributing extents of mutation pressure and natural selection on CUB can be evaluated using a neutrality plot [2,15]. As a result, the small regression coefficients, varying from 0.0522 to 0.1034, indicated that mutation pressure contributed less than 10.34%. Conversely, the remaining 89.66% of the bias was attributed to natural selection. Therefore, these results strongly underscored the dominant role of natural selection in the CUB pattern, but mutation pressure plays a minor role in determining the CUB of *HSP20* genes in the four cruciferous species. Similar results have been noted in seven Rosales species [37].

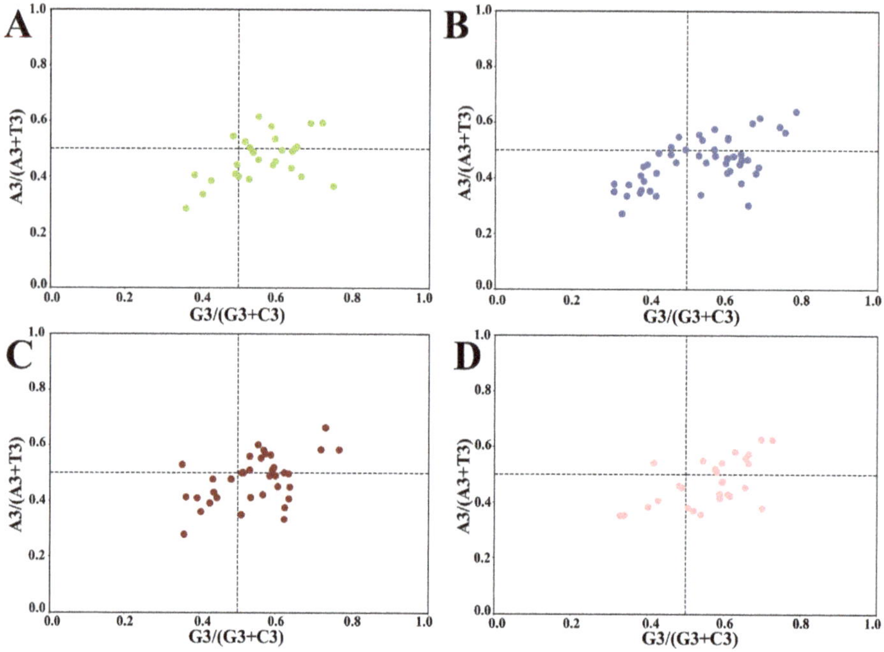

Figure 3. Parity rule 2 plot analysis of HSP20 genes in four species (GC bias on the x-axis and AT bias on the y-axis). The center point at 0.5 represents A = T and G = C, which means that there is no codon usage deviation between the two DNA strands. (**A**) *A. thaliana*; (**B**) *B. napus*; (**C**) *B. rapa*; and (**D**) *C. sativa*.

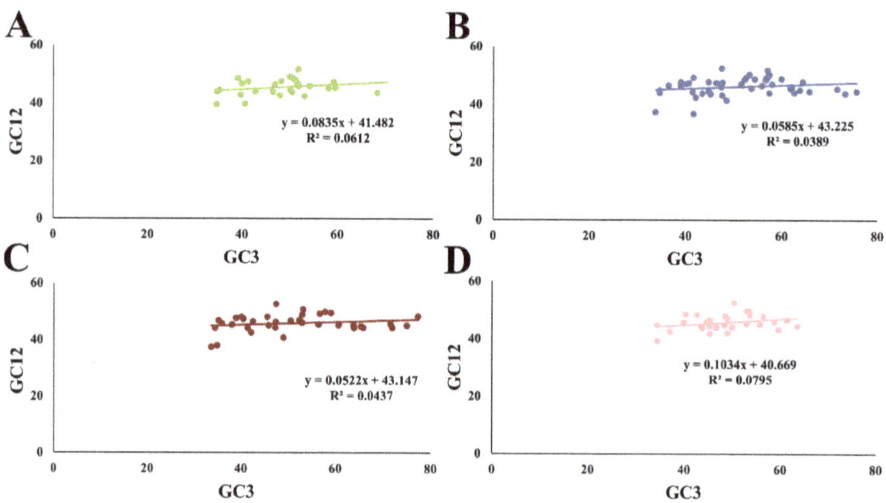

Figure 4. Neutrality plot between GC12 (the mean GC content at the first and second positions) and GC3 (GC content at the third codon position). (**A**) *A. thaliana*; (**B**) *B. napus*; (**C**) *B. rapa*; and (**D**) *C. sativa*.

2.6. Correlation Coefficient Analysis of Codon Usage Indices

It was observed that the CBI showed a highly significant correlation with the CAI and Fop indices in all species ($p < 0.001$) (Figure 5). Likewise, C3s showed a highly significant correlation with FOP, CBI, and CAI ($p < 0.001$), except for the C3s and CAI in *C. sativa* ($p < 0.05$). Intriguingly, a significant correlation between the GC content and GC3 in all species was evident ($p < 0.001$), indicating that the GC composition of *HSP20* genes could markedly affect GC3 content substantially. GC content also exhibited a correlation with GC1 and GC2 content in *A. thaliana, B. napus, B. rapa*, and *C. sativa*, confirming that the GC composition also could affect the GC1 and GC2 content to some extent (Figure 5A–D). Meanwhile, the GC content exhibited a highly significant correlation with T3s, C3s, and A3s across all species ($p < 0.001$). Except for the G3s in *A. thaliana* and *C. sativa*, those in the other two species correlated with the GC content ($p < 0.05$), implying that the GC composition impacted the base at the third position (Figure 5). We also found that GC3, A3s, T3s, C3s, and G3s were extremely significant in the four species ($p < 0.001$).

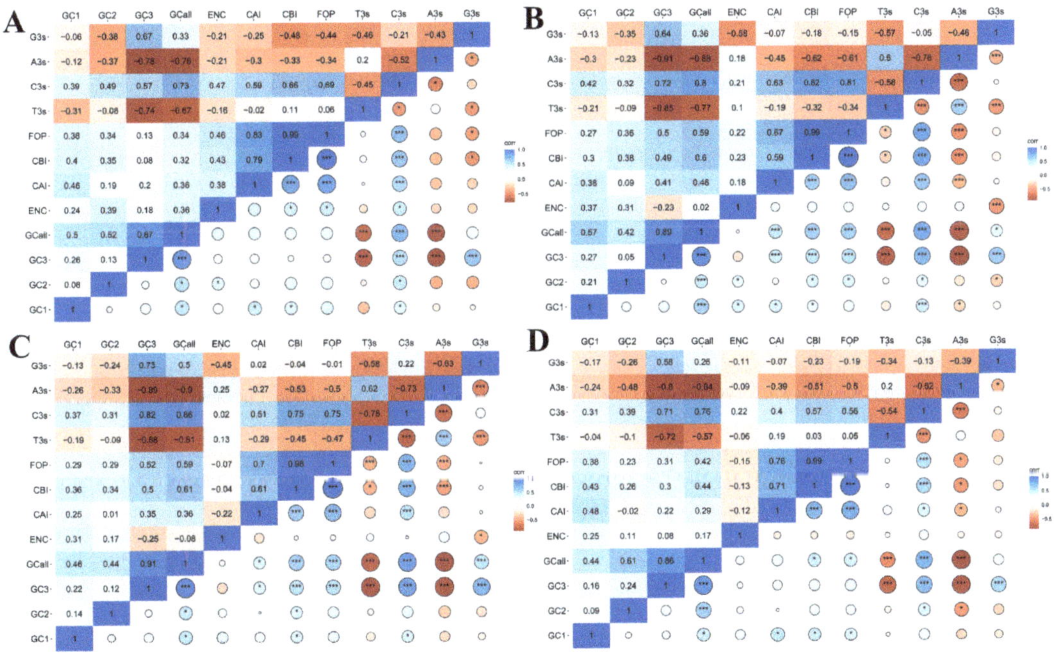

Figure 5. Pearson correlation analysis of codon usage indices in four cruciferous species. The color changing from red to blue represents an increasing correlation index. (**A**) *A. thaliana*; (**B**) *B. napus*; (**C**) *B. rapa*; and (**D**) *C. sativa*. * and *** represented statistical significance of $p < 0.05$ and $p < 0.001$, respectively.

2.7. Relationship between ENc and Gene Expression Level

FPKM is an indicator of the gene expression level, with a lower value denoting a lower gene expression level [38]. The FPKM values of 27 *HSP20* genes in *A. thaliana* revealed that the expression pattern of the *HSP20* genes was complex and varied under various conditions (Table S2). Some *HSP20* genes, like *AT5G12030*, were notably induced under drought and salinity stress compared with the control condition. We performed a Karl Pearson's correlation analysis of FPKM and ENc, and a negative correlation that ranged from −0.0320 to −3.3535 was observed, indicating that gene expression levels might be affected by the CUB (Figure 6).

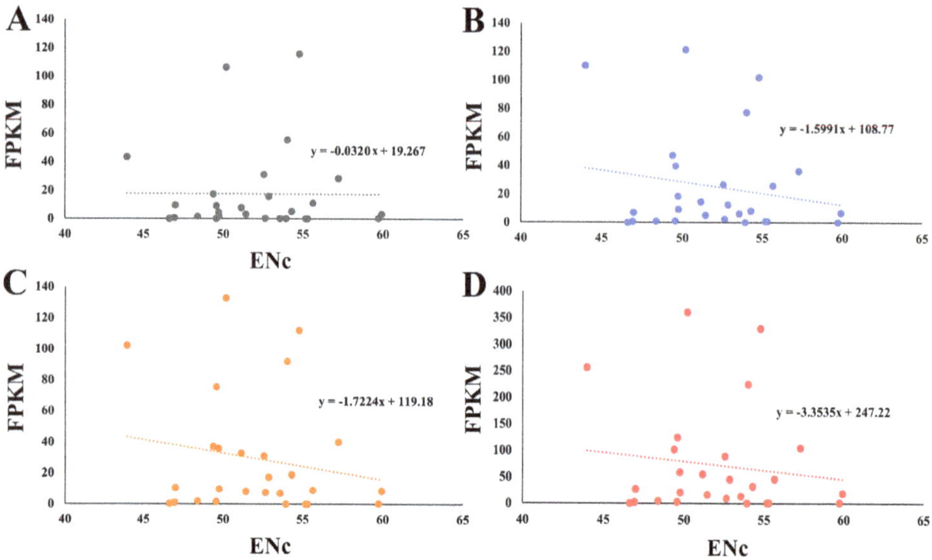

Figure 6. The relationship between FPKM and ENc. The FPKM values of *HSP20* genes under (**A**) normal, (**B**) salinity, and (**C**) drought conditions were applied for calculation. The FPKM values of *HSP20* genes used in (**D**) represented the sum total of those under normal, drought, and salt conditions.

2.8. Identification of Optimal Codons

The RSCU values for the *HSP20* genes in the four species were depicted in a stacked bar chart (Figure 7), with detailed RSCU values listed in Table S3. These stacked bars demonstrated high genetic conservation across species. After the removal of stop codons and codons with RSCU = 1, there were 29 (*A. thaliana*), 28 (*B. napus*), 27 (*B. rapa*), and 26 (*C. sativa*) high-frequency codons with RSCU over 1 in the four cruciferous species. A total of 12 codons satisfying RSCU > 1 and ΔRSCU ≥ 0.08 were grouped as optimal codons, of which 6, 6, 3, and 1 optimal codons were detected in *A. thaliana*, *C. sativa*, *B. rapa*, and *B. napus*, respectively. UAC was shared by *B. napus* and *B. rapa*. AAC and GGU were shared by *A. thaliana* and *C. sativa*, while AGA was shared by *B. rapa* and *C. sativa*. The remaining codons AGG, CCA, GAA, GAG, GCU, GUG, UCU, and UGA were used as the optimal codons for the individual species. The analysis revealed a preference for codons ending with A/T, as evidenced by 4 of 12 optimal codons ending with A, 3 each with T and G, and only 2 with C.

2.9. Comparison of Evolutionary Relationship and Cluster Analysis

The evolutionary relationship of *HSP20* genes was constructed on the basis of their CDSs using the NJ method. This resulting phylogenetic tree was highly supported, with more than 75% nodes obtained bootstrap over 90 BP (Figure 8). In the phylogenetic tree, the *HSP20* genes were clustered into various clades, reflecting their relationship. Genes in the same clades revealed a close relationship, while those in different clades were distant during evolution. As shown in Figure 8, 140 *HSP20* genes were classified into 13 subfamilies, including CI, CII, CIII, CV, CVI, CVII, CIV, ER, MI, MII, P, Po, and HSP20-like, containing 13, 3, 4, 2, 3, 9, 6, 3, 4, 4, 4, and 85 CDSs, respectively. Notably, our phylogenetic results indicate that *HSP20* genes from the same species were separated into different clades, suggesting that gene duplication leading to these different clades occurred before the species divergence. For instance, 5, 4, 2, and 1 *HSP20* genes from *A. thaliana*, *C. sativa*,

B. rapa, and *B. napus*, respectively, were included in subfamily CI, similar to the phylogenetic pattern observed in the *expansin* genes in the *Brassica* species [34].

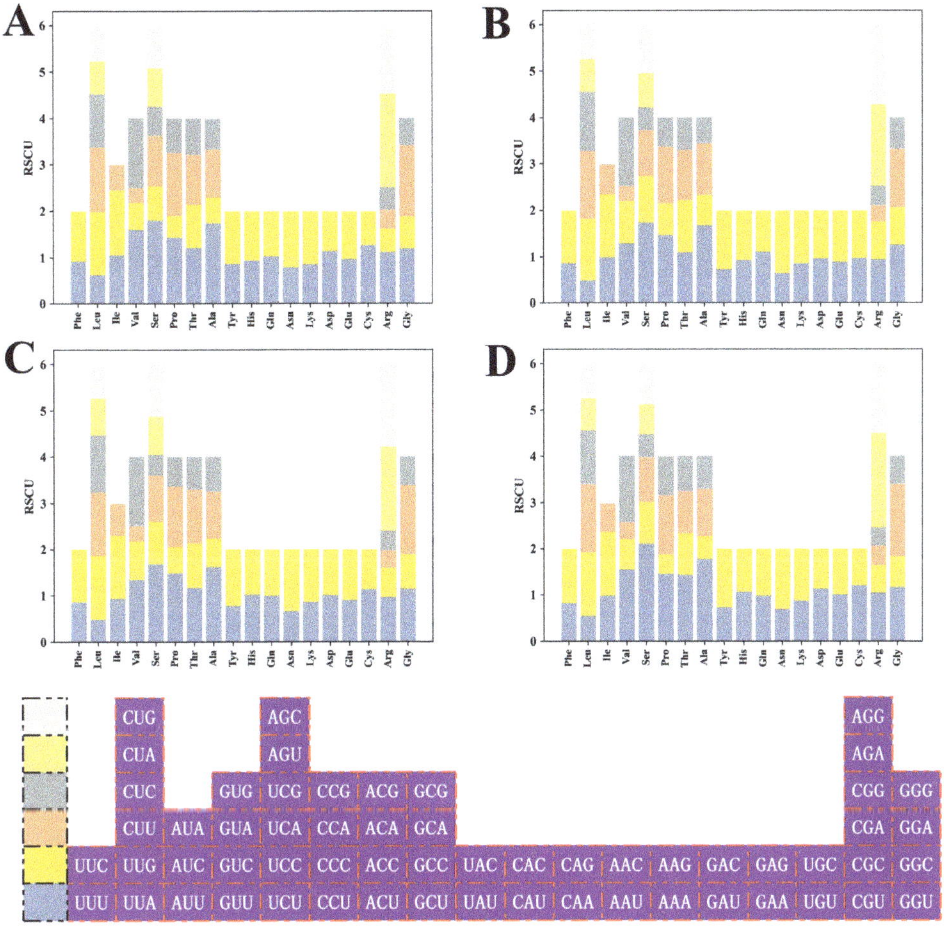

Figure 7. RSCU analysis of *HSP20* genes in four cruciferous species. (**A**) *A. thaliana*; (**B**) *B. napus*; (**C**) *B. rapa*; and (**D**) *C. sativa*.

Meanwhile, cluster analysis based on the indices provides statistical visualization to compare the codon usage similarity and has been widely used in plant CUB analysis [37,39]. The indices employed in previous studies included GC3 in *Ginkgo biloba* [39] and RSCU values in *Rosales* species [37]. In our study, RSCU values were used to construct the cluster tree, and the *HSP20* genes were grouped into various subfamilies rather than by species (Figure S1A), similar to the results in phylogenetic tree. The *HSP20* genes in the same subfamily had a close distance, and those with close genetic relationship shared a similar CUB. Furthermore, another cluster tree was created using four super-HSP20 genes, each representing a composite of the *HSP20* genes from the same species (Figure S1B). In this tree, *B. napus* and *B. rapa* formed a monophyletic group (Figure S1B), indicating that closer genetic relationships resulted in more similar CUB patterns among the species [4].

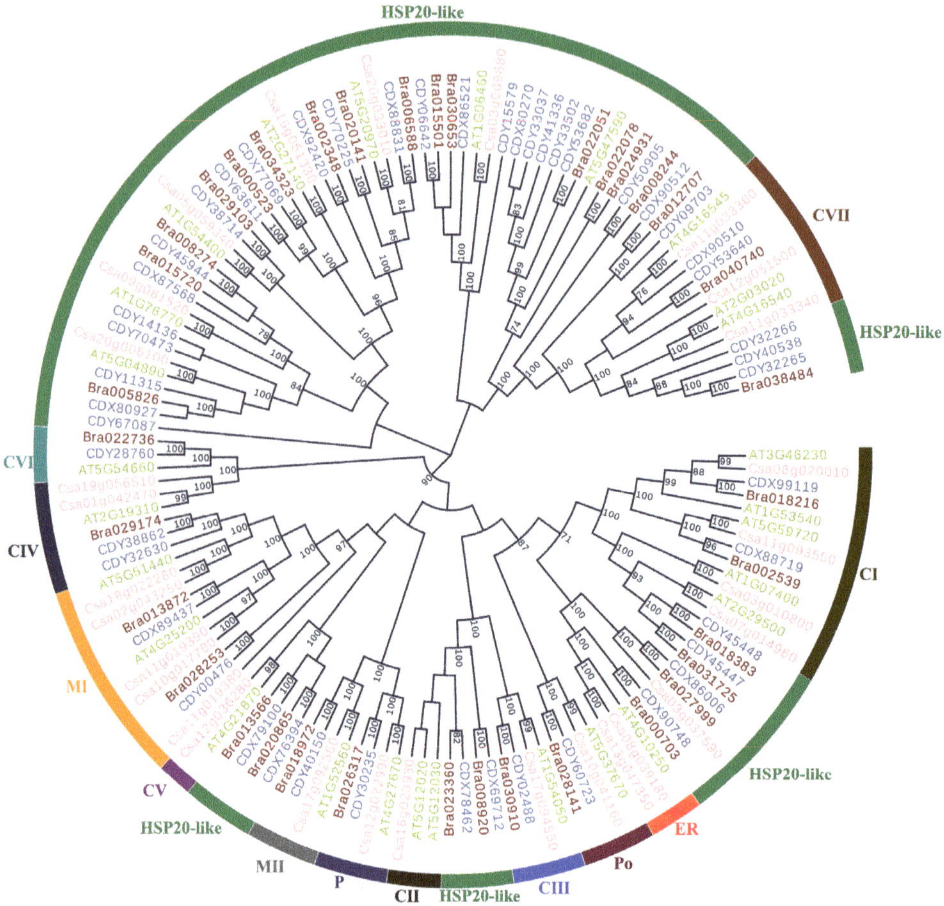

Figure 8. The phylogenetic analysis based on CDS of *HSP20* genes in four cruciferous species.

2.10. PCA of RSCU Values with Species and Subfamilies

In the phylogenetic tree, the clustering manner of *HSP20* genes was dependent on functional subfamilies rather than species (Figure 8). Genes within the same functional subfamilies exhibited close relationships, irrespective of species origin. We performed PCA to explore whether the CUB pattern of the *HSP20* genes was involved with evolutionary history, and hence, the PCA plots of the first and second principal components were generated against each other in terms of subfamilies and species, respectively. As depicted in Figure 9A, we observed that all dots divided by different species were dispersed with no clear clustering trend. However, in Figure 9B, the genes within the same functional subfamilies were close to each other, and the majority of genes from the 12 subfamilies, excluding those in the *HSP20*-like subfamilies, were clustered on the basis of subfamilies. The results strongly indicated that the CUB pattern of the *HSP20* genes was intensively linked to their evolutionary history, with genes that shared a close evolutionary relationship exhibiting similar codon usage bias.

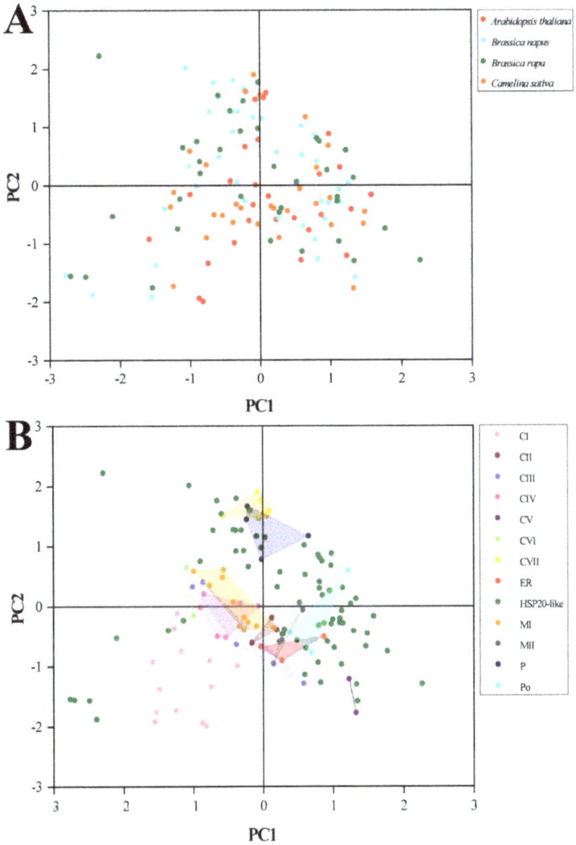

Figure 9. The PCA plots of first axis against second axis. (**A**) Points for *HSP20* genes classified by species. (**B**) Points for *HSP20* genes classified by subfamilies.

3. Discussion

The synonymy of the genetic code decreases the mutation of protein products, even though variation in the genetic code occurs within a gene [2]. Nevertheless, the CUB occurs ubiquitously and contributes to numerous biological processes, including the intensive production of exogenous proteins, RNA stability, environmental adaptability, viral infection, and evolutionary history [4,12,40,41]. Moreover, synonymous codons were believed to affect protein folding due to a translational pause caused by rare codons, and hence, the codon-specific structural properties were conducive to offering insights into the potential causes of conformational diseases [14]. Among the various biological factors shaping the CUB, mutation pressure and natural selection are probably major constituents of the evolutionary forces on the CUB in which natural selection would be more preponderant in the high-expression genes than in the low-expression genes on the condition that natural selection outweighs mutation pressure [42,43]. Given the continuous release of greenhouse gases [44], plants are expected to encounter various environmental stresses, especially temperature, drought, and oxidative stress. The *HSP20* genes are instrumental in plant responses to different environmental stresses, such as high temperature, drought, cold, and oxidative stress, because they can modulate their mRNA and (or) protein levels [21,38], and thus, it is attractive to carry out CUB analysis on *HSP20* genes to evaluate whether mutation pressure or natural selection exerts more influence on the CUB pattern. Nowadays, several CUB analyses are achieved at the genomic level due to the completion of plant genomes [4], which showed that the CUB pattern was associated with a genetic relationship. However, studies focused on the CUB of gene families remained scarce, which led to an insufficient understanding of CUB patterns.

In the present study, genome-wide mining was implemented, resulting in the identification of 140 *HSP20* genes in four species. The number of *HSP20* genes in four species is more than in fungi, algae, bryophytes, pteridophytes, and gymnosperms [45]. It was suggested that the *HSP20* genes experienced an initial expansion during the landing process of algae, followed by further expansion in seed plants and angiosperms [45]. On certain conditions, this greater expansion of the gene family was attributed to whole genome duplications, as seen in soybean [24]. Previous phylogenetic analyses succeeded in the classification of the *HSP20* family into distinct subfamilies, but their number varied across the plant lineage [24]. Comparative analyses of the genomes of *Solanum lycopersicum*, *Cucumis sativus*, and *Coix lacryma-jobi* identified 42, 30, and 32 *HSP20* family members, classified into 13, 11, and 13 subfamilies, respectively [22,23,25]. In the present study, the *HSP20* genes from different species that clustered into the same subfamilies showed closer relationships than those from the same species but different subfamilies. These findings suggest that the *HSP*20 genes in the four cruciferous species likely originated from a common ancestor before the divergence of these species, leading to the different subfamilies. Therefore, the second issue to be addressed is whether the CUB pattern of *HSP20* genes in the same subfamily is closer than that in the same species.

The mean GC content of the *HSP20* genes in these species ranged from 46.5% to 48.1%, indicating that their CDSs were AT-rich. The GC content varied during the evolutionary process, with the highest in algae and the lowest in dicots [4]. In the dinoflagellate *Alexandrium tamarense* CCMP 1598, the average GC content was 58.4% [2,46]. In addition, the GC contents of nuclei genes were higher than those of organelle genes [47]. Among GC1, GC2, and GC3 values, the highest content of GC1 and the lowest content of GC2 were obtained. Previous studies in *A. tamarense*, *Camellia sinensis* cultivars, and *Triticum aestivum* also exhibited the lowest GC2 content [2,47,48]. On the contrary, the lowest GC3 content was found in 18 *Oryza* species [2], implying the diversity of base composition in different plants. The *HSP20* genes preferred A- and T-ending codons due to the global GC3s percentage of 48.5%, consistent with previous studies in which dicot genes tend to end with A or T, while monocot genes preferred G- or C-ending codons [49]. Furthermore, the higher prevalence of T- and G-ending codons in the *HSP20* genes highlighted the inequality of bases in the degenerate codon position, facilitating the comparative analysis of CUB in the genus Bungarus [50]. Intriguingly, a significant correlation was found between overall GC content and the GC composition at three codon positions, suggesting that the CUB of the *HSP20* genes in these species was affected by both natural selection and mutational pressure, coinciding with the result of Li et al. [51] in *Amanita* species. Additionally, the majority of optimal codons in *HSP20* genes ended with A/T, paralleling the pattern observed in *MYB10* genes in Populus [13].

Various *HSP20* genes in *A. thaliana* exhibited complex expression patterns under different conditions, implying that they might have diverse biological functions. For instance, *AT5G12030* appeared to enhance drought and salt tolerance in *Arabidopsis*, probably due to its chaperone activity on citrate synthase [52]. In addition, we observed a negative correlation between FPKM and ENc, indicating that genes with higher expression exhibited a more pronounced CUB than those with lower expression [10]. Chakraborty et al. [50] also observed a significant correlation between expression levels and ENc in *Bungarus* species, highlighting the influence of CUB on gene expression. The classic theory linking the gene's expression level to its codon preference is that frequent codons are more adapted in highly expressed genes in order to match the tRNA abundances [53,54]. The relationship between CUB and tRNA abundance contributes to the accuracy and efficiency of translation. Firstly, the preferred codons are less likely to be recognized by mismatched tRNAs during translation, thereby reducing erroneous protein production [55]. Secondly, the preferred codons, aligning with abundant tRNAs, require shorter time for matching, thus improving the efficiency of ribosome utilization [56]. In summary, genes with a high CUB tend to exhibit higher expression levels than do genes with a low CUB, and the substitution of rare codons with preferred ones in the *HSP20* genes could enhance associated protein production.

In our study, the weak correlation between GC3 and GC12 indicated the greater contribution of natural selection than mutation pressure to the CUB in cruciferous species. Additionally, the dominant role of natural selection was further strengthened by the smooth slope of the regression line in the neutrality plot; ENc-GC3s plot together with PR2 bias, which agreed with the previous study on *Cuscuta australis* [10].

Based on the phylogenetic, clustering, and PCA results, our findings also supported the notion that the *HSP20* genes with a closer phylogenetic relationship exhibited a more similar CUB pattern. Although the *HSP20* genes within the same subfamilies were not derived from the same species, these members showed a close phylogenetic relationship together with close distance (Figure 7 and Figure S1). In addition, our results provided evidence that *B. napus* and *B. rapa* shared a more similar CUB pattern than *A. thaliana* and *C. sativa* (Figure S1). Therefore, our results, along with previous reports [4,57], supported that the CUB pattern was consistent with the genetic relationship at the gene level and (or) species level and provide supplementary insight into the genetic relationship derived from phylogenetic analysis [51].

4. Materials and Methods

4.1. Identification of HSP20 Genes

Following the methodology of Tian et al. [58], the proteomes of four cruciferous species were downloaded from the Ensembl plants database (https://plants.ensembl.org/index.html) (accessed on 1 June 2023) for use as local datasets. A Hidden Markov model (HMM) approach, employing the Pfam (http://pfam.xfam.org/) (accessed on 20 June 2023) profile of PF00011 as query, was utilized to identify HSP20 proteins. An *E*-value threshold of less than 1×10^{-5} was set. Subsequently, the candidates, which contained α-crystallin domain and had molecular weight ranging from 14–42 KD, remained. Moreover, BLASTP tool was performed to align the members from *B. napus*, *B. rapa*, and *C. sativa* with those from *A. thaliana* proteome with a criterion of *E*-value $< 1 \times 10^{-5}$. Finally, the protein sequences of the remaining HSP20 members were resubmitted once again to Pfam, SMART (https://smart.embl.de/) (accessed on 14 September 2023), and InterPro (https://www.ebi.ac.uk/interpro/) (accessed on 16 September 2023) to verify the conserved Hsp20 domain.

4.2. Codon Usage Bias Analysis

Complete CDSs of *HSP20* genes were extracted using TBtools-IIv2.052, ensuring that ATG acted as the start codon and that TGA, TAA, and TAG acted as the stop codons. The relative synonymous codon usage (RSCU), codon adaptation index (CAI), synonymous codon GC content, and 3rd base contents (A3s, T3s, G3s, and C3s) were determined using CodonW 1.4.2 [59]. Codon usage frequencies were calculated with the CUSP tool on the EMBOSS services (https://www.bioinformatics.nl/emboss-explorer/) (accessed on 11 October 2023), while the effective codon numbers were obtained using CHIPS, while the effective codon numbers were obtained using CHIPS.

4.3. ENc-GC3s Plot

To analyze the relationship between base composition and codon bias, an ENc-GC3s plot was created using GC3s content on the x-axis and ENc values on the y-axis. The expected ENc values were determined using the following equation [51]:

$$\text{ENc} = 2 + \text{GC3s} + 29/(\text{GC3s}^2 + (1 - \text{GC3s})^2)$$

The ENc values range from 20 to 61, with 35 usually considered as the threshold for assessing the strength of the CUB [60]. If the points in the ENc-GC3s plot lie on or near the expected curve, the CUB of a given gene is mostly affected by mutation pressure.

Conversely, the points situated below the expected curve suggest natural selection as the main contributor to the CUB of the given gene [61].

4.4. PR2-Plot Analysis

A PR2 Rule bias plot was generated with A3/(A3 + T3) on the x-axis and G3/(G3 + T3) on the y-axis. If the base content of one strand is A = T and G = C, it signifies that mutation pressure and natural selection would not affect both complementary DNA strands [59]. The center of the plot is set at (0.5, 0.5), where the distance and direction from the center point represent the degree and direction of PR2 bias, respectively. A longer distance away from the center equates to a stronger PR2 bias [7]. According to the proportion of four bases, the influence of base mutation on base variation could be inferred.

4.5. Neutrality Plot

A neutrality plot is a commonly assessed tool to determine the extent to which mutation pressure and natural selection play a role in CUB. To discern which factor was more influential, a neutrality plot was generated using Origin, with the GC3 on the x-axis against GC12 (the average value of GC1 and GC2) on the y-axis [7]. The slope of the regression line ranged from 0 to 1; 0 represents no difference in codon base usage across the 1st, 2nd, and 3rd positions of codons, while 1 means a predominant role of mutation pressure [62].

4.6. Correlation Analysis of Codon Usage Bias Indices

SPSS 26 software was utilized to analyze the significant differences in the paired CUB parameters of the *HSP20* genes. The parameters were compared in pairs, and the number of * represented the significance level, while the numerical magnitude represented the correlation strength.

4.7. The Relationship between CUB and Gene Expression Level

To evaluate the relationship between the CUB and the gene expression level, the FPKM (fragments per kilobase of exon model per million mapped fragments) and ENc values of the *HSP20* genes were applied as indicators of the gene expression level and the CUB, respectively. In our prior study, we obtained the transcriptome of *A. thaliana* under normal, drought, and salinity conditions [63]. Then, the analysis was conducted using SPSS software in which we plotted the ENc values on the x-axis and the FPKM values on the y-axis.

4.8. Determination of Optimal Codons

Optimal codons refer to those occurring with greater frequency in high-expression genes than in low-expression genes [47]. On the basis of the ENc values, the bottom and top five genes were collected and formed the high-expression and low-expression databases, respectively. Then, the optimal codon was defined using the following two criteria. First, a given codon should possess an RSCU value greater than 1 in both databases. Second, the RSCU value of a given codon in a high-expression database is 0.08 greater than that in a low-expression database [7].

4.9. Phylogenetic and Clustering Analysis

Phylogenetic analysis of the *HSP20* genes in the four species was constructed using their coding sequences (CDS) and the neighbor-joining (NJ) method. The NJ method employed the P-distance replacement model and included 1000 replicates, with the gaps removed [64]. The RSCU values of 59 codons (excluding the three stop codons, AUG, and UGG) of the four species were clustered using SPSS v26.0 software. The clustering approach adopts intergroup linkage, defining the distance between genes based on the square Euclidean distance of the RSCU values.

4.10. Principal Component Analysis

Principal component analysis (PCA), a multivariate statistical method, was utilized to explore the trends shaping the CUB of the *HSP20* genes [15]. Specifically, the RSCU values of 59 synonymous codons were represented as a 59-dimensional vector, which was then transformed into principal components. Subsequently, the PCA plot of the first two axes was constructed, effectively encapsulating most of the principal components shaping the CUB among the genes.

5. Conclusions

In this work, 140 *HSP20* genes were identified across four cruciferous species. The CUB of the *HSP20* genes was found to be relatively low. Both mutation pressure and natural selection were proposed to contribute to the CUB, with natural selection emerging as the main contributor according to various analyses. Meanwhile, the consistency between the genetic relationship and the CUB pattern provided the necessary evidence for understanding the evolutionary process and underscored the potential strategy for genetic transformation in the future.

Supplementary Materials: The following supporting information can be downloaded at https://www.mdpi.com/article/10.3390/plants13040468/s1. Supplementary Figure S1. The cluster trees based on RSCU values. (A) The cluster tree in terms of genes. (B) The cluster tree in terms of species. Supplementary Table S1. The information of *HSP20* genes in four cruciferious species. Supplementary Table S2. The FPKM values of *HSP20* genes in *A. thaliana* under various conditions. Supplementary Table S3. The RSCU values of codons in four cruciferous species.

Author Contributions: All authors are responsible for the paper. H.J. performed the experiments. H.J. and J.L. analyzed the data. Y.C., X.Y., C.L., and L.S. contributed to the analysis tool. J.Z. and H.L. wrote the paper and were responsible for the manuscript. All authors have read and agreed to the published version of the manuscript.

Funding: This work was partially funded by the National Natural Science Foundation of China (No. 32270410), the Innovation Program for Technology of Chengdu City (2022-YF05-01357-SN), and the Fundamental Research Funds for the Central Universities (No. 2682022ZTPY053).

Data Availability Statement: All data are available within this publication and Supplementary Material.

Conflicts of Interest: The authors declare no conflicts of interest.

Abbreviations

AROMO, aromaticity; CAI, codon adaptation index; CBI, codon bias index; CUB, codon usage bias; ENc, effective number of codons; Fop, frequency of optical codons; GC, guanine–cytosine; GC3s, GC content at the third position of synonymous codons; GRAVY, hydrophobicity; HSP20, heat shock protein 20; PCA, principal component analysis; pI, isoelectric point; PR2, parity rule 2; RSCU, relative synonymous codon usage; T3, C3, A3, G3, frequency of the nucleotides at the third position of codons.

References

1. Jia, J.; Xue, Q. Codon usage biases of transposable elements and host nuclear genes in *Arabidopsis thaliana* and *Oryza sativa*. *Genom. Proteom. Bioinform.* **2009**, *7*, 175–184. [CrossRef]
2. Chakraborty, S.; Yengkhom, S.; Uddin, A. Analysis of codon usage bias of chloroplast genes in *Oryza* species. *Planta* **2020**, *252*, 67. [CrossRef]
3. Tyagi, N.; Sardar, R.; Gupta, D. Natural selection plays a significant role in governing the codon usage bias in the novel SARS-CoV-2 variants of concern (VOC). *PeerJ* **2022**, *10*, e13562. [CrossRef]
4. Li, N.; Sun, M.-H.; Jiang, Z.-S.; Shu, H.-R.; Zhang, S.-Z. Genome-wide analysis of the synonymous codon usage patterns in apple. *J. Integr. Agric.* **2016**, *15*, 983–991. [CrossRef]
5. Murray, E.E.; Lotzer, J.; Eberle, M. Codon usage in plant genes. *Nucleic Acids Res.* **1989**, *17*, 477–498. [CrossRef]
6. Wang, L.; Xing, H.; Yuan, Y.; Wang, X.; Saeed, M.; Tao, J.; Feng, W.; Zhang, G.; Song, X.; Sun, X. Genome-wide analysis of codon usage bias in four sequenced cotton species. *PLoS ONE* **2018**, *13*, e0194372. [CrossRef] [PubMed]

7. Jiang, L.; Zhang, Q.; Xiao, S.; Si, F. Deep decoding of codon usage strategies and host adaption preferences of soybean mosaic virus. *Int. J. Biol. Macromol.* **2022**, *222*, 803–817. [CrossRef] [PubMed]
8. Xiao, G.; Zhou, J.; Huo, Z.; Wu, T.; Li, Y.; Li, Y.; Wang, Y.; Wang, M. The Shift in Synonymous Codon Usage Reveals Similar Genomic Variation during Domestication of Asian and African Rice. *Int. J. Mol. Sci.* **2022**, *23*, 2860. [CrossRef] [PubMed]
9. Henry, I.; Sharp, P.M. Predicting gene expression level from codon usage bias. *Mol. Biol. Evol.* **2007**, *24*, 10–12. [CrossRef] [PubMed]
10. Liu, X.Y.; Li, Y.; Ji, K.K.; Zhu, J.; Ling, P.; Zhou, T.; Fan, L.Y.; Xie, S.Q. Genome-wide codon usage pattern analysis reveals the correlation between codon usage bias and gene expression in Cuscuta australis. *Genomics* **2020**, *112*, 2695–2702. [CrossRef] [PubMed]
11. Jin, H.; Zhang, J.; Wang, Y.; Ge, W.; Jing, Y.; Cao, X.; Huo, Y.; Fu, Y. A codon-based live-cell biomonitoring system for assessing intracellular phenylalanine bioavailability in cyanobacteria. *Biosens. Bioelectron.* **2024**, *244*, 115792. [CrossRef] [PubMed]
12. Tyagi, S.; Kabade, P.G.; Gnanapragasam, N.; Singh, U.M.; Gurjar, A.K.S.; Rai, A.; Sinha, P.; Kumar, A.; Singh, V.K. Codon Usage Provide Insights into the Adaptation of Rice Genes under Stress Condition. *Int. J. Mol. Sci.* **2023**, *24*, 1098. [CrossRef]
13. Bu, Y.; Wu, X.; Sun, N.; Man, Y.; Jing, Y. Codon usage bias predicts the functional MYB10 gene in *Populus*. *J. Plant Physiol.* **2021**, *265*, 153491. [CrossRef] [PubMed]
14. Komar, A.A.; Samatova, E.; Rodnina, M.V. Translation Rates and Protein Folding. *J. Mol. Biol.* **2023**. [CrossRef] [PubMed]
15. Nasrullah, I.; Butt, A.M.; Tahir, S.; Idrees, M.; Tong, Y. Genomic analysis of codon usage shows influence of mutation pressure, natural selection, and host features on Marburg virus evolution. *BMC Evol. Biol.* **2015**, *15*, 174. [CrossRef]
16. Wu, P.; Xiao, W.; Luo, Y.; Xiong, Z.; Chen, X.; He, J.; Sha, A.; Gui, M.; Li, Q. Comprehensive analysis of codon bias in 13 *Ganoderma* mitochondrial genomes. *Front. Microbiol.* **2023**, *14*, 1170790. [CrossRef] [PubMed]
17. Zhou, M.; Li, X. Analysis of synonymous codon usage patterns in different plant mitochondrial genomes. *Mol. Biol. Rep.* **2009**, *36*, 2039–2046. [CrossRef] [PubMed]
18. Yang, J.; Zhang, B.; Gu, G.; Yuan, J.; Shen, S.; Jin, L.; Lin, Z.; Lin, J.; Xie, X. Genome-wide identification and expression analysis of the R2R3-MYB gene family in tobacco (*Nicotiana tabacum* L.). *BMC Genom.* **2022**, *23*, 432. [CrossRef]
19. Feng, P.M.; Chen, W.; Lin, H.; Chou, K.C. iHSP-PseRAAAC: Identifying the heat shock protein families using pseudo reduced amino acid alphabet composition. *Anal. Biochem.* **2013**, *442*, 118–125. [CrossRef]
20. Deng, Y.; Zheng, H.; Yan, Z.; Liao, D.; Li, C.; Zhou, J.; Liao, H. Full-Length Transcriptome Survey and Expression Analysis of Cassia obtusifolia to Discover Putative Genes Related to Aurantio-Obtusin Biosynthesis, Seed Formation and Development, and Stress Response. *Int. J. Mol. Sci.* **2018**, *19*, 2476. [CrossRef]
21. Zhong, L.; Zhou, W.; Wang, H.; Ding, S.; Lu, Q.; Wen, X.; Peng, L.; Zhang, L.; Lu, C. Chloroplast small heat shock protein HSP21 interacts with plastid nucleoid protein pTAC5 and is essential for chloroplast development in *Arabidopsis* under heat stress. *Plant Cell* **2013**, *25*, 2925–2943. [CrossRef]
22. Huang, J.; Hai, Z.; Wang, R.; Yu, Y.; Chen, X.; Liang, W.; Wang, H. Genome-wide analysis of HSP20 gene family and expression patterns under heat stress in cucumber (*Cucumis sativus* L.). *Front. Plant Sci.* **2022**, *13*, 968418. [CrossRef]
23. Hua, Y.; Liu, Q.; Zhai, Y.; Zhao, L.; Zhu, J.; Zhang, X.; Jia, Q.; Liang, Z.; Wang, D. Genome-wide analysis of the HSP20 gene family and its response to heat and drought stress in Coix (*Coix lacryma-jobi* L.). *BMC Genom.* **2023**, *24*, 478. [CrossRef]
24. Lopes-Caitar, V.S.; de Carvalho, M.C.; Darben, L.M.; Kuwahara, M.K.; Nepomuceno, A.L.; Dias, W.P.; Abdelnoor, R.V.; Marcelino-Guimaraes, F.C. Genome-wide analysis of the Hsp20 gene family in soybean: Comprehensive sequence, genomic organization and expression profile analysis under abiotic and biotic stresses. *BMC Genom.* **2013**, *14*, 577. [CrossRef] [PubMed]
25. Yu, J.; Cheng, Y.; Feng, K.; Ruan, M.; Ye, Q.; Wang, R.; Li, Z.; Zhou, G.; Yao, Z.; Yang, Y.; et al. Genome-Wide Identification and Expression Profiling of Tomato Hsp20 Gene Family in Response to Biotic and Abiotic Stresses. *Front. Plant Sci.* **2016**, *7*, 1215. [CrossRef]
26. Centurion, V.B.; Lacerda-Junior, G.V.; Duarte, A.W.F.; Silva, T.R.; Silva, L.J.; Rosa, L.H.; Oliveira, V.M. Dynamics of microbial stress responses driven by abiotic changes along a temporal gradient in Deception Island, Maritime Antarctica. *Sci. Total Environ.* **2021**, *758*, 143671. [CrossRef] [PubMed]
27. Mu, C.; Zhang, S.; Yu, G.; Chen, N.; Li, X.; Liu, H. Overexpression of small heat shock protein LimHSP16.45 in Arabidopsis enhances tolerance to abiotic stresses. *PLoS ONE* **2013**, *8*, e82264. [CrossRef]
28. Sun, X.; Sun, C.; Li, Z.; Hu, Q.; Han, L.; Luo, H. AsHSP17, a creeping bentgrass small heat shock protein modulates plant photosynthesis and ABA-dependent and independent signalling to attenuate plant response to abiotic stress. *Plant Cell Environ.* **2016**, *39*, 1320–1337. [CrossRef]
29. Arias, T.; Beilstein, M.A.; Tang, M.; McKain, M.R.; Pires, J.C. Diversification times among *Brassica* (Brassicaceae) crops suggest hybrid formation after 20 million years of divergence. *Am. J. Bot.* **2014**, *101*, 86–91. [CrossRef]
30. Fang, C.; Hamilton, J.P.; Vaillancourt, B.; Wang, Y.W.; Wood, J.C.; Deans, N.C.; Scroggs, T.; Carlton, L.; Mailloux, K.; Douches, D.S.; et al. Cold stress induces differential gene expression of retained homeologs in *Camelina sativa* cv Suneson. *Front. Plant Sci.* **2023**, *14*, 1271625. [CrossRef] [PubMed]
31. Kim, H.S.; Oh, J.M.; Luan, S.; Carlson, J.E.; Ahn, S.J. Cold stress causes rapid but differential changes in properties of plasma membrane H(+)-ATPase of camelina and rapeseed. *J. Plant Physiol.* **2013**, *170*, 828–837. [CrossRef]

32. Qin, Y.; Li, Q.; An, Q.; Li, D.; Huang, S.; Zhao, Y.; Chen, W.; Zhou, J.; Liao, H. A phenylalanine ammonia lyase from *Fritillaria unibracteata* promotes drought tolerance by regulating lignin biosynthesis and SA signaling pathway. *Int. J. Biol. Macromol.* **2022**, *213*, 574–588. [CrossRef]
33. Mazumdar, P.; Binti Othman, R.; Mebus, K.; Ramakrishnan, N.; Ann Harikrishna, J. Codon usage and codon pair patterns in non-grass monocot genomes. *Ann. Bot.* **2017**, *120*, 893–909. [CrossRef] [PubMed]
34. Li, K.; Ma, B.; Shen, J.; Zhao, S.; Ma, X.; Wang, Z.; Fan, Y.; Tang, Q.; Wei, D. The evolution of the expansin gene family in *Brassica* species. *Plant Physiol. Biochem.* **2021**, *167*, 630–638. [CrossRef] [PubMed]
35. Shen, Z.; Gan, Z.; Zhang, F.; Yi, X.; Zhang, J.; Wan, X. Analysis of codon usage patterns in citrus based on coding sequence data. *BMC Genom.* **2020**, *21*, 234. [CrossRef] [PubMed]
36. Chi, X.; Zhang, F.; Dong, Q.; Chen, S. Insights into Comparative Genomics, Codon Usage Bias, and Phylogenetic Relationship of Species from Biebersteiniaceae and Nitrariaceae Based on Complete Chloroplast Genomes. *Plants* **2020**, *9*, 1605. [CrossRef] [PubMed]
37. Zhang, Y.; Shen, Z.; Meng, X.; Zhang, L.; Liu, Z.; Liu, M.; Zhang, F.; Zhao, J. Codon usage patterns across seven *Rosales* species. *BMC Plant Biol.* **2022**, *22*, 65. [CrossRef] [PubMed]
38. Liao, H.; Quan, H.; Huang, B.; Ji, H.; Zhang, T.; Chen, J.; Zhou, J. Integrated transcriptomic and metabolomic analysis reveals the molecular basis of tissue-specific accumulation of bioactive steroidal alkaloids in *Fritillaria unibracteata*. *Phytochemistry* **2023**, *214*, 113831. [CrossRef]
39. He, B.; Dong, H.; Jiang, C.; Cao, F.; Tao, S.; Xu, L.A. Analysis of codon usage patterns in Ginkgo biloba reveals codon usage tendency from A/U-ending to G/C-ending. *Sci. Rep.* **2016**, *6*, 35927. [CrossRef] [PubMed]
40. Palanisamy, N.; Osman, N.; Ohnona, F.; Xu, H.T.; Brenner, B.; Mesplede, T.; Wainberg, M.A. Does antiretroviral treatment change HIV-1 codon usage patterns in its genes: A preliminary bioinformatics study. *AIDS Res. Ther.* **2017**, *14*, 2. [CrossRef]
41. Zhang, Y.; Gao, W.; Li, H.; Wang, Y.; Li, D.; Xue, C.; Liu, Z.; Liu, M.; Zhao, J. Genome-wide analysis of the bZIP gene family in Chinese jujube (*Ziziphus jujuba* Mill.). *BMC Genom.* **2020**, *21*, 483. [CrossRef]
42. Chamani Mohasses, F.; Solouki, M.; Ghareyazie, B.; Fahmideh, L.; Mohsenpour, M. Correlation between gene expression levels under drought stress and synonymous codon usage in rice plant by in-silico study. *PLoS ONE* **2020**, *15*, e0237334. [CrossRef]
43. Sharp, P.M.; Tuohy, T.M.; Mosurski, K.R. Codon usage in yeast: Cluster analysis clearly differentiates highly and lowly expressed genes. *Nucleic Acids Res.* **1986**, *14*, 5125–5143. [CrossRef]
44. Jiang, R.; Zou, M.; Qin, Y.; Tan, G.; Huang, S.; Quan, H.; Zhou, J.; Liao, H. Modeling of the Potential Geographical Distribution of Three Fritillaria Species under Climate Change. *Front. Plant Sci.* **2021**, *12*, 749838. [CrossRef]
45. Waters, E.R.; Vierling, E. Plant small heat shock proteins—Evolutionary and functional diversity. *New Phytol.* **2020**, *227*, 24–37. [CrossRef]
46. Hsiao, Y.Y.; Lin, C.H.; Liu, J.K.; Wong, T.Y.; Kuo, J. Analysis of Codon Usage Patterns in Toxic Dinoflagellate Alexandrium tamarense through Expressed Sequence Tag Data. *Comp. Funct. Genom.* **2010**, *2010*, 138538. [CrossRef] [PubMed]
47. Zhang, W.J.; Zhou, J.; Li, Z.F.; Wang, L.; Gu, X.; Zhong, Y. Comparative Analysis of Codon Usage Patterns among Mitochondrion, Chloroplast and Nuclear Genes in *Triticum aestivum* L. *J. Integr. Plant Biol.* **2007**, *49*, 246–254. [CrossRef]
48. Zhanjun, W.; Ziqi, W.; Zhaoxia, W.; Zulan, O.; Jie, L.; Qianwen, C.; Zhongdong, X.; Zhaoliang, Z. A comparative study of the evolution and codon usage bias in WOX gene family of three Camellia sinensis cultivars. *J. Nanjing For. Univ. (Nat. Sci. Ed.)* **2022**, *46*, 71–80. [CrossRef]
49. Kawabe, A.; Miyashita, N.T. Patterns of codon usage bias in three dicot and four monocot plant species. *Genes Genet. Syst.* **2003**, *78*, 343–352. [CrossRef] [PubMed]
50. Chakraborty, S.; Nag, D.; Mazumder, T.H.; Uddin, A. Codon usage pattern and prediction of gene expression level in *Bungarus* species. *Gene* **2017**, *604*, 48–60. [CrossRef] [PubMed]
51. Li, Q.; Luo, Y.; Sha, A.; Xiao, W.; Xiong, Z.; Chen, X.; He, J.; Peng, L.; Zou, L. Analysis of synonymous codon usage patterns in mitochondrial genomes of nine *Amanita* species. *Front. Microbiol.* **2023**, *14*, 1134228. [CrossRef]
52. Sun, W.; Bernard, C.; van de Cotte, B.; Van Montagu, M.; Verbruggen, N. At-HSP17.6A, encoding a small heat-shock protein in *Arabidopsis*, can enhance osmotolerance upon overexpression. *Plant J.* **2001**, *27*, 407–415. [CrossRef] [PubMed]
53. Duret, L.; Mouchiroud, D. Expression pattern and, surprisingly, gene length shape codon usage in *Caenorhabditis*, *Drosophila*, and *Arabidopsis*. *Proc. Natl. Acad. Sci. USA* **1999**, *96*, 4482–4487. [CrossRef] [PubMed]
54. Plotkin, J.B.; Kudla, G. Synonymous but not the same: The causes and consequences of codon bias. *Nat. Rev. Genet.* **2011**, *12*, 32–42. [CrossRef] [PubMed]
55. Drummond, D.A.; Wilke, C.O. Mistranslation-induced protein misfolding as a dominant constraint on coding-sequence evolution. *Cell* **2008**, *134*, 341–352. [CrossRef] [PubMed]
56. Cannarozzi, G.; Schraudolph, N.N.; Faty, M.; von Rohr, P.; Friberg, M.T.; Roth, A.C.; Gonnet, P.; Gonnet, G.; Barral, Y. A role for codon order in translation dynamics. *Cell* **2010**, *141*, 355–367. [CrossRef]
57. Xiong, B.; Wang, T.; Huang, S.; Liao, L.; Wang, X.; Deng, H.; Zhang, M.; He, J.; Sun, G.; He, S.; et al. Analysis of Codon Usage Bias in Xyloglucan Endotransglycosylase (XET) Genes. *Int. J. Mol. Sci.* **2023**, *24*, 6108. [CrossRef]
58. Tian, C.; Quan, H.; Jiang, R.; Zheng, Q.; Huang, S.; Tan, G.; Yan, C.; Zhou, J.; Liao, H. Differential roles of Cassia tora 1-deoxy-D-xylulose-5-phosphate synthase and 1-deoxy-D-xylulose-5-phosphate reductoisomerase in trade-off between plant growth and drought tolerance. *Front. Plant Sci.* **2023**, *14*, 1270396. [CrossRef]

59. Yang, J.; Chu, Q.; Meng, G.; Kong, W. The complete chloroplast genome sequences of three Broussonetia species and comparative analysis within the *Moraceae*. *PeerJ* **2022**, *10*, e14293. [CrossRef]
60. Jiang, W.; Lv, B.; Wu, X.; Wang, J.; Wu, G.; Shi, C.; Tang, X. Analysis of synonymous codon usage patterns in the edible fungus *Volvariella volvacea*. *Biotechnol. Appl. Biochem.* **2017**, *64*, 218–224. [CrossRef]
61. Li, G.; Pan, Z.; Gao, S.; He, Y.; Xia, Q.; Jin, Y.; Yao, H. Analysis of synonymous codon usage of chloroplast genome in *Porphyra umbilicalis*. *Genes Genom.* **2019**, *41*, 1173–1181. [CrossRef] [PubMed]
62. Akturk Dizman, Y. Codon usage bias analysis of the gene encoding NAD(+)-dependent DNA ligase protein of Invertebrate iridescent virus 6. *Arch. Microbiol.* **2023**, *205*, 352. [CrossRef] [PubMed]
63. Huang, S.; Huang, D.; Fu, L.; Qin, Y.; Tang, J.; Lv, S.; Liao, H.; Zhou, J. Characterization of Arabidopsis thaliana Hsp20 gene family and its expression analysis under drought and salt stress. *Life Sci. Res.* **2023**, *27*, 162–169.
64. Song, S.; Chen, A.; Zhu, J.; Yan, Z.; An, Q.; Zhou, J.; Liao, H.; Yu, Y. Structure basis of the caffeic acid O-methyltransferase from Ligusiticum chuanxiong to understand its selective mechanism. *Int. J. Biol. Macromol.* **2022**, *194*, 317–330. [CrossRef] [PubMed]

plants

MDPI

Article

Identification and Validation of the miR156 Family Involved in Drought Responses and Tolerance in Tea Plants (*Camellia sinensis* (L.) O. Kuntze)

Shengjing Wen [1], Chengzhe Zhou [1], Caiyun Tian [1], Niannian Yang [1], Cheng Zhang [1], Anru Zheng [1], Yixing Chen [1], Zhongxiong Lai [1,2] and Yuqiong Guo [1,3,*]

1 College of Horticulture, Fujian Agriculture and Forestry University, Fuzhou 350002, China; m18816476976@163.com (S.W.); chengzhechou@foxmail.com (C.Z.); cytian1997@foxmail.com (C.T.); niannian121@foxmail.com (N.Y.); zhangcheng9712@foxmail.com (C.Z.); zar527@foxmail.com (A.Z.); cyxing2001@163.com (Y.C.); laizx01@163.com (Z.L.)

2 Institute of Horticultural Biotechnology, Fujian Agriculture and Forestry University, Fuzhou 350002, China

3 Anxi College of Tea Science (College of Digital Economy), Fujian Agriculture and Forestry University, Quanzhou 362400, China

* Correspondence: guoyq828@163.com

Abstract: The microRNA156 (miR156) family, one of the first miRNA families discovered in plants, plays various important roles in plant growth and resistance to various abiotic stresses. Previously, miR156s were shown to respond to drought stress, but miR156s in tea plants (*Camellia sinensis* (L.) O. Kuntze) have not been comprehensively identified and analyzed. Herein, we identify 47 mature sequences and 28 precursor sequences in tea plants. Our evolutionary analysis and multiple sequence alignment revealed that csn-miR156s were highly conserved during evolution and that the rates of the csn-miR156 members' evolution were different. The precursor sequences formed typical and stable stem–loop structures. The prediction of *cis*-acting elements in the *CsMIR156s* promoter region showed that the *CsMIR156s* had diverse *cis*-acting elements; of these, 12 *CsMIR156s* were found to be drought-responsive elements. The results of reverse transcription quantitative PCR (RT-qPCR) testing showed that csn-miR156 family members respond to drought and demonstrate different expression patterns under the conditions of drought stress. This suggests that csn-miR156 family members may be significantly involved in the response of tea plants to drought stress. Csn-miR156f-2-5p knockdown significantly reduced the *Fv/Fm* value and chlorophyll content and led to the accumulation of more-reactive oxygen species and proline compared with the control. The results of target gene prediction showed that csn-miR156f-2-5p targeted SQUAMOSA promoter binding protein-like (*SPL*) genes. Further analyses showed that *CsSPL14* was targeted by csn-miR156f-2-5p, as confirmed through RT-qPCR, 5′ RLM-RACE, and antisense oligonucleotide validation. Our results demonstrate that csn-miR156f-2-5p and *CsSPL14* are involved in drought response and represent a new strategy for increasing drought tolerance via the breeding of tea plants.

Keywords: tea plant; miR156s; drought stress; SQUAMOSA promoter binding protein-like

1. Introduction

Drought stress is a devastating and unavoidable reality that is becoming increasingly prevalent in most parts of the world, and research data show that flash droughts occur frequently worldwide [1,2]. Drought not only causes serious agricultural production losses but also ecological damage, desertification, and soil erosion, among other things. As a result, drought has been recognized as an urgent global and environmental issue [3,4]. The tea plant (*Camellia sinensis* (L.) O. Kuntze) is an economically important crop grown all over the world. Drought stress is one of the most prominent natural challenges threatening major tea-producing areas in China; it remarkably affects the growth and development, yield, and quality of tea plants. Tea plants are sessile organisms that cannot escape drought when it occurs. Thus, to withstand drought stress, like other plants, tea plants display a series of interconnected response and defense mechanisms to safeguard their survival, which involve multiple processes. These may be morphological, physiological, or related to water stress sensing, stress signal transduction, and related gene expression regulation [5]. Recent studies have shown that (Z)-3-hexanol and eugenol can enhance the drought tolerance of tea plants by mediating glucosylation [6]. An exogenous application of 24-epibrassinolide can improve the drought resistance of tea plants by promoting the expression of genes involved in the biosynthesis of galactinol and abscisic acid [7]. In addition, several genes such as *CsUGT71A59*, *CsGSTU8*, and *CsSnRK2.5* have shown the ability to enhance drought tolerance by regulating the accumulation of reactive oxygen species and altering abscisic acid homeostasis [8–10]. However, the specifics of the tea plant's drought resistance mechanism are still unclear. Therefore, it is necessary to study said drought resistance mechanism and thereby improve the drought tolerance of tea plants.

MicroRNAs (miRNAs) are a class of 20–24-nucleotide (nt) endogenous non-coding RNAs in eukaryotes that regulate gene expression at the transcriptional, post-transcriptional, or translational level through complementary binding with target genes. As an important component of gene expression regulation, miRNAs play an important role in plant growth, development, and response to stress through the precise regulation of corresponding target genes [11,12]. Recently, the strategy of fine-tuning miRNAs based on the regulation of target gene transcription has been highlighted as an effective biotechnological method of improving tolerance to abiotic or biotic stresses in economically vital crops. Several studies have confirmed that many miRNAs, such as miR156 [13,14], miR166 [15], miR319 [16], and miR408 [17,18], can improve plant stress resistance to biotic or abiotic stresses by overexpressing or knocking out miRNAs. In apples, *MIR156a* overexpression weakens salt resistance [19]. The knock-down of miR166 can enhance abiotic stress resistance [20]. The overexpression of *MIR319b* and miR319 in the target simulated form (MIM319) demonstrated that miR319 positively regulates salt tolerance in switchgrass [21]. The drought resistance of Os-miR408 transgenic plants can be improved through changes in their leaf morphology [22]. miR156s are a family of small RNAs that were one of the first miRNAs discovered in Arabidopsis; they are highly conserved in plant species. In general, miR156s regulate the expression of the SQUAMOSA promoter binding protein-like (*SPL*) gene family through transcriptional cleavage or translation inhibition, thus inducing resistance to drought stress. *SPLs* are a family of plant-specific transcription factors with highly conserved SBP domains that play an essential role in plant growth and development and response to stress [14,23,24]. A recent study showed that the miR156-mediated silencing of *SPLs* improved drought stress resilience and promoted leaf gas exchange and abscisic acid sensitivity [25,26]. In alfalfa, miR156 was shown to improve drought resistance by downregulating *SPL13* expression; in the same study, the overexpression of miR156 led to stronger drought resistance [13]. Therefore, we used an analysis of the csn-miR156 family as the starting point of our study, progressing to explore the regulating effect of csn-miR156s on drought resistance.

In this study, 47 mature sequences and 28 precursor sequences in tea plants were identified based on published small-RNA sequencing results. Herein, we will provide a detailed overview of the evolution of miR156s and pre-miR156s in tea plants. We detected the expression level of csn-miR156s while being affected by different degrees of drought using RT-qPCR; then, csn-miR156f-2-5p was screened and knocked down. Compared to the control, the csn-miR156f-2-5p-knockdown tea plants showed lower tolerance to drought stress. In order to further explore the drought resistance mechanism of csn-miR156f-2-5p, we also predicted and analyzed the csn-miR156f-2-5p targets. The relationship between csn-miR156f-2-5p and its target genes was verified using reverse transcription quantitative PCR (RT-qPCR) and 5′RACE, followed by antisense oligonucleotide (AsODN) treatment in tea plants. To summarize, we explored the mechanisms mediated by the miR156-SPL regulatory module during drought in tea plants to provide a reference for breeding drought resistance in tea plants.

2. Results

2.1. Identification and Analysis of miR156 in Plants

To understand the distribution of the miR156 family in plants, a total of 447 mature sequences of miR156 from 54 species were downloaded from miRbase, and 47 csn-miR156 mature sequences were identified from published small-RNA sequencing results [27–31]. The miR156s were widely distributed in angiosperms, gymnosperms, and mosses, and the number of miR156s in angiosperms was the largest. The number of miR156 members in each species ranged from 1 to 47 (Figure 1A). To further investigate the evolutionary relationship between miR156s in the plants, we used *Physcomitrella patens* (3), *Selaginella moellendorffii* (4), *Oryza Sativa* (18), *Glycine max* (28), *Vitis vinifera* (9), *Arabidopsis thaliana* (13), *Medicago truncatula* (15), *Nicotiana tabacum* (10), *Brassica napus* (7), *Picea abies* (27), and *Camellia sinensis* (47) to construct a phylogenetic tree (Figure S1). The phylogenetic tree was divided into three clades. All the miR156s derived from the 5′ end of the precursor, such as csn-miR156a-1-5p, osa-miR156l-5p, mtr-miR156i-5p, ath-miR156c-5p, and so on, belonged to clade I and clade II. Clade III consisted of miR156s derived from the 3′ end of the precursor and csn-miR156i-1, csn-miR156j-1/2/3, and csn-miR156k-2.

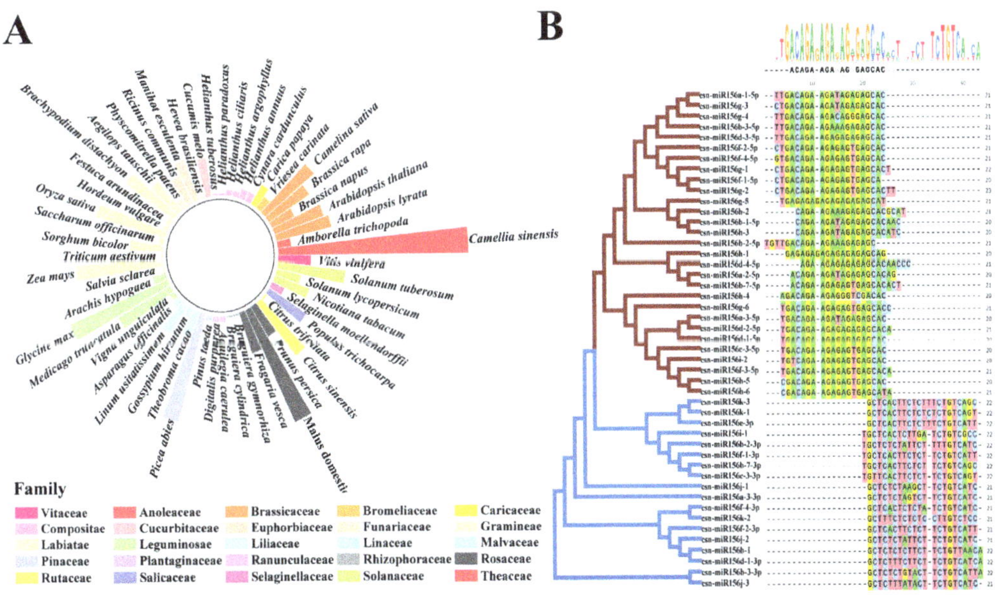

Figure 1. Statistics of plants' miR156 sequences alongside multiple sequence alignment of csn-miR156s. (**A**) Numeric distribution of miR156s in 55 plant species; the height of the column indicates their quantity. (**B**) Classification and multiple sequence alignment of miR156 family members in *Camellia sinensis*.

Alignment of the mature sequences of the miR156 family of the above 10 plants showed that miR156s were roughly divided into two groups based on whether they were derived from the 5′ or 3′ arms of the precursor (Figure S2). The results showed that most miR156 family members of different species were 21 nt, and a few were 20 nt or 22 nt. Except for a few base mutations, deletions, or insertions in the 12th, 15th, and 16th bases, the rest of the bases are highly conserved. Further comparison of the mature sequence of miR156 in tea plants showed that the csn-miR156s were divided into two types: the 5′-end sequence and the 3′-end sequence (Figure 1B). Except for the 1st, 12th, and 15th base differences at the 5′ end and the 5th, 13th, and 19th base differences at the 3′ end, the bases of the csn-miR156s were conserved.

2.2. Sequence Analysis and Chromosomal Localization of csn-MIR156s in the Tea Plant Genome

Based on published small-RNA sequencing results, 28 precursor sequences of the miR156 family were identified in the tea plants. The identified pre-miR156s were within 300 nt, meaning they were within the normal range of plant miRNA precursor length, and the sequences were conserved at the 5′ and 3′ ends (Figure S3). To determine the chromosomal locations of csn-miR156 family members, we aligned the precursor sequences of csn-miR156s (csn-MIR156s) with the 'Tieguanyin' genome (Figure 2A). The csn-MIR156s were located on chr01 (csn-MIR156a), chr02 (csn-MIR156b), chr06 (csn-MIR156c), chr07 (csn-MIR156d), chr10 (csn-MIR156e), chr14 (csn-MIR156f), and chr15 (csn-MIR156g). Csn-miR156a/b/c/d/e/f were named according to the position of the mature sequence on their corresponding precursor (Figure 2B). Among them, csn-miR156f-1-5p and csn-miR156f-2-5p were likely derived from csn-MIR156f-1 on chr14, while csn-miR156g/h/i/j/k did not correspond to the 28 precursor sequences.

2.3. Predicted Secondary Structure Analysis and Phylogenetic Analysis of csn-MIR156s in the Tea Plant Genome

In order to understand the secondary structure of csn-MIR156s, the RNAfold web server (http://rna.tbi.univie.ac.at/cgi-bin/RNAWebSuite/RNAfold.cgi; accessed on 25 July 2023) was used, with default parameters, to analyze the secondary structures of the csn-MIR156s. The results showed that almost all the csn-MIR156s formed typical and stable structures with a stem–loop hairpin (Figure S4). In addition, we also labeled the location of the mature sequence on the secondary structure of the precursor and found that the csn-MIR156s' secondary structures were highly stable in the region (arm) of the mature miRNA sequences, while the rest of the sequences (loop) were more unstable. The minimum folding free energy of the csn-MIR156s varied considerably, ranging from −95.60 to −13.40 kcal/mol. The minimum folding free energy was proportional to the sequence length of the csn-MIR156s.

To further examine the evolutionary relationship among the 28 csn-MIR156s, we used MEGA 10.0 software to analyze the evolution of the pre-miR156 sequences from *P. patens* (3), *S. moellendorffii* (4), *O. sativa* (12), *V. vinifera* (nine), *A. thaliana* (10), *G. max* (28), *M. truncatula* (10), *N. tabacum* (10), *C. sinensis* (28), *B. napus* (seven), and *P. abies* (27). The MIR156s were mainly divided into three branches. Specifically, 18 pab-MIR156s (pab-MIR156a/c/f/d/j/k/m/n/q/v/r/s/t/w/x/y/aa/ab/), 15 gam-MIR156 (gam-MIR156a/b/f/h/j/k/l/m/n/o/p/t/u/v/y), 14 csn-MIR156 (csn-MIR156a-1/a-2/b-1/b-3/b-6/c-3/d-1/d-2/d-3/f-3/f-4/f-6/g-1/g-2), 7 osa-MIR156 (osa-MIR156a/d/e/f/g/h/i), 4 vvi-MIR156s (vvi-MIR156b/f/g/i), 2 nta-MIR156 (nta-MIR156c/h), 8 mtr-MIR156 (mtr-MIR156 b/c/d/e/f/g/h/i), 8 ath-MIR156 (ath-MIR156a/c/d/e/f/h/i/j), 4 bna-MIR156 (bna-MIR156a/d/e/f), and 2 ppt-MIR156 (ppt-MIR156a/c) members were clustered into a broad branch. The second branch consisted of six csn-MIR156s (csn-MIR156b-2/b-4/b-5/d-4/f-5/f-7), three pab-MIR156s (pab-MIR156i/p/u), four gam-MIR156s (gam-MIR156i/q/r/w), five vvi-MIR156s (vvi-MIR156a/c/d/e/h), two osa-MIR156s (osa-MIR156k/l), eight nta-MIR156s (nta-MIR156a/b/d/e/f/g/i/j), one mtr-MIR156 (mtr-MIR156j), and one ath-MIR156 (ath-MIR156g). Eight csn-MIR156s (csn-MIR156a-3/a-4/b-4/c-1/c-2/e/f-1/f-2), six pab-MIR156s (pab-MIR156b/e/g/h/l/o/z), nine gam-MIR156s, three osa-MIR156s, one ath-MIR156s, three bna-MIR156s, and one ppt-MIR156 (ppt-MIR156b) were grouped into the third branch (Figure 3).

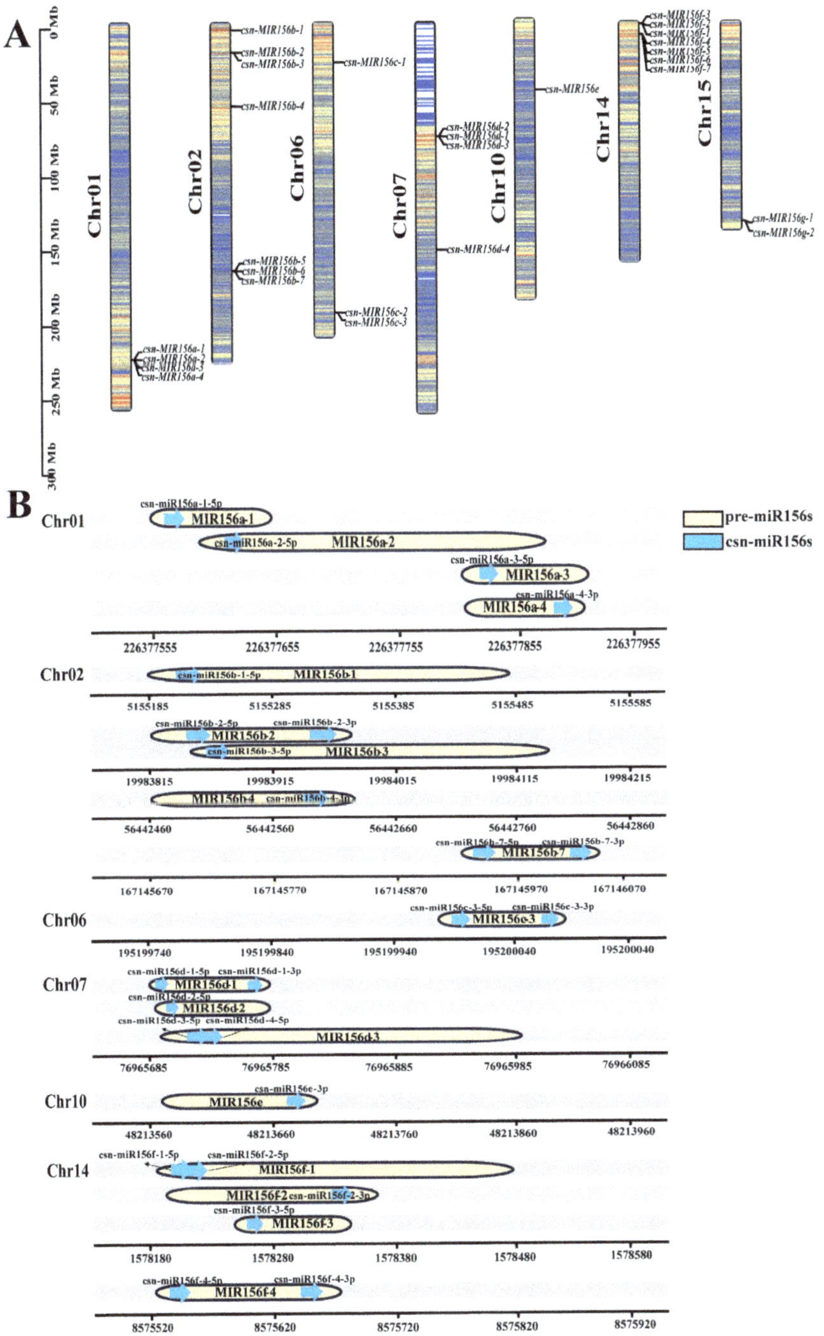

Figure 2. Location of the csn-MIR156 family in the 'Tieguanyin' genome. (**A**) Location of csn-miR156s' precursor sequences on chromosomes. The different colors in the chromosomes represent gene density. (**B**) The position of mature csn-miR156s on precursors. The blue arrow represents the mature sequence, and the yellow capsule represents the precursor sequence.

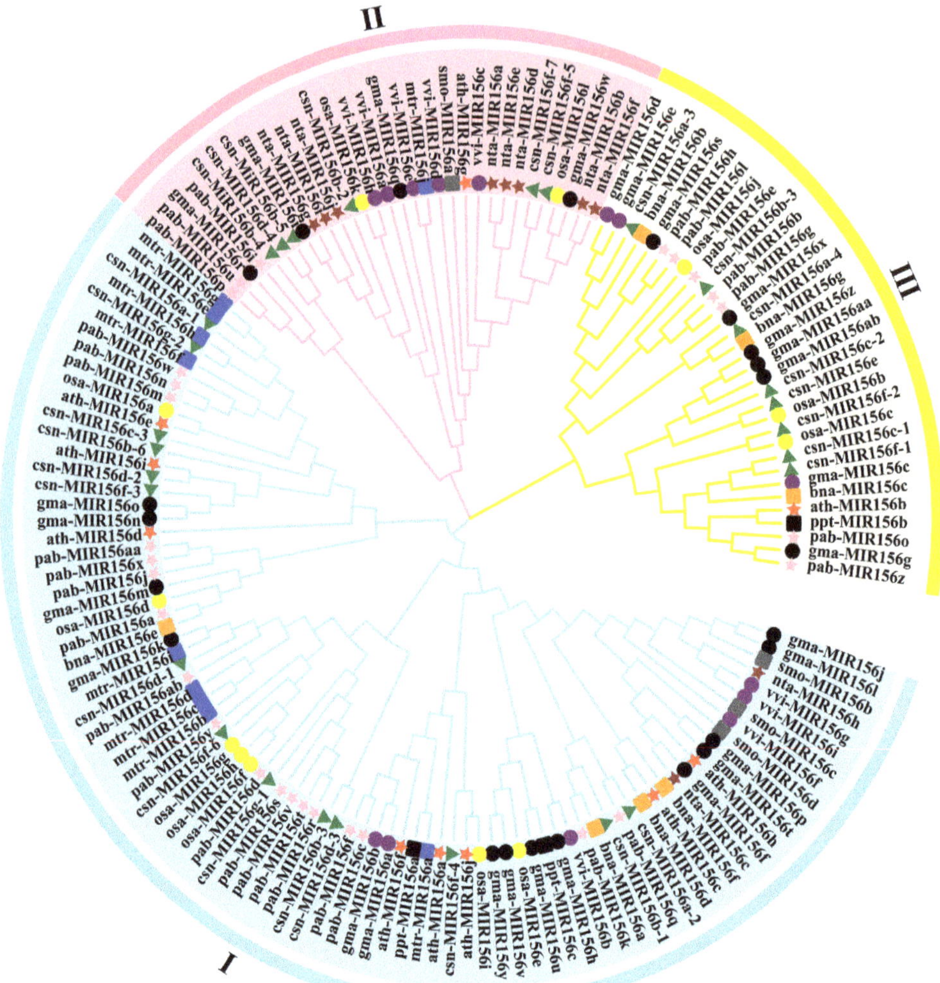

Figure 3. Phylogenetic tree of plants' MIR156 sequences. vvi. *Vitis vinifera*; smo. *Selaginella moellendorffii*; osa. *Oryza sativa*; csn. *Camellia sinensis*; ath. *Arabidopsis thaliana*; gma. *Glycine max*; mtr. *Medicago truncatula*; nta. *Nicotiana tabacum*; bna. *Brassica napus*; pab. *Picea abies*; ppt. *Physcomitrella patens*.

Further analysis revealed that the 28 miR156 precursors of the tea plant were present in three branches, as well as in *P. abies*, *A. thaliana*, *M. truncatula*, and *G. max*. Studies have shown that the evolutionary distance of miRNA families is not closely related to the genetic relationship between species [32–34]. In this study, members of csn-MIR156 and the MIR156s of monocotyledons, dicotyledons, gymnosperms, and ferns were widely distributed on the phylogenetic tree (Figure 3), suggesting that the csn-MIR156 family is a conserved and ancient miRNA family.

2.4. Analysis of Cis-Acting Elements in CsMIR156s Promoter Regions

Transcription initiation is a vital stage of gene expression, and promoters are an important regulatory component of gene transcription initiation. The identification and analysis of the *cis*-acting elements of miRNA promoters will help to further our understanding of the transcriptional activation mechanism of miRNA, as well as its role in the overall regulatory network [35]. Therefore, to identify the function of *CsMIR156s*, 2000 bp upstream regions of *CsMIR156s* in tea plants were used as putative promoter regions for the prediction of *cis*-acting elements. In 28 *CsmiR156s* promoters, in addition to the core promoter element TATA-box and common CAAT-box elements, a total of 52 specific promoter *cis*-acting elements were identified. (Table S1), which were divided into the following five categories: light-responsive, hormone-responsive, stress-responsive, biosynthesis- and metabolism-related *cis*-acting elements, and others (A-box, 3-AF3 binding sites, 60K protein binding sites, and AT-rich sequences) (Figure 4). Among them, the 'light-responsive' (44.7%) category was shared among the 28 *csn-MIR156s*, and six stress-response elements, namely, 'anaerobic-responsive element' (63.6%), 'drought-responsive element' (15.0%), 'defense and stress-responsive element' (7.5%), 'low-temperature-responsive element' (5.6%), 'anoxic-responsive element' (5.6%), and 'wound-responsive element' (2.8%), were identified. A total of 12 *CsMIR156s* (*CsMIR156a-1/a-2/a-3/a-4/b-3/c-1/c-2/c-3/d-1/d-2/d-3*) were identified as possessing drought-responsive elements (MBS).

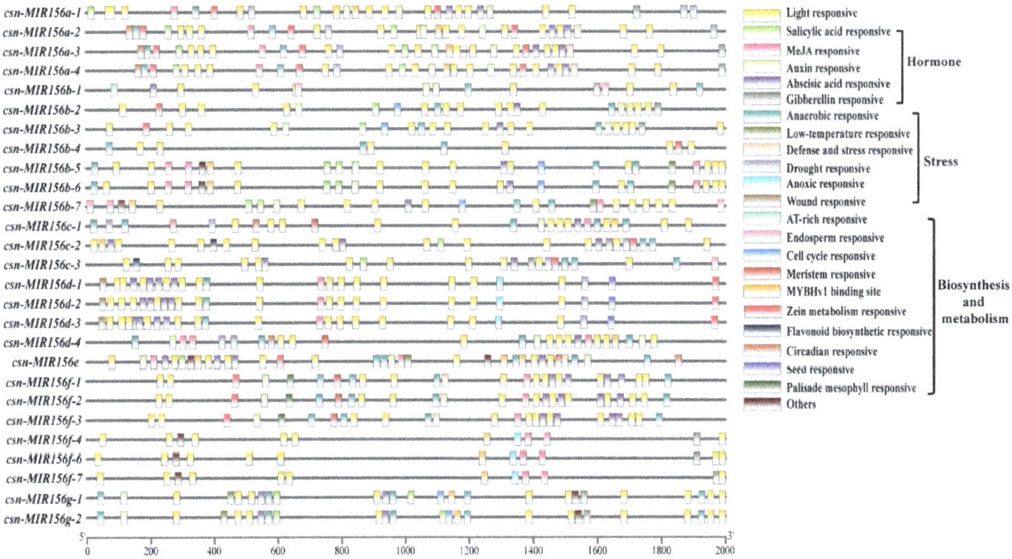

Figure 4. Promoter analysis of *CsMIR156* genes in *C. sinensis*.

2.5. Expression Analysis of csn-miR156s under Drought Stress

The length of miRNAs that guide AGO1 to cleave their target mRNAs in plants is generally 21 nt [12]. To explore the expression pattern of csn-miR156s under different drought degrees, nine 21 nt csn-miR156s that could localize on precursors (csn-miR156a-1-5p, csn-miR156a-4-3p, csn-miR156b-3-5p, csn-miR156b-7-5p, csn-miR156d-2-5p, csn-miR156f-2-5p, csn-miR156f-3-5p, csn-miR156f-4-5p, and csn-miR156f-4-3p) were chosen to detect expression patterns under drought stress via RT-qPCR. The expression patterns of the csn-miR156 members displayed a different trend in response to different degrees of drought. Specifically, the expression levels of miR156a-1-5p, miR156a-4-3p, miR156d-2-5p, and miR156f-3-5p showed a similar tendency. Their expression levels were upregulated compared with CK under T1, T2, and T3. The expression level of miR156f-4-5p increased significantly at T1 but decreased insignificantly at T2 and T3. Concurrently, the expression levels of miR156b-3-5p, miR156b-7-5p, miR156f-4-3p, and miR156f-2-5p decreased under different drought degrees compared with CK, and that of miR156f-2-5p declined continuously (Figure 5). Various studies have revealed that the expression levels of miR156s (mdmiR156a and ath-miR156) are continuously downregulated under abiotic stress, and miR156f-2-5p showed the same trend [19,36]. Therefore, miR156f-2-5p was selected for further analysis.

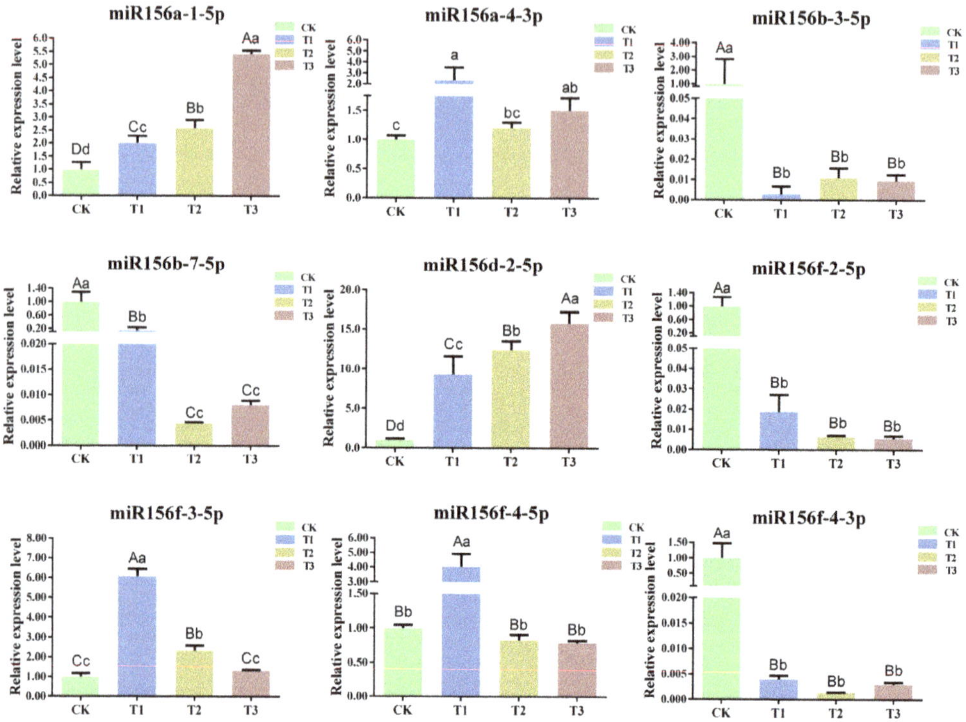

Figure 5. Expression patterns of csn-miR156s under different degrees of drought. CK: normal water supply; T1: mild drought stress, T2: moderate drought stress, and T3: severe drought stress. Data are the means of three independent replicates ± standard deviation (SD). Different lowercase letters indicate significant differences ($p < 0.05$); different uppercase letters indicate highly significant differences ($p < 0.01$).

2.6. csn-miR156f-2-5p Participates in Drought Stress Response in Tea Plants

As there is no existing efficient and stable genetic transformation system for the tea plant, we used a gene-specific antisense oligodeoxynucleotide (AsODN) suppression strategy to knockdown csn-miR156f-2-5p in order to further confirm that csn-miR156f-2-5p participates in drought tolerance in tea plants. Drought stress usually causes osmotic stress in plants, which leads to restricted plant development [4]. Therefore, we intended to simulate drought with PEG 6000; firstly, we observed the expression pattern of csn-miR156f-2-5p during a 15% (*w*/*v*) polyethyleneglycol (PEG) treatment of tea plants. In contrast with previous natural drought results, csn-miR156f-2-5p increased significantly at 4 h and 12 h after PEG treatment and then decreased at 24 h (Figure S5). Subsequently, AsODN and sODN solutions were injected into the leaves of hydroponic tea plants, respectively. After 24 h of incubation after the injection of the AsODN solution, we successfully knocked down the expression of csn-miR156f-2-5p relative to the control (Figure 6A). The tea plants of the csn-miR156f-2-5p-knockdown (AsODN) and control (sODN) tea plants were immersed in a 15% PEG solution to assess drought tolerance. After 24 h of drought, the csn-miR156f-2-5p-knockdown tea plants showed a more severe degree of wilting than the controls (Figure 6B). Diaminobenzidine (DAB) staining (Figure 6C) and nitroblue tetrazolium (NBT) staining (Figure 6D) showed that more H_2O_2 and O_2^- accumulated in the AsODN leaves than in the sODN leaves after 24 h of drought. Consistent with their phenotypes, chlorophyll fluorescence imaging demonstrated that the leaves of the csn-miR156f-2-5p-knockdown plants maintained a lower level of chlorophyll fluorescence (Figure 6E), and the *Fv/Fm* ratio was significantly decreased in the csn-miR156f-2-5p-knockdown tea plants compared with that of the controls (Figure 6F). In addition, the total chlorophyll content of the csn-miR156f-2-5p-knockdown plants was significantly lower than that of the control plants (Figure 6G); the proline content was significantly higher (Figure 6H), indicating that csn-miR156f-2-5p plays a role in tea plants' response to drought.

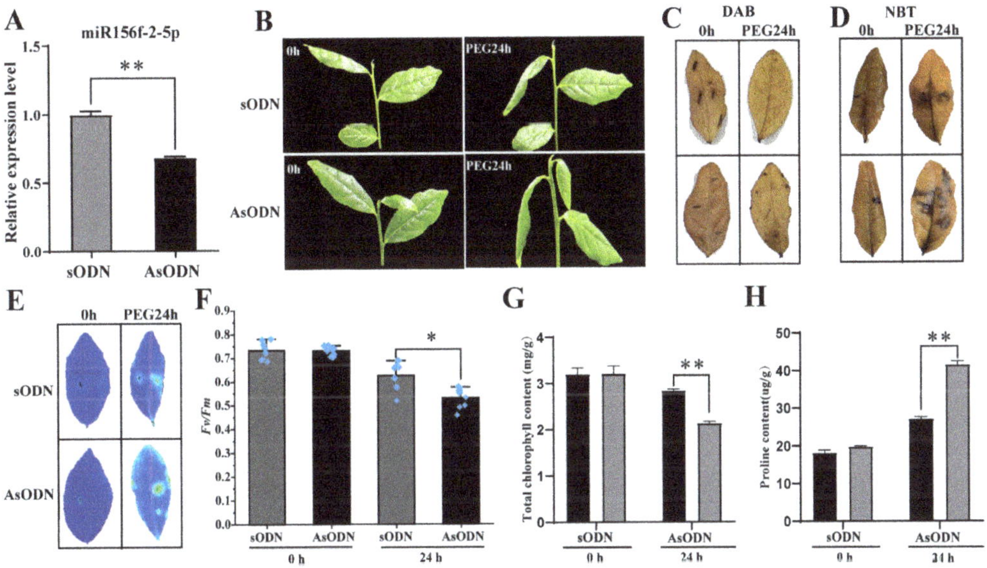

Figure 6. Csn-miR156f-2-5p plays a role in the drought tolerance of tea plants. (**A**) Expression levels of csn-miR156f-2-5p control (sODN) and csn-miR156f-2-5p-knockdown (AsODN) tea plants. (**B**) Phenotypes of the control (sODN) and csn-miR156f-2-5p-knockdown (AsODN) tea plants after 0 h and 24 h of being subjected to drought conditions. (**C**,**D**) DAB (**C**) and NBT (**D**) staining of csn-miR156f-2-5p-knockdown (AsODN) and sODN tea leaves subjected to 15% PEG at 0 h and 24 h. (**E**,**F**) Phenotype of damage (**E**) and *Fv/Fm* values (**F**) of csn-miR156f-2-5p-knockdown (AsODN) and sODN tea leaves subjected to 15% PEG after 0 h and 24 h that were used to assess drought stress resistance. Blue indicates a normal state of the photosynthetic apparatus, while green and yellow indicate damage to photosystem II due to drought. (**G**,**H**) Total chlorophyll (**G**) and proline content (**H**) of csn-miR156f-2-5p-knockdown (AsODN) and sODN tea leaves subjected to 15% PEG after 0 h and 24 h (* $p < 0.05$ and ** $p < 0.01$).

2.7. Prediction and Expression Analysis of csn-miR156f-2-5p Target Genes in Tea Plants

To further explore the mechanism of csn-miR156f-2-5p in the drought resistance of tea plants, the target genes of csn-miR156f-2-5p were predicted using the psRNATarget website (https://www.zhaolab.org/psRNATarget/; accessed on 3 August 2023), and the results showed that nine SPL genes were targeted by csn-miR156f-2-5p (Table S2). These *SPL* genes include two *SPL6* (CsTGY06G0000542 and CsTGY15G0001546), two *SPL12* (CsTGY06G0000121 and CsTGY10G0000091), one *SPL13A* (CsTGY15G0000933), one *SPL14* (CsTGY10G0002410) and three *SPL16* (CsTGY02G0000059, CsTGY02G0001323, and CsTGY10G0000081) genes (Table S3). In order to further understand the expression patterns of *CsSPLs*, the expression patterns of *CsSPLs* in different tissue parts of the tea plants (buds, flowers, stems, young leaves, old leaves, and roots) were analyzed (Figure S6). The results showed that the expression levels of *SPL* genes varied in different kinds of tissue, mainly in the buds, stems, and leaves. In particular, the expression levels of CsTGY06G0000121 (*SPL12-1*) and CsTGY10G0002410 (*SPL14*) were higher in buds and leaves. Subsequently, the expression patterns of nine *SPL* genes were explored under different degrees of drought (Figure 7). The expression levels of five *CsSPL* target genes (*SPL6-2*, *SPL12-1*, *SPL12-2*, *SPL16-1*, and *SPL16-3*) showed a similar upward trend, all of which were significantly upregulated compared with CK at T1 and T3. The expression levels of *SPL6-1* and *SPL13A* increased significantly at T1. In addition, compared with CK, the expression levels of *SPL14* and *SPL16-2* continuously increased from T1 to T3. In most cases, miRNAs inhibited the expression of target genes. The results indicated that nine *CsSPL* genes were responding to drought stress, among which *SPL14* and *SPL16-2* were negatively correlated with the expression of csn-miR156f-2-5p during subjection to different degrees of drought stress.

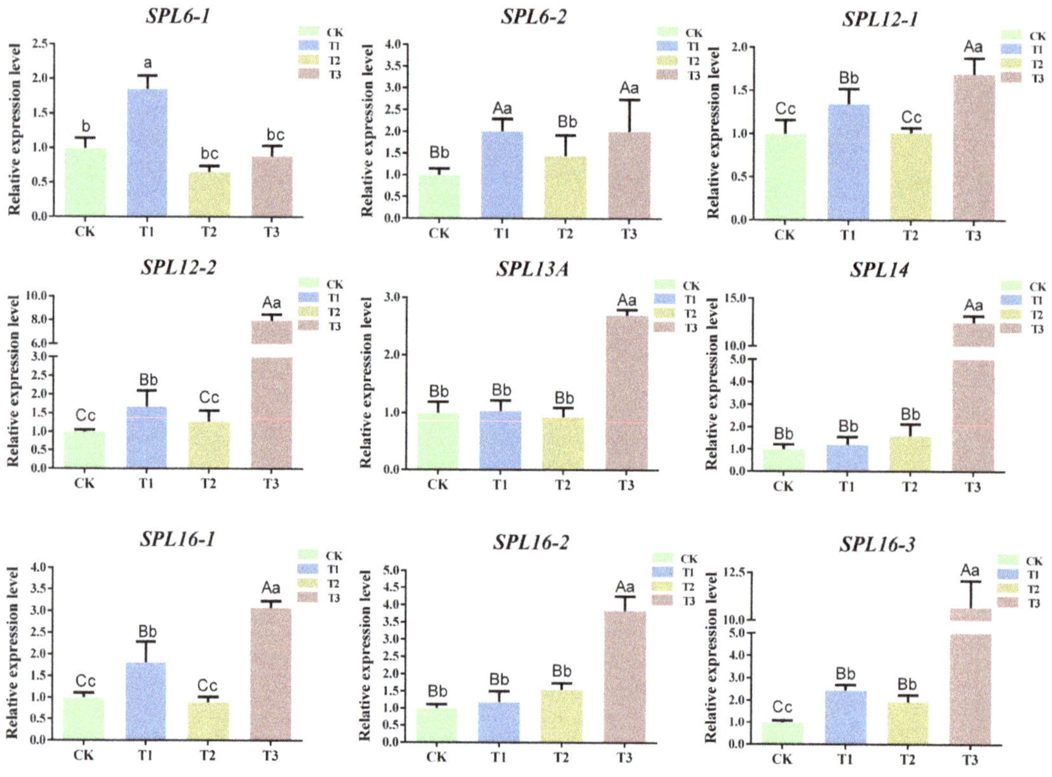

Figure 7. Expression patterns of targets during subjection to different degrees of drought. Data are the means of three independent replicates ± standard deviation (SD). Different lowercase letters indicate significant differences ($p < 0.05$); different uppercase letters indicate highly significant differences ($p < 0.01$).

2.8. *Csn-miR156f-2-5p Suppresses the Expression of CsSPL14 in Tea Plants*

Due to its low background expression level, only *CsSPL14* was successfully confirmed to contain cleaved sites. Multiple sequence alignment showed that the csn-miR156f-2-5p/*CsSPL14* duplex is cleaved between the 10th and 11th nucleotide from the 5′ end of csn-miR156f-2-5p (Figure 8A). These results confirmed that *CsSPL14* is the target gene of csn-miR156f-2-5p in tea plants. To explore the function of csn-miR156f-2-5p and its target gene *CsSPL14* in tea plants, we further verified the relationship between csn-miR156f-2-5p and *CsSPL14* expression. We incubated tender shoots of tea plants with either AsODN solution or sODN solution to knock down the expression of csn-miR156f-2-5p (Figure 8B).

With the extension of incubation time, the expression level of csn-miR156f-2-5p decreased significantly in contrast to that of CK (sODN) at 24 h. The knockdown of csn-miR156f-2-5p (miR156f-2-5p-KD) expression led to increased levels of *CsSPL14* expression compared with those of CK at 24 h after incubation, suggesting that *CsSPL14* can be suppressed by csn-miR156f-2-5p (Figure 8C). These results illustrated that csn-miR156f-2-5p may play a role in the drought resistance mechanism of tea plants by inhibiting the expression levels of *CsSPL14*.

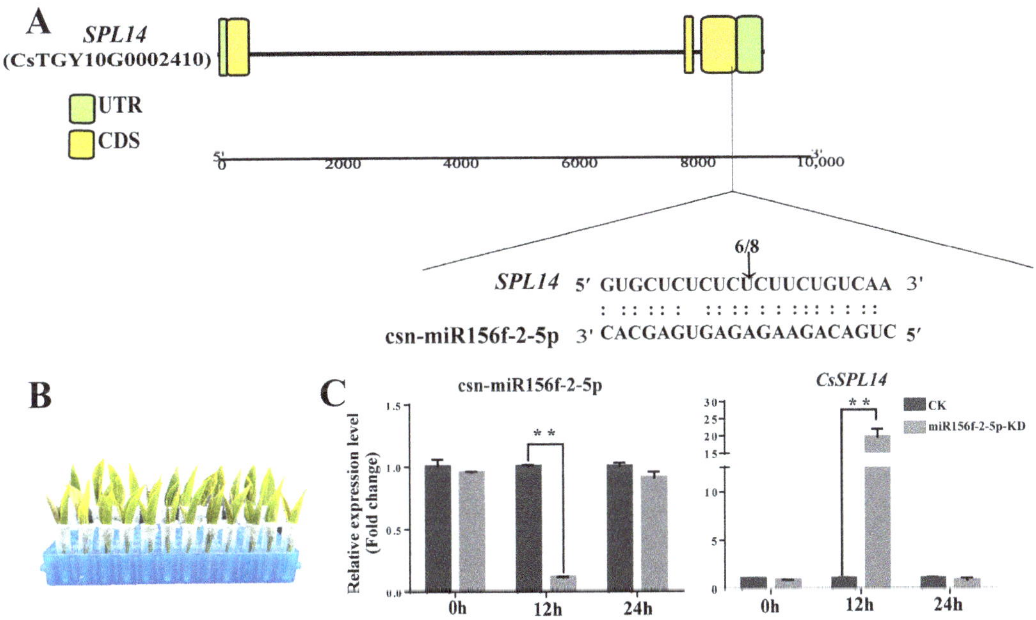

Figure 8. Verification of the relationship between csn-miR156f-2-5p and *CsSPL14*. (**A**) *CsSPL14* cleavage sites of csn-miR156f-2-5p, as identified using 5′RLM-RACE. (**B**) Knockdown of csn-miR156f-2-5p (miR156f-2-5p-KD) with AsODN and incubation, with a solution containing sense oligonucleotide (sODN) serving as a control. (**C**) RT-qPCR verification of csn-miR156f-2-5p knockdown (KD) and changes in the expression levels of *CsSPL14* after 0 h, 12 h, and 24 h of incubation (** $p < 0.01$).

3. Discussion

3.1. Evolutionary Characteristics of the csn-miR156s Family

miR156s were some of the first miRNAs to be discovered; they are highly conserved [37] and widely distributed in 56 plant species, including angiosperms, gymnosperms, ferns, and bryophytes (Figure 1A). In recent years, with the rapid development of plant-genome-sequencing technology and the increasing abundance of plant genomes, research on miR156 in different plant species has gradually grown in prominence. For example, studies of the miR156 family in *M. truncatula*, *Glycine max*, *A. thaliana*, and *O. sativa* have shown that miR156s are an ancient miRNA family that play an important role in the regulation of plant growth, developmental stage transitions, flowering time, and stress responses [38–42]. Numerous studies have shown that the miR156 family is highly conserved in monocotyledons and dicotyledons [34]. In the present study, the evolutionary characteristics of the csn-miR156 family were analyzed in tea plants. In accordance with previous studies [14], our multiple sequence alignment results showed that miR156s were highly conserved in different species (Figure 1B). Sequences of csn-miR156s were relatively conserved among different plant species, but they also showed a degree of diversity. Studies have shown that the evolutionary distance of miRNA families is not closely related to the genetic relationship between species [33,37]. On the evolutionary tree, members of the csn-miR156 family of tea plants are distributed across various species (Figure 3), indicating that the csn-miR156 family of plants is highly complex in terms of its evolution; the evolution rates of each member of the csn-miR156 family are also different. This high level of complexity may explain why the evolutionary distance of the csn-miR156 gene family has little to do with how closely related species are to each other. Gene polymorphism involves gene replication, mutation, deletion, and duplication [43]. The mature sequences of csn-miR156 from the 5′ or 3′ arm of the precursors have a few base mutations, deletions, or insertions in addition to their most base-conserved sequences. We speculate that the diversity of the csn-miR156s family may also be a consequence of random mutations, which result in sequence diversity among csn-miR156s.

Based on small-RNA sequencing results and the NCBI database, the number of mature sequences of csn-miR156s was determined to be 47, while the number of corresponding precursors was 28. In the miRBase database, the quantities of mature sequences and precursors of *A. thaliana*, *M. truncatula*, and *O. sativa* were also inconsistent. The quantities of their mature sequences were 15, 15, and 18, respectively, while those of the precursors were 10, 10, and 12, respectively. The *CsMIR156s* were located on chr01, chr02, chr06, chr07, chr10, chr14, and chr15. Csn-miR156a/b/c/d/e/f were named according to the position of the mature sequence on their corresponding precursors (Figure 2B), while csn-miR156g/h/i/j/k did not correspond to the 28 precursor sequences. We observed that there was no one-to-one correspondence between the number of mature sequences and the number of precursor sequences in the same species; moreover, precursor genes may undergo mutations such as base substitution or deletion during the cutting and processing process to produce one or more mature sequences. It is also possible that some precursors were not detected due to the use of inadequate genome-sequencing techniques.

3.2. Different csn-miR156 Members May Respond to Drought Stress with Different Strategies

Cis-regulatory elements are important for studying the transcriptional regulation of genes. The prediction of *cis*-acting elements in the *CsMIR156s* promoter of the tea plant showed the existence of *cis* elements responsive to drought (MBS) (Figure 4). As expected, the expression patterns of csn-miR156s were affected by drought stress. The expression patterns of miR156 family members were different during subjection to drought stress, among which csn-miR156a-1-5p, csn-miR156a-4-3p, csn-miR156f-3-5p, and csn-miR156d-2-5p showed a pattern of upregulation during drought stress, while csn-miR156b-3-5p, csn-miR156b-7-5p, csn-miR156f-2-5p, csn-miR156f-4-5p, and csn-miR156f-4-3p showed a pattern of downregulation (Figure 5). It has been reported that different members of the same miRNA family may play various modal or regulatory roles in resistance to drought stress [44,45]. This may be due to the fact that the different csn-miR156 family members are derived from different precursors with different *cis*-responsive elements. These results indicate that members of the csn-miR156 family have a multitude of different roles in the drought resistance mechanism of tea plants, and the regulation of these plants' genes during different degrees of drought may be complex and diverse. In fact, even within the same plant species, miRNAs can exhibit diverse responses to drought depending on specific conditions. For example, one study observed an increase in the expression level of miR398a/b in *M. truncatula* during drought stress [46], while another study reported a decrease in the expression level of this identical miRNA within the same plant species under similar drought conditions [47]. Indeed, the expression of csn-miR156f-2-5p in tea plants under PEG-simulated drought conditions was different from that found in a previous study. This difference may reflect different degrees of drought stress as well as the high sensitivity of some miRNAs to subtle differences in growth conditions. The differential expression of the same miRNA in the same plant species under drought conditions may be the result of different spatial–temporal behavior.

3.3. The csn-miR156 Family May Be Play a Role in the Drought Tolerance Mechanism by Cleaving CsSPLs

In the present study, we successfully knocked down csn-miR156f-2-5p and found that tea plants with csn-miR156f-2-5p knocked down might be more sensitive to drought (Figure 6). This is consistent with previous findings showing that plants with miR156silence are more susceptible to drought stress than the corresponding WT with regard to *Arabidopsis* and apples [48,49].MicroRNAs function by cleaving the corresponding mRNA or inhibiting translation to regulate the expression of their target genes [12]. *SPLs* are genes with crucial roles in plants' response to abiotic stress and their development, and they are the main target genes of miR156s [50]. Prediction of target genes revealed that csn-miR156f-2-5p targets *SPLs*, while RT-qPCR showed that *SPLs* respond to drought stress (Figure 7). This prediction was verified using 5′ RLM-RACE, which indicated that csn-miR156f-2-5p cleaved *CsSPL14* genes (Figure 8A). We also observed that the expression level of csn-miR156f-2-5p was inhibited by AsODN technology, and the expression level of *CsSPL14* was significantly increased (Figure 8B). The miR156ab-*SPL* module has been shown to improve drought resistance by accumulating auxin to maintain growth and by enhancing the activities of antioxidant enzymes in *Malus sieversii* [49]. Moreover, the drought tolerance of miR156-overexpressing and *SPL13*-silenced plants was significantly improved in alfalfa. Their survival rate and antioxidant enzyme activity were higher than those of the control [13]. In cassava, miR156-*MeSPL9* affected drought resistance by regulating the levels of protective metabolites and the jasmonic acid (JA) signaling pathway [51]. Therefore, we speculated that the cleaving of *CsSPL14* by csn-miR156f-2-5p could be an important mechanism involved in the mediation of drought resistance in tea plants (Figure 9).

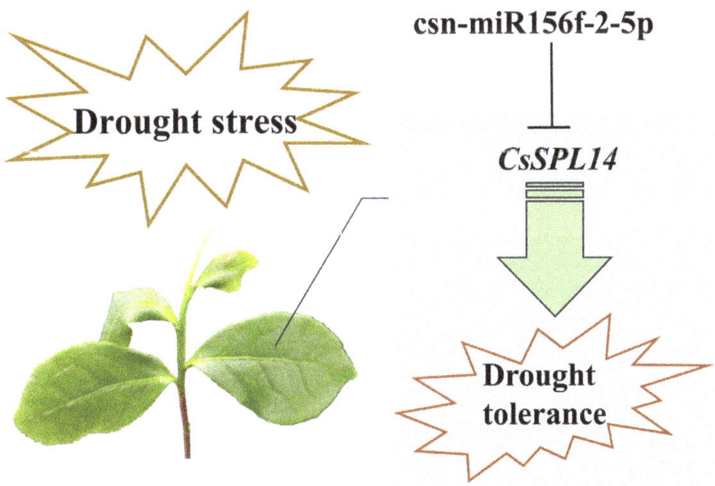

Figure 9. The potential drought tolerance mechanisms of csn-miR156f-2-5p-*CsSPL14* in tea plants.

Although it has been demonstrated that miR156-*SPL* modules are involved in plant responses to drought stresses, miR156-*SPL* combinations involving different *SPLs* may have different functions in different species. Therefore, it is necessary for us to further explore how the csn-miR156f-2-5p-*CsSPL14* module participates in the drought resistance mechanism of tea plants. In summary, our results demonstrate that the csn-miR156f-2-5p-*CsSPL14* module participates in drought tolerance in tea plants, providing a new research idea for improving tea plants' drought resistance and productivity in terms of both yield and quality.

4. Materials and Methods

4.1. Plant Materials

The experimental materials were specimens of *C. sinensis* 'Tieguanyin'. The experiment was conducted in the greenhouse of the College of Horticulture of Fujian Agriculture and Forestry University in June 2022 (119°14′ E, 26°05′ N). The daily temperature range of the greenhouse was 28–35 °C, and the atmospheric relative humidity was 50–75%. The soil field water capacity was 24%. Drought treatment was performed according to the method reported by Guo et al. [30]. The experimental design comprised treatment in four different forms: normal water supply (in which the soil water content was 80% ± 5% of the field water capacity; CK = 18.87%); mild drought stress treatment (wherein the soil water content was 60% ± 5% of field water capacity; T1 = 14.28%), moderate drought stress treatment (in which the soil water content was 40% ± 5% of the field water capacity; T2 = 9.96%); and severe drought stress (in which the soil water content was 20% ± 5% of the field water capacity; T3 = 5.90%). Each treatment was repeated 10 times. Tender *C. sinensis* leaves were fixed with liquid nitrogen and stored in a refrigerator at −80 °C until further analysis.

The relative water content of the soil was determined using the NY/T 52-1987 method. The relative water content of leaves was determined using the GB/T 8304-2002 method [45].

4.2. Bioinformatic Analyses of csn-miR156s and Target Gene Prediction

A total of 447 mature sequences of miR156 from 54 species were downloaded from miRbase, and we visualized the distribution of miR156 in 56 plant species using Origin (v2021) software. Some 47 mature sequences and 28 precursor sequences of csn-miR156s were obtained from the NCBI website (https://www.ncbi.nlm.nih.gov/; accessed on 18 June 2023) and named (Table S4). Subsequently, *P. patens*, *S. moellendorffii*, *O. Sativa*, *G. max*, *V. vinifera*, *A. thaliana*, *M. truncatula*, *N. tabacum*, *B. napus*, and *P. abies* members of the miR156 family were downloaded using the miRbase (Release 22.1) (http://www.mirbase.org/; accessed on 18 June 2023) database, which was used for subsequent analysis. DNAMAN 6.0 was used to analyze the csn-miR156 and pre-miR156 sequences. MEGA 10.0 software was used to construct a phylogenetic tree for the above sequences. The proximity neighbor-joining (NJ) method was implemented, and the Bootstrap coefficient was set to 1000. Evolution trees of csn-miR156s were visualized using Evolview (https://www.evolgenius.info/; accessed on 20 June 2023).

The secondary structure of csn-miR156 precursor sequences was predicted using RNAfold online website (http://rna.tbi.univie.ac.at/cgi-bin/RNAWebSuite/RNAfold.cgi; accessed on 25 July 2023). The 2000 bp upstream sequences of each csn-MIR156 and their *cis*-elements were predicted using the PlantCARE program (http://bioinformatics.psb.ugent.be/webtools/plantcare/html/; accessed on 28 July 2023).

Based on genome sequences of 'Tieguanyin' cultivars [52], the csn-miR156 targets were predicted using the psRNATarget (http://plantgrn.noble.org/psRNATarget/?function=3; accessed on 3 August 2023) online tool. The default parameters were applied, but the expected value was set to 2.0.

4.3. Suppression of csn-miR156f-2-5p Expression in Tea Leaves Treated with AsODN

Candidate AsODNs were selected using Soligo online website (http://sfold.wadsworth.org/cgi-bin/soligo.pl; accessed on 4 August 2023), using csn-miR156f-2-5p as the input sequence (Table S5). To knock down csn-miR156f-2-5p expression, 1 mL of 50 µM AsODN-csn-miR156f-2-5p solution was injected into the leaves of hydroponic tea plant seedlings, and seedlings injected with sODN were used as controls. After incubation for 24 h, tea plants were treated with 15% (*w*/*v*) polyethyleneglycol (PEG) 6000 for 24 h (sODN/AsODN) [6]. Three independent biological replicates were established for each treatment. The obtained leaves were immediately frozen in liquid nitrogen and stored at 80 °C for further analysis.

Using Soligo online website (http://sfold.wadsworth.org/cgi-bin/soligo.pl; accessed on 4 August 2023), the antisense oligonucleotides for csn-miR156f-2-5p were designed, with csn-MIR156f-1 being used as an input sequence (Table S5). To knock down csn-miR156f-2-

5p expression, one bud and two leaves of a freshly detached healthy plant of the variety 'Tieguanyin' were incubated in a 1.5 mL microcentrifuge tube that contained 50 μM of AsODN solution (containing 80 mM of sucrose solution) for various times (24 h and 48 h). Healthy tea branches incubated in sense ODNs (sODNs) together with 80 mM sucrose solution were used as controls [53]. The leaves were sampled at different time intervals to analyze gene expression levels. Three independent biological replicates were established for each treatment. The obtained leaves were immediately frozen in liquid nitrogen and stored at 80 °C for further analysis.

4.4. DAB and NBT Staining

Leaves were placed in a staining solution of 1 mg/mL of DAB or NBT (Solarbio, Beijing, China) prepared with 0.05 mol/L of sodium phosphate buffer (pH 7.5), evacuated for 5 min, and incubated in the dark at room temperature for 8 h before decolorization in 95% ethanol.

4.5. Determination of Fv/Fm, Chlorophyll, and Proline Content

The maximum efficiency of the photosystem II photochemistry (*Fv/Fm*) was measured using a chlorophyll fluorescence imaging system (IMAGING-PAM, Walz) after 30 min of dark adaptation and quantified using Origin (v2021) software. The chlorophyll content and proline content were determined using a chlorophyll content assay kit (HERUI, Fuzhou, China) and proline content kit (HERUI, Fuzhou, China), respectively.

4.6. RNA Extraction and RT-qPCR Analysis

Total RNA was extracted using Transzol UP (TransGen Biotech, Beijing, China). *CsSPL* genes and the first-strand cDNA of miRNA were synthesized using the TransScript® Uni One-Step gDNA Removal, cDNA Synthesis SuperMix (TransGen, Beijing, China), and TranScript® miRNA First-Strand cDNA Synthesis SuperMix (TransGen, Beijing, China) kits, respectively. The csn-miR156s, *CsSPLs*, and reference gene primers were designed using the Primer3 (https://bioinfo.ut.ee/primer3/; accessed on 5 August 2023) website (Table S5). RT-qPCR analysis of the csn-miR156s and their targets was conducted using TransStart® Tip Green qPCR SuperMix (TransGen, Beijing, China) via an ABI StepOnePlus™ Real-Time PCR System (Applied Biosystems, Waltham, MA). CsmiR222 and *CsGAPDH* were used as internal controls for the csn-miR156s and *CsSPLs*, respectively. The $2^{-\Delta\Delta Ct}$ method was used to calculate relative expression. Finally, Tbtools and Prism 5 software were used to draw heat maps and bar charts.

4.7. Cleavage Site Identification with Modified 5′ RLM-RACE

Experimental validation of the predicted targets was conducted using a FirstChoice™ RLM-RACE Kit (Thermo Fisher Scientific, Carlsbad, CA, USA). The target gene cleavage primers were designed using DNAMAN 6.0 (Table S5). Two rounds of nested PCR amplification were performed, using the 5′ RACE outer primer and inner primer alongside target-gene-specific outer primer and inner primers, respectively. The PCR products were purified using an EasyPure® Quick Gel Extraction Kit (TransGen, Beijing, China) and cloned into a Blunt Zero cloning vector (TransGen, Beijing, China); subsequently, single colonies of putative transformants of at least eight independent clones were sequenced.

4.8. Statistical Analysis

Data were analyzed using SPSS 25 software and are presented as the means ± SD of three independent biological replicates unless otherwise indicated. Significances were determined at $p < 0.05$ via one-way ANOVA analysis.

5. Conclusions

In this study, a total of 47 mature members and 28 precursor members of the miR156 family were identified in tea plants; they were found to be evolutionarily conservative and diverse. Moreover, our RT-qPCR analysis indicated that csn-miR156 family members are active in response to drought stress. In knocking down csn-miR156f-2-5p, we found that affected tea plants might be more sensitive to drought. The prediction and expression levels of target genes indicated that *CsSPLs* were involved in drought response. The 5′RACE and AsODN results demonstrated that csn-miR156f-2-5p cleaved and suppressed *CsSPL14*. Furthermore, AsODN confirmed that csn-miR156f-2-5p negatively regulated the expression of *CsSPL14*.

Supplementary Materials: The following supporting information can be downloaded at https://www.mdpi.com/article/10.3390/plants13020201/s1. Figure S1: Phylogenetic tree of plants' miR156 sequences. vvi, *Vitis vinifera*; smo, *Selaginella moellendorffii*; osa, *Oryza sativa*; csn, *Camellia sinensis*; ath, *Arabidopsis thaliana*; gma, *Glycine max*; mtr, *Medicago truncatula*; nta, *Nicotiana tabacum*; bna, *Brassica napus*; pab, *Picea abies*; ppt, *Physcomitrella patens*. Figure S2: Sequence alignment analysis of miR156s. Different highlight colors represent different homology levels of sequence. Black, ≥100%; red, ≥75%; green, ≥50%. Figure S3: Sequence alignment among *C. sinensis* pre-miR156s. Different highlight colors represent different homology levels of sequence. Black, ≥100%; red, ≥75%; green, ≥50%. Figure S4: Stem–loop structures of csn-MIR156s in *C. sinensis*. The mature miRNA sequences are marked with a black line. Figure S5: Expression patterns of the csn-miR156f-2-5p at 0 h, 2 h, 4 h, 12 h, and 24 h after 15% PEG treatment. Data are the mean of three independent replicates ± standard deviation (SD). Different lowercase letters indicate significant differences ($p < 0.05$); different uppercase letters indicate highly significant differences ($p < 0.01$). Figure S6: Tissue-specific gene expression patterns of nine *SPL* genes targeted by csn-miR156f-2-5p in buds, flowers, stems, leaves, and roots. Table S1: *Cis*-Acting regulatory element analysis of MIR156 promoters. Table S2: Summary of target genes corresponding to csn-miR156f-2-5p. Table S3: Identification of target genes corresponding to csn-miR156f-2-5p. Table S4: Identification of 47 mature miRNA sequences and 28 precursor sequences. Table S5: The gene primers presented in this paper.

Author Contributions: Conceptualization, S.W. and Y.G.; methodology, S.W. and C.Z. (Chengzhe Zhou); software, S.W. and C.Z. (Cheng Zhang); validation, S.W., C.T., N.Y., and A.Z.; formal analysis, S.W.; data curation, S.W. and Y.C.; writing—original draft preparation, S.W.; writing—review & editing, S.W., C.Z. (Chengzhe Zhou), and Y.G.; visualization, S.W.; supervision, Y.G. and Z.L.; funding acquisition, Y.G. All authors have read and agreed to the published version of the manuscript.

Funding: This research was funded by the Construction Project for Technological Innovation and Service System (K1520005A01), the Rural Revitalization Tea Industry Technical Service Project of Fujian Agriculture and Forestry University (11899170145), the "Double First-class" Scientific and Technological Innovation Capacity and Enhancement Cultivation Plan of Fujian Agriculture and Forestry University (KSYLP004), the Construction of Plateau Discipline of Fujian Province (102/71201801101), the Tea Industry Branch of Collaborative Innovation Institute of Fujian Agriculture and Forestry University (K1521015A), and the Special Fund for Science and Technology Innovation of Fujian Zhang Tianfu Tea Development Foundation (FJZTF01).

Data Availability Statement: Data is contained within the article or supplementary material.

Conflicts of Interest: The authors declare no conflicts of interest.

References

1. Christian, J.I.; Basara, J.B.; Hunt, E.D.; Otkin, J.A.; Furtado, J.C.; Mishra, V.; Xiao, X.; Randall, R.M. Global distribution, trends, and drivers of flash drought occurrence. *Nat. Commun.* **2021**, *12*, 6330. [CrossRef] [PubMed]
2. Ault, T.R. On the essentials of drought in a changing climate. *Science* **2020**, *368*, 256–260. [CrossRef] [PubMed]
3. Zhang, H.; Zhu, J.; Gong, Z.; Zhu, J. Abiotic stress responses in plants. *Nat. Rev. Genet.* **2022**, *23*, 104–119. [CrossRef] [PubMed]
4. Gupta, A.; Rico-Medina, A.; Caño-Delgado, A.I. The physiology of plant responses to drought. *Science* **2020**, *368*, 266–269. [CrossRef]
5. Nadarajah, K.; Kumar, I.S. Drought response in rice: The miRNA story. *Int. J. Mol. Sci.* **2019**, *20*, 3766. [CrossRef]
6. Jin, J.; Zhao, M.; Jing, T.; Wang, J.; Lu, M.; Pan, Y.; Du, W.; Zhao, C.; Bao, Z.; Zhao, W.; et al. (Z)-3-Hexenol integrates drought and cold stress signaling by activating abscisic acid glucosylation in tea plants. *Plant Physiol.* **2023**, *193*, 1491–1507. [CrossRef]
7. Zhang, Y.; Xiao, Y.; Zhang, Y.; Dong, Y.; Liu, Y.; Liu, L.; Wan, S.; He, J.; Yu, Y. Accumulation of galactinol and ABA is involved in exogenous EBR-induced drought tolerance in tea plants. *J. Agric. Food Chem.* **2022**, *70*, 13391–13403. [CrossRef]

8. Zhao, M.; Jin, J.; Wang, J.; Gao, T.; Luo, Y.; Jing, T.; Hu, Y.; Pan, Y.; Lu, M.; Schwab, W.; et al. Eugenol functions as a signal mediating cold and drought tolerance via *UGT71A59*-mediated glucosylation in tea plants. *Plant J.* **2022**, *109*, 1489–1506. [CrossRef]
9. Zhang, Y.; He, J.; Xiao, Y.; Zhang, Y.; Liu, Y.; Wan, S.; Liu, L.; Dong, Y.; Liu, H.; Yu, Y. *CsGSTU8*, a glutathione S-transferase from *Camellia sinensis*, is regulated by *CsWRKY48* and plays a positive role in drought tolerance. *Front. Plant Sci.* **2021**, *12*, 795919. [CrossRef]
10. Zhang, Y.; Wan, S.; Liu, X.; He, J.; Cheng, L.; Duan, M.; Liu, H.; Wang, W.; Yu, Y. Overexpression of *CsSnRK2.5* increases tolerance to drought stress in transgenic *Arabidopsis*. *Plant Physiol. Bioch.* **2020**, *150*, 162–170. [CrossRef]
11. Yu, Y.; Zhang, Y.; Chen, X.; Chen, Y. Plant noncoding RNAs: Hidden players in development and stress responses. *Annu. Rev. Cell Dev. Biol.* **2019**, *35*, 407–431. [CrossRef]
12. Song, X.; Li, Y.; Cao, X.; Qi, Y. MicroRNAs and their regulatory roles in plant-environment interactions. *Annu. Rev. Plant Biol.* **2019**, *70*, 489–525. [CrossRef]
13. Feyissa, B.A.; Arshad, M.; Gruber, M.Y.; Kohalmi, S.E.; Hannoufa, A. The interplay between miR156/SPL13 and DFR/WD40-1 regulate drought tolerance in alfalfa. *Bmc Plant Biol.* **2019**, *19*, 434. [CrossRef] [PubMed]
14. Wei, L.; Liu, J.; Huang, J.; Wang, C.; Zhang, L.; Feng, S. Genome-wide identification of miR156 and SPL family genes and phenotypic analysis of vegetative phase change in pepper (*Capsicum annuum* L.). *Gene* **2023**, *877*, 147542. [CrossRef] [PubMed]
15. Zhang, Q.L.; Su, L.Y.; Zhang, S.T.; Xu, X.P.; Chen, X.H.; Li, X.; Jiang, M.Q.; Huang, S.Q.; Chen, Y.K.; Zhang, Z.H.; et al. Analyses of microRNA166 gene structure, expression, and function during the early stage of somatic embryogenesis in *Dimocarpus longan* Lour. *Plant Physiol. Bioch.* **2020**, *147*, 205–214. [CrossRef] [PubMed]
16. Liu, L.; Chen, H.; Zhu, J.; Tao, L.; Wei, C. miR319a targeting of CsTCP10 plays an important role in defense against gray blight disease in tea plant (*Camellia sinensis*). *Tree Physiol.* **2022**, *42*, 1450–1462. [CrossRef]
17. Hajyzadeh, M.; Turktas, M.; Khawar, K.M.; Unver, T. miR408 overexpression causes increased drought tolerance in chickpea. *Gene* **2015**, *555*, 186–193. [CrossRef]
18. Guo, Y.; Wang, S.; Yu, K.; Wang, H.L.; Xu, H.; Song, C.; Zhao, Y.; Wen, J.; Fu, C.; Li, Y.; et al. Manipulating microRNA miR408 enhances both biomass yield and saccharification efficiency in poplar. *Nat. Commun.* **2023**, *14*, 4285. [CrossRef]
19. Ma, Y.; Xue, H.; Zhang, F.; Jiang, Q.; Yang, S.; Yue, P.; Wang, F.; Zhang, Y.; Li, L.; He, P.; et al. The miR156/SPL module regulates apple salt stress tolerance by activating MdWRKY100 expression. *Plant Biotechnol. J.* **2021**, *19*, 311–323. [CrossRef]
20. Li, Y.; Wang, X.; Guo, Q.; Zhang, X.; Zhou, L.; Zhang, Y.; Zhang, C. Conservation and diversity of miR166 family members from highbush blueberry (*Vaccinium corymbosum*) and their potential functions in abiotic stress. *Front. Genet.* **2022**, *13*, 919856. [CrossRef]
21. Liu, Y.; Li, D.; Yan, J.; Wang, K.; Luo, H.; Zhang, W. MiR319-mediated ethylene biosynthesis, signalling and salt stress response in switchgrass. *Plant Biotechnol. J.* **2019**, *17*, 2370–2383. [CrossRef] [PubMed]
22. Hang, N.; Shi, T.; Liu, Y.; Ye, W.; Taier, G.; Sun, Y.; Wang, K.; Zhang, W. Overexpression of Os-microRNA408 enhances drought tolerance in perennial ryegrass. *Physiol. Plantarum.* **2021**, *172*, 733–747. [CrossRef] [PubMed]
23. Gou, J.Y.; Felippes, F.F.; Liu, C.J.; Weigel, D.; Wang, J.W. Negative regulation of anthocyanin biosynthesis in *Arabidopsis* by a miR156-targeted SPL transcription factor. *Plant Cell* **2011**, *23*, 1512–1522. [CrossRef]
24. Stief, A.; Altmann, S.; Hoffmann, K.; Pant, B.D.; Scheible, W.R.; Baurle, I. *Arabidopsis* miR156 regulates tolerance to recurring environmental stress through SPL transcription factors. *Plant Cell* **2014**, *26*, 1792–1807. [CrossRef]
25. Visentin, I.; Pagliarani, C.; Deva, E.; Caracci, A.; Tureckova, V.; Novak, O.; Lovisolo, C.; Schubert, A.; Cardinale, F. A novel strigolactone-miR156 module controls stomatal behaviour during drought recovery. *Plant Cell Environ.* **2020**, *43*, 1613–1624. [CrossRef] [PubMed]
26. Brun, G. At the crossroads of strigolactones and abscisic acid pathways: A role for miR156. *Plant Cell Environ.* **2020**, *43*, 1609–1612. [CrossRef]
27. Jeyaraj, A.; Zhang, X.; Hou, Y.; Shangguan, M.; Gajjeraman, P.; Li, Y.; Wei, C. Genome-wide identification of conserved and novel microRNAs in one bud and two tender leaves of tea plant (*Camellia sinensis*) by small RNA sequencing, microarray-based hybridization and genome survey scaffold sequences. *Bmc Plant Biol.* **2017**, *17*, 212. [CrossRef]
28. Zhang, Y.; Zhu, X.; Chen, X.; Song, C.; Zou, Z.; Wang, Y.; Wang, M.; Fang, W.; Li, X. Identification and characterization of cold-responsive microRNAs in tea plant (*Camellia sinensis*) and their targets using high-throughput sequencing and degradome analysis. *Bmc Plant Biol.* **2014**, *14*, 271. [CrossRef]
29. Zhao, S.; Mi, X.; Guo, R.; Xia, X.; Liu, L.; An, Y.; Yan, X.; Wang, S.; Guo, L.; Wei, C. The biosynthesis of main taste compounds Is coordinately regulated by miRNAs and phytohormones in tea plant (*Camellia sinensis*). *J. Agric. Food Chem.* **2020**, *68*, 6221–6236. [CrossRef]
30. Guo, Y.; Zhao, S.; Zhu, C.; Chang, X.; Yue, C.; Wang, Z.; Lin, Y.; Lai, Z. Identification of drought-responsive miRNAs and physiological characterization of tea plant (*Camellia sinensis* L.) under drought stress. *Bmc Plant Biol.* **2017**, *17*, 211. [CrossRef]
31. Fan, K.; Fan, D.; Ding, Z.; Su, Y.; Wang, X. Cs-miR156 is involved in the nitrogen form regulation of catechins accumulation in tea plant (*Camellia sinensis* L.). *Plant Physiol. Bioch.* **2015**, *97*, 350–360. [CrossRef] [PubMed]
32. Axtell, M.J.; Meyers, B.C. Revisiting criteria for plant microRNA annotation in the era of big data. *Plant Cell* **2018**, *30*, 272–284. [CrossRef]
33. Lin, Y.; Zhang, Q.; Zeng, Y.; Chen, X.; Zhang, Z.; Chen, Y.; Lai, Z. Molecular evolution and spatiotemporal expression of longan miR166 family. *J. Hortic.* **2017**, *44*, 2285–2295.

34. Xie, Q.; Wang, X.; He, J.; Lan, T.; Zheng, J.; Li, Y.; Pan, J.; Lin, L.; Zhao, J.; Li, J.; et al. Distinct evolutionary profiles and functions of microRNA156 and microRNA529 in land plants. *Int. J. Mol. Sci.* **2021**, *22*, 11100. [CrossRef] [PubMed]
35. Pei, M.; Liu, H.; Wei, T.; Jin, H.; Yu, Y.; Ma, M.; Song, X.; Dai, R.; Guo, D. Identification, characterization, and verification of miR399 target gene in grape. *Hortic. Plant J.* **2023**, *10*, 91–102. [CrossRef]
36. Zhang, L.; Ding, H.; Jiang, H.L.; Wang, H.S.; Chen, K.X.; Duan, J.J.; Feng, S.J.; Wu, G. Regulation of cadmium tolerance and accumulation by miR156 in *Arabidopsis*. *Chemosphere* **2020**, *242*, 125168. [CrossRef] [PubMed]
37. Gramzow, L.; Theissen, G. Plant miRNA conservation and evolution. *Methods Mol. Biol.* **2019**, *1932*, 41–50.
38. Wang, H.; Lu, Z.; Xu, Y.; Kong, L.; Shi, J.; Liu, Y.; Fu, C.; Wang, X.; Wang, Z.Y.; Zhou, C.; et al. Genome-wide characterization of SPL family in *Medicago truncatula* reveals the novel roles of miR156/SPL module in spiky pod development. *Bmc Genom.* **2019**, *20*, 552. [CrossRef]
39. Naya, L.; Khan, G.A.; Sorin, C.; Hartmann, C.; Crespi, M.; Lelandais-Briere, C. Cleavage of a non-conserved target by a specific miR156 isoform in root apexes of *Medicago truncatula*. *Plant Signal Behav.* **2010**, *5*, 328–331. [CrossRef]
40. Ding, X.; Guo, J.; Lv, M.; Wang, H.; Sheng, Y.; Liu, Y.; Gai, J.; Yang, S. The miR156b-GmSPL2b module mediates male fertility regulation of cytoplasmic male sterility-based restorer line under high-temperature stress in soybean. *Plant Biotechnol. J.* **2023**, *21*, 1542–1559. [CrossRef]
41. Yin, H.; Hong, G.; Li, L.; Zhang, X.; Kong, Y.; Sun, Z.; Li, J.; Chen, J.; He, Y. miR156/SPL9 regulates reactive oxygen species accumulation and immune response in *Arabidopsis thaliana*. *Phytopathology* **2019**, *109*, 632–642. [CrossRef] [PubMed]
42. Wang, L.; Ming, L.; Liao, K.; Xia, C.; Sun, S.; Chang, Y.; Wang, H.; Fu, D.; Xu, C.; Wang, Z.; et al. Bract suppression regulated by the miR156/529-SPLs-NL1-PLA1 module is required for the transition from vegetative to reproductive branching in rice. *Mol. Plant* **2021**, *14*, 1168–1184. [CrossRef]
43. Ellis, J.; Dodds, P.; Pryor, T. Structure, function and evolution of plant disease resistance genes. *Curr. Opin. Plant Biol.* **2000**, *3*, 278–284. [CrossRef]
44. Eldem, V.; Celikkol, A.U.; Ozhuner, E.; Bakir, Y.; Uranbey, S.; Unver, T. Genome-wide identification of miRNAs responsive to drought in peach (*Prunus persica*) by high-throughput deep sequencing. *PLoS ONE* **2012**, *7*, e50298. [CrossRef] [PubMed]
45. Tian, C.; Zhou, C.; Zhu, C.; Chen, L.; Shi, B.; Lin, Y.; Lai, Z.; Guo, Y. Genome-wide investigation of the miR166 family provides new insights into Its involvement in the drought stress responses of tea plants (*Camellia sinensis* (L.) O. Kuntze). *Forests* **2022**, *13*, 628. [CrossRef]
46. Trindade, I.; Capitao, C.; Dalmay, T.; Fevereiro, M.P.; Santos, D.M. miR398 and miR408 are up-regulated in response to water deficit in *Medicago truncatula*. *Planta* **2010**, *231*, 705–716. [CrossRef] [PubMed]
47. Wang, T.; Chen, L.; Zhao, M.; Tian, Q.; Zhang, W.H. Identification of drought-responsive microRNAs in *Medicago truncatula* by genome-wide high-throughput sequencing. *Bmc Genom.* **2011**, *12*, 367. [CrossRef]
48. Chen, G.; Wang, Y.; Liu, X.; Duan, S.; Jiang, S.; Zhu, J.; Zhang, Y.; Hou, H. The mdmiR156n regulates drought tolerance and flavonoid synthesis in apple calli and *Arabidopsis*. *Int. J. Mol. Sci.* **2023**, *24*, 6049. [CrossRef]
49. Feng, C.; Zhang, X.; Du, B.; Xiao, Y.; Wang, Y.; Sun, Y.; Zhou, X.; Wang, C.; Liu, Y.; Li, T.H. MicroRNA156ab regulates apple plant growth and drought tolerance by targeting transcription factor MsSPL13. *Plant Physiol.* **2023**, *192*, 1836–1857. [CrossRef]
50. Cui, L.; Zheng, F.; Wang, J.; Zhang, C.; Xiao, F.; Ye, J.; Li, C.; Ye, Z.; Zhang, J. miR156a-targeted SBP-Box transcription factor SlSPL13 regulates inflorescence morphogenesis by directly activating SFT in tomato. *Plant Biotechnol. J.* **2020**, *18*, 1670–1682. [CrossRef]
51. Li, S.; Cheng, Z.; Li, Z.; Dong, S.; Yu, X.; Zhao, P.; Liao, W.; Yu, X.; Peng, M. MeSPL9 attenuates drought resistance by regulating JA signaling and protectant metabolite contents in cassava. *Theor. Appl. Genet.* **2022**, *135*, 817–832. [CrossRef] [PubMed]
52. Zhang, X.; Chen, S.; Shi, L.; Gong, D.; Zhang, S.; Zhao, Q.; Zhan, D.; Vasseur, L.; Wang, Y.; Yu, J.; et al. Haplotype-resolved genome assembly provides insights into evolutionary history of the tea plant *Camellia sinensis*. *Nat. Genet.* **2021**, *53*, 1250–1259. [CrossRef] [PubMed]
53. Wang, S.; Liu, S.; Liu, L.; Li, R.; Guo, R.; Xia, X.; Wei, C. miR477 targets the phenylalanine ammonia-lyase gene and enhances the susceptibility of the tea plant (*Camellia sinensis*) to disease during Pseudopestalotiopsis species infection. *Planta* **2020**, *251*, 59. [CrossRef] [PubMed]

plants

MDPI

Article

Analysis of the Genetic Diversity in Tea Plant Germplasm in Fujian Province Based on Restriction Site-Associated DNA Sequencing

Lele Jiang [1,†], Siyi Xie [2,†], Chengzhe Zhou [1,3], Caiyun Tian [1], Chen Zhu [4], Xiaomei You [5], Changsong Chen [5], Zhongxiong Lai [1,3] and Yuqiong Guo [1,6,*]

1 College of Horticulture, Fujian Agriculture and Forestry University, Fuzhou 350002, China; lelejiang0625@foxmail.com (L.J.); chengzhechou@foxmail.com (C.Z.); cytian1997@foxmail.com (C.T.); laizx01@163.com (Z.L.)
2 Key Laboratory of Tea Science of Ministry of Education, Hunan Agricultural University, Changsha 410128, China; siyixie979@foxmail.com
3 Institute of Horticultural Biotechnology, Fujian Agriculture and Forestry University, Fuzhou 350002, China
4 Guangdong Provincial Key Laboratory of Applied Botany & Key Laboratory of South China Agricultural Plant Molecular Analysis and Genetic Improvement, South China Botanical Garden, Chinese Academy of Sciences, No. 723 Xingke Road, Tianhe District, Guangzhou 510650, China; zhuchen19921118@foxmail.com
5 Tea Research Institute, Fujian Academy of Agricultural Sciences, No. 104 Pudang Road, Xindian Town, Jin'an District, Fuzhou 350012, China; yxm0593@163.com (X.Y.); ccs6536597@163.com (C.C.)
6 Anxi College of Tea Science (College of Digital Economy), Fujian Agriculture and Forestry University, Quanzhou 362400, China
* Correspondence: guoyq828@163.com
† These authors contributed equally to this work.

Abstract: Fujian province, an important tea-producing area in China, has abundant tea cultivars. To investigate the genetic relationships of tea plant cultivars in Fujian province and the characteristics of the tea plant varieties, a total of 70 tea cultivars from Fujian and other 12 provinces in China were subjected to restriction site-associated DNA sequencing (RAD-seq). A total of 60,258,975 single nucleotide polymorphism (SNP) sites were obtained. These 70 tea plant cultivars were divided into three groups based on analyzing the phylogenetic tree, principal component, and population structure. Selection pressure analysis indicated that nucleotide diversity was high in Southern China and genetically distinct from cultivars of Fujian tea plant cultivars, according to selection pressure analysis. The selected genes have significant enrichment in pathways associated with metabolism, photosynthesis, and respiration. There were ten characteristic volatiles screened by gas chromatography–mass spectrometry (GC–MS) coupled with multivariate statistical methods, among which the differences in the contents of methyl salicylate, 3-carene, cis-3-hexen-1-ol, (*E*)-4-hexen-1-ol, and 3-methylbutyraldehyde can be used as reference indicators of the geographical distribution of tea plants. Furthermore, a metabolome genome-wide association study (mGWAS) revealed that 438 candidate genes were related to the aroma metabolic pathway. Further analysis showed that 31 genes of all the selected genes were screened and revealed the reasons for the genetic differences in aroma among tea plant cultivars in Fujian and Southern China. These results reveal the genetic diversity in the Fujian tea plants as well as a theoretical basis for the conservation, development, and utilization of the Fujian highly aromatic tea plant cultivars.

Keywords: *Camellia sinensis*; genetic diversity; selection pressure; genome-wide association analysis

1. Introduction

The tea plant (*Camellia sinensis* (L.) O. Kuntze), an evergreen woody plant that originated in Southwest China, has been cultivated for approximately 3000 years and is now widely cultivated in more than 50 countries and regions around the world [1]. The majority of cultivated tea plants belong to the genus *Camellia* L., section *Thea* (L.) Dyer, in the family Theaceae and are categorized as one of two main varieties: *C. sinensis* var. *sinensis* (CSS) and *C. sinensis* var. *assamica* (Masters) Chang (CSA) [2]. The tea plants in China are mainly grown in Southwest China, Southern China, south of the Yangtze River, and north of the Yangtze River regions [3]. Fujian province is abundant in tea plant cultivars and tea varieties [4]. It is the birthplace of oolong tea, black tea, and white tea among the six tea categories, and the aroma for a relatively high proportion in the quality evaluation, especially in oolong tea [5,6]. Subsequent to its initial domestication, tea was further bred and cultivated to enhance certain organoleptic traits, primarily taste and aroma, as well as biotic and abiotic stress resistance properties [7]. Among these, aroma is an important indicator of the flavor quality of tea, determined by the genetic characteristics of tea plant cultivars. The components and contents of aroma are influenced by gene sequence, gene expression, and other factors, with gene sequence variation in different species being the constitutive factor that directly influences aroma [8–10]. The rapid and precise identification of genes that provide a favorable expression of traits is a high priority, and while genes related to tea plant type and leaf phenotype have been identified by researchers [11–13], fewer studies have been conducted to analyze the causes of aroma differences in different tea plant cultivars from the perspective of tea plant genetic diversity.

With the rapid development of high-throughput sequencing technology, reduced-representation genome sequencing (RRGS) has been widely used as an efficient marker development technology [14,15]. Amongst a significant amount of technology, thousands of SNP markers have been obtained using restriction site-associated DNA sequencing (RAD-seq) [16]. Additionally, with the development of mass spectrometry (MS) platforms and sequencing technology, metabolome-based genome-wide association studies (mGWAS) have successfully been applied to exploit natural genotypic variation and identify metabolic quantitative trait loci (mQTL) responsible for metabolomic variation in plants [17–21]. Based on RRGS data, the phylogenetic relationships, genetic diversity, and geographical distribution of tea plants have been revealed [11,22–24]. Analyses of genetic diversity and the identification of genes associated with excellent traits play an essential role in the breeding of superior varieties. Recently, based on mGWAS, the metabolic diversity of tea plants was revealed and the biosynthesis of tea plant flavonoids was further updated [17]. Genome-wide association analysis was conducted on 11 leaf-related traits of ancient tea plant populations in Yunnan and Guizhou, and four candidate genes that may be involved in regulating plant size, leaf color, and other related traits were screened [11]. Nevertheless, metabolite diversity among different tea plant cultivars and the identification of related genes remain to be explored.

In this study, the genetic relationships of 70 tea plant cultivars from Fujian and other 12 provinces in China were analyzed by RAD-seq to reveal the origin and evolutionary spread of tea plants in Fujian province. In addition, we also examined the volatiles of 70 tea plant cultivars, conducted mGWAS analysis of tea plant characteristic aroma traits based on SNP loci, and combined this with selection pressure analysis to investigate the causes of genetic differences in the aroma of different tea plant cultivars and accurate positioning key genes related to the aroma forming in the tea plant, which will provide scientific reference for the selection and breeding of tea plant cultivars with highly aromatic aromas [25].

2. Result

2.1. Overview of Sequencing Quality of 70 Tea Plant Cultivars

Seventy tea cultivars were used for the analysis of this study (Supplemental Table S1). The "Tieguanyin" genome (PRJCA003090) was selected as the reference genome by comparison of published genomic data [26]. A total of 1,686,990 restriction enzyme fragments were obtained by e-enzymatic digestion, and 1,135,399 effective restriction enzyme fragments were obtained. The effective restriction enzyme fragments were uniformly distributed in the genome, and the enzyme cutting exhibited a positive effect by using restriction endonuclease EcoR I, which can be used for library construction and sequencing (Figure 1a).

After sequencing, a total of 3.02 Gb reads of raw sequence data were obtained from 70 tea plant cultivars and 2.97 Gb of clean reads were valid after filtering (Supplemental Table S2). The average proportion of sequenced Q30 was 92%, the average of GC was 40%, and the sequence length of each sample was 150. The filtered high-quality reads were compared with the reference genome, and the comparison rate averaged 97.55%, which indicated that the distribution of base content was normal, and the quality of sequencing was high enough to meet the needs of subsequent analysis (Supplemental Table S3).

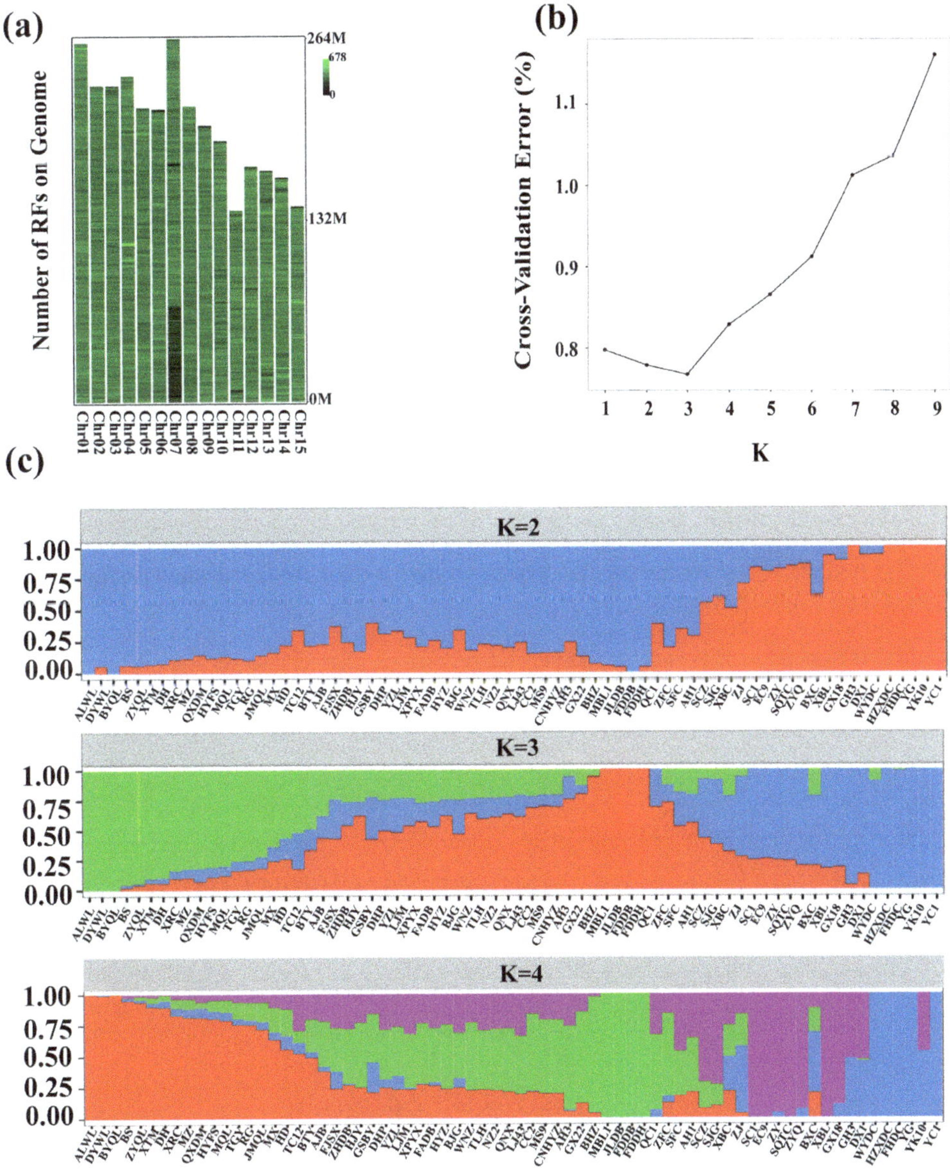

Figure 1. Population structure of 70 tea plant cultivars. (**a**) Effective-RF chromosome distribution map. (**b**) Cross-validation of K values. (**c**) Stacked diagram of tea plant community structure at K = 2, K = 3, and K = 4.

2.2. SNP Variation and Genetic Relationships Analysis of 70 Tea Plant Cultivars

A total of 60,258,975 SNP loci were detected in the 70 tea plant cultivars, of which 59,666,249 SNP loci were distributed on 15 chromosomes (Supplemental Table S4). The conversion-to-subversion ratio (Ts/Tv) was 1.8, with a higher conversion rate of 20.33% for C/T (12,724,288) (Supplemental Figure S1a). In addition, the number of insertion and deletion variants were detected as 1,009,010 and 1,528,058, respectively, with lengths mainly ranging from 1 to 8 bp and from 1 to 2 bp (Supplemental Figure S1b).

To illustrate the genetic relationship of 70 tea populations, population structure, principal component analysis (PCA), and a phylogenetic tree were constructed based on the SNPs. The values of cross-validation (CV) error were calculated for each K to select an optimal number of populations. The results show that the CV value reached its lowest value when K = 3, which indicated that the optimal number of populations should be three (Figure 1b,c). The results of the principal component analysis showed (Figure 2a) that some tea plants in Yunnan, Guangdong, and Guangxi provinces were distributed far away from other local tea plants. However, there was a severe overlap between Fujian province tea plants and the rest of the tea plant cultivars from other provinces. Therefore, further analysis of the overlapping areas (Figure 2b) revealed that group 1 was mainly composed of tea plant cultivars from Southern Fujian, Northern Fujian, and Taiwan, whereas group 2 was mainly composed of tea plant cultivars from Eastern Fujian, Northern Fujian, Zhejiang, Anhui, and Sichuan. The genetic diversity of 70 tea plant cultivars was also analyzed by conducting a phylogenetic tree. Based on the SNPs, 70 tea plant cultivars were classified into three subgroups (Figure 2c), which were in accordance with the results of the population structure and PCA. To further analyze the genetic relationship of tea plant cultivars in Fujian province, 34 tea plant cultivars in Fujian province were individually constructed into a phylogenetic tree. The results show (Figure 2d) that all the 34 tea plant cultivars were divided into two groups. Most of the tea plant cultivars from Southern Fujian and two from Northern Fujian were clustered into group I. All the tea plant cultivars from Eastern Fujian and most of the tea plant cultivars from Northern Fujian were clustered into one subgroup in group II. Meanwhile, the XBC from Western Fujian and BXC from Southern Fujian are closely related. The remaining tea plant cultivars from Southern Fujian are independent of other cultivars. All the above results are consistent with the results of the phylogenetic tree constructed from 70 tea plant cultivars.

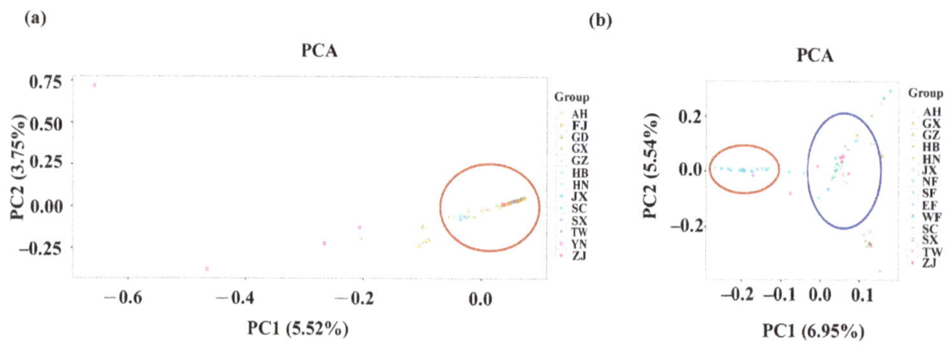

Figure 2. *Cont.*

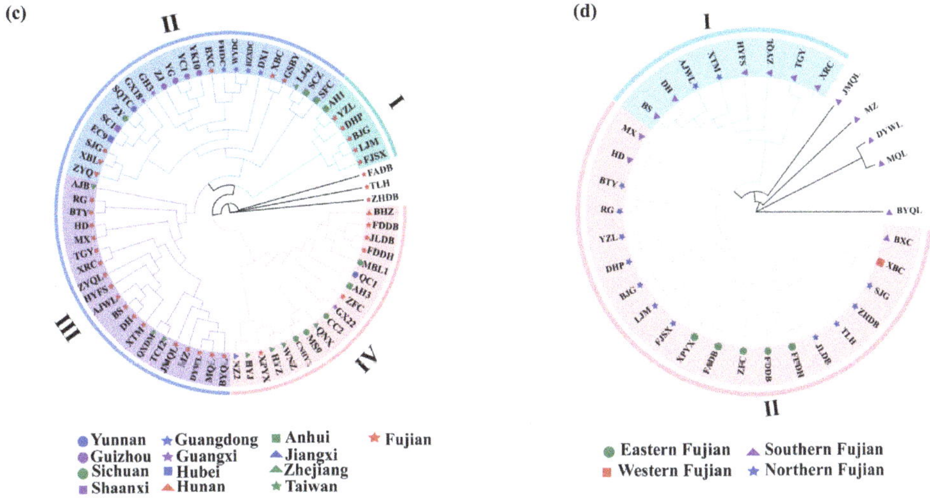

Figure 2. (**a**) Principal component analysis (PCA) of 70 tea plant cultivars; (**b**) The PCA of the overlapping segment in figure (**a**); (**c**) Phylogenetic tree of 70 tea plant cultivars; (**d**) Phylogenetic tree of tea plant cultivars in Fujian province. AH: Anhui; FJ: Fujian; GD: Guangdong; GX: Guangxi; GZ: Guizhou; HB: Hubei; HN: Hunan; JX: Jiangxi; SC: Sichuan; SX: Shaanxi; TW: Taiwan; YN: Yunnan; ZJ: Zhejiang; NF: Northern Fujian; EF: Eastern Fujian; SF: Southern Fujian; WF: Western Fujian.

2.3. Selective Pressure Analysis

Based on the filtered SNPs, nucleotide diversity (π) and genetic differentiation index (Fst) were calculated for Fujian, Eastern China excluding Fujian (EC-F), Southern China, Central China, and Western China (Supplemental Table S1) [27]. The π-value and Fst were both higher in Fujian and Southern China compared to the other three regions, indicating the greatest differentiation between them (Supplemental Figure S2). To explore the differences between tea plant cultivars in Fujian and other regions, the pairwise comparison of tea plant cultivars in Fujian and other regions was conducted for selection pressure analysis. There, 202, 158, 295, and 180 genes were screened in the selected regions from EC-F and Fujian (ECF), Southern China and Fujian (SCF), Central China and Fujian (CCF), and Western China and Fujian (WCF), respectively (Supplemental Figures S4 and S5).

To further explore the functions of the selected genes, these genes were annotated based on GO and KEGG databases. The GO enrichment analysis results show that the selected genes in the four subgroups were mainly enriched in the pathways related to respiration and photosynthesis and had the highest enrichment associated with cellular components (Supplemental Figure S6). KEGG enrichment analysis of the selected genes showed (Supplemental Figure S7) that the top 20 KEGG enrichment pathways were involved in phytohormone signaling, plant–pathogen interactions, phenylpropane biosynthesis, fatty acid biosynthesis, respiration, photosynthesis, etc. There were a large number of functionally enriched genes in respiration, photosynthesis, and other metabolism-related pathways in each subgroup, and the metabolic pathway had the largest number of enriched genes, which was consistent with the results of GO enrichment analysis. The selected genes related to the four subgroups of metabolic pathways were compared and analyzed together with the GO enrichment analysis results, and three genes (*Peroxidase 4-like*, *ATPA*, and *COX2*) were found to be selected in both ECF and SCF, among which *ATPA* and *COX2* were consistent with the GO annotation results in both regions. Two genes, *ND1* and *ND2*, were selected in both SCF and WCF. Three genes (*GLU3*, *HEXO3*, and *SAMDC*) were selected in CCF and ECF, and one gene (*PCAS-1*) was selected in CCF and WCF.

2.4. Overall Profile of Volatiles in 70 Tea Plant Cultivars

A total of 27 major volatiles were detected in the 70 tea leaves by GC–MS and classified based on the composition of the volatiles, including eight alcohols, eight aldehydes, eight hydrocarbons, and three lipids (Supplemental Table S5). The 20 volatiles were common to 70 tea plants, and the total volatile contents ranged from 688.26 to 125,924.90 μg/kg. To explore the differences in volatiles of tea plant cultivars under different geographical origins, the proportions of volatile content of each category were analyzed and it was found that the volatiles in all regions were mainly alcohols and carbon compounds (Supplemental Figure S8). A hierarchical clustering analysis (HCA) heatmap was conducted based on the volatile content of all tea plant cultivars, showing that 70 tea plant cultivars were clustered into two groups (Figure 3a). The overall volatile contents in the first category were lower than that in the second category, which mainly contains tea plant cultivars from Fujian, Sichuan, and Zhejiang, with the highest content of *cis*-3-hexen-1-ol and (*E,E*)-2,4-hexadienal. Tea plant cultivars from ten provinces (Anhui, Fujian, Zhejiang, Sichuan, Guizhou, Guangdong, Guangxi, Yunnan, Taiwan, and Shaanxi) were clustered into the second category, with a higher content of 3-carene and (*E*)-4-hexen-1-ol.

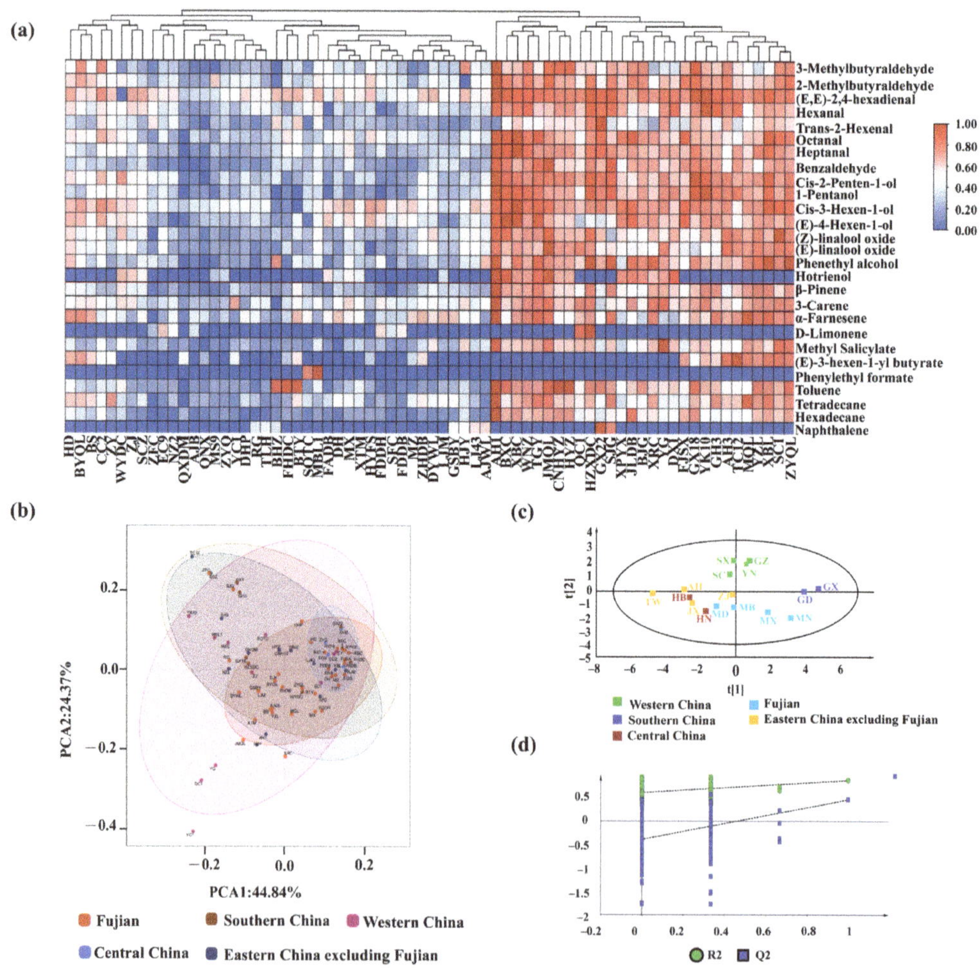

Figure 3. *Cont.*

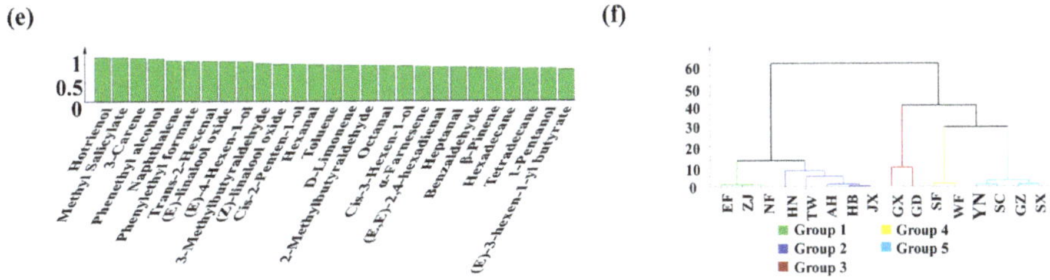

Figure 3. PCA of volatile and orthogonal projections to latent structures discriminant analysis (OPLS-DA) of volatiles in fresh tea leaves. (**a**) A hierarchical clustering analysis (HCA) heatmap of volatiles of 70 tea plants; (**b**) PCA of volatile in fresh tea leaves; (**c**) OPLS-DA score plot; (**d**) Permutation test plot; (**e**) The variable importance in the project (VIP) of 27 volatiles; (**f**) HCA plot (EF: Eastern Fujian; ZJ: Zhejiang; NF: Northern Fujian; HN: Hunan; TW: Taiwan; AH: Anhui; HB: Hubei; JX: Jiangxi; GX: Guangxi; GD: Guangdong; SF: Southern Fujian; WF: Western Fujian; YN: Yunnan; SC: Sichuan; GZ: Guangzhou; SX: Shaanxi).

2.5. Analysis of Characteristic Volatile in 70 Tea Plant Cultivars

The contributions of volatiles to tea aroma cannot completely depend on the amount of each volatile; therefore, we used odor activity value (OAV $\geq$ 1) to calculate the significance of the characteristic volatiles in 70 tea cultivars (Supplemental Table S6). The results show that a total of 15 volatiles revealed OAVs $\geq$ 1 in 70 tea cultivars, including seven aldehydes, four alcohols, three hydrocarbons, and one ester. Most of the aldehydes presented fresh and fruity aromas, among them, 3-methylbutyraldehyde, 2-methylbutyraldehyde, (*E*,*E*)-2,4-hexadienal, hexanal, octanal, and heptanal were the volatiles with OAVs $\geq$ 1 in more than 65 tea plant cultivars. *Cis*-3-hexen-1-ol and (*E*)-4-hexen-1-ol (grassy and fruity aromas) as alcohols and 3-carene (citrus-like and woody aromas) as a hydrocarbon were the volatiles both with OAVs $\geq$ 1 in most of tea plant cultivars. Of the esters, only methyl salicylate (peppermint aroma) was a volatile with OAV $\geq$ 1 in some tea plant cultivars.

To distinguish the geographically characterized volatiles in different tea plant cultivars, differential metabolites were screened by using PCA (Figure 3b) and orthogonal projections to latent structures discriminant analysis (OPLA-DA) (Figure 3c) simultaneously. It was found that both models could achieve good separation. The PCA results show that the tea plant cultivars from Yunnan were separated from other regions, and most tea plant cultivars from Fujian were clustered with Zhejiang, Anhui, and Sichuan. In the permutation test of validation of the OPLS-DA model (Figure 3d), $R^2Y = 0.708$ and $Q^2 = 0.0532$, indicating a good model fit. Subsequently, the variable importance in the project (VIP) was calculated (Figure 3e), and the volatiles with both VIP > 1 and OAVs $\geq$ 1 were selected as potential difference volatiles for further identification. A total of five components were identified as characteristic volatiles in 70 tea plant cultivars, including methyl salicylate, 3-carene, *cis*-3-hexen-1-ol, (*E*)-4-hexen-1-ol, and 3-methylbutyraldehyde. Ultimately, the HCA of characteristic volatiles showed that tea plant cultivars from different regions could be clearly distinguished based on the above five volatiles (Figure 3f).

2.6. MGWAS Analysis and Candidate Gene Screening

It revealed that the decay distance of genome-wide LD for the whole population was from 0 to 50 kb, where the R^2 dropped from 0.1 to 0.05 (Supplemental Figure S10a). Due to the faster rate of decay, we concluded that the correlation analysis was highly accurate. Ten characteristic volatiles (including α-farnesene, 3-carene, *cis*-2-penten-1-ol, (Z)-linalool oxide, (*E*,*E*)-2,4-hexadienal, (*E*)-4-hexen-1-ol, (*E*)-linalool oxide, hexanal, cis-3-hexen-1-ol, and methyl salicylate) from 70 tea cultivars were selected as phenotypic traits for subsequent mGWAS analysis. The skewness and kurtosis of the 10 traits were positive, and 3-carene had the highest kurtosis values and a high probability of data extremes (Table 1). Subsequently, association analysis was conducted between high-quality SNP loci and volatile phenotype data by using the MLM (K) model (Figure 4b1–b10). As a result (Figure 4a1–a10), α-farnesene and *cis*-2 pentan-1-ol were significantly associated with 963 and 56 markers and had the highest degree of association with Chr03_179317521 and Chr03_9258569, respectively. 3-carene (significantly associated with 2797 markers), (Z)-linalool oxide (significantly associated with 468 markers), (*E*,*E*)-2,4-hexadienal (significantly associated with 29 markers), (*E*)-4-hexen-1-ol (significantly associated with 1362 markers), and (*E*)-linalool oxide (significantly associated with 529 markers) had the highest degree of association with Chr14_122731139, Chr08_16489975, Chr07_118143306, Chr06_206575794, and Chr10_47770141, respectively. Hexanal was significantly associated with 298 markers and had the highest degree of association with Chr04_95153832, whereas *cis*-3-hexen-1-ol was significantly associated with 341 markers and had the highest degree of association with Chr01_95142047. Methyl salicylate was significantly associated with 1495 markers and had the highest degree of association with Chr05_33427085. Based on the SNP loci screened by mGWAS analysis and the reference genome, genes near 50 kb of significant loci on the reference genome were used as candidate genes. Subsequently, to screen the genes associated with aroma-related metabolic pathways, all candidate genes were subjected to GO and KEGG enrichment analysis, which showed that 38 aroma-related genes were found in α-farnesene, 154 for 3-carene, three for *cis*-2-penten-1-ol, 31 for (Z)-linalool oxide, one for (*E*,*E*)-2,4-hexadienal, 64 for (*E*)-4-hexen-1-ol, 41 for (*E*)-linalool oxide, 11 for hexanal, 16 for *cis*-3-hexen-1-ol, and 79 for methyl salicylate. The above 438 trait aroma-related genes were considered candidate genes.

Table 1. Characteristics of volatiles in fresh tea leaves.

Trait	Mean	Standard Deviation	Standard Error	Median	Min	Max	Range	Skew	Kyrtosis
α-farnesene	965.16	1710.24	204.41	300.42	5.5	10,144.3	10,138.79	3.45	13.75
3-Carene	2804.67	4639.92	554.58	1260.3	29.59	33,451.74	33,422.15	4.38	24.73
Cis-2-pentenol-1-ol	315.86	363.03	43.39	132.32	12.09	1681.63	1669.54	1.69	2.72
(Z)-linalool oxide	258.42	362.56	43.33	129.3	8.39	2141.99	2133.6	2.82	9.68
(*E*,*E*)-2,4-hexadienal	168.19	183.01	22.03	89.2	6.43	809.91	803.48	1.56	1.91
(*E*)-4-hexen-ol	879.54	1552.29	185.53	353.97	7.99	8209.08	8201.09	3.04	9.39
(*E*)-linalool oxide	615.06	920.17	109.98	272.16	19.98	6074.06	6054.08	3.06	15.51
Hexanal	184.14	244.71	29.25	73.66	3.89	1314.8	1337.91	2.23	5.88
Cis-3-hexen-1-ol	768.71	1054.45	126.03	350.32	1.61	6203.63	6202.02	2.57	8.55
Methyl salicylate	149.29	247.97	29.64	52.39	6.24	1585.38	1579.14	3.63	15.81

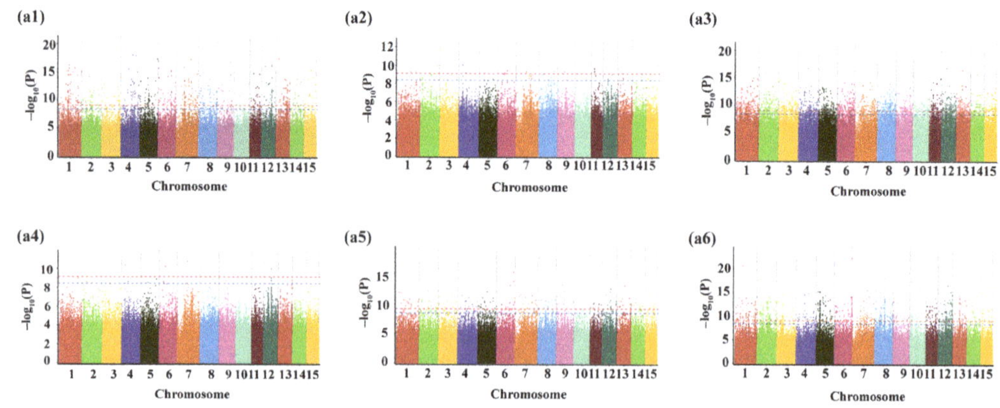

Figure 4. *Cont.*

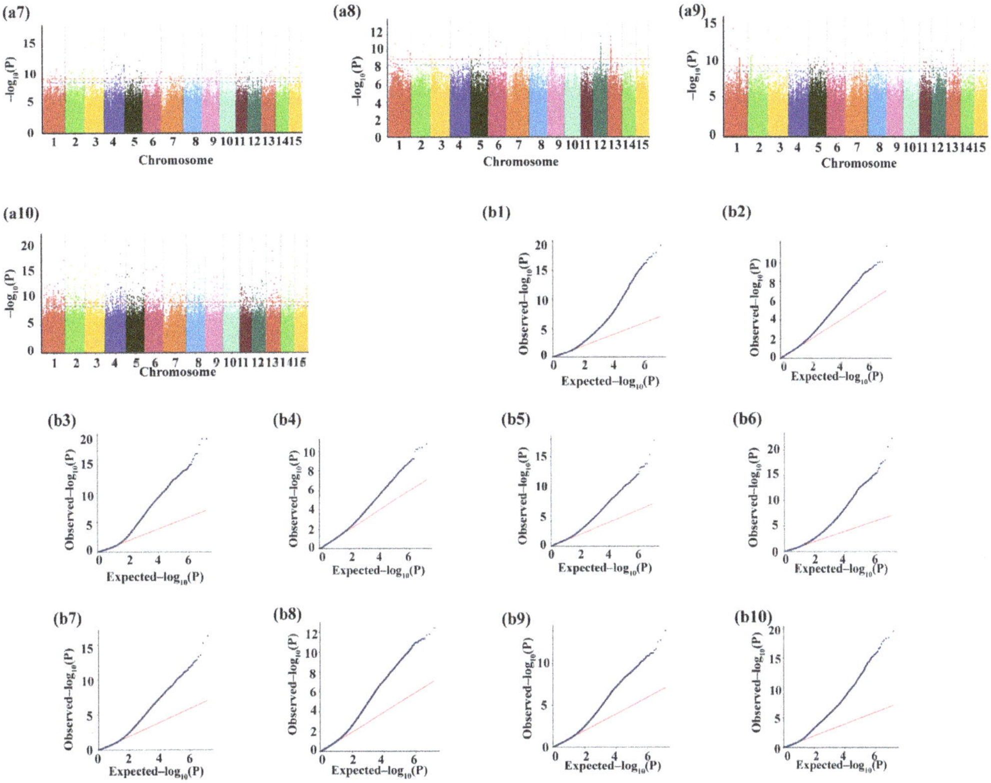

Figure 4. (**a1–a10**) Manhattan plots of α-farnesene association analysis (**a1**); *cis*-2-penten-1-ol association analysis (**a2**); 3-carene association analysis (**a3**); (Z)-linalool oxide association analysis (**a4**); (*E*,*E*)-2,4-hexadienal association analysis (**a5**); (*E*)-4-hexen-1-ol association analysis (**a6**); (*E*)-linalool oxide association analysis (**a7**); hexanal association analysis (**a8**); *cis*-3-hexen-1-ol association analysis (**a9**); methyl salicylate association analysis (**a10**); The blue and red dotted lines show significant associations between SNPs and metabolites value. (**b1–b10**) Quantile–Quantile (Q–Q) plots of the five traits in the same order with Manhattan plots showing the expected null distribution of *p*-value assuming no associations. The red line represents the predicted value and the blue dot represents the observed value, which can show the difference between the predicted value and the observed value.

2.7. Screening and Functional Analysis of Volatile-Associated Genes Subject to Selection Pressure

The analysis of selection pressure on tea plant cultivars in different regions showed that the nucleotide diversity in tea plant cultivars was high in Southern China, and the tea plant cultivars in Southern China were among the most distant from those in Fujian. Therefore, further analysis of the reasons for aroma differences between Fujian and Southern China during tea plant evolution was conducted. A total of 31 genes of all selected genes were found ($p < 0.05$); among them, five genes were significantly associated with α-farnesene, five for 3-carene, four for (*E*)-4-hexen-1-ol, eight for (*E*)-linalool oxide, and nine for methyl salicylate (Supplemental Figure S11). Furthermore, GO enrichment analysis of the above 31 genes revealed that these genes were mainly enriched in pathways related to metabolic and cells (Supplemental Figure S12). The results show that the ACT structural domain protein gene *ACR10*, LRR-like receptor serine/threonine protein kinase gene *At3g47570*, stress response protein NST1-like gene, pectin acetyl esterase gene *PAE8*, and 28S ribosomal protein genes were related to α-farnesene. Gag-Pol polyprotein gene, reactive oxygen species-related gene, O-fucosyltransferase gene *OFUT30*, WAT1 protein-related gene *At2g39510*, and pumilio domain-containing protein gene *APUM7* were related to 3-carene. AP-1 complex subunit gene *AP1M2*, solute carrier family 25 member gene *CIMG_07070*, receptor protein gene *RLP21*, and V-type proton ATPase gene were related to (*E*)-4-hexen-1-ol. Cytochrome C oxidase gene *COX2*, mitochondrial protein gene *AtMg00030*, ribosomal protein genes *RPS13* and *RPL5*, mitochondrial complex III subunit gene *MT-CYB*, WAT1 protein-related gene *At2g39510*, and pumilio domain-containing protein gene *APUM7* were related to (*E*)-linalool oxide. The genes related to methyl salicylate included chaperone protein genes, receptor-like protein 14 genes *RLP21* and *RLP14*, F-box/kelch-repeat protein gene *At1g60070*, nuclear pore complex protein gene *NUP98A*, E3 ubiquitin-protein ligase gene *RFI2*, and some unknown function genes.

3. Discussion

As a cash crop in China, the research of tea plant *cultivars* is the key to improve the tea quality [28]. Lou et al. studied the diversity in biochemical components of 76 tea varieties in Lishui, proving that tea plant *cultivars* have great selection potential and rich diversity [29]. In this study, we found that genes related to the quality and photosynthesis of tea plants were all subjected to selection during the evolutionary process, which is of great significance for breeding new tea plant varieties with high quality. Aroma plays a vital role in the quality of tea, and in addition to the tea manufacturing process, tea cultivars are the basis for the formation of tea aroma [8]. The differences in aroma of different tea plant varieties have been revealed and the candidate genes associated with volatiles were identified in this study, which will provide a reference for the establishment of highly aromatic core tea *plant cultivars* in Fujian.

3.1. Extensive Exchanges between the Tea Plant Cultivars in Fujian and Other Regions

Genetic diversity analysis has always been of great significance for the evaluation and utilization of crop germplasm, as well as for the study of origin and evolution [30]. Analysis of population structure demonstrated that tea cultivars in thirteen provinces were clustered into three groups (Figure 1b). However, most of the tea plant cultivars among the subgroups were genetically infiltrated, except for those in Northern Fujian, Yunnan, and Guangdong, indicating frequent genetic exchange among tea plants, which is consistent with the results of previous studies [2,26,31]. The Anxi in Southern Fujian and Fuding in Eastern Fujian tea populations are highly divergent, while the Wuyishan and Zhenghe in Northern Fujian have a high frequency of gene exchange with the Anxi in Southern Fujian and Fuding and Fuan in Eastern Fujian tea populations, which was confirmed by Nie et al. [32] These results will provide a reference for the study of molecular breeding of tea plant in Fujian Province. By analyzing the genetic diversity in oolong tea plant cultivars, it was found that Taiwan and Fujian oolong teas clustered into one taxon and there was a genotypic intersection between Southern Fujian and Northern Fujian tea plant cultivars [33]. Wuyi Mountain in Fujian province is an important mountain range in Eastern China, spanning Fujian, Guangdong, Jiangxi, and Zhejiang from north to south, which gradually changed from ocean to land during the Cretaceous Period of the Mesozoic Era, and then rose during the Ice Age at the end of the Quaternary Era, resulting in the creation of folded and fractured valleys between the mountains [34]. The geological age of Wuyi Mountain, the rocky nature of the soil and the presence of wild tea are a series of characteristics that suggest that the tea plant of Wuyi Mountain evolved from a homologous "isolated distribution" into the "Wuyi taxon" [35]. "Wuyi Caicha" occupies an important position in the taxonomy of tea plants, and TLH, SJG, BJG, and BTY were bred from it [36]. Therefore, we speculated that some tea plants in Fujian may have developed and evolved from northern Fujian due to geological changes [35,37].

3.2. Genes Related to Quality and Photosynthesis of Tea Plants May Be Selected during Evolutionary Processes

Tea quality has always been the focus of tea research. Accurate targeting of tea quality-related genes during tea plant evolution is significant for the cultivation of new high-quality tea plant varieties [38]. In this study, the annotation of selected genes by selection pressure analysis revealed that the gene *ACR10* was selected in all four populations (Supplemental Figure S6). The gene *ACR10* is involved in encoding ACR proteins, and all ACR proteins with multicopy ACT domains bind amino acids. For example, 3-phosphoglycerate dehydrogenase, which catalyzes the first step of serine synthesis, can bind serine; and branch acid translocase, which catalyzes the first two steps of phenylalanine synthesis, can bind phenylalanine. In other words, the terminal product amino acid can be bound to the ACT domain of its synthase, and the activity of the synthase can be regulated according to the concentration of the product amino acid, thus regulating the metabolism of the amino acid [39–41]. At present, 26 amino acids have been identified in tea leaves, which play an important role in the function and quality of tea leaves and their growth [42,43]. It was found that there were significant differences in amino acid contents among different groups of tea plant cultivars by analyzing the diversity in biochemical components of tea plant cultivars [44]. This may be the reason why tea plant cultivars from the Fujian region differ from those of other regions in terms of tea quality [45]. The selected differential genes in tea plant cultivars from ECF and CCF are mainly associated with enzyme activities that regulate plant growth and development, among which *SAMDC*-encoding S-adenosylmethionine decarboxylase was involved in caffeine biosynthesis. Caffeine plays an important role in plant defense against biotic stresses as a secondary metabolite of tea leaves, and its content, together with tea polyphenols, determines the suitability of *Camellia sinensis* for tea production and tea flavor [46,47]. Similar results were found in the study by Yang et al., suggesting that genes associated with tea leaf quality in tea plants were selected during long-term evolution [24].

In mainland China, the highly suitable area for tea planting covers about 16.83% of the continental area [48]. Studies have shown that climate change has important effects on plant growth and geographic distribution, resulting in large differences in genetic diversity, geographic distribution, community composition, and function among different tea plants [49]. Most of the selected genes were found to be involved in regulating photosynthesis and respiration in plants in this study (Supplemental Figure S6). As an important climatic condition, light is inseparable from the growth and development of plants. By analyzing the leaves of tea plants, from Guizhou, Sichuan, Hunan, Fujian, Zhejiang, and Guangdong provinces, which were planted in the same region, it was found that the differences between photosynthetic pigments and net photosynthetic rate were due to the genetic characteristics of the tea varieties [50]. Additionally, the geographic distribution of the tea plants was affected by the climate environment with regularity [51]. There was diversity in photosynthesis and physiological activity among different genetic populations of tea plants, and rich genetic variation in photosynthetic characteristics during adaptation to new ecological environments [52]. The climate in China is complex and diverse, and plant photosynthesis is susceptible to environmental stress; thus, variation in photosynthesis-related genes under selection pressure in different regions has led to differentiation among populations of tea plants [53]. Hence, we hypothesized that photosynthesis plays an important role in the evolution of tea plant propagation in terms of its adaptive capacity, resulting in genetic variation among populations of tea plants distributed in different regions with different genetic characteristics.

3.3. Selection-Pressured Volatile-Associated Genes May Be Markers for Screening Highly Aromatic Tea Plant Cultivars

Although the climate, soil conditions, and mountainous topography of both Fujian and Southern China are similar [54,55], there is a significant genetic gap between the two tea cultivars. It has been shown that there exists a high level of metabolite diversity in different tea populations and varieties, and that genetic background has a stronger effect on metabolite contents than environmental factors [56]. The 70 tea plant cultivars selected for this study were all planted in the same cultivation system; thus, it was inferred that the differences between varieties caused the differences in volatile fractions between tea cultivars in Fujian and South China. A total of 31 genes were screened by the association analysis of selection pressure and volatiles (Supplemental Figure S11), and some of them were associated with two volatiles. In this study, 3-carene, a volatile with a citrus-like aroma, was the main contributor to the volatiles from different tea plant cultivars. (Supplemental Table S6). Among the five genes significantly related to 3-carene, the Gag-pol polymer gene was involved in resistance regulation during plant growth [57,58]; The pumilio proteins may be involved in a wide range of post-transcriptional regulatory events, which are important for the rapid response of plants to changes in environmental conditions and throughout the developmental process [59]. Linalool and its oxides, α-farnesene, etc., as important aromatic compounds in the leaves of the highly aromatic tea plant variety HD were found by Wang et al. [8]. The *ACR10* gene is involved in encoding the acr protein, which binds to phenylalanine [60,61]. In addition, aromatic cyclic substances in tea are synthesized from phenylalanine via the shikimic acid pathway and are the main components of the fruity aroma of tea leaves. Cytochrome C oxidase is mainly involved in respiration, which lays the foundation for the formation of aroma and flavor by facilitating the transformation of the intrinsic qualities of tea leaves [62]. As described, all these volatile-associated genes subject to selection pressure were involved in regulating plant growth processes and associated with highly aromatic volatiles. Thus, these genes may have been involved in the formation of aroma during the evolution of tea plants and were key genes responsible for the formation of aroma differences between tea plant cultivars in Fujian and Southern China, which can be used as markers for screening highly aromatic tea plant cultivars.

4. Materials and Methods

4.1. Sample Collection

To investigate the origin of tea plant in Fujian Province and to explore the genes related to the flavor quality of tea plant, a total of 70 tea plant cultivars (Supplemental Table S1), which are representative species from different tea zones, including Western, Central, Southern, and EC-F were selected. All samples were collected from the tea germplasm nursery of the Tea Research Institute of Fujian Academy of Agricultural Sciences (119°34′ E, 27°13′ N) and the tea germplasm resource nursery with excellent characteristics in Fujian and Taiwan of Wuyixing Tea Industry Co., Ltd., Fujian, China (118°0′ E, 27°43′ N). In the spring of April 2021, a bud with two leaves of tea plants with good growth were taken as materials, and the samples were stored in the refrigerator at −80 °C for standby after solid sampling with liquid nitrogen.

4.2. DNA Extraction, Library Construction, Sequencing, and Data Processing

Genomic DNA was extracted from the 70 tea plants by using the CTAB method [63]. DNA purity and concentration were tested by using Qubit (Thermo Fisher Scientific, Waltham, MA, USA) and Nanodrop (Thermo Fisher Scientific, Waltham, MA, USA). Based on the published genomic data for comparison, the genome with the highest comparison rate was selected as the reference genome. As far as possible to test the efficiency of enzymatic digestion and analyze all markers, the number and coverage of enzymatic segments were counted by electron microdissection prediction of the reference genome using the restriction endonuclease EcoR I (G, AATTC). After that, a series of operations were conducted, including fragment end reparation, paired-end adapter ligation, polymerase chain reaction (PCR), and library construction [64]. Ultimately, sequencing was carried out using the PE 150 strategy on a Novaseq 6000 sequencer (Illumina, San Diego, CA, USA) at Kidio Biotechnology Co., Ltd. in Guangzhou, China. The sequencing original data from the Illumina platform were screened through FASTP software (version 0.18.0).

Reads with unknown nucleotides (N) $\geq$ 10%, Phred quality score $\leq$ 20 with more than 50% base content and those containing connectors were excluded [65]. Subsequently, the filtered reads were compared to the reference genome using the software BWA (version 0.7.12). Then, the alignment results were tagged and population SNPs were detected by using Picard (version 1.129) and GATK (version 4.0), respectively.

4.3. Population Structure, Principal Component, and Phylogenetic Tree Construction Analysis

Based on the SNPs obtained, the population structure of 70 tea plants was conducted by using Admixture (version 1.3.0), and the tested K was set from 1 to 9 [66]. According to the cross-validation method, the best genetic cluster K was determined and then the results graph was plotted with R language. PCA was conducted by using GCTA (version 1.92.2) based on the degree of SNP variation among tea plant samples, and PCA scatter plots were drawn using the R language. A phylogenetic tree was conducted by using MEGA (version 4.0) according to the maximum likelihood algorithm with 1000 bootstrap replicates.

4.4. Select Pressure Analysis

To identify potential selective regions, intra-population nucleotide diversity (π), neutrality test (Tajima's D), genetic differentiation index (Fst), and diversity difference ratio (π ratio) were analyzed using PopGenome (version 2.7.5) with a 100 kb sliding window and a step size of 10 kb [67,68]. Regions subject to selection tend to be chromosomal regions with low genetic diversity within populations, as well as regions with high rates of genetic differentiation between populations. This combination is based on the results of Fst, π, and π ratio, each of which is screened for significant regions by top 1% and top 5%. The regions, with decreasing heterozygosity ($-\log10\pi \geq$ threshold), significant differences between populations (Fst $\geq$ threshold), and diversity differences between populations (π ratio $\geq$ threshold) were obtained, and the intersection of these regions was located as potential selected regions, and the genes in these selected regions were identified. Ultimately, Venn diagram mapping, as well as GO and KEGG annotation of the selected genes, were conducted based on the selected genes, and subsequent analysis was conducted for variants such as SNPs and InDel in the coding regions of the candidate genes.

4.5. Analysis of Volatiles

The test tea samples were ground into tea powder of 40 mesh sizes after freeze-drying for 48. A total of 0.5 g of the powder was transferred to a 20 mL headspace vial and infused with 3 mL saturated NaCl solution. Subsequently, an internal standard of 5 μL 4-tert-butylcyclohexanol (30 mg·L^{-1}) was added into the vial, and then the vial was sealed immediately.

The method of volatile detection was adjusted by referring to the method of Wei Wang et al. [69]. GC conditions: 50 °C held for 5 min, then increased to 125 °C at a rate of 3 °C/min, then held for 2 min; and to 180 °C at a rate of 5 °C/min, then held for 3 min; and last to 230 °C at a rate of 15 °C/min, then held for 5 min. The carrier gas was helium (percentage purity > 99.999%) at a constant flow velocity of 1 mL/min. MC conditions: Electron ionization source; electron energy, 70 eV; ion source temperature, 230 °C; mass spectrometry transmission line temperature, 250 °C; mass scan range 45–500 AMU.

The volatiles were identified tentatively by comparing their mass spectra from the NIST11 standard reference database and further positively identified with retention indices, retention times, and the data available in the published literature. The content of each substance was calculated using 4-tert-butylcyclohexanol as the internal standard [70]. Odor activity value (OAV) is commonly used to evaluate the contributions of each volatile to the aroma of a sample. Numerically, OAV is equal to the ratio of the compound concentration (Ci) to the odor threshold (OTi) in water, and compounds with OAV > 1 are generally considered to greatly contribute to the aroma characteristics [70].

4.6. Genome-Wide Association Analysis

To identify SNP loci and candidate genes associated with tea plant aroma traits, variant quality control was conducted based on the above SNP loci. The raw marker loci were filtered by self-written Perl scripts combined with Plink software (version 2.0). Loci with non-double allelic loci, minor allele frequency < 0.05, deletion rate > 0.5, and heterozygosity rate > 0.8 were filtered. Phylogenetic tree construction, PCA, LD decay analysis, population structure analysis, and genetic relationship analysis were conducted based on filtered loci (Supplemental Figure S10). And based on the results of volatile detection, descriptive statistical analysis of the volatile content of different species was conducted using R language.

The above SNP data were used for genome-wide association analysis with the phenotypic trait data of volatiles of 70 tea plant cultivars. The mGWAS model was calculated by Gemma software (version 0.94.8), where the top five PCs of the PCA calculation results were used as the Q matrix of the corresponding model, and the genetic relationship matrix was used as the K matrix of the corresponding model. A total of four models, a general linear model (GLM), a generalized linear model with correction for PCA results (GLM (Q)), a mixed linear model with correction for genetic relationship results (MLM (K)), and a mixed linear model with correction for PCA and genetic relationship results (MLM (QK)) were used for pre-analysis in this study. The final analysis model was determined according to the deviation between the actual value and the expected value of the Q-Q plot.

5. Conclusions

In this study, the origin and evolution of tea plants in Fujian province were revealed by analyzing the genetic diversity and genetic relationship in tea plants. By analyzing the aroma components of tea plants, differences in the aroma of tea plant cultivars were identified and candidate genes associated with volatiles were mined. Based on selection pressure and association analysis, key genes causing differences in tea plant aroma in different regions were screened. All these aroma-related genes among different tea plant cultivars may be developed into a genetic marker to be used in the identification process of breeding.

Supplementary Materials: The following supporting information can be downloaded at: https://www.mdpi.com/article/10.3390/plants13010100/s1, Figure S1: (a) Statistics on the percentage of SNP types. (b) Insertion deletion variant loci statistics; Figure S2: Nucleotide diversity, neutrality test, and genetic index relationships of tea plants in various regions. FJ: Fujian; EC: Eastern China excluding Fujian (EC-F); WC: Western China; CC: Central China; SC: Southern China; Figure S3: (a–e) Nucleotide diversity in tea plants in different regions. (a) Fujian; (b) EC-F; (c) Southern China; (d) Central China; (e) Western China; (f–i) Genetic differentiation index of tea plants in different regions. (f) EC-F and Fujian (ECF); (g) Southern China and Fujian (SCF); (h) Central China and Fujian (CCF); (i) Western China and Fujian (WCF); Figure S4: Top 5% selected regions of the genome. (a1–a3) are ECF; (b1–b3) are SCF; (c1–c3) are CCF; (d1–d3) are WCF; Figure S5: A number of genes identified in the top 5% of selected genome regions. (a) ECF; (b) SCF; (c) CCF; (d) WCF; Figure S6: Top 20 GO annotations of selected genes in each region. (a) ECF; (b) SCF; (c) CCF; (d) WCF; Figure S7: Top 20 KEGG annotations of selected genes in each region. (a) ECF; (b) SCF; (c) CCF; (d) WCF; Figure S8: The proportion of volatile components in fresh tea leaves by region. (a) Western China; (b) Central China; (c) Southern China; (d) EC-F; (e) Fujian; Figure S9: The content of volatile components in fresh tea leaves; Figure S10: (a) Linkage disequilibrium; (b) Phylogenetic tree based on filtered SNPs; (c) Principal component analysis (PCA) based on filtered SNPs; Figure S11: Selective pressure volatile-associated genes; Figure S12: GO annotation of genes that were significantly associated with selected genes and traits. (a) α-farnesene; (b) 3-carene; (c) (*E*)-4-hexen-1-ol; (d) (*E*)-linalool oxide; (e) methyl salicylate. Table S1: Geographical origin distribution of 70 tea plant cultivars; Table S2: Sequencing data quality summary; Table S3: Filter before and after base information; Table S4: Number of SNPs on tea plant chromosomes; Table S5: Volatile content of fresh tea leaves; Table S6: OAV value of tea plant fresh leaf volatiles. Note: VC means volatile component.

Author Contributions: Conceptualization, L.J. and S.X.; methodology, L.J., S.X. and C.Z. (Chengzhe Zhou); software, L.J. and C.T.; validation, S.X. and C.Z. (Chen Zhu); formal analysis, L.J., C.T. and S.X.; resources, S.X., X.Y., C.C. and C.Z. (Chengzhe Zhou); data curation, L.J. and S.X.; writing—original draft preparation, L.J. and S.X.; writing—review and editing, Z.L.; supervision, Y.G. and Z.L.; project administration, Y.G.; funding acquisition, Y.G. All authors have read and agreed to the published version of the manuscript.

Funding: This work was supported by the Tea Industry Branch of Collaborative Innovation Institute (K1521015A), the Construction Project for Technological Innovation and Service System of Tea Industry Chain of Fujian Agriculture and Forestry University (K1520005A01), the "Double First-class" Scientific and Technological Innovation Capacity and Enhancement Cultivation Plan of Fujian Agriculture and Forestry University (KSYLP004), and the Special Fund for Science and Technology Innovation of Fujian Zhang Tianfu Tea Development Foundation (FJZTF01).

Data Availability Statement: All of the raw sequencing data were submitted to the National Genomics Data Center (NGDC) under project number PRJCA018106.

Conflicts of Interest: The authors declare no conflict of interest.

References

1. Chen, Z.; Chen, P. The rich resources of ancient tea plants are the best proof of the origin of tea plants in the world. *Agric. Archaeol.* **2007**, *5*, 257–267.
2. Wang, X.; Feng, H.; Chang, Y.; Ma, C.; Wang, L.; Hao, X.; Li, A.L.; Cheng, H.; Wang, L.; Cui, P.; et al. Population sequencing enhances understanding of tea plant evolution. *Nat. Commun.* **2020**, *11*, 4447. [CrossRef] [PubMed]
3. Yang, Y.; Liang, Y. *Chinese Asexual Tea Plant Varieties*; Shanghai Science & Technology Press: Shanghai, China, 2014; p. 273.
4. Zhang, W.; Sun, J.; Zhu, L.; Wu, Z. Progress on production and technology development of Fujian tea industry in 60 Years. *Acta Tea Sin.* **2019**, *60*, 51–56.
5. He, X.; Rao, Q.; Liu, L.; Fu, J. Microbial community and diversity in specialty teas from Fujian. *Fujian J. Agric. Sci.* **2020**, *35*, 28–37.
6. Yang, Z.Y.; Baldermann, S.; Watanabe, N. Recent studies of the volatile compounds in tea. *Food Res. Int.* **2013**, *53*, 585–599. [CrossRef]
7. Liang, Y.; Shi, M. Advances in tea plant genetics and breeding. *J. Tea Sci.* **2015**, *35*, 103–109.
8. Wang, P.J.; Yu, J.X.; Jin, S.; Chen, S.; Yue, C.; Wang, W.L.; Gao, S.L.; Cao, H.L.; Zheng, Y.C.; Gu, M.Y.; et al. Genetic basis of high aroma and stress tolerance in the oolong tea cultivar genome. *Hortic. Res.-Engl.* **2021**, *8*, 1–15. [CrossRef]
9. Chai, K.; Chen, S.; Wang, P.; Kong, W.; Ma, X.; Zhang, X. Multiomics analysis reveals the genetic basis of volatile terpenoid formation in oolong tea. *J. Agr. Food Chem.* **2023**, *71*, 19888–19899. [CrossRef]
10. He, S.; Wang, X.; Yin, M.; Ye, J.; Meng, J.; Zhou, L. Molecular identification of Lingyun Baihao wild and cultivated tea through genome-wide sequencing. *Genet. Resour. Crop Evol.* **2023**, *70*, 1407–1417. [CrossRef]
11. Lu, L.T.; Chen, H.F.; Wang, X.J.; Zhao, Y.C.; Yao, X.; Xiong, B.A.; Deng, Y.L.; Zhao, D.G. Genome-level diversification of eight ancient tea populations in the Guizhou and Yunnan regions identifies candidate genes for core agronomic traits. *Hortic. Res. Engl.* **2021**, *8*, 1–10. [CrossRef]
12. An, Y.L.; Mi, X.Z.; Zhao, S.Q.; Guo, R.; Xia, X.B.; Liu, S.R.; Wei, C.L. Revealing distinctions in genetic diversity and adaptive evolution between two varieties of *Camellia sinensis* by whole-genome resequencing. *Front. Plant Sci.* **2020**, *11*, 603819. [CrossRef] [PubMed]
13. Zhang, F.; Tian, W.L.; Cen, L.; Lv, L.T.; Zeng, X.F.; Chen, Y.L.; Zhao, Y.C.; Li, Y. Population structure analysis and genome-wide association study of tea (*Camellia sinensis* (L.) Kuntze) germplasm in Qiannan, China, based on SLAF-seq technology. *Phyton.-Int. J. Exp. Bot.* **2022**, *91*, 791–809. [CrossRef]
14. Davey, J.W.; Hohenlohe, P.A.; Etter, P.D.; Boone, J.Q.; Catchen, J.M.; Blaxter, M.L. Genome-wide genetic marker discovery and genotyping using next-generation sequencing. *Nat. Rev. Genet.* **2011**, *12*, 499–510. [CrossRef] [PubMed]
15. Wu, X.; Wang, M.; Zheng, X.; Zhang, R.; He, S.; Yan, Y. RAD-seq revealed the endangered mechanism of *Hydrocera triflora* (*Balsaminaceae*). *Guihaia* **2023**, *43*, 1414–1427.
16. Wang, Y.; Hu, Y.; Zhang, T. Current situation and prospect of RAD-seq technology in genome research. *Heredity* **2014**, *36*, 41–49.
17. Qiu, H.; Zhang, X.; Zhang, Y.; Jiang, X.; Ren, Y.; Gao, D.; Zhu, X.; Usadel, B.; Fernie, A.R.; Wen, W. Depicting the genetic and metabolic panorama of chemical diversity in the tea plant. *Plant Biotechnol. J.* **2023**. *online ahead of print*. [CrossRef]
18. Francisco, M.; Joseph, B.; Caligagan, H.; Li, B.; Corwin, J.A.; Lin, C.; Kerwin, R.E.; Burow, M.; Kliebenstein, D.J. Genome wide association mapping in *Arabidopsis thaliana* identifies novel genes involved in linking allyl glucosinolate to altered biomass and defense. *Front. Plant Sci.* **2016**, *7*, 1010. [CrossRef]
19. Slaten, M.L.; Yobi, A.; Bagaza, C.; Chan, Y.O.; Shrestha, V.; Holden, S.; Katz, E.; Kanstrup, C.; Lipka, A.E.; Kliebenstein, D.J.; et al. mGWAS uncovers Gln-Glucosinolate seed-specific interaction and its role in metabolic homeostasis. *Plant Physiol.* **2020**, *183*, 483–500. [CrossRef]

20. Wang, X.; Liu, S.J.; Zuo, H.; Zheng, W.K.; Zhang, S.S.; Huang, Y.; Pingcuo, G.S.; Ying, H.; Zhao, F.; Li, Y.R.; et al. Genomic basis of high-altitude adaptation in Tibetan prunus fruit trees. *Curr. Biol.* **2021**, *31*, 3848. [CrossRef]
21. Wen, W.; Li, D.; Li, X.; Gao, Y.; Li, W.; Li, H.; Liu, J.; Liu, H.; Chen, W.; Luo, J.; et al. Metabolome-based genome-wide association study of maize kernel leads to novel biochemical insights. *Nat. Commun.* **2014**, *5*, 3438. [CrossRef]
22. Lei, Y.; Yang, L.; Duan, S.; Ning, S.; Li, D.; Wang, Z.; Xiang, G.; Yang, L.; Wang, C.; Zhang, S.; et al. Whole-genome resequencing reveals the origin of tea in Lincang. *Front. Plant Sci.* **2022**, *13*, 984422. [CrossRef] [PubMed]
23. Zhao, Z.F.; Song, Q.F.; Bai, D.C.; Niu, S.Z.; He, Y.Q.; Qiao, D.H.; Chen, Z.W.; Li, C.Y.; Luo, J.; Li, F. Population structure analysis to explore genetic diversity and geographical distribution characteristics of cultivated-type tea plant in Guizhou Plateau. *BMC Plant Biol.* **2022**, *22*, 55. [CrossRef] [PubMed]
24. Yang, H.; Wei, C.; Liu, H.; Wu, J.; Li, Z.; Zhang, L.; Jian, J.; Li, Y.; Tai, Y.; Zhang, J.; et al. Genetic divergence between *Camellia sinensis* and its wild relatives revealed via genome-wide SNPs from RAD sequencing. *PLoS ONE* **2016**, *11*, e151424. [CrossRef] [PubMed]
25. Ma, J.; Yao, M.; Chen, L. Research progress on germplasm resources of tea plant. *J. Tea Sci.* **2015**, *35*, 11–16.
26. Zhang, X.; Chen, S.; Shi, L.; Gong, D.; Zhang, S.; Zhao, Q.; Zhan, D.; Vasseur, L.; Wang, Y.; Yu, J.; et al. Haplotype-resolved genome assembly provides insights into evolutionary history of the tea plant *Camellia sinensis*. *Nat. Genet.* **2021**, *53*, 1250–1259. [CrossRef]
27. Zhou, W. Geographical divisions in China. *Educ. Geogr.* **2011**, *39*, 2413–2416.
28. Wang, X.; Wang, L.; Hao, X.; Li, N.; Huang, J.; Ding, C.; Yang, Y. Retrospect and prospect of development and innovation of tea plant genetics and breeding in China. *J. Huazhong Agric. Univ.* **2022**, *41*, 1–8.
29. Hualou, Y.; Liu, Y.; He, W.; Ji, Q.; Zheng, S. Diversity analysis of biochemical components of 76 tea germplasm resources in Lishui city, Zhejiang. *J. Zhejiang Agric. Sci.* **2020**, *61*, 1990–1993.
30. Sun, J.; Xu, G.; Hu, Y.; Cheng, Y.; Wang, X.; Yang, J.; Yang, M.; Xie, D.; Dai, Z. Genome-wide assessment of genetic diversity and association mapping for salt tolerance traits in okra (*Abelmoschus esculentus* L. Moench) using genotyping-by-sequencing. *Sci. Hortic. Amst.* **2023**, *313*, 111922. [CrossRef]
31. Zhang, W.; Zhang, Y.; Qiu, H.; Guo, Y.; Wan, H.; Zhang, X.; Scossa, F.; Alseekh, S.; Zhang, Q.; Wang, P.; et al. Genome assembly of wild tea plant DASZ reveals pedigree and selection history of tea varieties. *Nat. Commun.* **2020**, *11*, 3719. [CrossRef]
32. Nie, C.; Chen, D.; LI, Y. Study on genetic diversity of tea plant from Fujian Province based on matK partial sequence. *J. Biol.* **2021**, *38*, 54–58.
33. Lin, Y.; Yu, W.; Zhou, L.; Fan, X.; Wang, F.; Wang, P.; Fang, W.; Cai, C.; Ye, N. Genetic diversity of oolong tea (*Camellia sinensis*) germplasms based on the nanofluidic array of single-nucleotide polymorphism (SNP) markers. *Tree Genet. Genomes.* **2019**, *16*, 3. [CrossRef]
34. Li, L.; Luo, S.; Wang, F.; Li, X.; Feng, H.; Shi, Y.; Ye, J.; Liu, F.; Zhao, J.; Li, S.; et al. Genetic analysis and marker development for Wuyi Tea(*Camellia sinensis*, Synonym: *Thea bohea* L.) based on GBS-SNP. *J. Tea Sci.* **2023**, *43*, 310–324.
35. Zhou, Y. The origin and development of Fujian tea plants. *Tea Fujian* **2013**, *35*, 52–53.
36. Yang, J.; Wang, R.; Kong, X.; Zheng, G.; Qiu, C. Genetic diversity and relationship of 94 tea germplasm. *Acta Tea Sin.* **2018**, *59*, 186–192.
37. Xia, E.; Tong, W.; Hou, Y.; An, Y.; Chen, L.; Wu, Q.; Liu, Y.; Yu, J.; Li, F.; Li, R.; et al. The reference genome of tea plant and resequencing of 81 diverse accessions provide insights into its genome evolution and adaptation. *Mol. Plant* **2020**, *13*, 1013–1026. [CrossRef] [PubMed]
38. Fang, W.; Lei, X.; Yang, B.; Wang, Y.; Ma, Y. Application of genome-wide association analysis in studying tea. *J. Huazhong Agric. Univ.* **2022**, *41*, 33–40.
39. Grant, G.A. The ACT domain: A small molecule binding domain and its role as a common regulatory element. *J. Biol. Chem.* **2006**, *281*, 33825–33829. [CrossRef]
40. Xu, J.; You, D.; Leng, P.; Ye, B. Allosteric regulation of a protein acetyltransferase in *Micromonospora aurantiaca* by the amino acids cysteine and arginine. *J. Biol. Chem.* **2014**, *289*, 27034–27045. [CrossRef]
41. Zhang, S.; Fitzpatrick, P.F. Identification of the allosteric site for phenylalanine in rat phenylalanine hydroxylase. *J. Biol. Chem.* **2016**, *291*, 7418–7425. [CrossRef]
42. Huang, R.; Wang, J.Y.; Yao, M.Z.; Ma, C.L.; Chen, L. Quantitative trait loci mapping for free amino acid content using an albino population and SNP markers provides insight into the genetic improvement of tea plants. *Hortic. Res.-Engl.* **2022**, *9*, uhab029. [CrossRef] [PubMed]
43. Fang, K.X.; Xia, Z.Q.; Li, H.J.; Jiang, X.H.; Qin, D.D.; Wang, Q.S.; Wang, Q.; Pan, C.D.; Li, B.; Wu, H.L. Genome-wide association analysis identified molecular markers associated with important tea flavor-related metabolites. *Hortic. Res.-Engl.* **2021**, *8*, 42. [CrossRef] [PubMed]
44. Huang, H.; Zhang, W.; Ao, C.; Cui, H.; Yu, J.; Zheng, X. Diversity analysis of biochemical components in local tea germplasm resources and screening of excellent tea varieties in Hangzhou. *Acta Agric. Jiangxi* **2019**, *31*, 21–25.
45. Yu, Z.; Yang, Z. Understanding different regulatory mechanisms of proteinaceous and non-proteinaceous amino acid formation in tea (*Camellia sinensis*) provides new insights into the safe and effective alteration of tea flavor and function. *Crit. Rev. Food Sci.* **2020**, *60*, 844–858. [CrossRef]
46. Xia, E.; Zhang, H.; Sheng, J.; Li, K.; Zhang, Q.; Kim, C.; Zhang, Y.; Liu, Y.; Zhu, T.; Li, W.; et al. The tea plant genome provides insights into tea flavor and independent evolution of caffeine biosynthesis. *Mol. Plant* **2017**, *10*, 866–877. [CrossRef] [PubMed]

47. Wang, Y.; Qian, W.; Li, N.; Hao, X.; Wang, L.; Xiao, B.; Wang, X.; Yang, Y. Metabolic changes of caffeine in tea plant (*Camellia sinensis* (L.) *O. Kuntze*) as defense response to *Colletotrichum fructicola*. *J. Agr. Food Chem.* **2016**, *64*, 6685–6693.
48. Tang, J.; Wang, P.; Youhao, E.; Ma, Y.; Wu, D.; Huo, Z. Chinese mainland tea planting climate suitability zoning. *J. Appl. Meteorol. Sci.* **2021**, *32*, 397–407.
49. Zhao, Y.C.; Zhao, M.Y.; Zhang, L.; Wang, C.Y.; Xu, Y.L. Predicting possible distribution of tea (*Camellia sinensis* L.) under climate change scenarios using maxent model in China. *Agriculture* **2021**, *11*, 1122. [CrossRef]
50. Lin, K.; Han, Z.; Song, Q.; Liang, S.; Zhou, Y. Photosynthetic physiological features and leaf functional traits of 15 tea cultivars (lines). *Southwest China J. Agric. Sci.* **2021**, *34*, 2370–2377.
51. Liu, S.; Duan, X.; Zhao, H.; Wei, J.; Guo, Y.; Yan, D. Biochemical diversity analysis of wild tea germplasms in Guizhou. *J. Plant Genet. Resour.* **2014**, *15*, 1255–1261.
52. Wang, F.; Chen, Y.; Wang, X.; You, Z.; Chen, C. Comparison of leaf functional and photosynthetic characteristics in different tea cultivars. *J. Tea Sci.* **2016**, *36*, 285–292.
53. Sun, Y.J.; Zhou, J.; Guo, J.S. Advances in the knowledge of adaptive mechanisms mediating abiotic stress responses in *Camellia sinensis*. *Front. Biosci.-Landmark Ed.* **2021**, *26*, 1714–1722. [CrossRef] [PubMed]
54. Zhang, Q.; Lin, F.; Li, X.; Hu, M. Major chemical components in rivers and their seaward fluxes in the southeastern coastal region of China. *Acta Oceanol. Sin.* **1985**, *05*, 561–566.
55. Li, S. Agricultural climate regionalization of China. *J. Nat. Resour.* **1987**, *01*, 71–83.
56. Yu, X.; Xiao, J.; Chen, S.; Yu, Y.; Ma, J.; Lin, Y.; Li, R.; Lin, J.; Fu, Z.; Zhou, Q.; et al. Metabolite signatures of diverse *Camellia sinensis* tea populations. *Nat. Commun.* **2020**, *11*, 5586. [CrossRef] [PubMed]
57. Sha, A.H.; Lin, X.H.; Huang, J.B.; Zhang, D.P. Analysis of DNA methylation related to rice adult plant resistance to bacterial blight based on methylation-sensitive AFLP (MSAP) analysis. *Mol. Genet. Genom.* **2005**, *273*, 484–490. [CrossRef] [PubMed]
58. Grandbastien, M. Activation of plant retrotransposons under stress conditions. *Trends Plant Sci.* **1998**, *3*, 181–187. [CrossRef]
59. Tam, P.P.; Barrette-Ng, I.H.; Simon, D.M.; Tam, M.W.; Ang, A.L.; Muench, D.G. The Puf family of RNA-binding proteins in plants: Phylogeny, structural modeling, activity and subcellular localization. *BMC Plant Biol.* **2010**, *10*, 44. [CrossRef]
60. Liu, Q. Computational identification and systematic analysis of the ACR gene family in Oryza sativa. *J. Plant Physiol.* **2006**, *163*, 445–451. [CrossRef]
61. Hayakawa, T.; Kudo, T.; Ito, T.; Takahashi, N.; Yamaya, T. ACT domain repeat protein 7, *ACR7*, interacts with a chaperone HSP18.0-CII in rice nuclei. *Plant Cell Physiol.* **2006**, *47*, 891–904. [CrossRef]
62. Li, Z.; Yang, W.; Ahammed, G.J.; Shen, C.; Yan, P.; Li, X.; Han, W. Developmental changes in carbon and nitrogen metabolism affect tea quality in different leaf position. *Plant Physiol. Bioch.* **2016**, *106*, 327–335. [CrossRef] [PubMed]
63. Tang, Y.; Guo, C.; Zhang, M.; Chen, C.; Chen, R. Application of ISSR markers to the genetic polymorphism of *Camellia sinensis*. *J. Fujian Agric. For. Univ. Nat. Sci. Ed.* **2007**, *36*, 51–55.
64. Wang, R.J.; Gao, X.F.; Yang, J.; Kong, X.R. Genome-wide association study to identify favorable snp allelic variations and candidate genes that control the timing of spring bud flush of tea (*Camellia sinensis*) using slaf-seq. *J. Agr. Food Chem.* **2019**, *67*, 10380–10391. [CrossRef] [PubMed]
65. Hu, Y.; Yu, Z.; Gao, X.; Liu, C.; Zhang, Y.; Smarda, P.; Guo, Q. Genetic diversity, population structure, and genome wide association analysis of ginkgo cultivars. *Hortic. Res.-Engl.* **2023**, *10*, uhad136. [CrossRef] [PubMed]
66. Lei, X.; Li, H.; Li, P.; Zhang, H.; Han, Z.; Yang, B.; Duan, Y.; Njeri, N.S.; Yang, D.; Zheng, J.; et al. Genome-wide association studies of Biluochun tea plant populations in Dongting mountain and comprehensive identification of candidate genes associated with core agronomic traits by four analysis models. *Plants* **2023**, *12*, 3719. [CrossRef] [PubMed]
67. Xu, X.; Liu, X.; Ge, S.; Jensen, J.D.; Hu, F.Y.; Li, X.; Dong, Y.; Gutenkunst, R.N.; Fang, L.; Huang, L.; et al. Resequencing 50 accessions of cultivated and wild rice yields markers for identifying agronomically important genes. *Nat. Biotechnol.* **2012**, *30*, 105–157. [CrossRef] [PubMed]
68. Pfeifer, B.; Wittelsbürger, U.; Ramos-Onsins, S.E.; Lercher, M.J. PopGenome: An efficient swiss army knife for population genomic analyses in R. *Mol. Biol. Evol.* **2014**, *31*, 1929–1936. [CrossRef]
69. Wang, W.; Huang, X.; Yi, G.; Xu, S.; Guo, Y. Analysis of volatile components in oolong tea by HS-GC/MS. *Nat. Prod. Res. Dev.* **2019**, *31*, 222–229.
70. Wang, P.; Wang, L.; Chen, J.; Zheng, T.; Chu, D.; Hu, W.; Ye, J.; He, H. Quantitative analysis of aroma components in tea by 4-tert-butylcyclohexanol as internal standard. *J. Fujian Agric. For. Univ. Nat. Sci. Ed.* **2021**, *50*, 198–205.

plants

MDPI

Article

Analysis of Characteristics in the Macro-Composition and Volatile Compounds of Understory Xiaobai White Tea

Mengcong Zhang [1], Chengzhe Zhou [1,2], Cheng Zhang [1], Kai Xu [1], Li Lu [1], Linjie Huang [1], Lixuan Zhang [1], Huang Li [1], Xuefang Zhu [3], Zhongxiong Lai [1,2] and Yuqiong Guo [4,5,*]

1 College of Horticulture, Fujian Agriculture and Forestry University, Fuzhou 350002, China; zmc.1998@foxmail.com (M.Z.); chengzhechou@foxmail.com (C.Z.); zhangcheng9742@foxmail.com (C.Z.); xukai97@foxmail.com (K.X.); lu18286159614@163.com (L.L.); huanglinjie1997@foxmail.com (L.H.); m15160004025@163.com (L.Z.); 13984128485@163.com (H.L.); laizx01@163.com (Z.L.)

2 Institute of Horticultural Biotechnology, Fujian Agriculture and Forestry University, Fuzhou 350002, China

3 Nanping Jianyang District Tea Development Center, Nanping 353000, China; 18950687671@163.com

4 Anxi College of Tea Science (College of Digital Economy), Fujian Agriculture and Forestry University, Quanzhou 362400, China

5 Tea Industry Research Institute, Fujian Agriculture and Forestry University, Fuzhou 350002, China

* Correspondence: guoyq@fafu.edu.cn

Abstract: Understory planting affects the growth environment of tea plants, regulating the tea plant growth and the formation of secondary metabolites, which in turn affects the flavor of Xiaobai white tea. The present research adopted biochemical composition determination, widely targeted volatilities (WTV) analysis, multivariate statistical analysis, and odor activity value (OAV) analysis to analyze the characteristics in the macro-composition and volatile compounds of understory white tea. The sensory evaluation results indicated that understory Xiaobai white tea (LWTs) was stronger than ordinary Xiaobai white tea (PWTs) in terms of the taste of smoothness, sweetness, and thickness as well as the aromas of the flower and sweet. Understory planting reduced light intensity and air temperature, increased air humidity, organic matter, total nitrogen, and available nitrogen contents, which improved the growth environment of tea plants. The phytochemical analysis showed that the water-extractable substances, caffeine, flavonoids, and soluble sugar contents of understory tea fresh-leaf (LF) were higher than those of ordinary fresh-leaf (PF). The phytochemical analysis showed that the free amino acids, theaflavins, thearubigins, water-extractable substances, and tea polyphenols contents of LWTs were significantly higher than those of PWTs, which may explain the higher smoothness, sweetness, and thickness scores of LWTs than those of PWTs. The 2-heptanol, 2-decane, damasone, and cedar alcohol contents were significantly higher in LWTs than in PWTs, which may result in stronger flowery and sweet aromas in LWTs than in PWTs. These results provide a firm experimental basis for the observed differences in the flavor of LWTs and PWTs.

Keywords: understory planting; environment factors; fresh-leaf; Xiaobai white tea; volatile compounds

1. Introduction

Tea is a shade plant, adapted to the understory of forests in its native habitat [1]. White tea is one of six kinds of tea in China and undergoes the fewest manufacturing processes, including prolonged withering and drying [2]. According to the different cultivars of tea trees from which fresh leaves are derived, white tea can be divided into Dabai, Xiaobai, and Shuixian white tea. Xiaobai white tea is one of the high-quality white teas in the Fujian Province, has a long history, with a slightly sweet and umami taste as well as a fresh odor, widely loved by consumers. The quality of Xiaobai white tea is determined by the fresh-leaf quality and the tea-making process, fresh-leaf is the foundation. Environmental factors, such as temperature, humidity, light, and rainfall were closely related to the growth and development of tea plants and the accumulation of fresh leaf components [3]. Tea plants grow well in conditions of 80% relative humidity and cloudy conditions (sunshine percentage ≤40%) [4,5]. Suitable effective accumulated temperatures (112.4 °C, with a minimum daily temperature of 5 °C), large temperature changes between the day and night, and scattered light, which contributes to the accumulation of aroma substances in tea, will increase the freshness and taste of tea to a certain extent, resulting in high-quality green tea [6]. Tea plants is an acidophilic plant, and the soil pH of 4.5~5.5 is suitable for its cultivation [7]. Soil type affects the exchange of water, air, and nutrients between the soil and the tea plant, which ultimately affects the growth and yield of the tea plant [8]. Nitrogen (N), phosphorus (P), and potassium (K) are the three main nutrients that regulate the growth of tea plants, and their availability directly affects the yield and quality of tea [9]. N and P are mainly involved in the synthesis of amino acids, vitamins, chlorophyll, and other substances, and directly regulate the content and proportion of biochemical components of the tea plants [10,11]. K helps to promote the growth and development of tea seedlings and the absorption capacity of tea roots, thus enhancing the resilience of tea plants [3].

Agroforestry system is a form of understory planting that involves planting trees alongside crops to improve the capture of temporal resources between species [12,13], thereby increasing resource use efficiency. Historically, agroforestry systems have been widely implemented as a way to achieve sustainability in agriculture [14,15]. To reduce the required labor force and protect the ecological environment, many places including Chongqing, Sichuan, Hubei, and Anhui planted Coptis chinensis Franch under natural forests using natural shade trees [16]. To overcome farmland shortage, understory *P. notoginseng* was developed as a new ecological planting model [17]. A related study investigated the production patterns of black oats grown under four different woodland types in southern Brazil and found that black oats grown under *P. rigida* were the most productive [18]. Agroforestry systems have been successfully practiced and researched on many species. The agroforestry pattern is also widely used in tea, many areas in Fujian Province have tried to plant tea under the forest. Using the existing forest land, the forest land is thinned with care without changing the stand structure, and tea plants are planted in the forest. Compared with ordinary planting methods, shade planting under the forest significantly changes the growing environment of tea. The chlorophyll content in the new shoots of tea plants in spring, summer, and autumn increased to varying degrees after shade tea plants [19]. Shao et al. [4] found in the study of the difference of metabolites processed into different tea types by preharvest shading that shading had a great effect on the composition of amino acids, flavonoids, and theaflavins in tea. The thickening of fence tissue and thinning of sponge tissue in tea fresh leaf cells in a shaded environment contributed to the accumulation of volatile metabolites, including volatile fatty acid derivatives and volatile phenylpropyl/phenyl esters (VPBs) [20,21], the accumulation may be related to upstream metabolism. Fang et al. [22] showed that preharvest shade covering helped to increase the total volatile compounds of steamed green tea. Understorey planting of tea plants improved soil organic matter, total nitrogen, total phosphorus, and total potassium content compared to ordinary tea plant gardens [23], which in turn improved the quality of tea plants.

To explore the influence of understory planting on the flavor quality of Xiaobai white tea, this study measured the main environmental factors of the understory Xiaobai white tea plantation and one ordinary Xiaobai white tea plantation. Then, the fresh leaves of the understory Xiaobai white tea plantation and ordinary Xiaobai white tea plantation were taken and processed into dry tea as experimental materials. The wide range of targeted volatile metabolomics (WTV) method, and odor activity values (OAV) method combined with stoichiometry were used to explore the effects of understory planting on the growth environment and fresh-leaf components of tea trees. Then screen the quality difference between LWTs and PWTs, which was aimed at providing a firm experimental basis for the promotion of understory tea planting mode.

2. Results and Discussion

2.1. Understory Planting Improves the Growing Environment for Tea Plants

Tea can achieve high quality only when it grows in the suitable environmental conditions of light temperature, humidity, soil, and fertilizer [3]. When the light intensity exceeds 3.7 J/cm^2·min, the photosynthetic rate of tea leaves will decline, and the optimal temperature for the growth and development of tea new shoots is at 20–25 °C, more than this temperature range, the new shoots growth rate slows down [24]. Hot and dry weather tends to make green tea bitter and astringent [25]. As shown in Figure 1A, the light intensity in the ordinary Xiaobai white tea garden (CK) was roughly twice as high as under shading. The average air temperature under shading was 1.27 °C lower than that in CK, and the average air humidity under shading was 5.09% RH higher than that in CK, respectively. These indicate that understory planting improves the growth environment of tea plants and the quality of tea. In addition, the nutrient status of the soil also has an important impact on the quality of tea, 11.0% nitrogen (N), 1.65% phosphorous (P), and 3.7% potassium (K) are reported to be necessary nutrients for tea [26]. As shown in Figure 1B, the organic matter, total nitrogen, and available nitrogen contents under shading were significantly higher than those in CK, other factors did not show a uniform pattern in the shaded and unshaded groups. In line with previous studies, Wen et al. [27] found that the soil nutrient content of the three intercropping systems was higher than that of the tea monoculture system, except for phosphorus [28]. The composition of the soil nutrients in the intercropping system is affected by many factors, such as the combination of different plant species, plant residues, and vertical structure [29]. Different forests have different soil nutrient requirements, which may be the reason why factors such as total phosphorus and total potassium in *Castanea henryi* understory Xiaobai white tea plantation (ZL), *Cunninghamia lanceolata* understory Xiaobai white tea plantation (SL), and *Phyllostachys edulis* understory Xiaobai white tea plantation (BL) did not show consistent regularity. In short, understory planting improves the growing environment of tea plants.

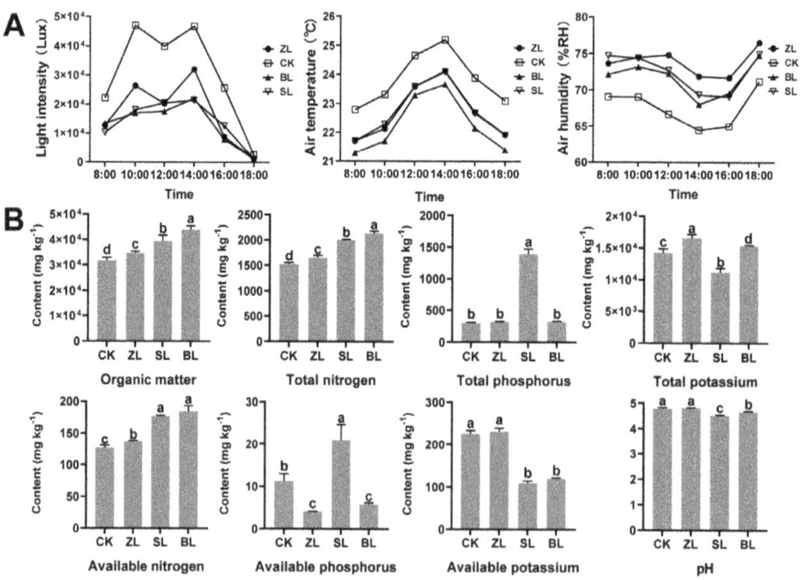

Figure 1. (**A**) Light intensity, air temperature, and air humidity, (**B**) the content of organic matter, total nitrogen, total phosphorus, total potassium, available nitrogen, available phosphorus, available potassium, and pH. The various superscripts show significant differences ($p < 0.05$).

2.2. Impact of Understory Planting on Fresh-Leaf Contents

To understand the effect of understory planting on the content of tea plants, the macroscopic components of tea fresh-leaf were determined and analyzed. The relative total water-extractable substances, caffeine, flavonoids, and soluble sugar contents in LF are significantly higher than those in PF, and the relative contents of free amino acids and thearubigins in LF are lower significantly than those in PF (Figure 2A); these provide a good foundation for the processing of dry tea. The growth of the tea plant is not only affected by its genetic factors, but also regulated by environmental factors such as light, temperature, water, and fertilizer, resulting in differences in the quantity and proportion of the components contained in the fresh-leaf [30]. Correlation analysis is a powerful tool for investigating the relationship between major environmental factors and the macroscopic components, the calculated results of the Pearson correlation coefficient between the two show that (Figure 2A), organic matter, total nitrogen, total phosphorus, available nitrogen, available phosphorus, available potassium, pH, and light intensity correlated closely with different macroscopic components ($r > 0.6$, $p < 0.05$), indicating that they are important factors underlying the macroscopic components of tea fresh-leaf.

Sponge tissue has a storage function, contains a large number of polyphenols and carbohydrates related to tea quality, and shading promotes the relative proportion of sponge tissue in leaves, which helps the accumulation of components contained in fresh-leaf [31]; this is consistent with the results of a negative correlation between water extract and light intensity (Figure 2B). The accumulation of free amino acid content in tea leaves under shade conditions is not due to increased synthesis but to the hydrolysis of chloroplast proteins [32]. Early shading can play a role in increasing the amino acid content of fresh leaves. This effect on the increase in amino acid content gradually weakens with the extension of shading time [33]; moreover, this is consistent with the fact that free amino acid have a significant negative correlation with organic matter, total nitrogen, and available nitrogen (Figures 1B and 2B). There was a significant positive correlation between the soil pH and free amino acids, which indicated that a proper increase in the pH value could improve the tea quality. However, it is not advisable to increase the soil pH above six, as that would be outside the fitness range of the tea tree [34]. When the soil pH value is greater than six, the growth of tea tree shoots is inhibited to a greater extent than nitrogen uptake, which reduces the yield of tea [35]. In the previous study, the long-term shading remarkably inhibited sugar metabolisms such as glycolysis, galactose metabolism, and pentose phosphate pathway in the leaves and roots of "Xiangfeicui" [36] to replenish the carbon backbone during shading, plants hydrolyze large amounts of polysaccharides into soluble sugars [37], which corresponded well with our findings. There was a significant positive correlation between soluble sugar and total nitrogen and available nitrogen, which was consistent with the report that the soluble sugar content decreased due to long-term nitrogen fertilizer loss [38]. The results suggested that the low content of available nitrogen in the soil led to the weakening of carbon assimilation in tea plants [39].

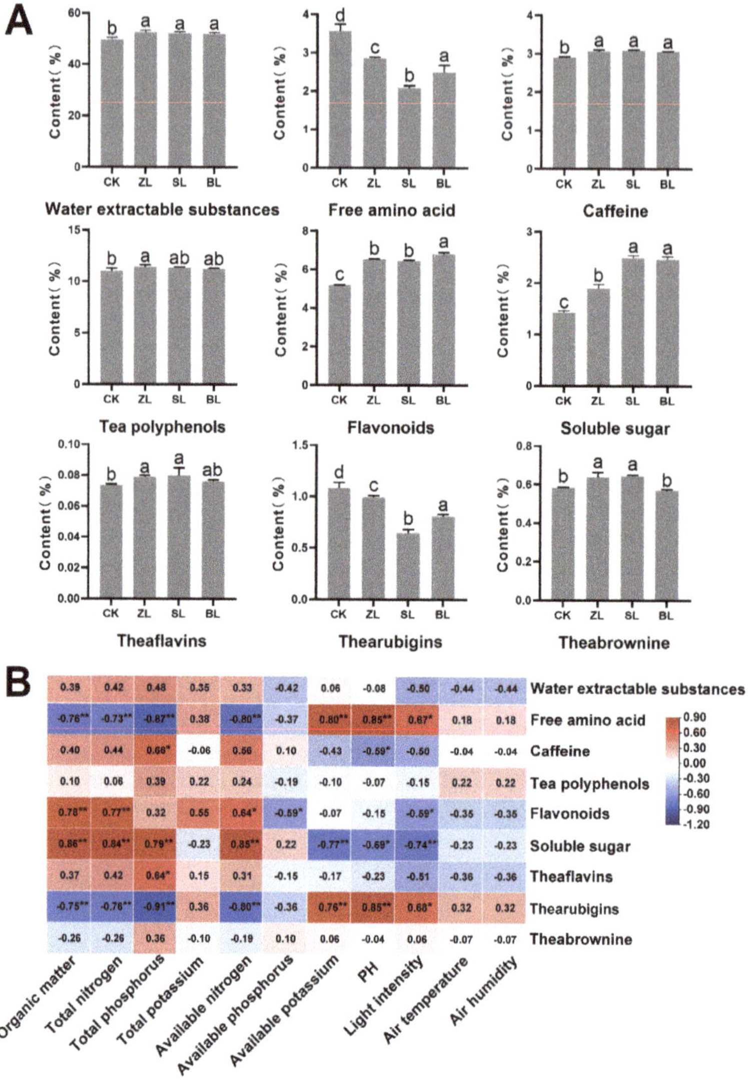

Figure 2. (**A**) The content of water-extractable substances, free amino acids, caffeine, tea polyphenols, flavonoids, soluble sugar, theaflavins, thearubigins, and theabrownines, (**B**) Correlation between macroscopic components and major environmental factors. The various superscripts show significant differences ($p < 0.05$). The symbols * and ** indicate statistical significance at $p < 0.05$ and $p < 0.01$, respectively. The sample information of fresh leaves is shown in Table 1.

Table 1. Fresh-leaf samples information.

Identifier	Source of Fresh Leaves	Picking Date	Picking Standards	Identifier
CK	ordinary tea plantation	25 April 2023	one bud and two leaves	CK-1 to 3
ZL	*Castanea henryi* understory tea plantation			ZL-1 to 3
SL	*Cunninghamia lanceolata* understory tea plantation			SL-1 to 3
BL	*Phyllostachys edulis* understory tea plantation			BL-1 to 3

2.3. Significant Difference in Aroma and Flavor between LWTs and PWTs

The quality of fresh tea leaves determines the quality of tea products. However, the quality-related metabolites in fresh leaves are significantly altered by the post-harvest processing of tea leaves. To more comprehensively analyze the effects of understory planting on the quality of tea products, also to enhance the representativeness of this study, repeated sampling was carried out. On 26 April, the *Castanea henryi* understory Xiaobai white tea samples (ZLL1) and the ordinary Xiaobai white tea samples (ZLP1) were obtained. On 30 April, the *Cunninghamia lanceolata* understory Xiaobai white tea samples (SLL1) and ordinary Xiaobai white tea (SLP1) were obtained. On 2 May, the *Phyllostachys edulis* understory Xiaobai white tea samples (BLL1) and ordinary Xiaobai white tea samples (BLP1) were obtained. On 7 May, 9 May, and 13 May, the *Castanea henryi* understory Xiaobai white tea samples (ZLL2) and ordinary Xiaobai white tea samples (ZLP2), the *Cunninghamia lanceolata* understory Xiaobai white tea samples (SLL2) and ordinary Xiaobai white tea samples (SLP2), the *Phyllostachys edulis* understory Xiaobai white tea samples (BLL2) and ordinary Xiaobai white tea samples (BLP2) were obtained again, respectively. Taken together, a total of 12 tea samples were obtained including ZLL1 and ZLP1, SLL1, and SLP1, BLL1, and BLP1, ZLL2, and ZLP2, SLL2, and SLP2, BLL2, and BLP2 (Table 2). All samples were processed in Zhangdun Town, Jianyang District, Nanping City using unified picking standards and processing techniques. The quality of the white tea was assessed according to the aroma and taste compounds and sensory properties. The sensory evaluation results showed that the appearance was in all samples, with the whole shoot and greenish auburn. In addition to ZL1, there were all LWTs in other groups, and PWTs in SL1 also had tippy. A yellow and bright liquor color was defined in all samples, the difference is that deeper of liquor color in LWTs than in PWTs (Figure 3A). LWTs were stronger than PWTs in terms of the taste of smoothness, sweetness, and thickness (Figure 3B), both LWTs and PWTs have a floral and sweet aroma, individual samples have a woody aroma, but the floral aroma of LWTs is stronger and more lasting than PWTs (Figure 3C). In general, LWTs has a more pleasant flavor than PWTs.

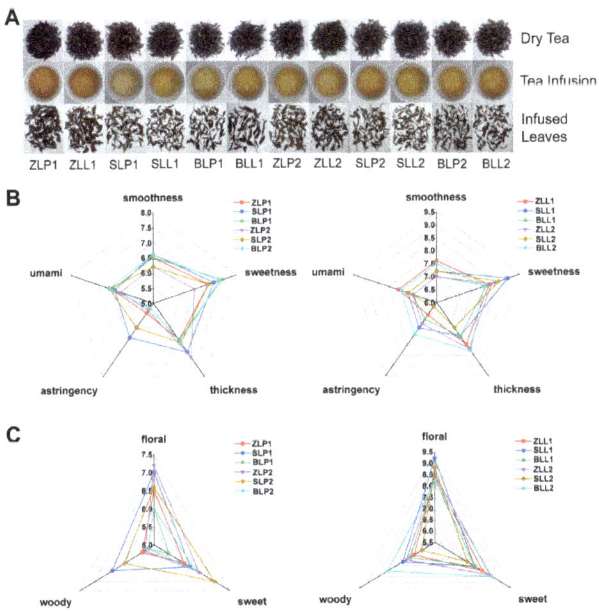

Figure 3. (**A**) The appearance and infusion colors of PWTs and LWTs; (**B**) Spider plots for the taste profiles of PWTs and LWTs; (**C**) Spider plots for the aroma profiles of PWTs and LWTs.

Table 2. Xiaobai white tea samples information.

Identifier	Source of Fresh Leaves	Production Date	Picking Standards	Identifier
ZL1	*Castanea henryi* understory tea plantation ordinary tea plantation	26–28 April 2023	one bud and two leaves	ZLL1-1 to 3 ZLP1-1 to 3
SL1	*Cunninghamia lanceolata* understory tea plantation ordinary tea plantation	30 April–2 May 2023		SLL1-1 to 3 SLP1-1 to 3
BL1	*Phyllostachys edulis understory tea plantation* ordinary tea plantation	2–5 May 2023		BLL1-1 to 3 BLP1-1 to 3
ZL2	*Castanea henryi* understory tea plantation ordinary tea plantation	7–9 May 2023		ZLL2-1 to 3 ZLP2-1 to 3
SL2	*Cunninghamia lanceolata* understory tea plantation ordinary tea plantation	9–12 May 2023		SLL2-1 to 3 SLP2-1 to 3
BL2	*Phyllostachys edulis* understory tea plantation ordinary tea plantation	13–15 May 2023		BLL2-1 to 3 BLP2-1 to 3

2.4. Analysis of Macro-Composition of LWTs and PWTs

The taste of tea is mainly determined by macro-composition, included among these, tea polyphenols enhance the bitterness of tea infusions, flavonoids supply the bitter taste, free amino acids are the source of the umami taste, soluble sugars are the main contributors to sweetness [40]. The relative total water-extractable substances, tea polyphenols, caffeine, and soluble sugar contents in LWTs are higher than those in PWTs and the relative contents of free amino acids contents in LWTs are lower than those in PWTs (Figure 4). This may be the reason for why LWTs has a stronger smoothness, sweetness, and thickness than PWTs in the taste. The macro-composition of fresh-leaf and tea are basically the same, this shows that the matrix of fresh-leaf contents contributes to the formation of good-quality tea. TRs, TBs, and TFs did not show uniform regular differences in the six groups of samples, which may be due to their unique woodland environment. During the withering process of white tea, on the one hand, the hydrolysis of large molecules within the withered leaves, starch, proteins, and other substances are hydrolyzed to monosaccharides, amino acids, and other water-soluble small molecules. On the other hand, along with the enhancement of the activity of enzymes such as polyphenol oxidase, the polyphenols, mainly catechins, undergo an oxidation and condensation reaction, which gives rise to oxidative products such as pigments and other products, and form the unique flavor of white tea.

To further identify the key macro-components that differentiate the tastes of LWTs and PWTs, orthogonal partial least squares–discriminant analysis (OPLS-DA) was performed on the macro-compositions (Figure 5A). Cross-validation analysis showed that the OPLS-DA models were reliable (Figure 5B). LWTs gathered in the first and fourth quadrants, and in PWTs, except for SLP1-3, the rest are clustered in the second and third quadrants (Figure 5A). The indication of the characteristic macro-composition helps to distinguish LWTs from PWTs. We can know from Figure 5C that the variable importance in projection (VIP) value of free amino acids, TFs, TRs, water-extractable substances, and tea polyphenols is greater than 1, and metabolites with VIP > 1 are generally considered to be significantly different. In the previous study [41], free amino acids, flavonoids, and TFs are the main metabolic pathways that improve the flavor of shade tea, which is basically consistent with this study.

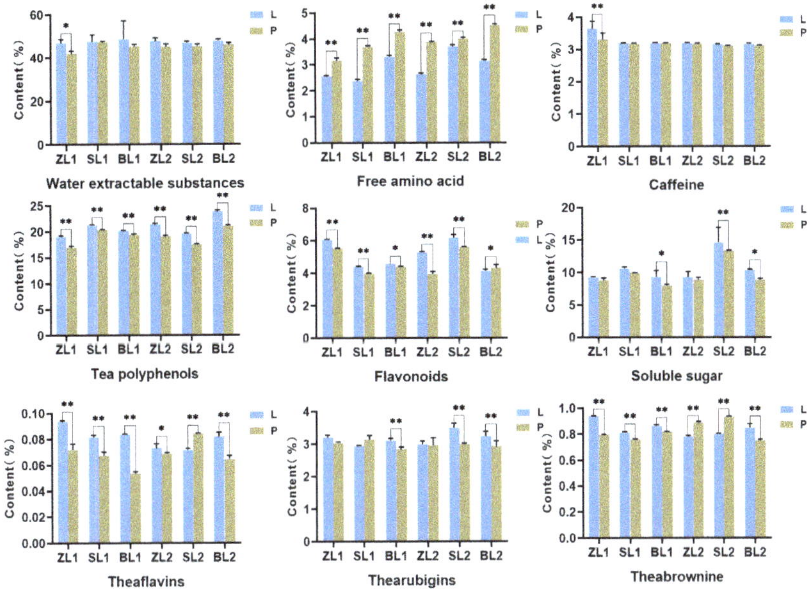

Figure 4. The content of water-extractable substances, free amino acids, caffeine, tea polyphenols, flavonoids, soluble sugar, theaflavins, thearubigins, and theabrownines. The symbols * and ** indicate statistical significance at $p < 0.05$ and $p < 0.01$, respectively. Undergrowth-picked was referred to as L, and ordinary-picked was called P.

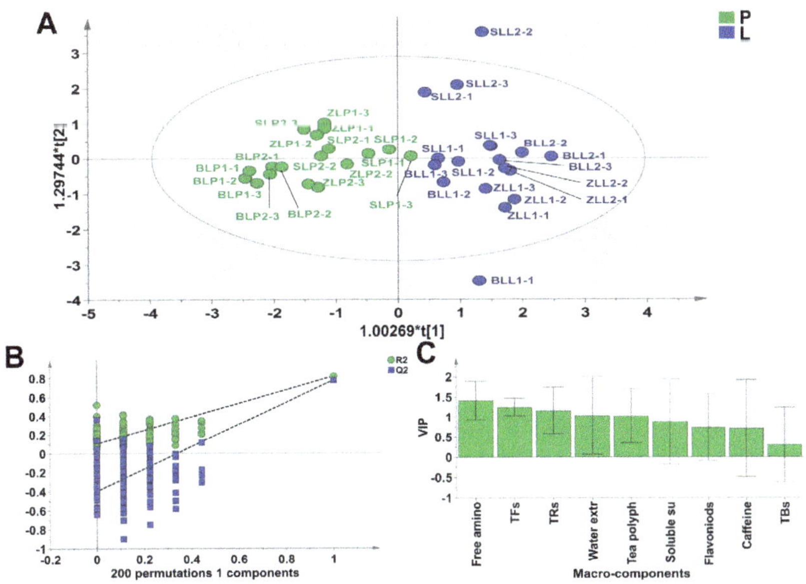

Figure 5. (**A**) The OPLS-DA scores for the non-volatile components; (**B**) Permutation test plots of the non-volatile components; (**C**) The variable importance in the project (VIP) of the non-volatile components. Undergrowth-picked was referred to as L, and ordinary-picked was called P.

2.5. Identification of VOCs in LWTs and PWTs

A Total of 220 Volatiles Were Detected

We used HS-SPME in combination with GC-MS/MS to determine the volatile content of LWTs and PWTs. The total ion flow (TIC) plot of the mixed sample is superimposed to determine the reproducibility of volatile extraction and detection, then the high stability of the instrument ensures data repeatability and reliability (Supplementary Figure S1). In total, 220 volatiles were detected, including 16 alcohols, 25 aromatic hydrocarbons, 8 phenols, 6 nitrogenous compounds, 15 aldehydes, 7 acids, 34 hydrocarbons, 28 ketones, 22 terpenes, 33 heterocyclic compounds, and 26 esters (Supplementary Table S1); the percentage of each substance is shown in the Figure 6A. Overall, the volatile metabolite composition in LWTs and PWTs is similar, but the exact content is slightly different (Supplementary Table S1). During the drying process, high temperatures produce the Maillard reaction, which causes amino acids and carbonyl compounds to polymerize and condense at room temperature or when heated, producing a large number of heterocyclic compounds [41,42], this may be the reason for the high content of volatile heterocyclic compounds in PWTs and LWTs.

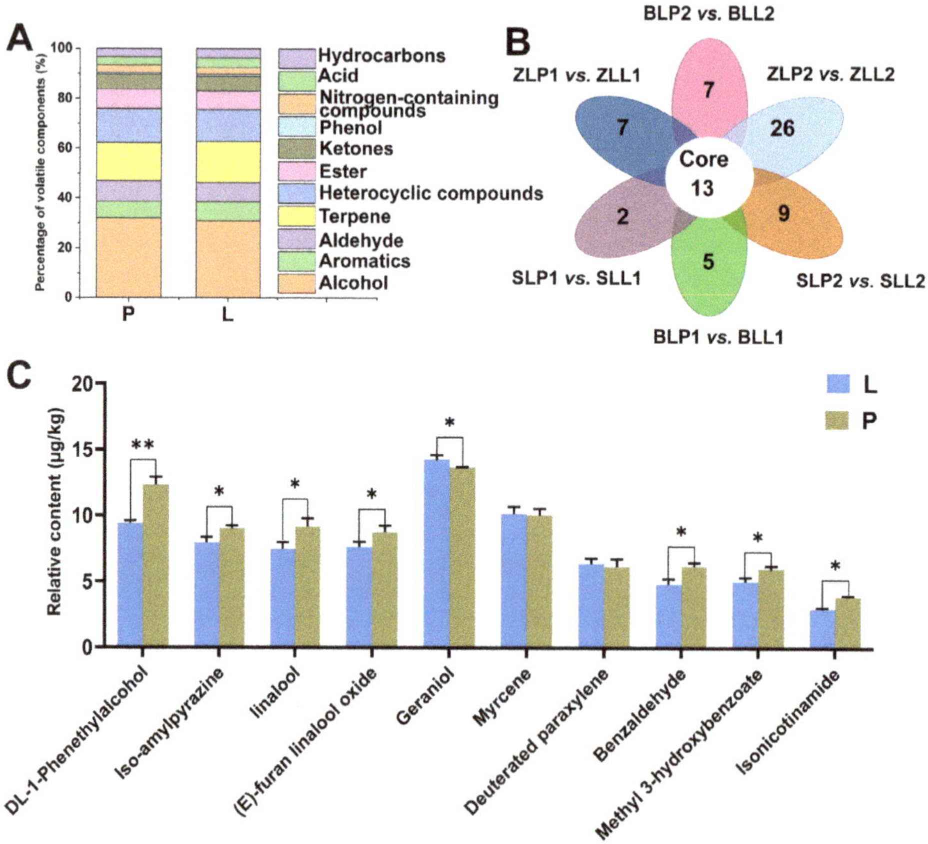

Figure 6. (**A**) The ratio of volatile components; (**B**) Venn diagram for different groups of key volatile components in BLP2 vs. BLL2, ZLP1 vs. ZLL1, ZLP2 vs. ZLL2, SLP1 vs. SLL1, SLP2 vs. SLL2, BLP1 vs. BLL1; (**C**) The content of top ten volatile components. Undergrowth-picked was referred to as L, and ordinary-picked was called P. The symbols * and ** indicate statistical significance at $p < 0.05$ and $p < 0.01$, respectively.

Among all VOCs, the top ten substances with relative content are DL-1-Phenethylalcohol, iso-amylpyrazine, linalool, (E)-furan linalool oxide, geraniol, myrcene, deuterated paraxylene, benzaldehyde, methyl 3-hydroxybenzoate, and isonicotinamide (Figure 6C). The relative content of geraniol in LWTs was the highest, accounting for an average of 10.11%, followed by myrcene (7.21%), DL-1-Phenethylalcohol (6.83%), iso-amylpyrazine (5.62%), and (E)-furan linalool oxide (5.42%), mainly displayed floral and green. The highest relative content in PWTs is geraniol, accounting for an average of 9.15%, followed by threuterol (8.28%), myrcene (6.70%), linalool (6.13%), and iso-amylpyrazine (6.03%), mainly displayed floral. The differences in the composition of these VOCs and differences in relative content may be responsible for the difference in the aroma of LWTs and PWTs.

2.6. *Screening for Differential VOCs in LWTs and PWTs*

2.6.1. Thirteen Important Differential VOCs with LWTs and PWTs Were Screened for

To investigate the VOCs responsible for the sensory differences between LWTs and PWTs, we used OPLS-DA further to analyze each of the six sets of samples. The OPLS-DA discriminant model is not overfitted, and the model is reliable, the LWTs and PWTs being significantly differentiated by VOCs (Supplementary Figure S2).

There were 52 (ZL1), 26 (SL1), 45 (BL1), 76 (ZL2), 55 (SL2), and 42 (BL2) compounds with VIP values greater than 1 and FC greater than 1.5 or FC less than 0.67 in the OPLS-DA model that was developed in this study. There are 13 important differential volatile metabolites shared by the six groups, namely 2-heptanol, cedrol, m-Xylene, β-isophorone, 2-Decanone, Jasmone, β-Damascenone, (+)-CUPARENE, (−)-Calamenene, 2-Aminopyridine, decyl formate, benzyl acetate, and cis-3-Hexenyl 2-methylbutanoate (Figure 6B), these are the important differentiating VOCs that distinguish LWTs from PWTs (Supplementary Table S2). Further analysis of the 13 VOCs showed that ketones compounds dominated the numbers, followed by esters. The highest VIP value (mean VIP value for the six groups of samples) was benzyl acetate, followed by decyl formate, (+)-CUPARENE, and (−)-Calamenene, which mainly display floral and herbal aromas. The benzyl acetate (jasmine, rose) relative content in LWTs (51.42 μg/kg on average, relative to internal standard) was almost twice that of the content in PWTs (98.61 μg/kg on average, relative to internal standard), which may contribute to a stronger flowery aroma for LWTs than PWTs.

Ketones generally have a "floral" and "woody" aroma [43]. In the previous study, the content of jasmone (jasmine) in tea was significantly affected by the different altitudes and seasons, at high altitudes with short sunshine hours and weak intensity, the content of jasmone in autumn tea is higher than that at low altitudes and summer [44], which is consistent with our study. Damasone, which has a flowery and fried potatoes aroma, was found to be the main contributor to the most to the aroma of Keemun black tea [45], and there is a downward trend with the increase in storage time in An tea [46]. 2-decane, which has a floral and citrus aroma, was found to be a characteristic volatile metabolite of beef bouillons. In the other study, 2-decone was produced in white tea after a certain period of aging [47]. The content of these ketones is higher in LWTs than in PWTs, which may be one of the reasons why LWTs have a higher flowery aroma than PWTs.

Esters, especially those with pleasant aromas, have been identified as key compounds that form the characteristic aroma of white tea [48]. The three esters of VIP > 1 were higher in LWTs than PWTs and mainly displayed "flowery" and "fruit" aromas. Benzyl acetate, as the characteristic aroma component of jasmine tea, is often used in cosmetics and foods to increase the aroma of jasmine and rose [49], which contributes to the flowery aroma of LWTs. Benzyl acetate is present in high volumes in aged dark teas; the reason is that, over time, β-D-glucosidase hydrolyzes the precursors of benzyl alcohol with glucosides to produce benzyl alcohol, which is further converted to benzaldehyde or benzyl acetate [50]. In many types of green tea, the aroma intensity of cis-3-Hexenyl 2-methylbutanoate (green) decreases with increasing storage temperature, which is greatly affected by the storage environment [51].

Alcohols usually have floral, sweet, and fruity aromas, which are the most critical contributor to the formation of white tea aromas [52]. Cedrol (woody, stale), identified as the main aromatic compound in aged white tea and Pu'er tea [53,54], showed the levels of LWTs were twice that of PWTs in almost all groups, which may be one of the reasons why LWTs have a higher woody aroma than PWTs. 2-heptanol has been identified as a key compound in the aroma of coffee and some tropical fruits. Some studies have shown that 2-heptanol has shallow olfactory threshold (OT) content in water (OT = 0.1) and is often described as "fresh fruity", "lemony", and "citrus" [55,56]. Terpenes often have spicy, fruity, and woody aromas, which are an essential source of tea aroma [56]. (+)-CUPARENE is a key differential volatile metabolite of roasted green tea at different altitudes [57], and its content is affected by tea grade [58]. Aroma components such as (−)-Calamenene (Herb, spice) are the main contributors to the formation of the irritating camphor fragrance characteristic of eagle tea [59]. Relevant studies have shown that rapid aging technology facilitates the production and accumulation of (−)-Calamenene [35]. In the previous study, volatile fatty acid derivatives and VPBs were significantly enhanced in dark-treated leaves [20]. As a volatile metabolite with VIP value > 1, the content of m-xylene (sweet, flowery) in LWTs was significantly higher than that of PWTs. It is greatly affected by different roasting degrees, and the content in medium-heat tea is significantly higher than that in low-heat tea [60].

The production of various key differential metabolites, such as damasone, 2-decane, benzyl acetate, and (−)-Calamenene, is related to or regulated by the aging process, and the content of these was significantly higher in LWTs than in PWTs. The results showed that long-term understory shade might regulate some metabolic mechanisms in the process of tea aging, and promote the formation of some flavor qualities of aged white tea. Aging gives white tea more pleasant aromas such as rose, sweet, and grassy aromas, and the overall taste is also improved, sweetness, thickness, smoothness, and acidity are enhanced, but bitterness and astringency are weakened [61]. In aged white tea, most ketones (such as 2-decaneone, 2-heptanone, and β-damascatone) and many esters (such as methyl salicylate) are on the rise, accompanied by some newly formed heterocyclic compounds [47], which is consistent with the distribution characteristics of VOCs in LWTs in this study. Understory shade planting may promote the formation of white tea aging flavor, and the specific related metabolic mechanism is a very meaningful research direction, which needs to be studied further.

2.6.2. 32. VOCs with OAV Greater Than 1 in LWTs and PWTs Were Screened for

The aroma of tea is determined by VOCs [62]. The composition of VOCs in white tea is very complex, and most of these compounds are odorless or have a weak odor, so they have little effect on the formation of white tea aroma. The characteristic flavor of the tea is not formed by a simple stacking of VOCs but relies heavily on the interaction between different VOCs, which is closely related not only to the content and aroma type of VOCs, but also to the odor threshold (OT) [63]. Therefore, screening the key active compounds of white tea aroma is essential to systematically reveal their important role in the formation of white tea aroma. Typically, OAVs are used to verify the contribution of active odor compounds and volatile organic compounds, which can well identify key aroma compounds [64–68].

The OAV was calculated for GC-MS test results, and there were 39 VOCs of OVA > 1 (Supplementary Table S3), including 6 alcohols, 2 aromatic hydrocarbons, 1 phenol, eight aldehydes, 5 terpenes, 3 acids, 10 ketones, 3 esters, and 1 heterocyclic compounds. Among them, there were 32 VOCs with OAV greater than 1 in LWTs and PWTs, which are the main contributors to the formation of the characteristic aroma of Xiaobai white tea (Supplementary Table S3). α-Terpineol (HY, TT) and methyl hexanoate (HY, XB, TT) only have OAV values > 1 in individual groups, which may be caused by special woodland environments. Combined with GC-MS-O and aroma recombination experiments, it was found that nalool, (E)-β-damascenone, camphene, 2-hexanone, phenylethyl alcohol, geraniol, and α-terpineol were the main contributors to the aroma of white tea [69], which is basically consistent with the results of this study.

2.6.3. Correlation Analysis between Aroma Profiles and Characteristic Volatile Metabolites of LWTs and PWTs

To further confirm the relationships between the volatiles and the aroma properties, PLSR analysis was applied to correlate the important volatiles (OAVs > 1) with the aroma properties (floral, woody, and sweet) in PWTs and LWTs (Figure 7). Geranylacetone, 3,5-OCTADIEN-2-ONE, β-CYCLOCITRAL, and Benzaldehyde were strongly correlated with the floral property. The contents of these substances in LWTs are higher than those in PWTs, which may account for the more pronounced floral aroma of the LWTs. The composition of sweet and woody aroma-related substances was different between PWTs and LWTs. However, LWTs scored higher than PWTs in sweet and woody properties, which may be due to the higher sum of OAV values of understory-related substances.

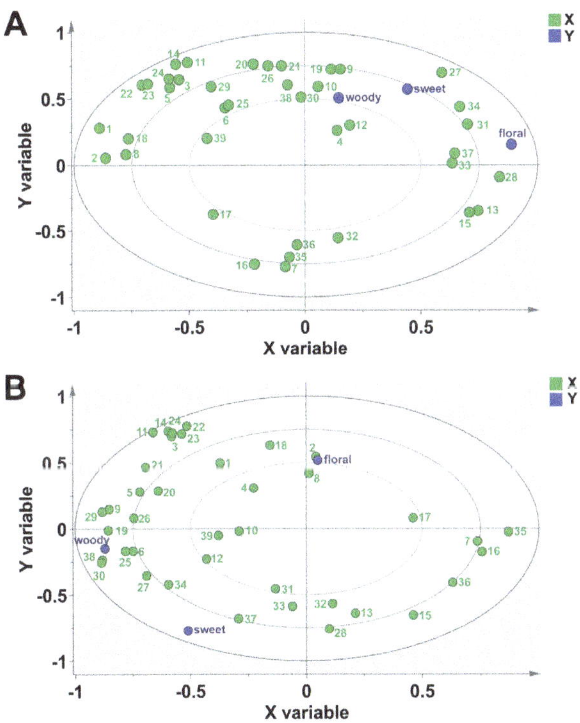

Figure 7. (**A**) PLSR analysis between the aroma properties and the characteristic volatile metabolites (OAVs > 1) of PWTs. X. 39 volatile metabolites (OAVs > 1); Y. aroma properties; (**B**) PLSR analysis between the aroma properties and the characteristic volatile metabolites (OAVs > 1) of LWTs. X. 39 volatile metabolites (OAVs > 1); Y. aroma properties. The serial numbers in the figure correspond to substances in the Supplementary Table S4.

2.6.4. Four Key Differential VOCs with LWTs and PWTs

Key differential volatile metabolites with VIP values greater than 1 are further combined with volatile metabolites with OAV values greater than 1, four substances, 2-heptanol, 2-decane, damasone, and cedar alcohol, were screened as characteristic VOCs that distinguished the difference in flavor quality between LWTs and PWTs, the content of these have a significant difference (Figure 8). It is worth noting that the relative content of cedar alcohol in HX, HY, and other groups is only PWTs (2.38–15.36 μg/kg), LWTs (3.70–66.93 μg/kg), while in TT it is as high as PWTs (45.18 μg/kg) and LWTs (374.91 μg/kg), which showed that different forest environment types had a significant effect on the VOCs content of Xiaobai white tea. It is of great significance to explore the optimal set of Xiaobai white tea forest environment types to improve the quality of LWTs and popularize the planting mode of LWTs.

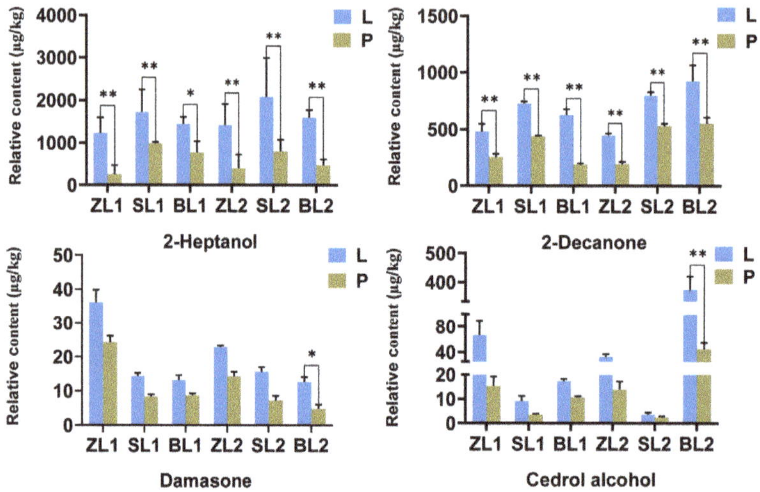

Figure 8. Four key differential volatile metabolites. The symbols * and ** indicate statistical significance at $p < 0.05$ and $p < 0.01$, respectively. Undergrowth-picked was referred to as L, and ordinary-picked was called P.

The analysis of content results showed that cedaryl alcohol and other substances were significantly affected by understory shading. Relevant studies have found, compared with the long-term light of 16 h, 20 h, and 24 h, 12 h light is more conducive to the accumulation of substances contained in white tea and the increase in volatile metabolites such as cedar alcohol (woody, floral), this is consistent with this study [69]. Although 2-heptanol (fresh, sweet, fruity) has been reported in oolong, green, and red raspberry leaf teas [67–69], it has been ignored in related studies on white tea. This may be that researchers focus more on compounds with higher content but ignore those with strong odors and low OT. In the recent study, different from previous studies on linalool as the largest contributor to the aroma of white tea, 2-heptanol was found to be the largest contributor to the aroma of GABA white tea for the first time [53]. Damasone (flowery, fried potatoes) has the highest OAV value among the four characteristic volatile metabolites, which is the key contributor to the flowery aroma of LWTs, but there are no studies on shade except for studies showing that its solubility in water is low and affected by storage time.

3. Materials and Methods

3.1. Chemicals

Hydrochloric acid, disodium hydrogen phosphate, potassium dihydrogen phosphate, ninhydrin, aluminum chloride, folinphenol, sodium bicarbonate, sulfuric acid, sodium carbonate, oxalic acid, ethyl acetate, and n-butanol were purchased from Sinopharm Chemical Reagent Co., Ltd. (Shanghai, China). The plant-soluble sugar content test kit was purchased from Suzhou Keming Biotechnology Co., Ltd. (Suzhou, Jiangsu, China). Sodium chloride, n-hexane (purity > 95%), and 3-hexanone-2,2,4,4-d4 (purity > 95%) were purchased from Sinopharm, Merck, and Sigma-Aldrich (St. Louis, MO, USA), respectively.

3.2. Materials

Fresh-leaf samples were taken from *Castanea henryi* understory Xiaobai white tea plantation (ZL), *Cunninghamia lanceolata* understory Xiaobai white tea plantation (SL), *Phyllostachys edulis* understory Xiaobai white tea plantation (BL), and one ordinary Xiaobai white tea plantation (CK) on the same hillside in Zhangdun Township, Jianyang District (118°46′ E, 27°47′ N). According to the five-point sampling method, five tea plants with good growth and no diseases and pests were selected for labeling. The light intensity, air humidity, and air temperature of the marker site were measured by 3415FQF photometric illuminance double radiometer (Spectrum, Stamford, CT, USA) and thermohygrometer every 2 h for 6 consecutive days from 8:00 to 18:00 on 23 April 2022, and the results were averaged. Sampling of fresh leaves of one bud and two leaves was gathered on 28 April 2022, undergrowth-picked was referred to as LF, and ordinary-picked was called PF. Soil samples were taken from 30 cm deep under the canopy of marked tea tree loci, mixed and air-dried for testing.

The Xiaobai white tea samples were produced in Jianyang District, Nanping City, China, from 26 April to 15 May 2023. On 26 April, the fresh leaves of the *Castanea henryi* understory tea plantation and the ordinary tea plantation were collected according to the five-point sampling method with the standard of one bud and two leaves, were withered immediately at 25 °C and 50% relative humidity for 48 h, then dried at 85 °C for 60 min, finally, the *Castanea henryi* understory Xiaobai white tea samples (ZLL1) and the ordinary Xiaobai white tea samples (ZLP1) were obtained. On 30 April, the processing of *Cunninghamia lanceolata* understory Xiaobai white tea samples (SLL1) and ordinary Xiaobai white tea (SLP1) were produced, with the same fresh leaf collecting method and processing technology as above. On 2 May, the processing of *Phyllostachys edulis* understory Xiaobai white tea samples (BLL1) and ordinary Xiaobai white tea samples (BLP1) were produced, with the same fresh leaf collecting method and processing technology as above. On 7 May, 9 May, and 13 May, the *Castanea henryi* understory Xiaobai white tea samples (ZLL2) and ordinary Xiaobai white tea samples (ZLP2), the *Cunninghamia lanceolata* understory Xiaobai white tea samples (SLL2) and ordinary Xiaobai white tea samples (SLP2), and the *Phyllostachys edulis* understory Xiaobai white tea samples (BLL2) and ordinary Xiaobai white tea samples (BLP2) were processed again, respectively. The second rounds of processing of fresh leaf picking method and processing technology as above. Taken together, a total of 12 tea samples were obtained including ZLL1 and ZLP1, SLL1, and SLP1, BLL1 and BLP1, ZLL2, and ZLP2, SLL2, and SLP2, and BLL2 and BLP2, once they were produced. Each sample was composed of three biological replicates, stored at 4 °C, and kept dry for further analysis.

3.3. Soil Chemical Analysis

Organic matter was measured using the oxidative thermal potassium dichromate oxidation method [70], the Kjeldahl digestion method was used to extract total nitrogen [71]. Total phosphorus and total potassium were determined using the sodium carbonate method and NaOH melt flamer, respectively [72]. In addition, available nitrogen was measured using the combination pH alkaline hydrolysable method. The hydrochloric acid and flame photometry analyzed the available phosphorus and available potassium, respectively [73]. The pH of soil was measured using the potentiometry [74].

3.4. Sensory Evaluation

Tea samples were evaluated and scored by six professional and trained sensory recognition panelists (three females and three males, 30 to 50 years old) from the Fujian Agriculture and Forestry University, who were professional assessors and trained according to the National Professional Standards for Tea Sensory Evaluation (Profession Code: 6–02-06-11, China) and had more than 10 years of descriptive sensory analysis experience with tea. Tea infusions were prepared according to national standards (GB/T 23776-2018) [75]: each tea sample (3 g) was brewed with 150 mL boiling water for 5 min and discarded, after which tea infusions labelled with a three-digit code were presented to each panelist in a randomized order. Then, the intensity values (0–10), taste descriptors (smoothness, sweetness, thickness, astringency, and umami), and aroma descriptors (floral, sweet, and woody) of each sample infusion were subjected to a sensory test by the six panelists. A scale from 0 to 10 (where 0 was none or no perception and 10 was extremely strong) as described in a previous study was used to symbolize intensity values [76].

3.5. Macro-Composition Quantification

Water extracts were measured using the suggested protocol from the Chinese National Standard, GB/T 8305-2013) [77], free amino acids were measured using the suggested protocol from the Chinese National Standard (GB/T 8314-2013) [78], tea polyphenols were measured using the suggested protocol from the Chinese National Standard (GB/T 8313-2018) with minor changes [79], caffeine was quantified using the UV spectrophotometric method based on the National Standard of China (GB/T 8312-2013) [80], soluble sugars were quantified using the plant-soluble sugar content test kit (Grace Biotechnology Co., Ltd., Suzhou, China) according to the manufacturer's instructions.

Quantification of total flavonoids content was according to an ultraviolet–visible spectrometry method reported by Ma et al. [81]. Briefly, the reaction contained 0.25 mL of the 70% (*v*/*v*) methanol extract or rutin standard solution and 1.25 mL of distilled water, followed by the addition of 0.75 mL 5% (*m*/*v*) sodium nitrite solution was mixed and rested for 5 min. Then, 0.15 mL of 10% (*m*/*v*) aluminum chloride solution was added and stood for 5 min before 0.5 mL of 1 M sodium hydroxide was added. The absorbance of the mixture was measured at 510 nm. The data were calculated using rutin as total flavonoids.

The total content of TBs, TFs, and TRs was determined using a previously discussed method [40]. Briefly, add 125 mL of boiling water to 3 g of tea powder, bathe in water for 10 min, filter, and cool to room temperature. The amount of 50 mL of ethyl acetate was pipetted in 50 mL of tea filtrate and shaken for 5 min, and the layers were separated after equilibration. A 0.8 mL portion of the ethyl acetate layer was pipetted in 10 mL test tubes and fixed volume with 95% ethanol (solution A). The 15 mL of $NaHCO_3$ solution (2.5%) was pipetted in 15 mL of the ethyl acetate layer and shaken for 30 s, testing to layering. A 1.6 mL ethyl acetate layer was pipetted in 10 mL test tubes and fixed volume with 95% ethanol (solution C). A 0.8 mL sample of portions of the aqueous layer was diluted to 10 mL with 2.4 mL distilled water, 0.8 mL saturated oxalic acid solution (10.2%, *m*/*v*), and fixed volume with 95% ethanol (solution D). A 25 mL amount of tea filtrate was pipetted and mixed with 25 mL of butyl alcohol. After shaking for 3 min, the layers were separated after equilibration. A 0.4 mL sample of the aqueous layer (second) was made to a volume of 10 mL with 0.4 mL of the saturated oxalic acid solution, 1.2 mL of distilled water, and 95% ethanol (solution B). The absorbance of solutions A, B, C, and D at 380 nm was measured spectrophotometrically using 95% ethanol as a blank. The results were calculated using the following formula:

$$\mathrm{TF_S} = \frac{E_c \times 2.25}{dried\ weight\ (\%)} \times 100\% \quad (1)$$

$$\mathrm{TR_S} = \frac{7.06 \times (2EA + 2ED - EC - 2EB)}{dried\ weight\ (\%)} \times 100\% \quad (2)$$

$$\mathrm{TB_S} = \frac{2EB \times 7.06}{dride\ weight\ (\%)} \times 100\% \quad (3)$$

3.6. Identification and Analysis of Volatiles

Volatiles were extracted from the samples using the headspace solid-phase microextraction (HS-SPME) method. After sampling, volatile analysis was performed using an Agilent Model 8890 GC equipped with a 30 m × 0.25 mm × 0.25 µm DB-5MS (5% phenyl-polymethylsiloxane) capillary column and a 7000 D mass spectrometer (Agilent, Santa Clara, CA, USA). The procedures of HS-SPME and gas chromatography–mass spectrometry

(GC-MS) analysis were performed, as previously described [40], with minor modifications. The temperature was programmed at 40 °C for 3.5 min, increasing to 100 °C at a rate of 10 °C/min, then to 180 °C at a rate of 7 °C/min, and finally to 280 °C at a rate of 25 °C/min and hold for 5 min. After that, VOCs were identified using the 7000 D MS (Agilent). The MS spectrometer operates in electron shock mode with an electron energy of 70 eV and a scanning range of 50–500 m/z. The temperatures of the ion source, quadrupole, and transfer line were set at 230, 150, and 280 °C, respectively. The more detailed GC-MS acquisition conditions are listed in Supplementary Table S5. The volatile peaks were identified by matching the National Institute of Standards and Technology (NIST) mass spectral database and retention index (RI, determined by n-alkane C7–C40). The chemical structure, name, and aroma of volatile organic compounds are from PubChem (https://pubchem.ncbi.nlm.nih.gov, accessed on 10 June 2023) and Good Scents Company Information System (http://www.thegoodscentscompany.com, accessed on 10 June 2023). The internal standard used was saturated NaCl solution (10 μL, 50 μg/mL), and the relative content of each volatile compound was calculated using the following formula:

$$C_i = \frac{\frac{A_i}{A_{is} \times m_{is}}}{m_i} \quad (4)$$

where *Ci* is the mass concentration of each component (μg kg^{-1}), m_{is} is the mass of the internal standard (μg), A_i, and A_{is} are the chromato-graphic peak area of each component and internal standard, respectively, and mi is the mass of the sample powder (kg).

OAV was calculated by dividing the calculated concentration of the volatile compound by its odor threshold in water and was used to evaluate the contributions of the volatile compounds to the aroma of tea samples. It is generally believed that OAV ≥ 1 of volatile compounds contributes to the flavor of samples [82]. The calculation formula for OAV value is:

$$OAV_i = \frac{C_i}{OT_i} \quad (5)$$

Note: *Ci* (μg/kg) was the content of the volatile compounds; O_{Ti} (μg/kg) was the aroma threshold of volatile components in water.

3.7. Statistical Analysis

All analyses were repeated three times. The statistical significance among the different flavors of Xiaobai white tea was determined by one-way analysis of variance (ANOVA) and Duncan's multiple range test using SPSS (Version 25.0, Armonk, NY, USA) [83]. TBtools (https://github.com/CJ-Chen/TBtools, accessed on 10 July 2023) was used for heat-map and hierarchical cluster analysis of the macro-Composition and volatile components. A radar plot was drawn using Origin (v2021, OriginLab Corporation, Northampton, MA, USA) [84]. OPLS-DA was performed using SIMCA software (version 13.0; Umetrics, Umea, Sweden) [85].

4. Conclusions

In this study, the macro-composition and volatile compounds of LWTs and PWTs were determined and analyzed by biochemical composition determination, widely targeted volatilities (WTV) analysis, multivariate statistical analysis, and odor activity value (OAV) analysis. The results show that understory planting improves the growth environment of tea plants, promotes the formation of the components of fresh-leaf, and provides a good material basis for the processing of dry tea. The free amino acids, TFs, TRs, water-extractable substances, tea polyphenols, 2-heptanol, 2-decane, damasone, and cedar alcohol were the key macro-composition and volatile compounds to distinguish LWTs and PWTs. This study will help to improve the quality of white tea and promote the application of the planting mode of Xiaobai white tea under forest.

Supplementary Materials: The following supporting information can be downloaded at: https://www.mdpi.com/article/10.3390/plants12244102/s1, Figure S1: Total ion chromatogram of mixed sample mass spectrometry analysis for quality control; Figure S2: Scatter plot of volatile components of OPLS-DA scores for six sets of samples (A), (C), (E), (G), (I), and (K); OPLS-DA cross-validation results of volatile components in six groups of samples (B), (D), (F), (H), (J), and (L). Table S1: The qualitative results for the VOCs in LWTs and PWT; Table S2: VIP and FC values of the 13 volatile components; Table S3: The VOCs of OVA is greater than 1 in all samples; Table S4: The serial number in the PLSR analysis corresponds to the VOCs; Table S5: The headspace solid-phase micro extraction (HS-SPME) and gas chromatography-mass spectrometry (GC-MS) conditions.

Author Contributions: Conceptualization, M.Z.; methodology, M.Z. and C.Z. (Chengzhe Zhou); software, M.Z. and K.X.; validation, C.Z. (Cheng Zhang) and L.L.; formal analysis, M.Z. and C.Z. (Chengzhe Zhou); resources, L.H., L.Z., H.L. and X.Z.; data curation, M.Z.; writing—original draft preparation, M.Z.; writing—review and editing, Z.L.; supervision, Y.G. and Z.L.; project administration, Y.G.; funding acquisition, Y.G. All authors have read and agreed to the published version of the manuscript.

Funding: This work was supported by the Tea Industry Technology Branch of Collaborative Innovation Institute (K1521015A),the Construction Project for Technological Innovation and Service System of Tea Industry Chain of Fujian Agriculture and Forestry University (K1520005A01), the "Double First-class" Scientific and Technological Innovation Capacity and Enhancement Cultivation Plan of Fujian Agriculture and Forestry University (KSYLP004), and the Special Fund for Science and Technology Innovation of Fujian Zhang Tianfu Tea Development Foundation (FJZTF01), the tea Industry Development Project of Zhangdun Town, Jianyang District (KH220027A).

Data Availability Statement: Data are contained within the article or Supplementary Materials.

Conflicts of Interest: The authors declare no conflict of interest.

References

1. Mohotti, A.J.; Lawlor, D.W. Diurnal variation of photosynthesis and photoinhibition in tea: Effects of irradiance and nitrogen supply during growth in the field. *J. Exp. Bot.* **2002**, *53*, 313–322. [CrossRef]
2. Chen, Q.; Shi, J.; Mu, B.; Chen, Z.; Dai, W.; Lin, Z. Metabolomics combined with proteomics provides a novel interpretation of the changes in nonvolatile compounds during white tea processing. *Food Chem.* **2020**, *332*, 127412. [CrossRef]
3. Ran, W.; Li, Q.; Hu, X.; Zhang, D.; Yu, Z.; Chen, Y.; Wang, M.; Ni, D. Comprehensive analysis of environmental factors on the quality of tea (*Camellia sinensis* var. *sinensis*) fresh leaves. *Sci. Hortic.* **2023**, *319*, 112177. [CrossRef]
4. Shao, C.; Deng, Z.; Liu, J.; Li, Y.; Zhang, C.; Yao, S.; Zuo, H.; Shi, Y.; Yuan, S.; Qin, L.; et al. Effects of Preharvest Shading on Dynamic Changes in Metabolites, Gene Expression, and Enzyme Activity of Three Tea Types during Processing. *J. Agric. Food Chem.* **2022**, *70*, 14544–14558. [CrossRef] [PubMed]
5. Santosh, K.C.; Long, L.; Zhang, Q.; Ni, K.; Ma, L.; Ruan, J. Effect of Interactions between Phosphorus and Light Intensity on Metabolite Compositions in Tea Cultivar Longjing43. *Int. J. Mol. Sci.* **2022**, *23*, 15194.
6. Wang, M.; Yang, J.; Li, J.; Zhou, X.; Xiao, Y.; Liao, Y.; Tang, J.; Dong, F.; Zeng, L. Effects of temperature and light on quality-related metabolites in tea [*Camellia sinensis* (L.) *Kuntze*] leaves. *Food Res. Int.* **2022**, *161*, 111882. [CrossRef] [PubMed]
7. Wang, Y.; Zhang, Q.; Li, J.; Lin, S.; Jia, X.; Zhang, Q.; Ye, J.; Wang, H.; Wu, Z. Study on the Effect of pH on Rhizosphere Soil Fertility and the Aroma Quality of Tea Trees and Their Interactions. *Agriculture* **2023**, *13*, 1739. [CrossRef]
8. Ye, J.; Chen, X.; Liu, G.; Jia, X.; Zhang, Q.; Zhu, C.; Wang, Y.; Wang, H.B. Effect of tea soil acidification on the diversity and function of fungi community. *J. Appl. Bot. Food Qual.* **2021**, *94*, 199–205. [CrossRef]
9. Tang, S.; Zhou, J.; Pan, W.; Sun, T.; Liu, M.; Tang, R.; Li, Z.; Ma, Q.; Wu, L. Effects of combined application of nitrogen, phosphorus, and potassium fertilizers on tea (*Camellia sinensis*) growth and fungal community. *Appl. Soil. Ecol.* **2023**, *181*, 104661. [CrossRef]
10. Chen, P.; Lin, S.; Liu, C.; Su, Y.; Cheng, H.; Shiau, J.; Chen, I. Correlation between nitrogen application to tea flushes and quality of green and black teas. *Sci. Hortic.* **2015**, *181*, 102–107. [CrossRef]
11. Lin, Z.; Qi, Y.; Chen, R.; Zhang, F.; Chen, L. Effects of phosphorus supply on the quality of green tea. *Food Chem.* **2012**, *130*, 908–914. [CrossRef]
12. Artru, S.; Garré, S.; Dupraz, C.; Hiel, M.; Blitz-Frayret, C.; Lassois, L. Impact of spatio-temporal shade dynamics on wheat growth and yield, perspectives for temperate agroforestry. *Eur. J. Agron.* **2017**, *82*, 60–70. [CrossRef]
13. Quinkenstein, A.; Wöllecke, J.; Böhm, C.; Grünewald, H.; Freese, D.; Schneider, B.U.; Hüttl, R.F. Ecological benefits of the alley cropping agroforestry system in sensitive regions of Europe. *Environ. Sci. Policy* **2009**, *12*, 1112–1121. [CrossRef]
14. Albrecht, A.; Kandji, S.T. Carbon sequestration in tropical agroforestry systems. *Agric. Ecosyst. Environ.* **2003**, *99*, 15–27. [CrossRef]
15. Torralba, M.; Fagerholm, N.; Burgess, P.J.; Moreno, G.; Plieninger, T. Do European agroforestry systems enhance biodiversity and ecosystem services? A meta-analysis. *Agric. Ecosyst. Environ.* **2016**, *230*, 150–161. [CrossRef]

16. Wang, Y.; Mo, Y.R.; Tan, J.; Wu, L.X.; Pan, Y.; Chen, X.D. Effects of growing *Coptis chinensis* Franch in the natural understory vs. under a manmade scaffold on its growth, alkaloid contents, and rhizosphere soil microenvironment. *Peer J.* **2022**, *10*, 13676. [CrossRef]
17. Huang, Z.; Shen, Z.; Liu, C.; Shi, H.; He, S.; Long, G.; Deng, W.; Yang, J.L.; Fan, W. Characteristics of heavy metal accumulation and risk assessment in understory Panax notoginseng planting system. *Environ. Geochem. Health* **2022**, *45*, 9029–9040. [CrossRef] [PubMed]
18. Sgarbossa, J.; Elli, E.F.; Schwerz, F.; Nardini, C.; de Cristo, E.; de Oliveira, D.; Caron, B.O. Morphology, growth and yield of black oats cultivated in agroforestry systems in southern Brazil. *Agric. Syst.* **2020**, *184*, 102911. [CrossRef]
19. Liu, J.; Yuan, D.; Si, H.; Pang, X.; Tang, X.; Yang, J. Effects of Shading on Ingredients of Tea Shoots in Different Seasons. *Southwest China J. Agric. Sci.* **2013**, *26*, 115–118.
20. Yang, Z.; Kobayashi, E.; Katsuno, T.; Asanuma, T.; Fujimori, T.; Ishikawa, T.; Tomomura, M.; Mochizuki, K.; Watase, T.; Nakamura, Y.; et al. Characterisation of volatile and non-volatile metabolites in etiolated leaves of tea (*Camellia sinensis*) plants in the dark. *Food Chem.* **2012**, *135*, 2268–2276. [CrossRef]
21. Shu, J.; Xu, Y. Microscopic and ultrastructural observations of tea tree leaves. *China Tea* **1990**, *6*, 18–19.
22. Fang, Q.; Jin, J.; Zhu, Y.; Ye, J.; Dong, J.; Liang, Y. Effect of shade treatmen and harvest season on the taste and aroma-related components in steamed green tea. *J. Tea* **2021**, *47*, 131–137.
23. Yang, X.; Leng, Y.; Zhou, Z.; Shang, H.; Ni, K.; Ma, L.; Yi, X.; Cai, Y.; Ji, L.; Ruan, J.; et al. Ecological management model for the improvement of soil fertility through the regulation of rare microbial taxa in tea (*Camellia sinensis* L.) plantation soils. *J. Environ. Manag.* **2022**, *308*, 114595. [CrossRef]
24. Duan, X. *Practical New Technology for Quality Tea Production*; Sichuan Education Publishing House: Chengdu, China, 2008; p. 63.
25. Li, X.; Ahammed, G.J.; Li, Z.; Zhang, L.; Wei, J.; Shen, C.; Yan, P.; Zhang, L.; Han, W. Brassinosteroids Improve Quality of Summer Tea (*Camellia sinensis* L.) by Balancing Biosynthesis of Polyphenols and Amino Acids. *Front. Plant Sci.* **2016**, *7*, 1304. [CrossRef]
26. Tongsiri, P.; Tseng, W.; Shen, Y.; Lai, H. Comparison of Soil Properties and Organic Components in Infusions According to Different Aerial Appearances of Tea Plantations in Central Taiwan. *Sustainability* **2020**, *12*, 4384. [CrossRef]
27. Wen, B.; Zhang, X.; Ren, S.; Duan, Y.; Zhang, Y.; Zhu, X.; Wang, Y. Characteristics of soil nutrients, heavy metals and tea quality in different intercropping patterns. *Agrofor. Syst.* **2020**, *94*, 963–974. [CrossRef]
28. De Costa, W.A.J.M.; Surenthran, P.; Attanayake, K.B. Tree-crop interactions in hedgerow intercropping with different tree species and tea in Sri Lanka: 2. Soil and plant nutrients. *Agrofor. Syst.* **2005**, *63*, 211–218. [CrossRef]
29. Zhang, G.; Chu, X.; Zhu, H.; Zou, D.; Li, L.; Du, L. The Response of Soil Nutrients and Microbial Community Structures in Long-Term Tea Plantations and Diverse Agroforestry Intercropping Systems. *Sustainability* **2021**, *13*, 7799. [CrossRef]
30. Huang, L.; Xu, K.; Zhou, C.; Shi, B.; Tian, C.; Lu, L.; Guo, Y. Quality Differences of Zhenghe White Tea from Different Altitudes. *Food Sci.* **2022**, *44*, 274–283.
31. Wang, S.; Yin, K.; Liang, G. Multivariate correlation analysis of leaf anatomy and quality traits of tea plants. *J. Tea Bus.* **2002**, *04*, 12–13.
32. Chen, Y.; Fu, X.; Mei, X.; Zhou, Y.; Cheng, S.; Zeng, L.; Dong, F.; Yang, Z. Proteolysis of chloroplast proteins is responsible for accumulation of free amino acids in dark-treated tea (*Camellia sinensis*) leaves. *J. Proteom.* **2017**, *157*, 10–17. [CrossRef]
33. Lu, A.; Ye, Y.; Liao, X.; Mao, S.; Yan, J.; Yang, L.; Zhang, Y.; Tong, H. Effects of Shading on Main Nitrogen Compounds in Fresh Leaves of Tea Plant. *Chin. Agric. Sci. Bull.* **2019**, *35*, 76–81.
34. Fu, H.; Li, H.; Yin, P.; Mei, H.; Li, J.; Zhou, P.; Wang, Y.; Ma, Q.; Jeyaraj, A.; Thangaraj, K.; et al. Integrated Application of Rapeseed Cake and Green Manure Enhances Soil Nutrients and Microbial Communities in Tea Garden Soil. *Sustainability* **2021**, *13*, 2967. [CrossRef]
35. Penn, C.J.; Camberato, J.J. A Critical Review on Soil Chemical Processes that Control How Soil pH Affects Phosphorus Availability to Plants. *Agriculture* **2019**, *9*, 120. [CrossRef]
36. Shao, C.; Jiao, H.; Chen, J.; Zhang, C.; Liu, J.; Chen, J.; Li, Y.; Huang, J.; Yang, B.; Liu, Z.; et al. Carbon and Nitrogen Metabolism Are Jointly Regulated During Shading in Roots and Leaves of Camellia Sinensis. *Front. Plant Sci.* **2022**, *13*, 894840. [CrossRef]
37. Du, Y.; Zhao, Q.; Chen, L.; Yao, X.; Zhang, W.; Zhang, B.; Xie, F. Effect of drought stress on sugar metabolism in leaves and roots of soybean seedlings. *Plant Physiol. Biochem.* **2020**, *146*, 1–12. [CrossRef] [PubMed]
38. Gong, X.; Li, L.; Qin, L.; Huang, Y.; Ye, Y.; Wang, M.; Wang, Y.; Xu, Y.; Luo, F.; Mei, H. Targeted Metabolomics Reveals Impact of N Application on Accumulation of Amino Acids, Flavonoids and Phytohormones in Tea Shoots under Soil Nutrition Deficiency Stress. *Forests* **2022**, *13*, 1629. [CrossRef]
39. Lin, Z.; Chen, C.; Zhong, Q.; Ruan, Q.; Chen, Z.; You, X.; Shan, R.; Li, X. The GC-TOF/MS-based Metabolomic analysis reveals altered metabolic profiles in nitrogen-deficient leaves and roots of tea plants (*Camellia sinensis*). *BMC Plant Biol.* **2021**, *21*, 506. [CrossRef] [PubMed]
40. Zhang, C.; Zhou, C.; Xu, K.; Tian, C.; Zhang, M.; Lu, L.; Zhu, C.; Lai, Z.; Guo, Y. A Comprehensive Investigation of Macro-Composition and Volatile Compounds in Spring-Picked and Autumn-Picked White Tea. *Foods* **2022**, *11*, 3628. [CrossRef]
41. Ho, C.; Zheng, X.; Li, S. Tea aroma formation. *Food Sci. Hum. Well.* **2015**, *4*, 9–27. [CrossRef]
42. Liu, Z.; Chen, F.; Sun, J.; Ni, L. Dynamic changes of volatile and phenolic components during the whole manufacturing process of Wuyi Rock tea (Rougui). *Food Chem.* **2021**, *367*, 130624. [CrossRef]

43. Lv, S.; Wu, Y.; Li, C.; Xu, Y.; Liu, L.; Meng, Q. Comparative Analysis of Pu-erh Tea and Fuzhuan Tea by Fully Automatic Headspace Solid-Phase Microextraction Coupled with Gas Chromatography-Mass Spectrometry and Chemometric Methods. *J. Agric. Food Chem.* **2014**, *62*, 1810–1818. [CrossRef]
44. Ni, Z.; Zhou, Z.; Liu, B.; Gao, F.; Li, L.; Wu, Q.; Deng, H.; Sun, Y. Study on Flavor Components in "Cloud-Mist Mountain Tea" Green Tea. *Sci. Technol. Food Ind.* **2021**, *42*, 214–221.
45. Wang, H. Study on the Characteristic Aroma Components of Keemun Black Tea. Master's Thesis, Shanghai Institute of Technology, Shanghai, China, 2017.
46. Shen, S.; Wu, H.; Li, T.; Sun, H.; Wang, Y.; Ning, J. Formation of aroma characteristics driven by volatile components during long-term storage of An tea. *Food Chem.* **2023**, *411*, 135487. [CrossRef]
47. Qi, D.; Miao, A.; Cao, J.; Wang, W.; Chen, W.; Pang, S.; He, X.; Ma, C. Study on the effects of rapid aging technology on the aroma quality of white tea using GC–MS combined with chemometrics: In comparison with natural aged and fresh white tea. *Food Chem.* **2018**, *265*, 189–199. [CrossRef] [PubMed]
48. Wang, L.; Cai, L.; Lin, Z.; Zhong, Q.; Lv, H.; Tan, J.; Guo, L. Analysis of Aroma Compounds in White Tea Using Headspace Solid-phase Micro-extraction and GC-MS. *J. Tea Sci.* **2010**, *30*, 115–123.
49. Lin, J.; Chen, Y.; Zhang, P.; Ren, M.; Xu, H.; Wang, X. A novel quality evaluation index and strategies to identify scenting quality of jasmine tea based on headspace volatiles analysis. *Food Sci. Biotechnol.* **2013**, *22*, 331–340. [CrossRef]
50. Nie, C.; Du, X.; Gao, Y.; He, L.; Wang, C.; Zhang, X.; Zhong, X. Comparison of different aroma-active compounds of Sichuan Dark brick tea (*Camellia sinensis*) and Sichuan Fuzhuan brick tea using gas chromatography–mass spectrometry (GC–MS) and aroma descriptive profile tests. *Eur. Food Res. Technol. = Z. Fur Lebensm. Unters. Forschung. A* **2019**, *245*, 1963–1979. [CrossRef]
51. Dai, Q.; Jin, H.; Gao, J.; Ning, J.; Yang, X.; Xia, T. Investigating Volatile Compounds' Contributions to the Stale Odor of Green Tea. *Int. J. Food Sci. Technol.* **2019**, *55*, 1606–1616. [CrossRef]
52. Li, Y.; Wu, T.; Deng, X.; Tian, D.; Ma, C.; Wang, X.; Li, Y.; Zhou, H. Characteristic aroma compounds in naturally withered and combined withered γ-aminobutyric acid white tea revealed by HS-SPME-GC-MS and relative odor activity value. *LWT* **2023**, *176*, 114467. [CrossRef]
53. Chen, Z.; Li, P.; Chen, X.; Yang, Y.; He, P.; Tu, Y. Effect of Compressed Processing on the Aroma of Aged White Tea. *Sci. Technol. Food Ind.* **2020**, *41*, 63–69.
54. Wang, Q.; Chen, D. Research on characteristic aroma material in Guangdong Chenxiang tea. *Guangdong Agric. Sci.* **2012**, *39*, 47–51.
55. El Hadi, M.A.; Zhang, F.; Wu, F.; Zhou, C.; Tao, J. Advances in Fruit Aroma Volatile Research. *Molecules* **2013**, *18*, 8200–8229. [CrossRef] [PubMed]
56. Zhou, Z.; Wang, X.; Luo, Z.; Fan, G.; Tian, C.; Tang, Q.; Pan, S. Identification and Analysis of Free and O-Glycoside-Bound Volatile Components in Wild *Rosa roxburghii* Juice. *Food Sci.* **2012**, *33*, 192–197.
57. Zhang, Y.; Zhu, Y.; Lv, H.; Huang, H.; Shao, C.; Peng, J.; Lin, Z. Quality composition analysis of roasted green tea at different altitudes. *Food Sci.* **2022**, *43*, 2257–2268.
58. Yin, X.; Fu, W.; Chen, Y.; Zhou, R.; Sun, W.; Ding, B.; Peng, X.; Gu, H. GC-MS-based untargeted metabolomics reveals the key volatile organic compounds for discriminating grades of Yichang big-leaf green tea. *LWT* **2022**, *171*, 114148. [CrossRef]
59. Zhang, L.; Wang, J.; Luo, L.; Zeng, L. Analysis of characteristic aroma components of eagle tea. *Food Sci.* **2019**, *40*, 220–228.
60. Xu, X. Study on the Difference of Roasting Degree, Grade and Regional Quality of Wuyi Cinnamon Tea Based on Metabolomics. Master's Thesis, Fujian Agriculture and Forestry University, Fuzhou, China, 2020.
61. Fan, F.; Huang, C.; Tong, Y.; Guo, H.; Zhou, S.; Ye, J.; Gong, S. Widely targeted metabolomics analysis of white peony teas with different storage time and association with sensory attributes. *Food Chem.* **2021**, *362*, 130257. [CrossRef]
62. Xue, J.; Guo, G.; Liu, P.; Chen, L.; Wang, W.; Zhang, J.; Yin, J.; Ni, D.; Engelhardt, U.; Jiang, H. Identification of aroma-active compounds responsible for the floral and sweet odors of Congou black teas using gas chromatography–mass spectrometry/olfactometry, odor activity value, and chemometrics. *J. Sci. Food Agric.* **2022**, *102*, 5399–5410. [CrossRef]
63. Guo, X.; Schwab, W.; Ho, C.; Song, C.; Wan, X. Characterization of the aroma profiles of oolong tea made from three tea cultivars by both GC–MS and GC-IMS. *Food Chem.* **2022**, *376*, 131933. [CrossRef] [PubMed]
64. Huang, Y.; Wan, J.; Wang, Z.; Sun, M.; Feng, T.; Ho, C.; Song, S. Variation of Volatile Compounds and Corresponding Aroma Profiles in Chinese Steamed Bread by Various Yeast Species Fermented at Different Times. *J. Agric. Food Chem.* **2022**, *70*, 3795–3806. [CrossRef]
65. Huang, G.; Li, Y.; Deng, X.; Su, D.; Shen, Y.; Li, Y.; Zhou, H. Analysis of aroma compounds of Yunnan white tea by four drying methods. *Sci. Technol. Food Ind.* **2022**, *18*, 283–292.
66. Yuan, Q.; Tu, M.; Gao, P.; Hu, C.; He, D. Comparative Analysis of Rapeseed Oils Prepared by Three Different Methods. *J. Oleo Sci.* **2020**, *69*, 1641–1648. [CrossRef] [PubMed]
67. Wang, W.; Huang, X.; Lin, Y.; Tang, R.; Guo, Y. Analysis of volatile compounds in Guanyin tea stem. *Trop. Crops* **2019**, *5*, 965–972.
68. Yu, S. Multi-Wavelength LED White Tea Quality Control and Its Production Line Research. Master's Thesis, Fujian Agriculture and Forestry University, Fuzhou, China, 2017.
69. Ni, H.; Jiang, Q.; Zhang, T.; Huang, G.; Li, L.; Chen, F. Characterization of the Aroma of an Instant White Tea Dried by Freeze Drying. *Molecules* **2020**, *25*, 3628. [CrossRef] [PubMed]

70. Roper, W.R.; Robarge, W.P.; Osmond, D.L.; Heitman, J.L. Comparing Four Methods of Measuring Soil Organic Matter in North Carolina Soils. *Soil. Sci. Soc. Am. J.* **2019**, *83*, 466–474. [CrossRef]
71. Keeney, D.R.; Nelson, D.W. *Methods of Soil Analysis. Part 2. Chemical and Microbiological Properties*; John Wiley & Sons: Hoboken, NJ, USA, 2015; Volume 5, pp. 643–698.
72. Lin, W.; Lin, M.; Zhou, H.; Wu, H.; Li, Z.; Lin, W. The effects of chemical and organic fertilizer usage on rhizosphere soil in tea orchards. *PLoS ONE* **2019**, *14*, e217018. [CrossRef] [PubMed]
73. Mózo, B.S.; Benites, A.C.; Sukma, M.; Paul, M.M. *Handbook of Soil Analysis: Mineralogical, Organic and Inorganic Methods*; Springer Science & Business Media: Berlin, Germany, 2016.
74. *HJ 962-2018*; Soil—Determination of pH—Potentiometry. China Environmental Science Press: Beijing, China, 2018.
75. *GB/T 23776-2018*; Methodology for Sensory Evaluation of Tea. Standards Press of China: Beijing, China, 2018.
76. Guo, X.; Ho, C.; Wan, X.; Zhu, H.; Liu, Q.; Wen, Z. Changes of volatile compounds and odor profiles in Wuyi rock tea during processing. *Food Chem.* **2021**, *341*, 128230. [CrossRef] [PubMed]
77. *GB/T 8305-2013*; Tea—Determination of Water Extracts Content. Standards Press of China: Beijing, China, 2013.
78. *GB/T 8314-2013*; Tea—Determination of Free Amino Acids Content. Standards Press of China: Beijing, China, 2013.
79. *GB/T 8313-2018*; Determination of Total Polyphenols and Catechins Content in Tea. Standards Press of China: Beijing, China, 2018.
80. *GB/T 8312-2013*; Tea—Determination of Caffeine Content. Standards Press of China: Beijing, China, 2013.
81. Ma, Z.; Li, S.; Zhang, M.; Jiang, S.; Xiao, Y. Light Intensity Affects Growth, Photosynthetic Capability, and Total Flavonoid Accumulation of Anoectochilus Plants. *HortScience* **2010**, *45*, 863–867. [CrossRef]
82. Li, K.; Wang, J.; Li, Y.; Li, Q. Characteristic aroma components of red pitaya wine. *Food Ferment. Ind.* **2019**, *45*, 217–223.
83. IBM Corp. *IBM SPSS Statistics for Windows*, Version 25.0; IBM Corp.: Armonk, NY, USA, 2017.
84. *Origin(Pro)*, version 2021; OriginLab Corporation: Northampton, MA, USA, 2021.
85. *SIMCA*, Version 13; Umetrics: Umea, Sweden, 2012.

plants

MDPI

Article

Transcriptome Analysis Reveals Fruit Quality Formation in *Actinidia eriantha* Benth

Peiyu Wang [1,2], Xin Feng [3], Jinlan Jiang [1,2], Peipei Yan [1,2], Zunwen Li [1,2], Weihong Luo [4], Yiting Chen [3,*] and Wei Ye [1,2,*]

1 Sanming Academy of Agricultural Sciences, Shaxian 365051, China; 17268144914@163.com (P.W.); 13616957010@163.com (J.J.); 18296793560@163.com (Z.L.)
2 The Key Laboratory of Crop Genetic Improvement and Innovative Utilization in Fujian Province (Mountain Area), Shaxian 365051, China
3 Fruit Tree Research Institute of Fujian Academy of Agricultural Sciences, Fuzhou 350002, China; fengxin1506@163.com
4 Institute of Horticultural Plant Bioengineering, Fujian Agriculture and Forestry University, Fuzhou 350002, China; L1224142990@163.com
* Correspondence: chyiting@163.com (Y.C.); yewei922@sina.com (W.Y.)

Abstract: *Actinidia chinensis* Planch. is a fruit tree originating from China that is abundant in the wild. Actinidia eriantha Benth. is a type of *A. chinensis* that has emerged in recent years. The shape of *A. eriantha* is an elongated oval, and the skin is covered with dense, non-shedding milk-white hairs. The mature fruit has flesh that is bright green in colour, and the fruit has a strong flavour and a grass-like smell. It is appreciated for its rich nutrient content and unique flavour. Vitamin C, sugar, and organic acids are key factors in the quality and flavour composition of *A. eriantha* but have not yet been systematically analysed. Therefore, we sequenced the transcriptome of *A. eriantha* at three developmental stages and labelled them S1, S2, and S3, and comparisons of S1 vs. S2, S1 vs. S3, and S2 vs. S3 revealed 1218, 4019, and 3759 upregulated differentially expressed genes and 1823, 3415, and 2226 downregulated differentially expressed genes, respectively. Furthermore, the upregulated differentially expressed genes included 213 core genes, and Gene Ontology enrichment analysis showed that they were enriched in hormones, sugars, organic acids, and many organic metabolic pathways. The downregulated differentially expressed genes included 207 core genes, which were enriched in the light signalling pathway. We further constructed the metabolic pathways of sugars, organic acids, and vitamin C in *A. eriantha* and identified the genes involved in vitamin C, sugar, and organic acid synthesis in *A. eriantha* fruits at different stages. During fruit development, the vitamin C content decreased, the carbohydrate compound content increased, and the organic acid content decreased. The gene expression patterns were closely related to the accumulation patterns of vitamin C, sugars, and organic acids in *A. eriantha*. The above results lay the foundation for the accumulation of vitamin C, sugars, and organic acids in *A. eriantha* and for understanding flavour formation in *A. eriantha*.

Keywords: *Actinidia eriantha* Benth; fruit quality; organic acid; vitamin C; sugars; transcriptome sequencing

1. Introduction

As a fruit tree native to China, *Actinidia eriantha* has become one of the most popular in recent years. *A. eriantha* is rich in nutrients, and the fruit contains a large amount of vitamin C (Vc), dietary fibre, and a variety of mineral nutrients [1]. In addition, *A. eriantha* contains approximately 500 to 1379 mg/100 g fresh weight (FW) of Vc; thus, it is not inferior to other varieties of kiwifruit [2]. However, more studies have been performed on the synthesis and metabolism of Vc in *Actinidia chinensis* and *Actinidia deliciosa* than in *A. eriantha*. In addition, *A. eriantha* is popular for its moderate sweetness and acidity and its unique flavour [3]. Sugar and acid components are important fruit quality components of research interest and have important effects on fruit's flavour. However, research on kiwifruit has focused on *A. chinensis* and *A. deliciosa*, which contain fructose, glucose, and sucrose [4], and quinic acid, malic acid, and citric acid as the primary acid components [5]. To date, the formation of unique *A. eriantha* flavours and the composition of sugars and organic acids have been less studied.

Sweetness is one of the primary factors used to measure fruit quality [6]. The content and composition ratio of sugar compounds such as sucrose, fructose, and glucose determine the sweetness and flavour of fruits [7,8]. For example, fructose is the main sugar substance in apples and pears [9,10], while sucrose is the main sugar substance in watermelon and cantaloupe [11,12]. Furthermore, previous studies have shown that different varieties of kiwifruit have different main sugar substances. In the fruit of kiwifruit 'Hort 16A', glucose is the most abundant, followed by fructose, while sucrose is the least abundant [13]. On the other hand, in the fruit of 'Mitsuko' kiwifruit, sucrose is the main sugar substance, and the contents of glucose and fructose are relatively low [14]. Sucrose, one of the key substances in plant metabolism, provides an important material foundation for plant growth and development. The synthesis of sucrose in plants is mainly regulated by sucrose synthase (*SUS*), which controls the synthesis of UDP-glucose for sucrose formation [15]. Subsequently, sucrose is converted into fructose and glucose by neutral invertase. Additionally, fructose and glucose can be phosphorylated by fructokinase and hexokinase, forming fructose-6-phosphate and glucose-6-phosphate [16].

Acidity is another important indicator that determines fruit quality [17]. The main organic acids in fruits are usually malic acid, citric acid, tartaric acid, and succinic acid. Additionally, different organic acids have varying levels of acidity, with malic acid being the strongest and tartaric acid being the weakest. The proportion of organic acids to sugar substances in fruit determines the balance of sweetness and acidity [18]. Furthermore, quinic acid and citric acid are the main organic acids found in kiwifruit 'Ganmi 6', yellow-fleshed kiwifruit 'Jinshi 1', and *A. chinensis* [19–21]. However, some studies also suggest that malic acid is a major component of organic acids in kiwifruit [22,23]. Moreover, malic acid and citric acid, as key intermediates in the tricarboxylic acid cycle, also participate in plant respiration and energy metabolism [24]. The types and levels of organic acids in fruits play a crucial role in determining the quality and unique flavour of horticultural crops.

Vc is an essential nutrient for the human body that cannot be synthesized endogenously and thus needs to be obtained from external sources. There are four pathways for the synthesis of Vc in plants, namely, the L-galactose, D-galactose, L-gulose, and myo-inositol pathways [25–28]. Additionally, plants have a Vc recycling pathway. Previous studies have shown that the majority of plants use the L-galactose and D-galactose pathways as the main routes of Vc synthesis. Kiwifruit also predominantly synthesizes Vc via these two pathways. However, there was significant variation in the Vc content among the different kiwifruit varieties. The fruits of *A. deliciosa* 'Hayward' contain 65.5 mg/100 g FW of Vc. The Vc content of the fruit of *A. deliciosa* ranged from 29 mg/100 g FW to 80 mg/100 g FW. The Vc content of the fruit of sanuki gold was more than three times that of *A. deliciosa* 'Hayward'. In *Actinidia arguta* fruit, the Vc content ranges from 37 to 185 mg/100 g FW [29,30].

As a technology to study the transcription of all genes in cells at the overall level, transcriptome sequencing has extensive applications in plant function research. Transcriptome sequencing was used to identify many genes related to flowering and fruiting in rabbit-eye blueberry (*Vaccinium ashei*) [31]. Using transcriptome sequencing, *Lycium ruthenicum* provides the basis for functional and molecular biology studies on its fruit colouration [32]. Transcriptome sequencing of different stages of fruit development helped to elucidate the molecular mechanism of *Pyrus bretschneideri* Rehd fruit ripening [33]. In addition, transcriptome techniques have been used in fruit trees such as *Prunus armeniaca* L. [34,35] and *Rosa roxburghii* [36]. *A. chinensis* Planch. is a perennial deciduous fruiting woody vine of *Actinidia, Actinidiaceae. A. eriantha* is a unique germplasm resource of kiwifruit in China, and more than 300 accessions have been recorded and evaluated. At present, most *A. eriantha* are found in the wild. *A. eriantha* 'White', *A. eriantha* 'Shanong 18', and *A. eriantha* 'Ganmi 6' are the dominant *A. eriantha*, and only one hybridized cultivar, *A. eriantha* 'Jinyan', has been registered. *A. eriantha* fruits are rich in Vc, minerals, and a variety of amino acids, with unique flavours [37]. Therefore, the synthesis and accumulation of vitamins, sugars, and organic acids during the growth and development process play a crucial role in the fruit quality and its unique flavour. Previous transcriptomic studies on kiwifruit have analysed its microsatellite markers for population genetics studies and the new formation mechanisms of the early maturity traits in kiwifruit [38]. However, systematic analysis and determination of the Vc, sugar, and organic acid contents in kiwifruit have not been reported. Therefore, this study used transcriptome sequencing technology to analyse samples from different stages of fruit growth and development of *A. eriantha* [39]. The transcriptome results showed that many differentially expressed genes (DEGs) were identifiable during the development of *A. eriantha* fruits. Similarly, we identified 213 and 207 core upregulated and downregulated genes among the DEGs. The Gene Ontology (GO) enrichment showed that the 213 core upregulated DEGs were mainly enriched in hormones, sugars, organic acids, and many organic matter metabolism pathways, and the 207 core downregulated DEGs showed a greatly reduced response to light during *A. eriantha* ripening. To further understand the fruit flavour formation of *A. eriantha,* we further constructed a synthesis pathway of sugar, organic acids, and Vc during fruit development in *A. eriantha*. Furthermore, we measured the contents of sugars, organic acids, and Vc during fruit development. Finally, we verified the relationship between genes and Vc, sugar, and organic acid using real-time fluorescence quantitative PCR (RT–qPCR). This study provides a reference for understanding the formation of thw unique flavours in *A. eriantha,* which can be used for genetic breeding and variety improvement.

2. Results

2.1. Transcriptomic Analysis of A. eriantha

Transcriptome sequencing was performed using nine samples of *A. eriantha* at different stages of growth and development (Figure 1A). The sequencing yielded 57.60 Gb of raw data, with a Qphred 30 (Q30, the probability of correct sequencing for one base is 99.99%) of all data greater than 93.35% (Supplemental Table S2). Further low-quality read data were removed using the Fsatp software with the default parameters and aligned with the *A. eriantha* genome, and the alignment rate was 82.37% to 84.11% (Supplemental Table S3). The results indicate a high quality of transcriptome data for sequencing. In addition, the results of principal component analysis (PCA) showed that the three biological replicates of the three samples clustered together and that principal component 1 (PC1) explained 89.8% of the observed variation, representing the difference between the samples (Figure 1B).

To gain insight into the dynamic changes in gene expression during *A. eriantha* ripening, we analysed the fragments per kilobase of exon model per million mapped fragments (FPKM) of data from three stages. We identified 1218, 4019, and 3759 upregulated differentially expressed genes and 1823, 3415, and 2226 distinct downregulated differentially expressed genes in S1 vs. S2, S1 vs. S3, and S2 vs. S3, respectively (Figure 2A–D). Moreover, further analysis of these DEGs revealed that the upregulated DEGs included 213 core genes (Figure 2E), and the downregulated DEGs included 207 core genes (Figure 2F). Therefore, we further performed a GO enrichment analysis on these core genes. The 213 core upregulated DEGs were mainly enriched in hormones, sugars, organic acids, and many organic matter metabolism pathways (Supplemental Table S4), indicating an important role in the accumulation and metabolism of hormones and organic matter during the maturation of *A. eriantha*. However, the GO enrichment of 207 core downregulated DEGs showed a greatly reduced response to light during *A. eriantha* ripening (Supplemental Table S5).

Figure 1. (**A**) The three developmental stages of *A. eriantha* fruit, S1: fruit at 180 days after flowering, S2: fruit at 210 days after flowering, S3: fruit at 240 days after flowering. (**B**) Principal component analysis of three fruit samples based on gene expression profiles.

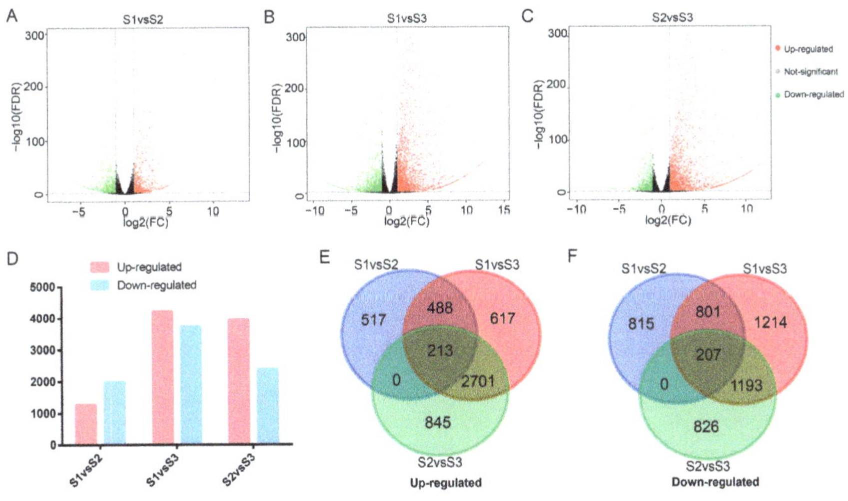

Figure 2. DEGs during *A. eriantha* maturation. (**A–C**) Volcano plot of DEGs between the three stages of *A. eriantha* ripening. (**D**) Number of DEGs. (**E**,**F**) Statistics of the core upregulated and downregulated DEGs.

2.2. Analysis of the Vc Synthesis Pathway in A. eriantha

Vc is one of the important factors in *A. eriantha* quality. To understand the dynamic changes in Vc content during fruit ripening, we constructed the Vc anabolic pathway (Figure 3A) of *A. eriantha*. The galactose and inositol pathways make limited contributions to Vc synthesis in plants. Therefore, in this paper, the expression profile of Vc synthesis-related genes in *A. eriantha* was constructed based on the L-galactose pathway, galacturonate pathway, and recycling pathway. The results showed that the expression patterns of genes in the L-galactose pathway, galacturonate pathway, and recycling pathway were consistent during the development of *A. eriantha*. They are highly expressed in early and middle fruit development, while gene expression decreases after fruit maturity. Furthermore, we determined the Vc content (Figure 3B) at the three stages of *A. eriantha* development. The results showed that the Vc content was significantly reduced after fruit maturation compared with early fruit development. However, the expression trends of five, two, and six genes in the L-galactose pathway, the galacturonic acid pathway, and the recycling pathway, respectively, showed inconsistencies with the pattern of Vc accumulation in Chinese kiwifruit (Figure 3A,B). Therefore, to further elucidate the correlation between Vc accumulation and gene expression in *A. eriantha,* we conducted RT–qPCR analysis of these genes in the above three pathways. In the L-galactose pathway, we screened five genes from the glucose-6-phosphate isomerase (*PGI*), mannose-6-phosphate isomerase (*PMI*), GDP-D-mannose pyrophosphorylase (*GMP*), and GDP-L-galactose phosphorylase (*GGP*) gene families (Figure 4A). Among the five genes, the *PGI* (*DTZ79_10g09380*) and *GMP* (*DTZ79_13g11340*) expression levels increased during fruit development. However, the expression of *PGI* (*DTZ79_10g12530*) and *PMI* (*DTZ79_02g10970*) decreased during fruit development, while *GGP* (*DTZ79_29g10040*) expression was decreased at S2 and peaked at S3. In addition, in the galacturonate pathway, the *D-galacturonate reductase* (*GalUR*) (*DTZ79_13g02680*) and *aldonolactonase* (*Alase*) (*DTZ79_03g07860*) expression levels decreased with fruit development (Figure 4B). In the Vc recycling pathway, we selected six genes (Figure 4C). The expression of *L-ascorbate peroxidase* (*APX*) (*DTZ79_25g10570*) peaked at S2, and the expression of *APX* (*DTZ79_05g07210*) increased with fruit ripening. Two genes of the L-ascorbate oxidase (*AO*) (*DTZ79_27g07430* and *DTZ79_27g07610*) and two genes of the *monodehydroascorbate reductase* (*MDHAR*) (*DTZ79_27g01630* and *DTZ79_27g11730*) decreased in expression as the fruit matured. The above results indicate that Vc in *A. eriantha* accumulates significantly at the early stages of fruit development while the accumulation pattern of Vc was basically consistent with the expression pattern of the gene.

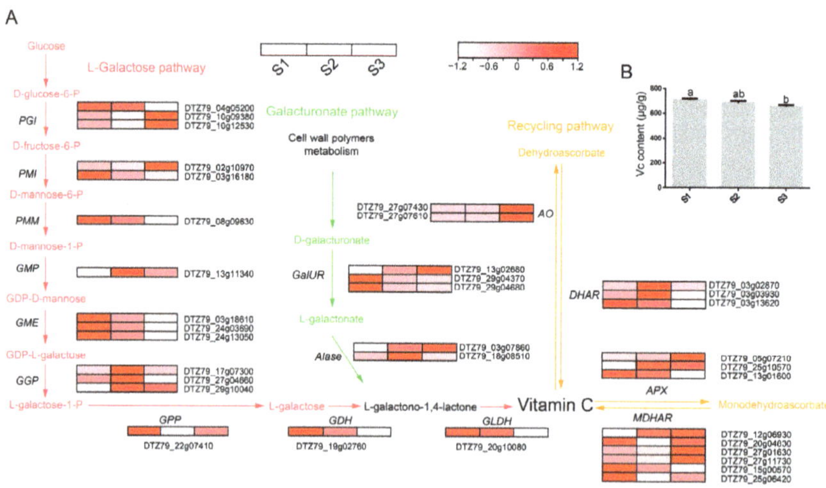

Figure 3. Schematic diagram of the *A. eriantha* Vc metabolic pathway. (**A**) Metabolite and gene expression profiles of Vc synthesis in *A. eriantha*. *AO*: L-ascorbate oxidase; *APX*: L-ascorbate peroxidase; *DHAR*: dehydroascorbate reductase; *GDH*: L-galactose dehydrogenase; *GGP*: GDP-L-galactose phosphorylase; *GLDH*: L-galactono-1,4-lactone dehydrogenase; *GME*: GDP-D-mannose-3,5-epimerase; *GMP*: GDP-D-mannose pyrophosphorylase; *GPP*: L-galactose-1-phosphate phosphatase; *MDHAR*: monodehydroascorbate reductase; *PGI*: glucose-6-phosphate isomerase; *PMI*: mannose-6-phosphate isomerase; *PMM*: phosphomannomutase, *Alase*: aldonolactonase; *GalUR*: D-galacturonate reductase. Blue arrows indicate the L-galactose pathway, green arrows indicate the galacturonate pathway, and yellow arrows indicate the recycling pathway. (**B**) Determination of Vc content during *A. eriantha* development. Differential analysis was performed using the LSD method with $p < 0.05$.

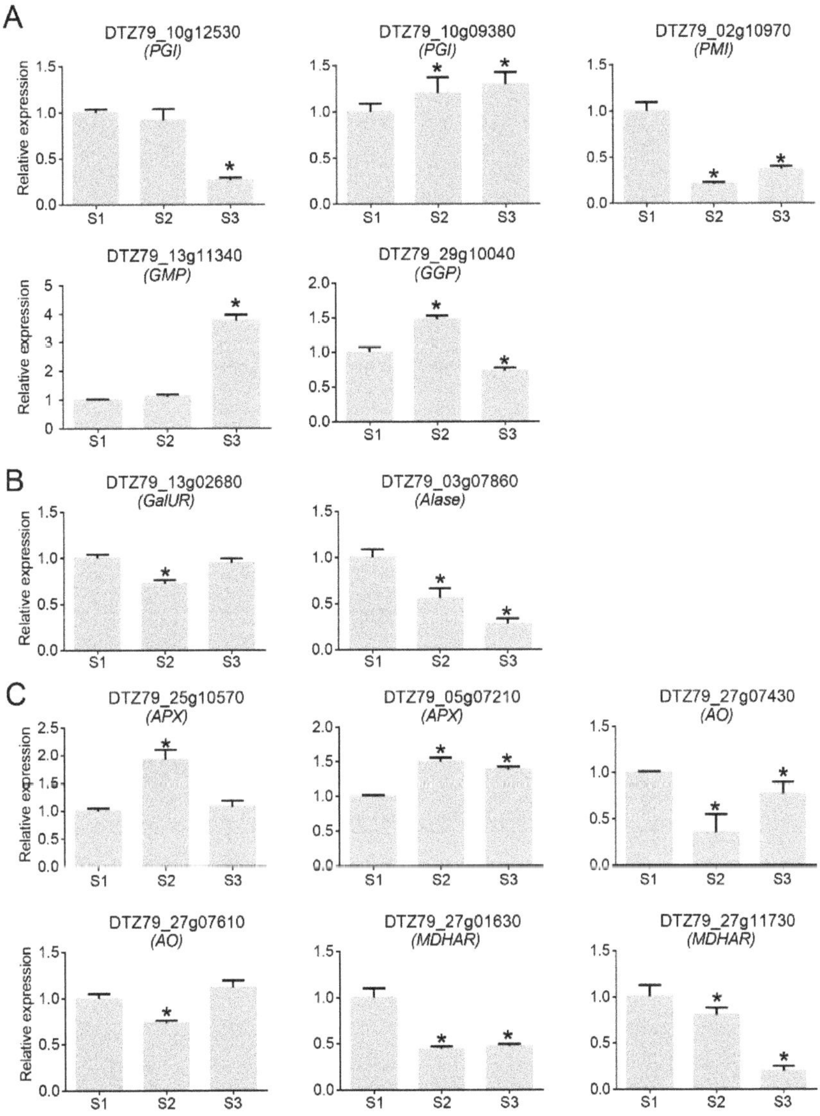

Figure 4. Analysis of expression patterns of related genes in different Vc synthesis pathways in *A. eriantha*. (**A**) L-galactose pathway, (**B**) galacturonate pathway, (**C**) recycling pathway. Asterisks indicate significant difference.

2.3. Analysis of the Sugar Metabolic Pathway in A. eriantha

Sugar compounds are one of the main factors of kiwifruit's flavour and fruit quality. To explore the anabolism of sugars in *A. eriantha*, several major sugar synthesis pathways were constructed in this study. The results showed that many genes involved in the synthesis of sucrose, fructose, and glucose were highly expressed during fruit ripening (Figure 5A). Among the 11 members of invertase (*A/N-INV*) that regulate the conversion of sucrose into fructose and glucose, 8 were highly expressed during fruit ripening. The above results indicate that the accumulation of sugar during the ripening process of *A. eriantha* shows a positive correlation with fruit ripening. In addition, we further analysed the contents of total sugar, sucrose, fructose, and glucose in the flesh of *A. eriantha* during the three stages of fruit ripening (Figure 5B). During the ripening process of *A. eriantha*, the contents of total sugar, sucrose, fructose, and glucose significantly increased and reached their peak at fruit maturity. To further validate the relationship between genes and sugar accumulation, we conducted RT–qPCR analysis of the expression patterns of six *A/N-INV* genes, two *hexokinase-like* (*HKL*) genes, and one *fucokinase-like* (*FRK*) gene during fruit development. Among the six *A/N-INV* genes, three members showed an increase in expression levels as the fruit matured, while three members showed a decrease in expression levels as the fruit matured. This result suggests that different members are involved in the synthesis of *A. eriantha* sugars at different stages. Moreover, the expression levels of *HKL* and *FRK* also decreased with fruit maturation (Figure 6). This suggests a decrease in the phosphorylation of sugars during the later stages of *A. eriantha* fruit development. The above results suggest that increased expression of some sugar synthesis genes and decreased expression of sugar phosphorylation genes contribute to sugar accumulation in *A. eriantha*.

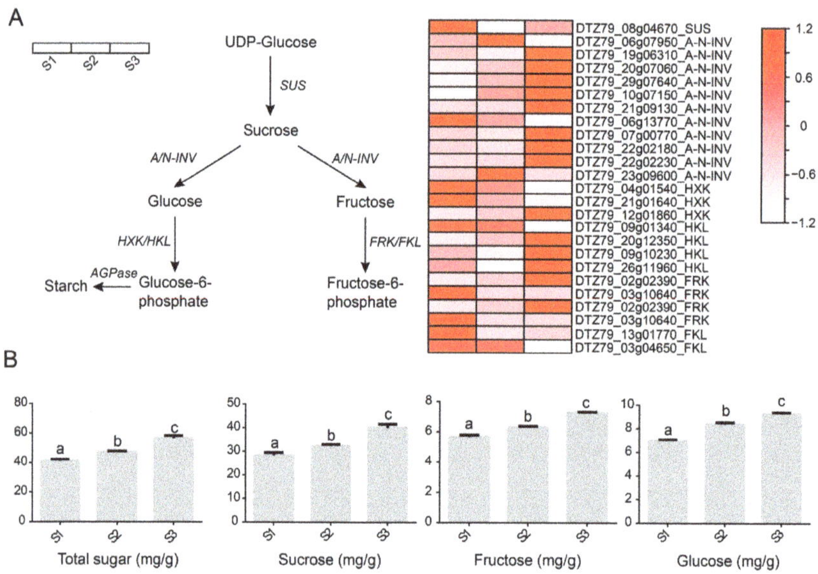

Figure 5. Schematic diagram of the *A. eriantha* glucose metabolism pathway. (**A**) Pathway of sucrose, fructose, and glucose metabolism in *A. eriantha*. *SUS*: sucrose synthase, *A/N-INVc*: invertase, *HXK*: hexokinase, *FRK*: fructokinase, *AGPase*: ADP-glucose pyrophosphorylase, *HKL*: hexokinase-like, *FKL*: fructokinase-like. (**B**) Determination of total sugar, sucrose, fructose, and glucose contents in *A. eriantha*. Letters indicate significant difference.

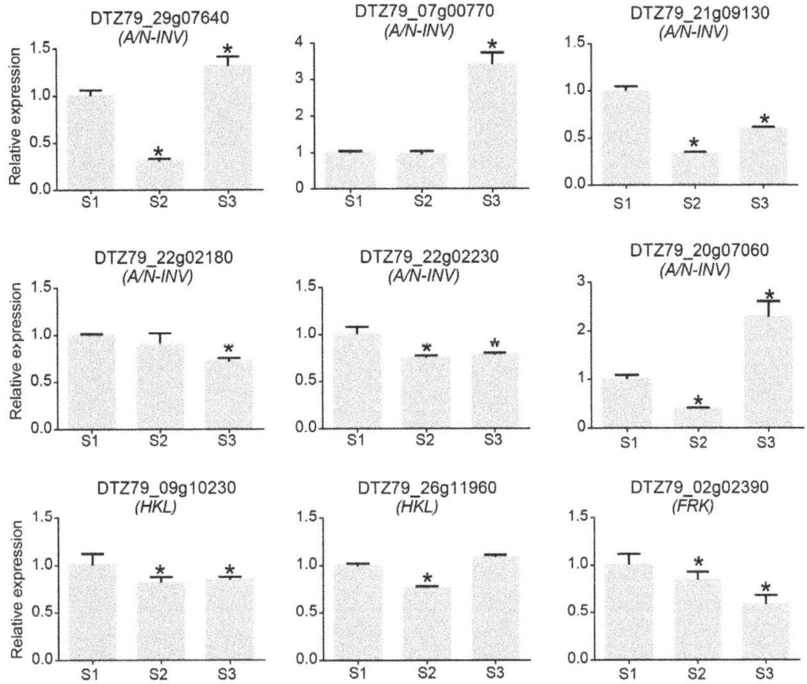

Figure 6. Quantitative analysis of genes related to glucose metabolism in *A. eriantha*. Asterisks indicate significant difference.

2.4. Analysis of the Organic Acid Metabolism Pathway in A. eriantha

Organic acids, as flavouring substances in fruits, play a crucial role in the formation of special flavours in fruits. To investigate the accumulation of organic acids in *A. eriantha,* we constructed a metabolic pathway of organic acids that showed a downward trend (Figure 7A). NAD-dependent malic enzyme (*NAD-ME*), as an important gene for the synthesis of malic acid, also has a high expression level during the ripening of *A. eriantha*. However, the accumulation of malic acid mainly depends on the transport of AL-activated malate transporter (*ALMT*) to the vacuoles for storage. However, the expression level of *ALMT* is relatively low during the maturation of *A. eriantha*. Therefore, the high expression of *NAD-ME* does not promote the accumulation of malic acid, and malic acid may be involved in other metabolic pathways. In addition, we further measured the contents of titratable acid, malic acid, and citric acid in the flesh of *A. eriantha* during the three stages of fruit ripening. The results showed that during the ripening process of *A. eriantha,* the content of organic acids significantly decreased (Figure 7B). The expression of *NAD-ME* differed from the accumulation pattern of organic acids. Therefore, according to transcriptome data, we selected two *NAD-ME* members with the largest expression differences during fruit ripening and carried out RT–qPCR analysis (Figure 7C). The results showed that the expression levels of these two *NAD-ME* genes significantly decreased during fruit ripening. The above results indicate that the content of organic acids decreases during the maturation process of *A. eriantha* and that the decrease in organic acid content is related to gene expression.

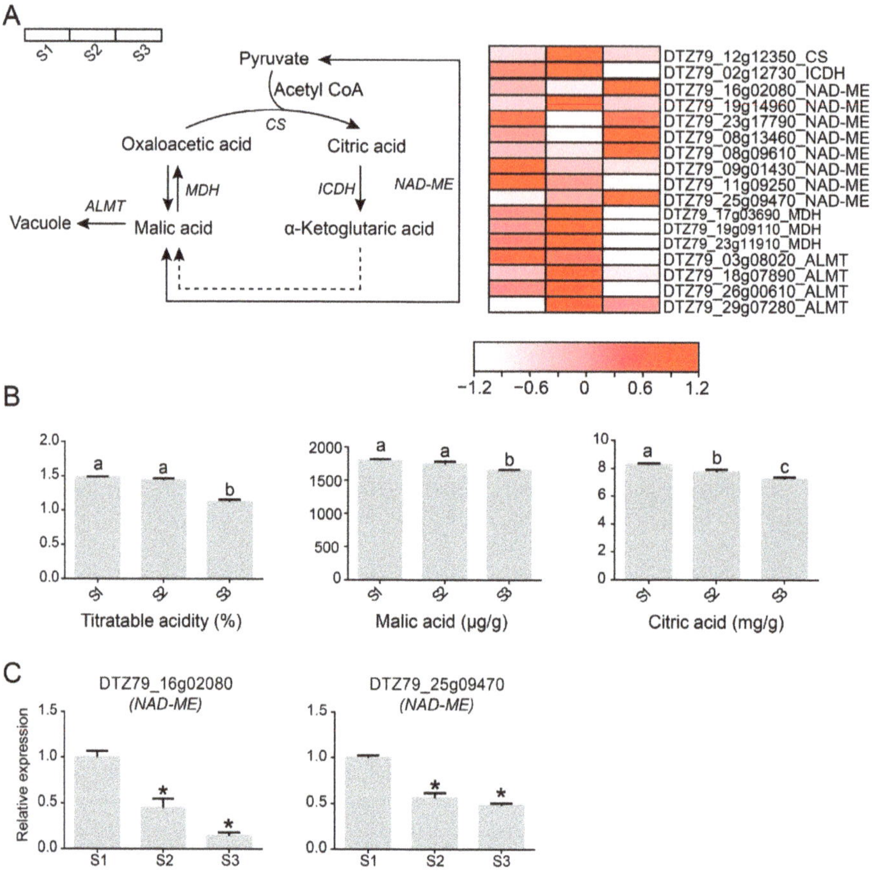

Figure 7. Schematic diagram of organic acid metabolism pathways in *A. eriantha.* (**A**) Metabolic pathways of malic acid and citric acid in *A. eriantha. CS*: citrate synthase, ICDH: isocitrate dehydrogenase, *ALMT*: AL-activated malate transporter, *MDH*: malate dehydrogenase, *NAD-ME*: malic enzyme. (**B**) Determination of titratable acid, malic acid, and citric acid in *A. eriantha.* Letters indicate significant difference. (**C**) Analysis of the expression patterns of two *NAD-ME* genes during the maturation process of *A. eriantha.* Asterisks indicate significant difference.

3. Discussion

Kiwifruit is a perennial fruiting woody vine originating from China. Although there are abundant wild resources of kiwifruit in China, there are significant differences in quality and shape between different varieties of kiwifruit. Therefore, it is very important to obtain a deeper understanding of the quality of kiwifruit and the formation of its unique flavour. *A. eriantha,* with a high Vc content, unique fruit flavour, and other *characteristics,* has become the most popular variety on the market together with *A. chinensis, A. deliciosa,* and *Actinidia arguta* [40].

Sugar and acid, as some of the most important qualities of fruits, have always been a research hotspot. During the development process of *A. eriantha* fruits, total sugar, sucrose, fructose, and glucose continue to accumulate. The accumulation pattern of sugar compounds in kiwifruit is similar to that in fruits such as dates, pears, and watermelons [41–43]. Furthermore, sucrose is the main carbohydrate compound in *A. eriantha,* with a proportion of approximately 80%, while the contents of fructose and glucose are similar to those in *A. arguta* [44]. The contents of sucrose, fructose, and glucose in *A. chinensis* 'Hort 16A' are similar [14]. The contents of titratable acid, malic acid, and citric acid in *A. eriantha* fruit decreased with ripening, and the content was the lowest at ripening. The content of citric acid is much higher than that of malic acid, which is also true for *A. chinensis* 'Hort 16A' and *A. arguta* [14,44]. In addition, the accumulation pattern of organic acids in kiwifruit is opposite to that in apples, peaches, plums, and loquats [41,45,46]. Therefore, even in different varieties of kiwifruit within the same species, there are significant differences in the accumulation patterns of sugar compounds and organic acids. The synthesis and metabolism of carbohydrate compounds and organic acids are key to fruit quality, and they are usually not regulated by a single gene. For example, previous studies on sugar and organic acids and related genes in fruits such as watermelon [11], peaches [42], and citrus [47] have shown that a single gene has a limited contribution to the overall accumulation of sugar and organic acids in the fruit. Therefore, we constructed a pathway for the synthesis of sugars and organic acids in *A. eriantha* and studied the expression profiles of genes involved in sugar and organic acid synthesis in *A. eriantha* during fruit development. Our results show that the genes involved in sugar and organic acid synthesis during *A. eriantha* fruit development have a basically similar expression trend with related sugars and organic acids. Furthermore, using transcriptome sequencing and RT–qPCR we found that the accumulation of organic acids in Trichotomis is directly related to organic acid synthesis genes. The three key gene families are the *CS* gene family, *NAD-ME* gene family and the *ALMT* gene family. *ALMT* and *NAD-ME* regulate the accumulation of malic acid in peaches, while *SUS* and *A/N-INV* participate in the accumulation of carbohydrates, which is similar to that in *A. eriantha* [48,49]. In the Chinese dwarf cherry (*CeraSUS humilis*, Bge.), key genes for the synthesis of organic acids and sugars are similar to those of *A. eriantha* [50]. In addition, similar studies in oriental melon [51], sweet cherry [52], and other fruit trees showed similar results to *A. eriantha*. Together, these results suggest that the accumulation patterns of sugars and organic acids may differ from those of other species, but the key genes are basically the same.

Vc is a necessary substance for the metabolism and growth of all Viridiplantae, and kiwifruit is one of the fruits with the highest Vc content on the market. This study shows that the content of Vc in *A. eriantha* fruit gradually decreases with fruit development. The accumulation pattern of Vc in *A. eriantha* is similar to that reported in previous studies on other kiwifruit species [53–55]. *A. eriantha* accumulates a large amount of Vc at the early stages of fruit development in the same way as apples [56] and peaches [57]. In addition, Vc is regulated by four linear and cyclic pathways. These pathways include dozens of enzymes that directly or indirectly participate in the metabolic reaction of Vc [26–28,58]. Furthermore, cloning and genetic transformation experiments were conducted on key genes in the Vc synthesis pathway, and no significant increase in Vc content was found [59–63], indicating that Vc is a multigene-, multinetwork-regulated substance. Subsequently, we constructed a pathway for the synthesis of Vc in *A. eriantha* and analysed the expression patterns of the key genes involved. The results showed that most of the genes involved in Vc synthesis were highly expressed at the early and middle stages of *A. eriantha* fruit development, and their expression levels decreased during fruit ripening. The specific expression of these genes is significantly correlated with fruit development, indicating that the development of *A. eriantha* fruit may be an important signal regulating gene expression. As a rate-limiting enzyme in the L-galactose pathway, *GGP* has a critical role in Vc synthesis in higher plants [64,65]. Numerous studies have shown that the expression level of *GGP* is closely related to the Vc content in plants and in kiwifruit [59,64,66–68]. The galacturonate pathway, as the second discovered Vc synthesis pathway in plant species, has also been shown to play a role in plant Vc synthesis [58,69]. Although it is not the main Vc synthesis pathway in plants, it is also a pathway for Vc in plants and is closely related to the synthesis of Vc [70,71]. In this study, the L-galactose pathway and the galacturonate pathway were constructed, and the key enzymes were found to be highly correlated with the Vc accumulation pattern in *A. eriantha*, reflecting the important role of the L-galactose pathway and the galacturonate pathway in the Vc synthesis of *A. eriantha* fruit. The pathway of vitamin synthesis and metabolism in higher plants is very complex. Understanding the accumulation and metabolism of Vc in *A. eriantha* can help us fully utilize the germplasm resources of kiwifruit. This study utilized high-throughput sequencing and molecular biology experiments to preliminarily explore the genes of *A. eriantha* and their function in Vc synthesis.

In conclusion, the transcriptome sequencing technique was used to identify gene expression during *A. eriantha* fruit development, providing important information for studying the dynamic changes in gene expression and helping identify key genes that regulate *A. eriantha* fruit development. In addition, the data can be used to identify specific functions of new genes and functional genomes, providing valuable information for the molecular breeding and germplasm resource development of kiwifruit and laying the foundation for genetic research on *A. eriantha*.

4. Materials and Methods

4.1. Plant Material

In this study, the wild *A. eriantha* materials were collected at the kiwifruit base of Xiamao Town, Shaxian District, the Sanming Research Institute of Agricultural Sciences. According to many years of cultivation observations, the whole *A. eriantha* fruit development process takes approximately 210 to 240 days. For this reason, the samples were collected at 180 days (S1), 210 days (S2), and 240 days (S3) after flowering; 10 to 20 fruits were picked each time, the flesh of the peeled fruits was taken as the experimental material, and the experiment was repeated 3 times. Samples were snap-frozen in liquid nitrogen and stored in a −80 °C refrigerator.

4.2. cDNA Library Construction and Transcriptome Sequencing

For more than 10 *A. eriantha* from the same period, the samples were ground, and the experiment was repeated with 3 biological replicates. The total RNA was subsequently extracted from the *A. eriantha* samples at different developmental stages using TRIzol™ RNA extraction reagent (Thermo Fisher, 15596026, Waltham, MA, USA) according to the instructions. The RNA concentration of the samples was measured using a NanoDrop 2000 (Thermo Fisher Scientific, Wilmington, DE, USA), and the RNAn integrity was further assessed using an RNA Nano 6000 from the Agilent 2100 bioanalyzer system (Agilent Technologies, Santa Clara, CA, USA). One microgram of RNA from each sample was used in the NEBNext Ultra™ RNA Library Prep Kit (New England BioLabs, Ltd., Ipswich, MA, USA) to construct cDNA libraries of samples from different stages of kiwifruit development. Finally, transcriptome sequencing was performed using double-end sequencing on the Illumina platform.

4.3. Transcriptome Data Analysis

Raw data from transcriptome sequencing were subjected to data quality control using the Fastp.0.23.4 software [72]. The filtered data were aligned with the kiwifruit genome sequence (https://kiwifruitgenome.org/, accessed on 23 August 2023). Subsequently, the results of HISAT2 [73] were compressed and sorted using the SAMtools.1.18 software [74]. Then, the FPKM value for each gene was calculated using the FeatureCounts script 2.16.0 [75]. Finally, we constructed the expression level of the genes in each sample into an expression matrix using the Trinity.2.15.1 software [76].

4.4. Differential Gene Analysis and Enrichment Analysis

Differential genes between the different samples were obtained using the DESeq2 R package calculation. DESeq2 provides statistical routines for determining differential expression in digital gene expression data using a model based on the negative binomial distribution. The resulting p values were adjusted using the Benjamini–Hochberg approach for controlling the false discovery rate. Genes with an adjusted p value < 0.01 and $|\log2 \text{ fold change}| > 1$ found using DESeq2 were assigned as differentially expressed. In addition, using the GO and KEGG databases, functional enrichment and channel significance enrichment were performed for differential genes between the selected samples using the ClusterProfiler R package 4.10.0 [77].

4.5. Vc, Sugar and Organic Acid Determination

An ascorbic acid (ASA) content assay kit (Solarbio® BC1230, Beijing, China) was used to determine the Vc content. For each sample, a total of 0.1 g of the sample was ground well, 1 mL of extract was added, the sample was centrifuged, the supernatant was removed, and the microplate reader wavelength was adjusted to 534 nm for index determination. The glucose content was tested using a Glucose Content Assay Kit (Solarbio® BC2505, Beijing, China), a 0.1 g tissue sample was weighed, 1 mL of distilled water was added, and the sample was ground in a 95 °C water bath for 10 min. After cooling, the sample was subjected to centrifugation, the supernatant was removed, and the microplate reader wavelength was adjusted to 505 nm for index measurement. For sucrose content detection, the Plant Sucrose Content Assay Kit (Solarbio® BC2465, Beijing, China) was utilized. First, 0.1 g of the sample was weighed and ground at normal temperature, followed by the addition of 0.5 mL of extract with proper grinding. The sample was then quickly transferred into a centrifuge tube, placed in an 80 °C water bath for 10 min, and, after cooling, the supernatant was removed. Then, 2 mg of the reagent was added and decolouring was conducted at 80 °C for 30 min, followed by the addition of 0.5 mL of extract. The supernatant was removed, and the enzyme wavelength was adjusted to 480 nm for index determination. For testing the fructose content, the Plant Tissue Fructose Content Assay Kit (Solarbio® BC2455, Beijing, China) was used. Approximately 0.1 g of the sample was weighed, and 1 mL of extract was added, followed by proper grinding. The sample was then placed in an 80 °C water bath for 10 min, and then was centrifuged. The supernatant was removed, followed by the addition of 2 mg of reagent and decolouration at 80 °C for 30 min. Then, 1 mL of extract was added followed by centrifugation at the standard temperature, with the microplate wavelength adjusted to 480 nm for index determination. The total sugar content test was performed using the Total Carbohydrate Content Assay Kit (Solarbio® BC2715, Beijing, China). For citric acid content detection, the Citric Acid (CA) Content Assay Kit (Solarbio® BC2155, Beijing, China) was used, with 0.1 g of tissue weighed, to which 1 mL of reagent was added for ice bath homogenization, followed by centrifugation. The supernatant was placed on ice, and the microplate reader was adjusted to 545 nm for index determination. For malic acid content detection, the Malic Acid Content Assay kit (Solarbio® BC5495, Beijing, China) was used, weighing out a 0.1 g tissue sample, adding 1 mL of extract, and conducting ice bath grinding of the sample. The sample was then centrifuged to produce the supernatant, and the microplate reader was adjusted to 450 nm for index determination. The titratable acid content was tested according to the international standard GB12293-90 《Fruit and vegetable products -Determination of titratable acidity》.

4.6. Quantitative Analysis

The total RNA was extracted at the 3 stages of kiwifruit development using the TRIzol™ RNA extraction kit (Thermo Fisher, 15596026, Waltham, MA, USA). The concentration of the RNA from each sample was determined using a NanoDrop 2000 (Thermo Scientific, MA, USA), and the integrity of the RNA was determined using gel electrophoresis. The total RNA was subsequently reverse-transcribed to quantify the desired cDNA using the PrimeScript™ RT reagent kit reverse transcription kit (TaKaRa, Tokyo, Japan) according to the instructions. The RT–qPCR was performed using an ABI QuantStudio 3-sequence detection system (Applied Biosystems, Waltham, MA, USA) using the SYBR Premix Ex Taq Kit (TaKaRa, Tokyo, Japan) with the following reaction conditions: 95 °C for 3 min, 40 cycles of 95 °C for 15 s, 56 °C for 30 s, and 72 °C for 20 s. The reference gene TCTB was used for normalization, the expression data were calculated using the 2ΔΔCt formula, and the difference was calculated using the SPSS 20.0 software (LSD method, $p < 0.05$). The primers used for quantification are listed in Supplemental Table S1.

Supplementary Materials: The following supporting information can be downloaded at https://www.mdpi.com/article/10.3390/plants12244079/s1. Table S1: Primers used in this study; Table S2: Statistical table of the sequencing data; Table S3: Statistics of the matching rate of *Actinidia eriantha* Benth. transcriptome data; Table S4: GO analysis of 213 core upregulated differential genes; Table S5: GO analysis of 207 core downregulated differential genes.

Author Contributions: Conceptualization, X.F.; methodology, J.J.; software, W.L.; validation, Z.L. and P.Y.; formal analysis, P.W. and W.Y.; investigation, P.W., Y.C. and X.F.; resources, Y.C.; data curation, P.W.; writing—original draft preparation, P.W.; writing—review and editing, P.W.; project administration, Y.C. All authors have read and agreed to the published version of the manuscript.

Funding: This research was supported in part by the Fujian Agricultural Guidance (Key) Project (no. 2021N0046) and the "14th Five-Year Plan" Fujian Province Seed Industry Innovation and Industrialization Project (no. zycxny2021010).

Data Availability Statement: All sequencing data, including those of the genome and transcriptome, can be found in the China National Center for Bioinformation database (PRJCA014460).

Conflicts of Interest: The authors declare no conflict of interest.

References

1. Liao, G.; He, Y.; Li, X.; Zhong, M.; Xu, X. Effects of bagging on fruit flavor quality and related gene expression of AsA synthesis in *Actinidia eriantha*. *Sci. Hortic.* **2019**, *256*, 108511. [CrossRef]
2. Du, G.; Li, M.; Ma, F.; Liang, D. Antioxidant capacity and the relationship with polyphenol and Vitamin C in Actinidia fruits–ScienceDirect. *Food Chem.* **2009**, *113*, 557–562. [CrossRef]
3. Huang, C.H.; Tao, J.J.; Liao, G.L.; Xie, M.; Xu, X.B. Dynamic changes of phenol and antioxidant capacity during fruit development of three *Actinidia* species (kiwifruit). *Sci. Hortic.-Amsterdam* **2020**, *273*, 109571. [CrossRef]
4. Hale, T.; Hassell, R.; Phillips, T. Taste Panel Perception of Sweetness as it Correlates to Refractometer Measurements and HPLC Analysis of Sugar Content. *Hortscience* **2004**, *39*, 879. [CrossRef]
5. Gang, W.; Tao, W.; Delin, P.; Zhanhui, J.; Jiyu, Z.; Zhongren, G. The Components and Contents of Organic Acid in Kiwifruit Fruits from Different Cultivars. *J. Agric.* **2017**, *7*, 81–84.
6. Ackermann, J.; Fischer, M.; Amado, R. Changes in sugars, acids, and amino acids during ripening and storage of apples (cv. *Glockenapfel*). *J. Agric. Food Chem.* **1992**, *40*, 1131–1134. [CrossRef]
7. Kroger, M.; Meister, K.; Kava, R. Low-calorie Sweeteners and Other Sugar Substitutes: A Review of the Safety Issues. *Compr. Rev. Food Sci. Food Saf.* **2010**, *5*, 35–47. [CrossRef]
8. Pangborn, R. Relative Taste Intensities of Selected Sugars and Organic Acids. *J. Food Sci.* **2006**, *28*, 726–733. [CrossRef]
9. Zhang, H.P.; Wu, J.Y.; Qin, G.H.; Yao, G.F.; Qi, K.J.; Wang, L.F.; Zhang, S.L. The role of sucrose-metabolizing enzymes in pear fruit that differ in sucrose accumulation. *Acta Physiol. Plant.* **2014**, *36*, 71–77. [CrossRef]
10. Wu, J.; Gao, H.; Zhao, L.; Liao, X.; Chen, F.; Wang, Z.; Hu, X. Chemical compositional characterization of some apple cultivars. *Food Chem.* **2007**, *103*, 88–93. [CrossRef]
11. Ren, Y.; Guo, S.; Zhang, J.; He, H.; Sun, H.; Tian, S.; Gong, G.; Zhang, H.; Levi, A.; Tadmor, Y.; et al. A Tonoplast Sugar Transporter Underlies a Sugar Accumulation QTL in Watermelon. *Plant Physiol.* **2018**, *176*, 836–850. [CrossRef] [PubMed]
12. Lingle, S.E.; Dunlap, J.R. Sucrose Metabolism in Netted Muskmelon Fruit during Development. *Plant Physiol.* **1987**, *84*, 386–389. [CrossRef]
13. Richardson, A.C.; Boldingh, H.L.; Mcatee, P.A.; Gunaseelan, K.; Luo, Z.; Atkinson, R.G.; David, K.M.; Burdon, J.N.; Schaffer, R.J. Fruit development of the diploid kiwifruit, *Actinidia chinensis* "Hort16A". *BMC Plant Biol.* **2011**, *11*, 182. [CrossRef]
14. Gook, J.; Kenji, K.; Kataoka, B. Physical and compositional characteristics of "Mitsuko" and local hardy kiwifruits in Japan. *Hortic. Environ. Biotechnol.* **2012**, *53*, 1–8. [CrossRef]
15. Koch, K. Sucrose metabolism: Regulatory mechanisms and pivotal roles in sugar sensing and plant development. *Curr. Opin. Plant Biol.* **2004**, *7*, 235–246. [CrossRef]
16. Rolland, F.; Baena-Gonzalez, E.; Sheen, J. Sugar Sensing and Signaling in Plants: Conserved and Novel Mechanisms. *Annu. Rev. Plant Biol.* **2006**, *57*, 675–709. [CrossRef] [PubMed]
17. Montero, T.M.; Mollá, E.M.; Esteban, R.M.; López-Andréu, F.J. Quality attributes of strawberry during ripening. *Sci. Hortic.* **1996**, *65*, 239–250. [CrossRef]
18. Giné Bordonaba, J.; Terry, L.A. Manipulating the taste-related composition of strawberry fruits (Fragaria × ananassa) from different cultivars using deficit irrigation. *Food Chem.* **2010**, *122*, 1020–1026. [CrossRef]
19. Jiang, Q.; Wang, Z.; Hu, G.; Yao, X. Genome-wide identification and characterization of AP2/ERF gene superfamily during flower development in Actinidia eriantha. *BMC Genomics* **2022**, *23*, 1–16. [CrossRef]
20. Wang, R.; Shu, P.; Zhang, C.; Zhang, J.; Chen, Y.; Zhang, Y.; Du, K.; Xie, Y.; Li, M.; Ma, T.; et al. Integrative analyses of metabolome and genome-wide transcriptome reveal the regulatory network governing flavor formation in kiwifruit (*Actinidia chinensis*). *New Phytol.* **2022**, *233*, 373–389. [CrossRef]

21. Jiang, Z.; Huang, Q.; Jia, D.; Zhong, M.; Tao, J.; Liao, G.; Huang, C.; Xu, X. Characterization of Organic Acid Metabolism and Expression of Related Genes During Fruit Development of *Actinidia eriantha* "Ganmi 6". *Plants* **2020**, *9*, 332. [CrossRef]
22. Macrae, E.A.; Bowen, J.H.; Stec, M.G.H. Maturation of kiwifruit (*Actinidia deliciosa* cv Hayward) from two orchards: Differences in composition of the tissue zones. *J. Sci. Food Agric.* **2010**, *47*, 401–416. [CrossRef]
23. Marsh, K.; Attanayake, S.; Walker, S.; Gunson, A.; Macrae, E.A. Acidity and taste in kiwifruit. *Postharvest Biol. Technol.* **2004**, *32*, 159–168. [CrossRef]
24. Etienne, A.M.G.; Lobit, P.D.M.; Bugaud, C. What controls fleshy fruit acidity? A review of malate and citrate accumulation in fruit cells. *J. Exp. Bot.* **2013**, *34*, 1451–1469. [CrossRef] [PubMed]
25. Agius, F.; Amaya, I.; Botella, M.A.; Valpuesta, V. Functional analysis of homologous and heterologous promoters in strawberry fruits using transient expression. *J. Exp. Bot.* **2005**, *56*, 37–46. [CrossRef] [PubMed]
26. Wolucka, B.A.; Van Montagu, M. GDP-Mannose 3′, 5′-Epimerase Forms GDP-L-gulose, a Putative Intermediate for the de Novo Biosynthesis of Vitamin C in Plants. *J. Biol. Chem.* **2003**, *278*, 47483–47490. [CrossRef] [PubMed]
27. Davey, M.W.; Montagu, M.; Inze, D.; Sanmartin, M.; Kanellis, A.; Smirnoff, N.; Benzie, I.; Strain, J.J.; Favell, D.; Fletcher, J. Review plant L-ascorbic acid: Chemistry, function, metabolism, bioavailability and effects of processing. *J. Sc. Food Agr.* **2000**, *80*, 825–860. [CrossRef]
28. Wheeler, G.L.; Jones, M.A.; Smirnoff, N. The biosynthetic pathway of vitamin C in higher plants. *Nature* **1998**, *393*, 365–369. [CrossRef]
29. Nishiyama, I.; Yamashita, Y.; Yamanaka, M.; Shimohashi, A.; Fukuda, T.; Oota, T. Varietal Difference in Vitamin C Content in the Fruit of Kiwifruit and Other Actinidia Species. *J. Agric. Food Chem.* **2004**, *52*, 5472–5475. [CrossRef]
30. Cai-Hong, Z.; Peng, Z.; Zheng-Wang, J.; Sheng-Mei, W.; Fei, H.; Li-Yun, X.U.; Hong-Wen, H. Dynamic Changes of Carbohydrate and Vitamin C in Fruits of Actinidia chinensis and *A. eriantha* during Growth Season. *Plant Sci. J.* **2011**, *4*, 319–330. [CrossRef]
31. Gao, X.; Wang, L.; Zhang, H.; Zhu, B.; Lv, G.; Xiao, J. Transcriptome analysis and identification of genes associated with floral transition and fruit development in rabbiteye blueberry (*Vaccinium ashei*). *PLoS ONE* **2021**, *16*, e259119. [CrossRef]
32. Qi, Y.; Wei, H.; Gu, W.; Shi, W.; Jiang, L.; Deng, L.; Liu, X. Transcriptome profiling provides insights into the fruit color development of wild Lycium ruthenicum Murr. from Qinghai-Tibet Plateau. *Protoplasma* **2021**, *258*, 33–43. [CrossRef] [PubMed]
33. Xie, M.; Huang, Y.; Zhang, Y.; Wang, X.; Yang, H.; Yu, O.; Dai, W.; Fang, C. Transcriptome profiling of fruit development and maturation in Chinese white pear (*Pyrus bretschneideri Rehd*). *BMC Genom.* **2013**, *14*, 823. [CrossRef] [PubMed]
34. Ester Garcia-Gomez, B.S.J.L. Identification of QTLs linked to fruit quality traits in apricot (*Prunus armeniaca* L.) and biological validation through gene expression analysis using qPCR. *Mol. Breed.* **2019**, *39*, S96–S101. [CrossRef]
35. García-Gómez, B.E.; Ruiz, D.; Salazar, J.A.; Rubio, M.; Martínez-Gómez, P. Analysis of Metabolites and Gene Expression Changes Relative to Apricot (*Prunus armeniaca* L.) Fruit Quality During Development and Ripening. *Front. Plant Sci.* **2020**, *11*, 1269. [CrossRef] [PubMed]
36. Su, L.; Zhang, T.; Wu, M.; Zhong, Y.; Cheng, Z.M. Transcriptome and Metabolome Reveal Sugar and Organic Acid Accumulation in Rosa roxburghii Fruit. *Plants* **2023**, *12*, 3036. [CrossRef] [PubMed]
37. Jia, H.; Tao, J.; Zhong, W.; Jiao, X.; Chen, S.; Wu, M.; Gao, Z.; Huang, C. Nutritional Component Analyses in Different Varieties of *Actinidia eriantha* Kiwifruit by Transcriptomic and Metabolomic Approaches. *Int. J. Mol. Sci.* **2022**, *6*, 217. [CrossRef]
38. Liao, G.; Liu, Q.; Xu, X.; He, Y.; Li, Y.; Wang, H.; Ye, B.; Huang, C.; Zhong, M.; Jia, D. Metabolome and Transcriptome Reveal Novel Formation Mechanism of Early Mature Trait in Kiwifruit. *Front. Plant Sci.* **2021**, *12*, 760496. [CrossRef]
39. Xiong, Y.; Yan, P.; Du, K.; Li, M.; Xie, Y.; Gao, P. Nutritional component analyses of kiwifruit in different development stages by metabolomic and transcriptomic approaches. *J. Sci. Food Agr.* **2020**, *100*, 2399–2409. [CrossRef]
40. Jeong, H.R.; Cho, H.S.; Cho, Y.S.; Kim, D.O. Changes in phenolics, soluble solids, vitamin C, and antioxidant capacity of various cultivars of hardy kiwifruits during cold storage. *Food Sci. Biotechnol.* **2020**, *29*, 1763–1770. [CrossRef]
41. Umer, M.J.; Bin Safdar, L.; Gebremeskel, H.; Zhao, S.; Yuan, P.; Zhu, H.; Kaseb, M.O.; Anees, M.; Lu, X.; He, N.; et al. Identification of key gene networks controlling organic acid and sugar metabolism during watermelon fruit development by integrating metabolic phenotypes and gene expression profiles. *Hortic. Res.* **2020**, *7*, 193. [CrossRef] [PubMed]
42. Peng, Q.; Wang, L.; Ogutu, C.; Liu, J.; Liu, L.; Mollah, M.; Han, Y. Functional Analysis Reveals the Regulatory Role of PpTST1 Encoding Tonoplast Sugar Transporter in Sugar Accumulation of Peach Fruit. *Int. J. Mol. Sci.* **2020**, *21*, 1112. [CrossRef] [PubMed]
43. Zhang, Y.; Fernie, A.R. On the role of the tricarboxylic acid cycle in plant productivity. *J. Integr. Plant Biol.* **2018**, *60*, 1199–1216. [CrossRef] [PubMed]
44. Huiqin, Z.; Ming, X.; Jinping, X.; Liqiu, Z.; Genhua, S. Characterization of fruit development of the diploid kiwifruit, *Actinidia eriantha* "White". *J. Fruit. Sci.* **2015**, *32*, 238–246. [CrossRef]
45. Famiani, F.; Battistelli, A.; Moscatello, S.; Cruzcastillo, J.G.; Walker, R.P. The organic acids that are accumulated in the flesh of fruits: Occurrence, metabolism and factors affecting their contents—A review. *Rev. Chapingo Ser. Hortic.* **2015**, *21*, 97–128. [CrossRef]
46. Yao, Y.X.; Ming, L.; Liu, Z.; You, C.X.; Wang, D.M.; Zhai, H.; Hao, Y.J. Molecular cloning of three malic acid related genes MdPEPC, MdVHA-A, MdcyME and their expression analysis in apple fruits. *Sci. Hortic.* **2009**, *122*, 404–408. [CrossRef]
47. Li, S.; Yin, X.; Xie, X.; Allan, A.C.; Ge, H.; Shen, S.; Chen, K. The Citrus transcription factor, CitERF13, regulates citric acid accumulation via a protein-protein interaction with the vacuolar proton pump, CitVHA-c4. *Sci. Rep.* **2016**, *6*, 20151. [CrossRef]

48. Zheng, B.; Zhao, L.; Jiang, X.; Cherono, S.; Liu, J.; Ogutu, C.; Ntini, C.; Zhang, X.; Han, Y. Assessment of organic acid accumulation and its related genes in peach. *Food Chem.* **2021**, *334*, 127567. [CrossRef]
49. Vimolmangkang, S.; Zheng, H.; Peng, Q.; Jiang, Q.; Wang, H.; Fang, T.; Liao, L.; Wang, L.; He, H.; Han, Y. Assessment of Sugar Components and Genes Involved in the Regulation of Sucrose Accumulation in Peach Fruit. *J. Agr. Food Chem.* **2016**, *64*, 6723–6729. [CrossRef]
50. Guo, C.; Wang, P.; Zhang, J.; Guo, X.; Mu, X.; Du, J. Organic acid metabolism in Chinese dwarf cherry [*Cerasus humilis* (Bge.) *Sok.] is controlled by a complex gene regulatory network. Front. Plant Sci.* **2022**, *13*, 982112. [CrossRef]
51. Cheng, H.; Kong, W.; Tang, T.; Ren, K.; Zhang, K.; Wei, H.; Lin, T. Identification of Key Gene Networks Controlling Soluble Sugar and Organic Acid Metabolism During Oriental Melon Fruit Development by Integrated Analysis of Metabolic and Transcriptomic Analyses. *Front. Plant Sci.* **2022**, *13*, 830517. [CrossRef] [PubMed]
52. Yang, H.; Tian, C.; Ji, S.; Ni, F.; Fan, X.; Yang, Y.; Sun, C.; Gong, H.; Zhang, A. Integrative analyses of metabolome and transcriptome reveals metabolomic variations and candidate genes involved in sweet cherry (*Prunus avium* L.) *fruit quality during development and ripening. PLoS ONE* **2021**, *16*, e260004. [CrossRef]
53. Bulley, S.M.; Rassam, M.; Hoser, D.; Otto, W.; Schünemann, N.; Wright, M.; Macrae, E.; Gleave, A.; Laing, W. Gene expression studies in kiwifruit and gene over-expression in Arabidopsis indicates that GDP-L-galactose guanyltransferase is a major control point of vitamin C biosynthesis. *J. Exp. Bot.* **2009**, *60*, 765–778. [CrossRef] [PubMed]
54. Shu, P.; Zhang, Z.; Wu, Y.; Chen, Y.; Li, K.; Deng, H.; Zhang, J.; Zhang, X.; Wang, J.; Liu, Z.; et al. A comprehensive metabolic map reveals major quality regulations in red-flesh kiwifruit (*Actinidia chinensis*). *New Phytol.* **2023**, *238*, 2064–2079. [CrossRef] [PubMed]
55. Li, M.; Ma, F.; Liang, D.; Li, J.; Wang, Y. Ascorbate Biosynthesis during Early Fruit Development Is the Main Reason for Its Accumulation in Kiwi. *PLoS ONE* **2010**, *5*, e14281. [CrossRef] [PubMed]
56. Li, M.; Chen, X.; Wang, P.; Ma, F. Ascorbic Acid Accumulation and Expression of Genes Involved in Its Biosynthesis and Recycling in Developing Apple Fruit. *J. Am. Soc. Hortic. Sci.* **2011**, *136*, 231–238. [CrossRef]
57. Imai, T.; Ban, Y.; Terakami, S.; Yamamoto, T.; Moriguchi, T. L-Ascorbate biosynthesis in peach: Cloning of six L-galactose pathway-related genes and their expression during peach fruit development. *Physiol. Plant.* **2010**, *136*, 139–149. [CrossRef]
58. Agius, F.; González-Lamothe, R.; Caballero, J.L.; Muñoz-Blanco, J.; Botella, M.A.; Valpuesta, V. Engineering increased vitamin C levels in plants by overexpression of a D-galacturonic acid reductase. *Nat. Biotechnol.* **2003**, *21*, 177–181. [CrossRef]
59. Bulley, S.; Wright, M.; Rommens, C.; Yan, H.; Rassam, M.; Lin-Wang, K.; Andre, C.; Brewster, D.; Karunairetnam, S.; Allan, A.C.; et al. Enhancing ascorbate in fruits and tubers through over-expression of the l-galactose pathway gene GDP-l-galactose phosphorylase. *Plant Biotechnol. J.* **2012**, *10*, 390–397. [CrossRef]
60. Qian, W.; Yu, C.; Qin, H.; Liu, X.; Zhang, A.; Johansen, I.E.; Wang, D. Molecular and functional analysis of phosphomannomutase (PMM) from higher plants and genetic evidence for the involvement of PMM in ascorbic acid biosynthesis in *Arabidopsis* and *Nicotiana benthamiana*. *Plant J.* **2007**, *49*, 399–413. [CrossRef]
61. Giovannoni, S.; Foster, R.; Rappe, M.; Epstein, S. New Cultivation Strategies Bring More Microbial Plankton Species into the Laboratory. *Oceanography* **2007**, *20*, 62–69. [CrossRef]
62. Gatzek, S.; Wheeler, G.L.; Smirnoff, N. Antisense suppression of l-galactose dehydrogenase in Arabidopsis thaliana provides evidence for its role in ascorbate synthesis and reveals light modulated l-galactose synthesis. *Plant J.* **2002**, *30*, 541–553. [CrossRef] [PubMed]
63. Keller, R.; Renz, F.S.A.; Kossmann, J. Antisense inhibition of the GDP-mannose pyrophosphorylase reduces the ascorbate content in transgenic plants leading to developmental changes during senescence. *Plant J.* **1999**, *19*, 131–141. [CrossRef]
64. Linster, C.L.; Gomez, T.A.; Christensen, K.C.; Adler, L.N.; Young, B.D.; Brenner, C.; Clarke, S.G. Arabidopsis VTC2 Encodes a GDP-l-Galactose Phosphorylase, the Last Unknown Enzyme in the Smirnoff-Wheeler Pathway to Ascorbic Acid in Plants. *J. Biol. Chem.* **2007**, *282*, 18879–18885. [CrossRef] [PubMed]
65. Laing, W.A.; Wright, M.A.; Cooney, J.; Bulley, S.M. The missing step of the l-galactose pathway of ascorbate biosynthesis in plants, an l-galactose guanyltransferase, increases leaf ascorbate content. *Proc. Natl. Acad. Sci. USA* **2007**, *104*, 9534–9539. [CrossRef] [PubMed]
66. Sun, Y.; Yang, M.; An, H. Expression of GDP-L-galactose pyrophosphatase and its relationship with ascorbate accumulation in Rosa roxburghii. *Acta Hortic. Sin.* **2014**, *41*, 1175–1182.
67. Mellidou, I.; Chagné, D.; Laing, W.A.; Keulemans, J.; Davey, M.W. Allelic Variation in Paralogs of GDP-l-Galactose Phosphorylase Is a Major Determinant of Vitamin C Concentrations in Apple Fruit. *Plant Physiol.* **2012**, *160*, 1613–1629. [CrossRef]
68. Liu, X.; Wu, R.; Bulley, S.; Zhong, C.; Li, D. Kiwifruit MYBS1-like and GBF3 transcription factors influence L-ascorbic acid biosynthesis by activating transcription of GDP-L-galactose phosphorylase 3. *New Phytol.* **2022**, *234*, 1782–1800. [CrossRef]
69. Cruz-Rus, E.; Botella, M.A.; Valpuesta, V.; Gomez-Jimenez, M.C. Analysis of genes involved in l-ascorbic acid biosynthesis during growth and ripening of grape berries. *J. Plant Physiol.* **2010**, *167*, 739–748. [CrossRef]
70. Li, M.J.; Liu, J.; Liang, D.; Guo, C.; Ma, F. The relationship between GalUR expression and ascorbate accumulation in Kiwifruit. *Acta Hortic. Sin.* **2011**, *38*, 1641–1649.
71. Li, M.J.; Ma, F.W.; Zhang, M.; Pu, F. Distribution and metabolism of ascorbic acid in apple fruits (*Malus domestica Borkh* cv. *Gala)*. *Plant Sci.* **2008**, *174*, 606–612. [CrossRef]

72. Chen, S. Ultrafast one-pass FASTQ data preprocessing, quality controland deduplication using fastp. *iMeta* **2023**, *2*, e107. [CrossRef]
73. Kim, D.; Ben, L.; Salzberg, S.L. HISAT: A fast spliced aligner with low memory requirements. *Nat. Methods* **2015**, *12*, 357–360. [CrossRef]
74. Li, H.; Handsaker, B.; Wysoker, A.; Fennell, T.; Ruan, J.; Homer, N.; Marth, G.; Abecasis, G.; Durbin, R. The Sequence Alignment/Map format and SAMtools. *Bioinformatics* **2009**, *25*, 2078–2079. [CrossRef]
75. Yang, L.; Smyth, G.K.; Wei, S. featureCounts: An efficient general purpose program for assigning sequence reads to genomic features. *Bioinformatics* **2014**, *30*, 923–930. [CrossRef]
76. Grabherr, M.G.; Haas, B.J.; Yassour, M.; Levin, J.Z.; Amit, I. Trinity: Reconstructing a full-length transcriptome without a genome from RNA-Seq data. *Nat. Biotechnol.* **2013**, *29*, 644–652. [CrossRef]
77. Yu, G.; Wang, L.; Han, L.; He, Q. clusterProfiler: An R package for Statistical Analysis and Visualization of Functional Profiles for Genes and Gene Clusters. *OMICS J. Integr. Biol.* **2012**, *16*, 284–287. [CrossRef]

plants

MDPI

Article

Transcriptome Analysis Reveals Candidate Genes Involved in Gibberellin-Induced Fruit Development in *Rosa roxburghii*

Xiaolong Huang [1,2,3], Xiaoai Wu [1], Guilian Sun [1,2,3], Yu Jiang [1] and Huiqing Yan [1,*]

1 School of Life Sciences, Guizhou Normal University, Guiyang 550001, China; huangxiaolong@gznu.edu.cn (X.H.); wuxiaoai1234@163.com (X.W.); sungl2022@163.com (G.S.); abc20090820@163.com (Y.J.)

2 Key Laboratory of Plant Physiology and Development Regulation, Guizhou Normal University, Guiyang 550001, China

3 Laboratory of State Forestry Administration on Biodiversity Conservation in Mountainous Karst Area of Southwestern China, Guizhou Normal University, Guiyang 550001, China

* Correspondence: yanhuiqing@gznu.edu.cn

Abstract: Gibberellins (GAs) play indispensable roles in the fruit development of horticultural plants. Unfortunately, the molecular basis behind GAs regulating fruit development in *R. roxburghii* remains obscure. Here, GA_3 spraying to *R. roxburghii* 'Guinong 5' at full-bloom promoted fruit size and weight, prickle development, seed abortion, ascorbic acid accumulation, and reduction in total soluble sugar. RNA-Seq analysis was conducted to generate 45.75 Gb clean reads from GA_3- and non-treated fruits at 120 days after pollination. We obtained 4275 unigenes belonging to differently expressed genes (DEGs). Gene ontology and the Kyoto Encyclopedia of Genes and Genomes displayed that carbon metabolism and oxidative phosphorylation were highly enriched. The increased critical genes of DEGs related to pentose phosphate, glycolysis/gluconeogenesis, and citrate cycle pathways might be essential for soluble sugar degradation. Analysis of DEGs implicated in ascorbate revealed the myoinositol pathway required to accumulate ascorbic acid. Finally, DEGs involved in endogenous phytohormones and transcription factors, including R2R3 MYB, bHLH, and WRKY, were determined. These findings indicated that GA_3-trigged morphological alterations might be related to the primary metabolites, hormone signaling, and transcription factors, providing potential candidate genes that could be guided to enhance the fruit development of *R. roxburghii* in practical approaches.

Keywords: *Rosa roxburghii*; gibberellin; transcriptome; fruit development

1. Introduction

Fruit development of horticultural plants is a refined reproductive and consecutive process divided into three stages: fruit set, the cell division or expansion stage, and fruit maturation [1]. The transition from flowering to fruit set upon fertilization in angiosperm refers to the initiation of fruit development. It begins with a period of cell division, and its high cell division activities often rapidly take place within a few weeks inside the ovary after anthesis and mitotic cells can still be found [2]. The second phase lasts a long time, mainly determining the final fruit size and weight. The last period is fruit thorough maturation and the fruit size grows very slightly at this stage. However, multiple storage products and secondary metabolites accumulate, thus forming different intrinsic and extrinsic qualities of fruits that are accompanied and could be manipulated by several phytohormones [3].

Gibberellins (GAs) are phytohormones and play critical roles in all the physiological periods of horticultural fruit development [4]. Abundant GAs are perceived as signals by the ovary or other fruit tissues after fertilization, resulting in fruit cell expansion [2]. For example, enhancing expression levels of genes in the gibberellin synthesis positively regulates the fruit set of tomato plants (*Solanum lycopersicum*) [4,5]. Similarly, endogenous GA accumulation is consistent with high cell division levels and tomato fruit expansion [6]. Moreover, GA application in the woodland strawberry (*Fragaria vesca*) promotes longitudinal fruit elongation, increasing the fruit size and weight [7]. Likewise, an early study proved that exogenous spraying of GAs increased the fruit weight of apples (*Malus* × *domestica*) and the wax layer thickness to decrease the fruit water-loss rate. It also lowers the proportion of asymmetrical fruit [8]. Transient overexpression of *PbGA20ox2* promoted the fruit development of pear trees (*Pyrus bretschneideri*) and delayed the drop of nonpollinated fruit [4]. Beyond that, GA treatment induced parthenocarpy (fertilization-independent fruit) in horticultural fruits, such as cucumbers and grapes [9,10]. All conclusions prove that GAs play essential roles in fruit development.

Recent studies focus on GA signaling and how it determines the biological functions in horticultural fruits [11]. When bioactive GAs are increased, a GA-binding pocket in GID1 (GIBBERELLININ SENSITIV EDWARF1) captures bioactive GAs. Simultaneously, DELLAs bind with the GID1, forming a GID1-GA-DELLA complex, and are subsequently degraded by the 26S proteasome, releasing repressed multiple transcription factors (TFs) to regulate transcripts of corresponding downstream genes, for example, MYB, bHLH, NAC, and TCP could be activated by GAs [12,13]. These candidate TFs exert their potential functions on fruit development [13]. In addition, exogenous GAs trigger diverse and complex crosstalk with other endogenous phytohormones. For example, GA-signaling repressor SlDELLA binds with auxin signaling components SlIAA9 and SlARF7 by yeast two-hybrid assays and coimmunoprecipitation, thereby repressing fruit initiation and inducing the parthenocarpy of tomatoes [2]. Arabidopsis DELLA also interacts with auxin-responsive factors (AtARF6, AtARF7, and AtARF8) to regulate hypocotyl elongation [14]. Gibberellin stimulates polar auxin transport and has a common transcriptome with auxin in the development of Populus wood [15]. Gibberellic acid (GA_3) application also results in the promotion of histidine kinase (*AHK*) in the cytokine signaling pathway, which positively regulates the initiation of axillary meristem in garlic (*Allium sativum*) [16]. Therefore, the signaling pathways or other phytohormones induced by GAs potentially influence the fruit development of horticultural plants.

Rosa roxburghii Tratt is also known as a chestnut rose belonging to the Rosaceae family. It is a deciduous, thick shrub that produces prickles on its stems and fruits [17]. Its fruits are widely consumed in China for their pleasant flavor and high nutrients. They are even dubbed the king of vitamin C, since they provide an excellent L-ascorbic acid source [18]. Some studies prove that phytohormones regulate the seedling growth and fruit quality of *R. roxburghii*. For example, a low concentration (1 mM) of methyl jasmonate (MeJA) elicited the root growth of *R. roxburghii* and promoted triterpenoid synthesis [19]. Exogenous GA treatment was performed on the fruits of *R. roxburghii* in vitro [20]. Unfortunately, information regarding the effect of GAs on fruit development in vivo and the molecular mechanisms behind it are still unavailable. Advances in high-throughput sequencing technologies enable the comprehensive analysis of deep mechanisms, providing an opportunity to understand the complex genetic networks and molecular basis underlying the effects of GAs [21–23]. However, extensive transcriptome alterations in response to exogenous GA_3 application in *R. roxburghii* fruits remain scarce.

In this current study, we used Illumina sequencing technology to conduct RNA-Seq transcriptome of *R. roxburghii* fruits with and without GA_3 application. Subsequent comparison of the global expression profiles of GA_3-treated and untreated *R. roxburghii* fruits allowed the identification of numerous GA_3-responsive genes. This study aims to explore these mechanisms through a robust transcriptome analysis of gibberellin-treated fruits, providing several potential candidate genes to promote the fruit quality of *R. roxburghii*.

2. Results

2.1. Exogenous Gibberellin Applications Result in Physiological Changes in R. roxburghii Fruits

To explore the roles of GA_3 on the fruit development of *R. roxburghii*, flowers of seven-year-old seedlings in full blossom were treated with GA_3 (200 mg/L) daily for 7 days. We observed that the fruit treated with GA_3 was more significant than the mock at both

60 days after pollination (DAP) or 120 DAP (Figure 1A). Accordingly, GA_3 induced a 13.61% increase in fresh fruit weight and a 21.79% increase in fruit shape index when fruits matured at 120 DAP, suggesting that GA_3 substantially positively regulated the size of *R. roxburghii* fruit cells (Figure 1B). Interestingly, the prickle length of the fruit increased by about a third, reaching 33.59%. In contrast, the seed number decreased by 31.23% after GA_3 treatment. Considering that total soluble sugars (TSS) (sucrose, glucose, and fructose) and L-ascorbic acid were critical qualities of *R. roxburghii* fruits, we measured their biochemical traits as shown in Figure 1B. GA_3 application resulted in a significant decrease in TSS but increased L-ascorbic acid concentration. Hence, GA_3 significantly influenced the fruit development of *R. roxburghii*, including regulating fruit weight and size, prickle, seed failure, and fruit quality.

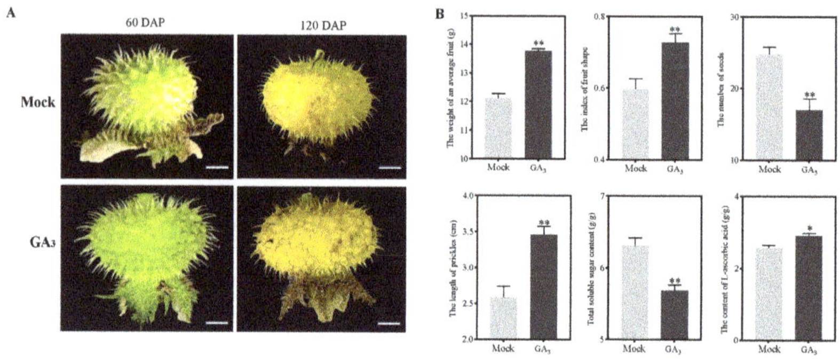

Figure 1. Exogenous gibberellin applications influenced the fruit development of *R. roxburghii*. (**A**) The morphology of fruits at 60 DAP (days after pollination) and 120 DAP after gibberellin (GA_3) spraying. Bar = 1 cm. (**B**) Effects of GA_3 on fruit weight, fruit shape index, seed number, prickle length, concentration of total soluble sugar, and L-ascorbic acid contents. Values were Mean ± S. D. from at least twenty fruits. The asterisk indicates a significant difference. "*": $p < 0.05$; "**": $p < 0.01$.

2.2. Transcriptome Sequencing and De Novo Assemble of *R. roxburghii*

To better explore the molecular mechanism underlying the physiological changes by GA_3, we used RNA-seq to profile the transcriptomes of GA_3-treated and non-treated fruits at 120 DAP. High-throughput sequencing generated a total of 45.75 Gb clean reads for six samples, and the percentage of Q30 bases ranged from 94.59% to 95.11% (Table S1). The clean reads were de novo assembled into transcripts, and 32,895 unigenes were obtained with an average length of 1294 bp (N50 = 1933). A length of 23,395 (71.11%) unigenes was over 500 bp, while those of 15,437 (46.92%) unigenes were over 1 Kb, suggesting that the assembly quality of the *R. roxburghii* transcriptome was satisfactory (Figure 2A). A total of 26,294 unigenes were annotated according to at least one database. Among them, 93.85% of (24,677) unigenes were annotated based on the National Center for Biotechnology Information (NCBI) non-redundant (Nr) database. In terms of the Nr annotations, 40.24% of the sequences had strong homology (E-value $< 1 \times 10^{-150}$), 11.96% and 20.71% of the annotated sequences showed homology ($1 \times 10^{-150} <$ E-value $< 1 \times 10^{-100}$), and ($1 \times 10^{-100} <$ E-value $< 1 \times 10^{-50}$) to available plant sequences, respectively (Figure 2B). The similarity distribution was comparable, with 41.56% of the sequences having similarities higher than 90%, while 48.66% of the sequences of *R. roxburghii* had similarities of 50–90% (Figure 2C). Regarding homolog species, Nr annotated 54.09% of unigenes which had top matches to sequences from woodland strawberries (*Fragaria vesca* subsp), followed by *Prunus persica* (2.03%), *Prunus avium* (1.48%), *Prunus mume* (1.15%), *Pyrus* × *bretschneideri* (0.99%), *Malus domestica* (0.88%), and others (39.17%), suggesting that *Fragaria vesca* was the outstanding species that shared significantly high homologous to *R. roxburghii* (Figure 2D).

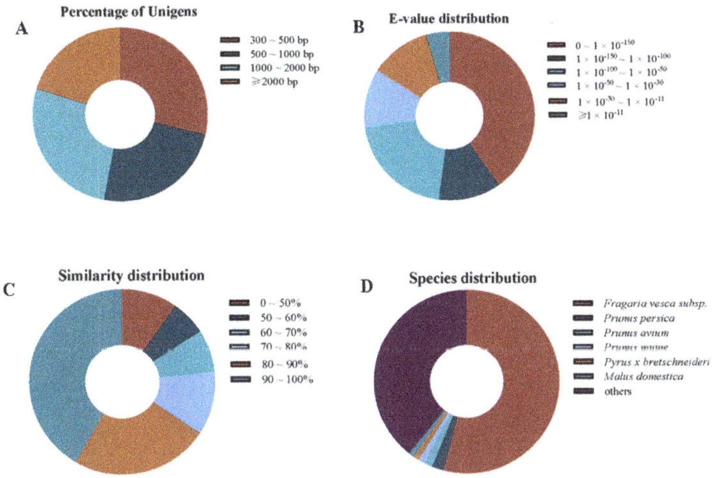

Figure 2. Characterization of the unigenes of *R. roxburghii* fruit transcriptome. (**A**) Statistics of the length distribution of assembled unigenes. (**B**) E-value distribution of the homology search of unigenes against the non-redundant (Nr) database. (**C**) Similarity distribution of the unigenes for each unique sequence. (**D**) Nr annotated species distribution similar to *R. roxburghii* fruit transcriptome.

2.3. Transcriptome-Scale Analysis of GA_3-Responsive DEGs in *R. roxburghii* Fruits

Differentially expressed genes (DEGs) were analyzed using the Fragments Per Kilobase of the exon model per million mapped fragments (FPKM) to assess the degree of overlap between GA_3- and non-treated fruits. A total of 4275 DEGs were detected, with 2782 up-regulated and 1513 down-regulated genes (Figure 3A). Among all DEGs, 89.47% (3825) unigenes were annotated using at least one database (Table S2).

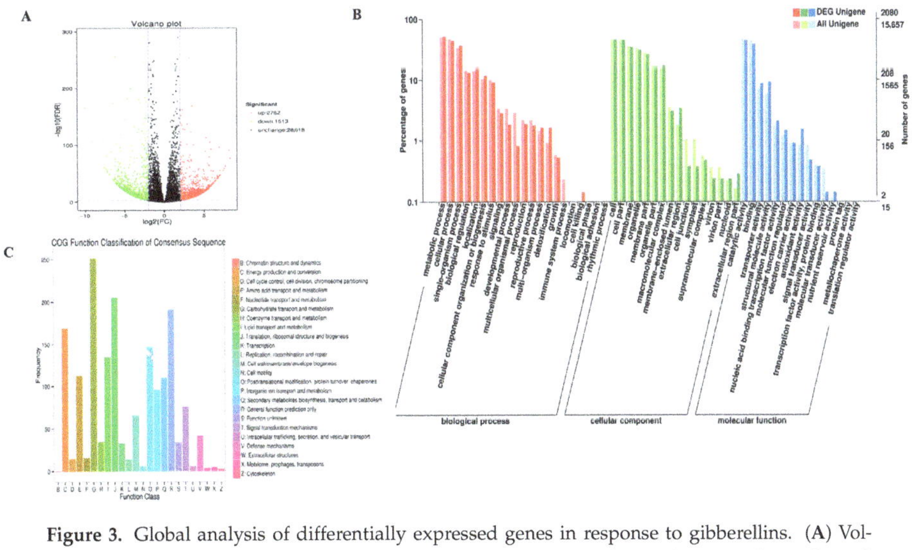

Figure 3. Global analysis of differentially expressed genes in response to gibberellins. (**A**) Volcano plot of DEGs expressed genes with the cutoff (|log2(fold change)| ≥ 2 and an adjusted *p*-value < 0.001). (**B**) Most significant GO functions (**C**) COG function annotations of DEGs expressed in *R. roxburghii* fruits.

Many DEGs annotated within the biological process category of Gene ontology (GO) were analyzed (Figure 3B). For instance, most genes were identified to play essential roles in metabolic, cellular, and single-organism processes, biological regulation, response to stimulus, and signaling. In addition, they were also found to be involved in other multiple biological processes, including developmental, multicellular organismal, and reproductive processes (Table S3). We also evaluated DEGs based on their cellular components and their molecular functions. Regarding the cellular component category, most DEGs fell within the cell part, membrane, organelle, macromolecular complex, and membrane-enclosed lumen. As for molecular function classification, the highest abundance of DEGs was related to catalytic activity and binding. At the same time, other fascinating groups, including transporter, structural molecule, and transcription factor activities, were clustered. Moreover, most DEGs also participated in molecular function regulator, electron carrier, antioxidant, and signal transducer activities. Clusters of Orthologous Groups of proteins (COG) assignment was performed to classify the functions of DEGs (Figure 3C). GA_3 treatment significantly regulated the expression of numerous genes encoding carbohydrate transport and metabolism, indicating their potential roles in the TSS accumulation in *R. roxburghii* fruits after GA_3 application. Other COG functions triggered by GA_3 were translation, ribosomal and biogenesis, energy production and conversion, and secondary metabolites biosynthesis.

The Kyoto Encyclopedia of Genes and Genomes (KEGG) enrichment analysis can identify the major biochemical and signal transduction pathways in which the DEGs were involved. The significant entries of DEGs were implicated in the carbon metabolism, oxidative phosphorylation, citrate cycle, glycolysis/gluconeogenesis, phenylpropanoid biosynthesis, and biosynthesis of amino acids (Figure 4A). DEGs associated with ascorbate, aldarate, tyrosine, tryptophan, starch, and sucrose metabolism were also enriched (Figure 4B).

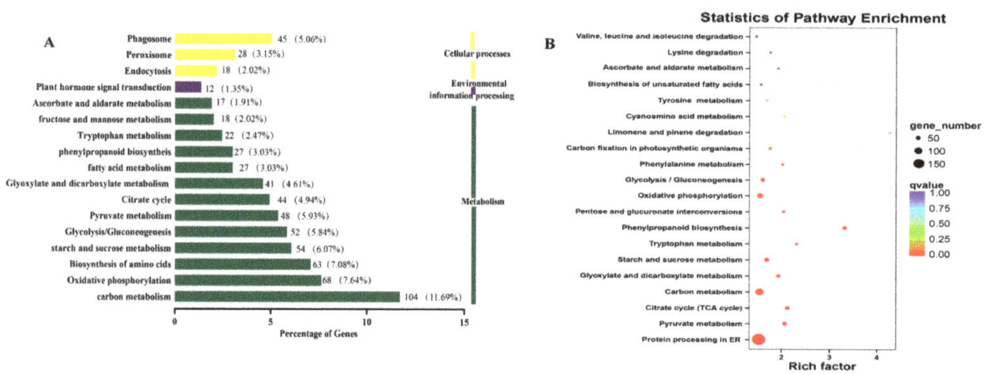

Figure 4. KEGG pathway enrichment of DEGs. (**A**) Statistical analysis of annotated genes in KEEG pathways, (**B**) scatterplot of KEEG pathway for DEGs.

2.4. DEGs Related to Primary Metabolism after GA_3 Application in R. roxburghii Fruits

Considering that the TSS accumulation of *R. roxburghii* fruits decreased significantly after exogenous GA_3 treatment, we determined the DEGs concerning primary metabolism. Their pathways are roughly outlined in Figure 5. The KEGG enrichments suggested that 52, 44, and 15 DEGs were associated with glycolysis/gluconeogenesis, citrate cycle (TCA), and the pentose phosphate pathway, respectively. Regarding the glycolysis/gluconeogenesis pathway, the expression profiles of DEGs displayed two clusters. Among these, 32.69% of DEGs (17) were significantly enhanced in fruits after gibberellin application. For example, two unigene-encoded phosphoenolpyruvate carboxykinase were significantly raised. Moreover, we found that unigenes encoding pyruvate kinase, pyruvate dehydrogenase E1 component, fructose–bisphosphate aldolase, glyceraldehyde-3-phosphate dehydrogenase, dihydrolipoyl dehydrogenase, and glyceraldehyde 3-phosphate dehydrogenase were also substantially increased, suggesting that these DEGs might play critical roles in decomposing soluble sugar. (Table S4). Moreover, about half of DEGs related to the pentose phosphate pathway showed a similar tendency after GA_3 treatment. Among these, the expression levels of unigenes annotated as 6-phosphogluconate dehydrogenase and transaldolase were promoted. Likewise, 43.19% of DEGs involved in the TCA pathway were significantly enhanced, especially ATP citrate (pro-S)-lyase, malate dehydrogenase, and phosphoenolpyruvate carboxykinase. Hence, some candidate crucial genes in the primary metabolic pathway by exogenous GA_3 in *R. roxburghii* fruits were identified, indicating that they might play crucial roles in degrading the total soluble sugar of *R. roxburghii* fruits.

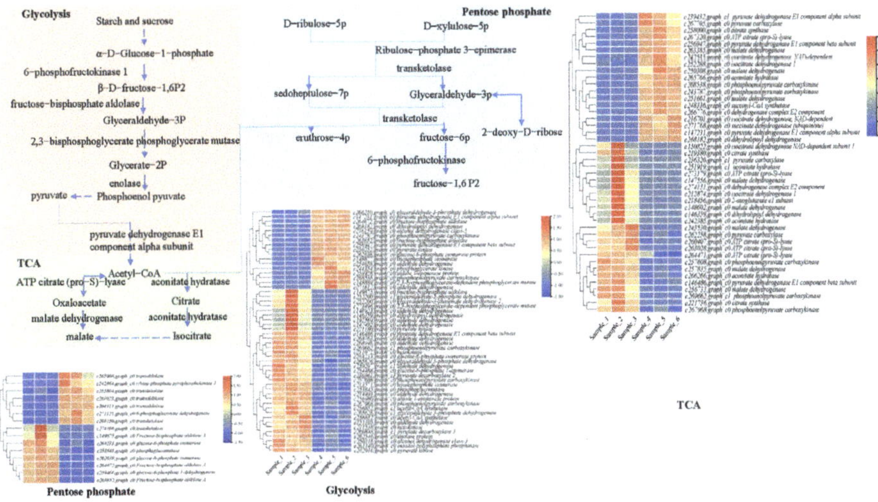

Figure 5. Expression profiles of DEGs related to primary metabolism of *R. roxburghii* fruit. The pathways of glycolysis/gluconeogenesis, citrate cycle (TCA), and pentose phosphate pathways were outlined and expression levels of candidate DEGs were listed using a heatmap. Bars represent the scale of the FPKM of each gene, as indicated by the red and blue rectangles.

2.5. Ascorbate and Hormone-Related Signaling Pathways upon Exogenous GA_3 Applications

The proposed L-ascorbic acid synthetic and recycling pathways are shown in Figure 6A. We identified and analyzed candidate DEGs implicated in ascorbate metabolism. The expression levels of DEGs encoding biosynthesized ascorbate were remarkably increased, mainly including UDP-glucose 6-dehydrogenase and inositol oxygenase (Figure 6B). L-ascorbic acid can be oxidized to ascorbate catalyzed by ascorbate oxidase and reversely by monodehydroascorbate reductase (NADH) activities. Despite the decrease in unigenes encoding these two enzymes, the reduction in unigenes annotated as L-ascorbate oxidase is more significant than the expression of monodehydroascorbate reductase, resulting in the accumulation of L-ascorbic acid in *R. roxburghii* fruits (Table S5).

To investigate the roles of other endogenous hormones in *R. roxburghii* fruits, genes associated with hormone signaling pathways were identified upon GA_3 treatment (Figure 6C). Among them, DEGs were mainly involved in five principal plant hormones: auxin, cytokinins (CK), salicylic acid (SA), abscisic acid (ABA), and ethylene (Table S6). DEGs encoding AUXIN/Indole-3-acetic acid (AUX/IAA), Small auxin-up RNA (SAUR), and Gretchen Hagen 3 (GH3) were found, supporting the crosstalk of the GA_3 signaling pathway with the auxin signaling pathway in *R. roxburghii*. Given that AUX/IAA proteins repressed auxin signaling, the decrease in DEGs annotated as AUX/IAA suggested that GA_3 spraying enhanced the auxin signaling pathway. Likewise, histidine-containing phosphotransferase protein 1-like (*RrAHP*) was stimulated to regulate cell division. By contrast, DEGs related to the signaling pathways of salicylic acid, abscisic acid, and ethylene were reduced, indicating that exogenous GA_3 may be conducive to alleviating the inhibitory effect of ABA and ethylene signal transduction that is commonly known to inhibit the growth of specific tissues and cells. These results indicated that exogenous GA_3 applications could affect other hormone-related pathways in *R. roxburghii*.

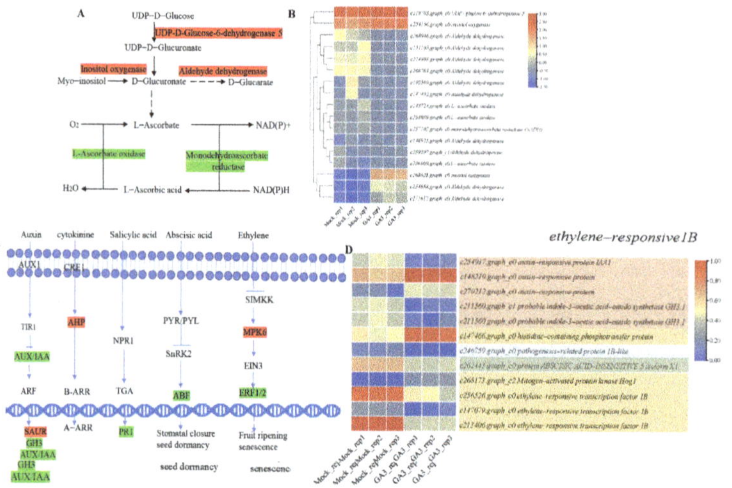

Figure 6. Expression profiles of DEGs related to ascorbate metabolism and hormone signaling of *R. roxburghii* fruit. (**A**) The DEGs involved in the ascorbate pathway are listed. (**B**) Heatmap of DEGs-influenced ascorbate. (**C**) The simple sketch of hormone signaling pathways. (**D**) Heatmap of DEGs implicated in endogenous hormone signaling pathways. The enzymes in the red shadow represent up-regulated, whereas those in the green shadow indicate down-regulated.

2.6. DEGs Related to Transcription Factors in R. roxburghii Fruits

The gene expression of transcription factors influenced by gibberellin in *R. roxburghii* fruits was investigated. A total of 577 candidate TFs were annotated in the transcriptome of *R. roxburghii* fruits. We obtained 24 TFs with significant differences (Table S7). TF members comprised the MYB family (MYB82 and R2R3 MYB), WRKY family (WRKY41 and WRKY75), bZIP family, MADS, NAC, and others. The expression levels of *TGA*$_3$, *C2H2*, *R2R3 MYB*, and *WRKY* were significantly increased (Figure 7). These TFs might play significant roles in regulating the development of *R. roxburghii* fruit, providing information for studying the functions of TFs in the promoting effect of GA_3 on the fruit quality of *R. roxburghii*.

2.7. Validation of DEGs by qRT-PCR

To verify the reliability of the transcriptome results, we randomly selected nine genes in the related pathways of *R. roxburghii* fruits for qRT-PCR validation, including unigenes involved in L-ascorbate acid, sugar, hormone signaling, and key transcriptional factors. Primers were listed (Table S8), and their expression levels were calculated using the $2^{-\Delta\Delta Ct}$ method. We further compared the expression data of the DEGs obtained by RNA-seq and qRT-PCR, and the results displayed a similar tendency (Figure 8A). The correlation between RNA-Seq (FPKM) and qPCR ($2^{-\Delta\Delta Ct}$) results for the nine DEGs was also calculated using $\log_2$fold variation measurements to produce a scatter plot. The qRT-PCR results of nine DEGs were significantly similar to the RNA-seq results (R^2 = 0.7598), indicating that our RNA-seq data were accurate and reproducible (Figure 8B).

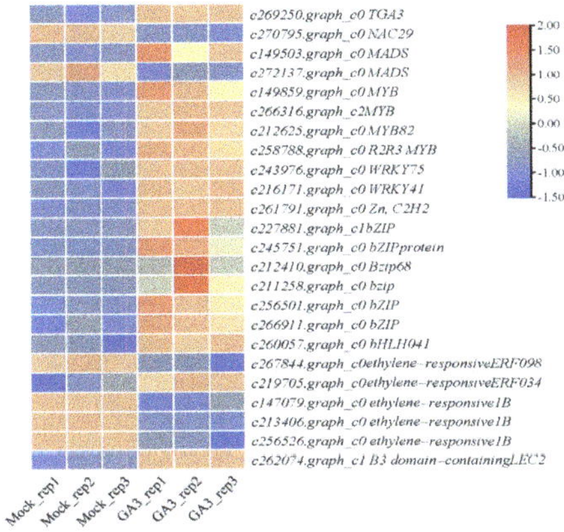

Figure 7. GA_3 promoted the expression levels of most transcription factors.

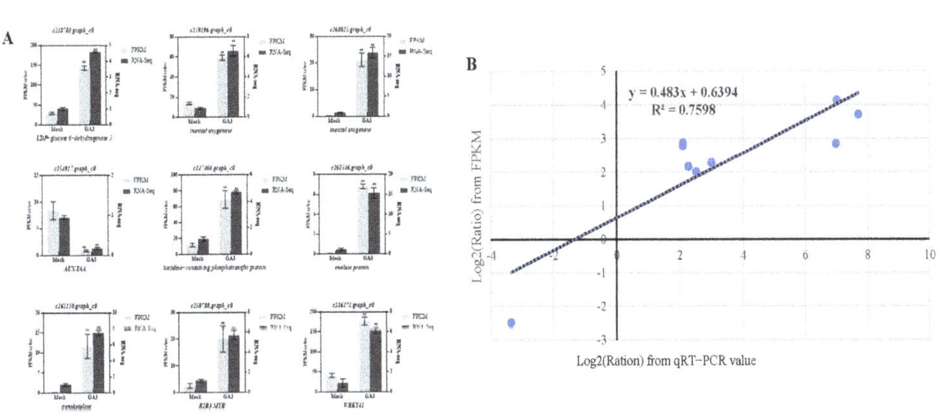

Figure 8. The qRT-PCR verification of DEGs. (**A**) The boxplot reveals the comparison of expression levels by RNA-seq and qRT-PCR. (**B**) Correlation analysis of qPCR ($2^{-\Delta\Delta Ct}$) and RNA-Seq results (RPKM). Log_2fold change of RNA-seq (y-axis) and qRT-PCR (x-axis) were listed. The asterisk indicates a significant difference. "**": $p < 0.01$.

3. Discussion

3.1. Exogenous Gibberellin Spraying Regulates Fruit Qualities of *R. roxburghii*

The GA application in *R. roxburghii* induced the apparent physiological divergences of the fruits, including promoting fruit size and weight, ascorbic acid accumulation, and prickle development. The increase in fruit size was consistent with other horticultural fruits, such as grapes, apples, and tomatoes [7]. Interestingly, the development of fruit prickle was also promoted by GA_3, the same as the functions of increased wax layer thickness by GA_3 treatment in apples, which could reduce fruit water loss rate [8]. Meanwhile, the increase in ascorbic acid and prickle length accompanied by the reduction in TSS concentration was noticed, which might be aroused by the flux allocation between primary and secondary metabolism [24]. Moreover, we observed that the seed number was significantly decreased, possibly due to fertilization failure, sharing the same phenomenon of parthenocarpy (fertilized-independent fruit) that could be induced by gibberellin [4]. Similarly, overexpression of GA biosynthesis enzymes could significantly decrease the seed number of tomatoes [25]. Likewise, CRISPR/Cas9-generated knockout GA biosynthesis enzyme mutants increased the seed numbers of soybeans (*Glycine max*), suggesting that GA_3 application negatively regulated seed set [26].

3.2. GA_3 Facilitated the Decomposition of Total Soluble Sugars by increasing Crucial Genes in Primary Metabolism

In this current study, six DGE libraries were constructed using RNA-Seq and used to screen DEGs after GA_3 treatment. A total of 26,294 unique sequences were obtained, of which 24,677 unigenes were annotated. The results of the present study would be helpful to clone coding domain sequences and analyze gene families in *R. roxburghii*. Multiple genes were altered during fruit setting after GA_3 treatment through RNA-Seq analysis. The roles of many annotated genes showed the same plant responses with GAs in other species. For example, *c241716.graph_c1* was annotated as an expansion, which was also increased in *Malus domestica* after the spray of gibberellin [27]. Moreover, GA_3 significantly decreases the transcript of genes encoding UDP-glycosyltransferase in *R. roxburghii*, as in *F. vesca* [28]. We analyzed GO and KEGG pathways and determined that 3825 DEGs participated in several pathways, yielding novel comprehensive insights into the GA_3 response in *R. roxburghii* fruits.

A few transcriptional events of GAs drive drastic shifts at the primary metabolites and developmental events [29]. We found that starch and sucrose metabolism pathways were significantly enriched, similar to the metabolic pathway that GAs induced during the fruit setting of triploid Loquat (*Eriobotrya japonica*) [21]. The RNA-seq data displayed that multiple DEGs involved in the glycolytic pathway, TCA, and pentose phosphate pathway were up-regulated, indicating that GAs accelerated the decomposing of storage sugars [30]. Spraying GA_3 reduced TSS contents in *R. roxburghii* fruits, mainly because of the enhanced DEGs in the primary metabolisms. A similar tendency was found that the sugar contents were also reduced in germinated seeds of *Fraxinus hupehensis* treated with GAs [30]. Most enzymes were encoded by gene families that displayed various expression levels in *R. roxburghii*. However, many enzymes might be increased after GA application by considering expressions of all genes in the same family [31]. For example, glyceraldehyde-3-phosphate dehydrogenase (GAPDH), a key enzyme in the glycolytic pathway that catalyzes the conversion of D-glyceraldehyde 3-phosphate to 1,3-diphosphoglycerate, was annotated by four genes. The increase in one unigene (*c266210.graph_c0*) is more extensive than the sum of the other four genes (*c252200.graph_c0*, *c252200.graph_c1*, *c265626.graph_c0*, and *c265626.graph_c1*) (Table S4). Therefore, GA_3 facilitated the oxidative decomposition of soluble sugars by increasing crucial genes in primary metabolism.

3.3. The Myoinositol Pathway Was Required for the Accumulation of Ascorbic Acid in R. roxburghii

The abundance of ascorbic acid was influenced by the biosynthesis and recycling pathways [32]. GA_3 spraying could significantly enhance the concentration of ascorbic acid. Moreover, the RNA-seq determined that DEGs implicated in the biosynthesis of L-ascorbate were all impressively enhanced, especially UDP-glucose 6-dehydrogenase and inositol oxygenase, possibly suggesting that L-galactose and myoinositol pathways were essential for the accumulation of ascorbic acid. Meanwhile, exogenous GA_3 substantially decreased the expression of L-ascorbate oxidase, increasing the reduced content and redox state. The results would help the understanding of the molecular basis regulating ascorbic acid in *R. roxburghii* fruits.

3.4. The Altered Hormonal Signaling and TF Induced by GA_3 Were Responsible for the Fruit Development of R. roxburghii

Transcriptome analyses illustrated that other hormone signaling pathways showed enhanced sensitivity to the application of GA_3. Five auxin-related DEGs were identified. *c246759.graph_c0* encoding (AUX/IAA), which could bind to auxin response factors, was significantly down-regulated (Table S6) [33]. An early study also proved that silencing of tomato *SlIAA9* led to the parthenocarpic fruit of Solanaceae [34]. Moreover, two genes encoding GH3 were decreased. Since GH3 was reported to catalyze the synthesis of indole-3-acetic acid (IAA)-amino acid conjugates, the reduction in GH3 indicated a high level of free IAA concentration [35]. During the early periods of fruit expansion, the transcript of GH3 was at a lower level, according to the high auxin concentration at the same development stage in apples [36]. Hence, GAs may play a role in increasing auxin signaling to regulate the fruit development of *R. roxburghii*. Similarly, Illumina HiSeq high-throughput sequencing revealed that auxin-responsive genes, playing the same roles of *c254917.graph_c0* in *R. roxburghii*, also increased in *F. vesca* when treated with 50 mg/L GA_3 [37]. In addition, histidine-containing phosphotransfer protein 1-like involved in the cytokinin signaling pathway was promoted after GA_3 treatment, indicating that GAs possibly induced fruit physiological changes via the cytokinin signal pathway [38]. ABA inhibits fruit growth in the early stages of strawberries, accompanied by low ABA levels in early development and a sharp increase during ripening [7]. Our study showed the decreased DEGs implicated in the ABA signaling pathway. Morever, some DEGs involved in other phytohormones were also determined. These results confirmed that auxin, ethylene, abscisic acid, and salicylic acid, which GAs can directly or indirectly trigger, might regulate fruit development, similar to the complex hormonal regulation in tomatoes [39].

Based on DGE analysis, the expression levels of the genes encoding several transcription factors were significantly changed during GA_3 treatment in *R. roxburghii*, including genes encoding NAC, MADS, bHLH, MYBs, and others. TFs play crucial roles in fruit qualities. For example, the development of *R. roxburghii* fruit prickles is determined by the complex of MYB-bHLH-WD40. Interestingly, the prickle length was significantly enhanced by GA application in our study. The same treatment also induced trichome formation by MYB (GLABROUS 1) in Arabidopsis [40]. RrGL1 (MYB), RrGL3/EGL3 (bHLH), and RrTTG1 (WD40) are responsible for prickle development [41,42]. The two former TFs are required for prickle elongation. In contrast, WD40 is necessary for inducing prickle initiation. Interestingly, The RNA-seq analyzed the transcripts of genes encoding *R2R3 MYB* and *bHLH041* that were significantly up-regulated, showing that the conserved domains with *RrGL1* and *RrGL3/RrEGL3* might play similar effects on prickle development. However, *RrTTG1* displayed no significant change in *R. roxburghii*, which explained why prickle only became longer but not numerous [17].

4. Materials and Methods

4.1. The Plant Materials

Seedlings of the *Rosa roxburghii* Tratt cultivar 'Guinong 5' were planted at the garden of Guizhou Normal University in Guiyang. The trees were 7 years old, in the adult phase. Flowers of *R. roxburghii* in full blossom were sprayed with 200 mg/L of GA_3 (Sigma-Aldrich, cat# Sigma G7645, St. Louis, MI, USA) daily for 7 days as treatment and distilled H_2O as a control on 15 May 2022 (Figure S1). We collected at least twenty fruits at 60 DAP and 120 DAP for morphology analysis. Some fresh fruits treated with GA_3 and control at 120 DAP were harvested from three separate seedings for measurement. The others for RNA extraction were snap-frozen immediately in nitrogen and stored in a −80 °C refrigerator. Transcriptome sequencing was performed with three biological replications.

4.2. Plant Biomass Yield and Biochemical Analysis of R. roxburghii

Physiological data of at least twenty fruits from three separate seedlings were recorded on the following traits: average fruit weight (g), fruit shape index (calculated as longitudinal diameter (cm)/horizontal diameter (cm)), seed number, and prickle length per each fruit. We weighed 100 g of fruits per treatment and homogenized it with 5000 rpm. Then, 2 g of the sample was collected, powdered with liquid nitrogen with 20 mL ethanol (80%, *v*/*v*), and sonicated for 15 min at 80 °C. After evaporation, the residue was dissolved with 2 mL of distilled water for HPLC analysis. The percentages (g/g) of total soluble sugar (TSS) and L-ascorbic acid concentration (g/g) were determined using high-performance liquid chromatography (HPLC) as described by Huang et al. [31].

4.3. RNA Extraction, Library Construction, and Sequencing of *R. roxburghii*

Total RNA was extracted from *R. roxburghii* fruits using the TRIzol reagent (Invitrogen, Carlsbad, CA, USA). RNA quality and quantity were determined using a NanoDrop spectrophotometer and Agilent 2100 Bioanalyzer for further transcriptome sequencing. RNA (1 μg) per sample was used for the preparations. Following the manufacturer's recommendations, sequencing libraries were generated using NEBNext®Ultra™ RNA Library Prep Kit for Illumina® (NEB, Ipswich, MA, USA). The library preparations were sequenced on an Illumina Hiseq 2000 platform. Clean data were obtained by removing reads containing adapter, ploy-N, and low-quality reads from raw data. All the RNA-seq raw data have been uploaded and are available in Genbank under accession number PRJNA1003688.

4.4. De Novo Assembles and Unigenes Annotation of *R. roxburghii*

Transcriptome assembly was accomplished based on the left. fq and right. fq using Trinity with min_kmer_cov set to 2 by default and all other parameters set to default [43]. Gene functions were annotated and based on BLASTx with cutoff E-value $\leq 1 \times 10^{-5}$ and HMMER $\leq 1 \times 10^{-10}$ [44] using Nr, GO [45], COG [46], and KEGG [47] databases.

4.5. Differential Expression Analysis of Unigenes Expressed in *R. roxburghii* Fruits

The expression level was estimated with RNA-Seq by Expectation Maximization (RSEM). FPKM values indicate the abundance of the corresponding unigene [48]. DESeq R package (1.10.1) was adopted to compare the GA_3 treatment with the reference group [49]. The resulting p values were adjusted using Benjamini and Hochberg's approach for controlling the false discovery rate (FDR). An adjusted p-value < 0.001 and $|\log_2(\text{fold change})| \geq 2$ were set as the threshold for determining the DEGs. GO and KEGG pathway analyses were conducted to interpret the DEGs.

4.6. Real-Time Quantitative PCR

Total RNAs from the *R. roxburghii* fruits were isolated, and RT-PCR Kit® (TaKaRa, Shiga, Japan) was performed using 2 μg of RNA and oligo dT-adaptor primer. Real-time quantitative PCR was performed in a LightCycler480 instrument (Roche, Basel, Switzerland). Each reaction contained 10 μL of SYBR green PCR master mix (TaKaRa, Shiga, Japan), 1.0 μL cDNA, 200 nM primers, and ddH_2O up to the final volume of 20 μL. Amplification was performed at 95 °C for 5 min, followed by 40 cycles at 95 °C for 20 s, 58 °C for 30 s, and 72 °C for 1 min. The expression levels relative to the RrActin were estimated using the $2^{-\Delta\Delta Ct}$ method. All samples were performed for three biological replicates.

4.7. Statistical Analysis

The results were statistically evaluated by one-way analysis of variance (ANOVA) followed by Tukey's test using Statistical Program for Social Sciences (SPSS) program version 20.0 (SPSS Inc., Chicago, IL, USA). Statistical differences with p values under 0.05 are considered significant, using an asterisk that indicates a significant difference. "*": $p < 0.05$; "**": $p < 0.01$.

5. Conclusions

Our study reveals a complex network of genes regulated by GA_3 treatment in *R. roxburghii* fruits. The enriched GO terms and KEGG pathways highlight the multi-faceted roles of GAs in promoting cell growth, accelerating fruit ripening, and modulating stress responses. These findings provide valuable insights into the molecular mechanisms of GA action in *R. roxburghii* fruits and provide a foundation for future studies to improve fruit quality and yield. Although significant DEGs have been identified regarding the roles of GAs in *R. roxburghii*, many problems remain to be studied. Efficient genetic transformation and plant regeneration systems are still lacking for *R. roxburghii*, limiting the functional validation of gene studies on relevant studies. It is worth constructing transgenetic systems in *R. roxburghii* in the future, which will undoubtedly promote further understanding of the molecular mechanisms underlying the effects of GAs on fruit development in *R. roxburghii*.

Supplementary Materials: The following supporting information can be downloaded at: https://www.mdpi.com/article/10.3390/plants12193425/s1, Figure S1. The experimental design map used in this study. Table S1. Statistics of the reads and RNA-seq mapping of *R. roxburghii* fruits; Table S2. All DEGs between GA_3- and non-treated fruits and their annotations were listed; Table S3. Top GO enriched functions of DEGs between GA_3- and non-treated fruits; Table S4. DEGs involved in the primary metabolism of *R. roxburghii* fruits; Table S5. DEGs associated with ascorbate metabolism of *R. roxburghii* fruits; Table S6. DEGs implicated in the endogenous hormone signaling pathway of *R. roxburghii* fruits; Table S7. The expression levels of DEGs encoding transcription factors following GA_3 treatment; Table S8. Primers used in qRT-PCR analysis.

Author Contributions: Conceptualization and visualization, X.H.; methodology, Y.J.; validation, X.W.; writing—review and editing, G.S.; funding acquisition, X.H. and H.Y. All authors have read and agreed to the published version of the manuscript.

Funding: This work was supported by grants from the National Natural Science Foundation of China (Grant No. 32160699 and 32060587). The Joint Fund of the National Natural Science Foundation of China and the Karst Science Research Center of Guizhou province (Grant No. U1812401). Guizhou Province Natural Science Foundation (QKHJC-ZK[2023]ZD026). Guizhou Normal University Qian Shi Xin Miao ([2022]18).

Data Availability Statement: The RNA-Seq of raw data is publicly available in the NCBI under number: PRJNA1003688 with the link: https://www.ncbi.nlm.nih.gov/sra/PRJNA1003688 accessed on 11 August 2023.

Conflicts of Interest: The authors declare no conflict of interest.

References

1. Azzi, L.; Deluche, C.; Gevaudant, F.; Frangne, N.; Delmas, F.; Hernould, M.; Chevalier, C. Fruit growth-related genes in tomato. *J. Exp. Bot.* **2015**, *66*, 1075–1086. [CrossRef] [PubMed]
2. Hu, J.; Israeli, A.; Ori, N.; Sun, T.P. The interaction between DELLA and ARF/IAA mediates crosstalk between gibberellin and auxin signaling to control fruit initiation in tomato. *Plant Cell* **2018**, *30*, 1710–1728. [CrossRef] [PubMed]
3. Cheniclet, C.; Rong, W.Y.; Causse, M.; Frangne, N.; Bolling, L.; Carde, J.P.; Renaudin, J.P. Cell expansion and endoreduplication show a large genetic variability in pericarp and contribute strongly to tomato fruit growth. *Plant Physiol.* **2005**, *139*, 1984–1994. [CrossRef] [PubMed]
4. Wang, H.; Wu, T.; Liu, J.; Cong, L.; Zhu, Y.; Zhai, R.; Yang, C.; Wang, Z.; Ma, F.; Xu, L. *PbGA20ox2* regulates fruit set and induces parthenocarpy by enhancing GA(4) content. *Front. Plant Sci.* **2020**, *11*, 113. [CrossRef] [PubMed]
5. Pattison, R.J.; Catalá, C. Evaluating auxin distribution in tomato (*Solanum lycopersicum*) through an analysis of the PIN and AUX/LAX gene families. *Plant J.* **2012**, *70*, 15. [CrossRef]
6. Zhao, X.; Muhammad, N.; Zhao, Z.; Yin, K.; Liu, Z.; Wang, L.; Luo, Z.; Wang, L.; Liu, M. Molecular regulation of fruit size in horticultural plants: A review. *Sci. Hortic.* **2021**, *288*, 110353. [CrossRef]
7. Liao, X.; Li, M.; Liu, B.; Yan, M.; Yu, X.; Zi, H.; Liu, R.; Yamamuro, C. Interlinked regulatory loops of ABA catabolism and biosynthesis coordinate fruit growth and ripening in woodland strawberry. *Proc. Natl. Acad. Sci. USA* **2018**, *115*, E11542–E11550. [CrossRef]
8. Liu, C.; Xiao, P.; Jiang, F.; Wang, S.; Liu, Z.; Song, G.; Li, W.; Lv, T.; Li, J.; Wang, D.; et al. Exogenous gibberellin treatment improves fruit quality in self-pollinated apple. *Plant Physiol. Biochem.* **2022**, *174*, 11–21. [CrossRef]
9. Lu, L.; Liang, J.; Zhu, X.; Xiao, K.; Li, T.; Hu, J. Auxin- and cytokinin-induced berries set in grapevine partly rely on enhanced gibberellin biosynthesis. *Tree Genet. Genomes* **2016**, *12*, 41. [CrossRef]
10. Qian, C.; Ren, N.; Wang, J.; Xu, Q.; Chen, X.; Qi, X. Effects of exogenous application of CPPU, NAA and GA(4+7) on parthenocarpy and fruit quality in cucumber (*Cucumis sativus* L.). *Food Chem.* **2018**, *243*, 4. [CrossRef]
11. Harada, T.; Kurahashi, W.; Yanai, M.; Wakasa, Y.; Satoh, T. Involvement of cell proliferation and cell enlargement in increasing the fruit size of *Malus* species. *Sci. Hortic.* **2005**, *105*, 447–456. [CrossRef]
12. Xie, J.; Tian, J.; Du, Q.; Chen, J.; Li, Y.; Yang, X.; Li, B.; Zhang, D. Association genetics and transcriptome analysis reveal a gibberellin-responsive pathway involved in regulating photosynthesis. *J. Exp. Bot.* **2016**, *67*, 3325–3338. [CrossRef] [PubMed]
13. Marin-de la Rosa, N.; Sotillo, B.; Miskolczi, P.; Gibbs, D.J.; Vicente, J.; Carbonero, P.; Onate-Sanchez, L.; Holdsworth, M.J.; Bhalerao, R.; Alabadi, D.; et al. Large-scale identification of gibberellin-related transcription factors defines group VII ethylene response factors as functional DELLA partners. *Plant Physiol.* **2014**, *166*, 1022–1032. [CrossRef]

14. Oh, E.; Zhu, J.Y.; Bai, M.Y.; Arenhart, R.A.; Sun, Y.; Wang, Z.Y. Cell elongation is regulated through a central circuit of interacting transcription factors in the Arabidopsis hypocotyl. *eLife* **2014**, *3*, e03031. [CrossRef] [PubMed]
15. Bjorklund, S.; Antti, H.; Uddestrand, I.; Moritz, T.; Sundberg, B. Cross-talk between gibberellin and auxin in development of *Populus* wood: Gibberellin stimulates polar auxin transport and has a common transcriptome with auxin. *Plant J.* **2007**, *52*, 499–511. [CrossRef] [PubMed]
16. Liu, H.; Wen, Y.; Cui, M.; Qi, X.; Deng, R.; Gao, J.; Cheng, Z. Histological, Physiological and transcriptomic analysis reveal gibberellin-induced axillary meristem formation in Garlic (*Allium sativum*). *Plants* **2020**, *9*, 970. [CrossRef]
17. Huang, X.; Yi, P.; Liu, Y.; Li, Q.; Jiang, Y.; Yi, Y.; Yan, H. RrTTG1 promotes fruit prickle development through an MBW complex in *Rosa roxburghii*. *Front. Plant Sci.* **2022**, *13*, 939270. [CrossRef]
18. Zhang, Y.; Bian, X.; Yan, G.; Sun, B.; Miao, W.; Huang, M.; Li, N.; Wu, J.L. Discovery of novel ascorbic acid derivatives and other metabolites in fruit of *Rosa roxburghii* Tratt through untargeted metabolomics and feature-based molecular networking. *Food Chem.* **2023**, *405*, 134807. [CrossRef]
19. Ning, Y.; Ding, Y.-k.; Chang, Y.-h.; Zhang, S.; An, H.-m.; Fu, Y.-j. Low concentration of MeJA-elicited the positive growth of *Rosa roxburghii* via balancing phytohormone signal transduction and triterpenoids synthesis. *Plant Growth Regul.* **2023**, *101*, 187–199. [CrossRef]
20. Zhang, S.; Li, L.; Lu, M.; An, H. Effects of three exogenous plant growth regulators on AsA accumulation and expression of genes involved in its metabolism in *Rosa roxburghii* fruit. *J. Agric. Biotechnol.* **2018**, *26*, 10.
21. Jiang, S.; Luo, J.; Xu, F.; Zhang, X. Transcriptome analysis reveals candidate genes involved in gibberellin-induced fruit setting in triploid Loquat (*Eriobotrya japonica*). *Front. Plant Sci.* **2016**, *7*, 1924. [CrossRef] [PubMed]
22. Cheng, C.; Jiao, C.; Singer, S.D.; Gao, M.; Xu, X.; Zhou, Y.; Li, Z.; Fei, Z.; Wang, Y.; Wang, X. Gibberellin-induced changes in the transcriptome of grapevine (*Vitis labrusca* × *V. vinifera*) cv. Kyoho flowers. *BMC Genom.* **2015**, *16*, 128. [CrossRef] [PubMed]
23. Zhao, L.; Li, M.; Ma, X.; Luo, D.; Zhou, Q.; Liu, W.; Liu, Z. Transcriptome analysis and identification of abscisic acid and gibberellin-related genes during seed development of alfalfa (*Medicago sativa* L.). *BMC Genom.* **2022**, *23*, 651. [CrossRef] [PubMed]
24. Shih, M.L.; Morgan, J.A. Metabolic flux analysis of secondary metabolism in plants. *Metab. Eng. Commun.* **2020**, *10*, e00123. [CrossRef]
25. Vriezen, W.H.; Feron, R.; Maretto, F.; Keijman, J.; Mariani, C. Changes in tomato ovary transcriptome demonstrate complex hormonal regulation of fruit set. *New Phytol.* **2008**, *177*, 60–76. [CrossRef]
26. Hu, D.; Li, X.; Yang, Z.; Liu, S.; Hao, D.; Chao, M.; Zhang, J.; Yang, H.; Su, X.; Jiang, M.; et al. Downregulation of a gibberellin 3beta-hydroxylase enhances photosynthesis and increases seed yield in soybean. *New Phytol.* **2022**, *235*, 502–517. [CrossRef]
27. Galimba, K.D.; Bullock, D.G.; Dardick, C.; Liu, Z.; Callahan, A.M. Gibberellic acid induced parthenocarpic 'Honeycrisp' apples (*Malus domestica*) exhibit reduced ovary width and lower acidity. *Hortic. Res.* **2019**, *6*, 41. [CrossRef]
28. Gu, T.; Jia, S.; Huang, X.; Wang, L.; Fu, W.; Huo, G.; Gan, L.; Ding, J.; Li, Y. Transcriptome and hormone analyses provide insights into hormonal regulation in strawberry ripening. *Planta* **2019**, *250*, 145–162. [CrossRef]
29. Wang, H.; Schauer, N.; Usadel, B.; Frasse, P.; Zouine, M.; Hernould, M.; Latché, A.; Pech, J.C.; Fernie, A.R.; Bouzayen, M. Regulatory features underlying pollination-dependent and -independent tomato fruit set revealed by transcript and primary metabolite profiling. *Plant Cell* **2009**, *21*, 25. [CrossRef]
30. Song, Q.; Cheng, S.; Chen, Z.; Nie, G.; Xu, F.; Zhang, J.; Zhou, M.; Zhang, W.; Liao, Y.; Ye, J. Comparative transcriptome analysis revealing the potential mechanism of seed germination stimulated by exogenous gibberellin in *Fraxinus hupehensis*. *BMC Plant Biol.* **2019**, *19*, 199. [CrossRef]
31. Gao, X.; Yuan, Y.; Liu, Z.; Liu, C.; Xin, H.; Zhang, Y.; Gai, S. Chilling and gibberellin acids hyperinduce beta-1,3-glucanases to reopen transport corridor and break endodormancy in tree peony (*Paeonia suffruticosa*). *Plant Physiol. Biochem.* **2021**, *167*, 771–784. [CrossRef] [PubMed]
32. Huang, M.; Xu, Q.; Deng, X.X. L-Ascorbic acid metabolism during fruit development in an ascorbate-rich fruit crop chestnut rose (*Rosa roxburghii* Tratt). *J. Plant Physiol.* **2014**, *171*, 12. [CrossRef] [PubMed]
33. Zhou, J.; Sittmann, J.; Guo, L.; Xiao, Y.; Huang, X.; Pulapaka, A.; Liu, Z. Gibberellin and auxin signaling genes RGA1 and ARF8 repress accessory fruit initiation in diploid strawberry. *Plant Physiol.* **2021**, *185*, 1059–1075. [CrossRef] [PubMed]
34. Wang, H.; Jones, B.; Li, Z.; Frasse, P.; Delalande, C.; Regad, F.; Chaabouni, S.; Latché, A.; Pech, J.-C.; Bouzayen, M. The Tomato Aux/IAA Transcription Factor IAA9 is involved in fruit development and leaf morphogenesis. *Plant Cell* **2005**, *17*, 2676–2692. [CrossRef]
35. Wu, B.; Sun, M.; Zhang, H.; Yang, D.; Lin, C.; Khan, I.; Wang, X.; Zhang, X.; Nie, G.; Feng, G.; et al. Transcriptome analysis revealed the regulation of gibberellin and the establishment of photosynthetic system promote rapid seed germination and early growth of seedling in pearl millet. *Biotechnol. Biofuels* **2021**, *14*, 94. [CrossRef]
36. Devoghalaere, F.; Doucen, T.; Guitton, B.; Keeling, J.; Payne, W.; Ling, T.J.; Ross, J.J.; Hallett, I.C.; Gunaseelan, K.; Dayatilake, G.A.; et al. A genomics approach to understanding the role of auxin in apple (*Malus* × *domestica*) fruit size control. *BMC Plant Biol.* **2012**, *12*, 7. [CrossRef]
37. Li, W.; Wang, B.; Sun, H.; Zhang, Z. Transcriptome profiling of runner formation induced by exogenous gibberellin in *Fragaria vesca*. *Sci. Hortic.* **2021**, *281*, 109966. [CrossRef]
38. Matsuo, S.; Kikuchi, K.; Fukuda, M.; Honda, I.; Imanishi, S. Roles and regulation of cytokinins in tomato fruit development. *J. Exp. Bot.* **2012**, *63*, 5569–5579. [CrossRef]

39. Min, D.; Li, Z.; Ai, W.; Li, J.; Zhou, J.; Zhang, X.; Mu, D.; Li, F.; Li, X.; Guo, Y. The Co-regulation of ethylene Biosynthesis and ascorbate-glutathione cycle by methy jasmonate contributes to aroma formation of tomato fruit during postharvest ripening. *J. Agric. Food Chem.* **2020**, *68*, 10822–10832. [CrossRef]
40. Perazza, D.; Vachon, G.; Herzog, M. Gibberellins promote trichome formation by up-regulating *GLABROUS1* in Arabidopsis. *Plant Physiol.* **1998**, *117*, 375–383. [CrossRef]
41. Yan, H.; Wu, Z.; Liu, Y.; Weng, Q.; Yi, Y.; Huang, X. Functional divergence of *RrGL3* and *RrEGL3* from *Rosa roxburghii* in mediating trichome development. *Plant Cell Tissue Organ Cult.* **2021**, *147*, 313–324. [CrossRef]
42. Huang, X.; Yan, H.; Zhai, L.; Yi, Y. GLABROUS1 from *Rosa roxburghii* Tratt regulates trichome formation by interacting with the GL3/EGL3 protein. *Gene* **2019**, *692*, 60–67. [CrossRef] [PubMed]
43. Grabherr, M.G.; Haas, B.J.; Yassour, M.; Levin, J.Z.; Thompson, D.A.; Amit, I.; Adiconis, X.; Fan, L.; Raychowdhury, R.; Zeng, Q.; et al. Full-length transcriptome assembly from RNA-Seq data without a reference genome. *Nat. Biotechnol.* **2011**, *29*, 644–652. [CrossRef]
44. Altschul, S.F.; Madden, T.L.; Schaffer, A.A.; Zhang, J.; Zhang, Z.; Miller, W.; Lipman, D.J. Gapped BLAST and PSI-BLAST: A new generation of protein database search programs. *Nucleic Acids Res.* **1997**, *25*, 3389–3402. [CrossRef] [PubMed]
45. Ashburner, M.; Ball, C.A.; Blake, J.A.; Botstein, D.; Butler, H.; Cherry, J.M.; Davis, A.P.; Dolinski, K.; Dwight, S.S.; Eppig, J.T.; et al. Gene ontology: Tool for the unification of biology. The Gene Ontology Consortium. *Nat. Genet.* **2000**, *25*, 25–29. [CrossRef]
46. Tatusov, R.L.; Galperin, M.Y.; Natale, D.A.; Koonin, E.V. The COG database: A tool for genome-scale analysis of protein functions and evolution. *Nucleic Acids Res.* **2000**, *28*, 33–36. [CrossRef]
47. Kanehisa, M.; Goto, S.; Kawashima, S.; Okuno, Y.; Hattori, M. The KEGG resource for deciphering the genome. *Nucleic Acids Res.* **2004**, *32*, D277–D280. [CrossRef]
48. Trapnell, C.; Williams, B.A.; Pertea, G.; Mortazavi, A.; Kwan, G.; van Baren, M.J.; Salzberg, S.L.; Wold, B.J.; Pachter, L. Transcript assembly and quantification by RNA-Seq reveals unannotated transcripts and isoform switching during cell differentiation. *Nat. Biotechnol.* **2010**, *28*, 511–515. [CrossRef]
49. Leng, N.; Dawson, J.A.; Thomson, J.A.; Ruotti, V.; Rissman, A.I.; Smits, B.M.; Haag, J.D.; Gould, M.N.; Stewart, R.M.; Kendziorski, C. EBSeq: An empirical Bayes hierarchical model for inference in RNA-seq experiments. *Bioinformatics* **2013**, *29*, 1035–1043. [CrossRef]

plants

Article

Transcriptome and Metabolome Reveal Sugar and Organic Acid Accumulation in *Rosa roxburghii* Fruit

Liyao Su, Tian Zhang, Min Wu, Yan Zhong and Zongming (Max) Cheng *

State Key Laboratory of Crop Genetics and Germplasm Enhancement, College of Horticulture, Nanjing Agricultural University, Nanjing 210095, China
* Correspondence: zcheng@utk.edu

Abstract: Sugars and organic acids significantly impact fruit sensory quality, but their accumulation patterns and regulatory mechanisms during the development of *Rosa roxburghii* fruit are still unclear. We utilized transcriptomics and metabolomics to investigate genes related to sugar and organic acid metabolism in *Rosa roxburghii*. Metabolomics data revealed that sucrose, glucose and fructose were the primary sugars, whereas citric acid and malic acid were the primary organic acids in *Rosa roxburghii* fruit. We constructed the metabolic pathways of major sugars and organic acids in *Rosa roxburghii* and identified five key genes involved in sugar and organic acid synthesis. In addition, we identified a module containing 132 transcription factors that was significantly associated with sucrose, citric acid and malic acid. Based on quantitative polymerase chain reaction (qPCR), we identified 13 transcription factors involved in sugar and organic acid metabolism, including the transcription factor *RrANL2* and the sucrose synthase gene *RrSUS3*. Further yeast one-hybrid and dual luciferase assays showed that *RrANL2* could bind to the promoter of *RrSUS3* to increase its expression. These results provide new insights into the metabolism of sugars and organic acids in *Rosa roxburghii* fruit.

Keywords: *Rosa roxburghii*; sugar; organic acid; transcriptome; metabolome; transcription factor

1. Introduction

Sweetness is an important factor that impacts the quality of most fruits [1]. The total amount of sugar compounds, such as sucrose, glucose and fructose, determines the sweetness of fruits, whereas differences in their relative ratios impact the unique flavor of each fruit [2,3]. For example, fructose and glucose are predominant in grapes [4], sucrose is the main sugar in watermelon [5] and muskmelon [6], and fructose is the main sugar in apples and pears [7,8]. Sucrose is one of the primary products of photosynthesis which is transported through the phloem to be stored in organs, including fruit [9]. As a key step in the sugar metabolic pathway, sucrose is converted to glucose and fructose by neutral converting enzymes. Subsequently, glucose and fructose are phosphorylated by hexokinase and fructokinase to synthesize glucose 6-phosphate (G6P) and fructose 6-phosphate (F6P). In addition, excess fructose and glucose are transported into vesicles for storage by sugar transport proteins [10].

Sourness is another factor that determines the quality and ripeness of fruit [11]. The sweet and sour taste of fruits depends primarily on the content and ratio of sugars and organic acids [12]. The accumulation of organic acids, such as malic acid, citric acid, succinic acid and tartaric acid, determines the acidity of fruits. Different kinds of acids have different acidities, which makes not only the total content of organic acids but also the relative ratios of different organic acids important for determining fruit acidity. In most fruits, malic and citric acids are the main constituents of organic acids [13]. As part of the tricarboxylic acid cycle, malic and citric acids are closely related to plant respiration and energy metabolism [14]. The impact of fruit organic acids and sugars on taste makes understanding key enzymes and regulatory factors, which impact their quantity, crucial for improving the quality of fruit and the economic value of horticultural crops.

Rosa roxburghii is an economically important fruit tree that is native to China, where it is mainly distributed in southwest regions such as Guizhou Province, Yunnan Province and Sichuan Province. *Rosa roxburghii* fruits have recently been found to contain very high vitamin C content (>2000 mg/100 g FW, fresh weight), in addition to other bioactive molecules such as vitamins, polyphenols, flavonoids, organic acids and Superoxide dismutase [15,16]. Despite these beneficial nutritional qualities, *Rosa roxburghii* fruit has a slightly sour and bitter taste, reducing its potential consumer appeal. At present, *Rosa roxburghii* fruit is primarily utilized in an array of processed products, including dry fruit, canned fruit and fruit drinks [15]. To improve the value of fresh *Rosa roxburghii* fruit, a better understanding of the dynamic changes of sugar and organic acid content during the course of its development is required. Although the metabolic processes associated with sugars and organic acids are well known in many plant species, they are poorly understood during *Rosa roxburghii* fruit development. A better understanding of these processes can lead to the development of *Rosa roxburghii* fruit with better consumer appeal, while retaining its beneficial nutritional qualities.

In this study, the dynamics of sugars and organic acids during *Rosa roxburghii* development were analyzed through metabolomic and transcriptomic analyses. Metabolic pathways for sugars and organic acids were constructed, and several key enzymes regulating sugar and organic acid synthesis were identified. We also conducted a focused investigation of the role of transcription factors in the accumulation of sugars and organic acids in *Rosa roxburghii*. Application of weighted gene co-expression analysis (WGCNA) led to the creation of modules that correlated transcription factors with sugars and organic acids. Within the modules significantly associated with sugar and organic acid accumulation, we identified many transcription factors and selected the differentially expressed transcription factors and key enzymes regulating sugar and organic acid synthesis in *Rosa roxburghii* for quantitative polymerase chain reaction validation. The regulatory relationships of transcription factors with sugar and organic acid synthesis enzymes were also verified by yeast one-hybrid and dual luciferase assays. This study provides a reference for understanding the mechanisms of sugar and organic acid accumulation in *Rosa roxburghii* fruits which can be applied to genetic breeding and quality improvement.

2. Results

2.1. Transcriptome and Metabolome Analysis

To study the dynamic changes of inclusion content and related genes in *Rosa roxburghii* fruits, we used transcriptomics combined with metabolomics to assay gene expression and metabolite content at five different stages of *Rosa roxburghii* fruit development (Figure 1A). Based on the OPLS-DA results, the metabolites that differed between tissues were initially identified from the obtained variable importance in projection (VIP) (VIP $\geq$1) of the multivariate analysis OPLS-DA model. We also combined this information with the differential gene expression data (fold change $\geq$ 2 and fold change $\leq$ 0.5) and performed KEGG analysis of differentially accumulated metabolites and differentially expressed genes in the five stages of *Rosa roxburghii* fruit development (Figure 1B–E). This analysis revealed changes in metabolites and genes associated with flavonoid biosynthesis, phenylpropanoid biosynthesis, ABC transporters, anthocyanin biosynthesis and cyanoamino acid metabolism during *Rosa roxburghii* development. Additionally, starch and sucrose metabolism, fructose and mannose metabolism, the citric acid cycle and the pyruvic acid cycle were also enriched at several time points. These results indicate that dynamic changes in sugars and organic acids take place during *Rosa roxburghii* fruit development.

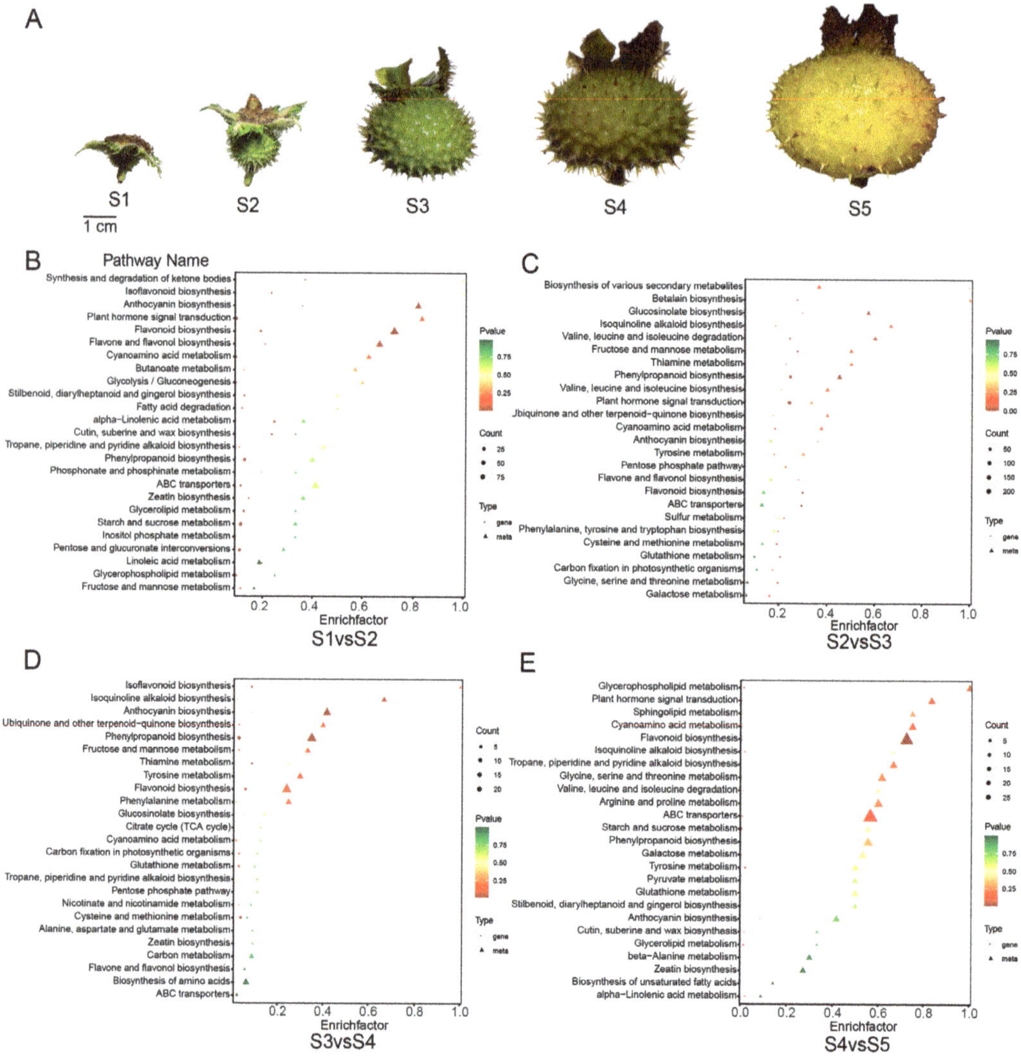

Figure 1. Analysis of differentially expressed genes and differentially accumulated metabolites. (**A**) The five developmental stages of *Rosa roxburghii* fruit. (**B–E**) KEGG enrichment of differentially accumulated metabolites and differentially expressed genes at different stages.

2.2. Dynamics of Sugar and Acid Content during Fruit Development of Rosa roxburghii

Sucrose, fructose and glucose are the primary sugars present in most plant species. We found changes in the relative levels of all three sugars during the course of *Rosa roxburghii* development. The levels of these sugars decreased from the young fruit stage to the fruit expansion stage and then increased dramatically at the fruit ripening stage (Figure 2A, Table S2). For subsequent analysis, we utilized malic acid, citric acid, tartaric acid and succinic acid as representative organic acids. We found that tartaric acid content was negligible compared with other acids in *Rosa roxburghii* fruit (Figure 2B). The content of succinic acid decreased during the development of *Rosa roxburghii* fruit, reaching its lowest level in ripe fruit, indicating it is not a primary source of sour flavor (Figure 2B, Table S2).

Malic acid and citric acid content, on the other hand, peaked in ripe *Rosa roxburghii* fruit. Taken together, these results demonstrate that glucose and fructose are the primary sugars in ripe *Rosa roxburghii* fruit, whereas malic acid and citric acid are the primary organic acids.

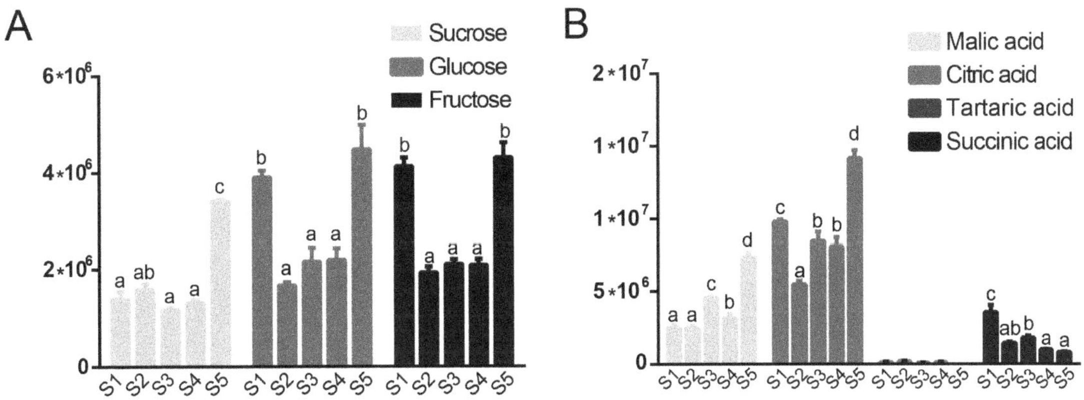

Figure 2. Sugar (**A**) and organic acid (**B**) content in five developmental stages of *Rosa roxburghii* fruit. a, b, c and d indicate significant difference (LSD, $p < 0.05$). Tartaric acid content was negligible compared with other acids. Therefore, we did not determine statistical significance.

2.3. Metabolic Pathways of Sugars and Organic Acids in Rosa roxburghii

The sugar and organic acid content of fruit are the two most important factors impacting its flavor. Having already identified the primary sugars and organic acids in *Rosa roxburghii* fruit, we sought to construct their metabolic pathways (Figure 3). Based on previous studies, we identified a total of 29 genes from the *Rosa roxburghii* genome that are likely involved in sugar metabolism, including members of the SUS, A N INVC, HXK, FRK, AGPase, AMY and BMY gene families (Table S3). SUS genes are critical for sucrose synthesis in plants, and evm.TU.chr4.1713 SUS3 expression was highest in the second fruit stage of *Rosa roxburghii*. However, another SUS gene (evm.TU.chr1.6623) was only highly expressed in S1 and S2. The A-N-INVG gene family members evm.TU.chr3.3618, evm.TU.chr5.1377 and evm.TU.chr6.3991 were highly expressed at the fruit ripening stage, and evm.TU.chr4.5684 and evm.TU.chr7.916 were highly expressed at the early stages of fruit development. In addition, members of the AGPase gene family were downregulated over the course of fruit development, reducing the conversion of sugar to starch. However, members of the AMY gene family were all upregulated during development, enhancing the degradation of starch to glucose. These findings indicate that sugar metabolism is regulated by multiple genes working together during *Rosa roxburghii* fruit development.

We identified 25 genes from seven gene families involved in organic acid metabolism in the *Rosa roxburghii* genome (Table S3). Similar to gene families related to sugar metabolism, gene families involved in organic acid metabolism also exhibited stage-specific expression profiles for different gene classes, including PK, ICDH and ME gene families. Most of these members were highly expressed at later stages of fruit development. However, the expression of MDH and ALMT gene family members was downregulated in the later stages of fruit development, whereas the expression of CS and PECP gene family members was upregulated at this same point. Analysis of these differences revealed a dynamic balance between citric acid and malic acid synthesis and metabolism in *Rosa roxburghii* fruits, with citric acid synthesis being induced later in development.

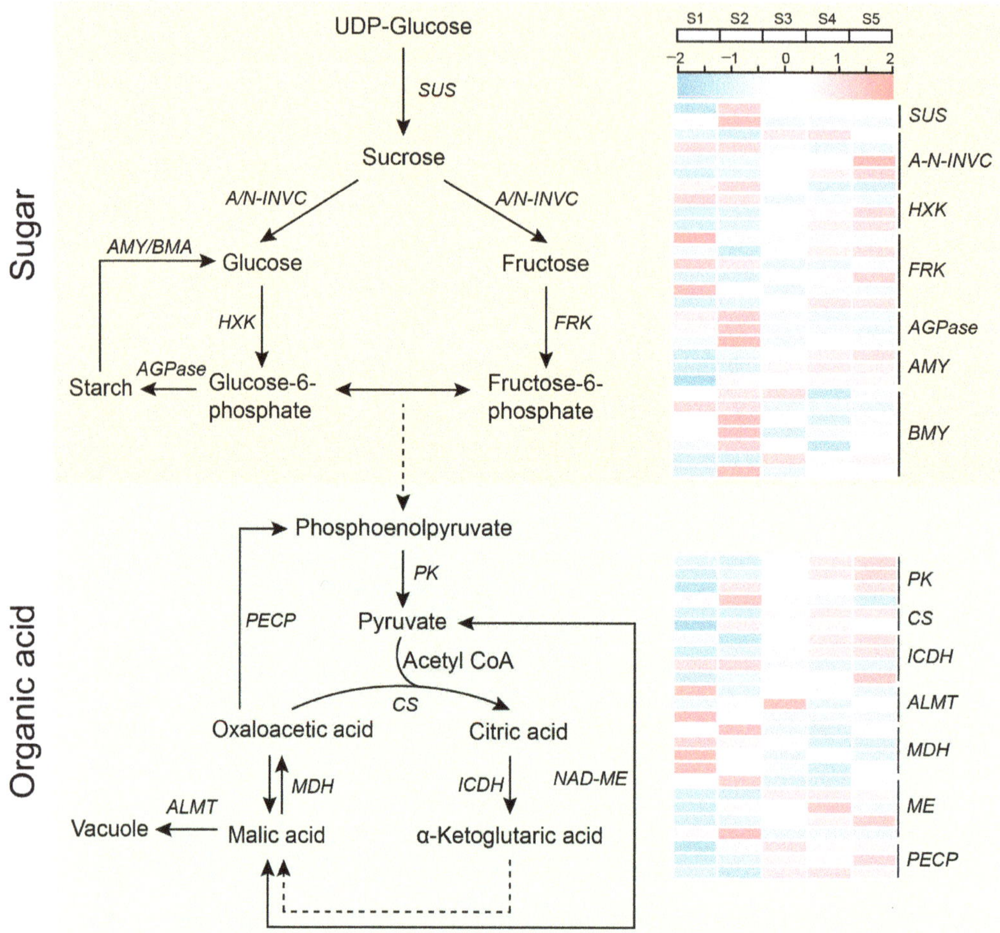

Figure 3. Sugar and organic acid biosynthetic pathways. SUS: sucrose synthase; A-N-INVC: invertase; HXK: hexokinase; FRK: fructokinase; AGPase: ADP-glucose pyrophosphorylase; AMY: α-amylase; BMY: β-amylase; PK: pyruvate kinase; CS: citrate synthase; ICDH: isocitrate dehydrogenase; ALMT: AL-activated malate transporter; MDH: malate dehydrogenase; ME: malic enzyme; PECP: phosphoenolpyruvate carboxylase.

To identify the most important genes involved in sugar and organic acid anabolism in *Rosa roxburghii* fruits, we analyzed the expression profiles of *Rosa roxburghii* sugar and acid metabolic pathway gene families at different developmental stages. This analysis indicated that SUS, A-N-INVG and AMY gene family members play key roles in sugar metabolism. The expression levels of evm.TU.chr4.1713 (*SUS3*), evm.TU.chr6.3991 (*A-N-INVC*), evm.TU.chr4.5028 (*AMY1*) and evm.TU.chr3.4947 (*AMY3*) consistently increased during fruit development, with 1.58, 2.33, 7.08 and 2.45-fold increases compared with the S1 stage. There were also 4.68 and 8.61-fold increases in the organic acid metabolic genes *CS2* (evm.TU.chr1.2040) and *ME1* (evm.TU.chr1.7029), respectively. We also performed qPCR analysis on the aforementioned genes, and the results were comparable to those of RNA-seq analysis. Additionally, all surveyed genes, except *AMY3*, were significantly expressed at the S5 stage (Figure 4). These results suggest that these genes have important roles in the synthesis of sugars and organic acids in *Rosa roxburghii* fruit.

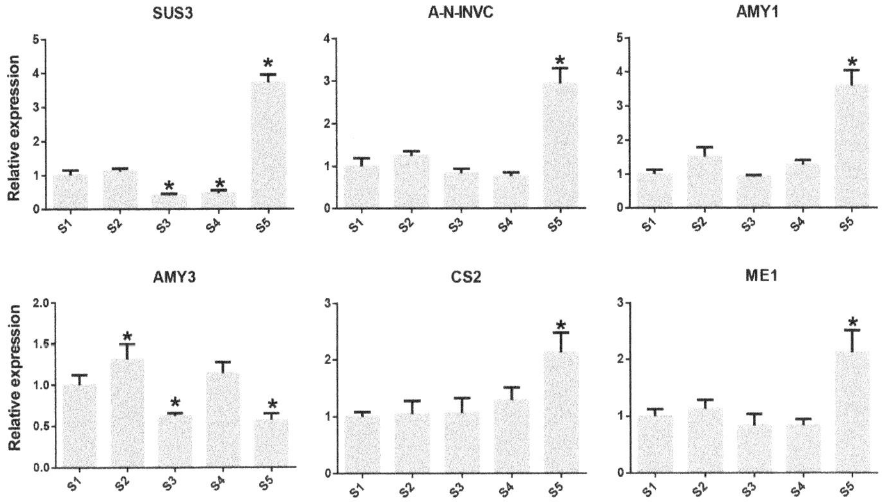

Figure 4. Relative expression of six genes involved in sugar and organic acid metabolism during *Rosa roxburghii* fruit development. Asterisk indicates significant difference (LSD, $p < 0.05$).

2.4. Co-Expression Network Construction and Transcription Factor Screening

To investigate the key transcription factors involved in regulating sugar and organic acid synthesis in *Rosa roxburghii*, we first identified a total of 2496 high confidence transcription factors (Table S4). Additionally, WGCNA was used to construct a network of key transcription factors using sugar and organic acid contents in *Rosa roxburghii* fruits as the phenotypic data. As shown in Figure 5A, all transcription factors were divided into nine modules. The brown module was significantly positively correlated with sucrose, citric acid and malic acid, with r2 values of 0.77, 0.86 and 0.74, respectively. Interestingly, no modules were significantly correlated with glucose or fructose. We also found that the yellow, green and blue modules were significantly negatively correlated with one or more different sugars or organic acids. These results suggest that transcription factors are closely related to the metabolism of *Rosa roxburghii* sugars and organic acids.

There was a total of 132 transcription factors included in the brown module (Table S5). After filtering to remove low-abundance transcripts (FPKM value > 10), 78 genes were retained and used to construct a network (Figure 5B). Within this network, a total of 59 transcription factors were upregulated over the course of *Rosa roxburghii* fruit development. These 59 transcription factors included NAC (11), WRKY (4), trihelix (4), bZIP (4), ERF (3) and others (Figure 5C). The gene with the largest fold difference in expression between S1 and S5 stages was evm.TU.chr6.2120 (LBD4, 202.87-fold), followed by evm.TU.chr3.5029 (HD-zip, 15.45-fold), evm.TU.chr1.1921 (NAC42, 9.69-fold) and evm.TU.chr1.5472 (ZIP6, 5.04-fold) (Table S6). We selected 14 genes (fold change $\geq$2) for qPCR analysis, including RrANL2, RrARF17, RrBEH4, RrbZIP6, RrGBF3, RrGRAS2, RrHD-zip17, RrHY5, RrLBD4, RrLZF1, RrNAC42, RrPIF8, RrWRKY3 and RrWRKY6. This analysis revealed expression changes that were similar to the RNA-seq data for all transcription factors, except PIF8, with detectible gene expression for all genes at S5 (Figure 6). Taken together, these results indicate that a large number of different transcription factors may be involved in regulating sugar and organic acid metabolism of *Rosa roxburghii* fruit.

2.5. Identification of Transcription Factors Involved in the Metabolism of Rosa roxburghii Sugars and Organic Acids

RrSUS3, *RrME1* and *RrCS2* are key genes regulating sucrose, malic acid and citric acid synthesis in *Rosa roxburghii* fruit. We therefore investigated the interactions between the above 14 transcription factors and the promoters of these three genes. Yeast one-hybrid experiments revealed that no transcription factors bound to the promoters of evm.TU.chr1.7029 (*RrME1*) and evm.TU.chr1.2040 (*RrCS2*), whereas *RrANL2* and *RrGBF3* bound to the promoter of evm.TU.chr4.1713 (*RrSUS3*) (Figure 7A).

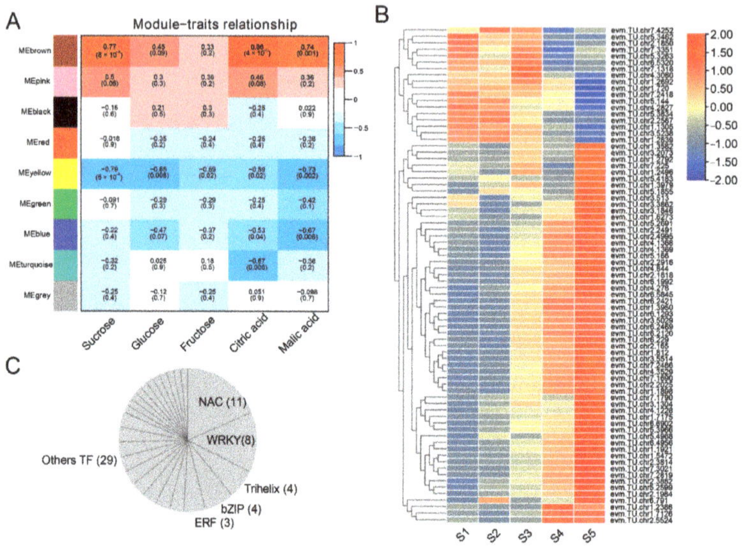

Figure 5. Weighted gene co-expression network analysis (WGCNA) of transcription factors. (**A**) Relationships of gene modules to sugars and organic acids. (**B**) Expression profiles of transcription factors with FPKM values > 10 in the brown module. (**C**) Classification of transcription factors that were upregulated during *Rosa roxburghii* fruit development.

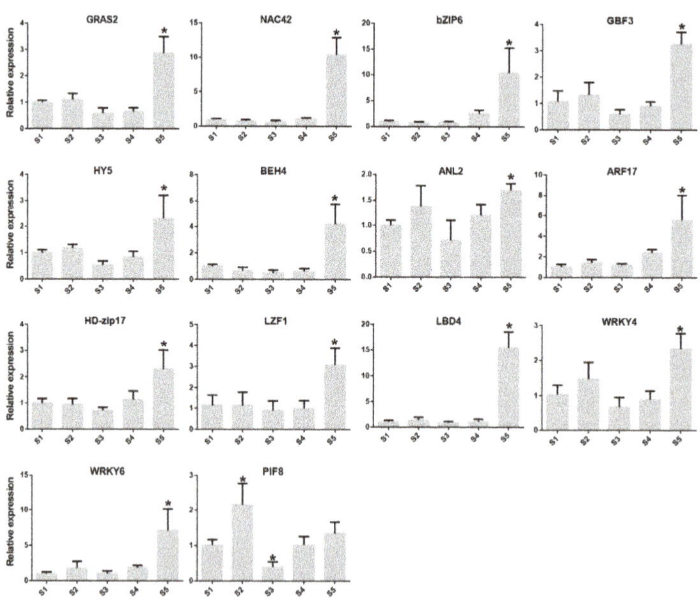

Figure 6. Relative expression of 14 transcription factors involved in sugar and organic acid metabolism. Asterisk indicates significant difference (LSD, $p < 0.05$).

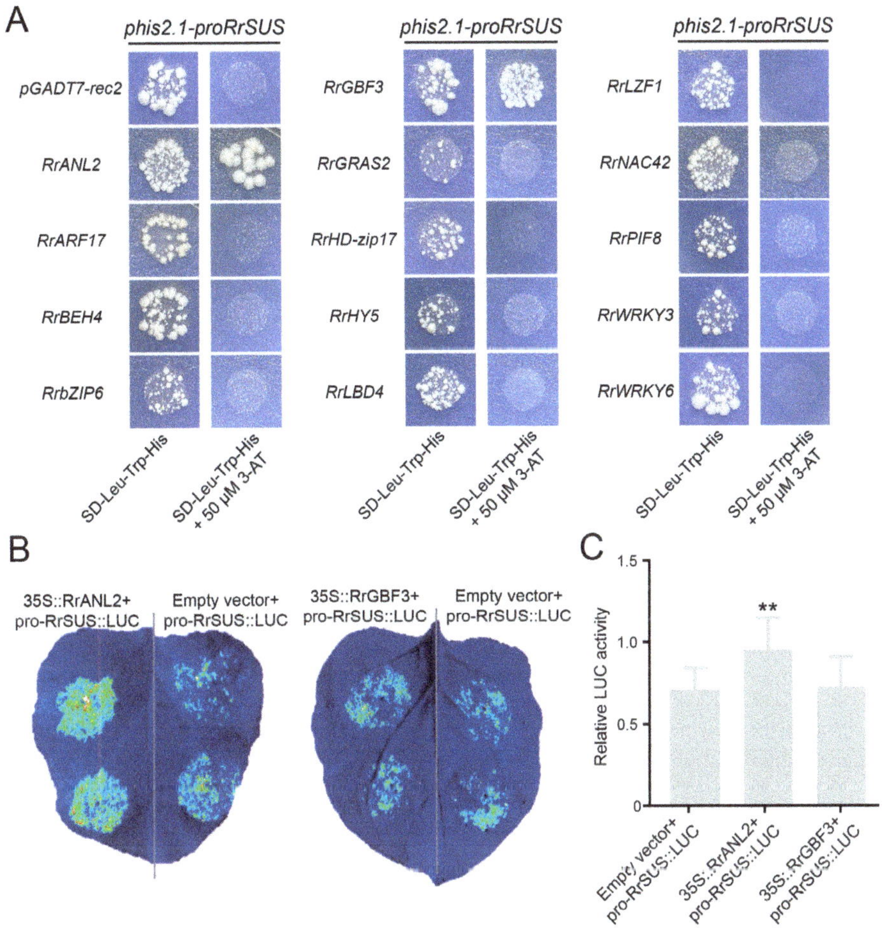

Figure 7. Validation of transcription factor activity. (**A**) Validation of the interaction of 14 transcription factors with the RrSUS promoter. The 3-AT concentration was 50 µM. (**B**,**C**) Transient expression in tobacco leaves to verify the LUC activity of transcription factors. Asterisk indicates significant difference (LSD, $p < 0.01$).

To further understand the interactions between transcription factors (*RrANL2* and *RrGBF3*) and *RrSUS*, we performed dual luciferase assays. This analysis showed that *RrANL2* could enhance the fluorescence signal of pro-RrSUS::LUC, whereas *RrGBF3* had no effect on the fluorescence signal of pro-RrSUS::LUC (Figure 7B,C). These results suggest that *RrANL2* impacts sucrose accumulation in *Rosa roxburghii* fruit by regulating the expression of *RrSUS3*.

3. Discussion

Fruit development is impacted by various environmental factors, genes and hormones, and the resulting sugars and organic acids produced during this process have a significant impact on fruit quality. In this study, transcriptomic and metabolomic data were used to construct the first networks that control sugars and organic acids in *Rosa roxburghii* fruit.

3.1. Accumulation Patterns of Sugars and Organic Acids in *Rosa roxburghii*

Organic acids usually accumulate significantly in the early stages of fruit development but decrease as development progresses. These changes are thought to be largely due to the increased synthesis of sugars during the later stages of fruit development [17]. In this study, we found that the main organic acids in *Rosa roxburghii* fruit were malic acid and citric acid, whose levels increased continuously during fruit development and peaked at fruit ripening. This was consistent with the accumulation pattern of organic acids in apples, peaches, plums, loquats and watermelons [18–20]. In addition, we found that sucrose, glucose and fructose all increased during *Rosa roxburghii* fruit ripening, but each had a distinct accumulation pattern. Glucose and fructose contents were found to be very high at the beginning of fruit development, but their content suddenly decreased midway through development, before accumulating again at fruit ripening. This pattern of changes was different from other more well-studied fruits, such as jujube, pear, watermelon and others, making it important to better understand the regulatory networks underlying changes in sugars and organic acids during *Rosa roxburghii* fruit development [19,21,22].

3.2. Regulatory Network of *Rosa roxburghii* Sugars and Organic Acids

The synthesis and catabolism of sugars and organic acids have significant impacts on the resulting fruit quality, and they are often regulated by multiple metabolic pathways. Previous studies have shown that the expression of a single gene does not have a significant effect on the overall sugar and organic acid accumulation in fruit. For example, studies have shown that perturbations in the sugar transporter protein *CiTST2* in watermelon [5], the sugar transporter protein PpTST1 in peach [21], the sugar transporter protein *CsPH8* in citrus fruit [23] and the ALMT gene family [13,24,25] often have minor effects on total sugar and organic acid content. We therefore constructed the metabolic pathways of sugar and organic acids in *Rosa roxburghii* and identified several genes that were significantly expressed at fruit ripening. Analysis of these pathways showed that genes involved in sugar and organic acid metabolism had highly similar expression trends, suggesting that the sugars and organic acids are regulated by complex networks.

Numerous studies have shown that transcription factors play important roles in the accumulation of sugars and organic acids in fruits. For example, the strawberry NAC transcription factor *FaRIF* promotes sugar accumulation in fruit [26], the grape transcription factor *WRKY22* regulates sucrose non-fermenting-1-related protein kinase 1 to promote sugar accumulation [27], and the transcription factor *AREB2* in apple acts on sugar transporter proteins and amylases to promote sugar accumulation [28]. Additionally, tomato *SlAREB1* promotes the accumulation of organic acids [29], strawberry *FaMYB5* participates in citric acid metabolism [30] and *CitERF13* regulates *CitVHA* to alter citric acid accumulation in citrus fruit [23]. We therefore further divided all transcription factors in the *Rosa roxburghii* genome into 10 modules by WGCNA, and found that the brown module, which contained 59 significantly expressed transcription factors, was significantly associated with sucrose, citric acid and malic acid.

3.3. Transcription Factor RrANL2 Regulates the Involvement of RrSUS3 in Sucrose Metabolism

We analyzed the expression of 14 transcription factors during *Rosa roxburghii* fruit development and found that most were consistent with the expression patterns of genes related to sucrose, malic acid and citric acid synthesis. Furthermore, expression of these transcription factors was consistent with changes in metabolite content, indicating the importance of transcription factors in sugar and organic acid metabolism in *Rosa roxburghii*. Subsequent yeast one-hybrid assays and dual luciferase assays showed that the transcription factor *RrANL2* can bind to the promoter of RrSUS3 to promote its expression, impacting sucrose levels and potentially affecting yield, fiber quality, fruit quality and growth rate [31–35]. Additionally, *ANL2* (ANTHOCYANINLESS2) belongs to the HD-ZIP IV subfamily [35] and functions in the development of plant root epidermal cells and the accumulation of anthocyanin [36–38]. *ANL2* has been shown to be involved in sucrose metabolism in plants, where it impacts the level of the sucrose transporter protein *AtSUC1* in Arabidopsis thaliana [39]. Additionally, sucrose acts as a molecular signal to regulate anthocyanin accumulation [38]. In this study, we found that *RrANL2* was an upstream regulator of *RrSUS3*, which provides new insights into the regulation of sugar metabolism in *Rosa roxburghii* fruit.

In this study, we integrated transcriptomic and metabolomic techniques to investigate the accumulation patterns of sugars and organic acids during *Rosa roxburghii* fruit development. During this process, genes associated with sugar and organic acid pathways were differentially expressed, with biosynthetic genes primarily increasing in expression during fruit development. Additionally, numerous transcription factors that may control sugar and organic acid metabolism in *Rosa roxburghii* were identified by WGCNA. Moreover, the transcription factor *RrANL2* was found to regulate the expression of RrSUS3, a key gene in sucrose synthesis. Taken together, these results partially elucidate the metabolism of sugars and organic acids in *Rosa roxburghii* and provide a biological basis for understanding their accumulation in *Rosa roxburghii* fruit.

4. Materials and Methods

4.1. Plant Materials

The *Rosa roxburghii* material used in this study was collected from the Baima Experimental Base of Nanjing Agricultural University, Jiangsu Province, China. Three biological replicates were collected for each sample 10, 40, 70, 100 and 130 days after flowering and named S1, S2, S3, S4 and S5. Samples were frozen in liquid nitrogen after collection and stored in a −80 °C freezer for later use.

4.2. Extraction and Analysis of Inclusions

Inclusions of samples from the five fruit stages were dissolved in 70% methanol and left overnight at 4 °C. After centrifugation at 10,000 rpm for 10 min, the supernatant was analyzed using a UPLC-ESI-MS/MS system (UPLC–SHIMADZU Nexera X2; MS–Applied Biosystems 4500 Q TRAP). Metabolites were considered differentially accumulated when they had an absolute Log2FC (fold change) ≥ 1 with a VIP ≥ 1. Metabolites were annotated and examined for enrichment using the KEGG compound database (http://www.kegg.jp/kegg/compound/, accessed on 11 February 2022) and KEGG pathway database (http://www.kegg.jp/kegg/pathway.html, accessed on 11 February 2022).

4.3. RNA-Seq Library Construction and Data Analysis

Total RNA was extracted from five stages of fruit using an RNAprep Pure Plant Plus Kit (Tiangen, Beijing, China). RNA concentration and purity were measured using a NanoDrop 2000 spectrophotometer (Thermo Fisher Scientific, Wilmington, DE, USA), and RNA integrity was assessed using an RNA Nano 6000 Assay Kit from the Agilent Bioanalyzer 2100 system (Agilent Technologies, Santa Clara, CA, USA). After quality assessment, 1 µg of RNA was utilized for RNA-seq library construction. Paired-end RNA-seq reads were generated using the Illumina NovaSeq 6000 platform. Raw RNA-seq data were then filtered with fastp [40] and mapped against the *Rosa roxburghii* genome by HISAT2 with default parameters [41]. We calculated the fragments per kilobase of transcript per million fragments mapped (FPKM) metric from featureCounts for each gene and identified differentially expressed genes with the DEseq2 R package, using a cutoff of adjusted p-value < 0.05. We then performed KEGG enrichment analysis of differentially expressed genes with clusterProfiler.

4.4. Identification of Key Transcription Factors by WGCNA

To understand the effects of transcription factors on sugars and organic acids during *Rosa roxburghii* fruit development, we identified transcription factors in the *Rosa roxburghii* genome by BLASTP [42] using all transcription factors of Arabidopsis thaliana as queries. The identified transcription factors were further analyzed via hmmscan to determine their conserved structural domains [43]. Additionally, we used WGCNA to construct a gene regulatory network for sugars and organic acids during *Rosa roxburghii* fruit development [44]. Using glucose, fructose, sucrose, malic acid and citric acid as phenotypic data, we normalized and iteratively examined expression data for all transcription factors.

4.5. Total RNA Extraction and qRT-PCR Analysis

Total RNA was extracted from five stages of *Rosa roxburghii* fruit using a Foregene Plant Total RNA Extraction kit (Chengdu, China). RNA was subsequently reverse-transcribed into cDNA by the PrimeScript RT Reagent Kit (TaKaRa, Beijing, China). The relative expression patterns of target genes during *Rosa roxburghii* fruit development were analyzed with the Hieff UNICON Universal Blue qPCR SYBR Green Master Mix (YEASEN, Shanghai, China). We utilized the expression of *Rosa roxburghii* RrUBQ as a reference gene for calculating relative expression by the $2^{-\Delta Ct}$ method and determined statistical significance using SPSS22. All primers used in this analysis are listed in Table S1.

4.6. Yeast One-Hybrid and Dual Luciferase Assays

We obtained total RNA and genomic DNA from *Rosa roxburghii* fruit using the Foregene Total Plant RNA Extraction Kit and Foregene Plant DNA Extraction Kit, respectively. cDNA was obtained by reverse transcription of RNA using the PrimeScript RT Reagent Kit (TaKaRa). The transcription factor sequences were amplified by PrimeSTAR HS DNA Polymerase (TaKaRa) and ligated into pGADT7 and PJX001 vectors. Additionally, the promoter sequence of each target gene was amplified from *Rosa roxburghii* genomic DNA and ligated into pHIS2.1 and pGreenII 0800-LUC vectors. All primers used in this analysis are listed in Table S1.

Yeast one-hybrid assays were used to determine the binding relationships between transcription factors and promoters. The pHIS2.1 vector carrying the transcription factor pGADT7 vector and the pHIS2.1 vector carrying the promoter sequence were transformed into the Y187 yeast strain by a modified lithium acetate method. The yeast strain was subsequently plated on SD-Leu-Trp solid agar medium (Coolaber, BeiJing, China) and incubated at 30 °C. After 3 days, single colonies of yeast were picked, suspended in ddH_2O and their concentration was adjusted to an OD600 of 0.05. Next, 5 μL of the yeast strain suspension was aspirated and transferred to SD/-Leu/-Trp/-His liquid medium (Coolaber, BeiJing, China) containing different concentrations of 3-AT (0, 25, 50, 75 and 100 μM) and grown for 4 days at 30 °C.

The regulatory relationship between transcription factors and promoters was further verified using dual luciferase assays. PJX001 vectors carrying transcription factors and pGreenII 0800-LUC vectors carrying promoter sequences were transformed into the Agrobacterium tumefaciens GV3101 receptor. The different combinations of transcription factors and promoters were then injected into tobacco leaves in a 9:1 ratio. After 3 days, tobacco leaves were assessed for LUC activity to determine the interactions between different promoters and transcription factors.

Supplementary Materials: The following supporting information can be downloaded at: https://www.mdpi.com/article/10.3390/plants12173036/s1, Table S1. Primers used in this study; Table S2. Sugar and organic acid contents of *Rosa roxburghii* during fruit ripening; Table S3. FPKM values of genes related to sugar and organic acid metabolism; Table S4. Transcription factor identification of *Rosa roxburghii*; Table S5. The FPKM value of transcription factor in MEbrown module; Table S6. Transcription factors expression fold change of S5 vs. S1.

Author Contributions: Z.C. and L.S. designed the research plan. T.Z., M.W. and Y.Z. performed the experiments. L.S. performed the data analyses and wrote the manuscript. All authors have read and agreed to the published version of the manuscript.

Funding: This research was supported in part by the Priority Academic Program Development of Jiangsu Higher Education Institutions and the State Key Laboratory of Crop Genetics and Germplasm

Enhancement (ZW201813). This study was supported by the high-performance computing platform of the Bioinformatics Center, Nanjing Agricultural University.

Data Availability Statement: All sequencing data, including that of the genome and transcriptome, can be found in the China National Center for Bioinformation database (PRJCA014460).

Conflicts of Interest: The authors declare that they have no conflict of interest.

References

1. Ackermann, J.; Fischer, M.; Amado, R. Changes in sugars, acids, and amino acids during ripening and storage of apples (cv. Glockenapfel). *J. Agric. Food Chem.* **1992**, *40*, 1131–1134. [CrossRef]
2. Kroger, M.; Meister, K.; Kava, R. Low-calorie Sweeteners and Other Sugar Substitutes: A Review of the Safety Issues. *Compr. Rev. Food Saf.* **2006**, *5*, 35–47. [CrossRef]
3. Pangborn, R. Relative Taste Intensities of Selected Sugars and Organic Acids. *J. Food Sci.* **2006**, *28*, 726–733. [CrossRef]
4. Shiraishi, M.; Fujishima, H.; Chijiwa, H. Evaluation of table grape genetic resources for sugar, organic acid, and amino acid composition of berries. *Euphytica* **2010**, *174*, 1–13. [CrossRef]
5. Ren, Y.; Guo, S.; Zhang, J.; He, H.; Sun, H.; Tian, S.; Gong, G.; Zhang, H.; Levi, A.; Tadmor, Y.; et al. A Tonoplast Sugar Transporter Underlies a Sugar Accumulation QTL in Watermelon. *Plant Physiol.* **2018**, *176*, 836–850. [CrossRef]
6. Lingle, S.E.; Dunlap, J.R. Sucrose Metabolism in Netted Muskmelon Fruit during Development. *Plant Physiol.* **1987**, *84*, 386–389. [CrossRef]
7. Wu, J.; Gao, H.; Zhao, L.; Liao, X.; Chen, F.; Wang, Z.; Hu, X. Chemical compositional characterization of some apple cultivars. *Food Chem.* **2007**, *103*, 88–93. [CrossRef]
8. Zhang, H.P.; Wu, J.Y.; Qin, G.H.; Yao, G.F.; Qi, K.J.; Wang, L.F.; Zhang, S.L. The role of sucrose-metabolizing enzymes in pear fruit that differ in sucrose accumulation. *Acta Physiol. Plant.* **2014**, *36*, 71–77. [CrossRef]
9. Ruan, Y. Sucrose Metabolism: Gateway to Diverse Carbon Use and Sugar Signaling. *Annu. Rev. Plant Biol.* **2014**, *65*, 33–67. [CrossRef]
10. Rolland, F.; Baena-Gonzalez, E.; Sheen, J. Sugar Sensing and Signaling in Plants: Conserved and Novel Mechanisms. *Annu. Rev. Plant Biol.* **2006**, *57*, 675–709. [CrossRef] [PubMed]
11. Montero, T.M.; Mollá, E.M.; Esteban, R.M.; López-Andréu, F.J. Quality attributes of strawberry during ripening. *Sci. Hortic.-Amst.* **1996**, *65*, 239–250. [CrossRef]
12. Giné Bordonaba, J.; Terry, L.A. Manipulating the taste-related composition of strawberry fruits (Fragaria × ananassa) from different cultivars using deficit irrigation. *Food Chem.* **2010**, *122*, 1020–1026. [CrossRef]
13. Etienne, A.; Génard, M.; Lobit, P.; Mbeguié-A-Mbéguié, D.; Bugaud, C. What controls fleshy fruit acidity? A review of malate and citrate accumulation in fruit cells. *J. Exp. Bot.* **2013**, *64*, 1451–1469. [CrossRef] [PubMed]
14. Zhang, Y.; Fernie, A.R. On the role of the tricarboxylic acid cycle in plant productivity. *J. Integr. Plant Biol.* **2018**, *60*, 1199–1216. [CrossRef]
15. Li, H.; Fang, W.; Wang, Z.; Chen, Y. Physicochemical, biological properties, and flavour profile of Rosa roxburghii Tratt, Pyracantha fortuneana, and Rosa laevigata Michx fruits: A comprehensive review. *Food Chem.* **2022**, *366*, 130509. [CrossRef]
16. Wang, L.; Lv, M.; An, J.; Fan, X.; Dong, M.; Zhang, S.; Wang, J.; Wang, Y.; Cai, Z.; Fu, Y. Botanical characteristics, phytochemistry and related biological activities of Rosa roxburghii Tratt fruit, and its potential use in functional foods: A review. *Food Funct.* **2021**, *12*, 1432–1451. [CrossRef]
17. Lombardo, V.A.; Osorio, S.; Borsani, J.; Lauxmann, M.A.; Bustamante, C.A.; Budde, C.O.; Andreo, C.S.; Lara, M.V.; Fernie, A.R.; Drincovich, M.F. Metabolic Profiling during Peach Fruit Development and Ripening Reveals the Metabolic Networks That Underpin Each Developmental Stage. *Plant Physiol.* **2011**, *157*, 1696–1710. [CrossRef]
18. Famiani, F.; Battistelli, A.; Moscatello, S.; Cruzcastillo, J.G.; Walker, R.P. The organic acids that are accumulated in the flesh of fruits: Occurrence, metabolism and factors affecting their contents—A review. *Rev. Chapingo Ser. Hortic.* **2015**, *21*, 97–128. [CrossRef]
19. Umer, M.J.; Bin Safdar, L.; Gebremeskel, H.; Zhao, S.; Yuan, P.; Zhu, H.; Kaseb, M.O.; Anees, M.; Lu, X.; He, N.; et al. Identification of key gene networks controlling organic acid and sugar metabolism during watermelon fruit development by integrating metabolic phenotypes and gene expression profiles. *Hortic. Res.* **2020**, *7*, 193. [CrossRef]
20. Yao, Y.X.; Ming, L.; Liu, Z.; You, C.X.; Wang, D.M.; Zhai, H.; Hao, Y.J. Molecular cloning of three malic acid related genes MdPEPC, MdVHA-A, MdcyME and their expression analysis in apple fruits. *Sci. Hortic.* **2009**, *122*, 404–408. [CrossRef]
21. Peng, Q.; Wang, L.; Ogutu, C.; Liu, J.; Liu, L.; Mollah, M.; Han, Y. Functional Analysis Reveals the Regulatory Role of PpTST1 Encoding Tonoplast Sugar Transporter in Sugar Accumulation of Peach Fruit. *Int. J. Mol. Sci.* **2020**, *21*, 1112. [CrossRef]
22. Zhang, C.; Bian, Y.; Hou, S.; Li, X. Sugar transport played a more important role than sugar biosynthesis in fruit sugar accumulation during Chinese jujube domestication. *Planta* **2018**, *248*, 1187–1199. [CrossRef] [PubMed]
23. Li, S.; Yin, X.; Xie, X.; Allan, A.C.; Ge, H.; Shen, S.; Chen, K. The Citrus transcription factor, CitERF13, regulates citric acid accumulation via a protein-protein interaction with the vacuolar proton pump, CitVHA-c4. *Sci. Rep.* **2016**, *6*, 20151. [CrossRef]
24. Angeli, A.; Baetz, U.; Francisco, R.; Zhang, J.; Chaves, M.M.; Regalado, A. The vacuolar channel VvALMT9 mediates malate and tartrate accumulation in berries of Vitis vinifera. *Planta* **2013**, *238*, 283–291. [CrossRef]

25. Etienne, C.; Rothan, C.; Moing, A.; Plomion, C.; Bodénès, C.; Svanella-Dumas, L.; Cosson, P.; Pronier, V.; Monet, R.; Dirlewanger, E. Candidate genes and QTLs for sugar and organic acid content in peach [*Prunus persica* (L.) Batsch]. *Theor. Appl. Genet.* **2002**, *105*, 145–159. [CrossRef]
26. Martín-Pizarro, C.; Vallarino, J.G.; Osorio, S.; Meco, V.; Urrutia, M.; Pillet, J.; Casañal, A.; Merchante, C.; Amaya, I.; Willmitzer, L.; et al. The NAC transcription factor FaRIF controls fruit ripening in strawberry. *Plant Cell* **2021**, *33*, 1574–1593. [CrossRef]
27. Huang, T.; Yu, D.; Wang, X. VvWRKY22 transcription factor interacts with VvSnRK1.1/VvSnRK1.2 and regulates sugar accumulation in grape. *Biochem. Bioph. Res. Commun.* **2021**, *554*, 193–198. [CrossRef]
28. Ma, Q.J.; Sun, M.H.; Lu, J.; Liu, Y.J.; Hu, D.G.; Hao, Y.J. Transcription Factor AREB2 Is Involved in Soluble Sugar Accumulation by Activating Sugar Transporter and Amylase Genes. *Plant Physiol.* **2017**, *174*, 2348–2362. [CrossRef]
29. Bastías, A.; López-Climent, M.; Valcárcel, M.; Rosello, S.; Gómez-Cadenas, A.; Casaretto, J.A. Modulation of organic acids and sugar content in tomato fruits by an abscisic acid-regulated transcription factor. *Physiol. Plant.* **2011**, *141*, 215–226. [CrossRef] [PubMed]
30. Liu, Y.; Zhu, L.; Yang, M.; Xie, X.; Sun, P.; Fang, C.; Zhao, J. R2R3-MYB transcription factor FaMYB5 is involved in citric acid metabolism in strawberry fruits. *J. Plant Physiol.* **2022**, *277*, 153789. [CrossRef] [PubMed]
31. Barrero-Sicilia, C.; Hernando-Amado, S.; González-Melendi, P.; Carbonero, P. Structure, expression profile and subcellular localisation of four different sucrose synthase genes from barley. *Planta* **2011**, *234*, 391–403. [CrossRef] [PubMed]
32. Poovaiah, C.R.; Mazarei, M.; Decker, S.R.; Turner, G.B.; Sykes, R.W.; Davis, M.F.; Stewart, C.N. Transgenic switchgrass (Panicum virgatum L.) biomass is increased by overexpression of switchgrass sucrose synthase (PvSUS1). *Biotechnol. J.* **2015**, *10*, 552–563. [CrossRef] [PubMed]
33. Qazi, H.A.; Paranjpe, S.; Bhargava, S. Stem sugar accumulation in sweet sorghum—Activity and expression of sucrose metabolizing enzymes and sucrose transporters. *J. Plant Physiol.* **2012**, *169*, 605–613. [CrossRef] [PubMed]
34. Ruan, Y. Signaling Role of Sucrose Metabolism in Development. *Mol. Plant* **2012**, *5*, 763–765. [CrossRef]
35. Nakamura, M.; Katsumata, H.; Abe, M.; Yabe, N.; Komeda, Y.; Yamamoto, K.T.; Takahashi, T. Characterization of the Class IV Homeodomain-Leucine Zipper Gene Family in Arabidopsis. *Plant Physiol.* **2006**, *141*, 1363–1375. [CrossRef]
36. Kubo, H.; Hayashi, K. Characterization of root cells of anl2 mutant in Arabidopsis thaliana. *Plant Sci.* **2011**, *180*, 679–685. [CrossRef] [PubMed]
37. Nadakuduti, S.S.; Pollard, M.; Kosma, D.K.; Allen, C., Jr.; Ohlrogge, J.B.; Barry, C.S. Pleiotropic Phenotypes of the sticky peel Mutant Provide New Insight into the Role of CUTIN DEFICIENT2 in Epidermal Cell Function in Tomato. *Plant Physiol.* **2012**, *159*, 945–960. [CrossRef]
38. Solfanelli, C.; Poggi, A.; Loreti, E.; Alpi, A.; Perata, P. Sucrose-Specific Induction of the Anthocyanin Biosynthetic Pathway in Arabidopsis. *Plant Physiol.* **2006**, *140*, 637–646. [CrossRef] [PubMed]
39. Sivitz, A.B.; Ward, R.J.M. Arabidopsis sucrose transporter AtSUC1 is important for pollen germination and sucrose-induced anthocyanin accumulation. *Plant Physiol.* **2008**, *147*, 92–100. [CrossRef] [PubMed]
40. Chen, S.; Zhou, Y.; Chen, Y.; Gu, J. fastp: An ultra-fast all-in-one FASTQ preprocessor. *Bioinformatics* **2018**, *34*, i884–i890. [CrossRef]
41. Kim, D.; Langmead, B.; Salzberg, S.L. HISAT: A fast spliced aligner with low memory requirements. *Nat. Methods* **2015**, *12*, 357–360. [CrossRef] [PubMed]
42. Altschul, S.F. Basic local alignment search tool (BLAST). *J. Mol. Biol.* **2012**, *215*, 403–410. [CrossRef] [PubMed]
43. Johnson, L.S.; Eddy, S.R.; Portugaly, E. Hidden Markov model speed heuristic and iterative HMM search procedure. *BMC Bioinform.* **2010**, *11*, 431. [CrossRef] [PubMed]
44. Langfelder, P.; Horvath, S. WGCNA: An R package for weighted correlation network analysis. *BMC Bioinform.* **2008**, *9*, 559. [CrossRef] [PubMed]

plants

MDPI

Article

Structural and Functional Analysis of the MADS-Box Genes Reveals Their Functions in Cold Stress Responses and Flower Development in Tea Plant (*Camellia sinensis*)

Juan Hu [1], Qianqian Chen [2], Atif Idrees [3], Wanjun Bi [1], Zhongxiong Lai [4,*] and Yun Sun [1,*]

1 Key Laboratory of Tea Science in Fujian Province, College of Horticulture, Fujian Agriculture and Forestry University, Fuzhou 350002, China; vicky232022@163.com (J.H.); biwanjun9701@126.com (W.B.)
2 College of Life Sciences, Fujian Agriculture and Forestry University, Fuzhou 350002, China; 3190515007@fafu.edu.cn
3 Guizhou Provincial Key Laboratory for Agricultural Pest Management of the Mountainous Region, Scientific Observing and Experimental Station of Crop Pest in Guiyang, Ministry of Agriculture and Rural Affairs, Institute of Entomology, Guizhou University, Guiyang 550025, China; atif_entomologist@yahoo.com
4 Institute of Horticultural Biotechnology, Fujian Agriculture and Forestry University, Fuzhou 350002, China
* Correspondence: laizx01@163.com (Z.L.); sunyun1125@126.com (Y.S.)

Abstract: MADS-box genes comprise a large family of transcription factors that play crucial roles in all aspects of plant growth and development. However, no detailed information on the evolutionary relationship and functional characterization of MADS-box genes is currently available for some representative lineages, such as the Camellia plant. In this study, 136 MADS-box genes were detected from a reference genome of the tea plant (*Camellia sinensis*) by employing a 569 bp HMM (Hidden Markov Model) developed using nucleotide sequencing including 73 type I and 63 type II genes. An additional twenty-seven genes were identified, with five MIKC-type genes. Truncated and/or inaccurate gene models were manually verified and curated to improve their functional characterization. Subsequently, phylogenetic relationships, chromosome locations, conserved motifs, gene structures, and gene expression profiles were systematically investigated. Tea plant MIKC genes were divided into all 14 major eudicot subfamilies, and no gene was found in Mβ. The expansion of MADS-box genes in the tea plant was mainly contributed by WGD/fragment and tandem duplications. The expression profiles of tea plant MADS-box genes in different tissues and seasons were analyzed, revealing widespread evolutionary conservation and genetic redundancy. The expression profiles linked to cold stress treatments suggested the wide involvement of MADS-box genes from the tea plant in response to low temperatures. Moreover, a floral 'ABCE' model was proposed in the tea plant and proved to be both conserved and ancient. Our analyses offer a detailed overview of MADS-box genes in the tea plant, allowing us to hypothesize the potential functions of unknown genes and providing a foundation for further functional characterizations.

Keywords: MADS-box; tea plant; cold stress; floral ABCE model; transcription factor

1. Introduction

The tea plant (*Camellia sinensis* (L.) O. Kuntze) belongs to the genus Camellia (Theaceae), which comprises more than 250 species [1,2]. The tea plant is a commercially important crop the leaves of which are used to produce tea for the world's three most widely consumed nonalcoholic beverages. Due to the highly repetitive, heterozygous, and relatively large (approximately 3G) genome of the tea plant, it is challenging to perform genome assembly on this crop. In recent years, with the revolution of next-generation sequencing technology and the efforts of researchers, the tea plant genome has developed from an initial draft genome (contig level) to the current reference genome (chromosome level) [3–6]. The increasing public availability of high-quality genome assembly and transcriptomes represent a great opportunity for elucidating the molecular basis underlying various biological processes and further accelerating tea plant research and breeding.

MADS-box is one of the most important transcription factor families and is widely distributed throughout plants. The initials of the four earliest discovered MADS-box genes constitute the most characteristic domain (MADS-domain) of this gene family, including *MINICHROMOSOME MAINTENANCE1* from *Saccharomyces cerevisiae* [7], *AGAMOUS* from *Arabidopsis thaliana* [8], *DEFICIENS* from *Antirrhinum majus* [9], and *SERUM RESPONSE FACTOR* from *Homo sapiens* [10]. Prior to the divergence of the plant and animal kingdoms, MADS-box genes were differentiated into two phylogenetic groups: type I (SRF-like or M-type) and type II (called MEF2-like or MIKC-type) [11,12]. Plant-specific MIKC-type genes encode a class of proteins comprising four characteristic domains (from N- to C-terminal): the MADS (M) domain, the intervening (I) domain, the keratin-like (K) domain, and the C-terminal (C) domain [13]. Type II MADS-box genes are long and contain five to eight exons, while type I genes have a relatively simple structure with no K domain and usually contain only one or two exons [14]. Type II genes are further divided into MIKC^C-type and MIKC*-type genes (also named Mδ in some studies) due to the variable intervening (I) domain [15]. MIKC^C proteins can be further subdivided into 14 distinct subfamilies based on their phylogenetic relationships [16,17]. Type I genes can be further divided into three subfamilies: Mα, Mβ, and Mγ [12,18].

MADS-box genes control all major aspects of life for land plants. The most famous MADS-box genes include four different classes of floral homeotic genes (A, B, C, E), which spell out the classic "ABCE" model of floral organ identity [19–21]. In *Arabidopsis*, Class A genes (*APETALA1* [*AP1*] and *APETALA2* [*AP2*]) alone determine sepal identity in whorl 1 [22,23]. Class A genes overlap with Class B genes *APETALA3* (*AP3*) and *PISTILLATA* (*PI*) to control petal development in whorl 2 [24,25]. Class B and Class C *AGAMOUS* (*AG*) genes together determine petals in whorl 3, and Class C genes alone specify carpel identity in whorl 4 [26]. Class E genes (*SEPALLATA1, 2, 3, 4* [*SEP1/2/3/4*; formerly named *AGL2/4/9/3*]) are involved in the specification of all four whorls of floral organs by redundantly interacting with genes of other classes to form heterotetramer protein complexes [27,28]. Generally speaking, floral organs in each of the four whorls are determined by combinations of MADS-box transcription factors with different homeotic functions, such as A-, B-, C-, and E-functions, to form homodimers and heterodimers and tetramers. With the exception of *AP2*, all ABCE homologous genes are MIKC^C-type genes. The ABCE model is very versatile for flowering plants; even the differentiated zygomorphic orchid flower can be primarily explained using a modified ABCE model [29,30].

In *Arabidopsis,* MADS-box transcription factors are also involved in regulating leaf morphogenesis; root, seed, and embryo development; and fruit ripening [31,32]. Several MIKC^C-type genes, such as *SHORT VEGETATIVE PHASE* (*SVP*)/*AGAMOUS-like* 24 (*AGL24*), *SUPPRESSOR OF OVEREXPRESSION OF CONSTANS1* (*SOC1*) and *FLOWERING LOCUS C* (*FLC*), are of great importance because they participate in flowering transition and flowering time by integrating various environmental signals [33,34]. Moreover, many studies on MADS-box genes of different subclades have revealed the functions of some plants in abiotic stress tolerance in different developmental processes [35]. For example, AP1, AP3, and AG homologs from tomatoes were shown to be induced by low temperatures and associated with the formation of abnormal flowers [36]. The rice *AGL12* ortholog, *OsMADS26*, acts as a regulator of stress-related responses such as drought and pathogen infections [37]. *LsMADS55 (AP1-like)* is responsible for heat-induced bolting and flowering in lettuce [38]. Several other MADS-box genes are implicated in plastic developmental processes (including vernalization, seed germination, seasonal growth cessation, and winter dormancy process) in response to seasonal changes caused by environmental conditions. For horticultural plants, these plant developmental processes and stress responses are strongly related to agronomic traits.

The prevalence and functional diversity of members of the MADS-box family have been comprehensively investigated in many angiosperms, including model plants *Arabidopsis*, *Oryza sativa*, and *Brachypodium disachyon*, as well as some commercially important crops (*Vitis vinifera*, *Triticum aestivum*, *Musa acuminata*, etc.) [39–41]. Although an overview of MADS-box genes in tea plants is available [42], there remains a lack of in-depth genome-wide identification and functional characterization for tea plant MADS-box genes in a well-resolved genome. In the present study, we identified 136 MADS-box gene members from the tea plant genome and transcriptome assembly by integrating several bioinformatics methods. Phylogenetic analysis, gene structures, conserved motifs, chromosomal locations, and syntenic relationships were also analyzed. Expression patterns of MADS-box genes linked to cold stress in different tissues during different seasons were further surveyed. Our results provide a detailed overview of the tea plant MADS-box gene family, which could facilitate future functional analyses of *CsMADS* genes in different biological processes.

2. Results

2.1. Identification of the MADS-Box Gene Family Based on Tea-Specific Nucleic Acid HMM and the Curation of Gene Models

A raw dataset of 132 non-redundant MADS-box genes was obtained through HMMER search employing downloaded HMM profiles as a query to search proteins containing the MADS-domain or K-box domain in the tea protein database (Table S1). In this dataset, forty-four genes were encoded for both the MADS-domain and K-box domain, seventy-nine genes for only the MADS-domain, and eight genes for only the K-box domain. In order to identify all genes that were missed in published annotations and lost their MADS-box domains or a large part of their sequences due to divergent evolution, a tea-specific nucleic acid HMM with a length of 569 bp was constructed based on the predicted mRNA nucleotide sequences of a well-aligned MADS-domain and K-box domain using hmmbuild from the HMMER v3 suite (Figure S1). The resulting HMM was adopted to probe the tea plant genome [3]. After removing sites with an e-value above 0.5, 169 genomic regions with candidate MADS domains were retained for subsequent manual verification and functional annotation. In total, 124 of 132 predicted proteins overlapped with candidate genomic regions, and the rest were excluded from posterior analyses, as it was further confirmed that only the k-box gene was preserved in these genes via screening extended genomic regions (Tables S2 and S3). For each nucleotide domain without counterparts of known genes, we performed repeated predictions with the Genewise software using genes of closely related species (grape or kiwifruit) or assembled transcripts of tea-reproductive organs. Of these, the proteins determined for twenty-seven genomic regions seemed to be functional, including five type II genes (three *AP1/FUL*-like genes, one *AGL15*, and one *FLC* gene) and twenty-two type I genes (Table S2).

In addition to mining novel gene loci, gene models of known genes were also evaluated and manually curated. A total of twenty-four known genes were truncated, among which sixteen genes only contained the MADS domain, and eight genes only contained the K-box domain. In four cases (*CSS0040885.1*, *CSS0001766.1*, *CSS0012111.1*, and *CSS0041561.1*), gene models were reconstructed based on transcriptome assembly. Four genes (*CSS0013866.1*, *CSS0009886.1*, *CSS0016920.1*, and *CSS0006612.1*) with only partial sequences were merged with nearby truncated genes (*CSS0033417.1*, *CSS0000825.1*, *CSS0032023.1*, and *CSS0043640.1*, respectively) to form a complete gene. Gene models of the 12 truncated genes could not be reconstructed, indicating that their functional MADS-box or K-box domains may have been lost during evolution (Supplementary Table S3). In two cases, candidates *CSS0032233.1* and *CSS0021126.1* allocated to non-chromosomal locations (Contig203 and Contig399) were discarded because their sequences were identical to those of *CSS0034367.1* and *CSS0012962.1* (Chr7 and Chr2), respectively. Ultimately, 136 full-length genes with high-confidence MADS domains were discovered in the tea plant genome, including 63 type II genes and 73 type I genes based on sequence similarity (Table S2).

2.2. Maximum Likelihood Phylogenetic Analysis of MADS-Box Genes

The maximum likelihood phylogeny of the type II MADS-box protein sequences of tea plant, grape, and *Arabidopsis* showed that 63 genes from the tea plant could be assigned into 14 major MIKC-type gene clades, with the members numbers described in square brackets: *AGAMOUS-LIKE6* (AGL6) [1], *SEPALLATA1* (*SEP1*) [3]/*SEP3* [2], *APETALA1* (*AP1*) [1]/*FUL* [6], *SUPPRESSOR OF OVEREXPRESSION OF CONSTANS1* (*SOC1*) [8], *B-sister*(BS) [1], *APETALA3* (*AP3*) [1]/*TOMATO MADS BOX GENE6*(TM6) [2]/*PISTILLATA* (*PI*) [2], *TM8* [1], *SHORT VEGETATIVE PHASE* (*SVP*) [4], *AGL12* [1], *AGL15* [2], *ANR1* [11], *AGAMOUS* (*AG*) [3]/*SEEDSTICK* (*AGL11*) [1], *FLOWERING LOCUS C* (*FLC*) [3], and *MIKC**(*M*δ) [10] (Figure 1). The *ANR1*, *MIKC**, and *SOC1* subfamilies had the largest numbers of *CsMADS* genes, while *AGL11*, *TM8*, and *AGL12* had the smallest number, with only one copy each. Except for the *TM8* and *TM6* subclades known from other flowering plants, all type II genes in the tea plant had orthologous genes from *Arabidopsis*. In addition, the numbers of AP3/TM6/PI, AP1/FUL1, SOC1, ANR1, and MIKC* subfamilies in the tea plant were much higher compared to their *Arabidopsis* and grape counterparts. Type I genes in the tea plant could be divided into two subclasses: Mα (39) and Mγ (34). No gene member was found in Mβ, which was similar to the results in grape and orchid [43,44]. Phylogenetic analysis of type I genes from grape, *Arabidopsis*, and the tea plant showed that many genes appeared to be unique to the tea plant, suggesting these genes may have undergone independent replication and evolution (Figure S2).

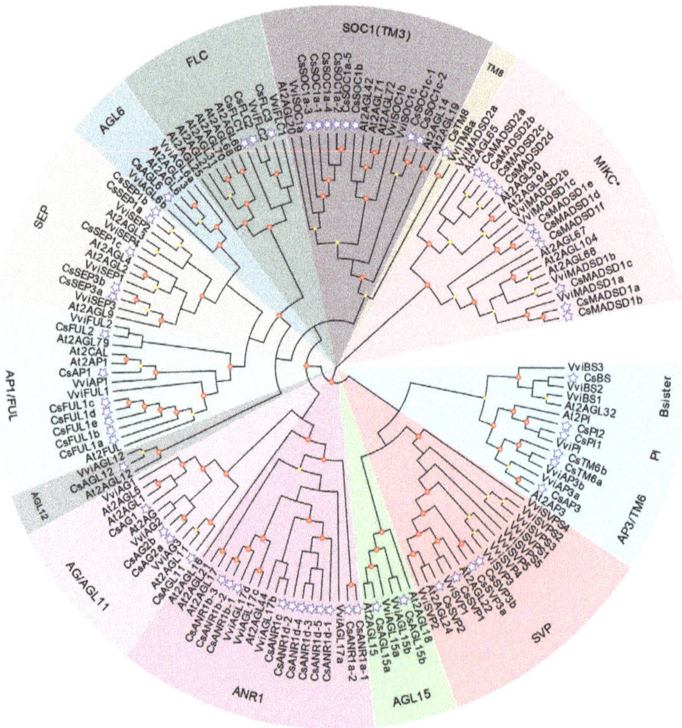

Figure 1. Phylogenetic relationship of type II MADS-box proteins. The maximum likelihood (ML) phylogenetic tree of MADS-box proteins from tea plant, *Arabidopsis*, and grape was constructed using FastTree with the GTR+CAT model. Tea plant genes are marked with blue stars. The subfamilies are presented in different colors. At, *A. thaliana*; Vvi, *Vitis vinifera*, Cs, *C. sinensis*.

2.3. Analysis of Genes Structure and Conserved Motifs

We analyzed the conserved motifs for all full-length MADS-box genes, as well as their exon-intron composition (Figure 2). Overall, the composition of motifs in the same subfamily was relatively conserved. Most of the MIKC^{C} genes shared six motifs (motifs 8, 1, 4, 6, 2, and 5), among which motif 1 represents the MADS-box domain. However, in one case of the AP3/TM6/PI subclade, *TM6-like* genes were characterized as having only three motifs (motifs 1, 2, and 8), indicating that their protein structures diverged from those of other B-class members during evolution. The distribution of specific conserved motifs varied among the different subfamilies. For example, all three members of the FLC subfamily contained no motif 6. Unlike MIKC^{C}, MIKC*-type genes contained fewer motifs (only three to four). Ten MIKC*-type genes with identifiable K domains could not be detected in pfam or prosite. Motif 3 and motif 9 were exclusive to Mα and Mγ, respectively. The structure of type II genes was longer and more complex than that of other genes due to the presence of multiple exons, while type I genes had only one or two exons. In contrast to the previous reports that nearly half of MIKC^{C} genes lacked introns in rice, *Arabidopsis*, and bamboo [45], homologs in the tea plant tended to retain more introns (about five to eight). Exceptions to this rule included two newly identified genes (*CsFUL1d* and *CsFUL1e*) and a known gene, *CsFUL1c*, whose intron numbers were extremely low, with only one intron each. Notably, we observed that a few genes (e.g., *CsFLC3*, *CsANR1a-1*, and *CsANR1a-2*) had extremely long introns (20 kb~60 kb). To date, such patterns have only been reported in a few species, such as bamboo and wheat [46,47].

Figure 2. Phylogenetic relationship, gene structure analysis, and motif compositions of MADS-box genes in the tea plant. (**A**) A total of 136 full-length sequences of CsMADS protein were used to build a neighbor-joining tree, with 1000 bootstrap replicates. (**B**) Conserved motif patterns were identified in tea plant MADS-box proteins using the MEME webserver. Different motifs are represented by different colors. (**C**) The exon–intron structures of *CsMADS*. Green blocks represent exons, and gray lines indicate introns. The size scale is indicated at the bottom.

2.4. Chromosomal Location and Gene Duplication Events of the MADS-Box Gene Family in Tea Plant

To explore the genomic distribution of the MADS-box gene family, we visualized the chromosomal positioning of the *CsMADS* genes according to genome annotation information (Figure 3). Tea MADS-box genes were unevenly spread across all chromosomes. Chr5 did not contain the MADS-box gene, while Chr2 and Chr11 contained the highest and lowest numbers of the gene, with twenty-seven and one, respectively. Owing to established gene nomenclature, type II genes from the same subfamily were clearly prone to duplicate within different chromosomes to realize full functionality. However, type I genes were often found to be evolutionarily closely related to each other, exhibiting a cluster in the physical map. For example, fifteen Mγ subfamily members were adjacent to Chr2 and five Mα members on Chr3. Moreover, eleven genes were still anchored to unattributed scaffold fragments, and five newly identified MIKC-type genes were located on Chr15, Chr13, Chr7, Chr3, and Chr10.

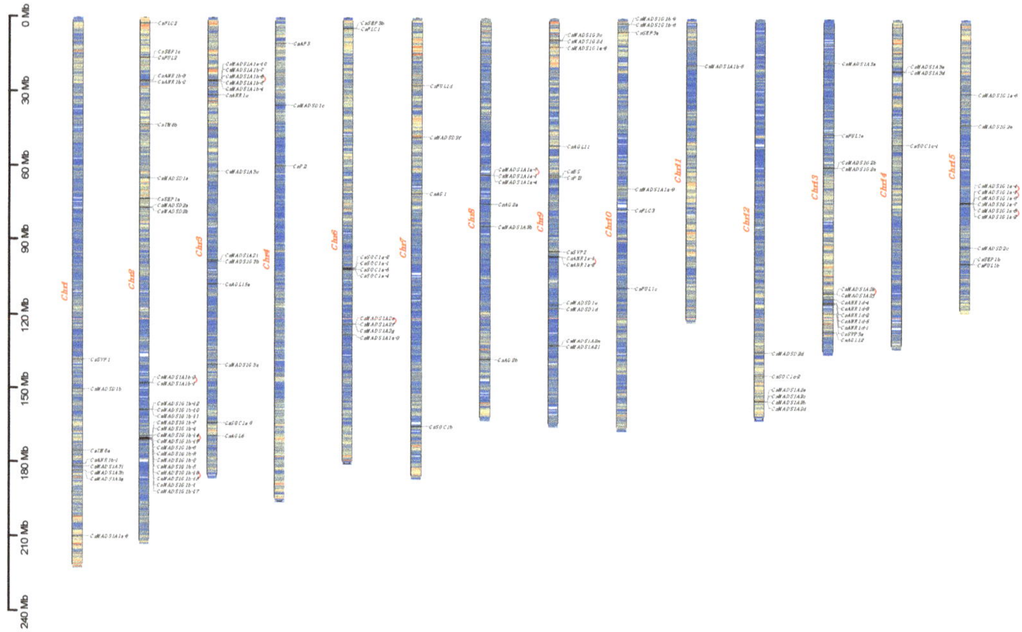

Figure 3. Chromosomal locations for the tea plant *MADS-box* gene family. Genes were mapped onto 14 chromosomes. The gene density of each chromosome was calculated with TBtools and is visualized by gradient colors from blue (low level) to red (high level). Tandem duplicates are linked with red curves.

To clarify the probable relationship between *CsMADS* genes and potential duplication events within the genome, both the occurrence of tandem duplication and large-scale segmental duplication during evolution were analyzed using a genome synteny analysis (Figure 4 and Table S4). A total of ten pairs of paralogous genes emerged from tandem duplications (TD), with nine of them from type I. Only one pair (*CsANR1a-1* and *CsANR1a-2*) was observed from type II, located within a 200 kb chromosomal region. Several ancient tandem arrangements were seemingly retained in the tea plant, including *CsSEP3b* and *CsFLC1* on Chr 1; *CsSEP3a*, *CsFLC3*, and CsFUL1c on Chr 10; *CsSEP1c* and *CsFUL2* on Chr 2; and *CsAGL6* and *CsSOC1a-3* on Chr 3. Such patterns have been previously reported in grapes, potatoes, peaches, and other species [48–50]. The tea genome contained 18 pairs with 34 *CsMADS* genes originating from WGD or segmental duplication. The largest number of MADS-box genes with collinearity was found on Chr 2 with five, followed by Chr 1 and Chr 9. Unlike tandem duplications, 88% (30/34) of segmental duplicated genes belonged to type II, while only two collinear gene pairs were from type I, namely, *CsMADS1A3i*/*CsMADS1A3h* and *CsMADS1G1b-8*/*CsMADSG1b-9*. Ultimately, segmental duplication played an important role in the expansion of type II genes, whereas tandem repeat duplication mainly contributed to type I gene expansion.

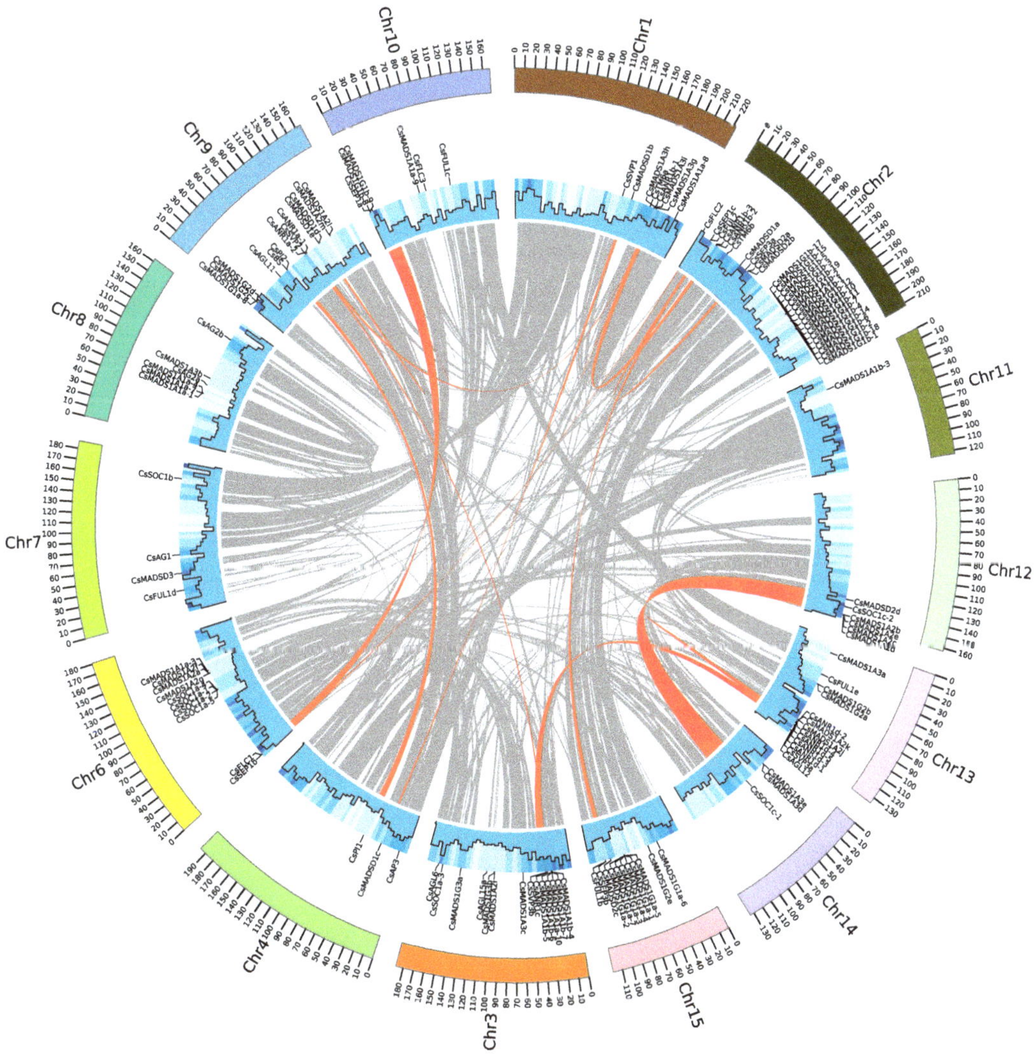

Figure 4. Synteny blocks of *CsMADS* genes in tea plant chromosomes. Gray lines indicate syntenic gene arrangements in the genome of the tea plant, while orange lines link gene pairs derived from segmental duplications or whole genome duplications.

To better understand the nature and extent of selection constraints acting on the tea plant MADS-box gene family, the non-synonymous (Ka) and synonymous (Ks) divergence values of 18 MADS-box gene pairs were calculated. For 12 segmental duplicated *CsMADS* gene pairs with valid Ks values, all Ka/Ks values of the corresponding MADS-box gene pairs were less than 1 (Table S5), suggesting that the tea plant MADS-box gene family might have experienced negative or purifying selective pressure during evolution [51,52]. This result shows that five gene pairs diverged about 27–43 million years ago (MYA), and four gene duplication events occurred around 100 MYA (Table S5).

2.5. MADS-Box Gene Expression Analysis among Different Tissues and Seasons

To explore the expression profiles of tea plant MADS-box genes, we analyzed the RNA-seq data of 136 gene members in different tissues and seasons. Log2-transformed expression values of type II and type I genes were visualized using a hierarchically clustered heatmap, (Figure 5A,B, Table S6). According to expression similarities, sixty-three type II genes were categorized into three major clusters (A-C) and eight expression modules (I-VIII) (Figure 5A). Cluster A was further assigned to expression modules I and II. The genes in this cluster showed flower-specific expression with the majority of putative floral homeotic genes clustered together. Gene expressions in module I (*CsAP1*, *CsAP3*, and *CsTM6a*) were also detected in vegetative tissues. All members from the SEP1/SEP3, PI, AGL11, AG, BS, and AGL6 clades were clustered in module II and showed exclusive expression in flowers. The overall expression levels of genes in cluster B were relatively low and could be further divided into III, IV, and V expression modules. Among them, module III comprised three *ANR1-like* (*CsANR1a-1*, *CsANR1a-2*, and *CsANR1c*) genes and one *FUL-like* (*CsFUL2*) gene. These *ANR1-like* genes were specifically expressed in roots, while *CsFUL2* was expressed during both root and flower development. In modules IV and V, most genes were undetectable or expressed at very low levels, including most members of the *ANR1* and *MIKC** subfamilies. However, two MIKC* genes (*CsMADSD1a* and *CsMADSD1b*) were specifically expressed in autumn flowers, and *CsTM8* in module IV was specifically expressed in the stems. Module VI included four genes from three subfamilies, *CsFUL1a*, *CsFLC2*, *CsFLC1*, and *CsSVP3a*, which exhibited extensive expression patterns throughout the annual cycle, with high transcript accumulation. Most of the *SOC1-like* genes were assigned to module VII, and together with two SVP members (*CsSVP2* and *CsSVP3b*), were widely expressed in non-floral tissues. The VIII expression module could be detected in all organs, but the overall expression level was low.

Generally, gene expression patterns from the same subfamily were relatively conserved, although expression levels of specific family members could vary across tissues. Notably, the expression patterns of five members of *AP1/FUL-like* genes (*CsAP1*, *CsFUL1a*, *CsFUL1b*, *CsFUL1c*, and *CsFUL2*) were very diverse and present in four expression modules. Transcript level changes of some genes specific to seasons were also observed in this study. For example, the expression of *CsAGL6* reached the highest level in spring flowers, followed by winter and autumn flowers. The same pattern was also detected in *CsAP1*, *CsFUL2*, *CsAGL11*, *CsBS*, and *CsSEP3a*, while *CsSEP1a-c* and *CsSEP3b* presented an opposite expression pattern. Notably, the transcript abundances of *CsFUL1a* and *CsFLC2* were significantly more upregulated in winter than in other seasons. In addition, *CsSVP3b* was preferentially expressed in the vegetative organs of autumn, while *CsSVP3a* tended to be expressed in winter and spring samples. The expression preferences of members from the same subfamily in different seasons and different tissues suggested expression differentiation and functional divergence after gene duplication. Most type I genes had very low or no detectable expression (Figure 5B). This result is consistent with previous studies showing that type I gene expression is restricted to specific cells and tissues, such as female gametophytes and developing seeds [53].

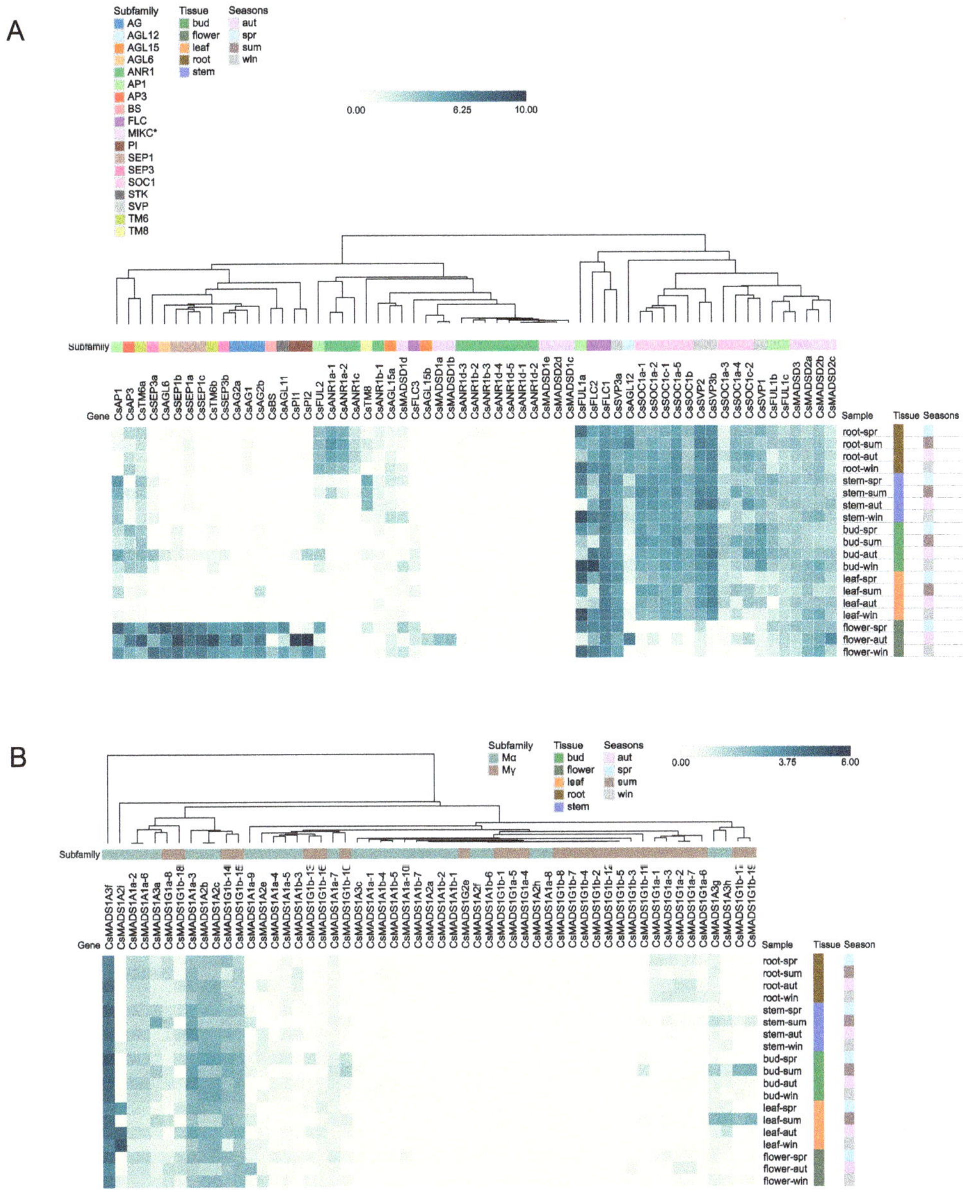

Figure 5. Expression patterns of tea plant MADS-box genes in different tissues (root, stem, bud, leaf, and flower) and seasons (spring, summer, autumn, and winter). (**A**) Clustering heatmap of type II *CsMADS* expression profiles. The three major expression groups are marked as A, B, and C, and roman numbers below/above the horizontal line represent further subdivided modules. (**B**) Clustering heatmap of type I C*sMADS* expression profiles. Genes and tissues are listed in supplementary Table S5. The TPM values were adopted for the relative gene expression levels and were normalized by genes (log2TPM). The color scale bar is provided at the top-left corner.

2.6. MADS-Box Gene Expression Analysis in Response to Cold Stress

To further study the role of MADS-box genes in response to cold stress, the expression profiles of MADS-box genes were obtained in tea leaves using transcriptome assembly (Tables S7 and S8). We assembled two datasets of low-temperature-related transcriptomes, one for short-term cold stress (Figure 6A) and the other for long-term cold acclimation and de-acclimation (Figure 6B). In the results, some MIKC-type genes showed cold-responsive expression patterns. The expression of *CsFUL1a* in one bud and two leaves was significantly up-regulated by low temperatures. Consistent with this result, *CsFUL1a* expression was remarkably increased during the 7 d chilling acclimation stage (CA2_7d) and then inhibited after recovering from cold acclimation. However, the expression trend of *CsFUL1b* was diametrically opposite. *CsTM6a* did not exhibit differential expression under cold treatment, but its expression level significantly increased after de-acclimation. *CsSVP1*, *CsMADSD1f*, and four *SOC1-like* members (*CsSOC1a-1*, *CsSOC1a-2*, *CsSOC1a-4*, and *CsSOC1a-5*) were also up-regulated during the de-acclimation stage. *CsFLC1* was slowly downregulated with continued cold acclimation, even after turning warm, while *CsFLC2* was up-regulated by cold stress. Except for the aforementioned genes that respond to low temperatures, the remaining genes suffered from an expression that was either too low or not sufficiently different from that of the control under cold stress.

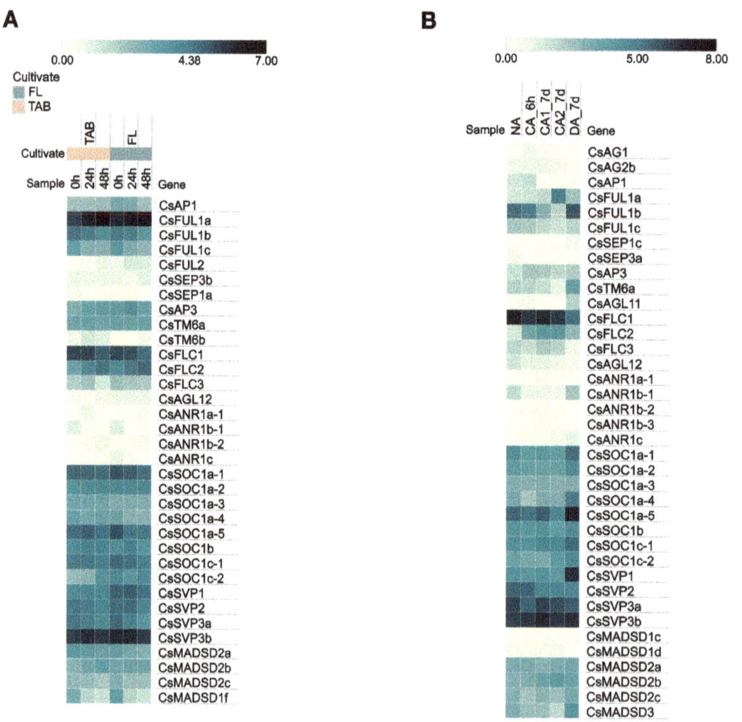

Figure 6. Expression analysis of *CsMADS* in response to cold stress. (**A**) Expression analysis of *CsMADS* in response to cold stress in fish leaf (FL) and two-leaf-and-one-bud (TAB) tissues. (**B**) Expression analysis of *CsMADS* in response to cold acclimation. The TPM values were adopted for the relative gene expression level and normalized by gene (log2TPM). The color scale bar is located at the top.

2.7. A Putative Floral ABCE Model for Tea Plant According to Expression Profiles

The expression patterns of twenty-one ABC(D)E model genes were detected in four floral organs and three developmental stages of flower based on transcriptome data (Figure 7A, Table S9). Among the five detected *AP1/FUL-like* gene members, only *CsAP1* and *CsFUL1b* were involved in both sepal and petal specification, and *CsFUL2* was expressed in sepals but at a low level. All five B-class genes (*CsAP3*, *CsPI1*, *CsPI2*, *CsTM6a*, and *CsTM6b*) from the tea plant displayed broad expression patterns. In addition to the expected second and third whorls, the expression domains of these B-class genes extended into the first and fourth whorls. For C-class genes, the transcripts of *CsAG2a* and *CsAG2b* strongly accumulated in pistils and stamens. Another *AG-like* gene, *CsAG1*, was preferentially expressed in stamens, indicating the stamen identity of the C-function. The E-class genes *CsSEP3a*, *CsSEP3b*, *CsSEP1b*, and *CsSEP1c* were expressed throughout four whorls and may represent candidate E-function genes. *CsSEP1a* may instead exert a partial E-function since it showed a relatively high expression in sepals but no expression in stamens. Notably, *CsAGL6* was mainly expressed in sepals and petals and may constitute the A-function together with *CsAP1* and *CsFUL1b*. A putative ABCE model was proposed according to the expression profiles and deduced functions of A-, B-, C-, and E-class genes in four whorls (Figure 7B). In total, 16 ABCE model-related genes were selected for RT-qPCR analysis to validate the reliability of the RNA-seq results (Figure 8). The RT-qPCR results strongly agreed with our RNA-seq results.

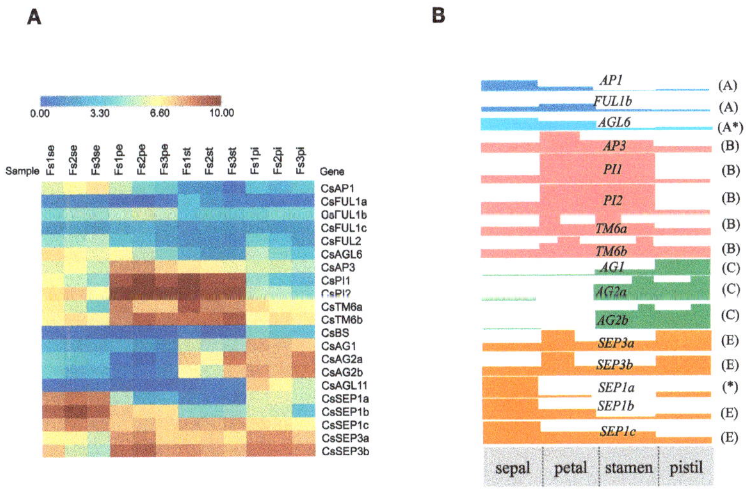

Figure 7. Expression profiles of ABC(D)E genes in the four whorls of the flower and a proposed floral ABCE model for the tea plant. (**A**) Heatmap of expression patterns of tea plant A-, B-, C-, D-, and E-class genes in floral organs (sepals, petals, pistils, and stamens) of three developmental stages. "FS" stands for flower stage, "se" for sepals, "pe" for petals, "pi" for pistils, and "st" for stamens (**B**) A floral model for the tea plant built according to the expression patterns described in Figure 7A and the expected functions of detected MADS-box genes. "*" represent that putative gene function was different from expectation due to expression split.

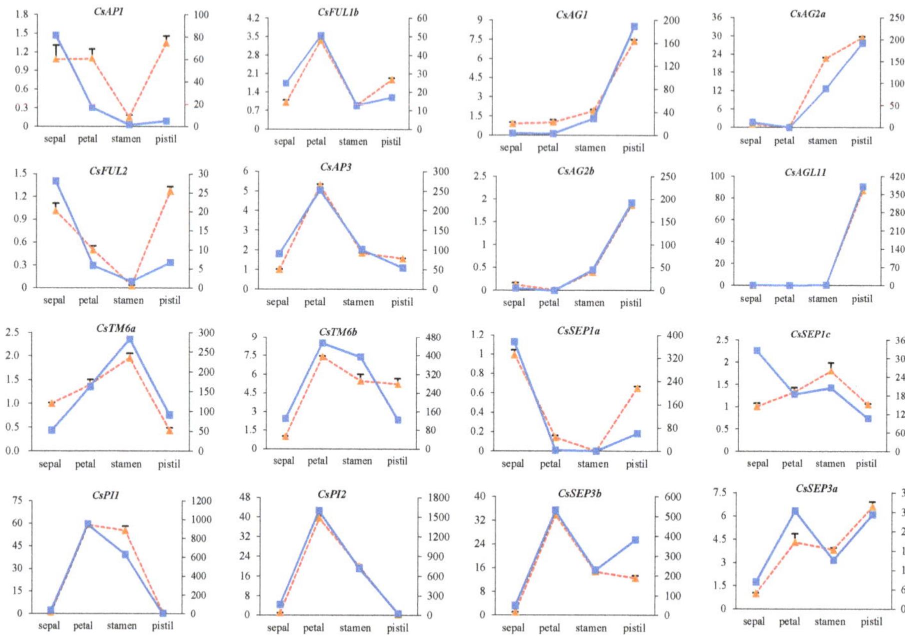

Figure 8. RT-qPCR verification of 16 ABCE model genes. The left y-axis represents the relative expression of the RT-qPCR results and the right y-axis stands for the FPKM value from the RNA-seq results. The blue solid line represents the TPM value, and the orange dashed line is the relative expression.

3. Discussion

MADS-box genes comprise a large family of transcription factors that are widely present in all flowering plants. Due to the availability of the recently published tea plant genome sequence, *CsMADS* genes can now be systematically identified and analyzed [42]. In this study, we searched for MADS-box genes in the tea plant genome using a self-built nucleic acid HMM, resulting in the identification of 136 family members. The total number of identified MADS-box genes was approximately two times that of a previous study [42]. Highly contiguous genome sequencing of the tea plant using a practical and delicate strategy led to a significant increase in gene numbers. A nucleic acid HMM search was sufficient to detect all potential MADS domains in genomic regions, including 124 known and 27 novel genes. Manual curation of the gene annotations greatly improved the accuracy of the gene model set for MADS-box sequences using assembled transcriptome datasets and homology predictions.

3.1. Evolution of the MADS-Box Family in Tea Plant

The tea plant presented a relatively high number of type II genes (63), with the number in many flowering plants ranging from 40 to 70 [54]. According to the phylogenetic analysis, tea plant type II MADS-box genes were distributed in all 14 major eudicot subfamilies, which is consistent with the classification and results for *Arabidopsis* and grape [18,55]. The presence of all relevant genes in these subfamilies reemphasized the completeness of our identification results. Different retention histories for duplicates in the different clades were observed in the tea plant, *Arabidopsis*, and grape (Figure 1). *TM6-like* and *TM8-like* genes are not found in *Arabidopsis*, but are present in tea plants and grapes. Previous findings indicated that independent gene loss occurred twice in the lineage leading to *Arabidopsis* [17,56]. Moreover, members of the AP3/TM6/PI, AP1/FUL, SOC1, ANR1, and MIKC* subfamilies were overrepresented in the tea plant, while *Arabidopsis* and grape contained more genes in the FLC and SVP clades. The presence of increased homologs from these clades might help fulfill specific functions, such as flower organization, flowering time, and root and gametophyte development.

Gene duplication events played a critical role in the expansion of the tea plant MADS-box gene, as observed in other species belonging to different taxonomic groups [57]. Based on genome synteny analysis, 30 paralogs composed of 53 genes were identified, of which 30 belonged to segmental duplications, and 23 were tandem duplication events. Our results showed that the expansion of type II MADS-box genes in the tea plant might have been primarily caused by WGD/segmental duplication, whereas type I genes were mostly derived from tandem duplications in individual chromosomes. Indeed, it has been suggested that several whole genome duplication events gave rise to most type II MADS-box genes, while type I genes are generally associated with smaller-scale duplication events [57]. The tea plant genome acquired an additional WGD after WGT-γ, which might have been responsible for the increased number of type II genes.

3.2. Extensively Conserved Function of MADS-Box Genes in Tea Plant Development

The gene expression patterns of most MADS-box genes in the tea plant showed extensive similarities with orthologous genes described in model plants such as *Arabidopsis* and related grape species, indicating that the functions of MADS-box genes were widely conserved among these species. For example, most tea plant MADS-box genes specifically expressed in flowers (*SEP1/SEP3-like*, *PI-like*, *AGL11-like*, *AG-like*, *BS-like*, and *AGL6-like* genes) were putative floral homeotic genes. Like their *Arabidopsis* and grape counterparts, *ANR1-like* genes in the tea plant presented the highest expression in roots. *CsAGL12* was mainly expressed in roots and flowers, which is consistent with the roles of the *AGL12* homologous gene (*AtXAL1*) in *Arabidopsis* in regulating root development and flowering transitions [58]. Tea plant *SOC1-like* genes showed expression patterns only in vegetative organs, which was consistent with the observations in *Arabidopsis* and grape, where the major expression domains of *SOC1-like* genes are not the floral organs [59]. Some *FLC-like* and *SVP-like* genes from the tea plant were ubiquitously expressed throughout the plant's life cycle in different tissues and might be involved in several developmental processes, as observed in *Arabidopsis*. The gene expression patterns were generally conserved among most subfamilies. However, diverse expression patterns of *AP1/FUL*-like genes were observed in the tea plant, suggesting that these homologous genes might be differentiated into multiple functions.

3.3. The Involvement of Tea Plant MADS-Box Genes in Cold Stress Responses

Previous research has indicated that some MADS-box genes play an important role in cold stress [60–63]. In this study, one *FUL1-like* gene *(CsFUL1a)* was strongly induced during the cold acclimation stage and showed significantly high expression in all winter tissues, which might be involved in the adaptation of tea plants to the severity of winter and flowering regulation. Apart from the *FUL1-like* gene (*TM4*), low temperatures can simultaneously increase the transcript levels of *TM5* (*SEP*), *TM6* (*DEF*), and *TAG1* (*AG*) in tomatoes. These genes may be involved in abnormal flower production at low temperatures [37,64,65]. The expression of these orthologs in tea plants exhibited no significant changes under cold stress, and *CsTM6a* was up-regulated after de-acclimation, indicating differential stress responses from tomatoes. Another gene strongly induced by cold acclimation was *CsFLC2*, the highest expression of which was detected in winter bud. Coincidently, four *SOC1-like* members (*CsSOC1a-1*, *CsSOC1a-2*, *CsSOC1a-4*, and *CsSOC1a-5*) preferentially expressed in summer were repressed under cold stress but up-regulated during the de-acclimation stage. The antagonistic expression patterns of *FLC* and *SOC1* fit well with the fact that *FLC* was a negative regulator of *SOC1* during flowering transitions. *SOC1* was involved in cross talk between cold response signaling and flowering regulation by directly repressing *C-Repeat Binding Factor* (*CBF*) genes, which could activate both *Cold Regulated* (*COR*) genes and FLC expression [66,67]. Moreover, in kiwifruit, SVP homologs from the SVP2 and SVP3 clades may act as suppressors during bud dormancy [68,69]. The expression of dormancy-associated *PpDAM5* and *PpDAM6* in peaches (*Prunus persica*) from the SVP clade was up-regulated in autumn and down-regulated by cold in winter [69,70]. Its ortholog, *CsSVP3b*, exhibited the highest expression in all autumn samples, and *CsSVP3a* was repressed by short-term cold stress, highlighting the potential involvement of these two genes in bud dormancy. Taken together, both seasonal expression analysis and cold treatment expression analysis suggested that these cold-responsive genes might play diverse roles in seasonal low-temperature-dependent developmental processes.

3.4. A Conserved and Ancient Floral 'ABCE' Model in Tea Plant

Four whorls of organs (sepal, petal, stamen, and pistil) from three different flower developmental stages were collected to investigate the regulatory relationships between ABCE homologs and floral organ development using RNA-seq and qRT-PCR (Figure 8). Our results demonstrate that the expression patterns of these genes largely agree with those of the classical floral 'ABCE' model. Previous studies have shown that *AGL6*-like genes formed a 'superclade' together with *SEP1*- and *SQUA*-like genes and shared similar functions with genes from these two sister clades in specifying floral meristem and organ identity [71]. In this study, a AGL6 homolog gene, *CsAGL6*, tended to act as an A-function candidate gene because its expression pattern was similar to the patterns of A-lineage genes. Five B-class genes were highly expressed in petals and stamens, indicating conserved functions in controlling the development of stamens and petals. C-class genes (*CsAG1*, *CsAG2a*, and *CsAG2b*) displayed high expression levels in the stamens and pistils, which is in line with its expected functions in specifying both male and female reproductive organs [20]. Five E-class genes fulfilled full or partial E-function activities that redundantly determined floral organ identity. Moreover, compared with eudicot model systems, a wider range of expression domains was observed in the tea plant. Notably, the expression domains of five B-class genes were expanded into sepals and pistils. This phenomenon is used to refer to the ancient 'ABCE' model mainly found in lineages that diverged early, such as *Aquilegia coerulea* [72], waterlilies [73], and magnoliids [74]. Interestingly, an expanded expression domain of B-function genes was also observed in other Camellia plants, *Camellia japonica* and *Camellia changii Ye*, which was thought to be responsible for petaloid phenotype and double flowers in these two species [75,76]. Thus, we speculate that this phenomenon may have occurred as early as in the ancestor of Camellia plants.

4. Materials and Methods

4.1. Identification of MADS-Box Genes and Curation of a Gene Model

The tea plant genome was retrieved from the TPIA website (http://tpia.teaplant.org/index.html) [77]. The Hidden Markov Model (HMM) of the SRF (MADS) family and (K-box) (PF00931 and PF01489) were downloaded from the Pfam website (http://pfam.sanger.ac.uk/ (accessed on 8 January 2021)) [78]. In addition, hmmsearch in the HMMER v3 software [79] was used to scan the tea plant protein sets for genes containing these two domains. A high-quality protein dataset was obtained through preliminary filtering based on the e-value and length. MAFFT v7.215 [80] was used for multiple sequence alignment of the conserved regions. The nucleic acid sequences corresponding to the aligned MADS-box and K-box domains were extracted and converted into a tea-specific nucleic acid HMM using the hmmbuild function of the HMMER software. This constructed HMM was used to scan the tea plant genome sequence by executing the "nhmmer" command to obtain sites containing this domain (Figure S1).

For the genome-wide candidate sites obtained in the last step, regions without corresponding mapped genes were considered potential new functional sites. Subsequently, Genewise [81] was used to screen the genomic regions near the new sites by inputting homologous genes from grape, kiwifruit, and other related species as a query. Gene sets of the floral organ transcriptome assembled in our laboratory (unpublished) were especially useful for the functional annotation of genes expressed in the flowers. The gene models of known genes without the MADS-domain or K-box domain were manually corrected using the same strategy. All full-length genes were retained for further analysis.

4.2. Sequence Alignment and Phylogenetic Analysis

To study the evolutionary relationships between the newly discovered MADS-domain genes in the tea plant, multiple sequence alignments were performed based on well-identified protein sequences of grape and *Arabidopsis* using MAFFT software [80] with 1000 bootstrap replicates. The MADS-box protein sequences in *Arabidopsis* and grape were obtained from the Plant TFDB website (http://planttfdb.gao-lab.org/index.php (accessed on 8 January 2020)) [82] and the research of *Grimplet* et al. [55], respectively. Maximum-likelihood phylogenetic trees were constructed using the maximum likelihood approach in FastTree v.2.1.10 software [83] with the GTR+CAT model. The phylogenetic tree was further refined using Evolview online tools (https://www.evolgenius.info/evolview/) [40].

4.3. Gene Nomenclature of Tea Plant MADS-Box Genes

To provide insight into the potential functional roles of relevant genes and facilitate future research, we renamed the MADS-box gene family members. For MIKC^C genes, we took the association and phylogenetic relationships of subfamilies into consideration and also referred to the gene nomenclature of orthologs in *Arabidopsis* and grape. Each gene begins with "*Cs*", the abbreviation of the species name *Camellia sinensis*, followed by the name of the most important *Arabidopsis* gene in this subfamily. In some clades where no *Arabidopsis* orthologs exist, we took the nomenclature of the grape as a reference, such as the *TM8* clade, with the exception of *TM6*-like genes, which are only found in eudicots and named after *AP3*-like genes in previous grape research. In cases where several subclades appear in a given subfamily, and previous studies have failed to describe them, if the subfamily name ends with a number, then we added lowercase letters (a, b, c, etc.) to distinguish each subfamily. If the subfamily could be further divided, we added a dash (-) and a number after the letter; if the subfamily name ends with a letter, we added a number to distinguish. For genes assigned to MIKC* and type I groups, we named them after the grape genes. The naming of Type I genes was largely performed the same as that of MIKC* genes, except that the number "1" was added after "MADS".

4.4. Characterization of the Gene Structure and Conserved Motifs of CsMADS

The available information for exons and introns was retrieved from the tea plant genome annotation file and then visualized using TBtools [84] with the full-length coding sequence (CDS) and genomic sequence. The online MEME suit software v 4.12.0 (http://meme-suite.org/tools/meme (accessed on 1 July 2015)) [85] was used to search conserved motifs in each observed CsMADS protein with the parameters as follows: maximum number of motifs, 10. The default values were employed for the other parameters. The motifs of MADS-box genes were annotated using the SMART program (http://smart.embl-heidelberg.de/ (accessed on 1 January 2009)).

4.5. Gene Mapping, Gene Duplication, Synteny, and Ka/Ks Substitutions of MADS-Box Genes in Tea Plant

According to the annotation file (GFF3) for the tea plant genome downloaded from the TPIA database (http://tpia.teaplant.org/index.html (accessed on 24 March 2020)), the positions of tea plant MADS-box genes on the chromosomes were determined, and Tbtools software [84] was used to map all MADS-box genes onto chromosomes. The Multiple Collinearity Scan toolkit X (MCScanX) [59] was applied to identify segmental and tandem duplications of *CsMADS* genes with default parameters, and the results of synteny were visualized using Circos software [60]. Synonymous (Ks) and nonsynonymous (Ka) substitution rates and their Ka/Ks ratios were estimated via the Simple Ka/Ks Calculator module in TBtools [84]. The divergence time was calculated according to the formula

T = Ks/2r, where λ refers to the clock-like rate (6.5×10^{-9} synonymous substitutions per site per year) [86].

4.6. Gene Expression Analysis of Tea Plant MADS-Box Genes

To explore the expression patterns of tea plant MADS-box genes, raw RNA-seq datasets were mined from the NCBI database under the projects PRJEB39502, PRJNA411886, and PRJNA387105 (Table S10). The project PRJEB39502 comprised nineteen tissue samples, including five tissues (root, stem, bud, leaf, and flower) in four seasons (flowers during summer excluded). The project PRJNA411886 comprised six treatments: FL0 h (control, 0 h), FL24 h (cold treatment, 24 h), and FL48h (cold treatment, 48 h) at the fish leaf phase; TAB0h (control, 0 h), TAB24h (cold treatment, 24 h), and TAB48 h (cold treatment, 48 h) at the two leaves and one bud phase. The temperature of the cold spell was set at 4 °C during the day and 2 °C at night. The project PRJNA387105 consisted of five treatments: CK (25~20 °C, control); CA1_6h (cold acclimated under 10 °C, 6 h); CA2_7d (cold acclimated under 10~4 °C, 7 days); CA2_7d (cold acclimated under 4~0 °C, 7 days); DA_7d (25~20 °C, 7 days). Detailed descriptions of these transcriptome datasets were published by Wang et al. [6], Hao et al. [87], and Li et al. [88], respectively. In addition, flower expression in the four whorls of the three developmental stages (the green bud stage, deglazing stage, and heyday stage) was based on unpublished transcriptome data from our laboratory (Table S10). The transcripts per million (TPM) values of tea plant MADS-box genes were calculated using a self-built script according to the HISAT2 and Stringtie pipeline [89]. The heatmaps were visualized via Morpheus (https://software.broadinstitute.org/morpheus) based on the transformed data of the log2 (TPM+1) values.

4.7. Plant Materials and Quantitative PCR Analysis

Tea flowers at the white bud stage were collected from tea bushes (*Camellia sinensis* var. Rougui) cultivated at the tea garden of Fujian Agriculture and Forestry University in the Fujian Province of China (26°04′ N, 119°14′ E) under natural conditions. Floral organs (sepals, petals, stamens, and pistils) were separated by hand and immediately inserted into liquid nitrogen for freezing and preservation. Floral organs from at least five randomly selected individuals were harvested as one replicate, and three biological replicates were used for each sample.

In this study, a polysaccharide and polyphenol kit (Tiangen Biotech, Beijing, China) was used to extract the total RNA of floral organs, according to the manufacturer's instructions. Total RNA was reverse-transcribed using EasyScript one-step gDNA Removal and cDNA Synthesis SuperMix (Transgen Biotech, Beijing, China). Genomic DNA was digested by DNase prior to reverse transcription, and the obtained cDNA template was diluted to a uniform concentration of about 100 ng/μL. Real-time PCR amplification was performed on a CFX96 Touch Deep Well system (Bio-rad, Hercules, CA, USA) using TransStart Top Green SuperMix (Tiangen Biotech, Beijing, China) with SBYR green fluorescent dye. The PCR mixture in a 20 μL volume contained 10 μL SuperMix, 0.4 μL forward primers, 0.4 μL reverse primers, 1 μL cDNA, and 8.2 μL ddH_2O. The thermal cycling conditions were set as follows: 95 °C for 30 s; 40 cycles at 95 °C for 5 s and 60 °C for 34 s; and 95 °C for 15 s, 60 °C for 1 min, and 95 °C for 15 s. A housekeeping gene, *GADPH*, was used as an internal control for data normalization. All reactions were run in technical triplicate and biological triplicate. The $2^{-\Delta\Delta CT}$ method (also called the comparative CT method) was used to calculate the relative gene expression levels. Details for the primers are provided in Table S11.

5. Conclusions

In this study, a tea-specific nucleotide acid HMM was developed to identify 136 MADS-box genes in the tea plant genome, including 63 genes of type II and 73 of type I. In total, 27 novel genes were identified in this study for the first time. In-depth artificial curation largely improved the accuracy of gene models based on assembled transcriptome datasets and homology predictions. Seasonal expression analysis and cold treatment expression analysis suggested these cold-responsive genes might play diverse roles in seasonal low-temperature-dependent developmental processes. Moreover, the expanded expression domain of 'ABCE' model genes indicated that the tea plant has a conserved and ancient ABCE floral model. Overall, this work contributes to our understanding of the evolutionary history of the MADS-box gene in the tea plant and provides perspectives on corresponding functional analyses.

Supplementary Materials: The following supporting information can be downloaded at: https://www.mdpi.com/article/10.3390/plants12162929/s1. Table S1: A preliminary protein set of tea plant MADS-box genes; Table S2: Known and novel MADS-box gene loci predicted using a nucleotide-based HMM; Table S3: The known genes from the preliminary MADS-box protein set deleted due to incomplete sequence data or assembly errors; Table S4: Duplication patterns of *CsMADS* genes identified in C. sinensis; Table S5: The Ka/Ks ratios and divergence time of the *CsMADS* gene pairs. Table S6: The TPM values of *CsMADS* genes expressed in different tissues and seasons; Table S7: The TPM values of *CsMADS* genes in response to cold stress in the fish leaf (FL) and the two-leaf-and-one-bud (TAB) tissues; Table S8: The TPM values of *CsMADS* genes during cold acclimation; Table S9: The TPM values of ABC(D)E-class MADS-box genes expressed in four whorls of tea plant; Table S10: Transcriptome datasets used in this study. Table S11: Primers used in this study for qRT-PCR purposes; Figure S1: Identification pipeline of the MADS-box gene family; Figure S2: A maximum likelihood (ML) phylogenetic tree of Type I MADS-box proteins from tea plant, *Arabidopsis*, and grape.

Author Contributions: Conceptualization, Y.S., Z.L. and J.H.; Formal analysis, J.H., A.I. and Q.C.; Methodology, J.H., A.I. and W.B.; Investigation, J.H. and Q.C.; Funding acquisition, Y.S.; Data curation, A.I., Q.C. and W.B.; Project administration, Y.S. and Z.L.; Software, A.I.; Resources, A.I.; Supervision, Y.S.; Writing—original draft, J.H.; Writing—review and editing, A.I., Y.S. and Z.L. All authors have read and agreed to the published version of the manuscript.

Funding: This work was financially supported by the China Agriculture Research System of MOF and MARA: The Earmarked Fund for China Agriculture Research System (grant number CARS-19); the Science and Technology Innovation Project of Fujian Agriculture and Forestry University (grant number CXZX2017177); and the Special Fund for Science and Technology Innovation of Fujian Zhang Tianfu Tea Development Foundation (grant number FJZTF01).

Data Availability Statement: All data in this study can be found in the manuscript or the supplementary materials.

Conflicts of Interest: The authors declare no conflict of interest.

References

1. Fan, Z.; Li, J.; Li, X.; Wu, B.; Wang, J.; Liu, Z.; Yin, H. Genome-wide transcriptome profiling provides insights into floral bud development of summer-flowering Camellia azalea. *Sci. Rep.* **2015**, *5*, 9729. [CrossRef]
2. Pereira, A.G.; Garcia-Perez, P.; Cassani, L.; Chamorro, F.; Cao, H.; Barba, F.J.; Simal-Gandara, J.; Prieto, M.A. Camellia japonica: A phytochemical perspective and current applications facing its industrial exploitation. *Food Chem. X* **2022**, *13*, 100258. [CrossRef]
3. Xia, E.; Tong, W.; Hou, Y.; An, Y.; Chen, L.; Wu, Q.; Liu, Y.; Yu, J.; Li, F.; Li, R.; et al. The Reference Genome of Tea Plant and Resequencing of 81 Diverse Accessions Provide Insights into Its Genome Evolution and Adaptation. *Mol. Plant* **2020**, *13*, 1013–1026. [CrossRef]
4. Zhang, Q.J.; Li, W.; Li, K.; Nan, H.; Shi, C.; Zhang, Y.; Dai, Z.Y.; Lin, Y.L.; Yang, X.L.; Tong, Y.; et al. The Chromosome-Level Reference Genome of Tea Tree Unveils Recent Bursts of Non-autonomous LTR Retrotransposons in Driving Genome Size Evolution. *Mol. Plant* **2020**, *13*, 935–938. [CrossRef]
5. Zhang, W.; Zhang, Y.; Qiu, H.; Guo, Y.; Wan, H.; Zhang, X.; Scossa, F.; Alseekh, S.; Zhang, Q.; Wang, P.; et al. Genome assembly of wild tea tree DASZ reveals pedigree and selection history of tea varieties. *Nat. Commun.* **2020**, *11*, 3719. [CrossRef] [PubMed]
6. Wang, X.; Feng, H.; Chang, Y.; Ma, C.; Wang, L.; Hao, X.; Li, A.; Cheng, H.; Wang, L.; Cui, P.; et al. Population sequencing enhances understanding of tea plant evolution. *Nat. Commun.* **2020**, *11*, 4447. [CrossRef] [PubMed]
7. Passmore, S.; Maine, G.T.; Elble, R.; Christ, C.; Tye, B.K. Saccharomyces cerevisiae protein involved in plasmid maintenance is necessary for mating of MAT alpha cells. *J. Mol. Biol.* **1988**, *204*, 593–606. [CrossRef]

8. Yanofsky, M.F.; Ma, H.; Bowman, J.L.; Drews, G.N.; Feldmann, K.A.; Meyerowitz, E.M. The protein encoded by the *Arabidopsis* homeotic gene agamous resembles transcription factors. *Nature* **1990**, *346*, 35–39. [CrossRef]
9. Sommer, H.; Beltrán, J.P.; Huijser, P.; Pape, H.; Lönnig, W.E.; Saedler, H.; Schwarz-Sommer, Z. Deficiens, a homeotic gene involved in the control of flower morphogenesis in *Antirrhinum majus*: The protein shows homology to transcription factors. *EMBO J.* **1990**, *9*, 605–613. [CrossRef]
10. Norman, C.; Runswick, M.; Pollock, R.; Treisman, R. Isolation and properties of cDNA clones encoding SRF, a transcription factor that binds to the c-fos serum response element. *Cell* **1988**, *55*, 989–1003. [CrossRef]
11. Martinez-Castilla, L.P.; Alvarez-Buylla, E.R. Adaptive evolution in the *Arabidopsis* MADS-box gene family inferred from its complete resolved phylogeny. *Proc. Natl. Acad. Sci. USA* **2003**, *100*, 13407–13412. [CrossRef]
12. Alvarez-Buylla, E.R.; Pelaz, S.; Liljegren, S.J.; Gold, S.E.; Burgeff, C.; Ditta, G.S.; Ribas de Pouplana, L.; Martinez-Castilla, L.; Yanofsky, M.F. An ancestral MADS-box gene duplication occurred before the divergence of plants and animals. *Proc. Natl. Acad. Sci. USA* **2000**, *97*, 5328–5333. [CrossRef]
13. Kaufmann, K.; Melzer, R.; Theissen, G. MIKC-type MADS-domain proteins: Structural modularity, protein interactions and network evolution in land plants. *Gene* **2005**, *347*, 183–198. [CrossRef]
14. Masiero, S.; Colombo, L.; Grini, P.E.; Schnittger, A.; Kater, M.M. The emerging importance of type I MADS box transcription factors for plant reproduction. *Plant Cell* **2011**, *23*, 865–872. [CrossRef]
15. Kofuji, R.; Sumikawa, N.; Yamasaki, M.; Kondo, K.; Ueda, K.; Ito, M.; Hasebe, M. Evolution and divergence of the MADS-box gene family based on genome-wide expression analyses. *Mol. Biol. Evol.* **2003**, *20*, 1963–1977. [CrossRef] [PubMed]
16. Arora, R.; Agarwal, P.; Ray, S.; Singh, A.K.; Singh, V.P.; Tyagi, A.K.; Kapoor, S. MADS-box gene family in rice: Genome-wide identification, organization and expression profiling during reproductive development and stress. *BMC Genom.* **2007**, *8*, 242. [CrossRef]
17. Becker, A. The major clades of MADS-box genes and their role in the development and evolution of flowering plants. *Mol. Phylogenetics Evol.* **2003**, *29*, 464–489. [CrossRef]
18. Parenicová, L.; de Folter, S.; Kieffer, M.; Horner, D.S.; Favalli, C.; Busscher, J.; Cook, H.E.; Ingram, R.M.; Kater, M.M.; Davies, B.; et al. Molecular and phylogenetic analyses of the complete MADS-box transcription factor family in *Arabidopsis*: New openings to the MADS world. *Plant Cell* **2003**, *15*, 1538–1551. [CrossRef] [PubMed]
19. Smyth, D.R. Evolution and genetic control of the floral ground plan. *New Phytol.* **2018**, *220*, 70–86. [CrossRef] [PubMed]
20. Causier, B.; Schwarz-Sommer, Z.; Davies, B. Floral organ identity: 20 years of ABCs. *Semin. Cell Dev. Biol.* **2010**, *21*, 73–79. [CrossRef]
21. Bowman, J.L.; Smyth, D.R.; Meyerowitz, E.M. The ABC model of flower development: Then and now. *Development* **2012**, *139*, 4095–4098. [CrossRef]
22. Jofuku, K.D.; den Boer, B.G.; Van Montagu, M.; Okamuro, J.K. Control of *Arabidopsis* flower and seed development by the homeotic gene *APETALA2*. *Plant Cell* **1994**, *6*, 1211–1225.
23. Mandel, M.A.; Gustafson-Brown, C.; Savidge, B.; Yanofsky, M.F. Molecular characterization of the *Arabidopsis* floral homeotic gene APETALA1. *Nature* **1992**, *360*, 273–277. [CrossRef] [PubMed]
24. Goto, K.; Meyerowitz, E.M. Function and regulation of the *Arabidopsis* floral homeotic gene *PISTILLATA*. *Genes Dev.* **1994**, *8*, 1548–1560. [CrossRef]
25. Jack, T.; Brockman, L.L.; Meyerowitz, E.M. The homeotic gene *APETALA3* of *Arabidopsis thaliana* encodes a MADS box and is expressed in petals and stamens. *Cell* **1992**, *68*, 683–697. [CrossRef] [PubMed]
26. Savidge, B.; Rounsley, S.D.; Yanofsky, M.F. Temporal Relationship between the Transcription of Two *Arabidopsis* MADS-Box Genes and the Floral Organ ldentity Genes. *Plant Cell* **1995**, *7*, 721–733.
27. Pelaz, S.; Ditta, G.; Baumann, E.; Wisman, E.; Yanofsky, M.F. B and C foral organ identity functions require *SEPALLATA* MADS-box genes. *Nature* **2000**, *405*, 200–203. [CrossRef] [PubMed]
28. Ditta, G.; Pinyopich, A.; Robles, P.; Pelaz, S.; Yanofsky, M.F. The *SEP4* gene of *Arabidopsis thaliana* functions in floral organ and meristem identity. *Curr. Biol.* **2004**, *14*, 1935–1940. [CrossRef]
29. Hsu, H.-F.; Hsu, W.-H.; Lee, Y.-I.; Mao, W.-T.; Yang, J.-Y.; Li, J.-Y.; Yang, C.-H. Model for perianth formation in orchids. *Nat. Plants* **2015**, *1*, 15046. [CrossRef]
30. Suetsugu, K.; Fukushima, K.; Makino, T.; Ikematsu, S.; Sakamoto, T.; Kimura, S. Transcriptomic heterochrony and completely cleistogamous flower development in the mycoheterotrophic orchid *Gastrodia*. *New Phytol.* **2023**, *237*, 323–338. [CrossRef]
31. Smaczniak, C.; Immink, R.G.; Angenent, G.C.; Kaufmann, K. Developmental and evolutionary diversity of plant MADS-domain factors: Insights from recent studies. *Development* **2012**, *139*, 3081–3098. [CrossRef]
32. Ferrándiz, C.; Liljegren, S.J.; Yanofsky, M.F. Negative regulation of the *SHATTERPROOF* genes by *FRUITFULL* during *Arabidopsis* fruit development. *Science* **2000**, *289*, 436–438. [CrossRef] [PubMed]
33. Moon, J.; Suh, S.S.; Lee, H.; Choi, K.R.; Hong, C.B.; Paek, N.C.; Kim, S.G.; Lee, I. The *SOC1* MADS-box gene integrates vernalization and gibberellin signals for flowering in *Arabidopsis*. *Plant J.* **2003**, *35*, 613–623. [CrossRef]
34. Lee, J.H.; Yoo, S.J.; Park, S.H.; Hwang, I.; Lee, J.S.; Ahn, J.H. Role of SVP in the control of flowering time by ambient temperature in *Arabidopsis*. *Genes Dev.* **2007**, *21*, 397–402. [CrossRef]

35. Castelán-Muñoz, N.; Herrera, J.; Cajero-Sánchez, W.; Arrizubieta, M.; Trejo, C.; García-Ponce, B.; Sánchez, M.P.; Álvarez-Buylla, E.R.; Garay-Arroyo, A. MADS-Box Genes Are Key Components of Genetic Regulatory Networks Involved in Abiotic Stress and Plastic Developmental Responses in Plants. *Front. Plant Sci.* **2019**, *10*, 853. [CrossRef]
36. Lozano, R.; Angosto, T.; Gomez, P.; Payan, C.; Capel, J.; Huijser, P.; Salinas, J.; Martinez-Zapater, J.M. Tomato flower abnormalities induced by low temperatures are associated with changes of expression of MADS-Box genes. *Plant Physiol.* **1998**, *117*, 91–100. [CrossRef]
37. Bartley, G.E.; Ishida, B.K. Developmental gene regulation during tomato fruit ripening and in-vitro sepal morphogenesis. *BMC Plant Biol.* **2003**, *3*, 4. [CrossRef] [PubMed]
38. Ning, K.; Han, Y.; Chen, Z.; Luo, C.; Wang, S.; Zhang, W.; Li, L.; Zhang, X.; Fan, S.; Wang, Q. Genome-wide analysis of MADS-box family genes during flower development in lettuce. *Plant Cell Environ.* **2019**, *42*, 1868–1881. [CrossRef] [PubMed]
39. Tian, Y.; Dong, Q.; Ji, Z.; Chi, F.; Cong, P.; Zhou, Z. Genome-wide identification and analysis of the MADS-box gene family in apple. *Gene* **2015**, *555*, 277–290. [CrossRef] [PubMed]
40. He, Z.; Zhang, H.; Gao, S.; Lercher, M.J.; Chen, W.H.; Hu, S. Evolview v2: An online visualization and management tool for customized and annotated phylogenetic trees. *Nucleic Acids Res.* **2016**, *44*, W236–W241. [CrossRef]
41. Schilling, S.; Pan, S.; Kennedy, A.; Melzer, R. MADS-box genes and crop domestication: The jack of all traits. *J. Exp. Bot.* **2018**, *69*, 1447–1469. [CrossRef] [PubMed]
42. Zhang, Z.B.; Jin, Y.J.; Wan, H.H.; Cheng, L.; Feng, Z.G. Genome-wide identification and expression analysis of the MADS-box transcription factor family in *Camellia sinensis*. *J. Appl. Genet.* **2021**, *62*, 249–264. [CrossRef] [PubMed]
43. Cai, J.; Liu, X.; Vanneste, K.; Proost, S.; Tsai, W.C.; Liu, K.W.; Chen, L.J.; He, Y.; Xu, Q.; Bian, C.; et al. The genome sequence of the orchid Phalaenopsis equestris. *Nat. Genet.* **2015**, *47*, 65–72. [CrossRef]
44. Lu, H.; Liu, Z.; Lan, S. Genome Sequencing Reveals the Role of MADS-box Gene Families in the Floral Morphology Evolution of Orchids. *Hortic. Plant J.* **2019**, *5*, 247–254. [CrossRef]
45. Cheng, Z.; Ge, W.; Li, L.; Hou, D.; Ma, Y.; Liu, J.; Bai, Q.; Li, X.; Mu, S.; Gao, J. Analysis of MADS-Box Gene Family Reveals Conservation in Floral Organ ABCDE Model of Moso Bamboo (*Phyllostachys edulis*). *Front. Plant Sci.* **2017**, *8*, 656. [CrossRef]
46. Zhang, Y.; Tang, D.; Lin, X.; Ding, M.; Tong, Z. Genome-wide identification of MADS-box family genes in moso bamboo (*Phyllostachys edulis*) and a functional analysis of *PeMADS5* in flowering. *BMC Plant Biol.* **2018**, *18*, 176. [CrossRef] [PubMed]
47. Schilling, S.; Kennedy, A.; Pan, S.; Jermiin, L.S.; Melzer, R. Genome-wide analysis of MIKC-type MADS-box genes in wheat: Pervasive duplications, functional conservation and putative neofunctionalization. *New Phytol.* **2020**, *225*, 511–529. [CrossRef]
48. Zhao, T.; Holmer, R.; de Bruijn, S.; Angenent, G.C.; van den Burg, H.A.; Schranz, M.E. Phylogenomic Synteny Network Analysis of MADS-Box Transcription Factor Genes Reveals Lineage-Specific Transpositions, Ancient Tandem Duplications, and Deep Positional Conservation. *Plant Cell* **2017**, *29*, 1278–1292. [CrossRef]
49. Gao, H.; Wang, Z.; Li, S.; Hou, M.; Zhou, Y.; Zhao, Y.; Li, G.; Zhao, H.; Ma, H. Genome-wide survey of potato MADS-box genes reveals that *StMADS1* and *StMADS13* are putative downstream targets of tuberigen StSP6A. *BMC Genom.* **2018**, *19*, 726. [CrossRef]
50. Wells, C.E.; Vendramin, E.; Jimenez Tarodo, S.; Verde, I.; Bielenberg, D.G. A genome-wide analysis of MADS-box genes in peach [*Prunus persica* (L.) Batsch]. *BMC Plant Biol.* **2015**, *15*, 41. [CrossRef]
51. Sharma, H.; Sharma, A.; Rajput, R.; Sidhu, S.; Dhillon, H.; Verma, P.C.; Pandey, A.; Upadhyay, S.K. Molecular Characterization, Evolutionary Analysis, and Expression Profiling of *BOR* Genes in Important Cereals. *Plants* **2022**, *11*, 911. [CrossRef]
52. Rathour, M.; Shumayla; Alok, A.; Upadhyay, S.K. Investigation of Roles of TaTALE Genes during Development and Stress Response in Bread Wheat. *Plants* **2022**, *11*, 587. [CrossRef] [PubMed]
53. Bemer, M.; Heijmans, K.; Airoldi, C.; Davies, B.; Angenent, G.C. An atlas of type I MADS box gene expression during female gametophyte and seed development in *Arabidopsis*. *Plant Physiol.* **2010**, *154*, 287–300. [CrossRef]
54. Gramzow, L.; Theissen, G. Phylogenomics of MADS-Box Genes in Plants-Two Opposing Life Styles in One Gene Family. *Biology* **2013**, *2*, 1150–1164. [CrossRef] [PubMed]
55. Grimplet, J.; Martinez-Zapater, J.M.; Carmona, M.J. Structural and functional annotation of the MADS-box transcription factor family in grapevine. *BMC Genom.* **2016**, *17*, 80. [CrossRef]
56. Kramer, E.M.; Su, H.J.; Wu, C.C.; Hu, J.M. A simplified explanation for the frameshift mutation that created a novel C-terminal motif in the *APETALA3* gene lineage. *BMC Evol. Biol.* **2006**, *6*, 30. [CrossRef]
57. Airoldi, C.A.; Davies, B. Gene duplication and the evolution of plant MADS-box transcription factors. *J. Genet. Genom.* **2012**, *39*, 157–165. [CrossRef] [PubMed]
58. Tapia-Lopez, R.; Garcia-Ponce, B.; Dubrovsky, J.G.; Garay-Arroyo, A.; Perez-Ruiz, R.V.; Kim, S.H.; Acevedo, F.; Pelaz, S.; Alvarez-Buylla, E.R. An AGAMOUS-related MADS-box gene, *XAL1* (*AGL12*), regulates root meristem cell proliferation and flowering transition in *Arabidopsis*. *Plant Physiol.* **2008**, *146*, 1182–1192. [CrossRef]
59. Diaz-Riquelme, J.; Lijavetzky, D.; Martinez-Zapater, J.M.; Carmona, M.J. Genome-wide analysis of MIKC^{C}-type MADS box genes in grapevine. *Plant Physiol.* **2009**, *149*, 354–369. [CrossRef] [PubMed]
60. Jia, J.; Zhao, P.; Cheng, L.; Yuan, G.; Yang, W.; Liu, S.; Chen, S.; Qi, D.; Liu, G.; Li, X. MADS-box family genes in sheepgrass and their involvement in abiotic stress responses. *BMC Plant Biol.* **2018**, *18*, 42. [CrossRef]

61. Shimada, S.; Ogawa, T.; Kitagawa, S.; Suzuki, T.; Ikari, C.; Shitsukawa, N.; Abe, T.; Kawahigashi, H.; Kikuchi, R.; Handa, H.; et al. A genetic network of flowering-time genes in wheat leaves, in which an *APETALA1/FRUITFULL-like* gene, *VRN1*, is upstream of *FLOWERING LOCUS T*. *Plant J.* **2009**, *58*, 668–681. [CrossRef] [PubMed]
62. Trevaskis, B.; Hemming, M.N.; Dennis, E.S.; Peacock, W.J. The molecular basis of vernalization-induced flowering in cereals. *Trends Plant Sci.* **2007**, *12*, 352–357. [CrossRef] [PubMed]
63. Hemming, M.N.; Peacock, W.J.; Dennis, E.S.; Trevaskis, B. Low-temperature and daylength cues are integrated to regulate *FLOWERING LOCUS T* in barley. *Plant Physiol.* **2008**, *147*, 355–366. [CrossRef]
64. Ishida, B.K.; Jenkins, S.M.; Say, B. Induction of AGAMOUS gene expression plays a key role in ripening of tomato sepals in vitro. *Plant Mol. Biol.* **1998**, *36*, 733–739. [CrossRef]
65. Vrebalov, J.; Ruezinsky, D.; Padmanabhan, V.; White, R.; Medrano, D.; Drake, R.; Schuch, W.; Giovannoni, J. A MADS-box gene necessary for fruit ripening at the tomato ripening-inhibitor (Rin) locus. *Science* **2002**, *296*, 343–346. [CrossRef]
66. Seo, E.; Lee, H.; Jeon, J.; Park, H.; Kim, J.; Noh, Y.-S.; Lee, I. Crosstalk between Cold Response and Flowering in *Arabidopsis* Is Mediated through the Flowering-Time Gene *SOC1* and Its Upstream Negative Regulator *FLC*. *Plant Cell* **2009**, *21*, 3185–3197. [CrossRef] [PubMed]
67. Sarah, J.; Gilmour, S.G.F.a.M.F.T. *Arabidopsis* transcriptional activators *CBF1*, *CBF2*, and *CBF3* have matching functional activities. *Plant Mol. Biol.* **2004**, *54*, 767–781.
68. Wu, R.M.; Walton, E.F.; Richardson, A.C.; Wood, M.; Hellens, R.P.; Varkonyi-Gasic, E. Conservation and divergence of four kiwifruit *SVP-like* MADS-box genes suggest distinct roles in kiwifruit bud dormancy and flowering. *J. Exp. Bot.* **2012**, *63*, 797–807. [CrossRef]
69. Liu, X.; Sun, Z.; Dong, W.; Wang, Z.; Zhang, L. Expansion and Functional Divergence of the *SHORT VEGETATIVE PHASE* (*SVP*) Genes in Eudicots. *Genome Biol. Evol.* **2018**, *10*, 3026–3037. [CrossRef]
70. Yamane, H.; Ooka, T.; Jotatsu, H.; Hosaka, Y.; Sasaki, R.; Tao, R. Expressional regulation of *PpDAM5* and *PpDAM6*, peach (*Prunus persica*) dormancy-associated MADS-box genes, by low temperature and dormancy-breaking reagent treatment. *J. Exp. Bot.* **2011**, *62*, 3481–3488. [CrossRef]
71. Litt, A.; Kramer, E.M. The ABC model and the diversification of floral organ identity. *Semin. Cell Dev. Biol.* **2010**, *21*, 129–137. [CrossRef]
72. Sharma, B.; Kramer, E.M. Aquilegia B gene homologs promote petaloidy of the sepals and maintenance of the C domain boundary. *Evodevo* **2017**, *8*, 22. [CrossRef]
73. Zhang, L.; Chen, F.; Zhang, X.; Li, Z.; Zhao, Y.; Lohaus, R.; Chang, X.; Dong, W.; Ho, S.Y.W.; Liu, X.; et al. The water lily genome and the early evolution of flowering plants. *Nature* **2020**, *577*, 79–84. [CrossRef]
74. Chanderbali, A.S.; Yoo, M.J.; Zahn, L.M.; Brockington, S.F.; Wall, P.K.; Gitzendanner, M.A.; Albert, V.A.; Leebens-Mack, J.; Altman, N.S.; Ma, H.; et al. Conservation and canalization of gene expression during angiosperm diversification accompany the origin and evolution of the flower. *Proc. Natl. Acad. Sci. USA* **2010**, *107*, 22570–22575. [CrossRef]
75. Gaopu, Z. The potential role of B-function gene involved in floral development for double flowers formation in *Camellia changii Ye*. *Afr. J. Biotechnol.* **2011**, *10*, 541–548. [CrossRef]
76. Li, X.; Li, J.; Fan, Z.; Liu, Z.; Tanaka, T.; Yin, H. Global gene expression defines faded whorl specification of double flower domestication in *Camellia*. *Sci. Rep.* **2017**, *7*, 3197. [CrossRef] [PubMed]
77. Xia, E.H.; Li, F.D.; Tong, W.; Li, P.H.; Wu, Q.; Zhao, H.J.; Ge, R.H.; Li, R.P.; Li, Y.Y.; Zhang, Z.Z.; et al. Tea Plant Information Archive: A comprehensive genomics and bioinformatics platform for tea plant. *Plant Biotechnol. J.* **2019**, *17*, 1938–1953. [CrossRef]
78. Finn, R.D.; Bateman, A.; Clements, J.; Coggill, P.; Eberhardt, R.Y.; Eddy, S.R.; Heger, A.; Hetherington, K.; Holm, L.; Mistry, J.; et al. Pfam: The protein families database. *Nucleic Acids Res.* **2014**, *42*, D222–D230. [CrossRef] [PubMed]
79. Finn, R.D.; Clements, J.; Eddy, S.R. HMMER web server: Interactive sequence similarity searching. *Nucleic Acids Res.* **2011**, *39*, W29–W37. [CrossRef]
80. Katoh, K.; Standley, D.M. MAFFT multiple sequence alignment software version 7: Improvements in performance and usability. *Mol. Biol. Evol.* **2013**, *30*, 772–780. [CrossRef] [PubMed]
81. Madeira, F.; Pearce, M.; Tivey, A.R.N.; Basutkar, P.; Lee, J.; Edbali, O.; Madhusoodanan, N.; Kolesnikov, A.; Lopez, R. Search and sequence analysis tools services from EMBL-EBI in 2022. *Nucleic Acids Res.* **2022**, *50*, W276–W279. [CrossRef]
82. Jin, J.; Tian, F.; Yang, D.C.; Meng, Y.Q.; Kong, L.; Luo, J.; Gao, G. PlantTFDB 4.0: Toward a central hub for transcription factors and regulatory interactions in plants. *Nucleic Acids Res.* **2017**, *45*, D1040–D1045. [CrossRef] [PubMed]
83. Price, M.N.; Dehal, P.S.; Arkin, A.P. FastTree: Computing large minimum evolution trees with profiles instead of a distance matrix. *Mol. Biol. Evol.* **2009**, *26*, 1641–1650. [CrossRef] [PubMed]
84. Chen, C.; Chen, H.; Zhang, Y.; Thomas, H.R.; Frank, M.H.; He, Y.; Xia, R. TBtools: An Integrative Toolkit Developed for Interactive Analyses of Big Biological Data. *Mol. Plant* **2020**, *13*, 1194–1202. [CrossRef]
85. Bailey, T.L.; Boden, M.; Buske, F.A.; Frith, M.; Grant, C.E.; Clementi, L.; Ren, J.; Li, W.W.; Noble, W.S. MEME SUITE: Tools for motif discovery and searching. *Nucleic Acids Res.* **2009**, *37*, W202–W208. [CrossRef]
86. Yue, C.; Chen, Q.; Hu, J.; Li, C.; Luo, L.; Zeng, L. Genome-Wide Identification and Characterization of *GARP* Transcription Factor Gene Family Members Reveal Their Diverse Functions in Tea Plant (*Camellia sinensis*). *Front. Plant Sci.* **2022**, *13*, 947072. [CrossRef]

87. Hao, X.; Tang, H.; Wang, B.; Yue, C.; Wang, L.; Zeng, J.; Yang, Y.; Wang, X. Integrative transcriptional and metabolic analyses provide insights into cold spell response mechanisms in young shoots of the tea plant. *Tree Physiol.* **2018**, *38*, 1655–1671. [CrossRef] [PubMed]
88. Li, Y.; Wang, X.; Ban, Q.; Zhu, X.; Jiang, C.; Wei, C.; Bennetzen, J.L. Comparative transcriptomic analysis reveals gene expression associated with cold adaptation in the tea plant *Camellia sinensis*. *BMC Genom.* **2019**, *20*, 624. [CrossRef]
89. Pertea, M.; Kim, D.; Pertea, G.M.; Leek, J.T.; Salzberg, S.L. Transcript-level expression analysis of RNA-seq experiments with HISAT, StringTie and Ballgown. *Nat. Protoc.* **2016**, *11*, 1650–1667. [CrossRef]

plants

MDPI

Article

Genome-Wide Identification of the *MPK* Gene Family and Expression Analysis under Low-Temperature Stress in the Banana

Zhengyang Fan [1], Bianbian Zhao [1], Ruilian Lai [1,2], Huan Wu [1], Liang Jia [1], Xiaobing Zhao [1], Jie Luo [1], Yuji Huang [1], Yukun Chen [1], Yuling Lin [1] and Zhongxiong Lai [1,*]

[1] Institute of Horticultural Biotechnology, Fujian Agriculture and Forestry University, Fuzhou 350002, China; ffzzyy789@163.com (Z.F.); zbb@fafu.edu.cn (B.Z.); lairl0618@163.com (R.L.); wuhuan980422@163.com (H.W.); kieraj@163.com (L.J.); zxb1344@126.com (X.Z.); luojie1998s@163.com (J.L.); yjhuang2004@163.com (Y.H.); cyk68@163.com (Y.C.); buliang84@163.com (Y.L.)

[2] Fruit Research Institute, Fujian Academy of Agricultural Sciences, Fuzhou 350013, China

* Correspondence: laizx01@163.com

Abstract: Mitogen-activated protein kinases (MAPKs and MPKs) are important in the process of resisting plant stress. In this study, 21, 12, 18, 16, and 10 *MPKs* were identified from *Musa acuminata*, *Musa balbisiana*, *Musa itinerans*, *Musa schizocarpa*, and *Musa textilis*, respectively. These *MPKs* were divided into Group A, B, C, and D. Phylogenetic analysis revealed that this difference in number was due to the gene shrinkage of the Group B subfamily of *Musa balbisiana* and *Musa textilis*. KEGG annotations revealed that K14512, which is involved in plant hormone signal transduction and the plant–pathogen interaction, was the most conserved pathway of the *MPKs*. The results of promoter cis-acting element prediction and *foc*TR4 (*Fusarium oxysporum* f. sp. *cubense* tropical race 4) transcriptome expression analysis preliminarily confirmed that *MPKs* were relevant to plant hormone and biotic stress, respectively. The expression of *MPKs* in Group A was significantly upregulated at 4 °C, and dramatically, the *MPKs* in the root were affected by low temperature. miR172, miR319, miR395, miR398, and miR399 may be the miRNAs that regulate *MPKs* during low-temperature stress, with miR172 being the most critical. miRNA prediction and qRT-PCR results indicated that miR172 may negatively regulate *MPKs*. Therefore, we deduced that *MPKs* might coordinate with miR172 to participate in the process of the resistance to low-temperature stress in the roots of the banana. This study will provide a theoretical basis for further analysis of the mechanism of *MPKs* under low-temperature stress of bananas, and this study could be applied to molecular breeding of bananas in the future.

Keywords: banana; MAPK; low temperature stress; miR172

1. Introduction

MAPK cascades are conserved in eukaryotes, which play an important role in plant development [1–3] and the stress-response mechanism [4–6]. Upstream signals of MPKs include mitogen-activated protein kinase kinase (MAPKK, MKK, and MEK) and mitogen-activated protein kinase kinase kinase (MAPKKK, MKKK, and MEKK) [7]. In general, MPKs relate to a variety of kinases with similar functions, which diversify the signal transduction pathways in plants to meet the needs of plant growth and development in complex environments. MKKKs, MKKs, and MPKs are phosphorylated in the signal transduction pathways. MPKs are activated and transferred to the nucleus or cytoplasm [8], where they then promote the expression of transcription factors or phosphorylate downstream substrates [9]. Eventually, these phosphorylation processes enable plant cells to respond to external signals.

Bananas are widely cultivated in tropical and subtropical regions as one of the main food crops in the world. However, bananas are sensitive to low temperatures, which indicates that high temperature and humidity conditions are better for their growth. Although researchers have conducted a large number of studies, the molecular mechanism underlying the cold resistance of bananas has not been completely understood [10–12]. AtMPK3/6 are negative regulators of the AtCBF-dependent cold response network in *Arabidopsis*. The AtICE1 protein has a capacity of being phosphorylated by AtMPK3/6 to inhibit downstream AtCBF expression, thereby affecting the tolerance of plants to low-temperature conditions [13]. Cascades of AtMEKK1-AtMKK2-AtMPK4 and AtMKK4/5-AtMPK3/6 play important roles in cold stress, in which a low temperature rapidly induces the expression of *AtMPK3/6* [14]. As a result of being phosphorylated by AtMPK3/6, the AtICE1 protein will not be able to regulate AtCBF effectively as a result. Eventually, the cold resistance of *Arabidopsis thaliana* is reduced, thereby showing how a negative regulatory network works [15]. Furthermore, the cascade of AtMEKK1-AtMKK2-AtMPK4 inhibits the phosphorylation of AtICE1 by AtMPK3/6, improving the stability of AtICE1, which will enhance the cold resistance of the plant as a result [16]. Interestingly, compared with AtMPK3/6, OsMPK3, which is present in rice, plays an opposite role under cold stress. Rice responds to cold stress through the OsMPK3-OsICE1-OsTPP1 pathway. OsICE1 activates OsTPP1 to promote the accumulation of trehalose in rice to resist low-temperature stress [17]. Therefore, it is significant for scientific research and agricultural production to analyze the mechanism of banana *MPKs* under cold stress. MAPK cascades have the capacity of improving the cold resistance of waxy corn [18]. Mel enhanced cold tolerance in the winter turnip rape crop (*Brassica rapa* L.), which depends on *MPK3/6* [19]. *SlMPK1/2* can interact with *SlBBX13* to improve cold tolerance in the tomato plant [20]. The COG1–OsSERL2 complex can promote the activation of OsMPK3 under low temperatures, thereby enhancing the cold resistance of japonica rice [21]. Previous studies have identified 25 members of the *MPK* gene family in the first version of the genome of *Musa acuminata* [22]; however, the differences and associations between the *MPKs* of different banana genomes have not been analyzed as of yet.

In this study, we analyzed the *MPKs* of five representative bananas. The *MPK* gene family was identified as *Musa acuminata*, *Musa balbisiana*, *Musa itinerans*, *Musa schizocarpa*, and *Musa textilis*. We then screened out a gene that is very critical to cold stress. In addition, the relationship between miR172 and *MPKs* was also researched. This study will provide the theoretical basis for further analysis of the mechanism of *MPKs* under cold stress in bananas, and this study may be applied to the molecular breeding of bananas in the future.

2. Results

2.1. Identification of the MPK Gene Family in Bananas

Based on the banana genome database, 21, 12, 18, 16, and 10 *MPK* family members were identified in *Musa acuminata*, *Musa balbisiana*, *Musa itinerans*, *Musa schizocarpa*, and *Musa textilis*, respectively. According to the location of these genes on chromosomes, they were named *MaMPK1-21*, *MbMPK1-12*, *MiMPK1-18*, *MsMPK1-16*, and *MtMPK1-10*, respectively. The number of amino acids, molecular weight, theoretical pI, instability index, and aliphatic index of the *MPKs* were between 366–1155, 41.68–130.81 KDa, 5.34–9.73, 28.07–51.77, and 99.1–78.41, respectively. GRAVY results revealed that these *MPKs* were hydrophilic proteins (Table S1).

2.2. Phylogenetic Analysis of the MPK Gene Family

To explore the relationship of these *MPKs* in bananas, MEGA-X was used to analyze the amino acid sequences and to construct the neighbor-joining (NJ) phylogenetic tree. According to the classification of *AtMPKs*, the phylogenetic tree was divided into groups A, B, C, and D (Figure 1).

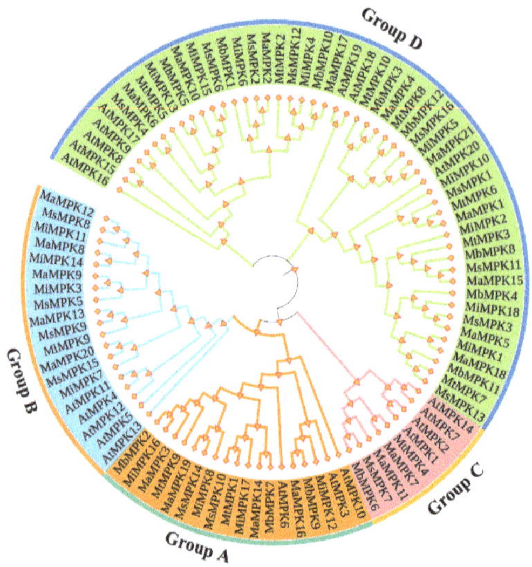

Figure 1. Phylogenetic analysis of the *MPK* gene family in *Arabidopsis thaliana* (At), *Musa acuminata* (Ma), *Musa balbisiana* (Mb), *Musa itinerans* (Mi), *Musa schizocarpa* (Ms), and *Musa textilis* (Mt).

Due to the difference in the number of *MPKs* in genomes, the number of subfamilies was counted according to the phylogenetic tree (Figure 2). It was clear that the quantity of *MPKs* was approximately equal in Group A, C, and D. Notably, no family members of *Musa balbisiana* and *Musa itinerans* were clustered in Group B. Therefore, the main reason for the difference in the amount of *MPKs* in bananas was the gene shrinkage of the Group B subfamily of *Musa balbisiana* and *Musa itinerans*.

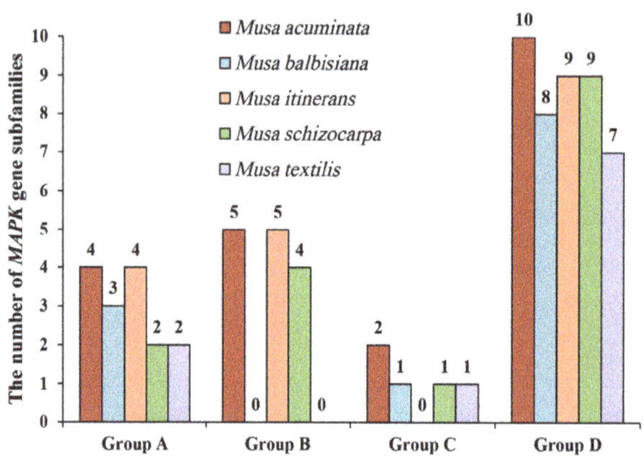

Figure 2. The number of *MPK* subfamily members.

2.3. Colinearity Analysis of the MPK Gene Family

Collinear analysis of the whole genome was used to understand gene duplication events of the *MPK* gene family. Meanwhile, unknown base content, GC content, and gene density data were supplemented as references. Unfortunately, the *Musa itinerans* and *Musa textilis* genomes were not assembled to the pseudochromosome level, meaning only *Musa acuminata*, *Musa balbisiana*, and *Musa schizocarpa* were analyzed (Figure 3). There were eight, six, and three segmental duplications and eight, zero, and three tandem duplications observed between the *MPK* family of *Musa acuminata*, *Musa balbisiana*, and *Musa schizocarpa*, respectively (Figure 3a,c,e). Interestingly, the 3–10 MB regions of Chr04 in *Musa acuminata* and *Musa schizocarpa* were discovered through eight and three times of segmental duplication, respectively (Figure 3b,f), but were not found in this region of *Musa balbisiana* (Figure 3d). The interspecific features of *MbMPKs* may come into being from long-term evolution and natural selection. These gene duplication events may refer to the functional differentiation and expansion of the *MPK* gene family in bananas.

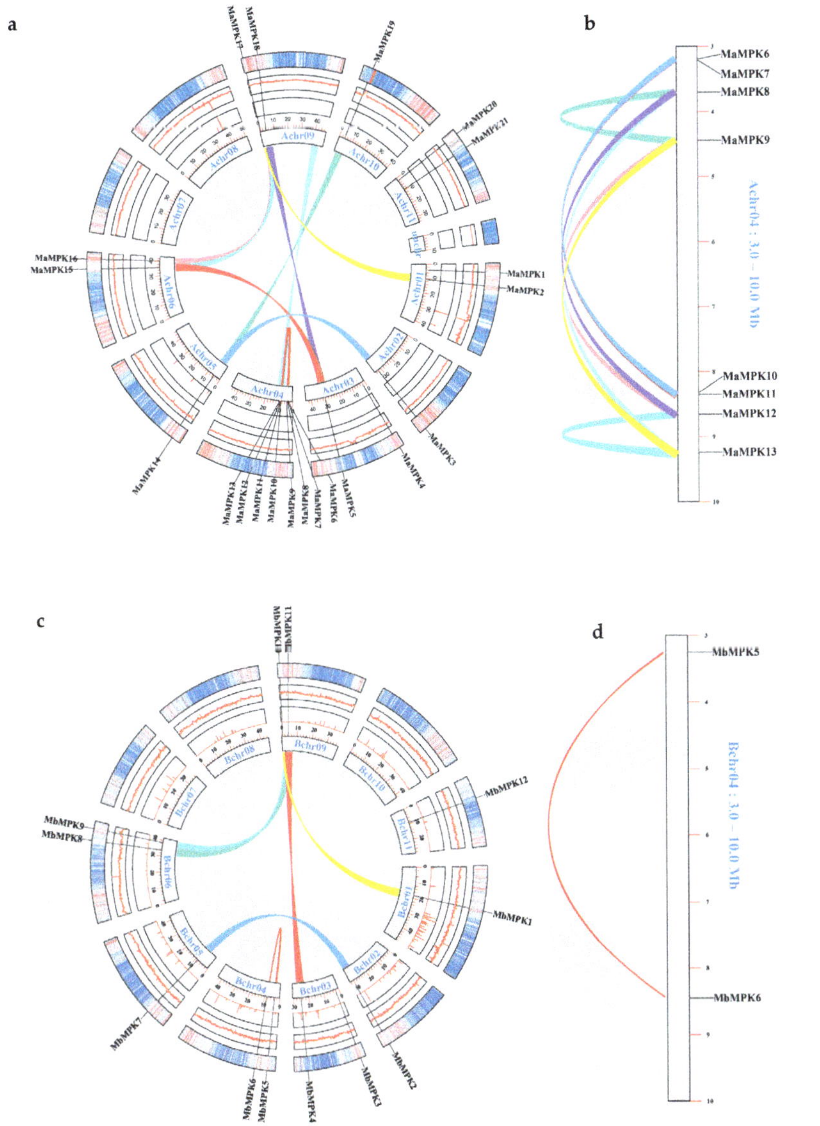

Figure 3. *Cont.*

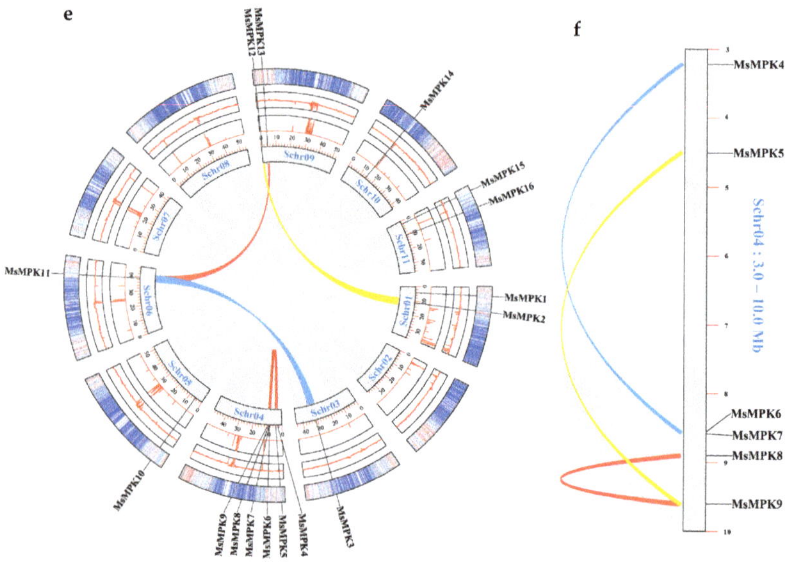

Figure 3. Collinearity analysis of the *MPK* gene family in bananas. The rings from the outside to the inside represent gene density, GC content, unknown base content, and pseudochromosomes. (**a**) Collinearity analysis of *Musa acuminata.* The colorful lines indicate the collinearity between the *MaMPK*s. (**b**) The 3–10 MB region of the Chr04 of *Musa acuminata.* (**c**) Collinearity analysis of *Musa balbisiana.* The colorful lines indicate the collinearity between the *MbMPK*s. (**d**) The 3–10 MB region of the Chr04 of *Musa balbisiana.* (**e**) Collinearity analysis of *Musa schizocarpa.* The colorful lines indicate the collinearity between the *MsMPK*s. (**f**) The 3–10 MB region of the Chr04 of *Musa schizocarpa.*

The results of collinearity analysis among *Musa acuminata, Musa balbisiana,* and *Musa schizocarpa* reveal that the positions of these *MPKs* on the chromosomes were also very similar (Figure 4), indicating that *MPKs* are quite conserved in bananas. Similarly, *MPKs* also have close genetic relationships in *Musa acuminata, Musa balbisiana,* and *Musa schizocarpa.* These results suggested that *MPKs* located in 3–10 MB of Chr04 may have evolved from the same gene.

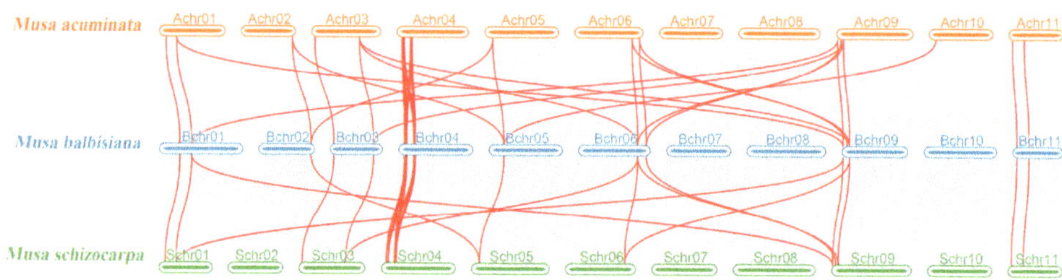

Figure 4. Collinearity analysis of the *MPK* gene family in *Musa acuminata, Musa balbisiana,* and *Musa schizocarpa.* The red lines mark the *MPK* gene pairs in bananas.

2.4. KEGG Annotation Analysis of the MPK Gene Family

To further verify the functional similarities and differences of the *MPK* gene family, functional annotations were made on their protein sequences (as shown in Figure 5). Most of the *MPKs* were annotated to K04371, K14512, K20535, K20536, and K20600, and others were not annotated. It is noteworthy that K14512 is the most conserved in bananas and existed in all species. K14512 is divided into three pathways, which plays an important role in plant growth and development. These pathways contains plant hormone signal transduction (map04075), plant–pathogen interaction (map04626), and MAPK signaling pathway–plant (map04016).

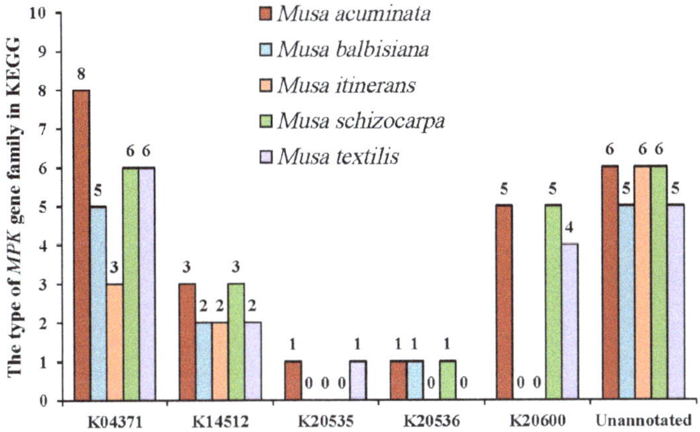

Figure 5. The type of *MPK* gene family in KEGG. One protein sequence may correspond to multiple entries.

2.5. Prediction Analysis of the MPKs Promoter Cis-Acting Elements

The results of promoter analysis showed that *MPKs* promoter cis-acting elements were related to plant hormones, the stress response, biological clocks, and plant development (Figure 6). In terms of plant hormones, it mainly contained abscisic acid, auxin, gibberellin, methyl jasmonate, and salicylic acid response elements, which, to some extent, confirmed the conclusion that *MPK* participated in plant hormone signal transduction pathways in KEGG. The stress response is mainly involved in anaerobic induction, stress defense, drought inducibility, low-temperature responsiveness, and wound responsiveness. Elements of cell cycle regulation and circadian control response are related to the biological clock regulation system. Finally, endosperm expression, light responsiveness, meristem expression, seed-specific regulation, and zein metabolism regulation elements are involved in plant development.

Overall, the most responsive element in the promoters was light responsiveness, while the least responsive element was wound responsiveness. This result suggested that *MPKs* may mainly rely on light, abscisic acid, and the MeJA metabolic pathways, which was found to be similar across different species of bananas. However, wound responsiveness was only found in the promoters of *MaMPK7* and *MiMPK2* (Figure 6a,b), but not in the others (Figure 6c–e). Seed-specific regulation did not exist in *Musa schizocarpa* (Figure 6d) but was found in all the others. In summary, the promoters of *MPKs* are both conserved and specific. A variety of promoter cis-acting elements could provide the *MPKs* with more functions.

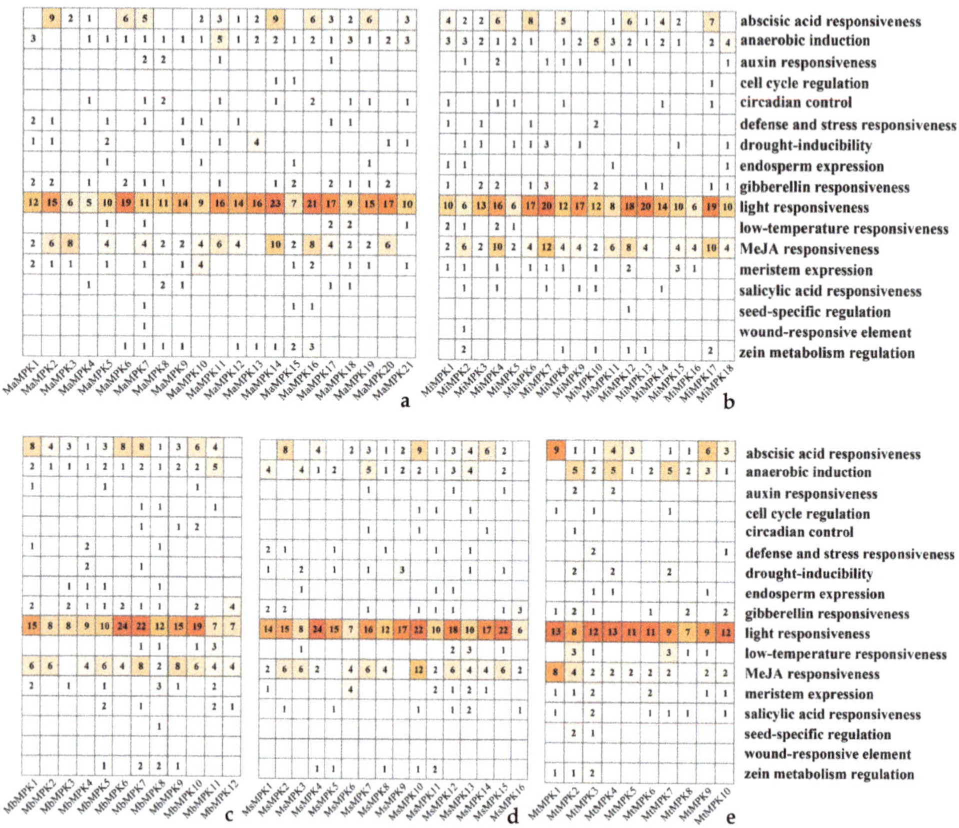

Figure 6. Prediction analysis of *MPKs* promoter cis-acting elements in *Musa acuminata* (**a**), *Musa itinerans* (**b**), *Musa balbisiana* (**c**), *Musa schizocarpa* (**d**), and *Musa textilis* (**e**). The color depth of the square represents the number of cis-acting elements.

2.6. Prediction of miRNA Regulation on MaMPKs

A total of 119 types of miRNAs regulating the *MPKs* were predicted by psRNATarge (Figure 7). Interestingly, sixteen miRNAs were found to regulate three genes simultaneously. *MaMPK3, MaMPK14,* and *MaMPK16* were regulated by miR5303; *MaMPK14, MaMPK16,* and *MaMPK19* were regulated by miR395; *MaMPK3, MaMPK14,* and *MaMPK19* were regulated by the other 14 miRNAs. Notably, the regulation of miR172 on *MaMPK3, MaMPK14,* and *MaMPK19* was found to be the most significant. Therefore, three miR172 family members were screened from the banana miRNA database for expression analysis under low-temperature stress (Figure 8). The results showed that mac-miR172a was significantly downregulated, while mac-miR172b and mac-miR172d were hardly affected by the low temperature. The expression level of mac-miR172a was about 4-fold higher than that of mac-miR172b and mac-miR172d at 28 °C. The expression of mac-miR172a was downregulated at 4 °C, indicating that mac-miR172a plays a major role in low-temperature stress. These results indicate that miR172 has a functional redundancy in bananas. In addition, miRNAs associated with cold stress, such as miR319, miR395, miR398, and miR399, were also predicted. This suggested that the regulatory mechanisms of miRNAs on *MaMPKs* were diverse.

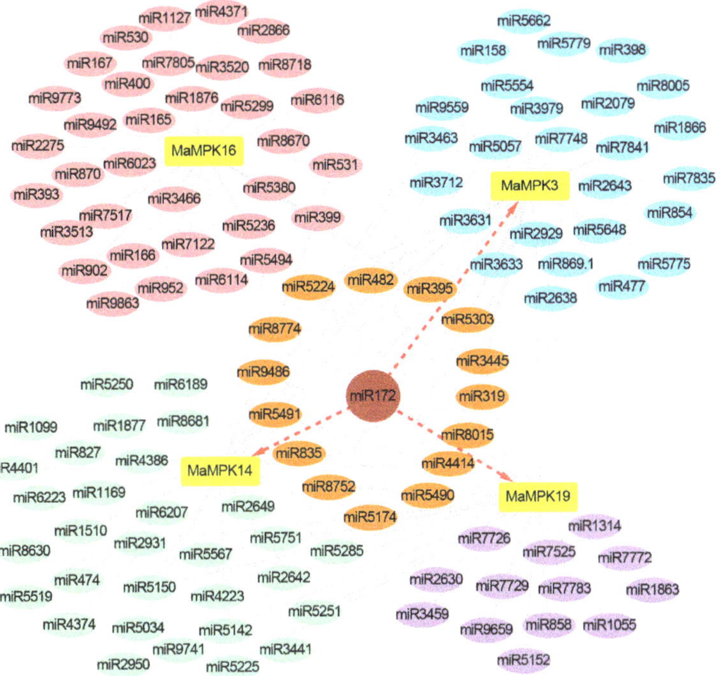

Figure 7. The interaction network between miRNAs and *MaMPKs*. The yellow rectangles represent genes, and the other ellipses represent miRNAs. The orange and red ellipses represent miRNAs that regulate three genes simultaneously. The pink, blue, green, and purple ellipses represent miRNAs that regulate one gene. This interaction network was visualized using Cytoscape.

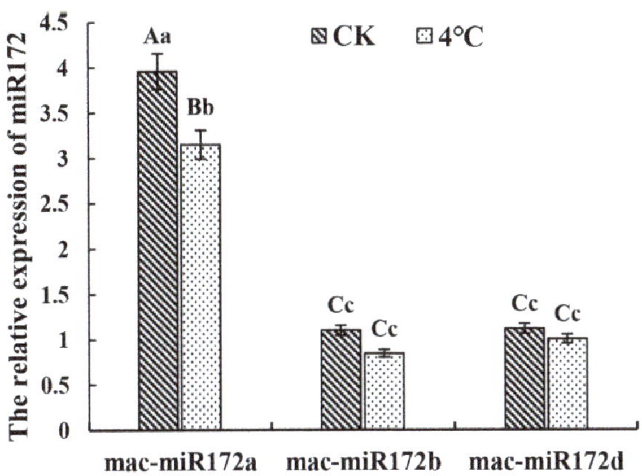

Figure 8. The expression of miR172 under low-temperature stress. Uppercase and lowercase letters indicate a significant difference at $p < 0.01$ and 0.05.

2.7. Expression Patterns of MPKs under Biotic and Abiotic Stress

The transcriptome FPKM data were used to analyze the expression patterns of *MPKs*, including pathogen stress and low-temperature stress (Figure 9). Thirteen genes were identified with low expression levels after 0, 1, 4, and 7 days of *Fusarium oxysporum* f. sp.

cubense tropical race 4 (*Foc*TR4) infection (Figure 9a), and their expression decreased with the extension of the infection time. The expression levels of the other *MPKs* were found to be higher than these 13 genes. Among them, the expression of *MaMPK8* and *MiMPK14* belonging to the Group B subfamily decreased most significantly. It is noteworthy that most of these *MPKs* are highly expressed. The expression of *MiMPK12*, *MaMPK16*, and *MbMPK9* increased most significantly following pathogen induction. *MsMPK11*, *MtMPK3*, *MiMPK2*, *MaMPK15*, and *MbMPK8* were downregulated from 0–1 d and upregulated from 1–14d. These results indicate that most *MPKs* can respond to *foc*TR4, especially *MiMPK12*, *MaMPK16*, and *MbMPK9*, and proves that *MPKs* play a crucial role in resisting pathogen stress in bananas. In addition, *MPKs* may rely on a large signal transduction network to either directly or indirectly regulate the expression of disease resistance genes to resist pathogen infection. These results also prove the correctness of the KEGG annotations.

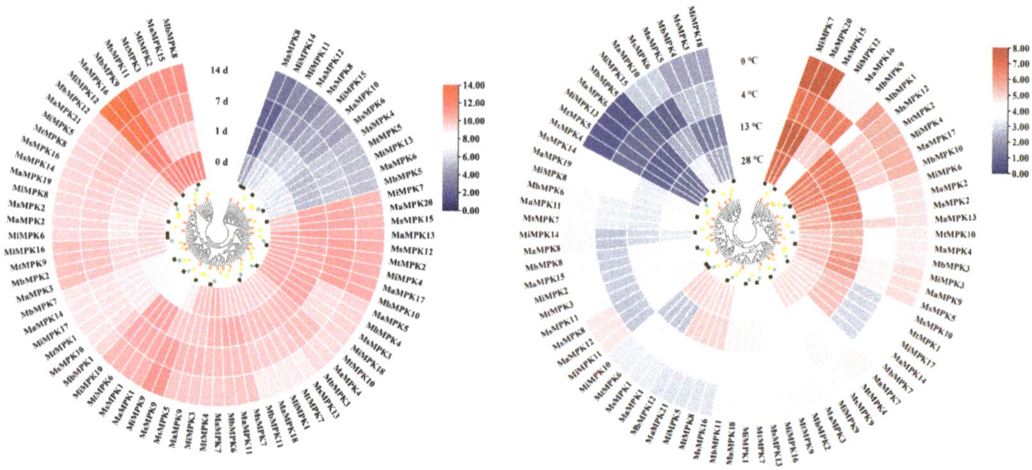

Figure 9. Expression patterns of *MPKs* under biotic and abiotic stress. (**a**) *Fusarium oxysporum* f. sp. *cubense* tropical race 4 (*Foc*TR4)-infected bananas at 0, 1, 4, and 7 days. (**b**) Bananas were treated at 0 °C, 4 °C, 13 °C, and 28 °C for 24 h. Red represents high expression and dark blue represents low expression.

On the whole, about half of the *MPKs* were highly expressed under low-temperature stress, and only a few were lowly expressed (Figure 9b). We found that several *MPKs* were highly expressed at each temperature, but also that the expression of some *MPKs* was induced to increase through low temperatures. For example, the expression levels of *MiMPK7*, *MaMPK20*, and *MsMPK15* were the highest at each temperature, while the expression levels of *MiMPK12*, *MaMPK16*, and *MbMPK9* were increased after low temperature induction. Under low-temperature stress, a few genes were differentially expressed from 28 °C to 13 °C. However, the expression of all genes was found to be significantly different from 13 °C to 4 °C, and the situation was also similar from 4 °C to 0 °C. Interestingly, while the expression levels of *MiMPK12*, *MaMPK16*, and *MbMPK9* were upregulated from 28 °C to 4 °C, they were downregulated at 0 °C in the high expression region. *MiMPK15*, *MaMPK10*, and *MsMPK6* were found to be downregulated and upregulated at 0 °C in the low expression region. From 13 °C to 4 °C, the expression levels of all genes at 4 °C were markedly different from 13 °C, which indicates that for the banana plant, the key temperature of low-temperature stress was 4 °C.

2.8. The Expression Analysis of Group A under Low-Temperature Stress

Phylogenetic tree, KEGG annotation analysis, and transcriptome results all suggested that Group A may participate in cold resistance, resulting in our focus on *MaMPK3*, *MaMPK14*, *MaMPK16*, and *MaMPK19*, which belong to Group A in *Musa acuminata*. The four genes in the genome of *Musa acuminata* was expressed in the roots, pseudostems, and leaves (Figures 10 and 11). These results showed that the expression levels of *MaMPK3* and *MaMPK19* were upregulated at 4 °C for 24 h. It also indicated that *MaMPK3* and *MaMPK19* were more sensitive to low temperatures. In contrast, the expression of *MaMPK14* did not change significantly in the pseudostems and leaves but was upregulated in the roots. The expression of *MaMPK16* decreased in the pseudostems and leaves while it was increased in the roots. The expression level in the roots was significantly enhanced. The expression levels of *MaMPK3* and *MaMPK19* were more intense compared to the other genes, which were upregulated about 20-fold and 35-fold compared with the control group, respectively. The expression levels of *MaMPK14* and *MaMPK16* in the roots increased about 4-fold. Together, these results indicate that the regulatory mechanisms of *MPKs* under low-temperature stress are diversified in bananas.

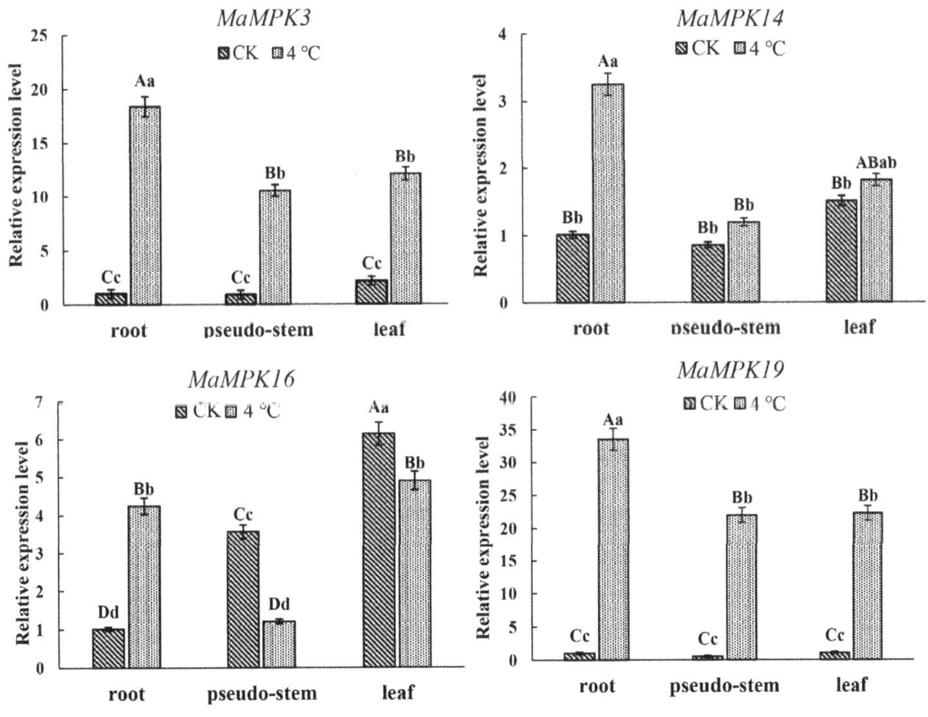

Figure 10. The expression analysis of Group A under low-temperature stress. Uppercase and lowercase letters indicate a significant difference at $p < 0.01$ and 0.05.

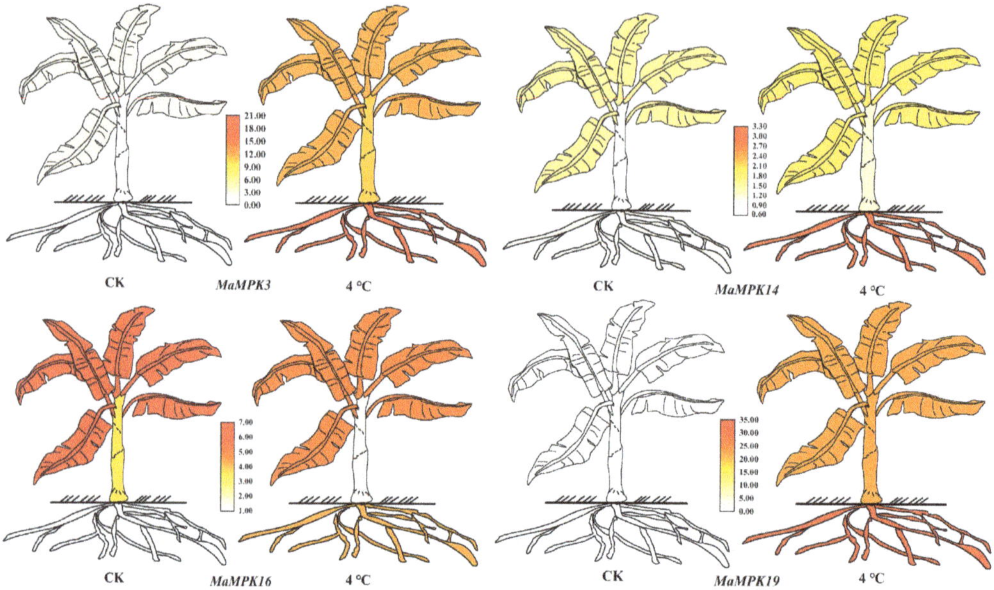

Figure 11. The expression of *MaMPKs* in different tissues. Red represents high expression and white represents low expression. These cartoon heat maps were visualized using TBtools.

3. Discussion

3.1. Interspecific Specificity and Genome Quality Together Determine the Number of MPKs in Bananas

An interesting phenomenon was that the number of Group B species was associated with the species themselves, while Group A, C, and D were almost unaffected by the species. The gene shrinkage of the Group B subfamily of *Musa balbisiana* and *Musa textilis* was the reason why their *MPKs* were significantly less than *Musa acuminata, Musa schizocarpa,* and *Musa itinerans*. Combined with the results of collinearity analysis, *Musa balbisiana* did not display many tandem duplications in Chr04: 3–10 MB, which was different from *Musa acuminata* and *Musa schizocarpa*. The results showed that the number of *MPKs* was contracted when *Musa balbisiana* experienced WGD (whole genome duplication) three times [23,24]. However, since the genomes of *Musa itinerans* and *Musa textilis* have not been assembled to the pseudochromosome level in this study [25,26], they were unable to verify the tandem duplication events in Chr04: 3–10 MB. This inference needs to be proven with the aid of more detailed genomic data of *Musa itinerans* and *Musa textilis*. Therefore, it is also an interesting research direction to further verify the differences of the *MPKs* in bananas through designing genetic complementation experiments for the genes in Chr04: 3–10 MB.

Genome-wide identification results showed that there were differences in the number of *MPKs* in bananas. These differences were closely related to the species, genome sequencing quality, and assembly quality. Five chromosomes achieved a complete telomere-to-telomere assembly of *Musa acuminata* for long-term improvement and optimization of the genome [27–29]. A high-quality genome typically indicates that more complete genes have been discovered, which may be the reason why the number of *MPKs* in *Musa acuminata* was found to be more than the others. The *Musa acuminata* and *Musa schizocarpa* genomes were assembled based on nanopore sequencing [30], which had higher continuity and integrity than the *Musa balbisiana* genomes. Overall, the assembly quality of *Musa acuminata, Musa balbisiana,* and *Musa schizocarpa* was better than of *Musa itinerans* and *Musa textilis*. According to the current genomic data, the number of *MPKs* is typically determined through interspecific specificity and genome quality. In future studies, more advanced genome sequencing techniques will be used to obtain high-quality genomes. It is an essential subject to completely eliminate the interference of sequencing and assembly techniques, which will not only benefit the comprehension of the influence of interspecific differences on the number of *MPKs* but will also facilitate the bioinformatics analysis and functional verification of the key genes. This will make outstanding contributions to the breeding of bananas using molecular biotechnology.

3.2. MPKs Retain the Most Critical Functions during Evolution in the Banana

A large number of studies have shown that Group A and B are involved in abiotic stress and biotic stress [31–33]. The number of K14152 pathways is roughly equal across all species of bananas, and all genes of K14152 belong to the Group A subfamily. Notably, the K14152 is closely associated with plant hormone signal transduction and the plant–pathogen interaction. In terms of plant immunity, the resistance of rice blast can be enhanced using the key immune component *OsWRKY31* of the OsMPK cascade [34]. *AtMPK3/6* can be phosphorylated through activating the extracellular ATP receptor P2K1, which improves the defense ability of plants to pathogens [35]. *AtMPK3/6* is involved in the regulation of plant immunity by SUMO (small ubiquitin-related modifier protein) and RNA-binding proteins [36]. In addition, the immune response of mechanosensory trichome cells can be activated through *AtMPK3/6* [37,38].

MPKs play a vital role in plant immunity depending on *MAPKKK3/MAPKKK5-MKK4/MKK5-MPK3/MPK6* [39,40]. The interaction between upstream and downstream genes forms a huge signal transduction network and regulates gene expression. *MPKs* associate with plant hormone synthesis, transport, and signal transduction [41]. The *MKK4/5-MPK6* pathway regulates the polar auxin transport in *Arabidopsis thaliana* [42]. *AtMPK3/6* can phosphorylate ACS in the ethylene synthesis pathway and participate in the process of response to abiotic stress or pathogen infection [43]. *LeMPK1* and *LeMPK2* are involved in the JA synthesis pathway to enhance tomato resistance to pests and diseases [44]. Similarly, *AtMPK3/6* participates in the immune response in the SA signal transduction pathway [45]. The promoter cis-acting elements showed that 76.62%, 25.97%, 88.31%, and 35.06% of the members contained the ABA, Aux, MeJA, and SA response elements, respectively; it indicated that *MaMPKs* were closely related to plant hormones. In conclusion, K14152 is the most conserved entry in *MaMPKs*, and it retains the plant hormone signal transduction pathway and the plant–pathogen interaction pathway under long-term evolution and natural selection. Promoter cis-acting element prediction and transcriptome data can preliminarily prove that *MPKs* play an important role in plant hormone signal transduction and in the plant–pathogen interaction in bananas.

3.3. MaMPKs Are Negatively Regulated by miR172 in Low–Temperature Stress

In this study, the expression levels of *MaMPK3*, *MaMPK14*, *MaMPK16*, and *MaMPK19* were significantly upregulated under low-temperature stress. In contrast, mac-miR172a was downregulated, but there was no significant change in the expression levels of mac-miR172b and mac-miR172d. This indicated that mac-miR172a negatively regulates *MaMPKs*. The ICE1-CBF-COR pathway has been widely studied in the cold resistance of bananas. It found that the cold resistance of Dajiao (*Musa* spp. ABB Group) was enhanced after *MpICE1* overexpression, while weakened after RNA interference with *MpMAPK3* [46,47]. It indicated that the MAPK cascade played a positive regulatory role in the cold resistance network of bananas.

A noteworthy phenomenon was that the expression of *MaMPKs* in the leaves was at its highest at 28 °C, and in the roots was highest at 4 °C. The roots were most markedly affected by low temperature. A previous study found that roots improved the ability of the stem cell niche to resist low-temperature stress through auxin accumulation and programmed cell death [48]. *AtMPK6* and its substrate EB1c (end-binding protein 1c) regulated the mitosis of root epidermal cells in *Arabidopsis* [49]. The RGF1–RGI1 ligand–receptor complex regulated downstream PLT1 and PLT2 through YDA-MKK4/5-MPK3/6 and then regulated root meristem development, which has been demonstrated by researchers from different perspectives [50,51]. The stem tip of *Arabidopsis thaliana* converts high-temperature signals into biochemical signals and transmits them to the roots to stimulate the response of the whole plant [52]. From what has been discussed above, *MaMPKs* expression is activated through cold stress and may mainly act on the roots. Therefore, it has been speculated that leaves may transmit low-temperature signals to the roots through the connection between aboveground and underground or the signal transduction pathways. In this way, the roots can provide nutrients or water for the overground to alleviate cold stress.

4. Materials and Methods

4.1. Plant Materials

The test-tube plantlet of 'Brazilian' (*Musa* AAA, Cavendish) was placed at 28 °C for 5 days and were then transplanted into the substrate. The cultural conditions were as follows: temperature 26 ± 2 °C, relative humidity 60–80%, light intensity 2500 Lux, and 12 h light/12 dark. Bananas were moved into the incubator at 4.0 ± 0.5 °C for 24 h, and the other conditions were the same as above. The leaves, pseudostems, and roots were quickly cut into pieces, frozen in liquid nitrogen, and then stored at −80 °C.

4.2. Identification of MPK Gene Family in Bananas

The genomic data of *Musa acuminata, Musa balbisiana, Musa itinerans, Musa schizocarpa,* and *Musa textilis* were downloaded from the Banana Genome Database [53] (https://banana-genome-hub.southgreen.fr/content/download, accessed on 15 April 2022). The gene and amino acid sequences of *AtMPKs* were downloaded from the TAIR database (https://www.arabidopsis.org/index.jsp, accessed on 21 April 2022). The candidate gene sequences of *MPKs* were screened using BLAST. The CDD database (https://www.ncbi.nlm.nih.gov/Structure/bwrpsb/bwrpsb.cgi, accessed on 22 April 2022) was used to screen the domains, and the genes containing the STKc_TEY_MAPK and STKc_TDY_MAPK domains were identified as the *MPK* gene family in bananas. Finally, Tbtools [54] was used to visualize the domains.

4.3. Protein Physicochemical Properties and Domain Analysis of MPKs

ExPAsy (https://web.expasy.org/protparam/, accessed on 28 April 2022) was used to predict the number of amino acids, molecular weight (Mw), theoretical pI, instability index, aliphatic index, and the grand average of hydropathicity (GRAVY) of the MPK proteins.

4.4. Phylogenetic Tree, Collinearity and Chromosome Localization Analysis

The phylogenetic tree containing *Arabidopsis thaliana, Musa acuminata, Musa balbisiana, Musa itinerans, Musa schizocarpa,* and *Musa textilis* was constructed through neighbor-joining of MEGA-X and beautified with iTOL (https://itol.embl.de/, accessed on 2 May 2022). Based on the gene location in the genomes, collinearity analysis was shown using Tbtools.

4.5. KEGG Annotation Analysis and Promoter Cis-Acting Element Prediction

The protein sequences of *MPKs* were uploaded to KofamKOALA (https://www.genome.jp/tools/kofamkoala/, accessed on 29 August 2022) to predict their KO numbers in KEGG (Kyoto Encyclopedia of Genes and Genomes). The upstream 2000 bp sequences of the CDS were extracted for promoter cis-acting element prediction with Plant CARE (http://bioinformatics.psb.ugent.be/webtools/plantcare/html/, accessed on 29 April 2022).

4.6. Expression Analysis of MPKs and Prediction of miRNA

FPKM (fragments per kilobase million) values of *MPKs* were extracted from *foc*TR4 [55] and low temperature [56] transcriptome data, which were displayed using Tbtools. The psRNATarget [57] (https://www.zhaolab.org/psRNATarget/, accessed on 12 November

2022) was used to predict the miRNA that regulates the *MPKs*, and then Cytoscape 3.8.2 was used to draw the interaction network between the miRNAs and genes.

4.7. RNA Extraction and qRT-PCR

Total RNA was extracted with the RNAprep Pure Plant Plus Kit (TIANGEN, Beijing, China). The RNA reverse transcription used the 1st strand cDNA synthesis superMix for qPCR (Yeasen, Shanghai, China). The primers for qRT-PCR have been listed in Table S2. The reaction system was as follows: 10 μL qPCR SYBR green master mix (Herui, Fuzhou, China), 0.4 μL forward primer (10 μM), 0.4 μL reverse primer (10 μM), 1 μL cDNA, and 8.2 μL ddH_2O. The CAC was used as a reference gene. The reverse transcription of miRNAs used the 1st strand cDNA synthesis Kit by stem-loop (Vazyme, Nanjing, China). The sequences of miR172 were derived from NCBI [58] (https://www.ncbi.nlm.nih.gov/geo/query/acc.cgi?acc=GSE77590, accessed on 12 December 2022). The miR172 reverse transcription and qRT-PCR primers have been displayed in Table S2. The reaction system was as follows: 10 μL 2 × miRNA Universal SYBR qPCR master mix (Vazyme, Nanjing, China), 0.4 μL specific primer (10 μM), 0.4 μL mQ primer R (10 μM), 1 μL cDNA, and 8.2 μL ddH_2O. The qPCR programs were as follows: 95 °C for 5min, 40 cycles of 95 °C for 10 s, and 60 °C for 30 s. The U6 was used as a reference gene. The relative expression levels of *MaMPKs* and miRNAs were calculated using the $2^{-\Delta\Delta Ct}$. SPSS26 was used to analyze the significance test between samples. The instrument used for this experiment was the Roche LightCycler 480.

Supplementary Materials: The following supporting information can be downloaded at: https://www.mdpi.com/article/10.3390/plants12162926/s1, Figure S1: Protein conserved domains of *MPKs* in bananas, Table S1: Basic information and protein physicochemical properties of the *MPK* family in bananas, Table S2: qRT-PCR and reverse transcription primers of *MPKs* and miRNAs, Table S3: annotation of *MPKs* in KEGG.

Author Contributions: Data curation, R.L., H.W., L.J., X.Z. and J.L.; Formal analysis, R.L.; Funding acquisition, Z.L.; Investigation, Z.F. and B.Z.; Methodology, Z.F. and Z.L.; Resources, B.Z.; Supervision, Y.H., Y.C., Y.L. and Z.L.; Visualization, Z.F.; Writing—original draft, Z.F.; Writing—review and editing, B.Z. and Z.L. All authors will be informed about each step of manuscript processing including submission, revision, revision reminder, etc., via emails from our system or assigned assistant editor. All authors have read and agreed to the published version of the manuscript.

Funding: This study was supported by the National Key R&D Program of China (2019YFD1000900), the Earmarked Fund for China Agricultural Research System (GARS-31-15), and the Plateau Discipline Construction Project of Fujian Province (102/71201801101).

Institutional Review Board Statement: Not applicable.

Informed Consent Statement: Not applicable.

Data Availability Statement: Not applicable.

Acknowledgments: We would like to thank the Institute of Horticultural Biotechnology of Fujian Agriculture and Forestry University for providing the conditions for this study.

Conflicts of Interest: The authors declare no conflict of interest.

References

1. Rahiman, N.; Akaberi, M.; Sahebkar, A.; Emami, S.A.; Tayarani-Najaran, Z. Protective effects of saffron and its active components against oxidative stress and apoptosis in endothelial cells. *Microvasc. Res.* **2018**, *118*, 82–89. [CrossRef]
2. Takahashi, Y.; Soyano, T.; Kosetsu, K.; Sasabe, M.; Machida, Y. HINKEL kinesin, ANP MAPKKKs and MKK6/ANQ MAPKK, which phosphorylates and activates MPK4 MAPK, constitute a pathway that is required for cytokinesis in Arabidopsis thaliana. *Plant Cell Physiol.* **2010**, *51*, 1766–1776. [CrossRef]
3. Xu, J.; Zhang, S. Mitogen-activated protein kinase cascades in signaling plant growth and development. *Trends Plant Sci.* **2015**, *20*, 56–64. [CrossRef] [PubMed]
4. Ma, H.; Chen, J.; Zhang, Z.; Ma, L.; Yang, Z.; Zhang, Q.; Li, X.; Xiao, J.; Wang, S. MAPK kinase 10.2 promotes disease resistance and drought tolerance by activating different MAPKs in rice. *Plant J.* **2017**, *92*, 557–570. [CrossRef] [PubMed]

5. Kumar, R.R.; Dubey, K.; Arora, K.; Dalal, M.; Rai, G.K.; Mishra, D.; Chaturvedi, K.K.; Rai, A.; Kumar, S.N.; Singh, B.; et al. Characterizing the putative mitogen-activated protein kinase (MAPK) and their protective role in oxidative stress tolerance and carbon assimilation in wheat under terminal heat stress. *Biotechnol. Rep.* **2021**, *29*, e00597. [CrossRef]
6. Shi, M.; Wang, Z.; Ma, Z.; Song, W.; Lu, W.; Xiao, K. Characterization onTaMPK14, an MAPK family gene of wheat, in modulating N-starvation response through regulating N uptake and ROS homeostasis. *Plant Cell Rep.* **2020**, *39*, 1285–1299. [CrossRef] [PubMed]
7. Su, Y.; Guo, A.; Huang, Y.; Wang, Y.; Hua, J. GhCIPK6aincreases salt tolerance in transgenic upland cotton by involving in ROS scavenging and MAPK signaling pathways. *BMC Plant Biol.* **2020**, *20*, 421. [CrossRef]
8. Wang, W.; Feng, B.; Zhou, J.-M.; Tang, D. Plant immune signaling: Advancing on two frontiers. *J. Integr. Plant Biol.* **2020**, *62*, 2–24. [CrossRef]
9. Wang, N.-N.; Li, Y.; Chen, Y.-H.; Lu, R.; Zhou, L.; Wang, Y.; Zheng, Y.; Li, X.-B. Phosphorylation of WRKY16 by MPK3-1 is essential for its transcriptional activity during fiber initiation and elongation in cotton (*Gossypium hirsutum*). *Plant Cell* **2021**, *33*, 2736–2752. [CrossRef]
10. Liu, W.; Cheng, C.; Chen, F.; Ni, S.; Lin, Y.; Lai, Z. High-throughput sequencing of small RNAs revealed the diversified cold-responsive pathways during cold stress in the wild banana (*Musa itinerans*). *BMC Plant Biol.* **2018**, *18*, 308. [CrossRef]
11. Sanchez Moreano, J.P.; Xu, X.; Aucapina Criollo, C.B.; Chen, X.; Lin, Y.; Munir, N.; Lai, Z. Genome-Wide Identification and Comprehensive Analyses of TCP Gene Family in Banana (*Musa* L.). *Trop. Plant Biol.* **2021**, *14*, 180–202. [CrossRef]
12. Han, Y.-C.; Fu, C.-C. Cold-inducible MaC2H2s are associated with cold stress response of banana fruit via regulating MaICE1. *Plant Cell Rep.* **2019**, *38*, 673–680. [CrossRef]
13. Li, H.; Ding, Y.; Shi, Y.; Zhang, X.; Zhang, S.; Gong, Z.; Yang, S. MPK3-and MPK6-Mediated ICE1 Phosphorylation Negatively Regulates ICE1 Stability and Freezing Tolerance in Arabidopsis. *Dev. Cell* **2017**, *43*, 630–642. [CrossRef] [PubMed]
14. Ding, Y.; Yang, S. Surviving and thriving: How plants perceive and respond to temperature stress. *Dev. Cell* **2022**, *57*, 947–958. [CrossRef] [PubMed]
15. Zhao, C.; Wang, P.; Si, T.; Hsu, C.-C.; Wang, L.; Zayed, O.; Yu, Z.; Zhu, Y.; Dong, J.; Tao, W.A.; et al. MAP Kinase Cascades Regulate the Cold Response by Modulating ICE1 Protein Stability. *Dev. Cell* **2017**, *43*, 618–629. [CrossRef] [PubMed]
16. Ye, K.; Li, H.; Ding, Y.; Shi, Y.; Song, C.; Gong, Z.; Yang, S. BRASSINOSTEROID-INSENSITIVE2 Negatively Regulates the Stability of Transcription Factor ICE1 in Response to Cold Stress in Arabidopsis. *Plant Cell* **2019**, *31*, 2682–2696. [CrossRef] [PubMed]
17. Zhang, Z.; Li, J.; Li, F.; Liu, H.; Yang, W.; Chong, K.; Xu, Y. OsMAPK3 Phosphorylates OsbHLH002/OsICE1 and Inhibits Its Ubiquitination to Activate OsTPP1 and Enhances Rice Chilling Tolerance. *Dev. Cell* **2017**, *43*, 731–743. [CrossRef] [PubMed]
18. Jiang, F.; Lv, S.; Zhang, Z.; Chen, Q.; Mai, J.; Wan, X.; Liu, P. Integrated metabolomics and transcriptomics analysis during seed germination of waxy corn under low temperature stress. *BMC Plant Biol.* **2023**, *23*, 190. [CrossRef]
19. Ma, C.; Pei, Z.-Q.; Bai, X.; Feng, J.-Y.; Zhang, L.; Fan, J.-R.; Wang, J.; Zhang, T.-G.; Zheng, S. Involvement of NO and Ca^{2+} in the enhancement of cold tolerance induced by melatonin in winter turnip rape (*Brassica rapa* L.). *Plant Physiol. Biochem.* **2022**, *190*, 262–276. [CrossRef] [PubMed]
20. Song, J.; Lin, R.; Tang, M.; Wang, L.; Fan, P.; Xia, X.; Yu, J.; Zhou, Y. SlMPK1-and SlMPK2-mediated SlBBX17 phosphorylation positively regulates CBF-dependent cold tolerance in tomato. *New Phytol.* **2023**, *239*, 1887–1902. [CrossRef] [PubMed]
21. Xia, C.; Liang, G.; Chong, K.; Xu, Y. The COG1-OsSERL2 complex senses cold to trigger signaling network for chilling tolerance in japonica rice. *Nat. Commun.* **2023**, *14*, 3104. [CrossRef] [PubMed]
22. Asif, M.H.; Lakhwani, D.; Pathak, S.; Bhambhani, S.; Bag, S.K.; Trivedi, P.K. Genome-wide identification and expression analysis of the mitogen-activated protein kinase gene family from banana suggest involvement of specific members in different stages of fruit ripening. *Funct. Integr. Genom.* **2014**, *14*, 161–175. [CrossRef]
23. Wang, Z.; Miao, H.; Liu, J.; Xu, B.; Yao, X.; Xu, C.; Zhao, S.; Fang, X.; Jia, C.; Wang, J.; et al. Musa balbisiana genome reveals subgenome evolution and functional divergence. *Nat. Plants* **2019**, *5*, 810–821. [CrossRef]
24. Van de Peer, Y.; Fawcett, J.A.; Proost, S.; Sterck, L.; Vandepoele, K. The flowering world: A tale of duplications. *Trends Plant Sci.* **2009**, *14*, 680–688. [CrossRef]
25. Wu, W.; Yang, Y.-L.; He, W.-M.; Rouard, M.; Li, W.-M.; Xu, M.; Roux, N.; Ge, X.-J. Whole genome sequencing of a banana wild relative Musa itinerans provides insights into lineage-specific diversification of the Musa genus. *Sci. Rep.* **2016**, *6*, 31586. [CrossRef]
26. Calano Galvez, L.; Lu Koh, R.B.; Cortez Barbosa, C.F.; Calundre Asunto, J.; Leonido Catalla, J.; Gomez Atienza, R.; Trinidad Costales, K.; Masinsin Aquino, V.; Zhang, D. Sequencing and de Novo Assembly of Abaca (*Musa textilis Nee*) var. Abuab Genome. *Genes* **2021**, *12*, 1202. [CrossRef]
27. D'Hont, A.; Denoeud, F.; Aury, J.-M.; Baurens, F.-C.; Carreel, F.; Garsmeur, O.; Noel, B.; Bocs, S.; Droc, G.; Rouard, M.; et al. The banana (*Musa acuminata*) genome and the evolution of monocotyledonous plants. *Nature* **2012**, *488*, 213–217. [CrossRef] [PubMed]
28. Martin, G.; Baurens, F.-C.; Droc, G.; Rouard, M.; Cenci, A.; Kilian, A.; Hastie, A.; Dolezel, J.; Aury, J.-M.; Alberti, A.; et al. Improvement of the banana "*Musa acuminata*" reference sequence using NGS data and semi-automated bioinformatics methods. *BMC Genom.* **2016**, *17*, 243. [CrossRef]
29. Belser, C.; Baurens, F.-C.; Noel, B.; Martin, G.; Cruaud, C.; Istace, B.; Yahiaoui, N.; Labadie, K.; Hribova, E.; Dolezel, J.; et al. Telomere-to-telomere gapless chromosomes of banana using nanopore sequencing. *Commun. Biol.* **2021**, *4*, 1047. [CrossRef]

30. Belser, C.; Istace, B.; Denis, E.; Dubarry, M.; Baurens, F.-C.; Falentin, C.; Genete, M.; Berrabah, W.; Chevre, A.-M.; Delourme, R.; et al. Chromosome-scale assemblies of plant genomes using nanopore long reads and optical maps. *Nat. Plants* **2018**, *4*, 879–887. [CrossRef] [PubMed]
31. Zhang, M.; Su, J.; Zhang, Y.; Xu, J.; Zhang, S. Conveying endogenous and exogenous signals: MAPK cascades in plant growth and defense. *Curr. Opin. Plant Biol.* **2018**, *45*, 1–10. [CrossRef] [PubMed]
32. Meng, X.; Zhang, S. MAPK Cascades in Plant Disease Resistance Signaling. In *Annual Review of Phytopathology*; VanAlfen, N.K., Ed.; Annual Reviews: Palo Alto, CA, USA, 2013; Volume 51, pp. 245–266.
33. Meng, X.; Xu, J.; He, Y.; Yang, K.-Y.; Mordorski, B.; Liu, Y.; Zhang, S. Phosphorylation of an ERF Transcription Factor by Arabidopsis MPK3/MPK6 Regulates Plant Defense Gene Induction and Fungal Resistance. *Plant Cell* **2013**, *25*, 1126–1142. [CrossRef]
34. Wang, S.; Han, S.; Zhou, X.; Zhao, C.; Guo, L.; Zhang, J.; Liu, F.; Huo, Q.; Zhao, W.; Guo, Z.; et al. Phosphorylation and ubiquitination of OsWRKY31 are integral to OsMKK10-2-mediated defense responses in rice. *Plant Cell* **2023**, *35*, 2391–2412. [CrossRef] [PubMed]
35. Kim, D.; Chen, D.; Ahsan, N.; Jorge, G.L.; Thelen, J.J.; Stacey, G. The Raf-like MAPKKK INTEGRIN-LINKED KINASE 5 regulates purinergic receptor-mediated innate immunity in Arabidopsis. *Plant Cell* **2023**, *35*, 1572–1592. [CrossRef] [PubMed]
36. Verma, V.; Srivastava, A.K.; Gough, C.; Campanaro, A.; Srivastava, M.; Morrell, R.; Joyce, J.; Bailey, M.; Zhang, C.; Krysan, P.J.; et al. SUMO enables substrate selectivity by mitogen-activated protein kinases to regulate immunity in plants. *Proc. Natl. Acad. Sci. USA* **2021**, *118*, e2021351118. [CrossRef] [PubMed]
37. Matsumura, M.; Nomoto, M.; Itaya, T.; Aratani, Y.; Iwamoto, M.; Matsuura, T.; Hayashi, Y.; Mori, T.; Skelly, M.J.; Yamamoto, Y.Y.; et al. Mechanosensory trichome cells evoke a mechanical stimuli-induced immune response in Arabidopsis thaliana. *Nat. Commun.* **2022**, *13*, 1216. [CrossRef] [PubMed]
38. Ai, G.; Li, T.; Zhu, H.; Dong, X.; Fu, X.; Xia, C.; Pan, W.; Jing, M.; Shen, D.; Xia, A.; et al. BPL3 binds the long non-coding RNA nalncFL7 to suppress FORKED-LIKE7 and modulate HAI1-mediated MPK3/6 dephosphorylation in plant immunity. *Plant Cell* **2023**, *35*, 598–616. [CrossRef] [PubMed]
39. Sun, T.; Zhang, Y. MAP kinase cascades in plant development and immune signaling. *EMBO Rep.* **2022**, *23*, e53817. [CrossRef] [PubMed]
40. Devendrakumar, K.T.; Li, X.; Zhang, Y. MAP kinase signalling: Interplays between plant PAMP- and effector-triggered immunity. *Cell. Mol. Life Sci.* **2018**, *75*, 2981–2989. [CrossRef] [PubMed]
41. Jiang, M.; Zhang, Y.; Li, P.; Jian, J.; Zhao, C.; Wen, G. Mitogen-Activated Protein Kinase and Substrate Identification in Plant Growth and Development. *Int. J. Mol. Sci.* **2022**, *23*, 2744. [CrossRef]
42. Zhang, M.; Wu, H.; Su, J.; Wang, H.; Zhu, Q.; Liu, Y.; Xu, J.; Lukowitz, W.; Zhang, S. Maternal control of embryogenesis by MPK6 and its upstream MKK4/MKK5 in Arabidopsis. *Plant J.* **2017**, *92*, 1005–1019. [CrossRef]
43. Zhang, M.; Zhang, S. Mitogen-activated protein kinase cascades in plant signaling. *J. Integr. Plant Biol.* **2022**, *64*, 301–341. [CrossRef] [PubMed]
44. Kandoth, P.K.; Ranf, S.; Pancholi, S.S.; Jayanty, S.; Walla, M.D.; Miller, W.; Howe, G.A.; Lincoln, D.E.; Stratmann, J.W. Tomato MAPKs LeMPK1, LeMPK2, and LeMPK3 function in the systemi n-mediated defense response against herbivorous insects. *Proc. Natl. Acad. Sci. USA* **2007**, *104*, 12205–12210. [CrossRef] [PubMed]
45. Wang, Y.; Schuck, S.; Wu, J.; Yang, P.; Doering, A.-C.; Zeier, J.; Tsuda, K. A MPK3/6-WRKY33-ALD1-Pipecolic Acid Regulatory Loop Contributes to Systemic Acquired Resistance. *Plant Cell* **2018**, *30*, 2480–2494. [CrossRef] [PubMed]
46. Dou, T.-X.; Hu, C.-H.; Sun, X.-X.; Shao, X.-H.; Wu, J.-H.; Ding, L.-J.; Gao, J.; He, W.-D.; Biswas, M.-K.; Yang, Q.-S.; et al. MpMYBS3 as a crucial transcription factor of cold signaling confers the cold tolerance of banana. *Plant Cell Tissue Organ Cult.* **2016**, *125*, 93–106. [CrossRef]
47. Gao, J.; Zhang, S.; He, W.-D.; Shao, X.-H.; Li, C.-Y.; Wei, Y.-R.; Deng, G.-M.; Kuang, R.-B.; Hu, C.-H.; Yi, G.-J.; et al. Comparative Phosphoproteomics Reveals an Important Role of MKK2 in Banana (*Musa* spp.) Cold Signal Network. *Sci. Rep.* **2017**, *7*, 40852. [CrossRef] [PubMed]
48. Hong, J.H.; Savina, M.; Du, J.; Devendran, A.; Ramakanth, K.K.; Tian, X.; Sim, W.S.; Mironova, V.V.; Xu, J. A Sacrifice-for-Survival Mechanism Protects Root Stem Cell Niche from Chilling Stress. *Cell* **2017**, *170*, 102–113. [CrossRef]
49. Komis, G.; Samajova, O.; Ovecka, M.; Samaj, J. Cell and Developmental Biology of Plant Mitogen-Activated Protein Kinases. In *Annual Review of Plant Biology*; Merchant, S.S., Ed.; Annual Reviews: Palo Alto, CA, USA, 2018; Volume 69, pp. 237–265.
50. Shao, Y.; Yu, X.; Xu, X.; Li, Y.; Yuan, W.; Xu, Y.; Mao, C.; Zhang, S.; Xu, J. The YDA-MKK4/M KK5-M PK3/MPK6 Cascade Functions Downstream of the RGF1-RGI Ligand-Receptor Pair in Regulating Mitotic Activity in Root Apical Meristem. *Mol. Plant* **2020**, *13*, 1608–1623. [CrossRef] [PubMed]
51. Lu, X.; Shi, H.; Ou, Y.; Cui, Y.; Chang, J.; Peng, L.; Gou, X.; He, K.; Li, J. RGF1-RGI1, a Peptide-Receptor Complex, Regulates Arabidopsis Root Meristem Development via a MAPK Signaling Cascade. *Mol. Plant* **2020**, *13*, 1594–1607. [CrossRef]
52. He, N.-Y.; Chen, L.-S.; Sun, A.-Z.; Zhao, Y.; Yin, S.-N.; Guo, F.-Q. A nitric oxide burst at the shoot apex triggers a heat-responsive pathway in Arabidopsis. *Nat. Plants* **2022**, *8*, 8–434. [CrossRef]
53. Droc, G.; Martin, G.; Guignon, V.; Summo, M.; Sempere, G.; Durant, E.; Soriano, A.; Baurens, F.-C.; Cenci, A.; Breton, C.; et al. The banana genome hub: A community database for genomics in the Musaceae. *Hortic. Res.* **2022**, *9*, uhac221. [CrossRef] [PubMed]

54. Chen, C.; Chen, H.; Zhang, Y.; Thomas, H.R.; Frank, M.H.; He, Y.; Xia, R. TBtools: An Integrative Toolkit Developed for Interactive Analyses of Big Biological Data. *Mol. Plant* **2020**, *13*, 1194–1202. [CrossRef] [PubMed]
55. Zhang, L.; Cenci, A.; Rouard, M.; Zhang, D.; Wang, Y.; Tang, W.; Zheng, S.-J. Transcriptomic analysis of resistant and susceptible banana corms in response to infection by Fusarium oxysporum f. sp. cubense tropical race 4. *Sci. Rep.* **2019**, *9*, 8199. [CrossRef]
56. Liu, W.; Cheng, C.; Lin, Y.; Xu, X.; Lai, Z. Genome-wide identification and characterization of mRNAs and lncRNAs involved in cold stress in the wild banana (*Musa itinerans*). *PLoS ONE* **2018**, *13*, e0200002. [CrossRef] [PubMed]
57. Dai, X.; Zhuang, Z.; Zhao, P.X. psRNATarget: A plant small RNA target analysis server (2017 release). *Nucleic Acids Res.* **2018**, *46*, W49–W54. [CrossRef]
58. Zhu, H.; Zhang, Y.; Tang, R.; Qu, H.; Duan, X.; Jiang, Y. Banana sRNAome and degradome identify microRNAs functioning in differential responses to temperature stress. *BMC Genom.* **2019**, *20*, 33. [CrossRef]

plants MDPI

Article

Characterization of *Carotenoid Cleavage Oxygenase* Genes in *Cerasus humilis* and Functional Analysis of *ChCCD1*

Chunzhen Cheng [1,†], Rui Yang [1,†], Lu Yin [1], Jianying Zhang [1], Limin Gao [2], Rong Lu [3], Yan Yang [1], Pengfei Wang [1], Xiaopeng Mu [1], Shuai Zhang [1], Bin Zhang [1] and Jiancheng Zhang [1,*]

1 College of Horticulture, Shanxi Agricultural University, Jinzhong 030801, China; ld0532cheng@sxau.edu.cn (C.C.); yangrui13934147962@163.com (R.Y.); yl18434764709@163.com (L.Y.); 18734448590@163.com (J.Z.); 18404968975@163.com (Y.Y.); 15110671026@163.com (X.M.); wwzs_1990@aliyun.com (S.Z.); jztgzhangbin@163.com (B.Z.)
2 Agricultural Technology Extension Service Center of Qianyang County, Baoji 721199, China; 13466896541@163.com
3 Rural Revitalization Bureau of Pu County, Linfen 041200, China; llrr0624@163.com
* Correspondence: zjcnd001@sxau.edu.cn
† These authors contributed equally to this work.

Abstract: Carotenoid cleavage oxygenases (CCOs) are key enzymes that function in degrading carotenoids into a variety of apocarotenoids and some other compounds. In this study, we performed genome-wide identification and characterization analysis of *CCO* genes in *Cerasus humilis*. Totally, nine *CCO* genes could be classified into six subfamilies, including carotenoid cleavage dioxygenase 1 (CCD1), CCD4, CCD7, CCD8, CCD-like and nine-*cis*-epoxycarotenoid dioxygenase (NCED), were identified. Results of gene expression analysis showed that *ChCCOs* exhibited diverse expression patterns in different organs and in fruits at different ripening stages. To investigate the roles of *ChCCOs* in carotenoids degradation, enzyme assays of the ChCCD1 and ChCCD4 were performed in *Escerichia coli* BL21(DE3) that can accumulate lycopene, β-carotene and zeaxanthin. The prokaryotic expressed ChCCD1 resulted in obvious degradation of lycopene, β-carotene and zeaxanthin, but ChCCD4 did not show similar functions. To further determine the cleaved volatile apocarotenoids of these two proteins, headspace gas chromatography/mass spectrometer analysis was performed. Results showed that ChCCD1 could cleave lycopene at 5, 6 and 5′, 6′ positions to produce 6-methy-5-hepten-2-one and could catalyze β-carotene at 9, 10 and 9′, 10′ positions to generate β-ionone. Our study will be helpful for clarifying the roles of *CCO* genes especially *ChCCD1* in regulating carotenoid degradation and apocarotenoid production in *C. humilis*.

Keywords: *Cerasus humilis*; carotenoid cleavage oxygenase; carotenoids; apocarotenoids; functional analysis

1. Introduction

Carotenoids, a subgroup of isoprenoids that typically contain 40 carbons and abundant conjugated double bounds, are the most conspicuous pigments and the most widely distributed secondary metabolites in plants [1,2]. They play vital roles in photosystem assembly, light harvesting and photoprotection and contribute greatly to the pigmentation, scents and flavors formation and stress responses of plants [1]. Carotenoids are important precursors of a large number of apocarotenoids and some other compounds, such as geranial, α-ionone, β-carotene, geranylacetone, farnesylacetone, pseudoionone, abscisic acids (ABA) and strigolactones (SL) [3]. It is noteworthy that these carotenoid-derived apocarotenoids have been proven to contribute greatly to diverse plant biological processes by acting as pigments, volatiles, signals and phytohormones, and so on [2].

Carotenoids can be cleaved into apocarotenoids by carotenoid cleavage oxygenases (CCOs), lipoxygenases, peroxidases and reactive oxygen species [2]. Among them, CCO-mediated carotenoid degradation is the main focus of attention for carotenoid degradation research. Depending on whether their substrates are epoxidized, CCOs can be divided into nine-cis-epoxycarotenoid dioxygenases (NCEDs) and carotenoid cleavage dioxygenases (CCDs) [4]. Ever since the first discovery of a maize *CCO* gene [5], a large number of plant *CCOs* have been identified and functionally studied. In the model plant Arabidopsis, the nine *AtCCO* members are divided into five subfamilies, including CCD1, CCD4, CCD7, CCD8 and NCED [4]. The classifications of CCOs in other plant species are mostly referred to as Arabidopsis or are based on sequence similarities, cleavage property, cleavage sites and substrate accessibility [3,6]. In addition to the five main CCO subfamilies, CCD-like, CCD2 and zaxinone synthase (ZAS) subfamily members were also identified in certain plant species [7–9].

Pieces of evidence have revealed that the functions of CCOs from different subfamilies varied a lot. For example, the CCD1, CCD2 and CCD4 subfamily members are reported to be involved in the biosynthesis of aroma-, flavor- and color formation-related apocarotenoids [10]; the CCD7 and CCD8 subfamily members are closely related to the SL biosynthesis [11]; the NCED members play a certain role in the growth and development of plants and are involved in ABA biosynthesis and [4] members from other subfamilies were found to be of diverse roles in the biosynthesis of some other carotenoids-derived compounds [6,12,13]. The functions of CCD1 and CCD4 subfamily members in carotenoid degradation and apocarotenoid production were the most widely studied in higher plants [14,15]. Research has revealed that CCD1 enzymes lack plastid localization peptides and are generally cytoplasmic [6]. Their cleaved products are usually responsible for the volatile formation [16,17], and their roles in controlling β-ionone generation have been reported in many plants [18–21]. Meanwhile, CCD4 enzymes are plastid-located, and their substrate specificity is generally higher than CCD1s [22]. Citrus CCD4 can cleave β-carotene, β,β-cryptoxanthin and zeaxanthin into apocarotenoids [23]. Grape VvCCD4 can cleave δ-carotene and lycopene to produce α-violonone and geranylacetone, respectively [24]. And Chrysanthemum CmCCD4a can cleave β-carotene into β-ionone, which is important for the flower pigmentation [25].

The Chinese dwarf cherry (*Cerasus humilis*) is a perennial woody fruit tree native to China [26]. Comparative genomic studies revealed that *C. humilis* shared a very close relationship with some fruit trees from the *Prunus* genus, such as *P. persia* and *P. armeniaca* [26]. The fruits of *C. humilis* contain a variety of carotenoids and their derived compounds [27]. Up until now, however, the *CCO* gene family in *C. humilis* has not been systematically studied. In this study, whole genome-wide identification and characterization of *C. humilis CCO* genes was performed. The expression patterns of *ChCCO* genes in different tissues and organs and in fruits at different ripening stages were studied using transcriptome data and quantitative real-time PCR analysis (qRT-PCR). Moreover, to uncover the functions of *ChCCO* genes in carotenoids degradation, *ChCCD1* and *ChCCD4* genes were cloned, inserted into prokaryotic expression vector pET-28a and subjected to enzyme assays in *Escerichia coli* BL21(DE3) that can accumulate lycopene, β-carotene and zeaxanthin. The results obtained in this study will provide a basis for understanding the roles of *CCO* genes in regulating carotenoid degradation and apocarotenoid production in *C. humilis*.

2. Results

2.1. Identification and Characterization of ChCCOs

Totally, we identified nine CCOs from the *C. humilis*. Phylogenetic analysis revealed that they can be classified into six subfamilies, including CCD-like, CCD1, CCD4, CCD7, CCD8 and NCED (Figure 1A). Among these subfamilies, the NCED subfamily consisted of three members (ChNCED1, ChNCED5 and ChNCED6), the CCD-like subfamily contained two members and the other subfamilies each contained only one member. Sequence similarity analysis revealed high similarities among ChNCEDs (Figure 1B). The similarity between *ChNCED1* and *ChNCED5* was about 71.51%, and their similarities with *ChNCED6* were about 58.63% and 59.37%, respectively. Similarities among their encoded proteins ranged from 66.16% to 78.29% (Figure 1C). Moreover, *ChCCD-like-a* and *ChCCD-like-b* shared a similarity of about 53.4%, and their encoded proteins shared a similarity of about 58.55%. Plant CCOs usually contained four conserved histidine active sites and three semi-conserved second shell glutamate residues [28]. Consistently, all the ChCCOs contained four conserved histidine and three conserved glutamate residues (Figure S1).

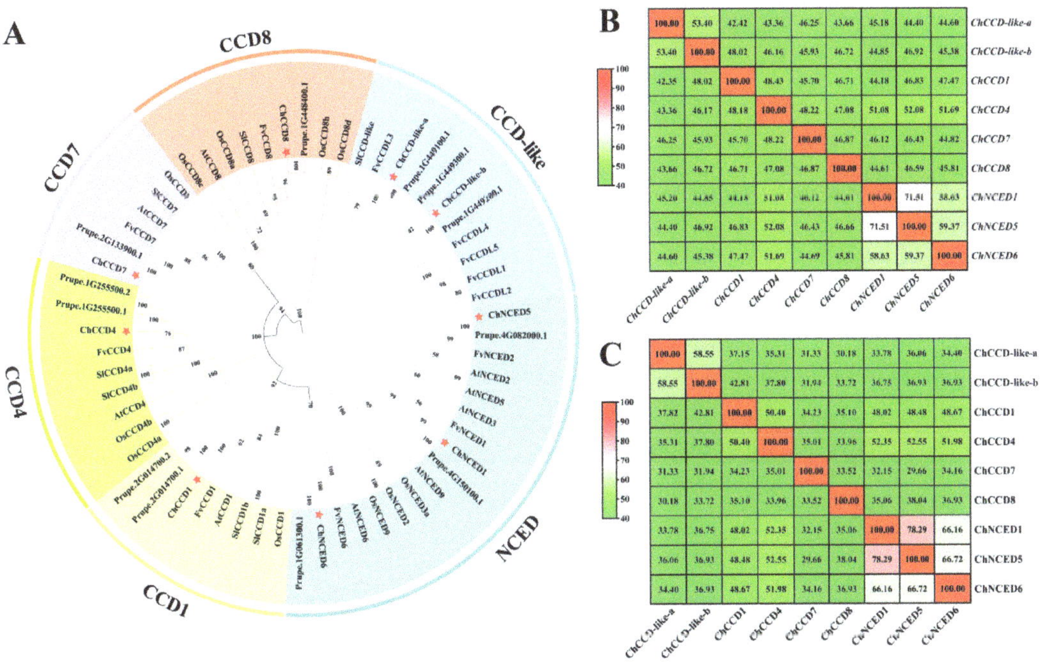

Figure 1. Phylogenetic analysis results (**A**) of CCO proteins from *Cerasus humilis* (Ch), *P. persica* (Pp), *Fragaria vesca* (Fv), *Solanum Lycopersicum* (Sl), *Oryza sativa* (Os) and *Arabidopsis thaliana* (At), and nucleotides (**B**) and proteins (**C**) similarity analysis results of ChCCOs. CCD1, CCD4, CCD-like, CCD8, CCD7 and NCED are six subfamilies of CCOs. Red stars in A represent *C. humilis* CCD members. In B and C: the redder the color, the higher the similarity; the greener the color, the lower the similarity.

The coding sequence (CDS) length of *ChCCOs* ranged from 547 bp (*ChCCD1*) to 702 bp (*ChCCD-like-a*). Their encoded proteins consisted of 547~702 amino acids with their molecular weight ranging from 61777.14 Da (ChCCD1) to 78364.26 Da (ChCCD-like-a) and their theoretical isoelectric point (pI) ranging from 5.44 (ChCCD-like-a) to 6.97 (ChNCED6) (Table 1). All the ChCCOs were predicted to be hydrophilic proteins (GRAVY < 0), and, except for ChCCD7, ChNCED1 and ChNCED5 (with instability index > 40), all other ChCCOs were stable proteins (Table 1). Subcellular localization analysis revealed that ChCCOs were mainly located in the cytoplasm and chloroplast.

2.2. Chromosome Location and Synteny Analysis of ChCCOs

Chromosome location analysis revealed that *ChCCOs* were located in three chromosomes of *C. humilis* (Table 1 and Figure S2), including five members (*ChCCD-like-a*, *ChCCD-like-b*, *ChCCD4*, *ChCCD8* and *ChNCED6*) in Chr1, two members (*ChNCED1* and *ChNCED5*) in Chr3 and two members (*ChCCD1* and *ChCCD7*) in Chr5. Synteny analysis revealed that *ChCCD-like-a* and *ChCCD-like-b* were tandem duplicated genes (Figure S2).

Table 1. Physiochemical prosperities and subcellular location analysis results of ChCCOs. Chr: chromosome; CDS: coding sequence; bp: base pair; Chr: chromosome; AA: amino acid; pI: isoelectric point; GRAVY: grand average of hydropathicity.

Gene Name	Chr	CDS/bp	Number of AA	Molecular Weight/Da	pI	Instability Index	GRAVY	Subcellular Location
ChCCD-like-a	Chr1	2109	702	78,364.26	5.44	36.82	−0.196	Chloroplast, Cytoplasm
ChCCD-like-b	Chr1	1677	558	62,470.54	5.49	28.40	−0.292	Cytoplasm
ChCCD1	Chr5	1644	547	61,777.14	6.05	34.76	−0.258	Cytoplasm
ChCCD4	Chr1	1794	597	65,717.9	6.10	38.51	−0.229	Chloroplast, Cytoplasm
ChCCD7	Chr5	1848	615	68,499.76	5.67	46.35	−0.219	Cytoplasm
ChCCD8	Chr1	1698	565	62,461.97	6.20	34.98	−0.270	Cytoplasm
ChNCED1	Chr3	1899	632	70,288.74	6.73	45.34	−0.435	Chloroplast, Cytoplasm, Nucleus
ChNCED5	Chr3	1854	617	68,027.19	6.45	46.67	−0.335	Cytoplasm
ChNCED6	Chr1	1818	605	66,921.84	6.97	38.59	−0.287	Chloroplast

2.3. Conserved Motifs in ChCCOs and Gene Structures of Their Corresponding Genes

Totally, we identified ten conserved motifs from the nine ChCCOs (Figure S3A). Among them, ChCCD-like-b, ChCCD1, ChCCD4, ChNCED1, ChNCED5 and ChNCED6 contained all the 10 conserved motifs. ChCCD-like-a did not contain Motif 1 and Motif 8, but it had two of Motif 10. ChCCD7 contained six motifs, including two of Motif 5 and one each of Motif 4, Motif 6, Motif 7 and Motif 8. And ChCCD8 only contained one each of Motif 2, Motif 3, Motif 4, Motif 6 and Motif 8.

Gene structure analysis results showed that, except *ChNCED1* and *ChNCED6*, all *ChCCOs* had introns (Figure S3B). The number of introns in *ChCCD-like-a* was the largest (19), followed by *ChCCD1* (14) and *ChCCD-like-b* (11). *ChCCD8* and *ChCCD7* had seven and five introns, respectively. And *ChCCD4* and *ChNCED5* both contained two introns.

2.4. Promoter Analysis of ChCCOs

The cis-acting elements in the promoter regions of *ChCCOs* were analyzed (Figure S4). Results showed that, in addition to the abundant light responsive, core promoter elements TATA-box and CAAT-box and growth and development-related elements, the *ChCCOs'* promoters also contained a variety of phytohormone- and stress-responsive elements.

Ten types of phytohormone-responsive elements involving six phytohormones (ABA, MeJA, auxin, gibberellin (GA), ethylene and salicylic acid (SA)) were identified from the promoters of *ChCCOs*. Notably, the ABA-responsive element ABRE was found in promoters of all *ChCCOs*. The *ChNCED6* promoter contained the largest number of the ABA-responsive element ABRE (18 in total), followed by *ChNCED5* (8). Except for *ChCCD4*, the promoters of all other *ChCCOs* contained MeJA-related cis-acting elements (TGACG-motif and CGTCA-motif). The promoters of *ChCCD-like-a, ChCCD-like-b, ChCCD1* and *ChNCED6* contained auxin-responsive elements. The promoters of *ChCCD-like-a, ChCCD-like-b, ChCCD4, ChNCED1* and *ChNCED6* contained GA-responsive elements. The promoters of *ChCCD4, ChCCD7, ChCCD8* and *ChNCED6* contained the ethylene-responsive element ERE, and the *ChCCD4, ChNCED1* and *ChNCED5* promoters contained the SA-responsive TCA-element.

Among the stress-responsive elements, the defense and stress-related element MYB was identified in the promoters of all *ChCCOs*. The *ChNCED5* promoter contained the largest number of MYB elements (a total of 7). However, except for *ChNCED5*, all the promoters of other *ChCCOs* contained the anaerobic inducible element ARE. Except for *ChNCED6*, all the promoters of *ChCCOs* contained the high-temperature response element STRE. Except for *ChNCED1* and *ChNCED6*, the promoters of all other *ChCCOs* contained drought-inducibility-related elements. Moreover, the low-temperature responsive element LTRS was found in the promoters of *ChCCD-like-a, ChCCD1* and *ChCCD8*.

Distributions of transcription factor binding sites (TFBS) on *ChCCOs'* promoters were also analyzed. Totally, binding sites for 43 TFs were identified in the *ChCCOs'* promoters (Figure S5). Among them, the total number of ERF binding sites identified in promoters of

ChCCOs was the largest (584), followed by bHLH (410), Dof (396) and BBR-BPC (378). In the promoters of *ChCCD1*, *ChCCD8*, *ChNCED1* and *ChCCD7*, binding sites for ERFs were found to be the most abundant, accounting for 127, 127, 126 and 103, respectively. The *ChCCD7* promoter contained 103 binding sites for BCR-BPCs. In the promoter of *ChCCD4*, the binding site for TCP was the most abundant (113). In the promoter of *ChCCD-like-a*, the binding site for MYB was the largest (45). The *ChCCD-like-b* promoter contained 45 Dof binding sites and 44 MYB binding sites. The bHLH binding sites were the most abundant in the promoters of *ChNCED6* (168) and *ChNCED5* (73), respectively.

2.5. Protein–Protein Interaction Analysis of ChCCOs

Based on the *P. persica* protein database, possible interacting proteins of ChCCOs were predicted. Results showed that all the ChCCOs were homologous proteins of *P. persica* CCOs (Figure 2). ChCCD-like-a, ChCCD-like-b, ChCCD1, ChCCD8, ChNCED1 and ChNCED5 were predicted to have the ability to interact with ABA2. ChCCD7 and ChCCD8 could interact with WD40, D27 (DWARF27) and MAX2 (MORE AXILLARY BRANCHING2). In addition, all the three ChNCEDs were predicted to be interacting proteins of MAX2. Moreover, ChCCD1, ChCCD4, ChCCD7 and all the three ChNCEDs were predicted to have the ability of interacting with CrtISO.

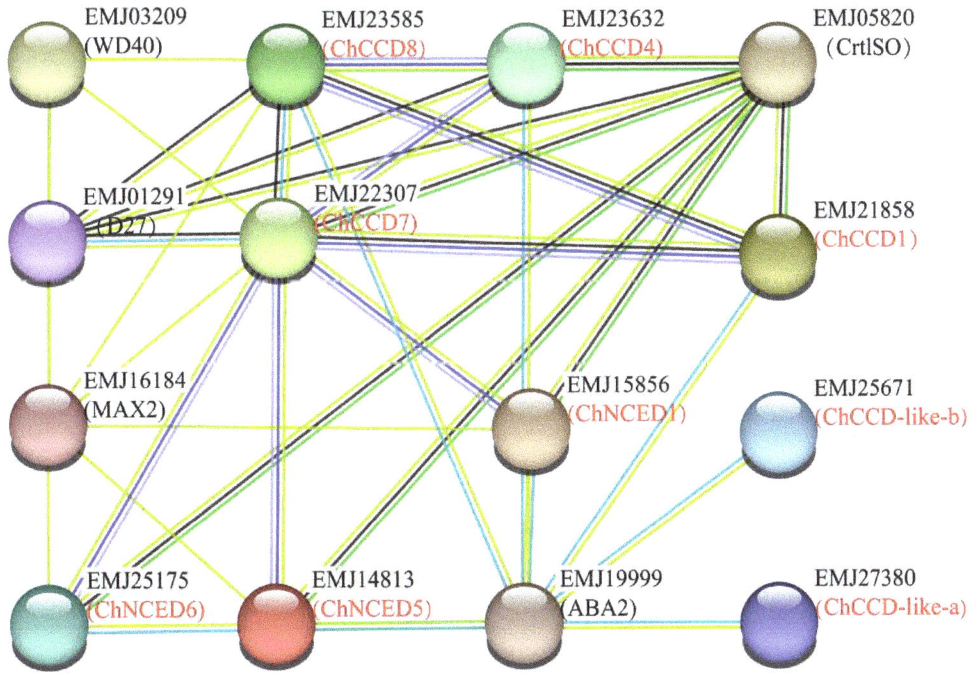

Figure 2. Protein–protein interaction network for ChCCOs based on the *Prunus persica* protein database.

2.6. Gene Expression Analysis of ChCCOs

According to the transcriptome data of five *C. humilis* organs, including fruit, leaf, kernel, rhizome and root, we found that the expression levels of *ChCCOs* in different parts varied a lot (Figure 3). Of the nine *ChCCOs*, only *ChCCD1*, *ChCCD4* and *ChNCED1* expressed in all the five organs; *ChCCD-like-a* and *ChCCD8* showed no expression in the kernel, and its expression in fruit, leaf, rhizome and root were all relatively low (FPKM < 2); *ChNCED5* showed expression in fruit, leaf and kernel; *ChCCD-like-b* and *ChCCD7* expressed in only rhizome and root; and the expression of *ChNCED6* was leaf-specific.

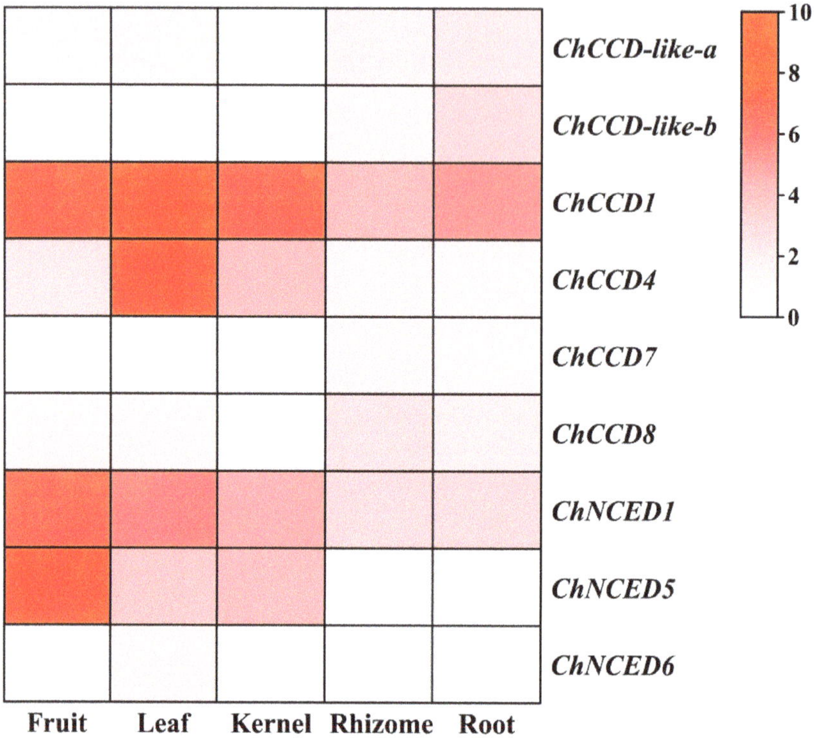

Figure 3. Heatmap for the transcriptome data-based expression analysis of *ChCCOs* in the fruit, leaf, kernel, rhizome and root of *C. humilis*. For heatmap drawing, $\log_2$(FPKM + 1) values of *ChCCO* genes were used. The redder the color, the higher the gene's expression, and white represents no expression.

Among the six fruit-expressing *ChCCOs* (*ChNCED5*, *ChCCD1*, *ChNCED1*, *ChCCD4*, *ChCCD8* and *ChCCD-like-a*), *ChNCED5* expressed the highest (FPKM > 400), followed by *ChCCD1* and *ChNCED1* (both with FPKM > 100). The expression level of *ChCCD4* ranked the fourth (with FPKM about 2) among the fruit-expressing *ChCCOs*. Although *ChCCD8* and *ChCCD-like-a* showed expression in fruit, their expression levels were very low. There were seven *ChCCOs* expressed in the leaf of *C. humilis*. Among them, the expression of *ChCCD1* and *ChCCD4* ranked top two, and their FPKM values were both higher than 400. *ChNCED1* and *ChNCED5* also expressed relatively high in the leaf. However, the expression levels of the other three leaf-expressing *ChCCOs* (*ChCCD-like-a*, *ChCCD8* and *ChNCED6*) were all very low (FPKM < 0.1). The expression levels of the four *ChCCOs* expressing in the kernel followed the order of *ChCCD1* > *ChNCED1* > *ChNCED5* > *ChCCD4*. And the FPKM value of *ChCCD1* in the kernel was more than 100. Except for *ChNCED5* and *ChNCED6*, all other *ChCCOs* showed expression in *C. humilis* rhizome and root with *ChCCD1* and *ChNCED1* both ranking top two.

The expression of the same *ChCCO* gene also showed obvious spatial differences. For example, *ChCCD1* expressed the highest in the leaf, followed by in the fruit and kernel; *ChCCD4* expressed the highest in the leaf, followed by in the kernel and fruit; the expression levels of *ChNCED1* and *ChNCED5* in the fruit were much higher than that in the other four organs.

The *C. humilis* fruits are rich in carotenoids and carotenoid-derived compounds [27]. To analyze the expression patterns of fruit-expressing *ChCCO* genes in fruits at different ripening stages, quantitative real-time PCR (qRT-PCR) analysis was performed (Figure 4). Results showed that the expression levels of *ChCCD-like-a* and *ChNCED1* increased sharply at 125 DAF (the color turning stage) but decreased at 135 DAF (the maturity stage). *ChCCD4* and *ChCCD8* exhibited a 'fall-rise-fall' expression change pattern during fruit ripening, and their expression levels both peaked at 125 DAF, followed by that in the fruit at 95 DAF. The expression of *ChCCD1* in fruits at 95 DAF and 110 DAF were significantly higher than that in fruits at 125 DAF and 135 DAF, and its expression at 135 DAF was found to be the lowest. However, the expression of *ChNCED5* in fruits at 125 DAF and 135 DAF was significantly higher than that in fruits at 95 DAF and 110 DAF, and its expression increased as fruit ripened. Except *ChCCD1* and *ChNCED5*, the relative expression levels of other fruit-expressing *ChCCOs* were all the highest at 125 DAF. Moreover, the expression levels of all the fruit-expressing *ChCCDs* (including *ChCCD-like-a*, *ChCCD1*, *ChCCD4* and *ChCCD8*) were the lowest at 135 DAF.

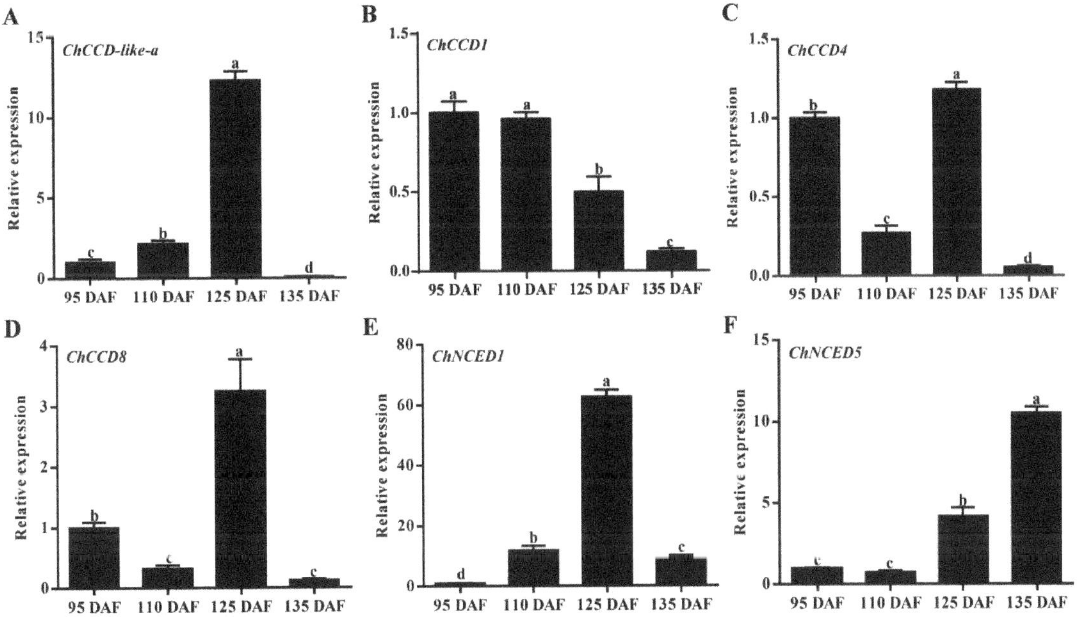

Figure 4. Quantitative real-time PCR analysis results of *ChCCOs* in fruits at four different ripening stages. (**A–F**) represents expression analysis result for *ChCCD-like-a*, *ChCCD1*, *ChCCD4*, *ChCCD8*, *ChNCED1* and *ChNCED5*, respectively. DAF: days after flowering. The different letters above the columns represent significant differences at $p < 0.05$ level.

2.7. Prokaryotic Expression and Enzyme Assay Analysis of ChCCD1 and ChCCD4 Proteins

The contributions of CCD1 and CCD4 in carotenoid degradation and apocarotenoid accumulation have been frequently demonstrated in many plant species [14,15]. To clarify the functions of ChCCD1 and ChCCD4, pET-ChCCD1 and pET-ChCCD4 prokaryotic expression vectors were constructed and individually transformed into *E. coli* BL21(DE3). After protein expression activation using IPTG, SDS-PAGE gel electrophoresis was used to detect the expression of ChCCD1 and ChCCD4 proteins. Results showed that *E. coli* BL21(DE3) carrying pET-ChCCD1 and pET-ChCCD4 could, respectively, express recombinant proteins with a molecular weight of about 61 kD and 65 kD (Figure S6), indicating that ChCCD1 and ChCCD4 proteins were correctly expressed.

To reveal the roles of the ChCCD1 and ChCCD4 in carotenoid degradation and apocarotenoid accumulation, pET-ChCCD1/pET-ChCCD4 vectors were introduced into *E. coli* BL21(DE3) together with plasmid pACCRT-EIB/pACCAR16ΔcrtX/pACCAR25ΔcrtX. Before IPTG addition, the color of bacterial cultures carrying pET-ChCCD1/ChCCD4 and pACCRT-EIB, pET-ChCCD1/ChCCD4 and pACCAR16ΔcrtX and pET-ChCCD1/ChCCD4 and pACCAR25ΔcrtX was red, yellow-orange and yellow-orange (Figure 5), respectively. After IPTG induction, bacterial cultures expressing recombinant ChCCD1 exhibited remarkable color changes, varying from a lighter color (cultures expressing pET-ChCCD1 and pACCRT-EIB/pACCAR25ΔcrtX) to almost white (cultures expressing pET-ChCCD1 and pACCAR16ΔcrtX). However, the ChCCD4 expression did not result in significant bacterial culture color change. These results indicated that ChCCD1 could cleave lycopene, β-carotene and zeaxanthin. ChCCD4, however, could not cleave these substrates as efficiently as ChCCD1.

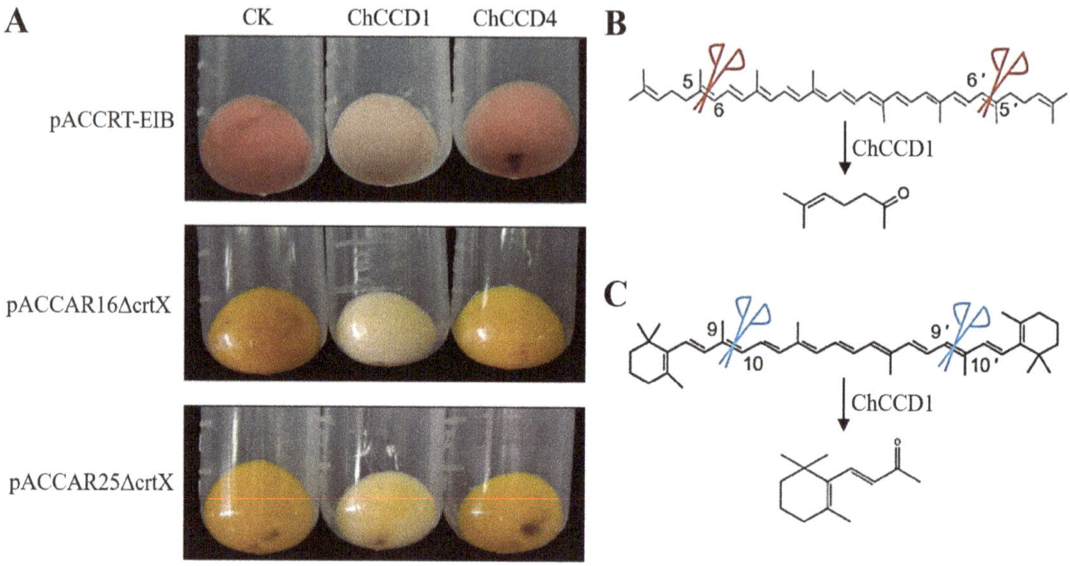

Figure 5. Functional analysis results of ChCCD1 and ChCCD4 proteins. (**A**) The influences of ChCCD1 and ChCCD4 expression on the color changes of *E. coli* strains that can accumulate lycopene (carrying pACCRT-EIB vector), β-carotene (carrying pACCAR16ΔcrtX vector) and zeaxanthin (carrying pACCAR25ΔcrtX vector); CK: control bacteria with no IPTG addition. (**B**) ChCCD1 can cleave lycopene into 6-methyl-5-heptene-2-one. (**C**) ChCCD1 can cleave β-carotene into β-ionone.

The volatile products of ChCCD1 and ChCCD4 in *E. coli* strains that can accumulate lycopene, β-carotene and zeaxanthin were further determined using headspace gas chromatography/mass spectrometer. Results showed that, after IPTG induction, among the volatiles released by the bacterial culture expressing pET-ChCCD1 and pACCRT-EIB, 6-methyl-5-heptene-2-one was detected (Figure 6). This indicated that ChCCD1 can cleave lycopene at 5, 6 and 5′, 6′ positions to produce 6-methyl-5-heptene-2-one. In the volatiles released by the bacterial culture expressing pET-ChCCD1 and pACCAR16ΔcrtX, β-ionone, an aroma substance produced by oxidative cleavage of β-carotene, was detected (Figure 6). This indicated that ChCCD1 can cleave β-carotene at 9, 10 and 9′, 10′ positions to produce β-ionone. No cleaved carotenoid products were detected in the volatiles released by other bacterial cultures.

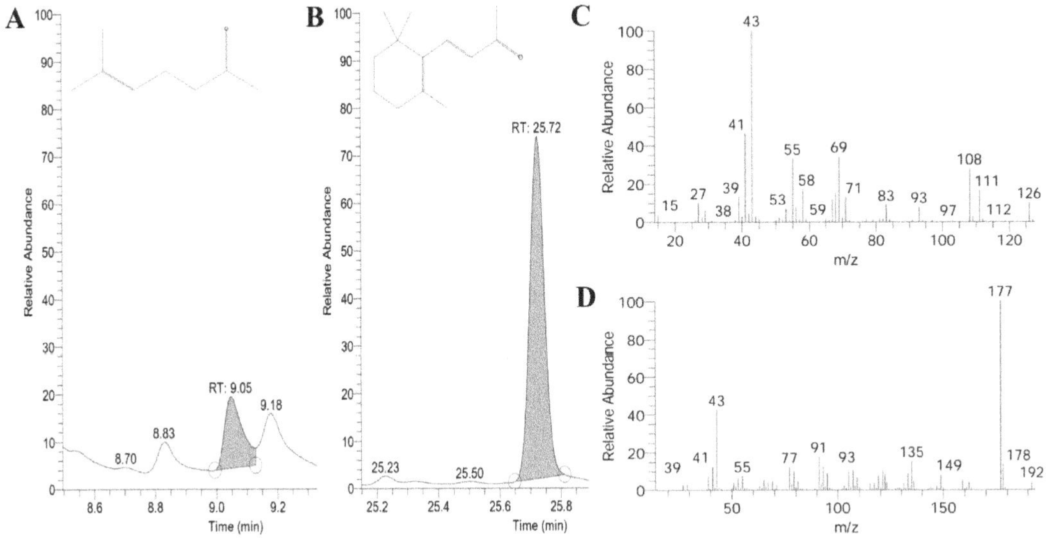

Figure 6. GC-MS detection results of the ChCCD1 cleaved volatile products of lycopene and β-carotene in *E. coli*. (**A**,**B**) for the cleaved volatile products of lycopene and β-carotene, respectively; (**C**,**D**) fragments pattern for 6-methy-5-hepten-2-one and β-ionone, respectively. Blue circles in (**A**) and (**B**) represent starting and ending time points of peak.

3. Discussion

In this study, for the first time, we performed whole genome-wide identification and characterization of the *CCO* genes in *C. humilis*. Totally, nine *ChCCOs* (including six *ChCCDs* and three *ChNCEDs*) belonging to six subfamilies were obtained. This classification was supported by their gene structures and conserved motifs in their encoded proteins. Most of the ChCCOs contained four histidine active sites, which might be closely related to their iron-binding abilities [28,29]. Subcellular localization analysis revealed that ChCCOs were mainly localized in cytoplasm and chloroplast, which was consistent with the CCOs from many other plant species [13,30–32]. Synteny analysis revealed that *ChCCD-like-a* and *ChCCD-like-b* were tandem duplicated genes, suggesting that the tandem duplication of CCD-like subfamily members contributed to the amplification of the *CCO* gene family in *C. humilis*.

Accumulated evidence demonstrated that *CCOs* might be involved in the plant responses to phytohormones and abiotic stresses [33–36]. In this study, we identified many phytohormone- and stress-responsive elements in the promoters of *ChCCOs*. The ABA-responsive element identified in the promoters of *CCD* genes from six Cucurbitaceae species was reported to be the most abundant among all the phytohormone-responsive elements [37]. Similarly, the abundance of an ABA-responsive element in the promoters of the litchi CCD1, CCD4, CCD7, CCD-like and NCED subfamily genes ranked the first among all the phytohormone-responsive elements [38]. Consistently, in our study, all the promoters of *ChCCOs* were predicted to contain the ABA-responsive element ABRE. There were nine, nine, nine and eight *ChCCOs* that contained MeJA-responsive elements, the anaerobic inducible element ARE, the high-temperature response element STRE and the drought-inducibility-related elements in their promoters, respectively. In addition, the promoters of *ChCCD-like-a*, *ChCCD1* and *ChCCD8* contained the low-temperature responsive element LTRS. These results suggested that *ChCCOs* might play roles in phytohormone and stress responses in *C. humilis*.

Transcription factors (TFs) play important roles in the biosynthesis of secondary metabolites including carotenoids. The expression of carotenoid metabolism-related genes has been continuously proven to be regulated by TFs [39,40]. In this study, we identified binding sites for 43 types of TFs in the promoters of *ChCCOs*, but the abundance of binding sites for different TFs varied a lot. For example, the binding site for ERF was the most abundant in the promoters of *ChCCD1*, *ChCCD7*, *ChCCD8* and *ChNCED1*; *ChCCD4* and *ChCCD-like-a* promoters had many binding sites for TCPs and MYBs, while *ChNCED6* and *ChNCED5* promoters were rich of bHLH binding sites. The distribution and abundance differences of TFBSs in their promoters suggested that the expression of *ChCCOs* might be regulated by different TFs.

The expression patterns of different *CCO* gene members varied a lot in different parts of plant species [3,37]. In this study, our transcriptome data-based gene expression analysis revealed that there were six, seven, four, seven and seven *ChCCOs* expressed in the fruit, leaf, kernel, rhizome and root, respectively. *ChCCD1*, *ChCCD4* and *ChNCED1* showed expression in all five organs, while the expression of *ChNCED6* was found to be leaf-specific. *ChCCD1* expressed much higher than other *ChCCOs* in the kernel, rhizome and root; *ChCCD1* and *ChCCD4* expressed highly in the leaf and *ChCCD1*, *ChNCED1* and *ChNCED5* showed high expression in the fruit. All this suggests that their roles in carotenoid degradation were spatially different in *C. humilis*. Moreover, our qRT-PCR analysis revealed that the expression patterns of *ChCCOs* in fruits at different ripening stages were also temporally different.

Our protein–protein interaction analysis also indicated that ChCCOs play different roles in *C. humilis*. Six ChCCOs (including ChCCD-like-a, ChCCD-like-b, ChCCD1, ChCCD8, ChNCED1 and ChNCED5) were predicted to interact with ABA2. In higher plants, ABA is derived from xanthophyll carotenoids via the C15 intermediate xanthoxin. ABA2, a xanthoxin dehydrogenase that catalyzes the conversion of xanthoxin to abscisic aldehyde, is a key protein function in ABA biosynthesis [41]. This indicates that these ChCCOs might play roles in ABA biosynthesis by interacting with ABA2. ChCCD7 and ChCCD8 were predicted to interact with WD40, D27 and MAX2. WD40 has been reported to function not only in the biosynthesis of flavonoids but also in the carotenoid-derived pigments [42]. D27 is a β-carotene isomerase that can catalyze the interconversion of all-trans- into 9-cis-β-carotene (the precursor of SLs) [43,44]. In saffron, CsD27-1 was found to be co-expressed with CCD7 and CCD8 in the mycorrhized roots [45]. MAX2 is a key regulatory gene in SL signal transduction [46]. The interactions of ChCCD7 and ChCCD8 with these proteins indicated that these two ChCCOs were involved in the carotenoid degradation and SL biosynthesis. All the three ChNCEDs were also predicted to be interacting proteins of MAX2, indicating that they also function in these processes. Additionally, ChCCD1, ChCCD4, ChCCD7 and the three ChNCEDs were identified to interact with CrtISO, a carotenoid isomerase that can catalyze the cis-to-trans isomerization of poly-cis-isomer of lycopene into all-trans lycopenes [45], indicating again that these ChCCOs function in carotenoids degradation.

The roles of CCD1 in regulating the formation of apocarotenoid volatiles [17–20,31], and the function of the *CCD4* gene in regulating the cleavage of carotenoids [47], have been confirmed in many plant species. The *Rosa damascene* RdCCD1 was reported to have the ability to cleave a variety of carotenoids at the 9, 10 and 9′, 10′ positions to produce a C14 dialdehyde and two C13 products, and it could also cleave lycopene at the 5, 6 and 5′, 6′ positions to produce 6-methyl-5-hepten-2-one [15]. The melon CmCCD1 could cleave a variety of carotenoids at 9, 10 and 9′, 10′ positions to generate several kinds of apocarotenoids [48]. In rice, CCD1 has been reported to have the ability to convert lycopene into volatiles, pseudoionone, 6-methyl-5-hepten-2-one and geranial [49]. In this study, we functionally analyzed the cleavage ability of ChCCD1 and ChCCD4 on lycopene and β-carotene and zeaxanthin by co-expressing them together with genes that can induce carotenoid accumulation in *E. coli*. Results showed that ChCCD1 can oxidize lycopene at 5, 6 and 5′, 6′ positions to produce 6-methyl-5-heptene-2-one and can cleave β-carotene at 9, 10 and 9′, 10′ positions to produce β-ionone. These results suggest that ChCCD1 plays a key role in the β-carotene accumulation in *C. humilis* and contributes greatly to the β-carotene-rich characteristics of *C. humilis* fruits [26,27]. Although ChCCD1 expression can also lead to the degradation of zeaxanthin, no volatile products were detected through GC/MS. This can be explained by the fact that the cleaved products of zeaxanthin by ChCCD1 are not volatiles. In addition, we also investigated the function of ChCCD4 in carotenoid degradation and apocarotenoid biosynthesis. Results showed that it did not show an obvious influence on the bacterial culture color change and volatiles release, which might be related to its higher substrate specificity [22].

4. Materials and Methods

4.1. Plant Materials

The two-year-old *C. humilis* cv. 'Jinou No. 1' materials were collected from the *C. humilis* resource nursery of Juxin Demonstration Park at Shanxi Agricultural University. At 95, 110, 125 and 135 days after flowering (DAF), fruits used for gene expression analysis were harvested, washed three times with sterile water, quick-frozen in liquid nitrogen and stored in a −80 °C freezer for further use.

4.2. Identification of Cerasus humilis CCO Genes

The *C. humilis* genome file [26] was provided by Dr. Pengfei Zhang from Shanxi Agricultural University. The *Prunus persica* CCO protein sequences were downloaded from NCBI (https://www.ncbi.nlm.nih.gov/, accessed on 3 March 2023) and used as queries to BLASTP against the *C. humilis* protein data to screen ChCCOs under the criterion of e-value $\leq 1 \times 10^{-5}$. Meanwhile, the RPE65 (retinal pigment epithelial membrane protein 65) domain Markov model (PF03055) downloaded from Pfam (http://pfam.xfam.org/, accessed on 3 March 2023) was used to search the putative CCO proteins in *C. humilis* using HMMER 3.0 under the criterion of e-value $\leq 1 \times 10^{-5}$. As some annotated *ChCCOs* do not have complete sequences, unigenes that were annotated as *CCOs* in our transcriptome data were subjected to gene cloning and sequencing confirmation to obtain their full-length CDSs. The obtained candidate CCOs were further subjected to conserved domain confirmation, and only proteins containing the RPE65 domain remained.

4.3. Bioinformatic Analysis of ChCCOs and Their Encoded Proteins

The physiochemical properties, subcellular localization and the existence of chloroplast transit peptide in ChCCOs were analyzed using ProtParam (https://web.expasy.org/protparam/, accessed on 5 March 2023), CELLO (http://cello.life.nctu.edu.tw/, accessed on 5 March 2023) and ChloroP (http://www.cbs.dtu.dk/services/ChloroP/, accessed on 5 March 2023), respectively. For gene structure analysis of *ChCCOs*, TBtools [50] was used. MEME (http://meme-suite.org/tools/meme, accessed on 5 March 2023) was applied to analyze the conservative motifs in each member of ChCCOs (the motif number was set as 10, and other parameters were set as default values). All the *ChCCO* genes were mapped to *C. humilis* chromosomes according to their location information, and synteny analysis was performed using MCscanX (Multiple Collinearity Scan Toolkit X version). MEGA7 software (Molecular Evolutionary Genetics Analysis Version 7.0) was applied for the multiple sequence alignments of CCO proteins from *C. humilis*, *P. persica*, *Fragaria vesca*, *Solanum Lycopersicum*, *Oryza sativa* and *Arabidopsis thaliana* and for phylogenetic tree construction by using the Neighbor Joining (NJ) method with default parameters (bootstrap = 1000). TBtools was used to extract the 2000 bp sequences upstream from the start codons (ATG) of *ChCCOs* from the *C. humilis* genome data. Additionally, the extracted sequences were used as promoter sequences of *ChCCOs* and were subjected to PlantCARE (http://bioinformatics.psb.ugent.be/webtools/plantcare/html/, accessed on 6 March 2023) and PlantTFDB (http://plantregmap.gao-lab.org/binding_site_prediction.php, accessed on 6 March 2023) for the cis-acting element and transcription factor binding sites (TFBS) analysis, respectively. Given the close relationship between *C. humilis* and *P. persica*, the interacting proteins of ChCCOs were predicted using STRING (https://cn.string-db.org/, accessed on 7 March 2023) based on the *P. persica* protein database.

4.4. Gene Expression Analysis

Based on our transcriptome data, the expression patterns of *ChCCO* genes in the fruit, leaf, kernel, rhizome and root were analyzed. TBtools was used for drawing the heatmap of their expression levels.

Trizol (Invitrogen, CA, USA) and the PrimeScript RT Master Mix (Perfect Real Time) kit (Takara, Dalian, China) were used for isolating the total RNA from *C. humilis* fruits and for biosynthesizing the complementary DNA (cDNA) used for quantitative real-time PCR, respectively. Gene-specific primers used for quantitative real-time PCR were designed using Vector NTI (Table 2). The expression of *ChCCO* genes in fruits at different ripening stages was investigated on an ABI 7500 real-time PCR system. The amplification system contained 2 μL cDNA, 0.8 μL each of the forward and reverse primers (10 μM), 0.4 μL ROX Reference DyeII, 10 μL SYBR solution and 6 μL ddH_2O. The reaction procedure was set as follows: pre-denaturation at 95 °C for 3 min; denaturation at 95 °C for 15 s; annealing at 58.5 °C for 30 s; extension at 72 °C for 15 s; 40 cycles. Three biological replications were made for each gene. By using *ChActin* as an internal reference gene, the relative expression of *ChCCOs* in different samples was calculated using the $2^{-\Delta\Delta CT}$ method.

Table 2. Information of primers used in this study. The nucleotide sequences underlined represent digestion site sequences of *Bam*HI (GGATCC) and *Xho*I (CTCGAG).

Gene Name	Forward Primer (5′–3′)	Reverse Primer (5′–3′)	Applications
ChActin	TTCAAAGACCAGCTCATCTGTGG	CAATGCCAGGGAACATAGTGGA	qRT-PCR
ChCCD-like-a	AAGTCAAGACCACCCTCTCCTCC	AACTCGTCTACGGGGCCAAAG	qRT-PCR
ChCCD1	ATGGCGGAGGTTGAAGATGAGG	TTGGTGGGAGGAGTTTCATCAAG	qRT-PCR
ChCCD1	CGGGATCCATGGCGGAGGTTGAAGATG	CCCTCGAGTTAGAGCTTTGCTTGTTCTTGC	Vector construction
ChCCD4	GGATGCCTTCTCTTCCTCTTTCC	CGCGAGCTTTTGTTGTTAGTGG	qRT-PCR
ChCCD4	CGGGATCCATGGATGCCTTCTCTTCCTC	CCCTCGAGCTACAACTTGTTGAGATCAC	Vector construction
ChCCD8	ATGGCTTCCATAGCATTTTCCG	AGCCACTATTGCCGCTCTCTCT	qRT-PCR
ChNCED1	CCTCTTCCTCTTCCAGCCCAA	GGCACTGGCTTTGAGGATTTAGA	qRT-PCR
ChNCED5	GCTCTTCCAAAAGCACCCAATT	GGAAGTGAAGAACCGAGGGAGAT	qRT-PCR

4.5. Expression of ChCCD1 and ChCCD4 in Escherichia coli

To clone the full-length CDSs of *ChCCD1* and *ChCCD4*, gene-specific primer pairs with *Bam*HI digestion site sequences (GGATCC) in the forward primers and *Xho*I digestion site sequences (CTCGAG) in the reverse primers were designed (Table 2). Amplified PCR products were double digested with *Bam*HI and *Xho*I and introduced into the prokaryotic expression vector pET28, which had been digested using the same two enzymes to generate pET-ChCCD1 and pET-ChCCD4 vectors. Then, vectors were transformed into *E. coli* BL21(DE3) and incubated at 37 °C with gentle shaking at 125 rpm till OD_{600} of 0.5. The expression of recombinant ChCCD1 or ChCCD4 proteins was induced by the addition of isopropyl β-D-thiogalactopyranoside (IPTG, with a final concentration of 0.5 mM), after which the cultures were grown at 37 °C for an additional 5 h. SDS-PAGE gel electrophoresis was applied to detect the protein expression.

4.6. Enzyme Assays In Vitro and Volatile Compounds Detection

pET-ChCCD1/pET-ChCCD4 vectors were transformed into *E. coli* BL21(DE3) together with plasmids pACCRT-EIB (carrying *crtE*, *crtB* and *crtI* genes), pACCAR16ΔcrtX (carrying *crtE*, *crtB*, *crtI* and *crtY* genes) and pACCAR25ΔcrtX (carrying *crtE*, *crtB*, *crtI*, *crtY* and *crtZ* genes) [51], respectively. After color observation and PCR detection using gene cloning primers for vector construction (Table 2), positive colonies respectively carrying pET-ChCCD1 and pACCRT-EIB, pET-ChCCD1 and pACCAR16ΔcrtX, pET-ChCCD1 and pACCAR25ΔcrtX, pET-ChCCD4 and pACCRT-EIB, pET-ChCCD4 and pACCAR16ΔcrtX and pET-ChCCD4 and pACCAR25ΔcrtX, were inoculated into an LB liquid medium containing appropriate antibiotics and incubated at 37 °C till OD_{600} of 0.6. After IPTG addition, bacterial cultures were grown at 37 °C for an additional 5 h and gently shaken at 125 rpm to induce protein expression. Bacterial cultures that were not treated with

IPTG were used as controls. The volatile compounds of bacterial cultures collected from the headspace were analyzed on a quadrupole GC/MS HP GCD (G1800A) coupled to an HP-5 silica capillary column (30 m × 0.25 mm) according to Huang et al. [14]. The oven temperature was held at 50 °C for 1 min and then increased to 200 °C at 4 °C/min intervals, with a helium flow rate of 1 mL/min. The EI-MS ionization voltage was 70 eV, and the ion source temperature was 280 °C. The mass range was recorded from 45 to 450 m/z, and spectra were evaluated with the Xcalibur software version 1.4.

5. Conclusions

In summary, for the first time, we identified and characterized the *CCO* gene family of *C. humilis*, investigated their expression patterns in different tissues and organs and in fruits at different ripening stages and functionally validated the roles of ChCCD1 in the degradation of lycopene, β-carotene and zeaxanthin. According to the results obtained in our study, it can be concluded that ChCCD1 plays a key role in carotenoid degradation and apocarotenoid accumulation in *C. humilis*.

Supplementary Materials: The following supporting information can be downloaded at: https://www.mdpi.com/article/10.3390/plants12112114/s1, Figure S1: Multiple sequence alignment results of ChCCOs. Red, yellow and green shading represent 100%, ≥75% and ≥50% amino acid sequence similarity, respectively; Figure S2: Chromosomal locations of *ChCCO* genes; Figure S3: Conserved motifs in ChCCOs proteins (A) and gene structures (B) of their corresponding genes; Figure S4: The identified cis-acting elements in *ChCCOs'* promoters. Red, yellow and green colors represent the high, moderate and low abundance of cis-acting elements in promoters, respectively. Figure S5: Heatmap for the transcription factor binding sites identified in promoters of *ChCCOs*. Red, yellow and green colors represent the high, moderate and low abundance of TFBS in promoters, respectively. Figure S6: SDS-PAGE gel electrophoresis detection results of ChCCD1 and ChCCD4 proteins. M: protein marker; 1–3: *E. coli* strains expressing ChCCD1; 4 and 5: *E. coli* strains carrying pET-ChCCD1 without IPTG induction; 6: *E. coli* strains carrying pET-ChCCD4 without IPTG induction; 7–9: *E. coli* strains expressing ChCCD4. Arrows represent target bands for recombinant ChCCD1 and ChCCD4.

Author Contributions: Conceptualization, C.C. and J.Z. (Jiancheng Zhang); methodology, R.Y., J.Z. (Jianying Zhang) and L.G.; software, R.Y., R.L., Y.Y., L.Y. and J.Z. (Jianying Zhang); validation, R.Y., Y.Y. and C.C.; formal analysis, J.Z. (Jianying Zhang), L.G. and R.Y.; investigation, J.Z. (Jianying Zhang), L.G. and R.Y.; resources, P.W., X.M., S.Z. and B.Z.; data curation, R.Y. and J.Z. (Jiancheng Zhang); writing—original draft preparation, C.C., R.Y., L.Y., L.G. and J.Z. (Jiancheng Zhang); writing—review and editing, C.C.; visualization, R.Y., L.G. and C.C.; supervision, J.Z. (Jiancheng Zhang); project administration, C.C. and J.Z. (Jiancheng Zhang); funding acquisition, C.C. and J.Z. (Jiancheng Zhang). All authors have read and agreed to the published version of the manuscript.

Funding: This work was supported by the Natural Science Basic Research Program of Shanxi Province (202203021211274), the earmarked fund for Modern Agro-industry Technology Research System of Shanxi Province (SXFRS-2022 and 2023CYJSTX07-02), the Fund for High-level Talents of Shanxi Agricultural University (2021XG010), and the Reward Fund for PhDs and Postdoctors of Shanxi Province (SXBYKY2022004).

Data Availability Statement: All data is available in this article.

Conflicts of Interest: The authors declare no conflict of interest.

References

1. Yuan, H.; Zhang, J.; Nageswaran, D.; Li, L. Carotenoid metabolism and regulation in horticultural crops. *Hortic. Res.* **2015**, *2*, 15036. [CrossRef]
2. Liang, M.H.; He, Y.J.; Liu, D.M.; Jiang, J.G. Regulation of carotenoid degradation and production of apocarotenoids in natural and engineered organisms. *Crit. Rev. Biotechnol.* **2021**, *41*, 513–534. [CrossRef]
3. Wang, Y.; Xu, J.; Liu, A. Identification of the carotenoid cleavage dioxygenase genes and functional analysis reveal *DoCCD1* is potentially involved in beta-ionone formation in *Dendrobium officinale*. *Front. Plant Sci.* **2022**, *13*, 967819. [CrossRef]
4. Tan, B.-C.; Joseph, L.M.; Deng, W.-T.; Liu, L.; Li, Q.-B.; Cline, K.; McCarty, D.R. Molecular characterization of the *Arabidopsis* 9-*cis* epoxycarotenoid dioxygenase gene family. *Plant J.* **2003**, *35*, 44–56. [CrossRef]

5. Schwartz, S.H.; Tan, B.C.; Gage, D.A.; Zeevaart, J.A.; McCarty, D.R. Specific oxidative cleavage of carotenoids by VP14 of maize. *Science* **1997**, *276*, 1872–1874. [CrossRef]
6. Dhar, M.K.; Mishra, S.; Bhat, A.; Chib, S.; Kaul, S. Plant carotenoid cleavage oxygenases: Structure-function relationships and role in development and metabolism. *Brief. Funct. Genom.* **2020**, *19*, 1–9. [CrossRef]
7. Wei, Y.; Wan, H.; Wu, Z.; Wang, R.; Ruan, M.; Ye, Q.; Li, Z.; Zhou, G.; Yao, Z.; Yang, Y. A Comprehensive analysis of carotenoid cleavage dioxygenases genes in *Solanum Lycopersicum*. *Plant Mol. Biol. Rep.* **2015**, *34*, 512–523. [CrossRef]
8. Fang, Q.; Li, Y.; Liu, B.; Meng, X.; Yang, Z.; Yang, S.; Bao, T.; Kimani, S.; Gao, X.; Wang, L. Cloning and functional characterization of a *carotenoid cleavage dioxygenase 2* gene in safranal and crocin biosynthesis from *Freesia hybrida*. *Plant Physiol. Biochem.* **2020**, *154*, 439–450. [CrossRef]
9. Wang, J.Y.; Haider, I.; Jamil, M.; Fiorilli, V.; Saito, Y.; Mi, J.; Baz, L.; Kountche, B.A.; Jia, K.P.; Guo, X.; et al. The apocarotenoid metabolite zaxinone regulates growth and strigolactone biosynthesis in rice. *Nat. Commun.* **2019**, *10*, 810. [CrossRef]
10. Frusciante, S.; Diretto, G.; Bruno, M.; Ferrante, P.; Pietrella, M.; Prado-Cabrero, A.; Rubio-Moraga, A.; Beyer, P.; Gomez-Gomez, L.; Al-Babili, S.; et al. Novel carotenoid cleavage dioxygenase catalyzes the first dedicated step in saffron crocin biosynthesis. *Proc. Natl. Acad. Sci. USA* **2014**, *111*, 12246–12251. [CrossRef]
11. Mishra, S.; Upadhyay, S.; Shukla, R.K. The role of strigolactones and their potential cross-talk under hostile ecological conditions in plants. *Front. Physiol.* **2016**, *7*, 691. [CrossRef]
12. Ahrazem, O.; Rubio-Moraga, A.; Berman, J.; Capell, T.; Christou, P.; Zhu, C.; Gomez-Gomez, L. The carotenoid cleavage dioxygenase CCD2 catalysing the synthesis of crocetin in spring crocuses and saffron is a plastidial enzyme. *New Phytol.* **2016**, *209*, 650–663. [CrossRef]
13. Yao, Y.; Jia, L.; Cheng, Y.; Ruan, M.; Ye, Q.; Wang, R.; Yao, Z.; Zhou, G.; Liu, J.; Yu, J.; et al. Evolutionary origin of the carotenoid cleavage oxygenase family in plants and expression of pepper genes in response to abiotic stresses. *Front. Plant Sci.* **2021**, *12*, 792832. [CrossRef]
14. Huang, F.C.; Molnar, P.; Schwab, W. Cloning and functional characterization of carotenoid cleavage dioxygenase 4 genes. *J. Exp. Bot.* **2009**, *60*, 3011–3022. [CrossRef]
15. Huang, F.C.; Horvath, G.; Molnar, P.; Turcsi, E.; Deli, J.; Schrader, J.; Sandmann, G.; Schmidt, H.; Schwab, W. Substrate promiscuity of RdCCD1, a carotenoid cleavage oxygenase from *Rosa damascena*. *Phytochemistry* **2009**, *70*, 457–464. [CrossRef]
16. Bouvier, F.; Suire, C.; Mutterer, J.; Camara, B. Oxidative remodeling of chromoplast carotenoids: Identification of the carotenoid dioxygenase *CsCCD* and *CsZCD* genes involved in Crocus secondary metabolite biogenesis. *Plant Cell* **2003**, *15*, 47–62. [CrossRef]
17. Auldridge, M.E.; Block, A.; Vogel, J.T.; Dabney-Smith, C.; Mila, I.; Bouzayen, M.; Magallanes-Lundback, M.; DellaPenna, D.; McCarty, D.R.; Klee, H.J. Characterization of three members of the Arabidopsis carotenoid cleavage dioxygenase family demonstrates the divergent roles of this multifunctional enzyme family. *Plant J.* **2006**, *45*, 982–993. [CrossRef]
18. Wei, S.; Hannoufa, A.; Soroka, J.; Xu, N.; Li, X.; Zebarjadi, A.; Gruber, M. Enhanced β-ionone emission in *Arabidopsis* over-expressing *AtCCD1* reduces feeding damage in vivo by the crucifer flea beetle. *Env. Entomol.* **2011**, *40*, 1622–1630. [CrossRef]
19. Simkin, A.J.; Schwartz, S.H.; Auldridge, M.; Taylor, M.G.; Klee, H.J. The tomato *carotenoid cleavage dioxygenase 1* genes contribute to the formation of the flavor volatiles β-ionone, pseudoionone, and geranylacetone. *Plant J.* **2004**, *40*, 882–892. [CrossRef]
20. Simkin, A.J.; Underwood, B.A.; Auldridge, M.; Loucas, H.M.; Shibuya, K.; Schmelz, E.; Clark, D.G.; Klee, H.J. Circadian regulation of the PhCCD1 carotenoid cleavage dioxygenase controls emission of β-ionone, a fragrance volatile of petunia flowers. *Plant Physiol.* **2004**, *136*, 3504–3514. [CrossRef]
21. Wang, J.; Wu, B.; Zhang, N.; Zhao, M.; Jing, T.; Wu, Y.; Hu, Y.; Yu, F.; Wan, X.; Schwab, W.; et al. Dehydration-induced carotenoid cleavage dioxygenase 1 reveals a novel route for β-ionone formation during tea (*Camellia sinensis*) withering. *J. Agric. Food. Chem.* **2020**, *68*, 10815–10821. [CrossRef] [PubMed]
22. Us-Camas, R.; Aguilar-Espinosa, M.; Rodriguez-Campos, J.; Vallejo-Cardona, A.A.; Carballo-Uicab, V.M.; Serrano-Posada, H.; Rivera-Madrid, R. Identifying *Bixa orellana* L. New carotenoid cleavage dioxygenases 1 and 4 potentially involved in bixin biosynthesis. *Front. Plant Sci.* **2022**, *13*, 829089. [CrossRef] [PubMed]
23. Latari, K.; Wust, F.; Hubner, M.; Schaub, P.; Beisel, K.G.; Matsubara, S.; Beyer, P.; Welsch, R. Tissue-specific apocarotenoid glycosylation contributes to carotenoid homeostasis in Arabidopsis leaves. *Plant Physiol.* **2015**, *168*, 1550–1562. [CrossRef] [PubMed]
24. Lashbrooke, J.G.; Young, P.R.; Dockrall, S.J.; Vasanth, K.; Vivier, M.A. Functional characterisation of three members of the *Vitis vinifera* L. carotenoid cleavage dioxygenase gene family. *BMC Plant Biol.* **2013**, *13*, 156. [CrossRef] [PubMed]
25. Ohmiya, A.; Kishimoto, S.; Aida, R.; Yoshioka, S.; Sumitomo, K. Carotenoid cleavage dioxygenase (CmCCD4a) contributes to white color formation in chrysanthemum petals. *Plant Physiol.* **2006**, *142*, 1193–1201. [CrossRef]
26. Wang, P.; Yi, S.; Mu, X.; Zhang, J.; Du, J. Chromosome-level genome assembly of *Cerasus humilis* using PacBio and Hi-C technologies. *Front. Genet.* **2020**, *11*, 956. [CrossRef]
27. Zhang, J.; Wang, P.; Hao, Y.; Yang, S.; Zheng, B.; Niu, Z.; Du, J. Overexpression of *ChPSY* gene from *Cerasus humilis* improved carotenoids synthesis in transgenic tomato. *Acta Hortic. Sin.* **2014**, *41*, 1563–1572. (In Chinese)
28. Harrison, P.J.; Bugg, T.D. Enzymology of the carotenoid cleavage dioxygenases: Reaction mechanisms, inhibition and biochemical roles. *Arch. Biochem. Biophys.* **2014**, *544*, 105–111. [CrossRef]
29. Poliakov, E.; Gentleman, S.; Cunningham, F.X., Jr.; Miller-Ihli, N.J.; Redmond, T.M. Key role of conserved histidines in recombinant mouse β-carotene 15,15′-monooxygenase-1 activity. *J. Biol. Chem.* **2005**, *280*, 29217–29223. [CrossRef]

30. Zhang, S.; Guo, Y.; Zhang, Y.; Guo, J.; Li, K.; Fu, W.; Jia, Z.; Li, W.; Tran, L.P.; Jia, K.P.; et al. Genome-wide identification, characterization and expression profiles of the CCD gene family in *Gossypium* species. *3 Biotech* **2021**, *11*, 249. [CrossRef]
31. Zhao, X.L.; Yang, Y.L.; Xia, H.X.; Li, Y. Genome-wide analysis of the carotenoid cleavage dioxygenases gene family in *Forsythia suspensa*: Expression profile and cold and drought stress responses. *Front. Plant Sci.* **2022**, *13*, 998911. [CrossRef] [PubMed]
32. Liu, D.; Qiu, C.; Lu, X.; Zeng, Y.; Zhang, C.; Li, T.; Zhu, G.; He, J.; Lin, Q. Cloning and prokaryotic expression of carotenoid cleavage dioxygenases from mulberry (*Morus notabilis*). *Evid. Based Complement Altern. Med.* **2022**, *2022*, 4811144. [CrossRef]
33. Su, W.; Zhang, C.; Feng, J.; Feng, A.; You, C.; Ren, Y.; Wang, D.; Sun, T.; Su, Y.; Xu, L.; et al. Genome-wide identification, characterization and expression analysis of the carotenoid cleavage oxygenase (*CCO*) gene family in *Saccharum*. *Plant Physiol. Biochem.* **2021**, *162*, 196–210. [CrossRef] [PubMed]
34. Wang, Y.; Ding, G.; Gu, T.; Ding, J.; Li, Y. Bioinformatic and expression analyses on carotenoid dioxygenase genes in fruit development and abiotic stress responses in *Fragaria vesca*. *Mol. Genet. Genom.* **2017**, *292*, 895–907. [CrossRef] [PubMed]
35. Chen, H.; Zuo, X.; Shao, H.; Fan, S.; Ma, J.; Zhang, D.; Zhao, C.; Yan, X.; Liu, X.; Han, M. Genome-wide analysis of carotenoid cleavage oxygenase genes and their responses to various phytohormones and abiotic stresses in apple (*Malus domestica*). *Plant Physiol. Biochem.* **2018**, *123*, 81–93. [CrossRef]
36. Zhao, J.; Li, J.; Zhang, J.; Chen, D.; Zhang, H.; Liu, C.; Qin, G. Genome-wide identification and expression analysis of the carotenoid cleavage oxygenase gene family in five Rosaceae species. *Plant Mol. Biol. Rep.* **2021**, *39*, 739–751. [CrossRef]
37. Cheng, D.; Wang, Z.; Li, S.; Zhao, J.; Wei, C.; Zhang, Y. Genome-wide identification of *CCD* gene family in six Cucurbitaceae species and its expression profiles in melon. *Genes* **2022**, *13*, 262. [CrossRef]
38. Yue, X.Q.; Zhang, Y.; Yang, C.K.; Li, J.G.; Rui, X.; Ding, F.; Hu, F.C.; Wang, X.H.; Ma, W.Q.; Zhou, K.B. Genome-wide identification and expression analysis of carotenoid cleavage oxygenase genes in Litchi (*Litchi chinensis* Sonn.). *BMC Plant Biol.* **2022**, *22*, 394. [CrossRef]
39. Zhang, J.; He, L.; Dong, J.; Zhao, C.; Wang, Y.; Tang, R.; Wang, W.; Ji, Z.; Cao, Q.; Xie, H.; et al. Integrated metabolic and transcriptional analysis reveals the role of carotenoid cleavage dioxygenase 4 (IbCCD4) in carotenoid accumulation in sweetpotato tuberous roots. *Biotechnol. Biofuels Bioprod.* **2023**, *16*, 45. [CrossRef]
40. Xiong, C.; Luo, D.; Lin, A.; Zhang, C.; Shan, L.; He, P.; Li, B.; Zhang, Q.; Hua, B.; Yuan, Z.; et al. A tomato B-box protein SlBBX20 modulates carotenoid biosynthesis by directly activating PHYTOENE SYNTHASE 1, and is targeted for 26S proteasome-mediated degradation. *New Phytol.* **2019**, *221*, 279–294. [CrossRef]
41. Gonzalez-Guzman, M.; Apostolova, N.; Belles, J.M.; Barrero, J.M.; Piqueras, P.; Ponce, M.R.; Micol, J.L.; Serrano, R.; Rodriguez, P.L. The short-chain alcohol dehydrogenase ABA2 catalyzes the conversion of xanthoxin to abscisic aldehyde. *Plant Cell* **2002**, *14*, 1833–1846. [CrossRef] [PubMed]
42. Meng, Y.; Wang, Z.; Wang, Y.; Wang, C.; Zhu, B.; Liu, H.; Ji, W.; Wen, J.; Chu, C.; Tadege, M.; et al. The MYB activator WHITE PETAL1 associates with MtTT8 and MtWD40-1 to regulate carotenoid-derived flower pigmentation in *Medicago truncatula*. *Plant Cell* **2019**, *31*, 2751–2767. [CrossRef] [PubMed]
43. Wang, Y.; Li, J. Branching in rice. *Curr. Opin. Plant Biol.* **2011**, *14*, 94–99. [CrossRef] [PubMed]
44. Lopez-Jimenez, A.J.; Morote, L.; Niza, E.; Mondejar, M.; Rubio-Moraga, A.; Diretto, G.; Ahrazem, O.; Gomez-Gomez, L. Subfunctionalization of D27 isomerase genes in Saffron. *Int. J. Mol. Sci.* **2022**, *23*, 10543. [CrossRef]
45. Giuliano, G.; Giliberto, L.; Rosati, C. Carotenoid isomerase: A tale of light and isomers. *Trends Plant Sci.* **2002**, *7*, 427–429. [CrossRef]
46. Dong, L.; Ishak, A.; Yu, J.; Zhao, R.; Zhao, L. Identification and functional analysis of three *MAX2* orthologs in chrysanthemum. *J. Integr. Plant Biol.* **2013**, *55*, 434–442. [CrossRef]
47. Ma, J.; Li, J.; Zhao, J.; Zhou, H.; Ren, F.; Wang, L.; Gu, C.; Liao, L.; Han, Y. Inactivation of a gene encoding carotenoid cleavage dioxygenase (CCD4) leads to carotenoid-based yellow coloration of fruit flesh and leaf midvein in peach. *Plant Mol. Biol. Rep.* **2013**, *32*, 246–257. [CrossRef]
48. Ibdah, M.; Azulay, Y.; Portnoy, V.; Wasserman, B.; Bar, E.; Meir, A.; Burger, Y.; Hirschberg, J.; Schaffer, A.A.; Katzir, N.; et al. Functional characterization of *CmCCD1*, a carotenoid cleavage dioxygenase from melon. *Phytochemistry* **2006**, *67*, 1579–1589. [CrossRef]
49. Ilg, A.; Beyer, P.; Al-Babili, S. Characterization of the rice carotenoid cleavage dioxygenase 1 reveals a novel route for geranial biosynthesis. *FEBS J.* **2009**, *276*, 736–747. [CrossRef]
50. Chen, C.; Chen, H.; Zhang, Y.; Thomas, H.R.; Frank, M.H.; He, Y.; Xia, R. TBtools: An integrative toolkit developed for interactive analyses of big biological data. *Mol. Plant* **2020**, *13*, 1194–1202. [CrossRef]
51. Misawa, N.; Satomi, Y.; Kondo, K.; Yokoyama, A.; Kajiwara, S.; Saito, T.; Ohtani, T.; Miki, W. Structure and functional analysis of a marine bacterial carotenoid biosynthesis gene cluster and astaxanthin biosynthetic pathway proposed at the gene level. *J. Bacteriol.* **1995**, *177*, 6575–6584. [CrossRef] [PubMed]

plants

MDPI

Article

Genome-Wide Identification, Expression and Stress Analysis of the GRAS Gene Family in *Phoebe bournei*

Jiarui Chang [1,†], Dunjin Fan [2,†], Shuoxian Lan [3], Shengze Cheng [2], Shipin Chen [2], Yuling Lin [3,*] and Shijiang Cao [2,4,*]

[1] International College, Fujian Agriculture and Forestry University, Fuzhou 350002, China; cherichang@163.com
[2] College of Forestry, Fujian Agriculture and Forestry University, Fuzhou 350002, China; fandunjin@foxmail.com (D.F.); csz231619@163.com (S.C.); fjcsp@126.com (S.C.)
[3] College of Horticulture, Fujian Agriculture and Forestry University, Fuzhou 350002, China
[4] Key Laboratory of Fujian Universities for Stress Physiology Ecology and Molecular Biology of Forest, Fuzhou 350002, China
* Correspondence: buliang84@163.com (Y.L.); 000q813040@fafu.edu.cn (S.C.)
† These authors contributed equally to this work.

Abstract: GRAS genes are important transcriptional regulators in plants that govern plant growth and development through enhancing plant hormones, biosynthesis, and signaling pathways. Drought and other abiotic factors may influence the defenses and growth of *Phoebe bournei*, which is a superb timber source for the construction industry and building exquisite furniture. Although genome-wide identification of the GRAS gene family has been completed in many species, that of most woody plants, particularly *P. bournei*, has not yet begun. We performed a genome-wide investigation of 56 *PbGRAS* genes, which are unequally distributed across 12 chromosomes. They are divided into nine subclades. Furthermore, these 56 *PbGRAS* genes have a substantial number of components related to abiotic stress responses or phytohormone transmission. Analysis using qRT-PCR showed that the expression of four *PbGRAS* genes, namely *PbGRAS7*, *PbGRAS10*, *PbGRAS14* and *PbGRAS16*, was differentially increased in response to drought, salt and temperature stresses, respectively. We hypothesize that they may help *P. bournei* to successfully resist harsh environmental disturbances. In this work, we conducted a comprehensive survey of the GRAS gene family in *P. bournei* plants, and the results provide an extensive and preliminary resource for further clarification of the molecular mechanisms of the GRAS gene family in *P. bournei* in response to abiotic stresses and forestry improvement.

Keywords: GRAS gene; *Phoebe bournei*; phylogenetic analysis; abiotic stress; expression analysis

1. Introduction

Throughout the life of a plant, temperature, water and salt stresses may recur over time and gradually affect its normal physiological activities. For example, salinity can cause oxidation, ionization, and osmosis, whereas heat and dryness can damage chloroplast membranes and promote chlorophyll breakdown, lowering photosynthetic efficiency [1]. These abiotic pressures may become more common in the natural environment as a result of climate change. Transcription factors (TFs) can precisely control the transcription of target genes via conjugation to deoxyribonucleic acid (DNA) sequences in response to unfavorable conditions [2–4]. TFs can help us understand the adaptive response networks that comprise many biological processes.

GRAS is an extremely significant TF that can regulate a variety of physiological processes in higher plants [5]. The GRAS gene family is named after the first three members to be discovered: Gibberellin Acid Insensitive (GAI), Repressor GA1 (RGA), and Scarecrow (SCR) [6,7]. The GRAS structural domain of a protein consists of an extremely volatile amino (N-) terminus region and a highly conserved carboxyl (C-) terminus region [8].

Leucine heptapeptide repeat I (LHRI), leucine heptapeptide repeat II (LHRII), VHIID, PFYRE, and SAW motifs make up the C terminus, while the N terminus is a disordered domain [9,10]. The main factors thought to influence the taxonomy of GRAS families are the sequences', structures', and phylogenetic relationships' diversity. Depending on the plant species tested, the number of subfamilies within the GRAS gene family can vary (DELLA, PAT1, LISCL, HAM, SCR, SHR, LAS, SCL3 in model plants and NSP1, NSP2, RAD1, SCLA, SCLB, RAM1, DL, SCL4/7, SCL32 in other plants) [10–13]. Take *Castanea mollissima*, for instance—the GRAS family is segmented into nine subfamilies, namely DELLA, PAT1, SCL3, SHR, SCR, LISCL, HAM, LAS and DLT [14]. The 150 GRAS genes in *G. hirsutum* were divided into 15 subfamilies: SCR (10), DLT (4), OS19 (4), LAS (2), SCL4/7 (4), OS4 (4), OS43 (2), DELLA (8), PAT1 (27), SHR (28), HAM (24), SCL3 (9), LISCL (20) and G_GRAS (4) [6]. The author also analyzed the evolution of algae and found that the GRASs originated from the period between algae and moss [6]. In the woody plant *Betula alba*, 265 GRAS genes were sorted into 17 groups using the analysis of 40 birch genes, 32 *A. thaliana* genes, 38 *O. sativa* genes, 52 tea tree genes, 59 *P. dactylifera* genes, and 44 *T. cocoa* genes [15]. Research such as this shows that within the plant kingdom, this family of genes is incredibly diverse. Additionally, various subfamilies may have distinct purposes during plant development due to variations in conserved structural domains. The SCR and SHR subfamilies operate in root and leaf growth—for example, SHR and SCR regulate bundle sheath cell development in *A. thaliana* and are also involved in asymmetric cell division and radial patterns in *A. thaliana* roots [16,17]. A recent study found that SCR-regulated SHR cells can autonomously inhibit formative pericyte division in moss plant leaf vein formation [18]. These results suggested that these genes are involved in regulating the direction of cell division and establishing genetic regulatory networks in land plants, both in flowering plants and bryophytes [18]. SCL is an essential subfamily involved in root cell elongation in the plant *A. thalian* by combining numerous signals and responses to different signaling/abiotic stresses (cold) as a co-activator in *G. hirsutum* [6,19]. These results suggest that the SCL subfamily may affect the regulation of plant signaling molecules. Meristem integrity in petunia stems is preserved by the HAM subfamily [20]. Moreover, the DELLA subfamily has different roles in the signaling of phytohormones such as gibberellins [21–23], which induce flower, stem and root growth in *A. thaliana* by regulating GA signaling [17]. In addition, DELLA genes also dominate plant tissue expansion [23,24]. Furthermore, similarly to the LAS subfamily's function affecting leaf and stem growth in *G. hirsutum*, the GV6 subfamily in *Solanum lycopersicum* also plays an essential role in the elongation of stems [6,25].

The GRAS family in various species of plants has been under extensive study in recent years, and not just at a surface level. In response to abiotic conditions such as drought, salt, and photo-oxidative stimuli, GRAS has been proven to be instrumental in numerous investigations [13]. Hydrogen peroxide (H_2O_2) accumulation is decreased in transgenic *O. sativa* due to OsGRAS23's favorable regulation of drought tolerance and oxidative stress tolerance by manipulating the activity of genes that engage in responding to stress [26]. GRAS1 in tobacco is implicated in signal transduction pathways that increase reactive oxygen species (ROS) levels in response to various stress treatments, suggesting a role in transcriptional regulation in response to stress [27,28]. Overexpression of the grape (*Vitis amurensis*) GRAS protein VaPAT1 enhanced abiotic stress tolerance in transgenic *A. thaliana* [29]. *Populus tremula*'s tolerance to drought and salt was improved by overexpressing the GRAS protein SCL7 [30]. Birch (*Betula alba*) demonstrated a salt-tolerant phenotype upon the overexpression of BpGRAS34, as well as increased ROS scavenging capabilities and proline content [15].

Phoebe bournei (Hemsl.) Yang, a member of the family Lauraceae, is one of the most economically and ecologically valuable timber species [31]. Below 1000 m in altitude, *Phoebe bournei* prefers to live in evergreen broad-leaved forests with healthy and well-drained soils [32]. Because of its straight trunk, large and dense crown, beautiful shape, excellent material and resistance to decay, fine grain structure, fragrant wood, and smooth and beautiful chipping, it is widely considered to be one of the best timber species for building construction, furniture making, craft carving, and ornamental purposes [33,34]. As major players in the molecular mechanisms of plant stress resistance, GRAS family members are responsible for a wide diversity of physiological procedures, covering root formation, fruit development, hormone signaling, stem cell maintenance, initiation of the axillary meristem in response to light, generation of male gametophytes, and response to adversity [17,20,35–37]. Despite the fact that genome-wide investigations of GRAS genes have been completed for a number of botanical species, no such analysis has been undertaken for *P. bournei*, and the role of GRAS proteins in abiotic strain responses remains unclear. In this work, we described 56 GRAS genes after conducting a genome-wide analysis of this gene family in *P. bournei*. Quantitative RT-PCR (qRT-PCR) was employed to examine the effects of exposure to drought, salt, and temperature stressors on a sub-set of *PbGRASs* genes.

In this research, we utilized the chromosome-level genomic information of *P. bournei* to systematically identify its GRAS family genes and analyzed the physicochemical nature of proteins and the structure and chromosomal distribution of the genes and established the evolutionary relationships, promoter cis-acting elements, and epistatic expression patterns of drought, salt and temperature stress genes in an attempt to offer a frame of reference for further research on the biological functions of *P. bournei* GRAS genes in the process of stress resistance. This research will be fundamental in elucidating the underlying molecular mechanisms of GRAS responsiveness to drought, salt and temperature stress, and will supply quality genes for genetically engineered breeding in *P. bournei* for improved resistance to either drought, salinity or high and cool temperatures.

2. Results

2.1. Phylogenetic Analysis of PbGRAS Gene

On the basis of the updated genome assembly, we identified 150 presumptive GRAS genes: 56 in *P. bournei*, 34 in *A. thaliana*, and 60 in *O. sativa*. The evolutionary interrelationships of the proteins of PbGRAS were structured using phylogenetic trees (Figure 1). Phylogenetic analysis revealed that these 150 GRAS proteins can be separated into nine groups (PAT1, SHR, LISCL, HAM, SCL4/7, LAS, SCL3, DELLA, and SCR). These results, as with those published for birch (*Betula platyphylla*), *Dendrobium chrysotoxum*, and soybean (*Glycine max*), demonstrated some variability in the number of GRAS genes in different subfamilies [15,38,39]. PbGRAS proteins are unevenly distributed in nine subfamilies, most of which belong to the DELLA subfamily (18 members), followed by the PAT1 subfamily (12 members). Only seven, six, five, three, two and one PbGRAS protein were detected in the HAM, SCR, SHR, SCL3, LISCL and LAS subfamilies, respectively. Several of the PbGRAS proteins were shown to have strong phylogenetic relationships with proteins from other species (bootstrap support $\geq$ 80), which could indicate that they are similar to GRAS proteins from *A. thaliana* or *O. sativa* serving the same purpose.

2.2. Chromosome Distribution and Synthesis Analysis of PbGRAS Gene

A total of 56 PbGRAS proteins were retrieved from the *P. bournei* genome databases and were characterized. Their physico-chemical properties were investigated further with the help of ExPasy (Table 1). The majority of these proteins have classical GRAS structural domains with around 400–700 amino acids (aa), reaching up to 777, while PbGRAS5 has a severely truncated GRAS domain with only 119 amino acids. The 56 PbGRASs had molecular weights ranging from 13.682 kDa to 84.836 kDa. Theoretical pI values were between 4.86 and 6.9 for the majority of PbGRAS proteins, with only a few pI values in the range of 7 to 9. This finding suggests that the majority of GRAS proteins are negatively charged, and only a minority are positively charged. The grand average of hydropathicity index for all PbGRAS proteins (from −0.555 to −0.014) indicates that all PbGRAS proteins are hydrophilic; these results are similar to those of GRAS proteins in *Betula platyphylla* [15]. The majority of the forecasted instability indices were above 40 (from

40.08 to 61.97), demonstrating that most PbGRAS proteins are unstable except for PbGRAS5 (39.37), PbGRAS35 (39.67), PbGRAS37 (35.3), and PbGRAS54 (32.34). Most of the PbGRASs do not have introns, suggesting that the sequence of PbGRASs is, to some extent, conserved.

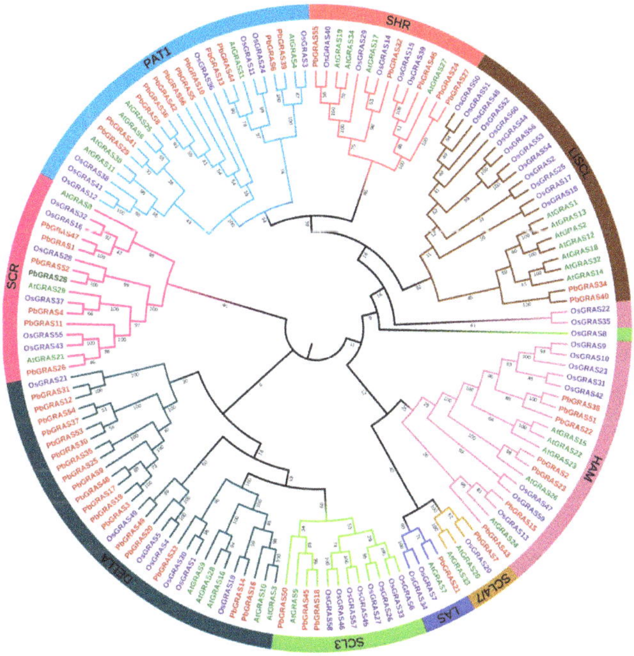

Figure 1. Phylogenetic tree of PbGRAS, AtGRAS and OsGRAS proteins. The different-colored arcs indicate subfamilies of the GRAS family. The tree was constructed using the 56 PbGRAs identified in *P. bournei*, the 34 AtGRAs identified in *A. thaliana* and the 60 OsGRAs identified in *O. sativa* by using MEGA11 with the bootstrap 1000 times.

Table 1. Characteristics of the GRAS proteins from *P. bournei*.

Member	Number of Amino Acid	Molecular Weight	Theoretical pI	Instability Index	Grand Average of Hydropathicity	Domain (aa)	Intron
PbGRAS1	684	76,352.99	5.91	60.92	−0.415	366	0
PbGRAS2	601	66,265.46	6.35	61.97	−0.149	364	0
PbGRAS3	522	59,304.67	5.94	48.53	−0.263	297	0
PbGRAS4	456	49,271.71	5.5	51.05	−0.132	356	0
PbGRAS5	119	13,682.77	8.46	39.37	−0.206	116	0
PbGRAS6	605	67,162.22	5.54	40.94	−0.315	369	1
PbGRAS7	406	44,804.64	5.47	58.55	−0.245	360	0
PbGRAS8	542	60,896.42	5.26	55.46	−0.52	329	1
PbGRAS9	407	45,766.31	7.25	46.25	−0.329	200	0
PbGRAS10	577	64,480.59	5.72	56.43	−0.418	369	1
PbGRAS11	578	64,135.67	5.51	51.57	−0.206	351	1
PbGRAS12	443	50,201.97	6.11	43.72	−0.368	366	0
PbGRAS13	433	48,160.99	9.95	49.35	−0.377	298	0
PbGRAS14	595	64,855.31	5.24	45.09	−0.236	366	1
PbGRAS15	515	56,248.95	5.28	44.28	−0.229	385	0
PbGRAS16	533	58,213.08	5.17	47.33	−0.178	358	0
PbGRAS17	595	66,650.92	7.87	47.27	−0.097	374	1

Table 1. *Cont.*

Member	Number of Amino Acid	Molecular Weight	Theoretical pI	Instability Index	Grand Average of Hydropathicity	Domain (aa)	Intron
PbGRAS18	723	80,619.74	5.5	51.06	−0.284	420	5
PbGRAS19	479	54,200.05	6.29	47.41	−0.242	263	0
PbGRAS20	489	54,185.93	6.3	53.76	−0.139	345	1
PbGRAS21	396	44,773.87	6.19	52.67	−0.24	365	0
PbGRAS22	755	82,162.03	5.91	59.13	−0.218	351	0
PbGRAS23	359	40,017.22	9.52	60.39	−0.023	341	0
PbGRAS24	414	46,394.48	5.14	45.58	−0.259	356	1
PbGRAS25	292	33,025.63	7.1	49.73	0.218	278	0
PbGRAS26	495	55,471.75	5.63	54.13	−0.385	351	1
PbGRAS27	406	45,917.81	5.28	54.2	−0.367	371	0
PbGRAS28	505	56,314.09	5.88	42.94	−0.129	354	0
PbGRAS29	426	46,516.11	5.27	53.99	−0.293	246	3
PbGRAS30	426	48,266.75	5.72	40.08	−0.014	346	0
PbGRAS31	452	51,198.4	6.23	47.47	−0.292	369	0
PbGRAS32	548	60,019.12	5.69	57.13	−0.52	400	0
PbGRAS33	547	59,484.68	5.62	46.5	−0.071	369	0
PbGRAS34	762	85,476.64	5.08	50.09	−0.496	372	0
PbGRAS35	593	67,285.73	5.14	39.67	−0.224	367	0
PbGRAS36	588	65,880.12	5.49	51.14	−0.443	355	0
PbGRAS37	499	56,222.09	4.86	35.3	−0.132	276	0
PbGRAS38	777	83,607.35	5.81	52.8	−0.168	358	0
PbGRAS39	582	64,216.62	5.01	48.6	−0.25	368	0
PbGRAS40	756	84,836.7	5.1	47.53	−0.555	372	0
PbGRAS41	586	65,363.6	5.7	48.7	−0.304	370	1
PbGRAS42	582	65,227.44	5.31	53.36	−0.448	369	0
PbGRAS43	545	61,817.03	4.91	46.94	−0.354	369	0
PbGRAS44	597	65,108.66	8.49	56.88	−0.344	367	0
PbGRAS45	474	53,276.12	5.76	52.93	−0.189	420	0
PbGRAS46	495	55,057.59	5.38	53.23	−0.343	381	0
PbGRAS47	652	73,122.66	6.4	60.59	−0.443	370	0
PbGRAS48	243	27,935.42	6.9	40.59	0.029	240	0
PbGRAS49	681	75,504.3	5.71	55.36	−0.333	364	1
PbGRAS50	412	46,004.99	7.22	50.54	−0.125	371	0
PbGRAS51	653	71,489.74	6.02	54.1	−0.236	363	0
PbGRAS52	460	50,659.31	5.91	46.48	−0.223	337	1
PbGRAS53	237	26,573	5.9	40.48	0.195	155	1
PbGRAS54	606	68,104.02	5.54	32.34	−0.098	372	1
PbGRAS55	453	50,294.48	5.66	47.16	−0.077	394	2
PbGRAS56	426	47,508.06	4.99	57.1	−0.507	142	2

The 56 *PbGRASs* identified were further localized on 12 *Phoebe bournei* chromosomes (Chr01 to Chr12) (Figure 2). Overall, the distribution of these 56 *PbGRASs* on the 12 chromosomes was heterogeneous. The distribution of *PbGRASs* on *P. bournei* chromosomes was variable and heterogeneous among chromosomal regions. No *PbGRASs* were detected on Chr06. Among the above chromosomes, the highest distribution of *PbGRASs* was on Chr08, with 10 *PbGRASs* (17.9%), followed by Chr07 (9, 16.1%), then Chr01 and Chr02 (6, 10.7%). Five *PbGRASs* (8.9%) were located on Chr05, four *PbGRASs* (7.1%) were located on Chr03, 04, and 09, three *PbGRASs* (5.3%) were located on Chr10 and 12, and two *PbGRASs* (3.6%) were located on Chr11. Based on previous studies, we hypothesized that no significant association or correlation was found between the amount of GRAS genes and chromosome length [5,40].

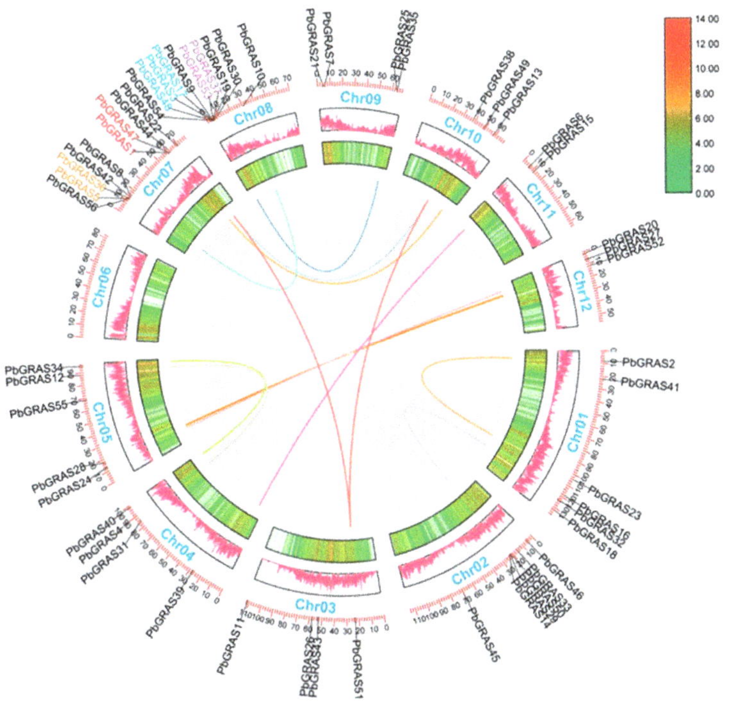

Figure 2. Genomic positions, duplication events and syntenic relationships of *PbGRAS* genes. The green and red blocks in the middle represent the gene density of each chromosome. Those genomes deriving from segmental replication events are linked by colored lines, and the gray lines denote synthetic blocks in the *P. bournei* genome. The genes in purple, blue, red, and yellow are those with tandem replication.

Tandem duplication events are defined as chromosomal regions that contain two or more genes within 200 kb and perform a critical task in the further generation of new functions in gene families [41]. Five tandem replication events were identified on Chr07 and Chr08, including *PbGRAS1/PbGRAS47*, *PbGRAS3/PbGRAS17*, *PbGRAS5/PbGRAS36*, *PbGRAS48/PbGRAS3*, and *PbGRAS53/PbGRAS37*, involving a total of nine *PbGRASs* (Figure 2). Except for *PbGRAS1* and *PbGRAS47* from the SCR subfamily, and *PbGRAS5* and *PbGRAS36* from the PAT1 subfamily, all genes that participate in tandem replication events belong to the DELLA subfamily, implying that the DELLA cluster, as the largest subfamily, plays a substantial role in the growing number of GRASs [7,42]. In addition, there are six fragment replicates situated at or near the terminal end of chromosomes 1, 2, 3, 4, 5, 7 and 10. To further investigate the gene duplications, we constructed a comparative tandem map of the three *PbGRAS* families (Figure 3). Among them, the intensity of association with *PbGRAS* genes was ranked from highest to lowest as follows: *O. sativa* (22), *A. thaliana* (18). The linear homology plots suggest that tandem replication events are the main mechanism for gene family expansion in plants, and more importantly, for gene family diversity [43,44].

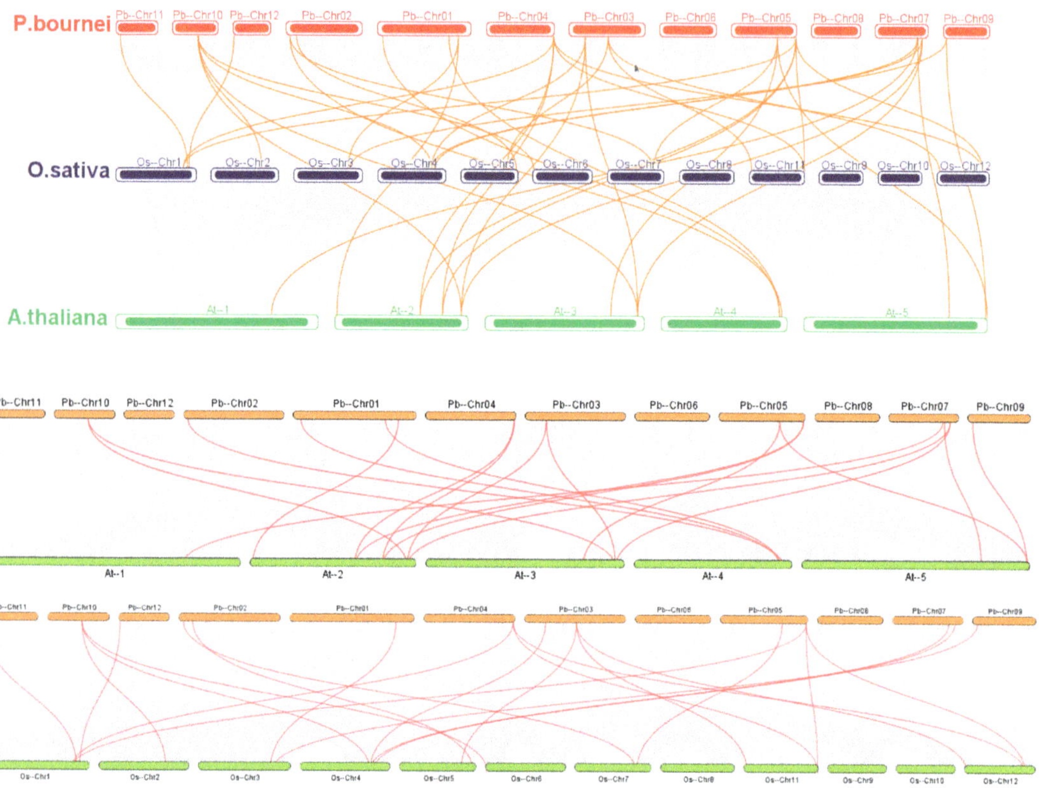

Figure 3. Analysis of homology between the *P. bournei* genome and plant genomes of *A. thaliana* and *O. sativa*. Gray lines symbolize pairs of genomes between homologous blocks, and red lines represent adjacent *PbGRAS* gene pairs. The orange lines emphasize the synthesized GRAS gene pairs in the three species.

2.3. Conserved Structural Domain Motif Analysis of PbGRASs

As a further exploration of the GRAS sequence characteristics and structure of *P. bournei*, we optimized the conserved motifs and intronic structures of *P. bournei* (Figure 4). To examine the structural specifics of the PbGRAS protein, we estimated 10 conserved motifs in the MEME program. GRAS proteins from the same subfamily contain comparable motif combinations in general, implying that GRAS proteins of the same subfamily likely have identical functions. Nearly all GRAS proteins included Motifs 1, 2, 3, 4, 5, 6, 7 and 10, demonstrating that the motifs are highly conserved and might have significant effects within the GRAS family. Motif 8 is only distributed in the PAT1 subfamily; Motif 9 is missing in the HAM subfamily except for *PbGRAS15*. *PbGRAS5* contains only Motif 3 and Motif 7, *PbGRAS23* contains only Motifs 2, 4, 6 and 7, and *PbGRAS43* contains only Motifs 1, 4, 5 and 7. Some motifs are distributed only in specific positions of the patterns. Motif 7, for instance, is consistently dispersed near the conclusion of the sequence, but Motif 5 is usually disseminated at the beginning. The intended uses of the majority of these preserved motifs are unknown. The variances in the order of appearance of these motifs across subfamilies imply that GRAS proteins from various subdivisions may have distinct purposes. Additionally, multiple genes within the same subfamily exhibit distinct motif distributions, implying that the tasks they perform may be unique as well. The tendency for similar motifs to cluster together in specific subgroups of GRAS proteins suggests the possibility of cross-functional similarities between these proteins. In addition, to determine the differences in gene structures, we compared the exon and intron structures of 56 *P. bournei* GRAS genes. Most of the GRAS genes have minimal or no introns, which was also evidenced in the present study with 67.86% of *PbGRAS* genes having no introns (Figure 4 and Table 1). Five and three introns were detected in *PbGRAS18* and *PbGRAS29*, respectively. This suggests that the family was relatively conservative in evolution.

Figure 4. Phylogenetic relationship, motif pattern, and gene structure of *PbGRAS* genes. (**a**) Phylogenetic tree of *PbGRAS*. (**b**) The motifs of PbGRAS. Motifs 1–10 are displayed in different colored boxes. The protein length can be estimated using the scale at the bottom. (**c**) Gene structure of *PbGRAS* genes. Green boxes indicate exons (CDS), black lines indicate introns, and yellow boxes indicate 5′ and 3′ untranslated regions.

2.4. Multiple Sequence Alignment and Cis-Acting Element Analysis of PbGRASs

Multiple sequence alignment of PbGRAS proteins showed a degree of conservation in the GRAS gene family. According to the output scores of multiple sequence alignment, there is a high similarity in the 320–360 and 460–510 regions, and it can be inferred that the function of these regions may be similar (Figure 5). Cis-acting components can affect the transcribed state of their linked genes since they are non-coding DNA regions in gene promoters. The occurrence of cis-acting elements in promoters might be related to the variety of gene-related processes and expression trends. We submitted base pair sequences from the 2 kb region upstream of the *PbGRAS* gene promoter to the PlantCARE database and found 19 cis-acting parts linked to reactions to environmental stress or phytohormone transduction, including nine stress-responsive elements and five phytohormone-related elements (Figure 6). As demonstrated in the picture, all *PbGRAS* genes have between seven and 11 cis-acting elements. Each GRAS gene has at a minimum one element associated with a strain response such as light responsiveness, essential for anaerobic induction, defense and stress responsiveness, low-temperature responsiveness, drought inducibility, anoxic specific inducibility, circadian control, and wound responsiveness, having a major effect in the generation of strain responses. Meanwhile, five phytohormone-related elements in the *PbGRAS* genes appear in most phytohormones; these include abscisic acid responsiveness, MeJA responsiveness, gibberellin responsiveness, auxin responsiveness, and salicylic acid responsiveness. Among them, we discovered that the photosensitive component is the richest cis-acting component, as it is extensively represented in the *PbGRAS* promoter region. These findings suggest that *PbGRASs* might be tolerant to non-biological stresses, such as light, salt, cold, and drought, and may be involved in plant evolutionary and developmental processes.

2.5. Heat Map of PbGRAS Gene Expression in Different Tissues

For analyzing the variation in *PbGRAS* exposure modes, we incorporated gene epistasis data from five tissues: leaf, root xylem, stem xylem, root bark, and stem bark (Figure 7). The outcomes displayed that *PbGRAS7* and *PbGRAS14* had the highest expression in the root xylem and stem xylem, and moderate expression in the root bark and stem bark; in particular, *PbGRAS7* reached maximum expression in stem xylem, which indicated that the performance of *PbGRA7* expression was closely related to the cultivation of the stem xylem. In addition, *PbGRAS10*, *PbGRAS41*, and *PbGRAS16* genes were also highly expressed in these four tissues, but their expression was lower in the root xylem and stem xylem compared with *PbGRAS7* and *PbGRAS14*. Meanwhile, the expression of *PbGRAS10* was comparatively higher in the root and stem bark. This is similar to the results in pepper, where members of the PAT1 subfamily, such as *CaGRAS30* and *CaGRAS34*, are expressed at higher concentrations in the stems than in other organs [3].

2.6. Abiotic Stress Experiments on the GRAS Gene Family of P. bournei

To investigate the responses of GRAS genes to abiotic stress, we tested the expression responsiveness of *P. bournei* genes to drought, salt, and temperature stress (Figure 8). The relatively high expression of *PbGRAS7*, *PbGRAS10*, *PbGRAS14* and *PbGRAS16* in different tissues can be seen more clearly in the gene expression heat map. In addition, cis-acting element analysis showed that *PbGRAS7*, *PbGRAS10*, *PbGRAS14* and *PbGRAS16* contain a large number of elements associated with drought and temperature stresses. Thus, we selected these four genes to validate the results further. The findings reveal that the levels of manifestation of GRAS genes were modulated in response to dryness, salt, and temperature. The expression of *PbGRAS10*, *PbGRAS14*, and *PbGRAS16* did not change significantly after 4 h of immersion with a nutrient solution containing 10% PEG, and only the expression of *PbGRAS7* was enhanced and was significantly higher at 6 h, about three times that of the control (Figure 8a). *PbGRAS7* may have a non-significant effect on *P. bournei*'s osmotic stress. When treated with 10% NaCl (Figure 8b), we discovered that *PbGRAS7* and *PbGRAS14* genes were repressed to different degrees. The expression of *PbGRAS16* increased somewhat after 4 h, while *PbGRAS10* increased significantly at 4 h after treatment and increased about 30-fold after 24 h compared with the expression at the beginning of treatment. One may presume that *PbGRAS7* may have a significant action in the tolerance of *P. bournei* to drought strain. Under the 10 °C treatment (Figure 8c), the expression of all genes increased to different degrees after 12 h. The expression of *PbGRAS10* and *PbGRAS16* was already increased at 4 h, and *PbGRAS7*, *PbGRAS14*, and *PbGRAS16* still maintained high expression levels after 24 h. The expression of *PbGRAS10* and *PbGRAS16* was increased at the beginning of the treatment. Among them, *PbGRAS10* expression was remarkably enhanced after 12 h and was about 25-fold compared with the control group. Under 40 °C treatment (Figure 8d), the expression of every gene was increased to various extents and remained high after 24 h. In particular, *PbGRAS10* and *PbGRAS16* expression increased about 5-fold after 24 h of treatment compared with the CK group. It can be postulated that *PbGRAS10* and *PbGRAS16* have significant effects on the plant's reactions to temperature stress.

Figure 5. The multiple sequence alignment in *PbGRAS7*, *PbGRAS10*, *PbGRAS14*, *PbGRAS16*, and *PbGRAS40* using Jalview software. The "*" under the conservation in the first line of the annotation indicates the highest similarity, and the number is the output score of the degree of conservatism. In sequence alignment, amino acid alignment of A, I, L, M, F, W, V, C is shown in blue, R, K in red, N, Q, S, T in green, E and D in magenta, G in orange, H, Y in cyan, and P in yellow.

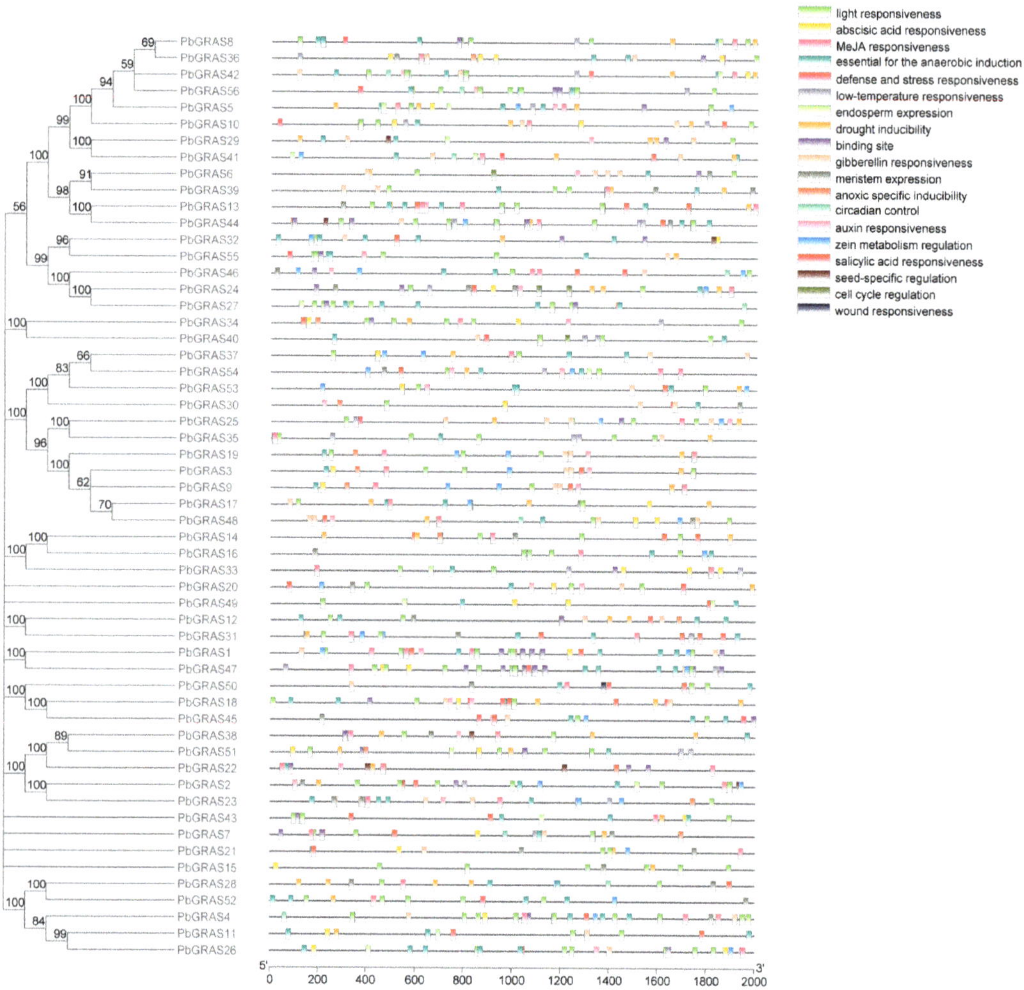

Figure 6. Predicted cis-acting elements in the promoter regions of *PbGRAS* genes. On the left is the ML phylogenetic tree (bootstrap replications: 1000) with branches labeled with bootstrap values. The one on the right is the promoter position (−2000 bp). The cis-acting regulatory elements in the promoter were categorized into 19 types with different colors. The lower axis denotes the quantity of each cis-acting element.

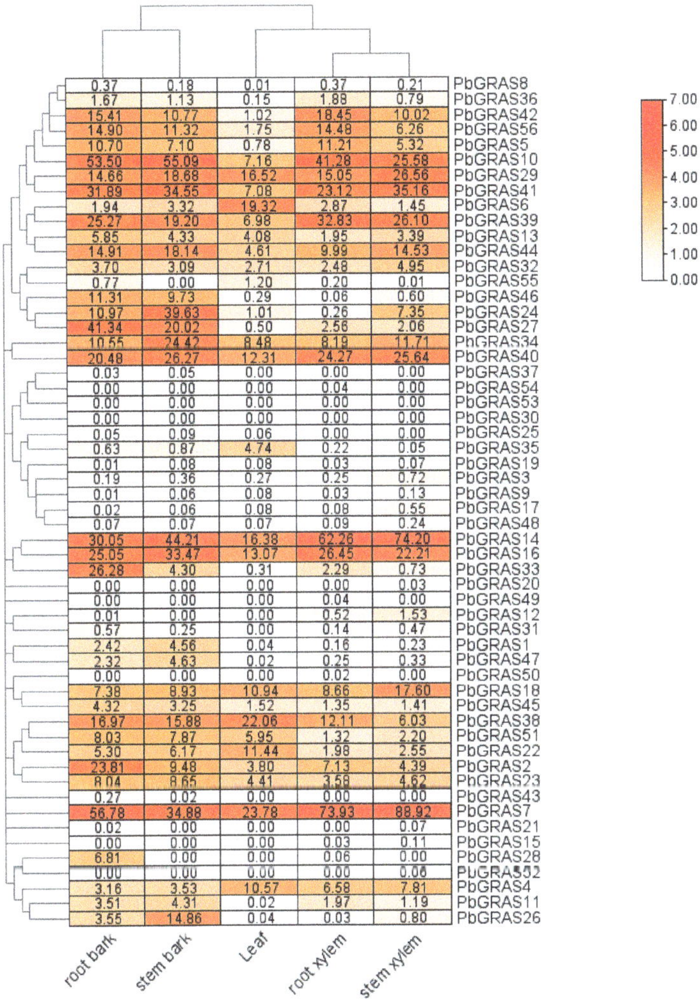

Figure 7. Expression profiling of *PbGRAS*. The expression of *PbGRAS* in five different tissues. The redder the color block, the higher the expression level, and the whiter the color, the lower the expression.

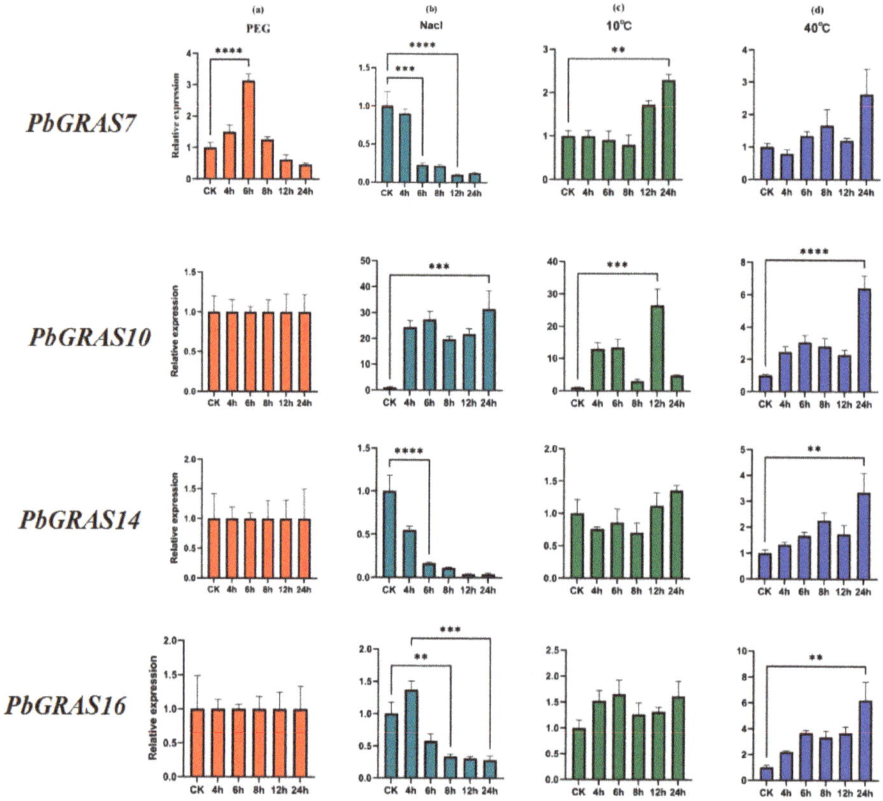

Figure 8. Expression of *PbGRAS* genes in *P. bournei* in response to drought, salt and temperature stresses was tested by means of qRT-PCR. (**a**) Relative gene expression levels at the same time points (4, 6, 8, 12, and 24 h) under treatments with a nutrient solution of 10% PEG simulating a drought environment. The control group was processed in distilled water. (**b**) Relative gene expression levels under the treatment with 10% NaCl nutrient solution immersion. The control group was processed in distilled water. (**c**) Relative gene expression levels at low temperature (10 °C) and for the control (25 °C). (**d**) Relative gene expression levels at high temperature (40 °C) and for the control (25 °C). (** $p < 0.01$, *** $p < 0.0005$, **** $p < 0.0001$).

3. Discussion

Plants primarily use gene regulation as an adaptive mechanism in response to environmental stressors [45]. The growth, differentiation, and stress responses of tissues and organs are tightly controlled by GRAS proteins [46]. The subtropical broadleaf evergreen tree *P. bournei* is extremely important, both economically and ecologically. Water scarcity and temperature variation severely affect *P. bournei* growth and wood accumulation. Jin and others found that *P. bournei* can regulate its own protective enzyme system and osmotic substances to adapt to drought under drought stress [47]. Wang showed that polyamines can increase soluble sugar and chlorophyll content under drought and improve the drought resistance of *P. bournei* [48]. However, there is a dearth of literature on the regulatory signaling pathways that help *P. bournei* to endure abiotic challenges such as drought. As GRAS genes have such a critical function in the signaling of *P. bournei*, it is crucial to comprehend their expression pattern. The correct identification of abiotic tolerance genes in *P. bournei* will be aided by a fundamental and thorough bioinformatic investigation of them.

Many plant species have been found to contain GRAS family genes, with 34 GRAS gene members in *A. thaliana* [17], 60 in *O. sativa* (*Oryza sativa*) [11], and 153 in wheat (*Triticum aestivum*) [49]. We undertook a genome-wide investigation of the GRAS gene family in *P. bournei* and identified 56 GRAS proteins, the majority of which have characteristic GRAS structural domains totaling roughly 360 aa, most of which share a conserved C-terminus (Table 1 and Figure 5), which is consistent with the study by Liu and Wang [50]. Most of the conserved sequences appeared in the 270–610 region. Given that intron loss in eukaryotes is massive due to selection pressure and that the GRAS group is a retrotransposon group that accelerates the generation of intron-less genes from bacterial to eukaryotic gene sets through both horizontal gene transfer and gene dimerization [51–53], it is possible that large and frequent gene duplication events have led to changes in the number of *PbGRAS* genes (38 out of 56, Table 1). This investigation found five simultaneous duplication events including a total of nine *PbGRASs*, and the majority of the genes implicated in these events belonged to the DELLA subfamily (Figure 2). However, some *PbGRASs* evolved different exon–intron outcomes, which may indicate that they evolved adaptive functions in order to adapt to their environment. Moreover, most GRAS proteins in Table 1 have pI values located at around 5~7, with a smaller possibility of interacting nonspecifically with acidic proteins [54], and the GRAS set may be engaged in protein–protein interfaces [55]. The findings of chromosomal localization reveal that the 56 *PbGRASs* identified were unevenly located on 12 chromosomes of *P. bournei* (Figure 2), with the highest distribution being on Chr08, accounting for 17.9% of the *PbGRASs*, while no *PbGRASs* were detected on Chr06. These outcomes are analogous to other research, such as where no *MdGRAS* geneswere found on chromosome 16 of *Malus domestica*, while chromosome 11 contained the most, accounting for 15.75% of the *MdGRAS* genes [56]. It is possible to attribute this to evolutionary processes such as gene duplication and chromosomal transfer. Chr08 has the largest amount of GRAS genes and may be one of the critical chromosomes for plants to deal with abiotic pressures. The addition of GRAS genes from *A. thaliana* and *O. sativa* expanded the dataset to 150 GRAS proteins divided into nine subfamilies after phylogenetic analysis (Figure 1). However, the conserved motifs in different subfamilies are different, which may be due to the subfamilies of PbGRASs performing different biological functions in the growth of *P. bournei*, as the motif arrangement was mostly similar among PbGRAS proteins of the same subfamily, suggesting that PbGRAS proteins of the same subfamily may have similar functions. As shown by To and Wang et al., GRAS proteins are stochastically assigned in the phylogenetic tree [7,57], but individuals from the same subfamily tend to perform comparable tasks [50].

Characterization of cis-acting elements demonstrated that *PbGRAS* genes may participate in multiple biotic procedures by regulating a variety of target genes associated with hormone induction, evolution, metabolism, and abiotic stresses (Figure 6). Among these cis-acting elements, those correlated to drought and temperature stress such as drought inducibility and low-temperature responsiveness are assigned to the promoter regions of *PbGRAS* genes, such as *PbGRAS10* and *PbGRAS14*. There are many elements in *PbGRAS14* that react to drought stress and salt stress, but fewer light response elements compared with other *PbGRAS* genes, which indicates that this gene may be less involved in light response and is the main gene used to cope with salt and drought stress. By comparing the other genes of the DELLA subfamily, there were relatively few cis-acting elements for light response, but more cis-acting elements for stress response, which indicated that the DELLA gene family may play a role in the stress response of *P. bournei*. Then, focusing on the stress experiment (Figure 8), it could be seen that the expression of *PbGRAS14* changed significantly under NaCl and 40 °C stress, which is also consistent with our analysis results. A similar example is the case for *CmGRAS* genes, which are more closely related to the reference species in Chinese chestnut and were found to have more cis-elements and showed higher expression modes at different phases of buds and fertile or aborted ovules, suggesting their significant roles in the evolution and functioning of the *CmGRAS* gene family [14]. Additionally, the results can be seen more clearly in the expression heat map of the genes; *PbGRAS7* and *PbGRAS14* are highly expressed in the root xylem and stem xylem, while *PbGRAS10* is relatively highly expressed in the root bark and stem bark. This likely relates to the fact that GRAS genes display a differential expression pattern across different tissues [13]. Consistent with some other observations, such as the expression of *NnuGRAS-05*, *NnuGRAS-10*, and *NnuGRAS-25* in lotus root bark compared to their expression in the leaf and petiole, members of the PAT1 subfamily seem to be more highly expressed in the root than in the leaf [58]. The expression levels of *CaGRAS30* and *CaGRAS34* in pepper were higher in stems than in other organs [3]. Almost all PAT1 subfamily members were highly expressed in eggplant stems [25]. This may be due to the fact that PAT1 subfamily members help to maintain rootstock growth and resist environmental disturbances.

Combining these findings and analysis, we selected four genes (*PbGRAS7*, *PbGRAS10*, *PbGRAS14*, and *PbGRAS16*) for validation of the stress treatments. By subjecting abiotic stress-treated *P. bournei* seedlings to qRT-PCR analysis, it was found that drought, salt and temperature stress conditions significantly induced the expression of the four *PbGRASs* in *P. bournei* plants (Figure 8). In particular, the expression of *PbGRAS10* significantly surged 30-fold and 25-fold after salt treatment and low-temperature treatment, respectively, *PbGRAS7* was expressed at a three times higher level than in the control after PEG treatment, and *PbGRAS10* and *PbGRAS16* increased approximately five-fold after high-temperature treatment compared with the control, which was consistent with the outcomes of cis-acting meta-analyses (Figure 6). In contrast, a low-temperature-responsive cis-element was predicted at the *PbGRAS14* promoter, but its expression level was comparable to the control condition. This, to some extent, reflects the fact that the accuracy of the cis-acting element prediction results needs to be further improved. Drought, salt, and temperature stress conditions significantly induced the expression of *PbGRASs* in *P. bournei* plants, suggesting that *PbGRASs* may have some degree of association with stress response. There have been other studies describing similar findings. In *O. sativa* (*Oryza sativa*), *OsGRAS23* in the SCL subfamily is highly sensitive to salinity stress, and the expression of this gene is up-regulated under drought stress [26,59]. Similar results were found in *Halostachys caspica*, where the expression of *HcSCL13* in the SCL subfamily was up-regulated under drought and salt treatments [30]. Salt stress in poplar (*Populus euphratica*) also regulates a stress response transcription factor, *PeSCL7* [60]. This provides more evidence that the SCL subfamily is crucial for adapting to adverse conditions. The GRAS family of grapevine is also regulated by abiotic stresses, for example, in that drought treatment upregulates the expression of *VviPAT3*, *VviPAT4*, *VviPAT6* and *VViPAT7* in seedlings [61], while salt treatment has an effect on *VviRGA5* in the DELLA subfamily [62]. Furthermore, the *Lycopersicon esculentum* gene *SlGRAS2*, which is homologous to *VviPAT4*, is highly sensitive to salt stress [51]. According to a recent report, GRAS genes play a key role in the regulation of salt stress [63], and thus, the role of GRAS in promoting plant growth and salt stress response has received widespread attention. However, further in-depth research is required to understand the function of GRAS genes in response to abiotic stressors in a wider range of plant species.

This study not only determined and examined the expression of *PbGRASs* using qRT-PCR, but it also systematically analyzed the GRAS gene family of *P. bournei*, suggesting that *PbGRASs* play an active part in the physiology of *P. bournei*'s resistance to drought, salt, high, and low temperature stresses. This study provides a basis for elucidating the molecular mechanisms of the *P. bournei* GRAS response to drought, salt, and high and low temperature stresses and posits ideas for the future improvement of *P. bournei* using genetic engineering methods.

4. Materials and Methods

4.1. Genome Data and Plant Material Source

The genome sequence data and annotation information of *P. bournei* were downloaded from the Sequence Archive of China National GeneBank Database (CNSA) with accession number CNP0002030 [64]. Genome sequence files of *A. thaliana* (version: Arabidopsis_thaliana.TAIR10.dna.toplevel.fa.gz, last modified: 17 February 2023 06:57, 35M) and *O. sativa* (version: Osativa_204_v7.0.fa.gz, last modified: 12 January 2014, 110.4M) were acquired from EnsemblPlents (https://plants.ensembl.org, accessed on 31 October 2022) and Phytozome v13 (https://phytozome-next.jgi.doe.gov, accessed on 31 January 2023), respectively. The RNA-seq data from different tissues of *P. bournei* were download from BioProject (https://www.ncbi.nlm.nih.gov/bioproject/, accessed on 12 March 2023) with accession number PRJNA628065. The plant materials were derived from 1-year-old *P. bournei* seedlings cultured in an artificial climate box under different treatments. *P. bournei* seedlings with the same growth potential for 1 year were selected for treatment, and the materials were divided into the control group and stress treatment group, with 30 strains in the treatment group and 3 strains in the control group. Every 2 strains in the treatment group were used as a biological replicate, and 3 groups of biological replicates were set in each time period. After various treatments, *P. bournei* leaf samples were collected and stored in liquid nitrogen at −80 °C for RNA extraction.

4.2. Identification and Physical and Chemical Property Analysis

The conserved domain of the *A. thaliana* GRAS gene family was downloaded from PlantTFDB (http://planttfdb.gao-lab.org/, accessed on 31 October 2022). A local BLASTp search was used to compare the conserved domain between *P. bournei* and *A. thaliana* to screen the candidate GRAS genes in *P. bournei* [39]. The repetitive results of the BLASTp search were removed. The identified protein sequences of GRAS genes from *P. bournei* were submitted to the NCBI for BLASTp to perform further searching. To further identify GRAS gene family members, the GRAS conserved domain HMM model (PF03514) was downloaded from the Pfam database (http://pfam.xfam.org/, accessed on 14 April 2023), using HMMER-3.2.1 (http://hmmer.org/download.html, accessed on 14 April 2023) with an e-value $<10^{-5}$ and other parameters by default. After the identification of GRAS genes in *P. bournei*, the online website ExPASy (https://web.expasy.org/prot-param/, accessed on 26 February 2023) was used to analyze the physical and chemical properties of the GRAS proteins that were identified [65].

4.3. Chromosomal Distribution and Gene Duplication of PbGRAS Genes

TBtools was used for grepping the location information of the *PbGRAS* genes from the genome (FASTA) file and the annotation (GFF) file of *P. bournei* [66]. The syntenic relationships of *PbGRAS* were determined using MCScanX with default parameters and plotted using Tbtools [67].

4.4. Collinearity Analysis of PbGRAS Genes

The syntenic relationships between *PbGRAS* genes and GRAS genes from *A. thaliana* and *O. sativa* were determined by using MCScanX software. Tbtools was used for visualization.

4.5. Phylogenetic Analysis

The sequences of GRAS proteins of *P. bournei*, *A. thaliana* and *O. sativa* were aligned using the Muscle program of MEGA11 with default settings for constructing maximum likelihood (bootstrap replications: 1000) phylogenetic tree [68,69]. iTOL (https://itol.embl.de/, accessed on 29 January 2023) was used to improve and beautify the phylogenetic tree.

4.6. Analysis of Conserved Motifs and Gene Structures

The protein sequence of *P. bournei* was identified using the online software MEME (http://meme-suite.org/, accessed on 8 March 2023), and the predicted value of the motif number was 10. The multiple GRAS protein sequences alignment was carried out using Jalview software (https://www.jalview.org/, accessed on 30 March 2023). We used the Batch CD-search (https://www.ncbi.nlm.nih.gov/Structure/bwrpsb/bwrpsb.cgi, accessed on 8 March 2023) with default parameters to detect the conserved domains of the PbGRAS protein.

4.7. Multiple Sequence Alignment and Promoter Cis-Element Analysis of PbGRAS Genes

Multiple sequence alignment of *PbGRAS* was performed using Jalview software (version: 2.11.2.6). To explore the cis-acting elements in the sequence, we extracted the upstream 2000 bp sequences from the *P. bournei* genome. The online software PlantCARE (https://bioinformatics.psb.ugent.be/webtools/plantcare/html/, accessed on 11 March 2023) was used to analyze the cis-acting regulatory elements in the promoter region of the *PbGRAS* genes. After selection and categorization, the data were visualized by means of TBtools software.

4.8. Treatment in Different Tissue

The fragments per kilobase of transcript per million fragments mapped (FPKM) transcriptomic data from five different tissues (leaf, root xylem, stem xylem, root bark and stem bark) were used to construct a expression profiling with TBtools.

Based on results of cis-element analysis and expression profiling of *PbGRAS*, we conducted different treatment. In the drought and salt stress experiment, 3 seedlings in the control group were soaked in distilled water after root washing, while those in the treatment group were soaked in a nutrient solution containing 10% PEG and 10% NaCl, respectively, and cultured in an artificial climate incubator with a temperature of 25 °C and humidity of 75%. The treatment group was sampled at 4 h, 6 h, 8 h, 12 h and 24 h, respectively, and the control group was sampled at 0h. In the temperature treatment (10 °C and 40 °C), the control group was incubated at room temperature, and the treatment group was incubated at 10 °C and 40 °C. The treatment group was sampled at 4 h, 6 h, 8 h, 12 h and 24 h, with 6 strains in each time period. The 40 °C and 10 °C control groups were sampled at 0 h.

4.9. qRT-PCR Analysis

An RNA extraction kit (HiPure Plant RNA Mini Kit from Magen, Shanghai, China) was used to extract RNA, while cDNA was synthesized using the Prime Script RT reagent Kit (Perfect Real Time from Takara, Dalian, China). Primer 3.0 software was used to design specific primers in the non-conserved region of the target gene, which were synthesized by Fuzhou Qingbaiwang Biotechnology Company. Real-time fluorescence quantitative analysis was used with cDNA template (1 μL), cDNA template SYBR Premix Ex TaqTM II (10 μL), specific primers (2 μL), and ddH_2O reaction program (7 μL): 95 °C for 30 s; 95 °C for 5 s; 60 °C for 30 s; 95 °C for 5 s; 60 °C for 60 s; and 30 s at 50 °C, with 40 cycles in total. The internal reference gene was *PbEF1α* (GenBank No. KX682032) [70]. The expression level of the target gene was calculated using the $2{-}\Delta\Delta Ct$ method, and the quantitative data were analyzed via t test using SPSS26 software. Finally, GraphPad Prism 9.0 was used to construct the graphs.

5. Conclusions

In this study, a total of 56 PbGRAS proteins were identified from the *P. bournei* genome and phylogenetically classified into nine subfamilies. Four GRAS genes, *PbGRAS7*, *PbGRAS10*, *PbGRAS14* and *PbGRAS16*, were induced by stress treatments such as drought, salt, and low and high temperature. The expression of each of the four genes varied under different treatments. These results suggest that *PbGRASs* may play a role in improving the tolerance of transgenic *P. bournei* to drought, salt, and high- and low-temperature stresses. This study lays the foundation for the further elucidation of the functions of *PbGRAS* family members, provides valuable information for further studies on the role of GRAS family genes in the development of abiotic stress resistance in *P. bournei*, and provides a basis for the improvement of *P. bournei*.

Author Contributions: Conceptualization, S.C. (Shijiang Cao), Y.L. and J.C.; methodology, D.F. and J.C.; software, D.F.; validation, J.C., D.F. and S.L.; investigation, J.C., S.L. and S.C. (Shengze Cheng); resources, S.C. (Shipin Chen); data curation, D.F.; writing—original draft preparation, J.C.; writing—review and editing, J.C., S.C. (Shijiang Cao) and Y.L. All authors have read and agreed to the published version of the manuscript.

Funding: This research was jointly funded by Fujian Agriculture and Forestry University Forestry peak discipline construction project (71201800739 to Shijiang Cao) and the 7th Project of Forest Seeding Breaking in Fujian Province (LZKG-202208 to Shipin Chen).

Data Availability Statement: Publicly available datasets were analyzed in this study. This data can be found here: The *P. bournei* genome data presented in this study are openly available in CNSA at https://db.cngb.org/search/project/CNP0002030/, accessed on 14 April 2023, CNP0002030. The *A. thaliana* genome data presented in this study are openly available in EnsemblPlents at https://ftp.ensemblgenomes.ebi.ac.uk/pub/plants/release-56/fasta/arabidopsis_thaliana/dna/, accessed on 14 April 2023, Arabidopsis_thaliana.TAIR10.dna.toplevel.fa.gz, 17 February 2023 06:57,35M. The *A. thaliana* genome data presented in this study are openly available in Phytozome v13 at https://phytozome-next.jgi.doe.gov, accessed on 14 April 2023, Osativa_323_v7.0.fa.gz. The *P. bournei* RNA-seq data presented in this study are openly available in https://www.ncbi.nlm.nih.gov/bioproject/, accessed on 14 April 2023, PRJNA628065.

Acknowledgments: We are grateful to the reviewers for their helpful comments on the original manuscript.

Conflicts of Interest: The authors declare that the research was conducted in the absence of any commercial or financial relationships that could be construed as a potential conflict of interest.

References

1. Bakshi, A.; Moin, M.; Kumar, M.U.; Reddy, A.B.M.; Ren, M.; Datla, R.; Siddiq, E.A.; Kirti, P.B. Ectopic Expression of Arabidopsis Target of Rapamycin (AtTOR) Improves Water-Use Efficiency and Yield Potential in Rice. *Sci. Rep.* **2017**, *7*, 42835. [CrossRef] [PubMed]
2. Riaño-Pachón, D.M.; Ruzicic, S.; Dreyer, I.; Mueller-Roeber, B. PlnTFDB: An Integrative Plant Transcription Factor Database. *BMC Bioinform.* **2007**, *8*, 42. [CrossRef]
3. Liu, B.; Sun, Y.; Xue, J.; Jia, X.; Li, R. Genome-Wide Characterization and Expression Analysis of GRAS Gene Family in Pepper (*Capsicum annuum* L.). *PeerJ* **2018**, *6*, e4796. [CrossRef] [PubMed]
4. Song, Y.; Xuan, A.; Bu, C.; Ci, D.; Tian, M.; Zhang, D. Osmotic Stress-Responsive Promoter Upstream Transcripts (PROMPTs) Act as Carriers of MYB Transcription Factors to Induce the Expression of Target Genes in *Populus simonii*. *Plant Biotechnol. J.* **2019**, *17*, 164–177. [CrossRef]
5. Liu, M.; Huang, L.; Ma, Z.; Sun, W.; Wu, Q.; Tang, Z.; Bu, T.; Li, C.; Chen, H. Genome-Wide Identification, Expression Analysis and Functional Study of the GRAS Gene Family in Tartary Buckwheat (*Fagopyrum tataricum*). *BMC Plant Biol.* **2019**, *19*, 342. [CrossRef] [PubMed]
6. Zhang, B.; Liu, J.; Yang, Z.E.; Chen, E.Y.; Zhang, C.J.; Zhang, X.Y.; Li, F.G. Genome-Wide Analysis of GRAS Transcription Factor Gene Family in *Gossypium hirsutum* L. *BMC Genom.* **2018**, *19*, 348. [CrossRef]
7. To, V.-T.; Shi, Q.; Zhang, Y.; Shi, J.; Shen, C.; Zhang, D.; Cai, W. Genome-Wide Analysis of the GRAS Gene Family in Barley (*Hordeum vulgare* L.). *Genes* **2020**, *11*, 553. [CrossRef]
8. Jaiswal, V.; Kakkar, M.; Kumari, P.; Zinta, G.; Gahlaut, V.; Kumar, S. Multifaceted Roles of GRAS Transcription Factors in Growth and Stress Responses in Plants. *iScience* **2022**, *25*, 105026. [CrossRef]
9. Sun, X.; Jones, W.T.; Rikkerink, E.H.A. GRAS Proteins: The Versatile Roles of Intrinsically Disordered Proteins in Plant Signalling. *Biochem. J.* **2012**, *442*, 1–12. [CrossRef]
10. Cenci, A.; Rouard, M. Evolutionary Analyses of GRAS Transcription Factors in Angiosperms. *Front. Plant Sci.* **2017**, *8*, 273. [CrossRef]
11. Tian, C.; Wan, P.; Sun, S.; Li, J.; Chen, M. Genome-Wide Analysis of the GRAS Gene Family in Rice and Arabidopsis. *Plant Mol. Biol.* **2004**, *54*, 519–532. [CrossRef] [PubMed]
12. Liu, X.; Widmer, A. Genome-Wide Comparative Analysis of the GRAS Gene Family in *Populus*, *Arabidopsis* and Rice. *Plant Mol. Biol. Rep.* **2014**, *32*, 1129–1145. [CrossRef]
13. Khan, Y.; Xiong, Z.; Zhang, H.; Liu, S.; Yaseen, T.; Hui, T. Expression and Roles of GRAS Gene Family in Plant Growth, Signal Transduction, Biotic and Abiotic Stress Resistance and Symbiosis Formation—A Review. *Plant Biol.* **2022**, *24*, 404–416. [CrossRef] [PubMed]
14. Yu, L.; Hui, C.; Huang, R.; Wang, D.; Fei, C.; Guo, C.; Zhang, J. Genome-Wide Identification, Evolution and Transcriptome Analysis of GRAS Gene Family in Chinese Chestnut (*Castanea mollissima*). *Front. Genet.* **2023**, *13*, 1080759. [CrossRef] [PubMed]

15. He, Z.; Tian, Z.; Zhang, Q.; Wang, Z.; Huang, R.; Xu, X.; Wang, Y.; Ji, X. Genome-Wide Identification, Expression and Salt Stress Tolerance Analysis of the GRAS Transcription Factor Family in Betula Platyphylla. *Front. Plant Sci.* **2022**, *13*, 1022076. [CrossRef] [PubMed]
16. Di Laurenzio, L.; Wysocka-Diller, J.; Malamy, J.E.; Pysh, L.; Helariutta, Y.; Freshour, G.; Hahn, M.G.; Feldmann, K.A.; Benfey, P.N. The SCARECROW Gene Regulates an Asymmetric Cell Division That Is Essential for Generating the Radial Organization of the Arabidopsis Root. *Cell* **1996**, *86*, 423–433. [CrossRef]
17. Lee, M.-H.; Kim, B.; Song, S.-K.; Heo, J.-O.; Yu, N.-I.; Lee, S.A.; Kim, M.; Kim, D.G.; Sohn, S.O.; Lim, C.E.; et al. Large-Scale Analysis of the GRAS Gene Family in Arabidopsis Thaliana. *Plant Mol. Biol.* **2008**, *67*, 659–670. [CrossRef]
18. Ishikawa, M.; Fujiwara, A.; Kosetsu, K.; Horiuchi, Y.; Kamamoto, N.; Umakawa, N.; Tamada, Y.; Zhang, L.; Matsushita, K.; Palfalvi, G.; et al. GRAS Transcription Factors Regulate Cell Division Planes in Moss Overriding the Default Rule. *Proc. Natl. Acad. Sci. USA* **2023**, *120*, e2210632120. [CrossRef]
19. Helariutta, Y.; Fukaki, H.; Wysocka-Diller, J.; Nakajima, K.; Jung, J.; Sena, G.; Hauser, M.-T.; Benfey, P.N. The SHORT-ROOT Gene Controls Radial Patterning of the Arabidopsis Root through Radial Signaling. *Cell* **2000**, *101*, 555–567. [CrossRef]
20. Stuurman, J.; Jäggi, F.; Kuhlemeier, C. Shoot Meristem Maintenance Is Controlled by a *GRAS*-Gene Mediated Signal from Differentiating Cells. *Genes Dev.* **2002**, *16*, 2213–2218. [CrossRef]
21. Peng, J.; Carol, P.; Richards, D.E.; King, K.E.; Cowling, R.J.; Murphy, G.P.; Harberd, N.P. The *Arabidopsis GAI* Gene Defines a Signaling Pathway That Negatively Regulates Gibberellin Responses. *Genes Dev.* **1997**, *11*, 3194–3205. [CrossRef] [PubMed]
22. Cheng, H.; Qin, L.; Lee, S.; Fu, X.; Richards, D.E.; Cao, D.; Luo, D.; Harberd, N.P.; Peng, J. Gibberellin Regulates *Arabidopsis* Floral Development via Suppression of DELLA Protein Function. *Development* **2004**, *131*, 1055–1064. [CrossRef] [PubMed]
23. Wild, M.; Davière, J.-M.; Cheminant, S.; Regnault, T.; Baumberger, N.; Heintz, D.; Baltz, R.; Genschik, P.; Achard, P. The Arabidopsis DELLA RGA-LIKE3 Is a Direct Target of MYC2 and Modulates Jasmonate Signaling Responses. *Plant Cell* **2012**, *24*, 3307–3319. [CrossRef] [PubMed]
24. Hou, X.; Lee, L.Y.C.; Xia, K.; Yan, Y.; Yu, H. DELLAs Modulate Jasmonate Signaling via Competitive Binding to JAZs. *Dev. Cell* **2010**, *19*, 884–894. [CrossRef]
25. Niu, Y.; Zhao, T.; Xu, X.; Li, J. Genome-Wide Identification and Characterization of GRAS Transcription Factors in Tomato (*Solanum lycopersicum*). *PeerJ* **2017**, *5*, e3955. [CrossRef]
26. Xu, K.; Chen, S.; Li, T.; Ma, X.; Liang, X.; Ding, X.; Liu, H.; Luo, L. OsGRAS23, a Rice GRAS Transcription Factor Gene, Is Involved in Drought Stress Response through Regulating Expression of Stress-Responsive Genes. *BMC Plant Biol.* **2015**, *15*, 141. [CrossRef]
27. Mayrose, M.; Ekengren, S.K.; Melech-Bonfil, S.; Martin, G.B.; Sessa, G. A Novel Link between Tomato GRAS Genes, Plant Disease Resistance and Mechanical Stress Response. *Mol. Plant Pathol.* **2006**, *7*, 593–604. [CrossRef]
28. Fode, B.; Siemsen, T.; Thurow, C.; Weigel, R.; Gatz, C. The *Arabidopsis* GRAS Protein SCL14 Interacts with Class II TGA Transcription Factors and Is Essential for the Activation of Stress-Inducible Promoters. *Plant Cell* **2008**, *20*, 3122–3135. [CrossRef]
29. Yuan, Y.; Fang, L.; Karungo, S.K.; Zhang, L.; Gao, Y.; Li, S.; Xin, H. Overexpression of VaPAT1, a GRAS Transcrip tion Factor from Vitis Amurensis, Confers Abiotic Stress Tolerance in *Arabidopsis*. *Plant Cell Rep.* **2016**, *35*, 655–666. [CrossRef]
30. Ma, H.-S.; Liang, D.; Shuai, P.; Xia, X.-L.; Yin, W.-L. The Salt- and Drought-Inducible Poplar GRAS Protein SCL7 Confers Salt and Drought Tolerance in *Arabidopsis thaliana*. *J. Exp. Bot.* **2010**, *61*, 4011–4019. [CrossRef]
31. He, Y.-H.; Liang, R.; Jiang, Y.; Sun, B. Research Progress of Precious Species Phoebe bournei and Its Development Strategies. *Guangxi For. Sci.* **2013**, *42*, 365–370. [CrossRef]
32. Han, S.; Han, X.; Li, Y.-B.; Zhang, Y.-T.; Zhang, J.-H.; Tong, Z.-K. ldentification of NF-Y Gene Family and Expres sion Analysis in Response to Drought Stress in *Phoebe bournei*. *Chin. J. Agric. Biotechnol.* **2022**, *30*, 1112–1127.
33. Wu, D.-R.; Zhu, Z.-D. Preliminary Study on Structure and Spatial Distribution Pattern of Phoebe. *Sci. Silvae Sin.* **2003**, *39*, 23–30.
34. Jiang, X.-M.; Xiao, F.-M.; Ye, J.-S.; Gong, B.; Liu, Z.-K. Geographic Variation and Estimation of Genetic Parameters of Seed and Growth Traits in Phoebe bournei Provenance at Seedling Stage. *Acta Agric. Univ. Jiangxiensis* **2008**, *150*, 666–670.
35. Dill, A.; Jung, H.-S.; Sun, T. The DELLA Motif Is Essential for Gibberellin-Induced Degradation of RGA. *Proc. Natl. Acad. Sci. USA* **2001**, *98*, 14162–14167. [CrossRef]
36. Greb, T.; Clarenz, O.; Schäfer, E.; Müller, D.; Herrero, R.; Schmitz, G.; Theres, K. Molecular Analysis of the *LATERAL SUPPRESSOR* Gene in *Arabidopsis* Reveals a Conserved Control Mechanism for Axillary Meristem Formation. *Genes Dev.* **2003**, *17*, 1175–1187. [CrossRef]
37. Li, X.; Qian, Q.; Fu, Z.; Wang, Y.; Xiong, G.; Zeng, D.; Wang, X.; Liu, X.; Teng, S.; Hiroshi, F.; et al. Control of Tillering in Rice. *Nature* **2003**, *422*, 618–621. [CrossRef]
38. Wang, T.-T.; Yu, T.-F.; Fu, J.-D.; Su, H.-G.; Chen, J.; Zhou, Y.-B.; Chen, M.; Guo, J.; Ma, Y.-Z.; Wei, W.-L.; et al. Genome-Wide Analysis of the GRAS Gene Family and Functional Identification of GmGRAS37 in Drought and Salt Tolerance. *Front. Plant Sci.* **2020**, *11*, 604690. [CrossRef]
39. Zhao, X.; Liu, D.-K.; Wang, Q.-Q.; Ke, S.; Li, Y.; Zhang, D.; Zheng, Q.; Zhang, C.; Liu, Z.-J.; Lan, S. Genome-Wide Identification and Expression Analysis of the GRAS Gene Family in *Dendrobium chrysotoxum*. *Front. Plant Sci.* **2022**, *13*, 1058287. [CrossRef]
40. Chen, Y.; Zhu, P.; Wu, S.; Lu, Y.; Sun, J.; Cao, Q.; Li, Z.; Xu, T. Identification and Expression Analysis of GRAS Transcription Factors in the Wild Relative of Sweet Potato Ipomoea Trifida. *BMC Genom.* **2019**, *20*, 911. [CrossRef]

41. Fan, Y.; Yan, J.; Lai, D.; Yang, H.; Xue, G.; He, A.; Guo, T.; Chen, L.; Cheng, X.; Xiang, D.; et al. Genome-Wide Identification, Expression Analysis, and Functional Study of the GRAS Transcription Factor Family and Its Response to Abiotic Stress in Sorghum [*Sorghum bicolor* (L.) Moench]. *BMC Genom.* **2021**, *22*, 509. [CrossRef] [PubMed]
42. Fan, Y.; Wei, X.; Lai, D.; Yang, H.; Feng, L.; Li, L.; Niu, K.; Chen, L.; Xiang, D.; Ruan, J.; et al. Genome-Wide Investigation of the GRAS Transcription Factor Family in Foxtail Millet (*Setaria italica* L.). *BMC Plant Biol.* **2021**, *21*, 508. [CrossRef]
43. Kotak, S.; Port, M.; Ganguli, A.; Bicker, F.; von Koskull-Döring, P. Characterization of C-Terminal Domains of *Arabidopsis* Heat Stress Transcription Factors (Hsfs) and Identification of a New Signature Combination of Plant Class A Hsfs with AHA and NES Motifs Essential for Activator Function and Intracellular Localization. *Plant J.* **2004**, *39*, 98–112. [CrossRef] [PubMed]
44. Liu, Y.; Wang, L.; Zhang, D.; Li, D. Expression Analysis of Segmentally Duplicated *ZmMPK3*-1 and *ZmMPK3*-2 Genes in Maize. *Plant Mol. Biol. Rep.* **2013**, *31*, 457–463. [CrossRef]
45. Lin, C.-W.; Huang, L.-Y.; Huang, C.-L.; Wang, Y.-C.; Lai, P.-H.; Wang, H.-V.; Chang, W.-C.; Chiang, T.-Y.; Huang, H.-J. Common Stress Transcriptome Analysis Reveals Functional and Genomic Architecture Differences Between Early and Delayed Response Genes. *Plant Cell Physiol.* **2017**, *58*, 546–559. [CrossRef]
46. Pysh, L.D.; Wysocka-Diller, J.W.; Camilleri, C.; Bouchez, D.; Benfey, P.N. The GRAS Gene Family in Arabidopsis: Sequence Characterization and Basic Expression Analysis of the SCARECROW-LIKE Genes. *Plant J.* **1999**, *18*, 111–119. [CrossRef] [PubMed]
47. Jin, Y.; Li, T.-H.; Wen, S.-Z.; Wang, B.; Xie, Y.-B. Growth and physiological characteristics of Phoebe bournei seedling under drought stress. *J. Cent. South Univ. For. Technol.* **2018**, *38*, 50–57. [CrossRef]
48. Wang, B. Studies on Photosynthesis and Physiology of Phoebe Bournei Seedlings under Drought Stress and Recover of Polyamine. Master's Dissertation, Central South University of Forestry and Technology, Changsha, China, 2019. Available online: https://kns.cnki.net/kcms2/article/abstract?v=3uoqIhG8C475KOm_zrgu4lQARvep2SAkEcTGK3Qt5VuzQzk0e7M1z79hmwnI7R4TyWQaG7VFWhQxxuF8z2IpxKazT8odri7D&uniplatform=NZKPT (accessed on 6 April 2022).
49. Li, Y.-F.; Yang, W.-L.; Gu, J.-J.; Zhang, A.-M.; Zhan, K.-H. Genome-widle ldentification and Characterization of the GRAS Gene Family in Bread Wheat (*Triticum aestivum* L.). *J. Triticeae Crops* **2019**, *39*, 549–559.
50. Liu, Y.; Wang, W. Characterization of the GRAS Gene Family Reveals Their Contribution to the High Adaptability of Wheat. *PeerJ* **2021**, *9*, e10811. [CrossRef] [PubMed]
51. Huang, W.; Xian, Z.; Kang, X.; Tang, N.; Li, Z. Genome-Wide Identification, Phylogeny and Expression Analysis of GRAS Gene Family in Tomato. *BMC Plant Biol.* **2015**, *15*, 209. [CrossRef]
52. Roy, S.W.; Penny, D. Patterns of Intron Loss and Gain in Plants: Intron Loss-Dominated Evolution and Genome-Wide Comparison of *O. Sativa* and *A. Thaliana*. *Mol. Biol. Evol.* **2006**, *24*, 171–181. [CrossRef] [PubMed]
53. Rogozin, I.B.; Carmel, L.; Csuros, M.; Koonin, E.V. Origin and Evolution of Spliceosomal Introns. *Biol. Direct* **2012**, *7*, 11. [CrossRef] [PubMed]
54. Takakura, Y.; Sofuku, K.; Tsunashima, M.; Kuwata, S. Lentiavidins: Novel Avidin-like Proteins with Low Isoelectric Points from Shiitake Mushroom (*Lentinula edodes*). *J. Biosci. Bioeng.* **2016**, *121*, 420–423. [CrossRef]
55. Jing, Z.; Qi, R.; Liu, C.; Ren, P. Study of Interactions between Metal Ions and Protein Model Compounds by Energy Decomposition Analyses and the AMOEBA Force Field. *J. Chem. Phys.* **2017**, *147*, 161733. [CrossRef]
56. Fan, S.; Zhang, D.; Gao, C.; Zhao, M.; Wu, H.; Li, Y.; Shen, Y.; Han, M. Identification, Classification, and Expression Analysis of GRAS Gene Family in *Malus domestica*. *Front. Physiol.* **2017**, *8*, 253. [CrossRef] [PubMed]
57. Wang, L.; Ding, X.; Gao, Y.; Yang, S. Genome-Wide Identification and Characterization of GRAS Genes in Soybean (*Glycine Max*). *BMC Plant Biol.* **2020**, *20*, 415. [CrossRef]
58. Wang, Y.; Shi, S.; Zhou, Y.; Zhou, Y.; Yang, J.; Tang, X. Genome-Wide Identification and Characterization of GRAS Transcription Factors in Sacred Lotus (*Nelumbo nucifera*). *PeerJ* **2016**, *4*, e2388. [CrossRef]
59. Czikkel, B.E.; Maxwell, D.P. NtGRAS1, a Novel Stress-Induced Member of the GRAS Family in Tobacco, Localizes to the Nucleus. *J. Plant Physiol.* **2007**, *164*, 1220–1230. [CrossRef]
60. Wang, H.-L.; Chen, J.; Tian, Q.; Wang, S.; Xia, X.; Yin, W. Identification and Validation of Reference Genes for *Populus Euphratica* Gene Expression Analysis during Abiotic Stresses by Quantitative Real-Time PCR. *Physiol. Plant.* **2014**, *152*, 529–545. [CrossRef]
61. Grimplet, J.; Agudelo-Romero, P.; Teixeira, R.T.; Martinez-Zapater, J.M.; Fortes, A.M. Structural and Functional Analysis of the GRAS Gene Family in Grapevine Indicates a Role of GRAS Proteins in the Control of Development and Stress Responses. *Front. Plant Sci.* **2016**, *7*, 353. [CrossRef]
62. Agudelo-Romero, P.; Erban, A.; Rego, C.; Carbonell-Bejerano, P.; Nascimento, T.; Sousa, L.; Martínez-Zapater, J.M.; Kopka, J.; Fortes, A.M. Transcriptome and Metabolome Reprogramming in *Vitis vinifera* Cv. Trincadeira Berries upon Infection with *Botrytis cinerea*. *J. Exp. Bot.* **2015**, *66*, 1769–1785. [CrossRef]
63. Harberd, N.P.; Belfield, E.; Yasumura, Y. The Angiosperm Gibberellin-GID1-DELLA Growth Regulatory Mecha nism: How an "Inhibitor of an Inhibitor" Enables Flexible Response to Fluctuating Environments. *Plant Cell* **2009**, *21*, 1328–1339. [CrossRef] [PubMed]
64. Han, X.; Zhang, J.; Han, S.; Chong, S.L.; Meng, G.; Song, M.; Wang, Y.; Zhou, S.; Liu, C.; Lou, L.; et al. The Chro mosome-Scale Genome of Phoebe Bournei Reveals Contrasting Fates of Terpene Synthase (TPS)-a and TPS-b Subfamilies. *Plant Commun.* **2022**, *3*, 100410. [CrossRef] [PubMed]
65. Chen, J.; Yan, Q.; Li, J.; Feng, L.; Zhang, Y.; Xu, J.; Xia, R.; Zeng, Z.; Liu, Y. The GRAS Gene Family and Its Roles in Seed Development in Litchi (*Litchi chinensis* Sonn). *BMC Plant Biol.* **2021**, *21*, 423. [CrossRef] [PubMed]

66. Chen, C.; Chen, H.; Zhang, Y.; Thomas, H.R.; Frank, M.H.; He, Y.; Xia, R. TBtools: An Integrative Toolkit Developed for Interactive Analyses of Big Biological Data. *Mol. Plant* **2020**, *13*, 1194–1202. [CrossRef] [PubMed]
67. Wang, Y.; Tang, H.; DeBarry, J.D.; Tan, X.; Li, J.; Wang, X.; Lee, T.H.; Jin, H.; Marler, B.; Guo, H.; et al. MCScanX: A Toolkit for Detection and Evolutionary Analysis of Gene Synteny and Collinearity. *Nucleic Acids Res.* **2012**, *40*, e49. [CrossRef]
68. Manuel, M. A New Semi-Subterranean Diving Beetle of the *Hydroporus Normandi*-Complex from South-Eastern France, with Notes on Other Taxa of the Complex (Coleoptera: Dytiscidae). *Zootaxa* **2013**, *3652*, 453. [CrossRef]
69. Tamura, K.; Stecher, G.; Kumar, S. MEGA11: Molecular Evolutionary Genetics Analysis Version 11. *Mol. Biol. Evol.* **2021**, *38*, 3022–3027. [CrossRef]
70. Zhang, J.; Zhu, Y.; Pan, Y.; Huang, H.; Li, C.; Li, G.; Tong, Z. Transcriptomic Profiling and Identification of Candidate Genes in Two Phoebe Bournei Ecotypes with Contrasting Cold Stress Responses. *Trees* **2018**, *32*, 1315–1333. [CrossRef]

plants

MDPI

Review

Exploiting *Brassica rapa* L. subsp. *pekinensis* Genome Research

Faujiah Nurhasanah Ritonga [1,2,†], Zeyu Gong [3,†], Yihui Zhang [1], Fengde Wang [1], Jianwei Gao [1], Cheng Li [1,*] and Jingjuan Li [1,*]

1 Shandong Key Laboratory of Bulk Open-field Vegetable Breeding, Ministry of Agriculture and Rural Affairs Key Laboratory of Huang Huai Protected Horticulture Engineering, Institute of Vegetables, Shandong Academy of Agricultural Sciences, Jinan 250100, China; faujiahnurhasanah@usu.ac.id (F.N.R.); zyh_0923@163.com (Y.Z.); wfengde@163.com (F.W.); jwg_738@163.com (J.G.)
2 Faculty of Forestry, Universitas Sumatera Utara, USU 2 Bekala Campus, Pancurbatu, Deli Serdang 20355, Indonesia
3 Shandong Provincial Key Laboratory of Plant Stress, College of Life Science, Shandong Normal University, Jinan 250358, China; s_f_gzy@foxmail.com
* Correspondence: lceye@126.com (C.L.); lijj0620@163.com (J.L.)
† These authors contributed equally to this work.

Abstract: Chinese cabbage, *Brassica rapa* L. subsp. *pekinensis* is a crucial and extensively consumed vegetable in the world, especially Eastern Asia. The market demand for this leafy vegetable increases year by year, resulting in multiple challenges for agricultural researchers worldwide. Multi-omic approaches and the integration of functional genomics helps us understand the relationships between Chinese cabbage genomes and phenotypes under specific physiological and environmental conditions. However, challenges exist in integrating multi-omics for the functional analysis of genes and for developing potential traits for Chinese cabbage improvement. However, the panomics platform allows for the integration of complex omics, enhancing our understanding of molecular regulator networks in Chinese cabbage agricultural traits. In addition, the agronomic features of Chinese cabbage are significantly impacted by the environment. The expression of these agricultural features is tightly regulated by a combination of signals from both the internal regulatory network and the external growth environment. To comprehend the molecular process of these characteristics, it is necessary to have a prior understanding of molecular breeding for the objective of enhancing quality. While the use of various approaches in Chinese cabbage is still in its early stages, recent research has shown that it has the potential to uncover new regulators both rapidly and effectively, leading to updated regulatory networks. In addition, the utilization of the efficient transformation technique in conjunction with gene editing using CRISPR/Cas9 will result in a reduction in time requirements and facilitate a more precise understanding of the role of the regulators. Numerous studies about Chinese cabbage have been conducted in the past two decades, but a comprehensive review about its genome still limited. This review provides a concise summary of the latest discoveries in genomic research related to Brassica and explores the potential future developments for this species.

Keywords: agricultural traits; Chinese cabbage; genome; omics; regulatory networks

1. Introduction

The Brassica crop species' are composed of three diploid species: *Brassica rapa* (AA), *Brassica nigra* (BB), and *Brassica oleracea* (CC). Additionally, there are three amphidiploid species: *Brassica juncea* (AABB), *Brassica napus* (AACC), and *Brassica carinata* (BBCC). These species collectively constitute the widely recognised 'triangle of U' paradigm [1]. *B. rapa* is regarded as a model plant for studying genome polyploidy in the context of evolutionary studies. Its genome is classified as AA. Previously, *Brassica rapa* L. was categorised as two separate species, namely *Brassica campestris* and *B. rapa*. These two species are capable of interbreeding. Regrettably, no molecular marker is currently available to differentiate between them. As a result, these two species have been merged into a single species, specifically *B. rapa*. *Brassica rapa* is cultivated globally as a significant vegetable and oilseed crop with commercial value. Due to frequent inter-species hybridization, nuclear genomes have undergone genetic recombination, resulting in the loss of species-specific DNA sequences [2].

Brassica rapa was discovered to exhibit a wide range of subspecies with distinct morphotypes, including turnip (subsp. *rapa*), yellow sarson (subsp. *trilocularis*) which is cultivated for seed oil, non-heading *B. rapa* (subsp. *chinensis*), and heading B. rapa (subsp. *pekinensis*) which is a leafy vegetable [3]. Brassica species have undergone a whole-genome triplication (WGT) event, which was absent in the model plant *Arabidopsis thaliana*. This event is crucial in the process of speciation and the development of different forms in Brassica plants [3,4]. WGT leads to an enlarged gene pool, which enables multicopy genes to develop distinct or novel roles, such as sub-functionalization or neofunctionalization [4].

Scientifically, Chinese cabbage, referred to as *Brassica rapa* L. subsp. *pekinensis*, is a crucial and extensively consumed vegetable in Eastern Asia [5]. Kimchi cabbage is a widely recognised term in Korea [6], but it is also referred to as napa cabbage [7,8], cool-weather Chinese cabbage [9], and headed Chinese cabbage [10]. Chinese cabbage is a type of vegetable that has large pale-green leaves and wide, white midribs, which come together to create a head [11]. Chinese cabbage possesses a rich nutritional profile, containing significant levels of antioxidants, phytochemicals, and mineral elements [12,13]. Additionally, it is known for its delightful flavour [12,13]. There is an increasing demand in the market for Chinese cabbage that possesses resilience to many types of stress, exhibits a delightful taste, has an attractive appearance, and is cost-effective. This has put significant pressure on agricultural researchers worldwide to develop high-quality varieties that meet these criteria [7].

Climate change has led to increased weather patterns, such as droughts, floods, and extreme temperatures, which are expected to cause huge crop losses and threaten worldwide food stability [7]. Pathogens are also a main cause of defective traits in plants. To improve food tolerance in biotic and abiotic environments, strategies have been developed to improve plant tolerance in *B. rapa* [6]. Introgression breeding and interspecific hybridization are widely used methods for genetic improvement, such as transferring clubroot tolerance from Chinese cabbage to canola [14] and vice versa [15]. These techniques enhance genetic diversity and improve the agronomic performance of novel cultivars. Inter-specific hybridization [16] and back-crossing [17] are also powerful tools for transferring favourable genes between species in plant breeding programs.

For nearly 100 years, traditional breeding has been successful for preventing starvation but has stagnated due to long breeding times and limited crossbreeding crops. A molecular-based selection strategy has emerged, replacing phenotype-based breeding. This has led to the widespread use of quantitative trait locus (QTL) and Marker Assisted Selection (MAS) in Chinese cabbage studies, as well as the potential application of Single Nucleotide Polymorphism (SNP) and InDel markers [18,19]. Before 2016, Chinese cabbage studies were primarily conducted in South Korea [20–22], but recent research has expanded to other countries, including China, and countries in Europe (Figure 1) [23–25]. In Figure 1, we present the co-occurrence map of Chinese cabbage studies to illustrate the updated studies of Chinese cabbage. This review elucidates the advanced genome research in Chinese cabbage and provides insights into the past, present, and future of Chinese cabbage breeding.

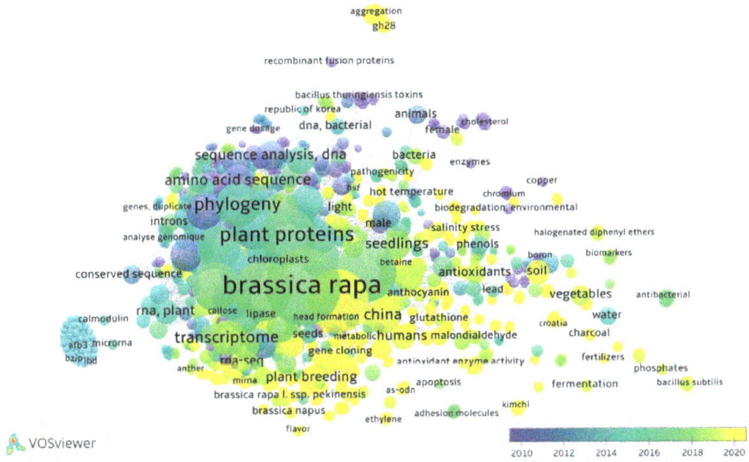

Figure 1. Term co-occurrence map of Chinese cabbage studies using VOSviewer software version 1.620 (https://www.vosviewer.com/). VOSviewer analysed text mining and bibliometric of scientific papers by observing the outputs of term (keyword) co-occurrence analysis. Studies about Chinese cabbage from 2010–2020 relate to biotic and abiotic stresses, head formation, and multi-omics (accessed: 30 September 2024).

2. The Field of Multi-Omics Technologies in Chinese Cabbage

Several intriguing omics technologies have arisen in recent decades. An essential stage in evaluating genome research of Chinese cabbage is to identify potential genes that are still limited to clarify the genetic basis of complex features in Chinese cabbage. In the past, this attempt encountered substantial obstacles because of a lack of relevant publications resulting from the limited accessibility of this research compared to other crops. An integrated panomics platform enables the incorporation of intricate omics to develop models for predicting complex characteristics. Integrating systems biology with multi-omics datasets can augment our comprehension of molecular regulator networks for Chinese cabbage management.

The omics-based methodologies have demonstrated their value in investigating the genetic and molecular mechanisms underlying crop development, specifically concerning alterations in mineral nutrients, metabolites, proteins, transcript levels, and DNA [26]. The introduction of next-generation sequencing (NGS) technologies has resulted in the efficient and quick production of large amounts of data for genomes, epigenomes, transcriptomes, proteomes, metabolomes, and phenomes [27]. This data can provide insights into gene functions and networks in the context of environmental and physiological stress responses [28]. To gather comprehensive information on the genomic data of Chinese cabbage, we extensively reviewed all relevant publications from various journals and publishers such as Elsevier, Springer Nature, MDPI, Frontiers, Wiley Online Library, Taylor and Francis, and others.

In this review, we present a comprehensive overview of the latest developments in omics investigations focused on Chinese cabbage.

2.1. Genomics

Genomics is the field of research that examines genes and genomes. It specifically investigates the structure, function, evolution, mapping, epigenomic, mutagenomic, and genome editing aspects [29]. Structural genomics is a genomics branch that focuses on identifying and utilising molecular markers in tagging and mapping genes of interest. These markers are then used in plant breeding programmes. There are two approaches used in structural genomics, namely PCR and non-PCR-based techniques. Unfortunately, the use of non-PCR-based approaches is restricted in Chinese cabbage investigations. In contrast, PCR-based approaches have been performed in Chinese cabbage for several decades [30,31].

Before multi-omics investigations, a PCR-based technique was used to detect certain traits of Chinese cabbage, such as microsatellite markers [30], anchor single sequence repeats (SSRs), Sequence-related Amplified Polymorphism (SRAP) [32,33], and single nucleotide polymorphisms (SNPs) [34]. The SNPs are genetic differences that occur at a single nucleotide position in the organism's genome. SNP analysis can be conducted by sequencing genomic PCR results from diverse individuals. A new Chinese cabbage map was constructed using 11 newly discovered polymorphic markers, in addition to 227 previously mapped marker loci. The updated map now has a total of 238 marker loci [30].

Microarray technology is used in plant molecular breeding to measure gene expression and detect specific DNA sequences, including SNPs. It involves binding thousands to millions of recognized nucleic acid fragments to a chip, that is then immersed with either DNA or RNA. This technique has been used to observe genes related to hormones and predict hybrid performance in Chinese cabbage. By using microarray analysis, it was a total of 4646 genes showed different expressions, following floral bud size which was related to hormones [35]. Furthermore, a heterotic prediction has been performed in headed Chinese cabbage to enhance breeding efficiency. This prediction is based on whole genome SNP markers and the phenotypic of parental inbred lines to determine hybrid performance [36].

Derived cleaved amplified polymorphic sequence (dCAPS) detects single base changes, but SNPs are unsuitable for CAPS markers. Kompetitive Allele-Specific PCR (KASP) marker genotyping technology allows accurate bi-allelic scoring of SNPs and specific loci across genomic DNA samples. KASP genotyping, combined with dCAPS, detects the retr02 allele in Chinese cabbage accessions. KASP is flexible, affordable, and efficient, making it beneficial for marker-assisted selection [37]. The study found that a single recessive nuclear gene, BraA09g066480.3C, controls the wax deficiency mutant cer1 phenotype in Chinese cabbage leaves. Using MutMap and KASP analysis, an SNP in the fourth exon of BraA09g066480.3C was assumed to cause a substitution of amino acid, resulting in protein changes [38].

The advent of NGS enabled the identification and usage of SNPs [34]. Eight gene-specific and gene-flanking markers for glucosinolates (GSLs) were identified in a combination of SSRs and recombinant inbred lines (RILs) derived from a cross between yellow sarson and Chinese cabbage. Additionally, 148 SSR markers were included in the preexisting ultra-dense genetic map of Chinese cabbage. The genetic map was utilised to conduct QTL mapping for GSLs, with the identification of loci responsible for the production of GSLs achieved by the application of gene-specific markers [33].

The first reference genome for Brassica species, Chiifu v1.5, was achieved using Chinese cabbage [29]. *B. rapa* Chiifu variant which originated from temperate regions, is more commonly used as a database source than *B. rapa* Kenshin which is from subtropical and tropical regions [39]. Sequencing technologies have led to the creation of two improved genome assembly versions of Chinese cabbage (Chiifu v2.0, v2.5, and v3.0) and an updated genome annotation version, namely Chiifu v3.5 [29,40–42]. Chiifu v1.5 and v2.5 were assembled using Illumina short reads, Chiifu v3.0 was generated using PacBio (Menlo Park, CA, USA), optical maps, and Hi-C technologies [29,40], and Chiifu v3.5 was updated using full-length PacBio sequencing [41]. Despite the considerable genomic studies performed on B. rapa, there is still a major absence of a complete genome annotation that encompasses gene structure, alternative splicing (AS) events, and non-coding genes in Chiifu v1.5, Chifuu v2.0, Chiifu v2.5, and Chifuu v3.0. Hence, Zhang, et al. [43] employed the single-molecular long-read method developed by PacBio to enhance gene models and generate the annotated genome version 3.5. The total number of full-length non-chimeric (FLNC) reads acquired was 753,041. These readings were then consolidated into 92,810 nonredundant consensus isoforms, which encompassed 48% of the genes that were annotated in the B. rapa reference genome annotation v3.1.

A study examined the effects of whole-genome triplication (WGT) on the variation within a species. This was completed by analysing the pan-genome of 16 newly assembled genomes and 2 previously published genomes using Illumina and PacBio sequencing technologies [43]. The availability of Chinese cabbage genome assemblies has been highly beneficial in comparative genomics and genetic breeding investigations of Brassica species.

Advances in sequencing technologies, like PacBio and Oxford Nanopore Technology (ONT), have revolutionized obtaining chromosome sequences, but the missing hundred sequences and thousand gaps still need to be minimized [3]. In addition, the Chinese cabbage Chiifu genome sequence enables the analysis of genome-wide identification, allowing for an understanding of multiple traits in Chinese cabbage. Several genome-wide identifications have been conducted in Chinese cabbage, namely TIR-NBS-LRR genes under TuMV infection [44], NRTs gene under low nitrate condition [45], microRNAs under *Plasmodiophora brassicae* infection [46], MYB transcription factors (TFs) in response to stresses in the male organ development stage [39], fork head-associated (FHA) for male sterility [47], expression and functional DELLA [48], and callose enzyme families [49].

Still, the demand for abiotic or biotic-resistant cultivars of Chinese cabbage is increasing, but traditional breeding methods are slow due to environmental conditions. The development of molecular biology techniques, particularly molecular markers, allows for the study of QTL and MAS, an effective breeding method [32]. To implement MAS, it is necessary to identify the specific gene responsible for a trait or the genetic region associated with that gene, which can be used for selective breeding purposes [11]. SSR markers together with MAS have been used in Chinese cabbage studies to improve quality, such as clubroot resistance in canola. This approach allows for the transfer of resistance genes to other cruciferous crops [14].

NGS-based bulked segregant analysis (BSA) is a valuable technique for identifying and mapping genes, particularly in crop breeding [50]. BSA is also known as a QTL mapping technique. It involves crossing a mutant to a wild type (WT) and identifying the mutants in the backcross population. This allows for rapid detection of increased homozygosity in a chromosomal region associated with the present phenotype of mutant [51]. Tipburn is a significant physiological disorder, causing 20% of phenotypic variance. Until now, most of the causes of tipburn cannot be successfully controlled by management precautions, suggesting a need for breeding tipburn resistant cultivars as the best long-term strategy in Chinese cabbage. The study identifies loci associated with tipburn resistance in the F2 population and the DH100 population using BSA and QTL mapping. Results indicate multiple genes contribute to tipburn disease resistance, with transgressive segregation patterns. High-quality bases were obtained from the DNA bulks of the tipburn susceptible and resistant populations, with 2,604,351 and 2,590,679 SNPs identified, respectively. It was also stated that the QTL gqbTRA06, containing *BrCRT2*, contributes to this issue where 51-bp deletion in the tipburn-susceptible line [51]. Breeding for tipburn resistance is the best long-term strategy. QTL has been studied for other traits like flowering time [18], yellow inner leaves [32], seed glucosinolates [33], leafy head formation [19], pathogen [10], and glossiness [52]. Representative breeding techniques are presented in Table 1. Chinese cabbage has a higher diversity of gene analogues than *A. thaliana*, with some domain genes being more abundant [53]. Despite the availability of high-quality genome sequences, studies about gene functions remain challenging [54]. A pangenome of Chinese cabbage was assembled to describe resistance gene analogues (RGAs), with significant conservation (93%) between the *B. rapa* and *B. napus*. It was found that presence–absence variation (PAV) affected 309 RGAs. Furthermore, a total of 223 missed RGAs were found in the reference genome. A total of 138 candidate RGAs were found within the recognized Chinese cabbage disease resistance QTL, with most under negative selection [10]. Whole-genome resequencing together with BSR-Seq and linkage analysis were also used to map and predict *BrCAO*, a gene that encodes chlorophyllide an oxygenase, a recessive nuclear gene causing a pale green mutant in Chinese cabbage [55].

Table 1. Representative techniques and trait improvements in Chinese cabbage genomic studies.

No	Technique	Trait Improvement	Gene Name	Reference
1	Bulk Segregant Analysis Sequencing (BSA-Seq) QTLmapping	*P. brassicae* resistance	Bra013275, Bra013299, Bra013336, Bra013339, Bra013341, and Bra013357	[50]
2	small RNA sequencing, degradome sequencing, and transcriptome sequencing	*P. brassicae resistance*	miRNAs	[46]
3	microarray	cytoplasmic male sterility	-	[35]
4	microarray		*BrMYBs*	[39]
5	Illumina HiSeq 2000 platform	male sterility	RALF genes	[56]
6	DNA-BSA, RNA-BSR, and QTL mapping	leaf yellowing	*BrHISN2*	[57]
7	computationally mined EST libraries	salt tolerance	miR168a and O-METHYLTRANSFERASE 1 (OMT1)	[58]
8	ONT ultralong-read sequencing and Hi-C technologies	evolution of centromeres	-	[3]
9	Genotyping, GWA mapping, bin-map construction, QTL analysis, BSA library construction and high-throughput DNA sequencing	-	*BrCRT2*, calreticulin gene	[51]
10	dCAPS and KASP	TMV resistance	TMV resistance gene	[37]
11	BSR-Seq, linkage analysis, and whole-genome resequencing	Pale green leaves color	*BraA10g007770.3C* (*BrCAO*)	[55]
12	BSA-Seq and KASP assays	Glossy trait	*BrWAX2*	[59]

Note: DNA-bulked segregant analysis (BSA), RNA-sequencing bulked segregant analysis (BSR), quantitative trait locus (QTL) mapping, Expressed Sequence Tag (EST), Oxford Nanopore Technology (ONT), derived cleaved amplified polymorphism sequence (dCAPS), Kompetitive Allele-Specific PCR (KASP), bulked segregant sequencing (BSA-Seq), Turnip mosaic virus (TMV).

Functional genomics and mutagenomics are used to study structural genomics and gene function. Authentic genome engineering is being used to improve crops, such as Chinese cabbage breeding, using tools like the CRISPR/Cas9 system [60] and transcription activator-like effector nuclease (TALEN) [61]. The CRISPR system allows for the induction of precisely targeted mutations in a plant genome using a binary construct commonly delivered into the *Agrobacterium tumefaciens* strain GV3101 [62]. Recently, CRISPR/Cas9 has begun to be widely used in Chinese cabbage breeding (Table 2) [57]. This system allows for targeted mutations in plant genomes, with studies showing high indel efficiency in flowering time [60,63,64]. A study found that *BraFLC2* and *BraFLC3* double knockout lines in the Kenshin background showed 97.7% and 100% indel efficiency, respectively, in the T0 generation [63]. Furthermore, high-quality protoplasts from leaf mesophylls are required for genome editing research [13]. Particle bombardment is also used for efficient genetic transformation of isolated microspores in Chinese cabbage, but low efficiency is a key issue affecting functional verification [65].

Table 2. Genome editing using CRISPR/Cas, transformation method, and the Agrobacterium strain that is used in Chinese cabbage.

No.	Gene Modified	Function	Agronomic Trait	Transformation Method	Agrobacterium Strain	References
1	*BraA1000785*	unknown function	unknown agronomic trait	Agrobacterium mediated transformation	GV3101; EHA105; LB4404; AGL1	[62]
2	*BrVRN1*	delayed flowering time	flowering time	Agrobacterium mediated transformation	GV3101	[60]
3	*BraFLC*	early flowering time	flowering time	Agrobacterium mediated transformation	LB4404	[63]
4	*BrFT1; BrFT2*	flowering	inflorescence; flowering	Agrobacterium mediated transformation	GV3101	[64]
5	*eIF(iso)4E*	TuMV resistance	TuMV resistance	Agrobacterium mediated transformation	GV3101	[66]
6	*BrLEAFY*	bolting time	bolting time	Agrobacterium mediated transformation	LB4404	[67]

Chinese cabbage, a major crop in Asia, produces over 50 megatons annually, accounting for 70% of the world's yield. The diploid genome has complex genetic control, making genetic and genomic analysis challenging. Mutagenesis is crucial for functional genomics studies and breeding materials, but fully characterized mutant resources are lacking [68]. EMS mutagenesis is a crucial breeding and genomic research tool, resulting in mutants with heritable morphological variations in leaf colour, shape, head, bolting, and fertility. Seed EMS mutagenesis generates large-scale mutant libraries, while microspore EMS mutagenesis facilitates rapid homozygous mutants [69]. Chinese cabbage mutants have been obtained using EMS mutagenesis including a male sterile mutant [70], non-heading flat leaves mutant [71], heading compact and wrinkled leaves mutant, and so on. Map-based cloning is often used to identify mutant causal genes involved in biosynthesis, such as the *BrWAX2* gene in cuticular wax biosynthesis [70], *BrCER60.A09* in cuticular wax biosynthesis [72], and OCTOPUS (*BrOPS*) gene in leafy head development [73]. Recently, a study combined MutMap, KASP genotyping, and map-based cloning to recognize a causal gene that is involved in GA biosynthesis and related to leafy head formation, namely *BraA09g001110.3C* [74].

2.2. Transcriptomics

NGS technology, including small RNA, degradome, and transcriptome sequencing, has been successfully applied to Chinese cabbage to identify miRNAs and their targets [46]. This technology allows for the resolution of relationships among subspecies and domestication history. A huge number of SNPs from across the nuclear genome can be used for resolving the relationships and genetic structure of Chinese cabbage [75]. Reference Qi, et al. [75] identified SNPs obtained from 126 accessions worldwide utilizing high-throughput transcriptome data. Extensive examination of over 31,000 single nucleotide polymorphisms (SNPs) in the *B. rapa* genome provided proof of five separate genetic groups and confirmed that *B. rapa* crops originated in Europe and Central Asia. It confirmed the well-established South Asian and East Asian *B. rapa* groupings and provided evidence that pak choi, Chinese cabbage, and yellow sarson are likely monophyletic groups. Conversely, the oil-type B. rapa subsp. oleifera and brown sarson, exhibited polyphyllenism. *B. rapa* was introduced to Asia approximately 2400 to 4100 years ago while Chinese cabbage originated some 1200 to 2100 years ago through the hybridization of pak choi and B. rapa from European-Central Asia. Substantial differences in the levels of founder effects were deduced among the several subspecies of *B. rapa*. Extant historical records documenting these crops are consistent with deductions.

The implementation of extensive transcriptome on long-time scales is conducted to encompass the primary stages of vegetative growth, including heading and non-heading [76].

A comparable investigation was undertaken to ascertain the causative genes in transcriptome analysis of leaf size differential [77]. Multiple studies utilising transcriptomic data have been published on various aspects of Chinese cabbage. These include investigations on yellow leafy head development [78], the transition that initiated the leaf heading between cabbage and Chinese cabbage [76], the colour difference between white and yellow Chinese cabbage inner-leaf [79], clubroot-resistant and clubroot-susceptible genotypes against *P. brassicae* infection [80], and leaf size [77].

Transcriptome analysis and SSRs have been used to identify candidate genes in Chinese cabbage, with 41 candidate genes identified across 10 chromosomes. Additionally, 341 SSRs were detected in 5 replicated types, namely single, double, triple, quintuple, and sextuple nucleotide replications [31]. Studies have also combined metabolomics and transcriptomics data to identify differentially expressed metabolites (DEMs) and differentially expressed genes (DEGs) and the correlation between DEMs and DEGs was might consistent or not consistent. The KEGG analysis is employed to validate the findings and identify multiple co-regulated pathways [81]. A study using the Illumina RNA-seq platform analysed transcriptome profiling in spring-sown Chinese cabbage. and the results were confirmed by RT-qPCR. KEGG pathway analysis revealed that there was an enrichment in pathways related to sucrose and starch metabolism, hormones, and signal transduction, suggesting potential trait candidates [82]. Moreover, a global transcriptome profiling of roots of clubroot-susceptible and clubroot-resistant Chinese cabbage revealed different gene expression in defence, transport, metabolism, and signal transduction, with 151 genes presenting different expression patterns between the two near-isogenic lines (NILs) [80].

Transcriptome studies in Chinese cabbage revealed the involvement of some TFs under stresses [83], glucosinolate pathways [84], and leafy head formation [78]. Chinese cabbage studies have identified MYB as the most prevalent TF family. Recently, *MYC2/3/4*, which are basic helix–loop–helix (bHLH) TFs, have been found to play a role in Chinese cabbage secondary metabolites. Additionally, *MYC2/3/4* activates the involvement of genes in various aspects of plant development. In particular, *BrMYC2* has been shown to reduce the length of roots and hypocotyls, as well as seed length and weight. It also promotes bolting time and significantly increases the total GS accumulation to confer the optimum resistance in response to the pathogenic fungus, *Sclerotinia sclerotiorum* [85]. The identified TFs in Chinese cabbage are presented in Table 3.

Table 3. Identified TFs and TF functions in Chinese cabbage.

No	TF Family	TF Names	Functions	Reference
1	MYB	*BrMYB28-1* *BrMYB28-2* *BrMYB28-3* *BrMYB29* *BrMYB29-1* *BrMYB34-1* *BrMYB34-2* *BrMYB34-3* *BrMYB34-4* *BrMYB51-1* *BrMYB34-2* *BrMYB34-3* *BrMYB122-1* *BrMYB122-2*	glucosinolate pathway	[84]
2	MYB	*BrMYB28* *BrMYB29*	glucosinolate pathway, temperature stresses	[86]
3	R2R3-MYB bHLH	*BrMYB90* *Bra037887*	anthocyanin biosynthesis	[17]

Table 3. *Cont.*

No	TF Family	TF Names	Functions	Reference
	AP2/ERF	*BrERF4*	drought and salt stress	[83]
	WRKY	*WRKY38*	pathogen	
4	MYB	*BrMYB1R25, BrMYB1R44, BrMYB1R70, BrMYB1R77, BrMYB1R171*, and *BrMYB1R 178*	cold stress	[39]
		BrMYB222, BrMYB118 and *BrMYB196*	salt stress	
		BrMYB118, BrMYB55, BrMYB168, and *BrMYB224*	drought stress	
		BrMYB7, BrMYB10, BrMYB55, BrMYB81, BrMYB98, BrMYB118, BrMYB168, BrMYB169, BrMYB196, BrMYB222, and *BrMYB224*	ABA	
		BrMYB81, BrMYB88, BrMYB98, Fusarium infectionBrMYB118, BrMYB163, BrMYB196, BrMYB200, BrMYB222, BrMYB224, and *BrMYB237*	JA treatment	
		BrMYB40, BrMYB58, BrMYB185, BrMYB118, and *BrMYB222*	Fusarium infection	
5	bHLH, MADS, MYB, MYB related, WRKY, NAC, AP2-EREBP	-	yellow leafy head	[78]
6	MYB	*BrMYB51*	*Pectobacterium carotovorum* infection	[87]
		BrMYB28.3, BrMYB51.1, and *BrMYB122.2*	glucosinolate content	
		BrMYB28s and *BrMYB34.3*	cold, drought, salt, and ABA stress	
7	bHLH	*MYC2/3/4*	growth and development	[85]
		BraA01gCHR11 and *BraA07gSCL1*	soluble sugar and GSL biosynthesis	[79]
8	MYB	*BraA09gMYB47* and *BraA06gMYB88*	cell size and proliferation	[77]

Note: Myeloblastosis (MYB).

2.3. Proteomics

Molecular proteomics is a technique employed to examine the total protein composition within an organism [88]. Proteomic analyses yield a substantial collection of data pertaining to the individual proteins implicated in particular biological reactions. Proteomic analysis has been employed in the study of Brassicaceae, specifically focusing on the processes of transmission of turnip mosaic virus over long distances in Brassica rapa L. ssp. Chinensis [89], fertility conversion in the thermo-sensitive male sterile line PK3-12S of winter rapeseed [90], and so on. Nevertheless, there is a scarcity of proteomic research particularly focused on Chinese cabbage. Highly sophisticated methods like SDS-PAGE, two-dimensional gel electrophoresis (2-DE), and two-dimensional differential gel electrophoresis (2D-DIGE) were devised and employed in gel-based approaches for protein separation [91,92]. Proteins extracted from the roots of B. rapa cv. For identification using mass spectrometry, 742 seedlings were subjected to 2-DE analysis. MALDI/TOF/MS analysis and BLASTp searches revealed the identification of nine differentially expressed genes (DEPs) from the roots of Brassica rapa L cv. Infection of 742 seedlings with plasmodiophora brassicae was conducted. Notably, all proteomic studies on Chinese cabbage primarily focus on P. brassicae infection. DEPs 3038 and 3129 were identified as proteins associated with salicylic acid (SA)-mediated systemic acquired resistance (SAR), DEPs 3414 and 2338 were associated with jasmonic acid (JA)/ethylene (ET)-mediated induced systemic resistance (ISR), and DEP 4043 was found to be involved in both SAR and ISR [93].

Isobaric tags for relative and absolute quantitation (iTRAQ) analysis are sophisticated methods used in quantitative proteomic research to effectively elucidate certain protein mechanisms [94]. The iTRAQ-based proteomic data revealed distinctive physiological and molecular alterations in the pre-intermediate stages of infection from P. brassicae in both the infected and control Chinese cabbage lines. The resistant/susceptible lines were found to have a total of 5003 proteins with varying abundances, which could be measured using dipeptide or polypeptide segments. Drawing from the differential proteome analysis results and pertinent previously published data, it is probable that Chinese cabbage either protected or opposed P. brassicae, as evidenced by the observed harm caused by the activation of cysteine, methionine, and tyrosine metabolism. Susceptible lines exhibited upregulation of glutamine synthetase cytosolic isozymes 1–5 and Arginine decarboxylase 2, indicating the involvement of these proteins in the response to P. brassicae poisoning [94]. Similarly, iTRAQ analysis performed by [95] revealed differential expression of 295 proteins involved in energy and lipid metabolism, plant defence, cell wall modification, and hormone biosynthesis and signalling during the secondary phase of the Chinese cabbage–P. brassicae interaction. Notably, the upregulation of proteins related to brassinosteroids (BR) metabolism was confirmed. Specifically, cycloartenol synthase (CAS1) and cytochrome P450 51G1 (CYP51G1), which play a significant role in BR biosynthesis, were increased after inoculation. This indicates that these proteins contribute to the development of clubroot tissue [95]. The proteomic findings shed fresh light on the development of germplasm resistance to P. brassicae and give a genetic foundation for breeding to enhance Chinese cabbage.

2.4. Metabolomics

Metabolomics is a comprehensive study of metabolites involved in biological system processes. However, metabolite profiling and variations in Chinese cabbage are still limited. Metabolomics and NGS have helped predict an initial metabolic network from an organism's genome sequence. Major biochemicals, such as GABA, acetate, and phenylalanine, differentiate cabbages grown in two regions. These differences suggest that environmental factors like climate and geology influence the levels of these metabolites in Chinese cabbage [20]. Similarly, a recent metabolomic study was performed using ultra-high-performance liquid chromatography–tandem mass spectrometry (UPLC-MS/MS) comparing two areas (Jiaozhou and Jinan). Chinese cabbage is rich in nutrients like amino acids, lipids, organic acids, phenolic acids, nucleotides, flavonoids, saccharides, glucosinolates, vitamins, and alcohols. Interestingly, some unique metabolites are recognized in Jiazhou and were absent in Jinan namely S-(Methyl)glutathione and nicotinic acid adenine dinucleotide. These two metabolite concentrations were elevated in Jiazhou, illustrating that they are Jiaozhou-specific metabolites of Chinese cabbage [96]. KEGG analysis showed that DAMs were significantly enriched in tryptophan metabolism and thiamine metabolism [96]. A study on Chinese cabbage in South Korea found that its taste variations depend on cultivation season, impacting kimchi production quality. The analysis revealed that glucose, fructose, as well as malic acid, exhibit a positive correlation with sweetness, while glucoerucin and glucobrassicin positively correlate with bitterness [97]. Compared to other cruciferous species, Chinese cabbage is the most sensitive in response to salt stress [98]. However, Chinese cabbage has been studied for its salt resistance, with several critical and potential metabolite candidates such as Coenzyme Q10, acetyl-CoA, pyruvic acid, nicotinic acid, and ATP that could reduce the salt stress level in the Chinese cabbage [81]. Qinghua, a cultivar with high salt tolerance, can improve photosynthetic fluorescence ability, promoting organic matter synthesis [99]. A study also explored the metabolic basis of soluble sugar and GSLs in white and yellow inner leaves of six Chinese cabbage cultivars. The genes *BraA05gAOP1*, *BraA04gAOP4*, *BraA03gHT7*, and *BraA01gHT4*, as well as the transcription factors *BraA01gCHR11* and *BraA07gSCL1* are responsible for the synthesis of glucosinolates and soluble sugars due to their higher relative expression in the yellow inner-leaf Chinese cabbage than that in the white inner-leaf Chinese cabbage based on qRT-PCR and transcriptome data. *BraA01gCHR11* and *BraA07gSCL1* may regulate soluble sugar and glucosinolate biosynthesis. The major metabolites found were aliphatic glucosinolate and two soluble sugars [79].

3. Explored Chinese Cabbage Traits

3.1. Glossy Traits in Chinese Cabbage

Plant cell walls maintain mechanical hardness and protect against damage from biotic and abiotic factors. Cuticular wax functions by covering the surface of the outer plant and plays a crucial role in protecting plants against stresses. Glossiness is an important quality-related trait of Chinese cabbage which is caused by abnormal cuticular wax accumulation [52]. The glossy phenotype is crucial for enhancing the genetic diversity of Chinese cabbage and developing vibrant green variants through breeding. Prior research has indicated that the glossy characteristics are a result of impaired wax biosynthesis [59,72], which exhibited a reduction in the overall levels of wax, alkanes, ketones, and alcohols in its leaves, whereas the concentration of aldehydes increased in comparison to the wild type.

A genetic study has revealed that *BrCER1* is the causative gene responsible for the glossy trait observed in the cer1 phenotype. This gene encodes aldehyde decarboxylase [38]. Furthermore, it has been discovered that *BrWAX2* is a potential gene associated with *CER1*. The glossy mutant lacks *CER1*. The *BrWAX2* gene was identified by map-based cloning and plays a role in the biosynthesis of cuticular wax. The glossy phenotype observed on the stem and leaf surface was a result of the dysfunction of *BrWAX2*, as well as the reduction of other gene expressions involved in the alkane-forming pathway [59].

Furthermore, a Chinese cabbage novel glossy mutant, known as SD369, has been discovered. The abundance of wax monomers with a carbon chain length exceeding 26 was dramatically reduced. Analysis of inheritance indicated that a single recessive gene, *BrWAX3*, controlled the glossy characteristic of SD369. The annotated genome of Chinese cabbage revealed the identification of an endoplasmic reticulum gene named *BrCER60*, which encodes a-ketoacyl-CoA synthase. This gene was found to be considerably downregulated in glossy plants' aerial organs [72].

Another gene, named *BrCER2*, has been found as a candidate responsible for the glossy appearance of Chinese cabbage, similar to *BrCER1* and *BrCER60*. The bioinformatics analysis discovered an insertion of a long interspersed nuclear element 1 (LINE-1) transposable element (BrLINE1-RUP) in the first exon of *BrCER2*, resulting in a premature termination codon. A mutation in *BrCER2* in Chinese cabbage that impairs its function causes a significant reduction in the levels of very long chain fatty acids. This, in turn, leads to the development of a glossy pistil, cauline leaf, and inflorescence stem, as well as flower [52].

3.2. Vernalization

Chinese cabbage requires vernalization for flowering, and the production of haploid and doubled haploid plants is used as a biotechnological tool to shorten the breeding process for new cultivars. Doubled haploid plants possess homozygosity at all loci and can serve as parental organisms for F1 hybrids through the utilization of androgenesis induced by isolated microspore cultures [100]. Reference [100] also confirmed significant genotypic influences on the haploid embryogenesis and plant regeneration capacity in isolated microspore culture. This highlights a critical need for selecting appropriate donor material to induce androgenesis. Both haploid and diploid plants are also useful for genetic research, mutagenesis, and genetic mapping procedures. Haploidization, particularly by androgenesis, is a commonly employed technique to expedite the breeding process, which is highly advantageous in the development of novel Chinese cabbage varieties.

Chinese cabbage cultivars predominantly consist of F1 hybrids with vegetative heterosis, which are harvested using self-incompatibility or cytoplasmic male sterility for commercial seed harvesting [11]. Cytoplasmic male sterility (CMS) is a maternally inherited trait in higher plants, causing aborted or infertile pollen grains. It is crucial for heterosis utilization in Chinese cabbage and hybrid seed production, primarily caused by mitochondrial genome mutations, not nuclear gene mutations [70].

The *BrABCG26* gene is responsible for male sterility, with mutants displaying degenerated stamen and no pollen. Mutations in the second and sixth exons of the *BrABCG26* gene resulted in a loss-of-function truncated protein, affecting stamen development and reduced expression in flower buds and anthers [70]. The msm1-1, msm1-2, and msm1-3 are three allelic male-sterile mutants triggered by the same nuclear gene and had microspores aborted and abnormal tapetums. Analysis using MutMap-based gene mapping and KASP revealed that three distinct SNPSs of *BraA09g012710.3C* (*BrACOS5*) were identified as the cause of male sterility. *BrACOS5* encodes an acyl-CoA synthetase involved in the production of sporopollenin and is particularly expressed in the anther. The acyl-CoA synthetase encodes the key enzyme sporopollenin biosynthesis. Consequently, the mutation of *BrACOS5* resulted in a reduction of enzyme activities and increased the content of fatty acid in mutants anther, resulting in male sterile mutants [101].

The FLOWERING LOCUS C (FLC) genes are a key repressor of flowering in Brassica rapa, with overexpression of natural antisense transcripts (NATs) significantly shortening plant growth cycles. The sequence and expression level variation of *BrFLC* genes is extensive. A non-required vernalization in a long NAT of *BrFLC2* overexpressing lines reduced growth cycles [102]. The double knockout lines of *BraFLC2* and *BraFLC3* show an early-flowering phenotype which is vernalization-free [63]. The FLOWERING LOCUS T (FT)-like clade of phosphatidyl ethanolamine binding proteins (PEBPs) associate with the basic leucine zipper (bZIP) TF FD to induce flowering and form complexes regulatory. *BrFT1/2* promotes flowering by upregulating floral meristem identity genes in long- or short-day photoperiods [64]. Identified genes related to flowering are presented in Table 4.

Table 4. List of genes related to flowering in Chinese cabbage.

No	Gene Name	Function	References
1	*BrVIN3.1; BrFLC1*	bolting time	[103]
2	*BrCRY1; BrVIN3; BrGNC*	flowering time	[18]
3	*HA6, VATG3, VHA-E, AHA8, BrEXT4* and *BrYLS9*	pollen development	[35]
4	*Brcer1*	bolting and flowering	[38]
5	*BrFHAs*	male sterility	[47]
6	GSL genes (*BraA10g005220, BraA09g063900, BraA03g033140* and *BraA05g038460*) endo-1,3-b-glucosidase genes (*BraA04g008040, BraA08g009700* and *BraA05g015330*)	pollen development	[49]
6	*BrSKS13*	pollen development	[104]
7	*BrFLC2* NATs	flowering	[102]
8	*ARG7; SAUR41, BSK11, PYR1, auxin-induced protein 15A*	bolting time	[82]
9	*BrFT1* and *BrFT2*	flowering; and inflorescence organogenesis	[64]
10	*BraFLC2* and *BraFLC3*	flowering	[63]
11	*BrLEAFY*	bolting time	[67]
12	*BrACOS5*	male sterility	[101]

Notes: H+-ATPase 6 (HA6), vacuolar ATP synthase G3 (VATG3), vacuolar H+-ATPase subunit E isoform (VHA-E), and Arabidopsis H+-ATPase 8 (AHA8), Extensin 4 (BrEXT4) and yellow-leaf-specific gene 9 (BrYLS9), glucan synthase-like (GSL).

3.3. Leafy Head and Leaf Size

Chinese cabbage has specialized leaf morphogenesis, named heading, which affects yield [73]. The leafy head formation is a specific process of leaf development that involves the growth of incurving leaves. Originating from non-heading plants around 6000 years ago, Chinese cabbage has two heading types: wrapped-over and joined-up. The degree of compactness of these Chinese cabbage varieties varies and they can be classified into loose, semi-heading, and fully-heading forms [105]. Wrapped-over refers to a plant that matures early, has a round-shaped head, and is well-suited for warmer regions. On the other hand, joined-up refers to a plant that matures later, has a longer-shaped head, and is fitted to cooler temperatures [11].

Leaf adaxial–abaxial polarity in Chinese cabbage is closely linked to leaf incurvature, which is essential for the formation of leafy heads. By identifying the specific genes that determine leaf ad-ab polarity and analysing their genetic variation, we can gain a better understanding of the underlying mechanisms behind the creation of leafy heads [31]. Leaf axial growth is also crucial during the heading stage, with no genes controlling leaf axial growth in Chinese cabbage. A previous study examined the characteristics of the mutant ic1, which exhibits inward curling, and identified *BrOPS* mutation. *BrOPS* controls the bending of leaves in a way that depends on the presence of brassinosteroids (BR). *BrOPS* and *BrBIN2* interact to regulate the phosphorylation of *BrBES1*, which in turn negatively controls the expression of the leaf polarity *BrAS1* TF. This interaction ultimately affects the curvature of leaves and the form of the heading in Chinese cabbage [73].

Hormones play a crucial role in the heading stage of yellow inner-leaf Chinese cabbage, affecting the expression levels of genes related to hormones like gibberellin, auxin, ethylene, cytokinin, BR, and abscisic acid, indicating the essential involvement of hormones in the heading stage [78]. Reference [78] also confirmed that several hormone metabolic pathways were identified as crucial to the developing process of leaves during the heading stage. Several differentially expressed genes (DEGs) were identified as participating in the hormonal pathways, including ko04075, ko00904, and ko00270. One group of genes exhibited differential expression in hormone biosynthesis and signal transduction pathways. As shown in Figure 2, auxin response factors (ARFs) interact with other leafy head genes (*WOX1*, *KAN2*, *AS1/2*, and so on). A previous study implied that BrAS1 correlates with BR in leafy head transformation on Chinese cabbage [73]. The transition to leaf heading is specialized leaf morphogenesis, but the transcriptional regulation mechanisms remain unclear. The ethylene pathway is particularly active in this stage, with upregulated auxin genes (*BrPIN3s*, *BrPIN4*, and *BrPIN7*) co-expressed with BR and ethylene pathways. This hormonal crosstalk may control PIN-related polar auxin transport. The study also identified a cluster of abaxial genes, including *BrEINs*, *BrARFs*, and *BrBIN2s*, interacting with hormones, the MAPK signalling pathway, microtubule dynamics, and the cytoskeleton. The cohesion among co-expression clusters offers novel insights into the interaction of distinct biological processes and suggests that ethylene, BR, and auxin might collaborate to regulate leafy head development [76].

Chinese cabbage plants have been found to have leafy heads due to an intricate hormone signalling network and abaxial–adaxial patterning pathways. Reference [74] discovered nhm4-1 and nhm4-2, two allelic non-heading mutants, and *BraA09g001440.3C* are responsible for mutant phenotypes. The concentration of gibberellic acid (GA) in the mutant leaves was noticeably decreased, and there was a tendency for the leaves to cluster together and form leafy heads after being treated with external GA_3. This indicates that gibberellin is also involved in the development of leafy heads in Chinese cabbage [74].

Leaf flattening is crucial for plant architecture and photosynthesis. An EMS-induced population yielded a non-heading mutant with flat-heading leaves (fg-1). Phytohormone analysis showed decreased IAA, ABA, JA, SA, and increased methyl IAA and trans-Zeatin levels. Transcriptome analysis showed decreased expression levels of *BrAUX1*, *BrLAXs*, and *BrPINs*, auxin transport, and responsive genes. Furthermore, the mutation also caused an up-regulated ABA-responsive gene transcript level. A significant reduction in BrIAMT1 transcripts might contribute to leaf epinastic growth. The expression levels of crucial ABA-responsive genes, such as *BrABF3*, were increased in the middle portions of the leaves, indicating that both auxin and ABA signalling pathways have significant functions in the regulation of leaf heading. We assumed that leaf heading in Chinese cabbage is controlled through hormone signalling and abaxial–adaxial patterning pathways as previously mentioned by [71]. Furthermore, a research study discovered that a mutant cwm, characterized by its compact and wrinkled leaves, was regulated by a solitary nuclear gene known as *Brcwm*. A single SNP (on exon 4 of *BraA07g021970.3C*) resulted in a proline to serine amino acid substitution. *BraA07g021970.3C* is a homologue of *AT3G55000*, which regulates a protein involved in the regulation of cortical microtubule organization [106]. The representative candidate genes related to leafy head formation in Chinese cabbage are presented in Table 5.

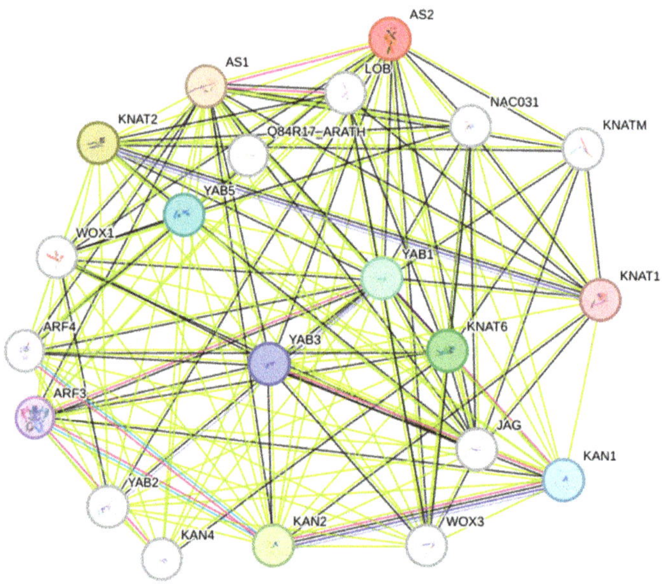

Figure 2. Interaction network analysis of genes related to leafy head formation in Chinese cabbage by using STRING. The interaction network has significantly more interactions than expected. This means that 21 *A. thaliana* proteins have more interactions among themselves, indicating the proteins are at least partially biologically connected as a group. The line colour is related to the type of interaction. The green line shows gene neighbourhood, the pink line means experimentally determined, the black line presents co-expression, the dark blue line indicates gene co-occurrence, and the blue line indicates protein homology. (For more interpretation of the colour codes in this figure legend, the reader is referred to the web version of this network analysis (https://string-db.org/cgi/network?taskId=bZ8uTOvikcqY&sessionId=bYgYPvAH4fiB (accessed: 30 September 2024)).

Moreover, the size of leaves is essential for the growth and production of plants, particularly in Brassica crops. The control of cell division and growth in leaves is determined by a combination of signals from both the internal regulatory network and the external environment. Gaining insight into the molecular mechanism of leaf size regulation is crucial for molecular breeding aimed at enhancing agricultural productivity. Recent research has discovered multiple genes and QTLs that play a crucial role in regulating leaf size [107]. These genes regulate growth by affecting pathways such as phytohormone production [108], transcription regulation, and short RNAs [107]. The primary contributors to leaf size in Chinese cabbage are cyclins *BraA09g010980.3C* (CYCB) and *BraA10g027420.3C* (CYCD). In addition, the *BraA09gMYB47* and *BraA06gMYB88* TFs are essential in determining the leaf size difference. However, the key regulators which regulate leaf size are phytohormone homeostasis or signalling. The cell numbers influence leaf size, with GA_1 and GA_3 playing crucial functions in early leaf growth, while ABA and IPA regulate cell proliferation throughout leaf growth [77]. Moreover, variations in light, ambient conditions, growth factors, and DELLA proteins influence the pace at which differentiated cells elongate during the cell division phase. Longer photoperiods promote the production of GA, leading to increased expression of the important GA synthesis genes GA20ox and GA3ox, thereby stimulating the production of active GAs in Arabidopsis [109]. While in peanut, *AhGA20ox* and *AhGA3ox* function in pod growth and development [110].

Table 5. Candidate genes for leafy head development in Chinese cabbage.

No	Gene Name	Function	Position	References
1	*BrETR2, BrERS2s, BrEIN3s, BrEBFs, BrEILs, BrACSs, BrACOs*, and *BrERFs*	leafy head development	-	[76]
2	*REV, PHB, PHV, ATHB8, AS1, AS2*	leaf ad-ab polarity controlling genes	adaxial determinants	[31]
	YAB1, YAB2, YAB3, YAB5, WOX1, and *WOX3*		middle domain determinants	
	KAN1, KAN2, KAN3, ARF3, and *ARF4*		abaxial determinants	
	AGO1, AGO7, AGO10, SGS3, RDR6, HYL1, DCL1, and *DCL4*,		small RNA pathway	
3	*BrCPS1*	non-heading trait		[74]
4	*BrOPS* *BrBIN2* *BrBES1* *BrAS1*	inward curling leaves		[73]
5	*BrCWM*	compact and wrinkled leaves		[106]
6	*BrCLV1*	non-heading trait		[111]

3.1. Leaf Colour

Leaf colour is crucial in crucifera crops, influencing nutritional and vitamin content. Chinese cabbage has various colours like pale green, green, and orange. Studies have explored how leaf colour affects indolic glucosinolates and pigment content [23], biomass [55], flavonoid biosynthesis [112], and soluble sugar and glucosinolate biosynthesis [79].

Chlorophyll, also known as Chl, plays a vital role in the process of plant greening and is influenced by a variety of genes. The yellowing of the inner leaves of Chinese cabbage is caused by the activation of the β-carotene production pathway, which involves the upregulation of genes such as GGPP and PSY, and the downregulation of genes such as *CrtL-e, NCED4*, and *DWARF-27*. Several genes involved in the synthesis of chlorophyll, including *NR, NIR, GS, GOCAT*, and *PROA*, were identified as being downregulated [78]. Chinese cabbage leaf yellowing is linked to chlorophyll metabolism and photosystem, with the enzyme *BrHISN2*, causing it. The *CBF2-BrHISN2* module enhances gene–environment interaction for breeding [57]. A mutation in pale green mutant (pgm) caused destructive chloroplast structure resulting in lower Chl content, higher Chl a/b ratio, and lower non-photochemical quenching. Genetic analysis reveals that the pale green phenotype is controlled by a recessive nuclear gene, *BrCAO*, which upregulates *BrCAO* expression and decreases CAO enzyme activity [55].

A yellow colour in Chinese cabbage seeds can be observed from seed stages, with brown-seeded and yellow-seeded heading varieties exhibiting different patterns. It was found that brown-seeded plants have higher phenolic compounds content than yellow-seeded plants, with large fragment deletion variations in the mapping region between brown-seeded and yellow-seeded. Proanthocyanidin (PA) is the main flavonoid affecting the colour of the seed coat. The TRANSPARENT TESTA GLABRA 1 (TTG1) is the candidate gene regulating seed coat colour due to large fragment deletions in the mapping region between the two varieties. A functional marker, Brsc-yettg1, was developed to detect TTG1 variation. However, *BrTTG1* is not tissue-specific and expression levels of four structural genes were higher in brown-seeded heading than yellow-seeded heading. The levels of expression of four structural genes, namely *BrDFR*, *BrANS*, *BrANR1*, and *BrANR2*, were higher in brown-seeded heading compared to yellow-seeded heading [113]. The seed coat colour of brown-seeded seeds varied from pale green to full brown at 32 DAF, while yellow-seeded seeds showed a transition from pale green to yellow at 26 DAF. Identified genes related to leaf colour are presented in Table 6.

Table 6. Identified genes related to leaf colour in Chinese cabbage.

No	Gene Name	Trait	Function	References
1	*GGPP*, *PSY*, *CrtL-e*, *NCED4*, and *DWARF-27*.	yellow inner leaves	β-carotene biosynthesis pathway	[78]
2	*BrHISN2*	leaf yellowing		[57]
3	*Bra037887* and *MYB90/PAP2*	leaf purple	anthocyanin biosynthesis	[17]
4	*BrTTG1*	seed coat color	proanthocyanidin/flavonoid biosyntesis	[113]
5	*BrF3′H*	purple coloration	flavonoid biosynthesis	[112]
6	*BrCAO*	pale green leaves	pale green variation trait	[55]

Flavonols and anthocyanins are the two major classes of flavonoids in Brassica rapa. The green variety of Chinese cabbage accumulates lower levels of quercetin, isorhamnetin, and cyanidin than the purple variety, illustrating that 3′-dihydroxylated flavonoids are not prevalent in the green variety. The majority of phenylpropanoid pathway gene expression patterns could not align with the patterns of flavonoid accumulation in both types (*BrPAL*, *BrC4H*, and *Br4CL*), with the exception of *BrPAL1.2*. However, it was observed that most early and late flavonoid biosynthesis genes exhibited significant levels of expression in the purple variety. Structural flavonoid metabolic profile features of purple Chinese cabbage are determined by the main enzyme, namely BrF3′H. The study also discovered that identical amino acid sequences of *BrF3′H*, which were obtained from two varieties, are highly conserved with F3′H genes from other species [112].

3.5. Biotic and Abiotic Stresses

Before the omics era, research conducted in the last twenty years has primarily examined the bacterial population, pesticide residues, parasite eggs, and heavy metals in Chinese cabbage due to food safety [114]. Chinese cabbage production is at risk due to abiotic and biotic stressors, climate change, and the proliferation of diseases. Research about glucosinolates (GSs) biosynthesis has been conducted in recent years [86]. GSs play a crucial role in plant–pathogen interactions and are a major secondary metabolite of the Brassicaceae family [85]. In addition, GSs and their hydrolysis products are recognized as crucial components in the interactions between plants and insects, as well as between plants and microorganisms [115]. Recent research has indicated that higher accumulation of GSs in the *BrMYC2* OE line confers the highest resistance to *S. sclerotiorum*. BrMYC2 decreased the lengths of roots and hypocotyls, seed length, and weight of Chinese cabbage. Additionally, it led to a faster bolting period and a much greater accumulation of total GSs. Elevated levels of GS accumulation in the BrMYC2OE line resulted in the greatest resistance to *S. sclerotiorum*, suggesting that overexpression may improve plant growth and development by increasing plant resistance to the fungal pathogen [85]. *Alternaria brassicifolii*, a unique species with weak pathogenicity, has been identified in *B. rapa* [116]. In addition, a first report of *Curvularia lunata* was found to cause leaf spot in Chinese cabbage [25].

Soft rot caused by *Pectobacterium carotovorum* and *Pectobacterium brasiliense* [80], and downy mildew caused by *Hyaloperonospora brassicae* [117], are the top three economically significant illnesses affecting Chinese cabbage, resulting in significant production and financial losses [118]. Polygalacturonase, pectate lyase, and cellulase are plant cell wall-degrading enzymes that are produced and released from the bacterial cytoplasm into the intercellular spaces of the plant tissue when *P. carotovorum* invades the host plant [119]. The symptoms of soft root in Chinese cabbage are characterized by a soft and watery texture, and a tan-coloured appearance [80].

A candidate pathogen-resistant WAK gene, BrWAK1, was identified and characterized, interacting with BrBAK1 to activate the mitogen-activated protein kinase (MAPK) cascade. The WAK gene, which provides disease resistance in Chinese cabbage, has been found for the first time, and its presence does not significantly influence plant biomass [117].

Plasmodiophora brassicae Woronin, an obligate parasite, is a major cause of clubroot, a soil-borne disease in Chinese cabbage [120], threatening the economic value of cruciferous crops [22,121]. MicroRNAs play a crucial role in post-transcriptional gene expression regulation. A total of 14 known and 10 potentially novel miRNAs were differentially expressed after *P. brassicae* treatment, with target genes mainly involved in seleno compound metabolism and plant hormone signal transduction [46]. Resistance genes to clubroot disease in *B. rapa* are located on different chromosomes. The Arabidopsis mutant rpp1 was more susceptible to clubroot disease, suggesting that deletion of rpp1 reduces plant resistance. *BrRPP1* interacts with *BrPLA1* to mediate the jasmonic acid signal transduction pathway in response to P. brassicae infection [121]. Interestingly, a study found that the causal agents of clubroot infection in *B. rapa* are robust effector-triggered immunity (ETI) and GA, as well as SA signalling [80]. Furthermore, *Variovorax* sp. YNA59 and *Lysinibacillus capsici* TT41 have been found to improve drought stress tolerance in Chinese cabbage. *Variovorax* sp. increased physiological parameters [122], while *L. capsici* enhanced drought stress tolerance by enhancing lactic acid levels [6].

Climate change has led to increased variable weather patterns affecting plants, with TFs and hormones involved in diverse plant reactions to stresses. Ethylene-responsive factors (ERFs) and ethylene or methyl jasmonate are involved in these reactions. *BrERF4* OE line improved tolerance to salt and drought stresses in Arabidopsis [83]. Reference [83] stated the *BrERF4* OE line showed green leaves after 4 days while WT showed more yellowing and began to die after 7 days of exposure to 300 mM NaCl. The study also suggests that reducing AtERF4 expression can improve salt and drought tolerance in BrERF4 OE lines by suppressing the signalling pathway mediated by AtERF4. This is supported by AtERF4's transcriptional repressor role, resulting in increased sensitivity to abiotic stresses and the down-regulation of salt-responsive genes. High similarity is found in the genetic structure of each subfamily (Pak-choi, Chinese cabbage, and Arabidopsis), with all CSP proteins containing highly conserved domains and HSE, which respond to some abiotic stress. Additionally, each *AtCSP1/2/4/5* have HSE, which responds to some abiotic stress [53]. Interestingly, numerous BrGRP genes exhibited abnormal expression in response to multiple stresses, suggesting potential resistance to multiple stresses (Table 7) [123].

miR168a may regulate OMT1, an essential melatonin biosynthesis gene in *B. rapa*, during salt stress. This regulates melatonin content, adjusting B. rapa's antioxidative potential, controlling osmolyte accumulation, and maintaining ion homeostasis, to improve salinity tolerance [58]. Under cold stress, a study on Chinese cabbage found that the C-repeat binding factor (CBF) response pathway could not be the primary mechanism for cold adaptation in the late-heading stage of deep yellow inner leaves. The study also found no significant differences in the expression of cold-response genes, suggesting that the CBF cold-response course is not the primary mechanism [78]. Abiotic stress, specifically Ca_2^+ deficient stress, causes tipburn, a collapse and necrosis of leaves. A candidate gene for tipburn resistance is *BrCRT2*, which showed higher expression levels during disease development [51].

Hydropriming boosts seed germination and early seedling growth in Chinese cabbage grown under salt stress by allowing seeds to absorb water without radicle appearance, promoting POD and CAT activities, and reducing MDA concentrations [8]. Similar to hydropriming, Nitrogen (N) is also crucial for seed germination, and deficiency is a significant environmental factor that leads to early senescence. Cotyledons provide an important N source during germination and early seedling development. Cotyledons are crucial for nitrogen during germination and seedling development. N deficiency enhances purine catabolism gene expression in Chinese cabbage cotyledons. Allopurinol, an inhibitor of xanthine dehydrogenase, reduces chlorophyll degradation and lowers fresh weight in seedlings [124]. Unfortunately, few studies have identified low N concentrations in Chinese cabbage. Nitrate transporters (NRTs) are essential for plant growth and development. The BrNRT1 gene family has a conserved structural pattern, with some genes' expression patterns specific to tissues. The promoters of BrNRT1 contain MFS elements, indicating the potential role of *BrNRT1* in NO_3^- transport regulation of Chinese cabbage [45].

Table 7. A (biotic) stress-related genes in Chinese cabbage.

No	Stress Type	Treatment	Gene Name	Reference
1	abiotic	low nitrate	*BrNRTs*	[45]
2	abiotic	drought stress		[122]
3	abiotic	heat stress		[9]
4	abiotic	NH_3 emission		[125]
5	abiotic	salinity stress		[126]
6	abiotic	drought and salt stress	*BrERF4*	[83]
7	abiotic and biotic	cold, heat stress and wound stress	*BrMYB28.3* and *BrMYB29.1*	[86]
8	abiotic	salt stress	*miR168a; OMT1*	[58]
9	abiotic	Ca^{2+} deficiency	*BrCRT2*	[51]
10	abiotic	cold stress	*BrCSPs*	[53]
11	abiotic	freezing stress cold stress	*BrCRG1, BrCRG 2, BrCRG 3* and *BrCRG 5* *BrCRG4* and *BrCRG6*	[127]
12	abiotic	calcium deficiency, Cd stress, salt stress	*BrP450s*	[5]
13	abiotic and biotic	high and low temperature, drought, and soft rot	*BrGRPs*	[123]
14	biotic	*P. brassicae*	*BrCRa*	[120]
15	biotic	*P. brassicae*	*BrRPP1*	[121]
16	biotic	*P. brassicae*	*Bra013275, Bra013299, Bra013336, Bra013339, Bra013341,* and *Bra013357*	[50]
17	biotic	*P. brassicae*	*miRNAs*	[46]
18	biotic	*Pectobacterium carotovorum*	*WRKY38*	[34]
19	biotic	*Hyaloperonospora brassicae*	*BrWAK1*	[117]
20	biotic	*Pseudomonas syringae*	*BrPR4*	[128]

4. Conclusions and Future Perspective

Multi-omics analysis plays a vital role in the identification of genetic processes, growth, development, and stress tolerance in Chinese cabbage. Diverse omics methodologies analyse gene networks and molecular pathways. Despite the development of several multi-omics studies on Chinese cabbage, several key traits remain underexplored such as leaf colour, glossy leaf, and especially nutrient efficiency. The pan-omics platform, which incorporates many omics approaches, can accurately forecast agriculturally significant characteristics for precise breeding and provide insights into the molecular regulatory networks involved in enhancing Chinese cabbage. The introduction of whole genome sequencing and other breeding techniques such as SNP, MAS, and KASP has facilitated the discovery of new regulators through mutant analysis and genome-wide identification, which is a novel approach in Chinese cabbage research. These techniques are expected to advance rapidly in the field of functional genomics in the future. In addition, the utilization of the efficient transformation technique in conjunction with gene editing using CRISPR/Cas9 will result in a reduction in time required and facilitate a more precise understanding of the role of the regulators. Integrating the regulatory networks of multiple discovered characteristics in Chinese cabbage is currently a significant concern. Thus, the advancement of molecular breeding techniques has enabled the exploration of features through the selection or manipulation of specific genes. High yield, resistance to stress, delectable flavour, and affordable price will all be achieved in the end.

Author Contributions: Conceptualization, F.N.R. and Z.G.; validation, Y.Z., F.W., J.G., C.L. and J.L.; investigation, F.N.R. and Z.G.; resources, F.N.R. and Z.G.; writing—original draft preparation, F.N.R. and Z.G.; writing—review and editing, F.N.R., Z.G., Y.Z., F.W., J.G., C.L. and J.L.; visualization, Y.Z., F.W., J.G., C.L. and J.L.; supervision, J.G., C.L. and J.L.; project administration, C.L. and J.L.; funding acquisition, Y.Z., F.W., J.G., C.L. and J.L. All authors have read and agreed to the published version of the manuscript.

Funding: This study was supported by the Natural Science Foundation of China (32172591); the Projects of 20 Rules for New Universities in Jinan, China (202228058); the Modern Agricultural Industrial Technology System Funding of Shandong Province, China (SDAIT-05); the Excellent Variety Project of Shandong Province (2023LZGC014); the Shandong Province Technology-based Small and Mediumsized Enterprise Innovation Capacity Improvement Project (2023TSGC0719); the Prospect of Shandong Seed Project, China (2022LZGC008); the Taishan Scholars Program of Shandong Province, China (tsqn201909167); the Formation Mechanism and Regulation of Fruit and Vegetable Quality in Sa-line-alkali Land (GYJ2023004).

Data Availability Statement: The original contributions presented in the study are included in the article, further inquiries can be directed to the corresponding author/s.

Conflicts of Interest: The authors declare no conflicts of interest.

References

1. Nagaharu, U. Genome Analysis in Brassica with Special Reference to the Experimental Formation of *B. Napus* and Peculiar Mode of Fertilization. *Jpn. J. Bot.* **1935**, *7*, 389–452.
2. Fareed, A.; Shindo, H.; Takahashi, H.; Nakamura, I. Analysis of PolA1 intron 19 and Ntag sequences reveals two ancestral lineages in the origin of the *Brassica rapa* complex and Chinese cabbage (*B. rapa* var. *pekinensis*). *J. Hortic. Sci. Biotechnol.* **2015**, *90*, 273–278. [CrossRef]
3. Zhang, L.; Liang, J.; Chen, H.; Zhang, Z.; Wu, J.; Wang, X. A near-complete genome assembly of *Brassica rapa* provides new insights into the evolution of centromeres. *Plant Biotechnol. J.* **2023**, *21*, 1022–1032. [CrossRef] [PubMed]
4. Azibi, T.; Hadj-Arab, H.; Lodé, M.; Ferreira de Carvalho, J.; Trotoux, G.; Nègre, S.; Gilet, M.M.; Boutte, J.; Lucas, J.; Vekemans, X.; et al. Impact of whole genome triplication on the evolutionary history and the functional dynamics of regulatory genes involved in Brassica self-incompatibility signalling pathway. *Plant Reprod.* **2020**, *33*, 43–58. [CrossRef] [PubMed]
5. Zhang, S.; Wu, Q.-R.; Zhang, H.-M.; Pei, Z.-M.; Gao, J.-W. Genome-wide identification and transcriptomic data exploring of the cytochrome P450 family in Chinese cabbage (*Brassica rapa* L. ssp. *pekinensis*). *J. Plant Interact.* **2021**, *16*, 136–155. [CrossRef]
6. Kim, M.; Lee, W.-K.; Kim, Y.-S.; Lim, C.-H.; Song, C.; Park, T.; Son, Y.; Son, Y.-M. Impact of thinning intensity on the diameter and height growth of *Larix kaempferi* stands in central Korea. *For. Sci Technol.* **2016**, *12*, 77–87. [CrossRef]
7. Kim, S.; Rho, H.Y.; Kim, S. The Effects of Climate Change on Heading Type Chinese Cabbage (*Brassica rapa* L. ssp. *Pekinensis*) Economic Production in South Korea. *Agronomy* **2022**, *12*, 3172. [CrossRef]
8. Yan, M. Hydro-priming increases seed germination and early seedling growth in two cultivars of Napa cabbage (*Brassica rapa* subsp. *pekinensis*) grown under salt stress. *J. Hortic. Sci. Biotechnol.* **2016**, *91*, 421–426. [CrossRef]
9. Quan, J.; Zheng, W.; Wu, M.; Shen, Z.; Tan, J.; Li, Z.; Zhu, B.; Hong, S.-B.; Zhao, Y.; Zhu, Z.; et al. Glycine Betaine and β-Aminobutyric Acid Mitigate the Detrimental Effects of Heat Stress on Chinese Cabbage (*Brassica rapa* L. ssp. *pekinensis*) Seedlings with Improved Photosynthetic Performance and Antioxidant System. *Plants* **2022**, *11*, 1213. [CrossRef] [PubMed]
10. Amas, J.C.; Bayer, P.E.; Hong Tan, W.; Tirnaz, S.; Thomas, W.J.W.; Edwards, D.; Batley, J. Comparative pangenome analyses provide insights into the evolution of *Brassica rapa* resistance gene analogues (RGAs). *Plant Biotechnol. J.* **2023**, *21*, 2100–2112. [CrossRef]
11. Okamoto, T.; Wei, X.; Mehraj, H.; Hossain, M.R.; Akter, A.; Miyaji, N.; Takada, Y.; Park, J.I.; Fujimoto, R.; Nou, I.S.; et al. Chinese cabbage (*Brassica rapa* L. var. *pekinensis*) breeding: Application of molecular technology. In *Advances in Plant Breeding Strategies: Vegetable Crops: Leaves, Flowerheads, Green Pods, Mushrooms and Truffles*; Springer International Publishing: Berlin/Heidelberg, Germany, 2021; Volume 10, pp. 59–94.

12. Wang, J.; Liu, Z.; Dou, J.; Lv, J.; Jin, N.; Jin, L.; Li, Z.; Zhang, B.; Tang, Z.; Yu, J. A Comparative Study on the Nutrients, Mineral Elements, and Antioxidant Compounds in Different Types of Cruciferous Vegetables. *Agronomy* **2022**, *12*, 3121. [CrossRef]
13. Sivanandhan, G.; Bae, S.; Sung, C.; Choi, S.-R.; Lee, G.-J.; Lim, Y.-P. Optimization of Protoplast Isolation from Leaf Mesophylls of Chinese Cabbage (*Brassica rapa* ssp. *pekinensis*) and Subsequent Transfection with a Binary Vector. *Plants* **2021**, *10*, 2636. [CrossRef] [PubMed]
14. Hirani, A.H.; Gao, F.; Liu, J.; Fu, G.; Wu, C.; Yuan, Y.; Li, W.; Hou, J.; Duncan, R.; Li, G. Transferring clubroot resistance from Chinese cabbage (*Brassica rapa*) to canola (*B. napus*). *Can. J. Plant Pathol.* **2016**, *38*, 82–90. [CrossRef]
15. Niemann, J.; Kaczmarek, J.; Książczyk, T.; Wojciechowski, A.; Jedryczka, M. Chinese cabbage (*Brassica rapa* ssp. *pekinensis*)—A valuable source of resistance to clubroot (*Plasmodiophora brassicae*). *Eur. J. Plant Pathol.* **2017**, *147*, 181–198. [CrossRef]
16. Xuan, S.X.; Shen, S.X.; Wang, Y.H.; Zhang, W.W.; Gu, A.X.; Feng, D.L. Genomic composition of microspore-derived plants from a Chinese cabbage (*Brassica rapa* ssp. *pekinensis*) × heading cabbage (*B. oleracea* var. *capitata*) allotriploid (ACC) hybrid based on genomic in situ hybridization. *J. Hortic. Sci. Biotechnol.* **2015**, *90*, 379–387. [CrossRef]
17. Xie, L.; Li, F.; Zhang, S.; Zhang, H.; Qian, W.; Li, P.; Zhang, S.; Sun, R. Mining for Candidate Genes in an Introgression Line by Using RNA Sequencing: The Anthocyanin Overaccumulation Phenotype in Brassica. *Front. Plant Sci.* **2016**, *7*, 1245. [CrossRef]
18. Liu, J.; Liu, B.; Cheng, F.; Liang, J.; Wang, X.; Wu, J. A High Density Linkage Map Facilitates QTL Mapping of Flowering Time in *Brassica rapa*. *Hortic. Plant J.* **2016**, *2*, 217–223. [CrossRef]
19. Yu, X.; Wang, H.; Zhong, W.; Bai, J.; Liu, P.; He, Y. QTL mapping of leafy heads by genome resequencing in the RIL population of *Brassica rapa*. *PLoS ONE* **2013**, *8*, e76059. [CrossRef]
20. Kim, J.; Jung, Y.; Song, B.; Bong, Y.-S.; Ryu, D.H.; Lee, K.-S.; Hwang, G.-S. Discrimination of cabbage (*Brassica rapa* ssp. pekinensis) cultivars grown in different geographical areas using 1H NMR-based metabolomics. *Food Chem.* **2013**, *137*, 68–75. [CrossRef]
21. Kim, Y.B.; Li, X.; Kim, S.-J.; Kim, H.H.; Lee, J.; Kim, H.; Park, S.U. MYB Transcription Factors Regulate Glucosinolate Biosynthesis in Different Organs of Chinese Cabbage (*Brassica rapa* ssp. *pekinensis*). *Molecules* **2013**, *18*, 8682–8695. [CrossRef]
22. Cho, K.H.; Park, S.H.; Kim, K.T.; Kim, S.; Kim, J.S.; Park, B.S.; Woo, J.G.; Lee, H.J. Mapping quantitative trait loci (QTL) for clubroot resistance in *Brassica rapa* L. *J. Hortic. Sci. Biotechnol.* **2012**, *87*, 325–333. [CrossRef]
23. Zhang, R.; Zhang, J.; Li, C.; Pan, Q.; Haq, S.U.; Mosa, F.W.A.; Fang, F.; Zhang, L.; Li, B. The Accumulation of Health-Promoting Nutrients from Representative Organs across Multiple Developmental Stages in Orange Chinese Cabbage. *Plants* **2023**, *12*, 2120. [CrossRef] [PubMed]
24. Yue, X.; Su, T.; Xin, X.; Li, P.; Wang, W.; Yu, Y.; Zhang, D.; Zhao, X.; Wang, J.; Sun, L.; et al. The Adaxial/Abaxial Patterning of Auxin and Auxin Gene in Leaf Veins Functions in Leafy Head Formation of Chinese Cabbage. *Front. Plant Sci.* **2022**, *13*, 918112. [CrossRef]
25. Wonglom, P.; Ito, S.; Sunpapao, A. First report of Curvularia lunata causing leaf spot of *Brassica rapa* subsp. *pekinensis* in Thailand. *New Dis. Rep.* **2018**, *38*, 15. [CrossRef]
26. Zemlyanskaya, E.V.; Dolgikh, V.A.; Levitsky, V.G.; Mironova, V. Transcriptional regulation in plants: Using omics data to crack the cis-regulatory code. *Curr. Opin. Plant Biol.* **2021**, *63*, 102058. [CrossRef]
27. Yang, Y.; Saand, M.A.; Huang, L.; Abdelaal, W.B.; Zhang, J.; Wu, Y.; Li, J.; Sirohi, M.H.; Wang, F. Applications of Multi-Omics Technologies for Crop Improvement. *Front. Plant Sci.* **2021**, *12*, 563953. [CrossRef]
28. Town, C.D.; Cheung, F.; Maiti, R.; Crabtree, J.; Haas, B.J.; Wortman, J.R.; Hine, E.E.; Althoff, R.; Arbogast, T.S.; Tallon, L.J.; et al. Comparative Genomics of *Brassica oleracea* and *Arabidopsis thaliana* Reveal Gene Loss, Fragmentation, and Dispersal after Polyploidy. *Plant Cell* **2006**, *18*, 1348–1359. [CrossRef]
29. Wang, X.; Wang, H.; Wang, J.; Sun, R.; Wu, J.; Liu, S.; Bai, Y.; Mun, J.H.; Bancroft, I.; Cheng, F.; et al. The genome of the mesopolyploid crop species *Brassica rapa*. *Nat. Genet.* **2011**, *43*, 1035–1039. [CrossRef]
30. Ge, Y.; Wang, T.; Wang, N.; Wang, Z.; Liang, C.; Ramchiary, N.; Choi, S.-R.; Lim, Y.P.; Piao, Z.Y. Genetic mapping and localization of quantitative trait loci for chlorophyll content in Chinese cabbage (*Brassica rapa* ssp. *pekinensis*). *Sci. Hortic.* **2012**, *147*, 42–48. [CrossRef]
31. Gao, Y.; Lu, Y.; Li, X.; Li, N.; Zhang, X.; Su, X.; Feng, D.; Liu, M.; Xuan, S.; Gu, A.; et al. Development and Application of SSR Markers Related to Genes Involved in Leaf Adaxial-Abaxial Polarity Establishment in Chinese Cabbage (*Brassica rapa* L. ssp. *pekinensis*). *Front. Genet.* **2020**, *11*, 00773. [CrossRef]
32. Li, Y.; Liu, Z.; Wang, Y.; Yang, N.; Xin, X.; Yang, S.; Feng, H. Identification of quantitative trait loci for yellow inner leaves in Chinese cabbage (*Brassica rapa* L. ssp. *pekinensis*) based on SSR and SRAP markers. *Sci. Hortic.* **2012**, *133*, 10–17. [CrossRef]
33. Hirani, A.H.; Geng, J.; Zhang, J.; Zelmer, C.D.; McVetty, P.B.E.; Daayf, F.; Li, G. Quantitative Trait Loci Mapping and Candidate Gene Identification for Seed Glucosinolates in *Brassica rapa* L. *Crop Sci.* **2016**, *56*, 942–956. [CrossRef]
34. Liu, M.-y.; Wu, F.; Ge, Y.-j.; Lu, Y.; Zhang, X.-m.; Wang, Y.-h.; Wang, Y.; Yan, J.-h.; Shen, S.-x.; Zhao, J.-j.; et al. Identification of soft rot resistance loci in *Brassica rapa* with SNP markers. *J. Integr. Agric.* **2022**, *21*, 2253–2263. [CrossRef]
35. Dong, X.; Kim, W.K.; Lim, Y.-P.; Kim, Y.-K.; Hur, Y. Ogura-CMS in Chinese cabbage (*Brassica rapa* ssp. *pekinensis*) causes delayed expression of many nuclear genes. *Plant Sci.* **2013**, *199–200*, 7–17. [CrossRef]
36. Yue, L.; Zhang, S.; Zhang, L.; Liu, Y.; Cheng, F.; Li, G.; Zhang, S.; Zhang, H.; Sun, R.; Li, F. Heterotic prediction of hybrid performance based on genome-wide SNP markers and the phenotype of parental inbred lines in heading Chinese cabbage (*Brassica rapa* L. ssp. *pekinensis*). *Sci. Hortic.* **2022**, *296*, 110907. [CrossRef]

37. Li, G.-L.; Qian, W.; Zhang, S.-J.; Zhang, S.-F.; Li, F.; Zhang, H.; Wu, J.; Wang, X.-W.; Sun, R.-F. Development of gene-based markers for the Turnip mosaic virus resistance gene retr02 in *Brassica rapa*. *Plant Breed.* **2016**, *135*, 466–470. [CrossRef]
38. Liu, C.; Song, G.; Wang, N.; Huang, S.; Gao, Y.; Fu, W.; Zhang, M.; Feng, H. A single SNP in Brcer1 results in wax deficiency in Chinese cabbage (*Brassica campestris* L. ssp. *pekinensis*). *Sci. Hortic.* **2021**, *282*, 110019. [CrossRef]
39. Saha, G.; Park, J.-I.; Ahmed, N.U.; Kayum, M.A.; Kang, K.-K.; Nou, I.-S. Characterization and expression profiling of MYB transcription factors against stresses and during male organ development in Chinese cabbage (*Brassica rapa* ssp. *pekinensis*). *Plant Physiol. Biochem.* **2016**, *104*, 200–215. [CrossRef]
40. Zhang, L.; Cai, X.; Wu, J.; Liu, M.; Grob, S.; Cheng, F.; Liang, J.; Cai, C.; Liu, Z.; Liu, B.; et al. Improved Brassica rapa reference genome by single-molecule sequencing and chromosome conformation capture technologies. *Hortic. Res.* **2018**, *5*, 50. [CrossRef]
41. Zhang, Z.; Guo, J.; Cai, X.; Li, Y.; Xi, X.; Lin, R.; Liang, J.; Wang, X.; Wu, J. Improved Reference Genome Annotation of Brassica rapa by Pacific Biosciences RNA Sequencing. *Front. Plant Sci.* **2022**, *13*, 841618. [CrossRef]
42. Chen, H.; Wang, T.; He, X.; Cai, X.; Lin, R.; Liang, J.; Wu, J.; King, G.; Wang, X. BRAD V3.0: An upgraded Brassicaceae database. *Nucleic Acids Res.* **2021**, *50*, gkab1057. [CrossRef] [PubMed]
43. Cai, X.; Chang, L.; Zhang, T.; Chen, H.; Zhang, L.; Lin, R.; Liang, J.; Wu, J.; Freeling, M.; Wang, X. Impacts of allopolyploidization and structural variation on intraspecific diversification in *Brassica rapa*. *Genome Biol.* **2021**, *22*, 166. [CrossRef] [PubMed]
44. Lv, S.; Zhang, C.; Tang, J.; Li, Y.; Wang, Z.; Jiang, D.; Hou, X. Genome-wide analysis and identification of TIR-NBS-LRR genes in Chinese cabbage (*Brassica rapa* ssp. *pekinensis*) reveal expression patterns to TuMV infection. *Physiol. Mol. Plant Pathol.* **2015**, *90*, 89–97. [CrossRef]
45. Zhang, Y.; Ritonga, F.N.; Zhang, S.; Wang, F.; Li, J.; Gao, J. Genome-Wide Identification of the NRT1 Family Members and Their Expression under Low-Nitrate Conditions in Chinese Cabbage (*Brassica rapa* L. ssp. *pekinensis*). *Plants* **2023**, *12*, 3882. [CrossRef]
46. Wei, X.; Xu, W.; Yuan, Y.; Yao, Q.; Zhao, Y.; Wang, Z.; Jiang, W.; Zhang, X. Genome-wide Investigation of microRNAs and Their Targets in *Brassica rapa* ssp. *pekinensis* Root with *Plasmodiophora brassicae* Infection. *Hortic. Plant J.* **2016**, *2*, 209–216. [CrossRef]
47. Shi, F.; Liu, Z.; Pang, Z.; Feng, H. The role of the FHA domain-containing protein family in pollen development of *Brassica rapa* L. *Sci. Hortic.* **2021**, *288*, 110339. [CrossRef]
48. Yuan, Y.; Wang, L.; Zhao, Q.; Liu, C.; Fu, X.; Li, X.; Qiu, M.; Li, J.; Zhang, Y.; Li, C.; et al. Genome-Wide Identification, Expression Analysis and Functional Study of DELLA Genes in Chinese Cabbage (*Brassica rapa* L. ssp. *pekinensis*). *FBL* **2024**, *29*, 5198. [CrossRef]
49. Pu, Y.; Hou, L.; Guo, Y.; Ullah, I.; Yang, Y.; Yue, Y. Genome-wide analysis of the callose enzyme families of fertile and sterile flower buds of the Chinese cabbage (*Brassica rapa* L. ssp. *pekinensis*). *FEBS Open Bio* **2019**, *9*, 1432–1449. [CrossRef]
50. Zhang, H.; Liu, X.; Zhou, J.; Strelkov, S.E.; Fredua-Agyeman, R.; Zhang, S.; Li, F.; Li, G.; Wu, J.; Sun, R.; et al. Identification of Clubroot (*Plasmodiophora brassicae*) Resistance Loci in Chinese Cabbage (*Brassica rapa* ssp. *pekinensis*) with Recessive Character. *Genes* **2024**, *15*, 274. [CrossRef]
51. Su, T.; Li, P.; Wang, H.; Wang, W.; Zhao, X.; Yu, Y.; Zhang, D.; Yu, S.; Zhang, F. Natural variation in a calreticulin gene causes reduced resistance to Ca2+ deficiency-induced tipburn in Chinese cabbage (*Brassica rapa* ssp. *pekinensis*). *Plant Cell Environ.* **2019**, *42*, 3044–3060. [CrossRef]
52. Li, B.; Yue, Z.; Ding, X.; Zhao, Y.; Lei, J.; Zang, Y.; Hu, Q.; Tao, P. A BrLINE1-RUP insertion in BrCER2 alters cuticular wax biosynthesis in Chinese cabbage (*Brassica rapa* L. ssp. pekinensis). *Front. Plant Sci.* **2023**, *14*, 1212528. [CrossRef] [PubMed]
53. Huang, F.; Wang, J.; Tang, J.; Hou, X. Identification, evolution and functional inference on the cold-shock domain protein family in Pak-choi (*Brassica rapa* ssp. chinensis) and Chinese cabbage (*Brassica rapa* ssp. pekinensis). *J. Plant Interact.* **2019**, *14*, 232–241. [CrossRef]
54. Sun, X.; Li, X.; Lu, Y.; Wang, S.; Zhang, X.; Zhang, K.; Su, X.; Liu, M.; Feng, D.; Luo, S.; et al. Construction of a high-density mutant population of Chinese cabbage facilitates the genetic dissection of agronomic traits. *Mol. Plant.* **2022**, *15*, 913–924. [CrossRef] [PubMed]
55. Zhao, Y.; Huang, S.; Wang, N.; Zhang, Y.; Ren, J.; Zhao, Y.; Feng, H. Identification of a biomass unaffected pale green mutant gene in Chinese cabbage (*Brassica rapa* L. ssp. *pekinensis*). *Sci. Rep.* **2022**, *12*, 7731. [CrossRef]
56. Shi, F.; Zhou, X.; Liu, Z.; Feng, H. Rapid alkalinization factor (RALF) genes are related to genic male sterility in Chinese cabbage (*Brassica rapa* L. ssp. *pekinensis*). *Sci. Hortic.* **2017**, *225*, 480–489. [CrossRef]
57. Su, T.; Wang, W.; Li, P.; Xin, X.; Yu, Y.; Zhao, X.; Zhang, D.; Yu, S.; Zhang, F. Natural variations of BrHISN2 provide a genetic basis for growth-flavour trade-off in different *Brassica rapa* subspecies. *New Phytol.* **2021**, *231*, 2186–2199. [CrossRef]
58. Shamloo-Dashtpagerdi, R.; Lindlöf, A.; Tahmasebi, S. Evidence that miR168a contributes to salinity tolerance of *Brassica rapa* L. via mediating melatonin biosynthesis. *Physiol. Plant.* **2022**, *174*, e13790. [CrossRef] [PubMed]
59. Yang, S.; Liu, H.; Wei, X.; Zhao, Y.; Wang, Z.; Su, H.; Zhao, X.; Tian, B.; Zhang, X.W.; Yuan, Y. BrWAX2 plays an essential role in cuticular wax biosynthesis in Chinese cabbage (*Brassica rapa* L. ssp. *pekinensis*). *Theor. Appl. Genet.* **2022**, *135*, 693–707. [CrossRef] [PubMed]
60. Hong, J.K.; Suh, E.J.; Park, S.R.; Park, J.; Lee, Y.-H. Multiplex CRISPR/Cas9 Mutagenesis of BrVRN1 Delays Flowering Time in Chinese Cabbage (*Brassica rapa* L. ssp. *pekinensis*). *Agriculture* **2021**, *11*, 1286. [CrossRef]
61. Joung, J.K.; Sander, J.D. TALENs: A widely applicable technology for targeted genome editing. *Nat. Rev. Mol. Cell Biol.* **2013**, *14*, 49–55. [CrossRef]

62. Li, X.; Li, H.; Zhao, Y.; Zong, P.; Zhan, Z.; Piao, Z. Establishment of A Simple and Efficient Agrobacterium-mediated Genetic Transformation System to Chinese Cabbage (*Brassica rapa* L. ssp. *pekinensis*). *Hortic. Plant J.* **2021**, *7*, 117–128. [CrossRef]
63. Jeong, S.Y.; Ahn, H.; Ryu, J.; Oh, Y.; Sivanandhan, G.; Won, K.-H.; Park, Y.D.; Kim, J.-S.; Kim, H.; Lim, Y.P.; et al. Generation of early-flowering Chinese cabbage (*Brassica rapa* spp. *pekinensis*) through CRISPR/Cas9-mediated genome editing. *Plant Biotechnol. Rep.* **2019**, *13*, 491–499. [CrossRef]
64. Lee, A.; Jung, H.; Park, H.J.; Jo, S.H.; Jung, M.; Kim, Y.S.; Cho, H.S. Their C-termini divide Brassica rapa FT-like proteins into FD-interacting and FD-independent proteins that have different effects on the floral transition. *Front. Plant Sci.* **2022**, *13*, 1091563. [CrossRef] [PubMed]
65. Liu, Y.; Zhang, S.; Zhang, S.; Zhang, H.; Li, G.; Sun, R.; Li, F. Efficient transformation of the isolated microspores of Chinese cabbage (*Brassica rapa* L. ssp. *pekinensis*) by particle bombardment. *Plant Methods* **2024**, *20*, 17. [CrossRef]
66. Lee, Y.R.; Siddique, M.I.; Kim, D.S.; Lee, E.S.; Han, K.; Kim, S.G.; Lee, H.E. CRISPR/Cas9-mediated gene editing to confer turnip mosaic virus (TuMV) resistance in Chinese cabbage (*Brassica rapa*). *Hortic. Res.* **2023**, *10*, uhad078. [CrossRef]
67. Shin, Y.H.; Park, Y.D. CRISPR/Cas9-Mediated Mutagenesis of BrLEAFY Delays the Bolting Time in Chinese Cabbage (*Brassica rapa* L. ssp. *pekinensis*). *Int. J. Mol. Sci.* **2022**, *24*, 541. [CrossRef]
68. Guo, Y. Mutant resource for Chinese cabbage. *Nat. Food* **2022**, *3*, 188. [CrossRef]
69. Gao, Y.; Qu, G.; Huang, S.; Liu, Z.; Zhang, M.; Fu, W.; Ren, J.; Feng, H. Comparison between Germinated Seed and Isolated Microspore EMS Mutagenesis in Chinese Cabbage (*Brassica rapa* L. ssp. *pekinensis*). *Horticulturae* **2022**, *8*, 232. [CrossRef]
70. Xu, J.; Liao, R.; Xue, M.; Shang, S.; Zhou, M.; Liu, Z.; Feng, H.; Huang, S. Mutations in BrABCG26, encoding an ATP-binding cassette transporter, are responsible for male sterility in Chinese cabbage (*Brassica rapa* L. ssp. pekinensis). *Theor. Appl. Genet.* **2024**, *137*, 63. [CrossRef]
71. Li, J.; Zhang, X.; Lu, Y.; Feng, D.; Gu, A.; Wang, S.; Wu, F.; Su, X.; Chen, X.; Li, X.; et al. Characterization of Non-heading Mutation in Heading Chinese Cabbage (*Brassica rapa* L. ssp. *pekinensis*). *Front. Plant Sci.* **2019**, *10*, 00112. [CrossRef]
72. Yang, S.; Tang, H.; Wei, X.; Zhao, Y.; Wang, Z.; Su, H.; Niu, L.; Yuan, Y.; Zhang, X. BrWAX3, Encoding a β-ketoacyl-CoA Synthase, Plays an Essential Role in Cuticular Wax Biosynthesis in Chinese Cabbage. *Int. J. Mol. Sci.* **2022**, *23*, 10938. [CrossRef] [PubMed]
73. Zhang, X.; Ma, W.; Liu, M.; Li, X.; Li, J.; Lu, Y.; Li, G.; Zhang, S.; Feng, D.; Wang, Y.; et al. OCTOPUS regulates BIN2 to control leaf curvature in Chinese cabbage. *Proc. Natl. Acad. Sci. USA* **2022**, *119*, e2208978119. [CrossRef] [PubMed]
74. Gao, Y.; Qu, G.; Huang, S.; Liu, Z.; Fu, W.; Zhang, M.; Feng, H. BrCPS1 function in leafy head formation was verified by two allelic mutations in Chinese cabbage (*Brassica rapa* L. ssp. *pekinensis*). *Front. Plant Sci.* **2022**, *2426*, 889798. [CrossRef] [PubMed]
75. Qi, X.; An, H.; Ragsdale, A.P.; Hall, T.E.; Gutenkunst, R.N.; Chris Pires, J.; Barker, M.S. Genomic inferences of domestication events are corroborated by written records in *Brassica rapa*. *Mol. Ecol.* **2017**, *26*, 3373–3388. [CrossRef]
76. Zhang, K.; Yang, Y.; Wu, J.; Liang, J.; Chen, S.; Zhang, L.; Lv, H.; Yin, X.; Zhang, X.; Zhang, Y.; et al. A cluster of transcripts identifies a transition stage initiating leafy head growth in heading morphotypes of Brassica. *Plant J.* **2022**, *110*, 688–706. [CrossRef]
77. Wang, L.; Zhang, S.; Zhang, Y.; Li, J.; Zhang, Y.; Zhou, D.; Li, C.; He, L.; Li, H.; Wang, F.; et al. Integrative analysis of physiology, biochemistry and transcriptome reveals the mechanism of leaf size formation in Chinese cabbage (*Brassica rapa* L. ssp. *pekinensis*). *Front. Plant Sci.* **2023**, *14*, 1183398. [CrossRef]
78. Li, Y.; Fan, Y.; Jiao, Y.; Wu, J.; Zhang, Z.; Yu, X.; Ma, Y. Transcriptome profiling of yellow leafy head development during the heading stage in Chinese cabbage (*Brassica rapa* subsp. *pekinensis*). *Physiol. Plant.* **2019**, *165*, 800–813. [CrossRef] [PubMed]
79. Wang, L.; Zhang, S.; Li, J.; Zhang, Y.; Zhou, D.; Li, C.; He, L.; Li, H.; Wang, F.; Gao, J. Identification of key genes controlling soluble sugar and glucosinolate biosynthesis in Chinese cabbage by integrating metabolome and genome-wide transcriptome analysis. *Front. Plant Sci.* **2022**, *13*, 1043489. [CrossRef]
80. Li, X.; Ren, X.; Ibrahim, E.; Kong, H.; Wang, M.; Xia, J.; Wang, H.; Shou, L.; Zhou, T.; Li, B.; et al. Response of Chinese cabbage (*Brassica rapa* subsp. *pekinensis*) to bacterial soft rot infection by change of soil microbial community in root zone. *Front. Microbiol.* **2024**, *15*, 1401896. [CrossRef]
81. He, L.; Li, J.; Shi, L.; Zhao, Q.; Wu, Z.; Zeng, S.; Li, Y.; Zhang, W.; Wang, F.-D.; Zhang, S.; et al. Exogenous metabolites spray, which identified from metabolomics analysis and transcriptomic analysis, can improve salt tolerance of Chinese cabbages (*Brassica rapa* L. ssp *pekinensis*)*. *J. Plant Interact.* **2021**, *16*, 452–461. [CrossRef]
82. Wei, X.; Rahim, M.A.; Zhao, Y.; Yang, S.; Wang, Z.; Su, H.; Li, L.; Niu, L.; Harun-Ur-Rashid, M.; Yuan, Y.; et al. Comparative Transcriptome Analysis of Early- and Late-Bolting Traits in Chinese Cabbage (*Brassica rapa*). *Front. Genet.* **2021**, *12*, 590830. [CrossRef] [PubMed]
83. Seo, Y.J.; Park, J.-B.; Cho, Y.-J.; Jung, C.; Seo, H.S.; Park, S.-K.; Nahm, B.H.; Song, J.T. Overexpression of the Ethylene-Responsive Factor Gene BrERF4 from *Brassica rapa* Increases Tolerance to Salt and Drought in Arabidopsis Plants. *Mol. Cells* **2010**, *30*, 271–278. [CrossRef] [PubMed]
84. Kumari, S.; Jo, J.S.; Choi, H.S.; Lee, J.G.; Lee, S.I.; Jeong, M.-J.; Kim, J.A. Molecular Characterization and Expression Analysis of MYB Transcription Factors Involved in the Glucosinolate Pathway in Chinese Cabbage (*Brassica rapa* ssp. *pekinensis*). *Agronomy* **2019**, *9*, 807. [CrossRef]
85. Teng, Z.; Zheng, W.; Yu, Y.; Hong, S.-B.; Zhu, Z.; Zang, Y. Effects of BrMYC2/3/4 on Plant Development, Glucosinolate Metabolism, and Sclerotinia sclerotiorum Resistance in Transgenic *Arabidopsis thaliana*. *Front. Plant Sci.* **2021**, *12*, 707054. [CrossRef]

86. Baskar, V.; Park, S.W. Molecular characterization of BrMYB28 and BrMYB29 paralogous transcription factors involved in the regulation of aliphatic glucosinolate profiles in *Brassica rapa* ssp. *pekinensis*. *C. R. Biol.* **2015**, *338*, 434–442. [CrossRef]
87. Seo, M.-S.; Jin, M.; Sohn, S.-H.; Kim, J.S. Expression profiles of BrMYB transcription factors related to glucosinolate biosynthesis and stress response in eight subspecies of *Brassica rapa*. *FEBS Open Bio* **2017**, *7*, 1646–1659. [CrossRef]
88. Liu, Y.; Lu, S.; Liu, K.; Wang, S.; Huang, L.; Guo, L. Proteomics: A powerful tool to study plant responses to biotic stress. *Plant Methods* **2019**, *15*, 135. [CrossRef] [PubMed]
89. Liu, C.; Sun, G.S.; Zhang, R.J.; Lv, S.W.; Gao, L.; Gao, L.W.; Liu, T.K.; Xiao, D.; Hou, X.L.; Zhang, C.W. Proteomic analysis provides integrated insight into mechanisms of Turnip mosaic virus long distance movement in Brassica rapa. *Biol. Plant.* **2019**, *63*, 164–173. [CrossRef]
90. Mi, W.B.; Fang, Y.; Liu, Z.G.; Xu, C.M.; Liu, G.Y.; Zou, Y.; Xu, M.X.; Zheng, G.Q.; Cao, X.D.; Fang, X.L. Differential proteomics analysis of fertility transformation of the winter rape thermo-sensitive sterile line PK3-12S (*Brassica rapa* L.). *Acta. Agro. Sin.* **2020**, *46*, 1507–1516.
91. Brunelle, J.L.; Green, R. One-dimensional SDS-polyacrylamide gel electrophoresis (1D SDS-PAGE). *Methods Enzymol.* **2014**, *541*, 151–159. [CrossRef]
92. Minden, J.S. Chapter 6—Two-Dimensional Difference Gel Electrophoresis (2D DIGE). In *Methods in Cell Biology*; Conn, P.M., Ed.; Academic Press: Cambridge, MA, USA, 2012; Volume 112, pp. 111–141.
93. Ji, R.; Wang, Y.; Wang, X.; Liu, Y.; Shen, X.; Feng, H. Proteomic analysis of the interaction between Plasmodiophora brassicae and Chinese cabbage (*Brassica rapa* L. ssp. *Pekinensis*) at the initial infection stage. *Sci. Hortic.* **2018**, *233*, 386–393. [CrossRef]
94. Lan, M.; Li, G.; Hu, J.; Yang, H.; Zhang, L.; Xu, X.; Liu, J.; He, J.; Sun, R. iTRAQ-based quantitative analysis reveals proteomic changes in Chinese cabbage (*Brassica rapa* L.) in response to Plasmodiophora brassicae infection. *Sci. Rep.* **2019**, *9*, 12058. [CrossRef] [PubMed]
95. Su, T.; Yu, S.; Wang, W.; Li, P.; Zhang, F.; Yu, Y.; Zhang, D.; Zhao, X. iTRAQ analysis of protein profile during the secondary stage of infection of Plasmodiophora brassicae in Chinese cabbage (*Brassica rapa* subsp. *pekinensis*). *J. Plant Pathol.* **2018**, *100*, 533–542. [CrossRef]
96. Li, J.; Qiu, M.; Ritonga, F.N.; Wang, F.; Zhou, D.; Li, C.; Li, H.; Zhang, Y.; Gao, J. Metabolite Profiling and Comparative Metabolomics Analysis of Jiaozhou Chinese Cabbage (*Brassica rapa* L. ssp. *pekinensis*) Planted in Different Areas. *Front. Biosci. (Landmark Ed.)* **2023**, *28*, 345. [CrossRef]
97. Ju, W.; Park, S.J.; Lee, M.J.; Park, S.H.; Min, S.G.; Ku, K.-M. Seasonal variation of metabolites in Kimchi cabbage: Utilizing metabolomics based machine learning for cultivation season and taste discrimination. *Hortic. Environ. Biotechnol.* **2024**, 1–16. [CrossRef]
98. Pavlović, I.; Mlinarić, S.; Tarkowská, D.; Oklestkova, J.; Novák, O.; Lepeduš, H.; Bok, V.V.; Brkanac, S.R.; Strnad, M.; Salopek-Sondi, B. Early Brassica Crops Responses to Salinity Stress: A Comparative Analysis Between Chinese Cabbage, White Cabbage, and Kale. *Front. Plant Sci.* **2019**, *10*, 450. [CrossRef]
99. Li, N.; Zhang, Z.; Gao, S.; Lv, Y.; Chen, Z.; Cao, B.; Xu, K. Different responses of two Chinese cabbage (*Brassica rapa* L. ssp. *pekinensis*) cultivars in photosynthetic characteristics and chloroplast ultrastructure to salt and alkali stress. *Planta* **2021**, *254*, 102. [CrossRef]
100. Adamus, A.; Szklarczyk, M.; Kiełkowska, A. Haploid and Doubled Haploid Plant Production in *Brassica rapa* L. subsp. *pekinensis* Via Microspore Culture. In *Hot Topics, Apiaceae, Brassicaceae, Solanaceae*; Seguí-Simarro, J.M., Ed.; Humana Press: Totowa, NJ, USA, 2021; Volume 2, pp. 181–198.
101. Zou, J.; Dong, S.; Fang, B.; Zhao, Y.; Song, G.; Xin, Y.; Huang, S.; Feng, H. BrACOS5 mutations induced male sterility via impeding pollen exine formation in Chinese cabbage (*Brassica rapa* L. ssp. *pekinensis*). *Theor. Appl. Genet.* **2023**, *136*, 6. [CrossRef]
102. Li, X.; Zhang, S.; Bai, J.; He, Y. Tuning growth cycles of Brassica crops via natural antisense transcripts of BrFLC. *Plant Biotechnol. J.* **2016**, *14*, 905–914. [CrossRef]
103. Su, T.; Wang, W.; Li, P.; Zhang, B.; Li, P.; Xin, X.; Sun, H.; Yu, Y.; Zhang, D.; Zhao, X.; et al. A Genomic Variation Map Provides Insights into the Genetic Basis of Spring Chinese Cabbage (*Brassica rapa* ssp. *pekinensis*) Selection. *Mol. Plant.* **2018**, *11*, 1360–1376. [CrossRef]
104. Ji, R.; Ge, W.; Wang, H.; Zhao, Y.; Feng, H. BrSKS13, a multiple-allele-inherited male sterility-related gene in Chinese cabbage (*Brassica rapa* L. ssp. *pekinensis*), affects pollen development and pollination/fertilization process. *Gene* **2019**, *696*, 113–121. [CrossRef] [PubMed]
105. Opeña, R.T.; Kuo, C.G.; Yoon, J.Y. *Breeding and Seed Production of Chinese Cabbage in the Tropics and Subtropics*; Asian Vegetable Research and Development Center (AVRDC): Taiwan, China, 1988.
106. Wu, Y.; Xin, Y.; Zou, J.; Huang, S.; Wang, C.; Feng, H. BrCWM Mutation Disrupted Leaf Flattening in Chinese Cabbage (*Brassica rapa* L. ssp. *pekinensis*). *Int. J. Mol. Sci.* **2023**, *24*, 5225. [CrossRef] [PubMed]
107. Karamat, U.; Sun, X.; Li, N.; Zhao, J. Genetic regulators of leaf size in Brassica crops. *Hortic. Res.* **2021**, *8*, 91. [CrossRef]
108. Ritonga, F.N.; Zhou, D.; Zhang, Y.; Song, R.; Li, C.; Li, J.; Gao, J. The Roles of Gibberellins in Regulating Leaf Development. *Plants* **2023**, *12*, 1243. [CrossRef]
109. Li, Q.; Li, J.; Zhang, L.; Pan, C.; Yang, N.; Sun, K.; He, C. Gibberellins are required for dimorphic flower development in *Viola philippica*. *Plant Sci.* **2021**, *303*, 110749. [CrossRef]

110. Sun, J.; Zhang, X.; Fu, C.; Ahmad, N.; Zhao, C.; Hou, L.; Naeem, M.; Pan, J.; Wang, X.; Zhao, S. Genome-wide identification and expression analysis of GA20ox and GA3ox genes during pod development in peanut. *PeerJ* **2023**, *11*, e16279. [CrossRef]
111. Zhao, Y.; Liu, C.; Fang, B.; Huang, S.; Wang, N.; Tan, C.; Ren, J.; Feng, H. Identification of a Leafy Head Formation Related Gene in Chinese Cabbage (*Brassica rapa* L. ssp. *pekinensis*). *Horticulturae* **2022**, *8*, 86. [CrossRef]
112. Park, S.; Lee, H.; Min, M.K.; Ha, J.; Song, J.; Lim, C.J.; Oh, J.; Lee, S.B.; Lee, J.-Y.; Kim, B.-G. Functional Characterization of BrF3'H, Which Determines the Typical Flavonoid Profile of Purple Chinese Cabbage. *Front. Plant Sci.* **2021**, *12*, 793589. [CrossRef]
113. Ren, Y.; He, Q.; Ma, X.; Zhang, L. Characteristics of Color Development in Seeds of Brown- and Yellow-Seeded Heading Chinese Cabbage and Molecular Analysis of Brsc, the Candidate Gene Controlling Seed Coat Color. *Front. Plant Sci.* **2017**, *8*, 1410. [CrossRef]
114. Jeong, J.-W.; Park, K.-J.; Lim, J.-H.; Sung, J.-M. Analysis of hazard on fresh and salted baechus (nappa cabbage, *Brassica rapa* L. subsp. *pekinensis*) in Korea. *J. Korean Soc. Appl. Biol. Chem.* **2013**, *56*, 69–76. [CrossRef]
115. Ji, R.; Lei, J.; Chen, I.W.; Sang, W.; Yang, S.; Fang, J.; Zhu-Salzman, K. Cytochrome P450s CYP380C6 and CYP380C9 in green peach aphid facilitate its adaptation to indole glucosinolate-mediated plant defense. *Pest Manag. Sci.* **2021**, *77*, 148–158. [CrossRef] [PubMed]
116. Deng, J.X.; Li, M.J.; Paul, N.C.; Oo, M.M.; Lee, H.B.; Oh, S.-K.; Yu, S.H. Alternaria brassicifolii sp. nov. Isolated from *Brassica rapa* subsp. *pekinensis* in Korea. *Mycobiology* **2018**, *46*, 172–176. [CrossRef] [PubMed]
117. Zhang, B.; Su, T.; Xin, X.; Li, P.; Wang, J.; Wang, W.; Yu, Y.; Zhao, X.; Zhang, D.; Li, D.; et al. Wall-associated kinase BrWAK1 confers resistance to downy mildew in *Brassica rapa*. *Plant Biotechnol. J.* **2023**, *21*, 2125–2139. [CrossRef] [PubMed]
118. Liu, M.; Wu, F.; Wang, S.; Lu, Y.; Chen, X.; Wang, Y.; Gu, A.; Zhao, J.; Shen, S. Comparative transcriptome analysis reveals defense responses against soft rot in Chinese cabbage. *Hortic. Res.* **2019**, *6*, 68. [CrossRef]
119. Abbott, D.W.; Boraston, A.B. Structural biology of pectin degradation by Enterobacteriaceae. *Microbiol. Mol. Biol. Rev.* **2008**, *72*, 301–316, table of contents. [CrossRef]
120. Liu, X.; Strelkov, S.E.; Sun, R.; Hwang, S.-F.; Fredua-Agyeman, R.; Li, F.; Zhang, S.; Li, G.; Zhang, S.; Zhang, H. Histopathology of the Plasmodiophora brassicae-Chinese Cabbage Interaction in Hosts Carrying Different Sources of Resistance. *Front. Plant Sci.* **2022**, *12*, 783550. [CrossRef]
121. Ge, W.; Lv, M.; Feng, H.; Wang, X.; Zhang, B.; Li, K.; Zhang, J.; Zou, J.; Ji, R. Analysis of the role of BrRPP1 gene in Chinese cabbage infected by *Plasmodiophora brassicae*. *Front. Plant Sci.* **2023**, *14*, 1082395. [CrossRef] [PubMed]
122. Kang, S.-M.; Adhikari, A.; Kim, Y.-N.; Lee, I.-J. Comparative Evaluation of Paclobutrazol and *Rhizobacterium Variovorax* sp. YNA59 Effects on Mitigating Drought Stress in Chinese Cabbage (*Brassica rapa* ssp. *pekinensis*). *Microbiol. Res.* **2023**, *14*, 1210–1224. [CrossRef]
123. Lu, X.; Cheng, Y.; Gao, M.; Li, M.; Xu, X. Molecular Characterization, Expression Pattern and Function Analysis of Glycine-Rich Protein Genes Under Stresses in Chinese Cabbage (*Brassica rapa* L. ssp. *pekinensis*). *Front. Genet.* **2020**, *11*, 774. [CrossRef]
124. Yi, S.Y.; Lee, M.; Jeevan Rameneni, J.; Lu, L.; Kaur, C.; Lim, Y.P. Xanthine-derived metabolites enhance chlorophyll degradation in cotyledons and seedling growth during nitrogen deficient condition in Brassica rapa. *Plant Signal. Behav.* **2021**, *16*, 1913309. [CrossRef]
125. Kang, Y.-G.; Chun, J.-H.; Lee, J.-H.; Park, S.-Y.; Luyima, D.; Oh, T.-K.; Yun, Y.-U. Effects of Varying Rates of Nitrogen and Biochar pH on NH3 Emissions and Agronomic Performance of Chinese Cabbage (*Brassica rapa* ssp. *pekinensis*). *Agronomy* **2022**, *12*, 61. [CrossRef]
126. Linić, I.; Šamec, D.; Grúz, J.; Vujčić Bok, V.; Strnad, M.; Salopek-Sondi, B. Involvement of Phenolic Acids in Short-Term Adaptation to Salinity Stress is Species-Specific among Brassicaceae. *Plants* **2019**, *8*, 155. [CrossRef] [PubMed]
127. Thamilarasan, S.K.; Park, J.-I.; Jung, H.-J.; Chung, M.-Y.; Cho, Y.-G.; Nou, I.-S. Expression profiling and characterization of cold, freezing-related genes from *Brassica rapa* cultivars. *Isr. J. Plant Sci.* **2016**, *63*, 105–115. [CrossRef]
128. Chung, S.Y.; Lee, K.A.; Oh, K.J.; Cho, T.J. Molecular characterization of a PR4 gene in Chinese cabbage. *Integr. Biosci.* **2005**, *9*, 239–244. [CrossRef]

plants

MDPI

Review

Multi-Omics Research Accelerates the Clarification of the Formation Mechanism and the Influence of Leaf Color Variation in Tea (*Camellia sinensis*) Plants

Yan-Gen Fan [1,†], Ting-Ting Zhao [1,†], Qin-Zeng Xiang [1], Xiao-Yang Han [1], Shu-Sen Yang [2], Li-Xia Zhang [1,*] and Li-Jun Ren [1,*]

[1] College of Horticulture Science and Engineering, Shandong Agricultural University, Tai'an 271018, China; fanyangen@sdau.edu.cn (Y.-G.F.); datingsdau@126.com (T.-T.Z.); qzshine@sdau.edu.cn (Q.-Z.X.); hanxy@sdau.edu.cn (X.-Y.H.)

[2] Yipinming Tea Planting Farmers Specialized Cooperative, Longnan 746400, China; leegenvan@163.com

* Correspondence: zlx_sdau@163.com (L.-X.Z.); renlijun403@163.com (L.-J.R.)

† These authors contributed equally to this work.

Abstract: Tea is a popular beverage with characteristic functional and flavor qualities, known to be rich in bioactive metabolites such as tea polyphenols and theanine. Recently, tea varieties with variations in leaf color have been widely used in agriculture production due to their potential advantages in terms of tea quality. Numerous studies have used genome, transcriptome, metabolome, proteome, and lipidome methods to uncover the causes of leaf color variations and investigate their impacts on the accumulation of crucial bioactive metabolites in tea plants. Through a comprehensive review of various omics investigations, we note that decreased expression levels of critical genes in the biosynthesis of chlorophyll and carotenoids, activated chlorophyll degradation, and an impaired photosynthetic chain function are related to the chlorina phenotype in tea plants. For purple-leaf tea, increased expression levels of late biosynthetic genes in the flavonoid synthesis pathway and anthocyanin transport genes are the major and common causes of purple coloration. We have also summarized the influence of leaf color variation on amino acid, polyphenol, and lipid contents and put forward possible causes of these metabolic changes. Finally, this review further proposes the research demands in this field in the future.

Keywords: albinism; etiolation; transcriptome; metabolome; anthocyanin

1. Introduction

The tea plant (*Camellia sinensis*) is a widespread crop with high economic and health potential. Tea plant buds and leaves can be processed into popular non-alcoholic beverages [1]. The primary reasons for tea's popularity are its heightened metabolites and rich flavors. Among these metabolites, catechins are an expansive class of polyphenols making up 12–24% of the dry weight of young tea leaves, imparting an astringent flavor [2]. Regarding human health, catechins have characteristic antioxidative, anti-inflammatory, and other important biological activities [3]. Theanine (γ-glutamylethylamide), a non-proteinogenic amino acid, is specifically synthesized in tea plants. Generally, theanine is the predominant free amino acid in tea leaves and makes up 1–2% of its dry weight [2]. Theanine has conventional health benefits, including antioxidation and neuroprotection [4]. When drinking tea, theanine is the main flavorful substance related to its sweetness and umami, which can neutralize the astringent flavors of polyphenols and the bitter taste resulting from caffeine [2]. Additionally, tea contains other important functional metabolites, including chlorophylls, carotenoids, lipids, and anthocyanins [5]. The content and ratio of these compounds contribute significantly to the economic and nutritional value of tea plants.

Typically, green-leaf tea varieties are most commonly planted and used to make tea. Recently, increasing numbers of tea mutants with variable leaf colors, including albinism, etiolation, and purple hues, have been used in tea production due to their characterized accumulation of certain bioactive substances and increased economic value [6]. As albinistic or etiolated leaves often lead to a limited photosynthetic capacity and an impaired plant survivability, chlorina tea varieties should adopt special survival strategies. In agricultural practices, certain tea cultivars with periodic chlorina behavior are widely utilized because of their good balance between quality and environmental adaptability [6–10]. For periodic chlorina tea varieties, an albinistic or etiolated phenotype is induced according to specific developmental stages or ecological conditions. For instance, chlorina leaves are observed in *Anjibaicha* (Alternative names: *Baiye No1*, *White leaf No.1*, and *Anji white 1*) and *Huabai 1* under low temperatures [11,12], and high-density light is required for the etiolated phenotype in other varieties such as *Huangjinya* [9,13,14]. When these environmental factors are not satisfied, these varieties will exhibit green leaves to enhance their environmental adaptability [14,15]. In other periodic chlorina tea varieties, leaf re-greening is not modulated by ecological factors but by foliar age [16]. Aside from periodic chlorina cultivars, some varieties with variegated leaves have also been chosen for tea cultivation. In variegated-leaf varieties, a percentage of green tissue is retained on the chlorina leaves, also offering the tea plants an acceptable environmental adaptability [6,17,18]. Alongside changes in leaf color, there are drastic and complex influences on the accumulation of various important functional components in chlorina tea [6,8,9].

Tea varieties possessing purple leaves represent another emerging popular tea type [19]. Compared to traditional green-leaf cultivars, purple-leaf tea is rich in purple-colored bioactive anthocyanins [20–22]. Anthocyanins are polyphenols and possess powerful antioxidative and anti-inflammatory capacities [23]. The high-level accumulation of anthocyanins in purple tea hinders the accumulation of other metabolites, predominantly flavonoids [19]. Clearly, enhanced anthocyanin accumulation will greatly alter the flavor of the tea, not just its appearance [19].

As a crop with its leaves acting as the primary economically valuable components, variations in tea leaf color often lead to positive changes in nutritional value and flavor quality. Exploring the metabolites that are associated with leaf color formation and quality changes has theoretical value and important practical value. Therefore, tea is an ideal model plant for investigating the relationship between leaf color and metabolism regulation. Due to the lack of effective gene identification and functional confirmation tools, the cloning of mutated genes for tea leaf color variations is relatively undefined. Fortunately, due to the recent rapid progress of multi-omics technology and its extensive application to tea studies, several high-quality genomes of tea plants have been documented and used to guide physiological and molecular analysis of various important traits [1,24–27]. To date, over 70 articles have reported the utilization of genome, transcriptome, metabolome, proteome, and lipidome approaches alone or in combination to uncover the color variations in tea. This review will summarize these recent advances and propose challenges for further research.

2. Multi-Omics Approaches Further Our Understanding of Leaf Color Variation in Tea

2.1. Physiological Mechanisms of Albinism or Etiolation in Tea

Chlorophylls and carotenoids are the two primary pigment types in plant leaves. The biosynthesis and degradation of chlorophylls and carotenoids are complex processes with multiple steps, and chloroplast development is a delicate process. Therefore, from a genetic point of view, the mutations causing chlorina phenotypes in the leaves are very rich [28].

As in all albinistic and etiolated tea mutants, varying degrees of reduced chlorophyll a, chlorophyll b, and total chlorophyll content are the most direct factors causing the leaf color to become lighter (Table 1) [12,16,29–34]. The ratio of chlorophyll a/b is down-regulated in *Anjibaicha* [10,35], *Huangjinya* [32,36,37], and *Yanlingyinbiancha* [18] but up-regulated in *Baijiguan* [38], *Xiangfeihuangye* [39], and *Huangjinju* [13]. Additionally, the content of carotenoids in chlorina leaves is also always lower as compared to green-leaf controls [10,13,14,16,18,29,30,32–34,39–42]. There are exceptions, however: for example, the chlorina tea shoot (HY) of *Danzicha* accumulates 1.84 times higher levels of carotenoids than the green shoot from the same tree [43]. The ratios of chlorophylls to carotenoids are decreased in all chlorina tea plants except in the HY shoot of *Danzicha* [10,13,14,16,18,29,30,32–34,39–42]. These findings suggest that the absolute and relative decreases in total chlorophyll content across most chlorina tea plants are much higher than for carotenoids, and the resulting leaf color appears yellowish. As chlorophyll and carotenoid accumulation are highly interdependent, it is challenging to determine whether the reduction in chlorophyll or carotenoid content is the fundamental cause of most chlorina tea plants' formation [44,45].

Table 1. A summary of multi-omics-generated molecular evidence related to leaf color variations in tea.

Tea Varieties	Omics Approaches	Potential Molecular Mechanisms	References
Anjibaicha (Alternative names: *Baiye No1*, *White leaf No.1*, and *Anji white 1*)	Succinyl-proteome	Photosynthetic chain: The succinylation levels of PsbS and light-harvesting complex LHCA4 are down-regulated; the succinylation level of LHCB4 is up-regulated.	[46]
	Proteome and acetyl-proteome	Photosynthetic chain: lower abundance of LHCB1, LHCB2, LHCB3, LHCB4, LHCB5, LHCB7, LHCA1, LHCA2, LHCA3, LHCA4, PsbC, PsbD, PsbO, PsbP, PsbQ, PsbR, PsbS, Psb27, PsaE, PsaG, PsaL, PsaN, PetA, PetC, PetE, PetF, PetH, ATPA, ATPB, ATPD, ATPE, ATPG; a lower acetylation level of LHCA1.	[8]
	Pangenome	Chlorophyll synthesis: The *GluRS/EARS* gene in *Anjibaicha* showed a loss of anti-codon recognition domains; SVs were revealed in *ChlD*. Chlorophyll degradation: 1 bp deletion is revealed in *NOL/NYC1*.	[24]
	Whole-transcriptome	Chlorophyll synthesis: down-regulated expression of *POR* (two alleles), *CLH1*, *PBGD/HMBS/HEMC*; up-regulated expression of *COX15*. Photosynthetic chain: down-regulated expression of *CAB7*, *CAB21*, *LHCA4*, *CAB40* (three alleles), *CAB13*, *LHCB5*, *PetA* (two alleles), *PsbA* (*TEA_001460*), *PsaB*, *PsbP1*, *ATPA*, *PsaH*, ATPI, *PetB*, *Psb28*; up-regulated expression of *PsbA* (*TEA_001460*).	[35]
Huangjinya	Proteome	Chlorophyll synthesis: higher abundance of GluRS/EARS, MgCh/ChlH, and HMBS; lower abundance of POR. Photosynthetic chain: lower abundance of Photosystem Q(B) protein.	[36]
	Metabolome and proteome	Chlorophyll synthesis: higher abundance of GluRS/EARS and UROD/HEME; lower abundance of POR and CAO. Chlorophyll degradation: higher abundance of PAO/ACD1 Photosynthetic chain: lower abundance of LHCA1, LHCA3, LHCB1, LHCB2, LHCB3, LHCB4, LHCB5, LHCB6, PsaD, PsaF, PsaL, PsaN, PsbA, PsbD, PsbE, PsbO, PsbP, PsbQ; higher abundance of PsbS, FNR, ATPB. Carotenoid synthesis: lower abundance of ZEP; higher abundance of PSY.	[9]
	Pangenome	Chlorophyll synthesis: The *GluRS/EARS* gene in *Huangjinya* showed a loss of anti-codon recognition domains, which may inhibit chlorophyll synthesis; SVs were detected in *CAO*, *CHLP*, and *GluTR*. Carotenoid synthesis: mutated amino acids in $CYP97A3^{HJY}$ and elevated expression of the $CYP97A3^{HJY}$ allele.	[24]
	Transcriptome	Carotenoid synthesis: down-regulated expression of *PSY*, *PDS*, *ZDS*, *LCYE*, *LCYB*, *CHY*, *ZEP*, and *VDE*.	[47]

Table 1. *Cont.*

Tea Varieties	Omics Approaches	Potential Molecular Mechanisms	References
Baijiguan	BSR-seq	Photosynthetic chain: co-down-regulated expression of *LHCA3*, three *LHCB1* alleles (*TEA001863, TEA001868, TEA030368*), two *LHCB3* alleles (*TEA017256, TEA021966*), and *LHCB4* in bulked groups and parents.	[29]
	Genome: Genotyping by sequencing and BSA-seq	Chlorophyll synthesis: a non-synonymous polymorphism (G1199A) in the magnesium chelatase I subunit (*CsChlI*).	[38]
HY1	Proteome	Photosynthetic chain: lower abundance of LHCA3, LHCA4, LHCB1, LHCB2, LHCB6, PsbC, PsBO, PsbS, PsaB, PsaC, PsaD, PsaF, PetA, PetH.	[17]
HY2	Proteome	Photosynthetic chain: lower abundance of LHCB2, PsbA, PsbD, PsbC, PsbB, PsbQ, PsaA, PsaB, PsaD, PsaH, PetB, PetA, beta F-type ATPase.	[17]
HY	Transcriptome	Chlorophyll degradation: the activation of *SGR* and *CLH*.	[43]
Xiangfeihuangye	Transcriptome, translatome, and metabolome	Chlorophyll synthesis: up-regulation of HY5 in EL inhibited the expression of *GluTR/HEMA* and *POR*.	[39]
Huabai 1	Transcriptome	Photosynthetic chain: down-regulated expression of light-harvesting complex II (*LHCII*) chlorophyll-a/b-binding protein. Chlorophyll degradation: up-regulated expression of *SGR*.	[12]
Albinistic branch of Huangshan	Transcriptome	Chlorophyll synthesis: down-regulated expression of four *CHLP* alleles (*TEA027589, TEA019124*, BGI_novel_G007262, *TEA016514*) and one POR allele (*TEA014780*); up-regulated expression of one *CHLP* allele (*TEA009538*), one *POR* allele (*TEA027994*), and one *CLH* allele (*TEA027808*). Photosynthetic chain: down-regulated expression of three *LHCB1* alleles (*TEA019232, TEA030366, TEA030368*), *PsbC*, five *PsbB* alleles (*TEA028468, TEA011113, TEA032780*, TEA018797, *BGI_novel_G013475*), *PsbP, Psb28, PsaA*, five *ATPD* alleles (*TEA030038, TEA004696, TEA002611, BGI_novel_G006800, BGI_novel_004911*), two ATPA alleles (*TEA019276, BGI_novel_G009498*), *ATPE*.	[48]
Huangjinju	Transcriptome	Chlorophyll synthesis: up-regulated expression of *POR*. Photosynthetic chain: down-regulated expression of *LHCA2, LHCA4, LHCB1, LHCB2, LHCB6*.	[13]
Yanlingyinbiancha	Transcriptome	Chlorophyll synthesis: down-regulated expression of *UROS/HEMD, PPOX, ChlH/GUN5, MgPEC/CRD1, DVR/PCB2*, and *CAO*. Chlorophyll degradation: down-regulated expression of *NOL/NYC1, HCAR, CLH1*, and *ACD2*. Photosynthetic chain: fifty-five DEGs involved in photosynthetic complexes were found to be down-regulated. Carotenoid synthesis: down-regulated expression of *Z-ISO, ZDS, ZEP, LUT2, NCED4*.	[18]
Yanling Huayecha	Transcriptome	Chlorophyll synthesis: down-regulated expression of *PPOX*. Photosynthetic chain: down-regulated expression of *LHCB6* and *FdC2*. Thylakoid membrane structure: down-regulated expression of *SCY1*.	[40]

Table 1. *Cont.*

Tea Varieties	Omics Approaches	Potential Molecular Mechanisms	References
Menghai Huangye	Transcriptome	Chlorophyll synthesis: four genes related to chlorophyll synthesis (*HEME2* and *POR*). Photosynthetic chain: ten genes related to photosynthesis (*LHCA* and *LHCB*) are down-regulated.	[41]
Zhonghuang 3	Transcriptome	Chlorophyll metabolism: down-regulated expression of *GluTR/HEMA3* and *CLH4*.	[49]
Zhonghuang 3	Transcriptome	Chlorophyll synthesis: down-regulated expression of *GluTR/HEMA*, *GSA-AM/HEML*, UROD/HEME, *HEMF/CPOX*, *DVR* (*CSS0009780*), and *CHLP*; up-regulated expression of *PBGS/HEMB*, *DVR* (*CSS0011936*), and *CLH*. Chlorophyll degradation: down-regulated expression of *NOL/NYC1* (*CSS0031926*) and *SGR* (*CSS0030812*); up-regulated expression of *NOL/NYC1* (*CSS0015127*), *SGR* (*CSS0050352*), and *SGRL* (*CSS0004139* and *CSS0036450*). Carotenoid synthesis: up-regulated expression of *Z-ISO*, CRTISO (*CSS0027469*, *CSS0033902*, and *CSS0044870*), *NCED1*, *NCED2*; down-regulated expression of *LCYB* and *NXS*.	[14]
Zhonghuang 2	Transcriptome	Transcripts encoding enzymes such as those functioning in early enzymatic steps, from the formation of glutamate 1-semialdehyde to protoporphyrin IX, showed lower levels. Critical enzymes for converting Mg-protoporphyrin IX into chlorophyll were also inhibited.	[42]
Koganemidori	Transcriptome	Chlorophyll synthesis: down-regulated expression of *POR*, *CAO*, and *ChlG*. Chlorophyll degradation: up-regulated expression of *CLH*. Transcriptional regulation: two homologs of *GLK* were significantly down-regulated.	[50]
Huangjinshuixian	Transcriptome and metabolome	Chlorophyll degradation: down-regulated expression of *SGR*. Carotenoid synthesis: the expression of *DXS* and *GGPPS* was significantly down-regulated. Transcriptional regulation: *PIFs* related to chlorophyll biosynthesis were significantly suppressed.	[16]
Huangyu	Transcriptome and metabolome	Chlorophyll synthesis: down-regulated expression of *UROD/HEME*, *MgCh/ChlH*, and *CAO*. Chlorophyll degradation: up-regulated expression of *CLH*. Photosynthetic chain: down-regulated expression of three *LHCII* genes (*CSS0013089*, *CSS0017825*, and *CSS0039893*)	[30]
Huangkui	Transcriptome	Transcriptional regulation: the transcriptional expression of *CsRVE1* increased during seasonal greening and was tightly correlated with increases in the expression of genes involved in light harvesting (*LHCB*) and chlorophyll biosynthesis (*MgCh/ChlH*, *GluTR/HEMA1*, and *CAO*).	[51]
Fuhuang 1	Transcriptome	Chlorophyll synthesis: down-regulated expression of *CAO*. Chlorophyll degradation: down-regulated expression of *NOL/NYC1* and *SGR*. Photosynthetic chain: down-regulated expression of *LHCA2*, *LHCA4*, *LHCB1*, *LHCB3*. Carotenoid synthesis: down-regulated expression of *LCYE*, *ZEP*, *NCED*; up-regulated expression of *PSY*, *PDS*, *VDE*.	[52]

Table 1. *Cont.*

Tea Varieties	Omics Approaches	Potential Molecular Mechanisms	References
Fuhuang 2	Transcriptome	Chlorophyll synthesis: down-regulated expression of *GluTR/HEMA*. Chlorophyll degradation: down-regulated expression of *CLH*. Photosynthetic chain: down-regulated expression of *PsbB, PetC, ATPF1B, LCHBs, LCHAs*. Carotenoid synthesis: down-regulated expression of *NCED*.	[53]

Targeted metabolomic methods are often used to analyze the changes in carotenoid composition and contents in chlorina leaves. In tea plants, lutein is the most prevailing carotenoid component, followed by neoxanthin, violaxanthin, cryptoxanthin, α-carotene, β-carotene, and zeaxanthin [10]. Compared to green-leaf varieties, the levels of lutein, neoxanthin, and violaxanthin in the chlorina leaves of *Anjibaicha* and *Huangjinya* are dramatically reduced, while the levels of α-carotene, β-carotene, and zeaxanthin are increased [10,32,37]. A different effect of the chlorina phenotype on carotenoid accumulation is observed in *Huangjinshuixian, Yujinxiang*, and *Huangyu*. In these three varieties, the contents of lutein, carotene, neoxanthin, and cryptoxanthin are all down-regulated [16,30,33]. The levels of β-carotene in *Zhonghuang 2* are also reduced [42]. An enhanced accumulation of zeaxanthin is consistently observed in *Zhonghuang 2, Yujinxiang*, and *Huangyu*, but not in *Huangjinshuixian* [16,30,33,42]. These findings indicate that the metabolic flow allocation in the carotenoid pathway varies greatly in different chlorina tea plants. This is logical as carotenoids are important photo-protective pigments in plants. Blocking photosynthetic electron transfer in chlorina leaves is more prone to producing excess reactive oxygen species and radicals under high light or UV stress. The efficacy of different carotenoids in scavenging various reactive oxygen species varies greatly [54]. For example, β-carotene has a high activity in $^1O_2$ quenching, while lutein and α-carotene exhibit a much lower efficacy [55]. For radical scavenging and UV protection, zeaxanthin is more efficient than β-carotene [56,57]. Albino leaves may enhance the metabolic flow toward more highly active carotenoid components to enhance antioxidative abilities when the total carotenoid content decreases.

2.2. Molecular Mechanisms of Leaf Chlorosis in Tea Plant

Cloning the mutated gene responsible for albinism or etiolation in tea varieties is the most important, but still challenging, approach to studying the formation of leaf chlorosis. To date, only two studies have identified highly reliable candidate genes for leaf chlorosis in tea plants via omics techniques [24,38]. Using genotyping-by-sequencing, Zhang et al. constructed a genetic map of the full-sibling population of *Baijiguan* and *Longjing43* and successfully identified the major effect QTL linked to the total chlorophyll contents [38]. Using bulked segregant analysis sequencing (BSA-Seq)-assisted genetic mapping, a nonsynonymous mutation in the magnesium chelatase I subunit encoding the gene *CsChlI* blocks the conversion of protoporphyrin IX (Proto IX) into MgP IX throughout chlorophyll synthesis, impacting leaf coloring [38]. In many annual plants, similar nonsynonymous mutations are reported to be responsible for the reduced function of ChlI and chlorophyll accumulation [58–61]. A common feature of these mutants and *Baijiguan* is that the degree of leaf yellowing is gene-dosage-dependent [58–60].

In another study, a pan-genome analysis of 22 tea accessions revealed various genomic deletions in the *glutamyl-tRNA synthetase* (*GluRS/EARS*) gene of both *Anjibaicha* and *Huangjinya* [24]. GluRS/EARS catalyzes the formation of L-Glu-tRNA, which is considered the first step of chlorophyll biosynthesis in higher plants [62]. The truncated GluRS/EARS protein may limit chlorophyll synthesis and contribute to the formation of the chlorina phenotype in these two varieties. Additionally, structural variations (SVs) have been identified in another three chlorophyll synthesis genes (*chlorophyllide a oxygenase, CAO*; *geranylgeraniol reductase, CHLP*; *glutamyl-tRNA reductase, GluTR*) and one (*magne-*

sium chelatase D subunit, ChlD) chlorophyll synthesis gene in *Huangjinya* and *Anjibaicha*, respectively. These findings suggest the presence of multiple blocked genetic sites in the chlorophyll synthesis pathways of *Anjibaicha* and *Huangjinya* [24]. Additionally, the genomic region of the chlorophyll-degradation-related gene *chlorophyll b reductase* (*NOL*) has a 1-base-pair deletion in *Anjibaicha*, suggesting that the chlorophyll degradation may also have changed [24,63,64]. Comparative genomic analysis identified a genomic variation in the *CYP97A3* gene causing a 20-amino-acid alteration that may affect its ability to catalyze zeaxanthin synthesis of lutein [24,65]. However, the true functions of the aforementioned candidate genes identified using either genetic mapping or pan-genome analysis remain unclear in vivo.

Given the difficulty of identifying mutated genes at the genomic level, much of the research in this field uses metabolomes, proteomes, and transcriptomes to indirectly investigate the molecular mechanisms underlying leaf chlorosis in tea. A joint assessment of proteomes and metabolomes revealed one potential blocked site of chlorophyll synthesis in etiolated leaves of *Huangjinya*: from protochlorophyllide (Pchlide) into chlorophyllide a (Chlide a). The significantly lowered protein abundance of protochlorophyllide oxidoreductase (POR) in the etiolated leaves was considered to be responsible for the excessive accumulation of the substrate (MgP IX) in the previous step [9]. While the metabolic flow from coproporphyrinogen III (Coprogen III) to Proto IX aligns with the trend of the excessive accumulation of substrates and a reduced product content, the abundance of protein (coproporphyrinogen III oxidase, COPX/HEMF) mediating this catalytic reaction does not change in etiolated leaves [9]. The alterations in the content of Coprogen III and Proto IX are more likely influenced by other metabolic events. In addition, a dramatically lower level of CAO may further decrease the content of chlorophyll b (Figure 1) [9]. Pheophorbide a oxygenase (PAO/ACD1)-mediated chlorophyll breakdown is essential for the loss of green pigments in plant leaves [66]. Overactivated PAO/ACD1 in *Huangjinya* may enhance the chlorina phenotype. In the etiolated leaves of *Huangjinya*, proteome analysis demonstrates that the function of the photosynthetic chain is greatly damaged [9,36]. Fan et al. reported that most protein members of Photosystem I (PSI), PSII, and plastid quinone pool (PQ) are significantly down-regulated in yellowish leaves [9]. In another study, differentially expressed protein (DEP) analysis also indicates an impaired photosynthetic chain, but the DEPs identified are less repetitive and reduced in number [36].

Proteome, acetyl-proteome, and succinyl-proteome analysis in *Anjibaicha* uncovered an enrichment of differentially expressed or modified proteins in the photosynthetic pathway during the whitening and re-greening processes of *Anjibaicha*, suggesting that the destruction and reconstruction of the photosynthetic chain function is closely associated with the change in leaf color [8,46,67]. In support of this, transcriptomic analysis also identified enriched down-regulated differentially expressed genes (DEGs) in the photosynthetic chain and chlorophyll metabolism, respectively, including eight and six genes [35]. Among them, the antenna protein LHCA4 exhibits a consistently lower abundance in etiolated leaves at the transcriptional, protein, and protein modification levels [8,35,46].

In most other albinistic or etiolated varieties, the evidence at the protein or transcriptome level supports one or more of the processes of chlorophyll metabolism, the photosynthetic chain, and carotenoid synthesis being impacted in chlorosis leaves (Table 1). Understandably, blocked biosynthesis of chlorophyll and carotenoids and activated chlorophyll degradation contribute to the lightened color of the leaves [68,69]. While not all genes in the biosynthetic pathways of chlorophyll and carotenoids are uniformly down-regulated (Table 1), the reduced expression of crucial upstream genes may explain the decreased accumulation of total chlorophyll or carotenoids in *Anjibaicha* (*POR* and *PBGD/HMBS/HEMC*) [35] and the albinistic branches of *Huangshan* (*CHLP* and *POR*) [48], *Huangjinshuixian* (*DXS* and *GGPPS*) [16], and *Zhonghuang 3* (*GluTR/HEMA, GSA-AM/HEML, UROD/HEME, HEMF/CPOX*, and *CHLP*) [14].

In plants, the biosynthesis and degradation of chlorophyll are highly coordinated with the structural and functional integrity of the photosynthetic chain [28]. The LHC proteins associated with the antenna complex are responsible for light harvesting and Chl a/b binding [70]. Incompletely assembled photosystems will hinder the localization of antenna proteins to the thylakoids and will cause reduced contents of LHC-bound chlorophyll [71]. As free chlorophyll is photosensitive, the absence of LHCB1 and LHCB2 can lead to reduced chlorophyll accumulation [72,73]. In various chlorina tea varieties, the functional modules (antenna complex [8,9,13,29,30,40,41,48,52,53], photosystems [8,9,12,17,48,53], quinone pools [36], cytochrome b6/f complexes [9,40], and ATPase complexes [8,9,48,53]) of the photosynthetic chain have been differentially impaired (Table 1). The reduction in antenna protein abundance and the photooxidative stress triggered by photosynthetic electron transfer defects may be common factors leading to or exacerbating leaf chlorosis in these tea variations [74].

Reduced chlorophyll content will disrupt photosynthetic chain stability. Light-induced thylakoid biogenesis is a prerequisite for the assembly of the photosynthetic chain, which is highly coordinated with the reduction of Pchlide into chlorophyllide (Chlide) in the penultimate step of chlorophyll biosynthesis [73]. In addition, the protein stability of the LHCII members is modulated by chlorophyll [64]. In *Arabidopsis cao1* mutant chloroplasts, down-regulated chlorophyll b promotes the proteolysis metabolism of the LHCII proteins [75]. Therefore, it is still difficult to distinguish whether a reduced chlorophyll content or an impaired photosynthetic chain is more likely to cause leaf chlorosis in tea according to the available omics results.

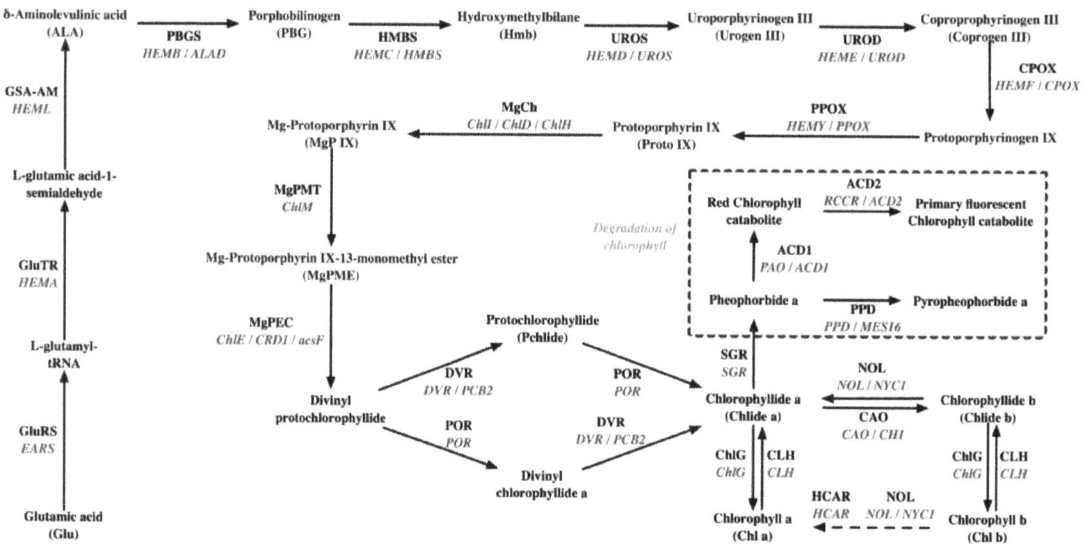

Figure 1. Proteins and genes involved in chlorophyll synthesis pathway. Next to the arrow, the word above represents the protein name, and the corresponding gene name is below. GluRS, glutamyl-tRNA synthetase; GluTR, glutamyl-tRNA reductase; GSA-AM, glutamate-1-semialdehyde 2,1-aminomutase; PBGS, porphobilinogen synthase; HMBS, hydroxymethylbilane synthase; UROS, uroporphyrinogen-III synthase; UROD, uroporphyrinogen decarboxylase; CPOX, coproporphyrinogen III oxidase; PPOX, protoporphyrinogen III oxidase; MgCh, magnesium chelatase; MgPMT, magnesium-protoporphyrin O-methyltransferase; MgPEC, magnesium-protoporphyrin IX monomethyl ester (oxidative) cyclase; DVR, divinyl chlorophyllide a 8-vinyl-reductase; POR, protochlorophyllide reductase; NOL, chlorophyll(ide) b reductase; CAO, chlorophyllide a oxygenase; ChlG, chlorophyll/bacteriochlorophyll a synthase; CLH, chlorophyllase; HCAR, hydroxymethyl chlorophyll a reductase; SGR, magnesium dechelatase; PAO/ACD1, pheophorbide a oxygenase; ACD2, red chlorophyll catabolite reductase; PPD, pheophorbidase.

Multi-omics technologies have identified potential transcriptional factors involved in regulating the chlorophyll metabolism in chlorina tea plants [16,39,50,51]. Both transcriptome and translatome analyses reveal that *ELONGATED HYPOCOTYL 5* (*HY5*) (Figure 1), a bZIP-type transcription factor (TF) involved in photomorphogenesis, is up-regulated in *Xiangfeihuangye* [39]. HY5 serves a conserved role in negatively regulating chlorophyll synthesis [76]. Gene expression analysis has uncovered that *HY5* is negatively correlated with genes associated with light harvesting and chlorophyll biosynthesis in tea plants. The expression of HY5 is inhibited, thereby relieving its inhibition of chlorophyll synthesis during the seasonal re-greening of *Xiangfeihuangye* [39]. During the seasonal re-greening of *Huangkui*, the positive regulation of genes related to light harvesting and chlorophyll biosynthesis due to increased *REVEILLE1* (*RVE1*) expression has been confirmed by transcriptome findings and other molecular biological results [51]. In the future, it will be necessary to determine whether there are genomic variations in these TF genes and characterize their genetic contributions to the formation of chlorina phenotypes in tea plants.

2.3. Physiological Mechanisms of Purple Leaf Coloration in Tea

Two main types of purple coloring behavior have been documented in tea plants: at the developmental stage and due to ecological-factor-induced purple pigment accumulation [19]. In purple varieties such as *Zijuan* [77], *Zixin* [78], and *Wuyiqizhong18* [79], purple pigments accumulate in the young leaves and stems but not in mature tissues. In some ecologically dependent varieties, such as *Longjing43* [80] and *TRFK 306* [81], the biosynthesis of purple pigments is triggered by environmental stress, such as temperature, water, and light.

Numerous studies have determined that water-soluble anthocyanins are the primary purple pigments in tea [19]. As illustrated in Table 2, glycosylated or acylated derivatives of cyanidin, delphinidin, petunidin, pelargonidin, peonidin, and malvidin can be detected in purple tea leaves [77,78,80,82–85]. Among these metabolites, only cyanidin-3-O-galactoside is the major anthocyanin composition in nearly all the studied purple tea varieties [77,78,80,82–85]. Apart from cyanidin-3-O-galactoside, delphinidin-3-O-galactoside, cyanidin-3-O-glucoside, cyanidin-3-rutinoside, and delphinidin-3-O-glucoside are another four anthocyanin components in tea identified with the second highest frequency in various environments and different varieties (Figure 2) [81,83,84].

Table 2. A summary of anthocyanin composition in purple leaves of tea.

Tea Varieties	Sampling Location	Measurement Technique	Anthocyanin Composition	References
Zijuan	Tea garden of South China Agricultural University, Guangzhou, China	HPLC	Major anthocyanin compositions: cyanidin-3-*O*-galactoside and delphinidin-3-*O*-galactoside.	[77]
	Tea garden of the Institute of Tea Science, Yunnan Province Academy of Agricultural Sciences (Menghai, China)	Non-targeted metabolomics approach: UHPLC–Orbitrap–MS/MS	Cyanidin 3-diglucoside 5-glucoside, cyanidin 3-*O*-(6-*O*-*p*-coumaroyl) glucoside, cyanidin 3-sambubioside, cyanidin 3-(6″-acetylglucoside)-5-glucoside, delphinidin 3-(6-*p*-coumaroyl) galactoside, delphinidin-3-*O*-arabinoside, pelargonidin 3-sophoroside 5-glucoside, pelargonidin 3-coumarylglucoside-5-acetylglucoside, pelargonidin 3-rhamnoside 5-glucoside; compared with *Yunkang*, the contents of cyanidin 3-diglucoside 5-glucoside and pelargonidin 3-sophoroside 5-glucoside are most increased in *Zijuan*.	[86]

Table 2. *Cont.*

Tea Varieties	Sampling Location	Measurement Technique	Anthocyanin Composition	References
Zijuan	Pu'er City Institute of Tea Science, Yunnan Province	UPLC–ESI–MS/MS metabolomic analysis	Specific metabolites: petunidin 3-O-glucoside, peonidin 3-O-glucoside chloride, peonidin 3-O-glucoside, peonidin O-hexoside, malvidin 3-O-glucoside (oenin), petunidin 3,5-O-diglucoside. Marker metabolites: cyanidin 3-O-galactoside, cyanidin 3-O-glucoside (Kuromanin), delphinidin 3-O-glucoside (Mirtillin), pelargonidin 3-O-glucoside	[87]
	Changsha, Hunan, China	UPLC–ESI–MS/MS metabolomic analysis	Major anthocyanin compositions: cyanidin-3-ogalactoside, delphinidin-3-O-galactoside, and petunidin-3-O-galactoside	[88]
	Dechang Fabrication Base of Shucheng County in Anhui Province, China	LC–TOF–MS	Cyanidin-3-O-galactoside, Cyanidin 3-O-(6-O-p-coumaroyl) galactoside, Delphinidin 3-O-(6-O-p-coumaroyl) galactoside, Delphinidin-3-O-galactoside.	[89]
Zijuan Ziyan and *Chuanzi* (ZZ)	Muchuan County, Sichuan Province, China	Targeted UPLC–ESI–MS/MS analysis	A total of 22 anthocyanins with a content ≥1 μg/g (DW) were detected in *Chuanzi*, *Ziyan*, and/or *Zijuan* and these included 6 cyanidins, 7 delphinidins, 5 pelargonidins, 2 peonidins, and 2 petunidins. In addition, 23 anthocyanins with a concentration of <1 μg/g were also detected.	[82]
Ziyan	Planted in plastic pots	HPLC	Delphinidin, cyanidin, and pelargonidin.	[90]
Hongyecha, *Zijuan*, *9803*, *Hongyafoshou*	Changsha, Hunan, China	UPLC–DAD–QTOF–MS	Cyanidin-(E)-p-coumaroylgalactoside, cyanidin-3-O-galactoside, delphinidin-3-O-galactoside, delphinidin-(Z)-p-coumaroylgalactoside, delphinidin-(E)-p-coumaroylgalactoside, pelargonidin-O-hexose, and pelargonidin-O-dihexose.	[91]
Jinmingzao	Tea plantation of Wuqu in Fuan City, Fujian Province, China	Widely targeted metabolomics: UPLC–ESI–MS/MS	Cyanidin 3-O-glucoside, cyanidin 3-O-galactoside, cyanidin 3-rutinoside, cyanidin chloride, delphinidin 3-O-glucoside, peonidin 3-O-glucoside chloride (most affected).	[83]
Zikui	South Campus of Guizhou University, Huaxi District, Guiyang City, Guizhou Province, China	ESI–QTRAP–MS/MS	Cyanidin 3-O-galactoside, cyanidin 3-O-glucosid, petunidin 3-O-glucoside.	[84]
Longjing43	Tea Research Institute, Chinese Academy of Agricultural Sciences, Hangzhou, China	LC–MS/MS	Delphinidin-hexose-coumaroyl showed the greatest increase.	[80]
TRFK 306	Tea Research Institute (TRI), Kericho County, Kenya	HPLC	Malvidin 3-glucoside, peonidin 3-glucoside, pelargonidin 3,5-O-diglucoside, cyanidin 3-O-glucoside, cyanidin 3-O-galactoside, cyanidin 3-O-rutinoside.	[81]

Table 2. *Cont.*

Tea Varieties	Sampling Location	Measurement Technique	Anthocyanin Composition	References
9 tea cultivars possessing purple leaves	Wuxi Institute of Tea Varieties in Wuxi City, Jiangsu Province, China	Widely targeted metabolomics: UPLC–ESI–MS/MS	Thirty-three anthocyanins were identified, and delphinidin 3-O-galactoside and cyanidin 3-O-galactoside were found to be the most abundant in PTLs.	[85]
Unknown	Experimental tea farm (IHBT-269) of CSIR—Institute of Himalayan Bioresource Technology, HP, India	UHPLC	3-O-alpha-l-arabinopyranosylproantho cyanidin A5′ and 3,3′-Di-O-galloylprocyanidin B.	[92]

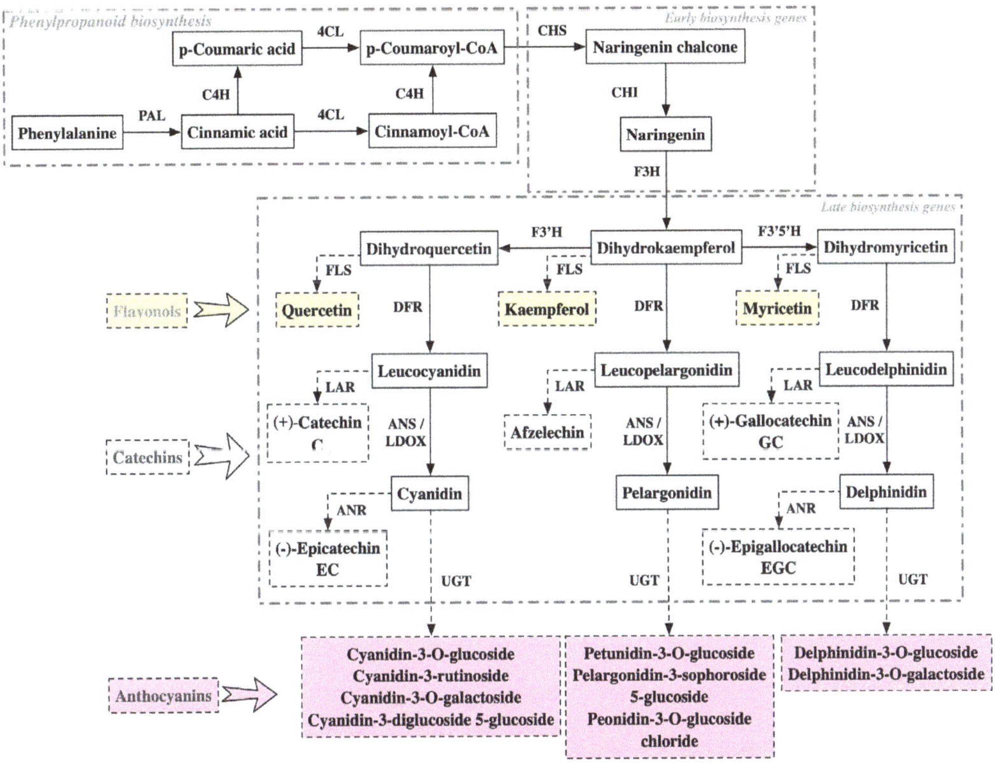

Figure 2. The synthesis process of flavonoids and anthocyanins. The yellow textboxes represent the flavonols, the pale green textboxes represent the catechins, and the purple textboxes represent the decorated anthocyanins. PAL, phenylalanine ammonia lyase; C4H, cinnamate-4-hydroxylase; 4CL, 4-coumarate: CoA ligase; CHS chalcone synthase; CHI chalcone isomerase; F3H, flavanone 3-hydroxylase; F3′H, flavonol 3′-hydroxylase; F3′5′H, flavonol 3′5′-hydroxylase; FLS, flavonol synthase; DFR, dihydroflavonol 4-reductase; LAR, leucoanthocyanidin reductase; ANS/LDOX, leucoanthocyanidin dioxygenase; ANR, anthocyanin reductase; UGT, UDP-glucose: flavonol-3-O-glycosyltransferase.

Besides the genetic factors, environmental variations also influence the anthocyanin composition [93]. For instance, in the most studied purple variety, *Zijuan*, delphinidin-3-O-galactoside is detected at high levels in five different geographical environments but not in 'Pu'er' [77,82,87–89,91]. The components of anthocyanin identified also depend on the detection strategy. In the study conducted by Tan et al., a total of 45 differentially accumulated anthocyanins were detected using targeted UPLC–ESI–MS/MS analysis in *Zijuan*, *Ziyan*, and *Chuanzi* [82]. This study identified the most diverse types of anthocyanins in *Zijuan* [82]. In addition, both of the two most abundant anthocyanins, cyanidin-3-O-galactoside and delphinidin-3-O-galactoside, cannot be detected in the steamed tea of *Zijuan*, suggesting that tea processing also influences the output of anthocyanin composition analysis substantially [82,86].

2.4. Molecular Mechanisms of Purple Coloration in Tea Leaves

Multi-omics studies have uncovered a range of differentially expressed genes or differential alternative splicing genes linked to anthocyanin biosynthesis, anthocyanin transportation, anthocyanin degradation, and transcriptional regulation (Table 3). In flowering plants, MYB TFs assemble into a well-known ternary complex MBW with members of the WD40 and bHLH families to modulate the expression of various structural genes involved in anthocyanin biosynthesis [94,95]. The activation of R2R3-MYB transcription factors, including *AN1/MYB75* [82,96], *MYB114* [24], and *MYB90* [84], is documented in most studies related to purple tea, suggesting that the MBW complex may possess a common and conserved transcriptional regulatory role in tea anthocyanin synthesis. Consistently, the molecular experimental evidence has verified this hypothesis. Performing the overexpression of *MYB90* from *Zikui* in the calluses of tobacco can strongly induce anthocyanin synthesis [84]. Moreover, Sun et al. reported that AN1/MYB75 could physically interact with bHLH TFs (GL3 and EGL3) to recruit the WD40 protein TTG1 to induce a large amount of anthocyanin accumulation in *Nicotiana benthamiana* leaves [77]. At the genomic level, a recent study has identified a 148 bp LTR/*Gspsy* insertion into the upstream region of *MYB114* in *Zijuan*, *Jinguanyin*, and *Jinmingzao*, which provides potential genomic clues supporting the high expression of the *MYB* gene in these purple varieties [24].

Table 3. A summary of multi-omics-generated molecular evidence related to purple leaves in tea.

Tea Varieties	Omics Approaches	Potential Molecular Mechanisms	References
Zijuan	Transcriptome	Transcriptional regulation: Activation of the R2R3-MYB transcription factor (TF) *anthocyanin1* (*CsAN1*) and the bHLH TF *CsGL3*; CsAN1 interacts with bHLH TFs (CsGL3 and CsEGL3) and recruits a WD-repeat protein CsTTG1 to form the MYB-bHLH-WDR (MBW) complex that regulates anthocyanin accumulation. Late biosynthetic genes (LBGs): activation of *CsF3′H*, *CsF3′5′H*, *CsDFR1*, *CsDFR2*, *CsANS1/LDOX1*, *CsANS2/LDOX2*, and *CsANS3/LDOX3*. Metabolic substrate competition: activation of *CsLAR1*, *CsLAR2*, and *CsLAR3*, which encode enzymes for catechin biosynthesis, was highly expressed in red foliage.	[77]
	Transcriptome, proteome	Phenylpropanoid metabolism: significantly increased expression of three *PALs* (*CSA016076, 022024, 022025*); significantly decreased expression of *4CL* (*CSA001434*). Early biosynthesis genes (EBGs): significantly increased expression of *CHS* (*CSA029775*); significantly decreased expression of *CHI* (*CSA008261*). LBGs: significantly increased expression of two *DFRs* (*CSA003949, XLOC_010242*), one *ANS/LDOX* (*CSA011508*), six *UGT75L12*/13 (*CSA005544, 005545, 010001, 036671, 036672, 029026*), and two *UGT94P1* (*CSA007394, 008750*); significantly decreased expression of *F3′5′H* (CSA031792), *ANS/LDOX* (*CSA035767*), two *UGT75L12s* (*CSA008693, 028873*), and two *UGT94P1s* (*CSA005965, 026000*). Metabolic substrate competition: significantly increased expression of two *LARs* (*CSA014943, XLOC_016774*).	[87]

Table 3. *Cont.*

Tea Varieties	Omics Approaches	Potential Molecular Mechanisms	References
Zijuan	Transcriptome	Phenylpropanoid metabolism: activation of *C4H*. LBGs: activation of *ANS/LDOX*, *UGT*. Chlorophyll degradation: activation of *CLH1*.	[97]
	Full-length transcriptome	Alternative splicing (AS) events identified in transcriptional regulation (*MYB113-1*), phenylpropanoid metabolism (*C4H1*, *PAL2*), LBGs (*UDP75L122*), and metabolic substrate competition (*FLS1*).	[98]
	Proteome	EBGs: increased abundance of CHS and CHI. LBGs: increased abundance of DFR, ANS/LDOX, and UGT. Anthocyanin transportation: increased abundance of ABC transporter B8.	[20]
	Transcriptome	Transcriptional regulation: Most of the members belonging to the MYB, WRKY, AP2, GRF, bZIP, and MYC groups had a higher expression in *Zijuan*. LBGs: significantly increased expression of *F3′5′H* (*CSS0022212.1*), *ANS/LDOX* (*CSS0010687.1*), *3GT* (*anthocyanidin 3-O-glucosyltransferase*, *CSS0024320.1*), *3AT* (*cyanidin-3-O-glucoside 6″-O-acyltransferase*, *CSS0015285.1*). Metabolic substrate competition: significantly decreased expression of *LAR* (*CSS0009063*.1). Anthocyanin degradation: *polyphenol oxidase* (*PPO*, *CSS0002951.1*), showed negative correlation with the three anthocyanins, especially delphinidin and delargonidin.	[89]
Chuanzi (ZZ)	Transcriptome	Transcriptional regulation: significantly increased expression of the well-known MYB transcription factor *CsAN1/CsMYB75* (*CSS0030514*). LBGs: significantly increased expression of *CsANSs/LDOXs* (*CSS0010687*, *CSS0018498* and *CSS0046216*), *CsUGT94P1* (*CSS0011196*), and the *anthocyanin O-methyltransferase* gene (*CsAOMT*, *CSS0015915*). Anthocyanin transportation: significantly increased expression of *CsGSTF1* (*CSS0022086*) and three other GST candidate genes (*CSS0031248*, *CSS0026690*, and *CSS0018634*) tightly linked to CsGSTF1. Metabolic substrate competition: down-regulated expression of *LARs* (*CSS0028235* and *CSS0009063*) and *ANRs* (*CSS0005927*, and *CSS0033195*).	[82]
Zijuan, *Jinguanyin* and *Jinmingzao*	Pangenome	Read depth of the LTR insertion region in the promoter of *CsMYB114* among a set of representative purple-leaf cultivars (*'ZJ'*, *'JMZ'*, and *'JGY'*) and tea cultivars with green leaves (*'FDDB'*, *'BHZ'*, and *'GH3H'*)	[24]
Ziyan	Transcriptome	Transcriptional regulation: UV-A induces the expression of the regulatory gene *TT8*; UV-AB induces the expression of the regulatory genes *EGL1* and *TT2*. LBGs: UV-A induces the expression of *F3H*, *F3′5′H*, *DFR*, and *ANS/LDOX*; UV-AB induces the expression of *F3′5′H*, *DFR*, *ANS/LDOX*, and *UGT*. Metabolic substrate competition: UV radiation repressed the expression levels of *LAR*, *ANR*, and *FLS*, resulting in reduced ANR activity and a metabolic flux shift towards anthocyanin biosynthesis.	[90]

Table 3. *Cont.*

Tea Varieties	Omics Approaches	Potential Molecular Mechanisms	References
Wuyiqizhong18	cDNA-AFLP	EBGs: increased expression of *CHS*. LBGs: increased expression of *AT* (*TDF #3341_2f*) and *UGT* (*TDF #2421_1d* and *TDF #2411_1f*).	[79]
	Proteome	EBGs: increased abundance of CHS and CHI. Metabolic substrate competition: increased abundance of FLS.	[99]
Jinmingzao	Transcriptome	Phenylpropanoid metabolism: activation of *PAL*, *C4H*, and *4CL*. LBGs: activation of *DFR*, *ANS/LDOX*, and *UGT* (*TEA004632* and *TEA004632*) genes.	[83]
Longjing43	Transcriptome	Transcriptional regulation: activation of *MYB75*. LBGs: activation of *ANS/LDOX* and *3-GT*. Anthocyanin transportation: activation of genes involved in anthocyanin transportation (*GST*, *glutathione S-transferase*).	[96]
	Transcriptome	Phenylpropanoid metabolism: activation of *PAL* and *C4H* by high temperature and/or light levels in summer. EBGs: activation of *CHI* and *CHS* by high temperature and/or light levels in summer. LBGs: activation of *ANR*, *ANS/LDOX*, and *DFR* by high temperature and/or light levels in summer. Metabolic substrate competition: activation of *FLS* and *LAR* by high temperature and/or light levels in summer.	[80]
TRFK 306	Transcriptome	Transcriptional regulation: transcripts encoding pathway regulators of the MYB–bHLH–WD40 (MBW) complex were repressed, possibly contributing to the suppression of late biosynthetic genes of the pathway during the dry season. Anthocyanin transportation: suppression of anthocyanin transport genes could be linked to reduced accumulation of anthocyanin in the vacuole during the dry season.	[81]
Zikui	Transcriptome	Transcriptional regulation: *CsMYB90* showed strong correlations with petunidin 3-O-glucoside, cyanidin 3-O-galactoside, and cyanidin 3-O-glucosid. LBGs: activation of two *F3′H* genes and two *ANS/LDOX* genes. Anthocyanin degradation: three negatively correlated *PPO* (*polyphenol oxidase*) genes with anthocyanin accumulation.	[84]
Hongyecha, Zijuan, 9803, Hongyafoshou	Transcriptome	Phenylpropanoid metabolism: activation of *4CL*. LBGs: activation of *ANS/LDOX* and *UGT*.	[88]

Anthocyanin biosynthesis begins with phenylalanine and is catalyzed by a series of enzymes involved in phenylpropanoid metabolism, the early flavonoid biosynthesis process, and the late flavonoid biosynthesis process [77]. The transcriptome and proteome data reveal a consistent up-regulation of the late biosynthesis gene (LBG) *anthocyanin synthase* (*ANS*) in all examined purple tea leaves (Table 3). ANS/LDOX catalyzes colorless leucoanthocyanins to form anthocyanidin, operating as the penultimate step of anthocyanin biosynthesis [100]. Potential binding sites of MYB transcription factors (TFs) are identified in the promoter regions of *ANS/LDOX* genes. Yeast one-hybrid assays have verified that the MYB TF AN1 can directly bind to the promoter of ANS/LDOX genes, indicating that ANS/LDOX plays a vital role in the MBW-regulated anthocyanin synthesis pathway [77].

Additionally, the up-regulation of certain other genes classified as LBGs, including flavonoid-3′-hydroxylase (F3′H), flavonoid 3′,5′-hydroxylase (F3′5′H), dihydroflavonol-4-reductase (DFR), and UDPG-flavonoid glucosyltransferase (UGT) are identified in the purple leaves of some tea varieties. However, the specific gene members identified vary from study to study (Table 3) [77,78,80,82–85]. A similar gene expression change behavior is observed for the genes involved in the substrate competition pathways in anthocyanin biosynthesis (Table 3). In particular, flavonoid synthase (FLS), leucoanthocyanidin reductase (LAR), and anthocyanidin reductase (ANR) are responsible for catalyzing delphinidin, leucodelphinidin, and delphinidin to produce flavonols, catechin, and epicatechin, respectively [77]. The deactivation of one or more metabolic substrate competition pathway in anthocyanin biosynthesis is differentially observed in various purple-leaf varieties [80,82,89,90,99].

Additionally, the multi-omics data have identified several specific auxiliary mechanisms of anthocyanin accumulation in the purple leaves of various tea varieties. For example, in *Zijuan* and *Zikui*, the down-regulated expression of genes associated with anthocyanin degradation (*Polyphenol oxidase, PPO*) could contribute to anthocyanin accumulation (Table 3) [84,89]. The activated transportation of anthocyanins into the vacuoles can limit their degradation and help to increase their accumulation [101,102]. *Zijuan* and *Chuanzi* each have an up-regulated anthocyanin transporter protein, ABC transporter B8 [20] and glutathione S-transferase, respectively [82]. The gene expression changes in these anthocyanin transporters in response to ecological factors are consistent with the change trends in the anthocyanin content [81,96]. Therefore, protecting anthocyanin from degradation may be an essential synergistic mechanism to increase the accumulation of anthocyanins in purple tea leaves.

3. Metabolic Reprogramming by Leaf Color Variations in Tea Plants

3.1. Influence of Chlorina Variations on Amino Acid Metabolism

In the albinistic or etiolated leaves of chlorina tea varieties, the formation of certain flavors is closely related to the changes in tea theanine and polyphenol levels. Generally, the content of total free amino acids is much higher in chlorina leaves than in green leaves from re-greening or green-leaf plants (Figure 3) [32,40,43,48,49,52,53,103]. As the most abundant free amino acid in tea leaves, alteration in the absolute content of theanine contributes heavily to the change in the total free amino acid [32,40,43,48,49,52,53,103]. Additionally, glutamic acid, glutamine, and aspartic acid are three other free amino acids with high levels after theanine in tea leaves. The accumulation of these three amino acids is also dramatically enhanced in chlorina leaves (Figure 3) [32,40,43,48,49,52,53,103].

However, not all chlorina variations demonstrate an elevated accumulation of free amino acids. For instance, the levels of total free amino acids in the etiolated leaves of *Huangjinju* only increase slightly compared to re-greening leaves, and the differences do not reach the significance level [13]. In another chlorina variety, *Baijiguan*, the levels of free amino acids in the etiolated leaves are not always higher than that in the green-leaf control, *Longjin43*, and the conclusions drawn can even be the opposite [104]. In addition, no significant differences are observed in the total amino acids between the etiolated and green half-siblings generated from *Baijiguan* and *Longjin43* [15]. One possible explanation for these findings, which contradict our understanding that chlorina enhances the accumulation of free amino acids, especially theanine, is that the change in amino acid accumulation in some green-leaf varieties is more drastic than that in chlorina varieties under specific environments. One piece of evidence that supports this hypothesis is that the levels of theanine in the etiolated variety *Yujinxiang* are higher and lower than that in *Shuchazao* under non-shaded and shaded conditions, respectively [33].

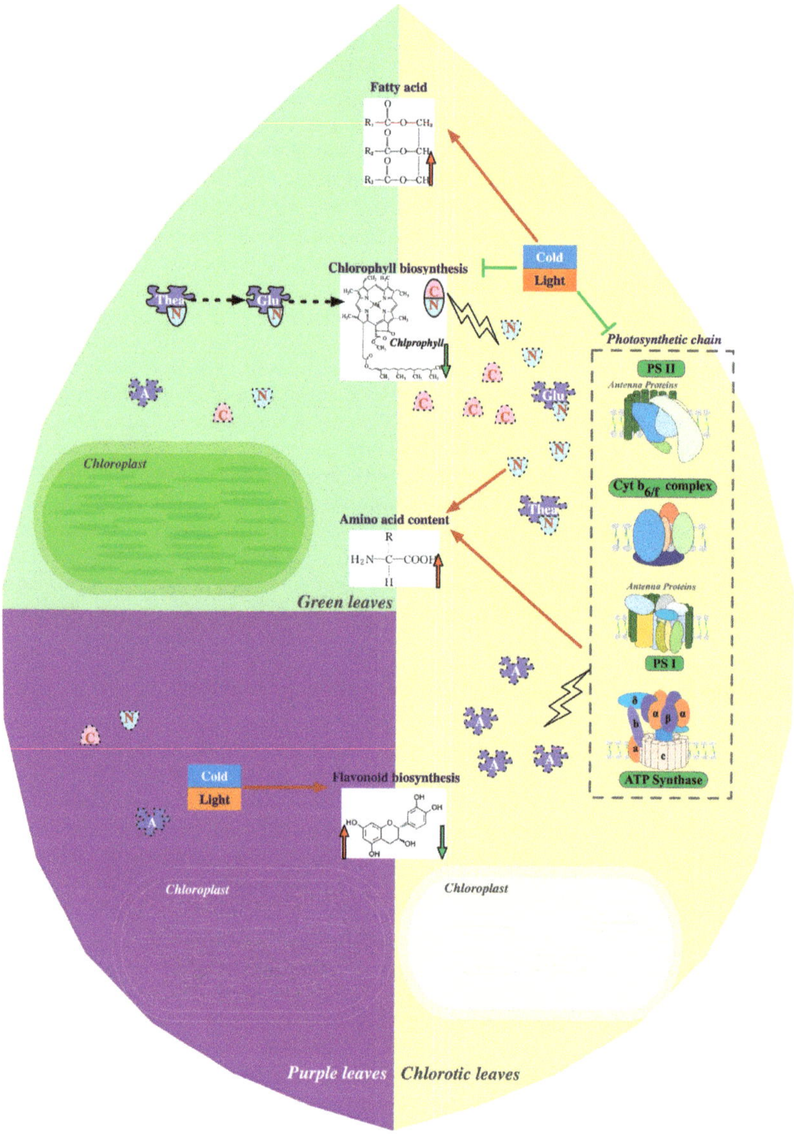

Figure 3. Red/green frameless arrows represent a promoting/inhibitory effect, while red/green framed arrows represent an increase/decrease in substances. Substances surrounded with a dashed outline are in a free state. Compared to green varieties, chlorotic varieties are hindered in chlorophyll synthesis and degradation under strong light or low-temperature conditions, resulting in imbalanced carbon (C) and nitrogen (N) metabolism and reduced flavonoid content in the chlorotic leaves. Excess free nitrogen promotes the accumulation of free amino acids in chlorotic leaves. On the other hand, the loss of chloroplast structure and degradation of photosynthetic-chain-related proteins further increase the content of free amino acids in chlorotic leaves. In addition, the synthesis of fatty acids has also increased in chlorotic varieties. In purple varieties, low temperature or strong light enhances the expression of flavonoid-pathway-related genes, promoting the synthesis of more anthocyanins.

The aforementioned clues indicate that the changes in theanine content and total free amino acids are consistent in most cases [32,40,43,48,49,52,53,103]. Therefore, many studies mainly focus on the molecular mechanisms underlying the differences in theanine content among different leaf color materials. The current multi-omics evidence mainly provides three plausible explanations for why leaf chlorosis can elevate the theanine levels in albinistic or etiolated tea varieties. The first possible explanation may be that the expression levels of the genes involved in theanine synthesis or degradation are activated or repressed (Figure 3). Glutamate and ethylamine serve as precursors of theanine biosynthesis in tea plants, and this enzyme reaction is catalyzed by theanine synthetase (TS), a member of the glutamine synthetases (GSs) [2]. The other GS members may also have a certain TS activity [105]. In *Anjibiacha*, *HY*, and *Zhonghuang 3*, transcriptome analysis has demonstrated that the expression levels of some TS or GS genes are positively correlated with the theanine content, suggesting enhanced TS activity in these varieties [14,43,48,52,106]. In addition, the up-regulation of glutamate synthase (GOGAT), glutamate dehydrogenase (GDH), alanine decarboxylase (AlaDC), and alanine transaminase (ALT) in chlorina leaves may increase the abundance of glutamate and ethylamine and promote theanine biosynthesis [42,49,52,106]. In *Zhonghuang 3*, the mRNA abundance of two genes involved in theanine hydrolysis, *PDX2* and its homolog *glutamine amidotransferase 5* (*GGP5*), is significantly down-regulated in the etiolated leaves, potentially influencing theanine accumulation [49,107].

Theanine is not only the most abundant free amino acid in the tea plant but also acts as the primary nitrogen storage source. A second possible reason for an increased theanine content is reduced nitrogen consumption in chlorina leaves. In higher plants, the theanine synthesis substrate glutamate plays a central hub role in nitrogen metabolism and is an important link connecting the carbon and nitrogen metabolic balance [108]. Importantly, the photosynthetic pigment chlorophyll's biosynthesis depends on glutamate as the nitrogen source. A much lower chlorophyll content in chlorina tea leaves may decrease the nitrogen consumption, contributing to a higher glutamate accumulation. Another indirect piece of evidence supporting this hypothesis is the observation of an increased NH_4^+ content in chlorina leaves [16,32]. As the biosynthesis of glutamate is a primary pathway that consumes NH_4^+, the accumulation of ammonium ions also supports repressed nitrogen utilization in albino leaves [32]. Based on this, we speculate that the reduction in the use of glutamic acid, an important organic nitrogen source, causes more nitrogen to be stored in the form of theanine in chlorina tea leaves.

Thirdly, the activation of protein degradation may be a potential cause of highly accumulated free amino acids, including theanine, in chlorina tea leaves [16,18,32,50,52]. Via transcriptome analysis, a group of up-regulated genes involved in the ubiquitin–proteasome system has been identified in chlorina tea leaves [16,18,32,50,52]. Specifically, the expression levels of associated genes encoding E3 ubiquitin ligase, ubiquitin, damaged DNA binding protein, and ubiquitin-specific protease are enhanced to varying degrees in *Yanlingyinbiancha* and *Fuhuang 1* [18,52]. In *Koganemidori*, the autophagy pathway, critical to organelle degradation, is also up-regulated, indicating that abnormally developed chloroplasts may be directed to the recycling system [50]. The aforementioned significant decrease in protein abundance of the photosynthetic chain may be attributed to the combined effects of protein degradation and chloroplast recycling. Accordingly, more free amino acids will be released.

3.2. Effects of Chlorina Variation on Flavonoid Metabolism

Tea polyphenols are an expansive group of flavonoids, including flavanols, flavanones, flavonols, leucoanthocyanidins, and anthocyanins [109]. In non-purple tea, flavanols represent the most abundant flavonoids (accounting for 60–80%) and are mainly composed of catechins.

Metabolome analysis reveals that the total flavonoids in chlorina varieties, including *Anjibaicha*, *Huangjinya*, *HY*, *Xiangfeihuangye*, *Huangjinju*, and *Zhonghuang 3*, are down-regulated [10,13,37,39,43,49]. At the molecular level, down-regulated expression levels of genes involved in phenylpropanoid biosynthesis and early biosynthesis pathways (Figure 2) may limit the metabolic flow to flavonoid biosynthesis. For instance, the transcript abundances of *PAL* in *Zhonghuang 3*, *CHI* in *Zhonghuang 3* and *Yujinxiang*, *CHS* in *HY* and *Yujingxiang*, and *F3H* in *Xiangfeihuangye* and *Yujingxiang* 1 are positively related to flavonoid contents. In *Huangjinya*, decreased levels of PAL and 4CL proteins have been detected using proteome analysis [36]. Genes in the phenylpropanoid metabolic pathway are not only regulated at the transcriptional level but also in terms of protein modification in *Anjibaicha* [8,10].

Because flavonoids account for over 10% of the dry matter in tea, flavonoid biosynthesis consumes a large number of photosynthetic products. Impaired photosynthesis in chlorina leaves may result in an insufficient carbon skeleton supply for flavonoid synthesis [37,39]. One compelling piece of evidence that supports this hypothesis is that decreased tea catechins are observed in the albinistic branches of normal tea plants but the expression levels of genes in the associated synthesis pathway are unaffected [48]. The lack of a carbon source may be a common cause for the limited polyphenol synthesis capacity in chlorina leaves.

There are some exceptions where the flavonoid content in chlorina varieties unexpectedly increases [19]. In *Yanlingyinbiancha*, a significantly higher catechin content is detected in albinistic tissues but not green tissues. The expression levels of genes involved in phenylpropanoid biosynthesis (*4CL*) and the early biosynthesis pathways (*CHI*) are up-regulated, consistent with elevated flavonoid contents. On the other hand, the up-regulation of *sucrose synthase* in the albinistic tissues of variegated leaves suggests that it may be able to supplement carbon sources from the surrounding green tissues [18]. Therefore, this counterexample supports the high-level expression of upstream structural genes and sufficient carbon sources, which are determinants for ensuring the normal synthesis of flavonoids in tea.

Comparative omics examinations show that differentially accumulated flavonoid compositions between chlorina and green leaves are complex and vary with the tested tea variety and growing environment [15,33,37,39,43]. For instance, the relative flavonol content in the albinistic leaves of *Anjibaicha* exhibits opposite trends across two independent studies, suggesting that the accumulation of certain flavonoids may be more impacted by environmental factors [10,103]. In most chlorina tea varieties, epigallocatechin (EGC) is a flavonoid component with roughly consistent content changes relative to the total flavonoid contents [10,13,15,43,50]. Causally, EGC and its gallate derivative (EGCG) account for most catechins and are more sensitive to changes in the supply levels of carbon sources.

At the molecular level, the expression changes in LBGs, such as *ANS*, *ANR*, and *FLS*, can only explain the changes in metabolic flux toward different subclasses of flavonoids (mainly catechins and flavonols) but cannot explain the changes in the contents of a single component within them [10,13,18,37,39,43,49]. One hypothesis is that the resistance of chlorina leaves to light stress is reduced, which may activate a stronger oxidative stress defense response and promote the accumulation of flavonoids with high antioxidant activity. In plants, some flavonoids containing dihydroxylated B-rings (e.g., quercetin) are accumulated in response to intense sunlight [110]. Also, in the chlorina leaves of both *Huangjinya* and *Yujinxiang*, the contrarian accumulation of quercetin supports the idea of an adaptive synthesis of flavonoids [33,37].

3.3. Effects of Chlorina Variations on Fatty Acid Metabolism

Leaf chlorosis also greatly influences the accumulation of lipids in tea [39]. In plants, lipids are critical backup energy storage substances [111]. Given that the accumulation of photosynthetic products is considerably damaged in chlorina leaves, these mutated tea variations may invoke fatty acids to elevate the supply of carbon sources by enhancing the activities of critical enzymes in lipid reuse. Transcriptional analysis has demonstrated that several genes involved in the β-oxidation of fatty acids are up-regulated in the albinistic tissues of *Yanlingyinbiancha*. Among these up-regulated DEGs related to fatty acid oxidation, *acyl-CoA oxidase* (*ACX*) and *peroxisomal 3-ketoacyl-CoA thiolase* (*PKT*) are responsible for the first and final steps of the β-oxidation process in plants [18,112,113], supporting the idea of enhanced energy release with fatty acids in albinistic tissues [18].

Another critical biological function of lipids is as the main components of chloroplast membranes [114]. Monogalactosyldiacylglycerol (MGDG) and digalactosyldiacylglycerol (DGDG) represent the two primary fatty acids in the lipid bilayer of the thylakoid membrane. An adequate ratio of MGDG to DGDG is necessary to maintain a stable thylakoid membrane [115]. In both *Arabidopsis* and soybeans, the MGDG:DGDG ratios are both approximately 2 [115,116]. And the composition of lipids in the chloroplast membrane is highly dynamic and can be modulated rapidly in response to environmental changes [116,117]. While in the etiolated leaves of *Baijiguan*, the ratio of MGDG to DGDG is relatively low, potentially supporting an observed damaged thylakoid membrane [111,118]. When subjected to shade treatment, the MGDG:DGDG ratio exhibits a trend of slightly increasing first and then decreasing. Although the differences among these shaded groups are not significant, the rapid response of the MGDG:DGDG ratio to shading may partially explain the improvement in chloroplast stability during the early re-greening stage [118]. In addition, more polar lipids are accumulated in the etiolated leaves of *Baijiguan* compared to re-greening leaves [118]. Similarly, the polar lipids involved in cell membrane assembly are significantly up-regulated in another etiolated mutant, *Huangjinshuixian*, compared to its green-leaf parent [16]. The findings of these two comparative analyses contradict the idea of chloroplasts acting as lipid synthesis locations, as the chloroplast function is more robust in the green leaves. According to limited information, we speculate that increased polar lipids may be a survival strategy for chlorina plants under normal light conditions. While reducing the light intensity down-regulates the expression of genes associated with polar lipid accumulation [118], the recovery of chloroplast function is insufficient to enhance polar lipid synthesis.

Epidermal wax is another crucial lipid type used by plant leaves to adapt to environmental stress [119,120]. In response to UV-B stress, *HY5* is activated and promotes the biosynthesis of epidermal wax by repressing the expression of a related negative regulator, *DEWAX*, in *Arabidopsis*. *HY5*-mutated or *DEWAX*-overexpressing plants are more sensitive to UV-B irradiation stress, supporting the idea of physical wax barrier protection in UV resistance [121]. Compared to green-leaf tea varieties, significantly increased epidermal wax is detected in *Huangjinshuixian* [15]. At the transcriptional level, increased expression of the *long-chain acyl-CoA synthetase* gene supports the activation of epidermal wax biosynthesis in the albinistic tissues of *Yanlingyinbiancha* [18]. These findings suggest that chlorina tea plants may adjust their lipid metabolism to enhance their adaptability to environmental stress.

3.4. Metabolic Changes in Flavonoids Other Than Anthocyanins in Purple Tea

As the primary accessory pigments, anthocyanins are the most well-documented flavonoids in purple tea. Along with enhanced anthocyanin biosynthesis, the levels of other flavonoid compositions have undergone substantial alterations in purple tea [21,78,86,89,91]. Unlike the significant increase in anthocyanins, the changing trend in the content of other flavonoids in purple tea relies on the control chosen for comparative analysis. For example, it is challenging to obtain similar findings when comparing the same purple variety with different green-leaf varieties. For catechins, the purple leaves of *Zijuan* exhibit significantly lower levels of C, GC, and EGCG and similar levels of EC, EGC, and ECG when compared to the green-leaf variety *Yunkang* [89]. Compared to another green-leaf tea variety, *Fudingdabai*, no significant difference in the contents of EC and significantly decreased EGCG contents are also observed in *Zijuan* [82]. However, the trends in the relative content change in EGC and ECG are quite different between these two different comparison group selection strategies [82,89]. The differences in differential metabolites are easy to understand because the accumulation behaviors of flavonoids across different green-leaf controls may also be different. Nevertheless, the consistent result between these two studies is that the total catechin in *Zijuan* is reduced compared to in *Fudingdabai* and *Yunkang* [82,89]. At the molecular level, the significantly lower level of *LAR* (Figure 2) in *Zijuan* may explain the decrease in metabolic flow toward catechin biosynthesis (Table 3) [89].

The levels of these non-purple flavonoids are also compared for two other groups of tea varieties, namely *Jinmingzao* compared to *Huangdan* and *Zikui* compared to *N61* [83,84]. The total catechin levels are higher in *Jinmingzao* and lower in *Zikui* when compared to the corresponding green-leaf variety [83,84]. Furthermore, the proanthocyanidin content is much lower in *Zikui* [84]. However, the opposite result is reported in another study, where monomeric catechin is lowered in purple tea varieties but the proanthocyanidin content increases [91]. In another comprehensive comparative study of nine purple tea varieties and three green-leaf tea varieties, a much higher level of catechins is more likely to be detected in purple tea [85]. These inconsistent results suggest the importance of selecting appropriate green-leaf controls when exploring metabolic alterations in purple leaves.

There are several studies comparing the metabolic differences between tender purple leaves and purple faded mature leaves [78,99,122]. In the purple tea variety *Wuyiqizhong18*, the contents of both total polyphenols and total catechins significantly decreased as the leaves matured and faded [99]. All five detected monomeric catechins were dramatically lowered in mature green leaves [99]. Higher expression levels of the EBGs in flavonoid synthesis in tender leaves support the idea of an elevated metabolic flux toward polyphenol accumulation [99]. Whereas, in the purple leaves of *Zijuan*, the abundance of EGCG is significantly lower, and the contents of EC, ECG, and EGC are much higher than those in green leaves [122]. According to the findings, the content alterations in the non-purple flavonoid components in diverse purple leaves do not exhibit a stable trend, potentially due to the highly complex regulation of genetic and environmental factors in this process.

4. Conclusions and Future Prospects

Leaf color variations give tea a unique appearance and nutritional quality. Under the current paradigm of homogenization in tea cultivars, varieties with leaf color variations typically have a higher economic value. Therefore, exploring the molecular mechanisms underlying leaf color formation will assist in the identification of mutation genes and also accelerate the utilization of leaf color variation varieties for tea breeding enhancement.

Using a systematic review of numerous omics investigations related to leaf color variation, we have concluded that the decrease in chlorophyll content and the accumulation of anthocyanin were the major common reasons for the color decoration of leaves in chlorina and purple-leaf tea varieties, respectively. In most chlorina tea varieties, the content and composition of carotenoids are altered, indicating that both photosynthetic activities and photooxidation protection capabilities are impaired in albinistic or etiolated leaves. The multi-omics results have also shown that decreased expression levels of critical genes in the biosynthesis of chlorophyll and carotenoids, activated chlorophyll degradation, and an impaired photosynthetic chain function are associated with the chlorina phenotype in tea plants. Decreased chlorophyll content not only impairs the structure and function of the chloroplasts but also leads to the degradation of the photosynthetic proteins that bind the plastid membrane. A decrease in the abundance of photosynthetic chain proteins also limits the content of bound chlorophyll and negatively influences its accumulation, as free chlorophyll is phytotoxic and unstable. Moreover, we have summarized that this tea plant chlorosis phenotype tends to result in the up-regulation of free amino acids and polar lipids and the inhibited accumulation of polyphenols. We have also proposed the possible causes of the changes in these bioactive metabolites in chlorina tea leaves. Overall, these metabolic changes in the chlorotic tissues embody a balance between leaf development and environmental adaptation, which fortunately endows the tea with a better flavor quality.

For purple-leaf tea varieties, the composition of purple-colored anthocyanin is affected by the tested environment and variety. The MBW complex plays a conserved upstream regulatory role in anthocyanin accumulation. The enhanced expression of LBG genes, especially those involved in the final two steps of anthocyanin synthesis, and positive changes in the genes related to the stable storage of anthocyanins play crucial roles in the formation of purple tea leaves. The decreased expression of genes associated with anthocyanin degradation may promote the accumulation of anthocyanins in some purple-leaf tea varieties. Additionally, heterosis of the purple-leaf phenotype is observed in the offspring of *Ziyan* and *Zijuan*, indicating the underlying genetic variations. The purple phenotype influences the accumulation of other flavonoids, which are also discussed in this paper.

While the results on the formation mechanisms and impact of the leaf color variation in tea can be identified according to certain macro patterns, the details of these physiological and molecular patterns vary greatly across different studies. The results indirectly suggest that these studies are likely limited to demonstrating the secondary outputs of leaf color variation, and the root causes of changes in leaf color or characteristic metabolite content remain poorly understood. Therefore, more in-depth studies must be carried out to uncover the mechanisms of leaf color formation in tea. (1) The function of candidate genes in tea plants should be verified using a transient transformation technology, such as a virus-induced gene-silencing system [123,124]. (2) Comparative omics research should select parents of leaf color variation materials as controls, reducing the interference from environmental conditions or developmental stages. (3) Genetic mapping and comparative genomics have been effective in identifying mutated genes in tea plants. Therefore, more efforts should focus on dissecting the genetic basis of leaf color variation and cloning mutated genes. (4) The influence of ecological or developmental factors on the metabolic flow allocation and the foundational molecular mechanisms should be explored. Through the above efforts, we will truly comprehend the mechanism of tea leaf color formation, explore its characteristic qualities more efficiently and accurately, and enhance the breeding utilization efficiency of these specific germplasms.

Author Contributions: L.-J.R. and L.-X.Z. conceived the present work; L.-J.R., Y.-G.F. and T.-T.Z. drafted the manuscript; Q.-Z.X., X.-Y.H. and S.-S.Y. retrieved the relevant literature; L.-X.Z. revised the manuscript. All authors have read and agreed to the published version of the manuscript.

Funding: This work was financially supported by the Lugan Technology Collaboration Special Project of the Science and Technology Planning Project of Gansu Province, China (Grant No. 22CX8NK246) and the Special Funds of Taishan Scholar Project of Shandong Province, China (Grant No. tsqnz20231202).

Data Availability Statement: All the data are contained within the article.

Conflicts of Interest: The authors declare no conflicts of interest.

References

1. Xia, E.H.; Tong, W.; Wu, Q.; Wei, S.; Zhao, J.; Zhang, Z.Z.; Wei, C.L.; Wan, X.C. Tea plant genomics: Achievements, challenges and perspectives. *Hortic. Res.* **2020**, *7*, 7. [CrossRef] [PubMed]
2. Wei, C.L.; Yang, H.; Wang, S.B.; Zhao, J.; Liu, C.; Gao, L.p.; Xia, E.H.; Lu, Y.; Tai, Y.L.; She, G.B.; et al. Draft genome sequence of *Camellia sinensis* var. *sinensis* provides insights into the evolution of the tea genome and tea quality. *Proc. Natl. Acad. Sci. USA* **2018**, *115*, E4151–E4158. [CrossRef] [PubMed]
3. Chen, Y.; Cheng, S.; Dai, J.G.; Wang, L.; Xu, Y.; Peng, X.Y.; Xie, X.F.; Peng, C. Molecular mechanisms and applications of tea polyphenols: A narrative review. *J. Food Biochem.* **2021**, *45*, e13910. [CrossRef] [PubMed]
4. Li, M.Y.; Liu, H.Y.; Wu, D.T.; Kenaan, A.; Geng, F.; Li, H.B.; Gunaratne, A.; Li, H.; Gan, R.Y. L-theanine: A unique functional amino acid in tea (*Camellia sinensis* L.) with multiple health benefits and food applications. *Front. Nutr.* **2022**, *9*, 853846. [CrossRef]
5. Samynathan, R.; Thiruvengadam, M.; Nile, S.H.; Shariati, M.A.; Rebezov, M.; Mishra, R.K.; Venkidasamy, B.; Periyasamy, S.; Chung, I.-M.; Pateiro, M.; et al. Recent insights on tea metabolites, their biosynthesis and chemo-preventing effects: A review. *Crit. Rev. Food Sci. Nutr.* **2023**, *63*, 3130–3149. [CrossRef]
6. Liu, D.D.; Mei, J.F.; Wang, J.Y.; Tang, R.J.; Chen, L.; Ma, C.L. Research Progress on Albino Trait of Tea Plant. *China Tea* **2020**, *42*, 24–35. [CrossRef]
7. Wang, K.R.; Liang, Y.R.; Li, M.; Zhang, L.J. Backbone parents of albino tea plants and teir germplasm traits. *J. Tea* **2015**, *41*, 130–132,136. [CrossRef]
8. Xu, Y.X.; Chen, W.; Ma, C.L.; Shen, S.Y.; Zhou, Y.Y.; Zhou, L.Q.; Chen, L. Proteome and acetyl-proteome profiling of *Camellia sinensis* cv. 'Anji Baicha' during periodic albinism reveals alterations in photosynthetic and secondary metabolite biosynthetic pathways. *Front. Plant Sci.* **2017**, *8*, 147. [CrossRef]

9. Fan, Y.G.; Zhao, X.X.; Wang, H.Y.; Tian, Y.Y.; Xiang, Q.Z.; Zhanf, L.X. Effects of light intensity on metabolism of light-harvesting pigment and photosynthetic system in *Camellia sinensis* L. cultivar 'Huangjinya'. *Environ. Exp. Bot.* **2019**, *166*, 103796. [CrossRef]
10. Zhang, Q.F.; Li, C.L.; Jiao, Z.X.; Ruan, J.Y.; Liu, M.Y. Integration of metabolomics and transcriptomics reveal the mechanism underlying accumulation of flavonols in albino tea leaves. *Molecules* **2022**, *27*, 5792. [CrossRef] [PubMed]
11. Li, Q.; Huang, J.N.; Liu, S.Q.; Li, J.; Yang, X.H.; Liu, Y.S.; Liu, Z.H. Proteomic analysis of young leaves at three developmental stages in an albino tea cultivar. *Proteome Sci.* **2011**, *9*, 44. [CrossRef]
12. Ma, Q.P.; Li, H.; Zou, Z.W.; Arkorful, E.; Lv, Q.R.; Zhou, Q.Q.; Chen, X.; Sun, K.; Li, X.H. Transcriptomic analyses identify albino-associated genes of a novel albino tea germplasm 'Huabai 1'. *Hortic. Res.* **2018**, *5*, 54. [CrossRef]
13. Jiang, X.F.; Zhao, H.; Guo, F.; Shi, X.P.; Ye, C.; Yang, P.X.; Liu, B.Y.; Ni, D.J. Transcriptomic analysis reveals mechanism of light-sensitive albinism in tea plant *Camellia sinensis* 'Huangjinju'. *BMC Plant Biol.* **2020**, *20*, 216. [CrossRef]
14. Liu, S.C.; Xu, Y.F.; Liu, Y.B.; Zhao, X.; Wei, J.; Lin, K.Q.; Liu, Y.; Yan, D.H. Metabolomic and transcriptomic analyses reveal the biosynthetic mechanisms of pigments and main taste compounds in an albino tea cultivar. *Physiol. Plant* **2023**, *175*, e13933. [CrossRef] [PubMed]
15. Chen, M.J.; Kong, X.G.; Zhang, Y.; Wang, S.Y.; Zhou, H.W.; Fang, D.S.; Yue, W.J.; Chen, C.S. Metabolomic profiling in combination with data association analysis provide insights about potential metabolic regulation networks among non-volatile and volatile metabolites in *Camellia sinensis* cv *Baijiguan*. *Plants* **2022**, *11*, 2557. [CrossRef] [PubMed]
16. Zheng, Y.C.; Wang, P.J.; Chen, X.J.; Yue, C.; Guo, Y.C.; Yang, J.F.; Sun, Y.; Ye, N.X. Integrated transcriptomics and metabolomics provide novel insight into changes in specialized metabolites in an albino tea cultivar (*Camellia sinensis* (L.) *O. Kuntz*). *Plant Physiol. Biochem.* **2021**, *160*, 27–36. [CrossRef]
17. Ma, C.Y.; Cao, J.X.; Li, J.K.; Zhou, B.; Tang, J.C.; Miao, A.Q. Phenotypic, histological and proteomic analyses reveal multiple differences associated with chloroplast development in yellow and variegated variants from *Camellia sinensis*. *Sci. Rep.* **2016**, *6*, 33369. [CrossRef] [PubMed]
18. Gao, X.Z.; Zhang, C.Y.; Lu, C.; Wang, M.H.; Xie, N.C.; Chen, J.J.; Li, Y.F.; Chen, J.H.; Shen, C.W. Disruption of photomorphogenesis leads to abnormal chloroplast development and leaf variegation in *Camellia sinensis*. *Front. Plant Sci.* **2021**, *12*, 720800. [CrossRef]
19. Li, X.X.; Li, Z.Y.; Zhu, W.; Wang, Y.Q.; Liang, Y.R.; Wang, K.R.; Ye, J.H.; Lu, J.L.; Zheng, X.Q. Anthocyanin metabolism and its differential regulation in purple tea (*Camellia sinensis*). *Plant Physiol. Biochem.* **2023**, *201*, 107875. [CrossRef]
20. Wang, L.; Pan, D.; Liang, M.; Abubakar, Y.S.; Li, J.; Lin, J.; Chen, S.P.; Chen, W. Regulation of anthocyanin biosynthesis in purple leaves of *Zijuan* Tea (*Camellia sinensis* var. *kitamura*). *Int. J. Mol. Sci.* **2017**, *18*, 833. [CrossRef]
21. Tan, L.Q.; Yang, C.J.; Zhou, B.; Wang, L.B.; Zou, Y.; Chen, W.; Xia, T.; Tang, Q. Inheritance and quantitative trait loci analyses of the anthocyanins and catechins of *Camellia sinensis* cultivar 'Ziyan' with dark-purple leaves. *Physiol. Plant* **2020**, *170*, 109–119. [CrossRef]
22. Chen, X.J.; Wang, P.J.; Wei, M.X.; Lin, X.Y.; Gu, M.Y.; Fang, W.P.; Zheng, Y.C.; Zhao, F.; Jin, S.; Ye, N.X. Lipidomics analysis unravels changes from flavor precursors in different processing treatments of purple-leaf tea. *J. Sci. Food Agric.* **2022**, *102*, 3730–3741. [CrossRef] [PubMed]
23. Shu, C.; Wu, S.Y.; Li, H.K.; Tian, J.L. Health benefits of anthocyanin-containing foods, beverages, and supplements have unpredictable relation to gastrointestinal microbiota: A systematic review and meta-analysis of random clinical trials. *Nutr. Res.* **2023**, *116*, 48–59. [CrossRef] [PubMed]
24. Chen, S.; Wang, P.J.; Kong, W.L.; Chai, K.; Zhang, S.C.; Yu, J.X.; Wang, Y.B.; Jiang, M.W.; Lei, W.L.; Chen, X.; et al. Gene mining and genomics-assisted breeding empowered by the pangenome of tea plant *Camellia sinensis*. *Nat. Plants* **2023**, *9*, 1986–1999. [CrossRef] [PubMed]
25. Zhang, X.T.; Chen, S.; Shi, L.Q.; Gong, D.P.; Zhang, S.C.; Zhao, Q.; Zhan, D.L.; Vasseur, L.; Wang, Y.B.; Yu, J.X.; et al. Haplotype-resolved genome assembly provides insights into evolutionary history of the tea plant *Camellia sinensis*. *Nat. Genet.* **2021**, *53*, 1250–1259. [CrossRef] [PubMed]
26. Wang, X.C.; Feng, H.; Chang, Y.X.; Ma, C.L.; Wang, L.Y.; Hao, X.Y.; Li, A.L.; Cheng, H.; Wang, L.; Cui, P.; et al. Population sequencing enhances understanding of tea plant evolution. *Nat. Commun.* **2020**, *11*, 4447. [CrossRef] [PubMed]
27. Zhang, W.Y.; Zhang, Y.J.; Qiu, H.J.; Guo, Y.F.; Wan, H.L.; Zhang, X.L.; Scossa, F.; Alseekh, S.; Zhang, Q.H.; Wang, P.; et al. Genome assembly of wild tea tree DASZ reveals pedigree and selection history of tea varieties. *Nat. Commun.* **2020**, *11*, 3719. [CrossRef]
28. Wang, P.; Grimm, B. Connecting chlorophyll metabolism with accumulation of the photosynthetic apparatus. *Trends Plant Sci.* **2021**, *26*, 484–495. [CrossRef]
29. Wang, J.Y.; Chen, J.D.; Wang, S.L.; Chen, L.; Ma, C.L.; Yao, M.Z. Repressed Gene Expression of photosynthetic antenna proteins associated with yellow leaf variation as revealed by bulked segregant RNA-seq in tea plant *Camellia sinensis*. *J. Agric. Food Chem.* **2020**, *68*, 8068–8079. [CrossRef]
30. Mei, X.; Zhang, K.K.; Lin, Y.G.; Su, H.F.; Lin, C.Y.; Chen, B.Y.; Yang, H.J.; Zhang, L.Y. Metabolic and Transcriptomic profiling reveals etiolated mechanism in huangyu tea (*Camellia sinensis*) leaves. *Int. J. Mol. Sci.* **2022**, *23*, 15044. [CrossRef]
31. Li, C.F.; Yao, M.Z.; Ma, C.L.; Ma, J.Q.; Jin, J.Q.; Chen, L. Differential metabolic profiles during the albescent stages of 'Anji Baicha' (*Camellia sinensis*). *PLoS ONE* **2015**, *10*, e0139996. [CrossRef] [PubMed]
32. Zhang, Q.F.; Liu, M.Y.; Ruan, J.Y. Integrated transcriptome and metabolic analyses reveals novel insights into free amino acid metabolism in *Huangjinya* tea cultivar. *Front. Plant Sci.* **2017**, *8*, 291. [CrossRef] [PubMed]

33. Liu, G.F.; Han, Z.X.; Feng, L.; Gao, L.P.; Gao, M.J.; Gruber, M.Y.; Zhang, Z.L.; Xia, T.; Wan, X.C.; Wei, S. Metabolic flux redirection and transcriptomic reprogramming in the albino tea cultivar 'Yu-Jin-Xiang' with an emphasis on catechin production. *Sci. Rep.* **2017**, *7*, 45062. [CrossRef] [PubMed]
34. Wu, Q.J.; Chen, Z.D.; Sun, W.J.; Deng, T.T.; Chen, M.J. De novo sequencing of the leaf transcriptome reveals complex light-responsive regulatory networks in *Camellia sinensis* cv. *Baijiguan*. *Front. Plant Sci.* **2016**, *7*, 332. [CrossRef] [PubMed]
35. Xu, C.B.; Li, J.L.; Wang, H.L.; Liu, H.J.; Yu, Z.H.; Zhao, Z. Whole-transcriptome sequencing reveals a cerna regulatory network associated with the process of periodic albinism under low temperature in *Baiye No. 1* (*Camellia sinensis*). *Int. J. Mol. Sci.* **2023**, *24*, 7162. [CrossRef]
36. Fang, D.; Shi, Y.Z.; Liu, M.Y.; Fan, K.; Zhang, Q.F.; Ruan, J.Y. iTRAQ-based quantitative proteomics analysis reveals the mechanism underlying the weakening of carbon metabolism in chlorotic tea leaves. *Int. J. Mol. Sci.* **2018**, *19*, 3943. [CrossRef]
37. Zhang, Q.F.; Liu, M.Y.; Ruan, J.Y. Metabolomics analysis reveals the metabolic and functional roles of flavonoids in light-sensitive tea leaves. *BMC Plant Biol.* **2017**, *17*, 64. [CrossRef]
38. Zhang, C.Y.; Liu, H.R.; Wang, J.Y.; Li, Y.Y.; Liu, D.D.; Ye, Y.Y.; Huang, R.; Li, S.J.; Chen, L.; Chen, J.D.; et al. A key mutation in magnesium chelatase I subunit leads to a chlorophyll-deficient mutant of tea (*Camellia sinensis*). *J. Exp. Bot.* **2023**, erad430. [CrossRef]
39. Zhang, C.Y.; Liu, G.Z.; Chen, J.J.; Xie, N.C.; Huang, J.A.; Shen, C.W. Translational landscape and metabolic characteristics of the etiolated tea plant (*Camellia sinensis*). *Sci. Hortic.* **2022**, *303*, 111193. [CrossRef]
40. Xie, N.C.; Zhang, C.Y.; Zhou, P.Q.; Gao, X.Z.; Wang, M.H.; Tian, S.H.; Lu, C.; Wang, K.B.; Shen, C.W. Transcriptomic analyses reveal variegation-induced metabolic changes leading to high L-theanine levels in albino sectors of variegated tea (*Camellia sinensis*). *Plant Physiol. Biochem.* **2021**, *169*, 29–39. [CrossRef]
41. Liu, Y.F.; Pang, D.D.; Jiang, H.B.; Chen, C.L.; Sun, Y.N.; Tian, Y.P.; Xu, Y.; Song, W.X.; Chen, L.B. Identifying key genes involved in yellow leaf variation in 'Menghai Huangye' based on biochemical and transcriptomic analysis. *Funct. Integr. Genom.* **2022**, *22*, 251–260. [CrossRef]
42. Wang, L.; Yue, C.; Cao, H.L.; Zhou, Y.H.; Zeng, J.M.; Yang, Y.J.; Wang, X.C. Biochemical and transcriptome analyses of a novel chlorophyll-deficient chlorina tea plant cultivar. *BMC Plant Biol.* **2014**, *14*, 352. [CrossRef] [PubMed]
43. Li, Y.Y.; Zhang, C.Y.; Ma, C.L.; Chen, L.; Yao, M.Z. Transcriptome and biochemical analyses of a chlorophyll-deficient bud mutant of tea plant (*Camellia sinensis*). *Int. J. Mol. Sci.* **2023**, *24*, 15070. [CrossRef] [PubMed]
44. Qin, G.J.; Gu, H.Y.; Ma, L.G.; Peng, Y.B.; Deng, X.W.; Chen, Z.L.; Qu, L.J. Disruption of *phytoene desaturase* gene results in albino and dwarf phenotypes in *Arabidopsis* by impairing chlorophyll, carotenoid, and gibberellin biosynthesis. *Cell Res.* **2007**, *17*, 471–482. [CrossRef] [PubMed]
45. Hertle, A.P.; García-Cerdán, J.G.; Armbruster, U.; Shih, R.; Lee, J.J.; Wong, W.; Niyogi, K.K. A Sec14 domain protein is required for photoautotrophic growth and chloroplast vesicle formation in *Arabidopsis thaliana*. *Proc. Natl. Acad. Sci. USA* **2020**, *117*, 9101–9111. [CrossRef] [PubMed]
46. Xu, Y.X.; Shen, C.J.; Ma, J.Q.; Chen, W.; Mao, J.; Zhou, Y.Y.; Chen, L. Quantitative succinyl-proteome profiling of *Camellia sinensis* cv. '*Anji Baicha*' during periodic albinism. *Sci. Rep.* **2017**, *7*, 1873. [CrossRef] [PubMed]
47. Song, L.B.; Ma, Q.P.; Zou, Z.W.; Sun, K.; Yao, Y.T.; Tao, J.H.; Kaleri, N.A.; Li, X.H. Molecular link between leaf coloration and gene expression of flavonoid and carotenoid biosynthesis in *Camellia sinensis* cultivar 'Huangjinya'. *Front. Plant Sci.* **2017**, *8*, 803. [CrossRef]
48. Lu, M.Q.; Han, J.Y.; Zhu, B.Y.; Jia, H.Y.; Yang, T.Y.; Wang, R.J.; Deng, W.W.; Zhang, Z.Z. Significantly increased amino acid accumulation in a novel albino branch of the tea plant (*Camellia sinensis*). *Planta* **2019**, *249*, 363–376. [CrossRef] [PubMed]
49. Xu, Y.X.; Yang, L.; Lei, Y.S.; Ju, R.N.; Miao, S.G.; Jin, S.H. Integrated transcriptome and amino acid profile analyses reveal novel insights into differential accumulation of theanine in green and yellow tea cultivars. *Tree Physiol.* **2022**, *42*, 1501–1516. [CrossRef]
50. Yamashita, H.; Kambe, Y.; Ohshio, M.; Kunihiro, A.; Tanaka, Y.; Suzuki, T.; Nakamura, Y.; Morita, A.; Ikka, T. Integrated metabolome and transcriptome analyses reveal etiolation-induced metabolic changes leading to high amino acid contents in a light-sensitive Japanese albino tea cultivar. *Front. Plant Sci.* **2021**, *11*, 611140. [CrossRef]
51. Liu, X.Y.; Cao, J.J.; Cheng, X.; Zhu, W.F.; Sun, Y.; Wan, X.C.; Liu, L.L. CsRVE1 promotes seasonal greening of albino *Camellia sinensis* cv. *Huangkui* by activating chlorophyll biosynthesis. *Tree Physiol.* **2023**, *43*, 1432–1443. [CrossRef]
52. Lin, X.Y.; Wang, P.J.; Yang, R.X.; Zheng, Y.C.; Chen, X.M.; Zhang, L.; Xian, S.S.; Xing, Y.N. The albino mechanism of a new high theanine tea cultivar *Fuhuang 1*. *Sci. Agric. Sin.* **2022**, *55*, 22. [CrossRef]
53. Lin, X.Y.; Shao, S.X.; Wang, P.J.; Yang, R.X.; Zheng, Y.C.; Chen, X.M.; Zhang, L.; Ye, N.X. The albino mechanism of a new high theanine tea cultivar *Fuhuang 2*. *Chin. J. Biotechnol.* **2022**, *38*, 3956–3972. [CrossRef]
54. Sandmann, G. Antioxidant protection from UV- and light-stress related to carotenoid structures. *Antioxidants* **2019**, *8*, 219. [CrossRef] [PubMed]
55. Di Mascio, P.; Kaiser, S.; Sies, H. Lycopene as the most efficient biological carotenoid singlet oxygen quencher. *Arch. Biochem. Biophys.* **1989**, *274*, 532–538. [CrossRef] [PubMed]
56. Boey Peng, L.; Akihiko, N.; Junji, T.; Kazunobu, T.; Tetsuya, S.; Kozo, T. Antioxidant activity of xanthophylls on peroxyl radical-mediated phospholipid peroxidation. *Biochim. Biophys. Acta (BBA)—Lipids Lipid Metab.* **1992**, *1126*, 178–184. [CrossRef]

57. Woodall, A.A.; Britton, G.; Jackson, M.J. Carotenoids and protection of phospholipids in solution or in liposomes against oxidation by peroxyl radicals: Relationship between carotenoid structure and protective ability. *Biochim. Biophys. Acta (BBA)—Gen. Subj.* **1997**, *1336*, 575–586. [CrossRef]
58. Du, H.Y.; Qi, M.Z.; Cui, X.P.; Cui, Y.M.; Yang, H.; Zhang, J.Y.; Ma, Y.J.; Zhang, S.S.; Zhang, X.; Yu, D.Y. Proteomic and functional analysis of soybean chlorophyll-deficient mutant *cd1* and the underlying gene encoding the CHLI subunit of Mg-chelatase. *Mol. Breed.* **2018**, *38*, 71. [CrossRef]
59. Satou, M.; Enoki, H.; Oikawa, A.; Ohta, D.; Saito, K.; Hachiya, T.; Sakakibara, H.; Kusano, M.; Fukushima, A.; Saito, K.; et al. Integrated analysis of transcriptome and metabolome of *Arabidopsis* albino or pale green mutants with disrupted nuclear-encoded chloroplast proteins. *Plant Mol. Biol.* **2014**, *85*, 411–428. [CrossRef]
60. Zhang, H.; Liu, L.L.; Cai, M.H.; Zhu, S.S.; Zhao, J.Y.; Zheng, T.H.; Xu, X.Y.; Zeng, Z.Q.; Niu, J.; Jiang, L.; et al. A point mutation of magnesium chelatase *OsCHLI* gene dampens the interaction between CHLI and CHLD subunits in rice. *Plant Mol. Biol. Rep.* **2015**, *33*, 1975–1987. [CrossRef]
61. Hansson, A.; Kannangara, C.G.; von Wettstein, D.; Hansson, M. Molecular basis for semidominance of missense mutations in the XANTHA-H (42-kDa) subunit of magnesium chelatase. *Proc. Natl. Acad. Sci. USA* **1999**, *96*, 1744–1749. [CrossRef] [PubMed]
62. Schön, A.; Krupp, G.; Gough, S.; Berry-Lowe, S.; Kannangara, C.G.; Söll, D. The RNA required in the first step of chlorophyll biosynthesis is a chloroplast glutamate tRNA. *Nature* **1986**, *322*, 281–284. [CrossRef] [PubMed]
63. Teng, K.; Tan, P.H.; Guan, J.; Dong, D.; Liu, L.U.; Guo, Y.D.; Guo, W.E.; Yuesen, Y.; Fan, X.F.; Wu, J.Y. Functional characterization of the chlorophyll b reductase gene *NYC1* associated with chlorophyll degradation and photosynthesis in *Zoysia japonica*. *Environ. Exp. Bot.* **2021**, *191*, 104607. [CrossRef]
64. Sato, R.; Ito, H.; Tanaka, A. Chlorophyll b degradation by chlorophyll b reductase under high-light conditions. *Photosynth. Res.* **2015**, *126*, 249–259. [CrossRef]
65. Niu, G.Q.; Guo, Q.; Wang, J.; Zhao, S.; He, Y.K.; Liu, L. Structural basis for plant lutein biosynthesis from α-carotene. *Proc. Natl. Acad. Sci. USA* **2020**, *117*, 14150–14157. [CrossRef] [PubMed]
66. Pružinská, A.; Tanner, G.; Anders, I.; Roca, M.; Hörtensteiner, S. Chlorophyll breakdown: Pheophorbide a oxygenase is a Rieske-type iron–sulfur protein, encoded by the *accelerated cell death 1* gene. *Proc. Natl. Acad. Sci. USA* **2003**, *100*, 15259–15264. [CrossRef] [PubMed]
67. Zhang, B.; Chen, Z.T.; Sun, Q.X.; Liu, J.X. Proteome-wide analyses reveal diverse functions of protein acetylation and succinylation modifications in fast growing stolons of bermudagrass (*Cynodon dactylon* L.). *BMC Plant Biol.* **2022**, *22*, 503. [CrossRef]
68. Kobayashi, K.; Masuda, T. Chapter Five—Transcriptional control for the chlorophyll metabolism. In *Advances in Botanical Research*; Grimm, B., Ed.; Academic Press: Cambridge, MA, USA, 2019; Volume 91, pp. 133–161.
69. Sathasivam, R.; Radhakrishnan, R.; Kim, J.K.; Park, S.U. An update on biosynthesis and regulation of carotenoids in plants. *S. Afr. J. Bot.* **2021**, *140*, 290–302. [CrossRef]
70. Engelken, J.; Brinkmann, H.; Adamska, I. Taxonomic distribution and origins of the extended LHC (light-harvesting complex) antenna protein superfamily. *BMC Evol. Biol.* **2010**, *10*, 233. [CrossRef]
71. Takabayashi, A.; Kurihara, K.; Kuwano, M.; Kasahara, Y.; Tanaka, R.; Tanaka, A. The oligomeric states of the photosystems and the light-harvesting complexes in the Chl *b*-less mutant. *Plant Cell Physiol.* **2011**, *52*, 2103–2114. [CrossRef] [PubMed]
72. Pietrzykowska, M.; Suorsa, M.; Semchonok, D.A.; Tikkanen, M.; Boekema, E.J.; Aro, E.-M.; Jansson, S. The light-harvesting chlorophyll a/b binding proteins Lhcb1 and Lhcb2 play complementary roles during state transitions in *Arabidopsis*. *Plant Cell* **2014**, *26*, 3646–3660. [CrossRef]
73. Wietrzynski, W.; Engel, B.D. Chlorophyll biogenesis sees the light. *Nat. Plants* **2021**, *7*, 380–381. [CrossRef]
74. Fu, C.X.; Tang, X.Y.; Rui, R.L.; Chen, Y.F.; Zhong, W.G. Pigment photobleach and its component changes in indica varieties leaves under high light. *Chin. J. Rice Sci.* **1993**, *7*, 83–87.
75. Nick, S.; Meurer, J.; Soll, J.; Ankele, E. Nucleus-Encoded Light-harvesting chlorophyll a/b proteins are imported normally into chlorophyll b-free chloroplasts of *Arabidopsis*. *Mol. Plant* **2013**, *6*, 860–871. [CrossRef]
76. Kindgren, P.; Norén, L.; Barajas López, J.d.D.; Shaikhali, J.; Strand, Å. Interplay between HEAT SHOCK PROTEIN 90 and HY5 controls *PhANG* expression in response to the GUN5 plastid signal. *Mol. Plant* **2012**, *5*, 901–913. [CrossRef]
77. Sun, B.M.; Zhu, Z.S.; Cao, P.R.; Chen, H.; Chen, C.M.; Zhou, X.; Mao, Y.H.; Lei, J.J.; Jiang, Y.P.; Meng, W.; et al. Purple foliage coloration in tea (*Camellia sinensis* L.) arises from activation of the R2R3-MYB transcription factor CsAN1. *Sci. Rep.* **2016**, *6*, 32534. [CrossRef] [PubMed]
78. Shen, J.Z.; Zou, Z.W.; Zhang, X.Z.; Zhou, L.; Wang, Y.H.; Fang, W.P.; Zhu, X.J. Metabolic analyses reveal different mechanisms of leaf color change in two purple-leaf tea plant (*Camellia sinensis* L.) cultivars. *Hortic. Res.* **2018**, *5*, 7. [CrossRef] [PubMed]
79. Zhou, Q.Q.; Sun, W.J.; Lai, Z.X. Differential expression of genes in purple-shoot tea tender leaves and mature leaves during leaf growth. *J. Sci. Food Agric.* **2016**, *96*, 1982–1989. [CrossRef]
80. Zhang, Q.F.; Hu, J.H.; Liu, M.Y.; Shi, Y.Z.; De Vos, R.C.H.; Ruan, J.Y. Stimulated biosynthesis of delphinidin-related anthocyanins in tea shoots reducing the quality of green tea in summer. *J. Sci. Food Agric.* **2020**, *100*, 1505–1514. [CrossRef]
81. Maritim, T.K.; Korir, R.K.; Nyabundi, K.W.; Wachira, F.N.; Kamunya, S.M.; Muoki, R.C. Molecular regulation of anthocyanin discoloration under water stress and high solar irradiance in pluckable shoots of purple tea cultivar. *Planta* **2021**, *254*, 85. [CrossRef]

82. Tan, L.Q.; Zhang, P.; Cui, D.; Yang, X.; Zhang, D.Y.; Yang, Y.; Chen, W.; Tang, D.D.; Tang, Q.; Li, P.W. Multi-omics analysis revealed anthocyanin accumulation differences in purple tea plants 'Ziyan', 'Zijuan' and their dark-purple hybrid. *Sci. Hortic.* **2023**, *321*, 112275. [CrossRef]
83. Chen, X.J.; Wang, P.J.; Zheng, Y.C.; Gu, M.Y.; Lin, X.Y.; Wang, S.Y.; Jin, S.; Ye, N.X. Comparison of metabolome and transcriptome of flavonoid biosynthesis pathway in a purple-leaf tea germplasm *Jinmingzao* and a green-leaf tea germplasm huangdan reveals their relationship with genetic mechanisms of color formation. *Int. J. Mol. Sci.* **2020**, *21*, 4167. [CrossRef]
84. Cai, J.; Lv, L.T.; Zeng, X.F.; Zhang, F.; Chen, Y.L.; Tian, W.L.; Li, J.R.; Li, X.Y.; Li, Y. Integrative analysis of metabolomics and transcriptomics reveals molecular mechanisms of anthocyanin metabolism in the *Zikui* tea plant (*Camellia sinensis* cv. *Zikui*). *Int. J. Mol. Sci.* **2022**, *23*, 4780. [CrossRef]
85. Shi, J.; Simal-Gandara, J.; Mei, J.F.; Ma, W.J.; Peng, Q.H.; Shi, Y.L.; Xu, Q.; Lin, Z.; Lv, H.P. Insight into the pigmented anthocyanins and the major potential co-pigmented flavonoids in purple-coloured leaf teas. *Food Chem.* **2021**, *363*, 130278. [CrossRef]
86. Li, M.W.; Shen, Y.; Ling, T.J.; Ho, C.T.; Li, D.X.; Guo, H.M.; Xie, Z.W. Analysis of Differentiated chemical components between *Zijuan* purple tea and Yunkang green tea by UHPLC-Orbitrap-MS/MS combined with chemometrics. *Foods* **2021**, *10*, 1070. [CrossRef] [PubMed]
87. Liu, Z.W.; Shi, X.Y.; Duan, S.M.; Nian, B.; Chen, L.J.; Zhang, G.H.; Lv, C.Y.; Ma, Y.; Zhao, M. Multiomics analysis of the mechanisms behind flavonoid differences between purple and green tender shoots of *Camellia sinensis* var. *assamica*. *G3-Genes Genom. Genet.* **2022**, *13*, jkac297. [CrossRef] [PubMed]
88. Huang, F.Y.; Duan, J.H.; Lei, Y.; Kang, Y.K.; Luo, Y.; Chen, Y.Y.; Ding, D.; Li, S.J. Metabolomic and transcriptomic analyses reveal a MYB gene, *CsAN1*, involved in anthocyanins accumulation separation in F1 between 'Zijuan' (*Camellia sinensis* var. *assamica*) and 'Fudingdabaicha' (*C. sinensis* var. *sinensis*) tea plants. *Front. Plant Sci.* **2022**, *13*, 1008588. [CrossRef] [PubMed]
89. Mei, Y.; Xie, H.; Liu, S.R.; Zhu, J.Y.; Zhao, S.Q.; Wei, C.L. Metabolites and transcriptional profiling analysis reveal the molecular mechanisms of the anthocyanin metabolism in the "*Zijuan*" tea plant (*Camellia sinensis* var. *assamica*). *J. Agric. Food Chem.* **2021**, *69*, 414–427. [CrossRef]
90. Li, W.; Tan, L.Q.; Zou, Y.; Tan, X.Q.; Huang, J.C.; Chen, W.; Tang, Q. The Effects of ultraviolet A/B treatments on anthocyanin accumulation and gene expression in dark-purple tea cultivar 'Ziyan' (*Camellia sinensis*). *Moleculars* **2020**, *25*, 354. [CrossRef]
91. Zhu, M.Z.; Zhou, F.; Ran, L.S.; Li, Y.L.; Tan, B.; Wang, K.B.; Huang, J.A.; Liu, Z.H. Metabolic profiling and gene expression analyses of purple-leaf formation in tea cultivars (*Camellia sinensis* var. *sinensis* and var. *assamica*). *Front. Plant Sci.* **2021**, *12*, 606962. [CrossRef]
92. Kumari, M.; Thakur, S.; Kumar, A.; Joshi, R.; Kumar, P.; Shankar, R.; Kumar, R. Regulation of color transition in purple tea (*Camellia sinensis*). *Planta* **2019**, *251*, 35. [CrossRef]
93. Li, Z.; Ahammed, G.J. Plant stress response and adaptation via anthocyanins: A review. *Plant Stress* **2023**, *10*, 100230. [CrossRef]
94. Schaart, J.G.; Dubos, C.; Romero De La Fuente, I.; van Houwelingen, A.M.M.L.; de Vos, R.C.H.; Jonker, H.H.; Xu, W.; Routaboul, J.-M.; Lepiniec, L.; Bovy, A.C. Identification and characterization of MYB-bHLH-WD40 regulatory complexes controlling proanthocyanidin biosynthesis in strawberry (*Fragaria × ananassa*) fruits. *New Phytol.* **2013**, *197*, 454–467. [CrossRef]
95. Cui, D.L.; Zhao, S.X.; Xu, H.N.; Allan, A.C.; Zhang, X.D.; Fan, L.; Chen, L.M.; Su, J.; Shu, Q.; Li, K.Z. The interaction of MYB, bHLH and WD40 transcription factors in red pear (*Pyrus pyrifolia*) peel. *Plant Mol. Biol.* **2021**, *106*, 407–417. [CrossRef]
96. Wei, K.; Zhang, Y.Z.; Wu, L.Y.; Li, H.L.; Ruan, L.; Bai, P.X.; Zhang, C.C.; Zhang, F.; Xu, L.Y.; Wang, L.Y.; et al. Gene expression analysis of bud and leaf color in tea. *Plant Physiol. Biochem.* **2016**, *107*, 310–318. [CrossRef]
97. Wang, X.W.; Liu, B.Y.; Zhao, Q.S.; Sun, X.M.; Li, Y.Y.; Duan, Z.F.; Miao, X.L.; Luo, S.; Li, J.B. Genomic variance and transcriptional comparisons reveal the mechanisms of leaf color affecting palatability and stressed defense in tea plant. *Genes* **2019**, *10*, 929. [CrossRef] [PubMed]
98. Chen, L.J.; Shi, X.Y.; Nian, B.; Duan, S.M.; Jiang, B.; Wang, X.H.; Lv, C.Y.; Zhang, G.H.; Ma, Y.; Zhao, M. Alternative splicing regulation of anthocyanin biosynthesis in *Camellia sinensis* var. *assamica* unveiled by PacBio Iso-Seq. *G3-Genes Genom. Genet.* **2020**, *10*, 2713–2723. [CrossRef] [PubMed]
99. Zhou, Q.Q.; Chen, Z.D.; Lee, J.W.; Li, X.H.; Sun, W.J. Proteomic analysis of tea plants (*Camellia sinensis*) with purple young shoots during leaf development. *PLoS ONE* **2017**, *12*, e0177816. [CrossRef] [PubMed]
100. Wilmouth, R.C.; Turnbull, J.J.; Welford, R.W.D.; Clifton, I.J.; Prescott, A.G.; Schofield, C.J. Structure and mechanism of anthocyanidin synthase from *Arabidopsis thaliana*. *Structure* **2002**, *10*, 93–103. [CrossRef] [PubMed]
101. Marrs, K.A.; Alfenito, M.R.; Lloyd, A.M.; Walbot, V. A glutathione S-transferase involved in vacuolar transfer encoded by the maize gene *Bronze-2*. *Nature* **1995**, *375*, 397–400. [CrossRef] [PubMed]
102. Braidot, E.; Zancani, M.; Petrussa, E.; Peresson, C.; Bertolini, A.; Patui, S.; Macrì, F.; Vianello, A. Transport and accumulation of flavonoids in grapevine (*Vitis vinifera* L.). *Plant. Signal. Behav.* **2008**, *3*, 626–632. [CrossRef]
103. Zeng, C.Z.; Lin, H.Y.; Liu, Z.X.; Liu, Z.H. Analysis of young shoots of 'Anji Baicha' (*Camellia sinensis*) at three developmental stages using nontargeted LC-MS-Based metabolomics. *J. Food Sci.* **2019**, *84*, 1746–1757. [CrossRef]
104. Huang, R.; Wang, J.Y.; Yao, M.Z.; Ma, C.L.; Chen, L. Quantitative trait loci mapping for free amino acid content using an albino population and SNP markers provides insight into the genetic improvement of tea plants. *Hortic. Res.* **2022**, *9*, uhab029. [CrossRef]
105. Yu, Y.; Kou, X.B.; Gao, R.S.; Chen, X.F.; Zhao, Z.; Mei, H.L.; Li, J.J.; Jeyaraj, A.; Thangaraj, K.; Periakaruppan, R.; et al. Glutamine synthetases play a vital role in high accumulation of theanine in tender shoots of albino tea germplasm "Huabai 1". *J. Agric. Food Chem.* **2021**, *69*, 13904–13915. [CrossRef]

106. Li, C.F.; Xu, Y.X.; Ma, J.Q.; Jin, J.Q.; Huang, D.J.; Yao, M.Z.; Ma, C.L.; Chen, L. Biochemical and transcriptomic analyses reveal different metabolite biosynthesis profiles among three color and developmental stages in 'Anji Baicha' (*Camellia sinensis*). *BMC Plant Biol.* **2016**, *16*, 195. [CrossRef]
107. Fu, X.M.; Cheng, S.H.; Liao, Y.Y.; Xu, X.L.; Wang, X.C.; Hao, X.Y.; Xu, P.; Dong, F.; Yang, Z.Y. Characterization of l-theanine hydrolase in vitro and subcellular distribution of its specific product ethylamine in tea (*Camellia sinensis*). *J. Agric. Food Chem.* **2020**, *68*, 10842–10851. [CrossRef]
108. Liao, H.S.; Chung, Y.H.; Hsieh, M.H. Glutamate: A multifunctional amino acid in plants. *Plant Sci.* **2022**, *318*, 111238. [CrossRef] [PubMed]
109. Shen, N.; Wang, T.F.; Gan, Q.; Liu, S.; Wang, L.; Jin, B. Plant flavonoids: Classification, distribution, biosynthesis, and antioxidant activity. *Food Chem.* **2022**, *383*, 132531. [CrossRef] [PubMed]
110. Tattini, M.; Galardi, C.; Pinelli, P.; Massai, R.; Remorini, D.; Agati, G. Differential accumulation of flavonoids and hydroxycinnamates in leaves of *Ligustrum vulgare* under excess light and drought stress. *New Phytol.* **2004**, *163*, 547–561. [CrossRef] [PubMed]
111. Fujii, S.; Wada, H.; Kobayashi, K. Role of Galactolipids in plastid differentiation before and after light exposure. *Plants* **2019**, *8*, 357. [CrossRef]
112. Pye, V.E.; Christensen, C.E.; Dyer, J.H.; Arent, S.; Henriksen, A. Peroxisomal plant 3-ketoacyl-coa thiolase structure and activity are regulated by a sensitive redox switch. *J. Biol. Chem.* **2010**, *285*, 24078–24088. [CrossRef] [PubMed]
113. Arent, S.; Pye, V.E.; Henriksen, A. Structure and function of plant acyl-CoA oxidases. *Plant Physiol. Biochem.* **2008**, *46*, 292–301. [CrossRef] [PubMed]
114. Kim, H.U. Lipid Metabolism in Plants. *Plants* **2020**, *9*, 871. [CrossRef] [PubMed]
115. Yu, C.W.; Lin, Y.T.; Li, H.M. Increased ratio of galactolipid MGDG: DGDG induces jasmonic acid overproduction and changes chloroplast shape. *New Phytol.* **2020**, *228*, 1327–1335. [CrossRef] [PubMed]
116. Narasimhan, R.; Wang, G.; Li, M.; Roth, M.; Welti, R.; Wang, X. Differential changes in galactolipid and phospholipid species in soybean leaves and roots under nitrogen deficiency and after nodulation. *Phytochemistry* **2013**, *96*, 81–91. [CrossRef] [PubMed]
117. Xu, D.Y.; Lam, S.M.; Zuo, J.H.; Yuan, S.Z.; Lv, J.Y.; Shi, J.Y.; Gao, L.P.; Chen, B.; Sui, Y.; Shui, G.H.; et al. Lipidomics reveals the difference of membrane lipid catabolism between chilling injury sensitive and non-sensitive green bell pepper in response to chilling. *Postharvest Biol. Technol.* **2021**, *182*, 111714. [CrossRef]
118. Zhou, Z.; Chen, M.j.; Wu, Q.j.; Zeng, W.; Chen, Z.d.; Sun, W.j. Combined analysis of lipidomics and transcriptomics revealed the key pathways and genes of lipids in light-sensitive albino tea plant (*Camellia sinensis* cv. *Baijiguan*). *Front. Plant Sci.* **2022**, *13*. [CrossRef]
119. Yang, X.Y.; Lu, M.Q.; Wang, Y.F.; Wang, Y.R.; Liu, Z.J.; Chen, S. Response mechanism of plants to drought stress. *Horticulturae* **2021**, *7*, 50. [CrossRef]
120. Bhanot, V.; Fadanavis, S.V.; Panwar, J. Revisiting the architecture, biosynthesis and functional aspects of the plant cuticle: There is more scope. *Environ. Exp. Bot.* **2021**, *183*, 104364. [CrossRef]
121. Saini, P.; Bhatia, S.; Mahajan, M.; Kaushik, A.; Sahu, S.K.; Kumar, A.; Satbhai, S.B.; Patel, M.K.; Saxena, S.; Chaurasia, O.P.; et al. ELONGATED HYPOCOTYL5 negatively regulates *DECREASE WAX BIOSYNTHESIS* to increase survival during UV-B stress. *Plant Physiol.* **2020**, *184*, 2091–2106. [CrossRef]
122. Song, S.S.; Tao, Y.; Gao, L.H.; Liang, H.L.; Tang, D.S.; Lin, J.; Wang, Y.C.; Gmitter, F.G.; Li, C.F. An integrated metabolome and transcriptome analysis reveal the regulation mechanisms of flavonoid biosynthesis in a purple tea plant cultivar. *Front. Plant Sci.* **2022**, *13*, 880227. [CrossRef] [PubMed]
123. Yang, W.; Chen, X.Y.; Chen, J.H.; Zheng, P.; Liu, S.Q.; Tan, X.D.; Sun, B.M. Virus-induced gene silencing in the tea plant (*Camellia sinensis*). *Plants* **2023**, *12*, 3162. [CrossRef] [PubMed]
124. Li, G.D.; Li, Y.; Yao, X.Z.; Lu, L.T. Establishment of a virus-induced gene-silencing (VIGS) system in tea plant and its use in the functional analysis of *CsTCS1*. *Int. J. Mol. Sci.* **2023**, *24*, 392. [CrossRef] [PubMed]

MDPI AG
Grosspeteranlage 5
4052 Basel
Switzerland
Tel.: +41 61 683 77 34

Plants Editorial Office
E-mail: plants@mdpi.com
www.mdpi.com/journal/plants

www.ingramcontent.com/pod-product-compliance
Lightning Source LLC
Chambersburg PA
CBHW061503220926
51850CB00004B/160
9783725862764